Thomas W. Wieting
May, 1963

AMERICAN MATHEMATICAL SOCIETY
COLLOQUIUM PUBLICATIONS, VOLUME XV

LINEAR TRANSFORMATIONS IN HILBERT SPACE
AND THEIR APPLICATIONS TO ANALYSIS

BY

MARSHALL HARVEY STONE
ASSOCIATE PROFESSOR IN YALE UNIVERSITY

NEW YORK
PUBLISHED BY THE
AMERICAN MATHEMATICAL SOCIETY
501 WEST 116TH STREET
1932

Reprinted 1958

OFFSET
PRINTED BY

SPAULDING-MOSS COMPANY
BOSTON, MASSACHUSETTS, U.S.A.

FOREWORD

When I began in the summer of 1928 to study the theory of linear transformations in Hilbert space, it was my intention to present the results in some of the current journals. Various circumstances have led, however, to the preparation of this volume, in which I have attempted to include not only my own independent contributions but also a substantial portion of the existing material bearing on the subject. The lack of English-language works dealing with the theory of Hilbert space appeared to be an adequate reason for planning a detailed treatment which would start with the foundations and carry the development as far as possible in every direction. It has accordingly been my object to provide a treatise of this character, with serious claims to completeness—one which will be, if my hopes are realized, a useful handbook both for the student and for the investigator. In carrying out this plan, I have confined myself strictly to the theory of Hilbert space, arbitrarily excluding any reference to the various related and similar theories of other types of space. The adoption of this course was merely an unfortunate consequence of the necessity of keeping the length of this book within reasonable limits, and does not imply any personal view or judgment concerning the interest and importance of other theories. I should be the first to urge the reader to consult discussions of such cognate topics.* Considerations of space have also

---

* For mathematical and bibliographical information, see Volterra, *The Theory of Functionals and Integro-Differential Equations,* London, 1930; —Hildebrandt, Bulletin of the American Mathematical Society, 37 (1931), pp. 185–212. There are several books in preparation which will be of interest in this connection: Mrs. Pell-Wheeler will contribute a volume to this Collo-

led to the omission of two chapters for which provision had originally been made. One of these, dealing with groups of transformations in Hilbert space, was completed in May, 1930, and will appear separately in the course of time. The other, outlining the remarkable applications of the abstract theory developed in the present volume to modern atomic physics, was never written. Although this chapter would have been purely expository in nature, I regret that it could not be included, since the point of view which it would have set forth in some detail has already proved so fruitful in the study of the atom and promises still more profound results in the future.*

The existence of the splendid article by Hellinger and Toeplitz in the Encyklopädie der Mathematischen Wissenschaften, $2^{3^{2}}$, has spared me the exacting labor of preparing a bibliography of the voluminous literature prior to 1924. I have taken some pains, however, to give references to the major contributions since that date.† While I have concerned myself very little

---

quium Series (compare the synopsis of her Colloquium Lectures, Bulletin of the American Mathematical Society, 33 (1927), pp. 664-665); in the Introduction to his book, *Les espaces abstraits,* Paris, 1928, Fréchet has promised a second volume dealing with functional analysis; and, I am informed, the researches of E. H. Moore and his school, which have long been practically unavailable to the mathematical public, will soon be presented in book form. A French translation of Banach's *Teorja Operacyj, Tom. I: Operacje linjowe,* Warszawa, 1931, is about to appear.

* See Dirac, *The Principles of Quantum Mechanics,* Oxford, 1930;—Weyl, *Gruppentheorie und Quantenmechanik,* Leipzig, 1928 (first edition), and 1931 (second edition), translated into English by H. P. Robertson under the title *The Theory of Groups and Quantum Mechanics,* New York, 1932;—and two papers of J. v. Neumann in the Göttinger Nachrichten, 1927, together with his recent book, *Mathematische Grundlagen der Quantenmechanik,* Berlin, 1932.

† The recent paper of J. v. Neumann, Annals of Mathematics, (2) 33 (1932), pp. 294-310, came to my attention too late to be cited. This paper throws a great deal of light on Theorems 2.9, 2.10, 2.26, 8.18, 9.5, 10.10 and Definition 8.3. For instance, the hypothesis of Theorem 2.26 can be weakened to read "*If T is a transformation whose domain is $\mathfrak{H}$ and if its adjoint T^* has domain everywhere dense in $\mathfrak{H}$*"; an independent proof of the modified theorem was communicated to me by Professor J. D. Tamarkin.

with questions of history or of priority, I wish to acknowledge in the most cordial spirit my scientific debt to J. v. Neumann. The initial impetus of my interest came from reading some of v. Neumann's early and still incomplete work, which was described in the Göttinger Nachrichten, 1927, pp. 1–55, foot-notes 12 and 27, to which I had access, but which was never published. Thereafter, I worked independently, the results announced in the Proceedings of the National Academy of Sciences, 15 (1929), pp. 198–200, pp. 423–425, and 16 (1930), pp. 172–175, being obtained without further knowledge of his progress along the same or similar lines. I have been only too glad to improve the final presentation of my own investigations by the continual use of v. Neumann's various memoirs on the theory of transformations in Hilbert space. While it is scarcely necessary to point out that this recent work is a natural continuation of that begun by Hilbert and his school, I wish to emphasize the important rôle played by the contributions of F. Riesz in preparing the ground for a successful consideration of non-bounded transformations. The concepts which Riesz developed in his book, *Les systèmes d'équations linéaires à une infinité d'inconnues,* Paris, 1913, marked the introduction of a new point of view and of new methods, without which progress might well have been retarded; their influence can be traced throughout the development of the theory given in these pages.

With a view to making the book useful as a work of reference, I have adopted the practice of stating all important definitions and theorems in italics and numbering them serially by chapters: thus Theorem 5.12 is the twelfth theorem in Chapter V, Definition 8.2 the second definition in Chapter VIII. At the same time, I have avoided, so far as possible, any elaborate system of cross-references in either the text or the foot-notes, so that the reader need not correlate a mass of widely scattered material when he is interested in a particular topic. The table of contents and the index are designed to serve as adequate guides to the various subjects treated. In order to compress the material into the compass of six hundred odd pages, it has been necessary to employ as concise a style as is consistent with completeness and clarity of statement, and to omit numerous comments, however illuminating, which will doubtless suggest themselves to the reader as simple ccrollaries or special cases of the general theory.

It is a great pleasure to express my gratitude to the many friends who have encouraged or aided me in the task of preparing this book. I wish, above all, to thank Professor J. D. Tamarkin of Brown University, whose interest fostered the project from its inception and whose patient criticism, freely and unselfishly given, has guided it to maturity. More formal thanks are owed to the administrators of the Milton Fund of Harvard University, who, during my connection with the University, granted moneys for the preparation of the manuscript; and to the Committee on Colloquium Publications of the American Mathematical Society, who have generously honored my work by applying to it the policy of publishing material which has not been presented in the form of Colloquium Lectures before the Society.

New Haven, Connecticut, M. H. Stone.
April, 1932.

CONTENTS

CHAPTER I

ABSTRACT HILBERT SPACE AND ITS REALIZATIONS

§ 1. The Concept of Space

The word "space" has gradually acquired a mathematical significance so broad that it is virtually equivalent to the word "class", as used in logic. Historically, the reason for this development is to be found in the recognition that many classes which are of special importance in mathematics enjoy internal properties analogous to the familiar ones of Euclidean space. For example, the class of all continuous real functions of a real variable defined on a given closed interval can be treated as a metric space, the distance between two functions of the class being the maximum numerical value of their difference. Another class or space of peculiar importance for analysis is the class of all real functions of a real variable on a given closed interval which are Lebesgue-measurable and have Lebesgue-integrable squares; in this space, the distance between two elements or points of the space may be defined as the square root of the integral of the square of their difference. In each of these two cases, the distance between two elements or points of the space has many of the properties of distance in Euclidean geometry. Spaces such as those just described are frequently referred to as "function-spaces" or "spaces of infinitely many dimensions". The systematic study of the possible internal relationships with which a general class or space may be endowed and of the various types of space characterized by such internal properties was begun by Fréchet and Hausdorff; to the efforts of these mathematicians and their numerous followers we

1

now owe an independent and fertile mathematical discipline.*
The geometrical point of view which pervades and illuminates these researches will be adopted in the treatment of topics considered in the present work.

Of the great variety of general spaces, we shall be concerned primarily with a single one, abstract Hilbert space. If we were to introduce Hilbert space by successive generalizations from real n-dimensional Euclidean space, we should proceed as follows: we should first extend Euclidean space to the complex domain, obtaining a type of space known as n-dimensional unitary space; and we should then generalize the unitary spaces so as to reach an analogous type of space with a denumerable infinity of dimensions. Since we are interested almost exclusively in Hilbert space in these pages, we shall adopt a quite different course. We shall proceed directly to the definition of Hilbert space, reserving such comments as we need to make concerning unitary spaces until a little later. The abstract characterization of Hilbert space and the discussion of its realizations constitute the subject of this chapter.

§ 2. Abstract Hilbert Space

We shall define Hilbert space as a collection of elements of entirely unspecified nature which satisfies the five postulates given below. Particular examples of spaces satisfying these postulates were first introduced and systematically studied by Hilbert, whose name has since been attached quite appropriately to all such spaces. The postulational treatment was carried through only recently by J. v. Neumann.†

---

* Fréchet, Rendiconti di Palermo, 22 (1906), pp. 1–74. — Hausdorff, *Grundzüge der Mengenlehre*, first edition, Leipzig, 1914. — Tychonoff and Vedenisoff, Bulletin des Sciences Mathématiques, (2) 50 (1926), pp. 15–27. — Fréchet, *Les espaces abstraits,* Paris, 1928.

† Hilbert, *Grundzüge einer allgemeinen Theorie der linearen Integralgleichungen,* Leipzig, 1912. This material appeared in six articles in the Göttingen Nachrichten, 1904–10. — J. v. Neumann, Göttinger Nachrichten, 1927, pp. 14–18; Mathematische Annalen, 102 (1929), pp. 64–66.

DEFINITION 1.1. *A class $\mathfrak{H}$ of elements $f, g, \cdots$ is called a Hilbert space if it satisfies the following postulates:*

POSTULATE A. *$\mathfrak{H}$ is a linear space; that is,*

(1) *there exists a commutative and associative operation, denoted by $+$, applicable to every pair f, g of elements of $\mathfrak{H}$, with the property that $f+g$ is also an element of $\mathfrak{H}$;*

(2) *there exists a distributive and associative operation, denoted by $\cdot$, applicable to every pair (a, f), where a is a complex number and f is an element of $\mathfrak{H}$, with the properties that $1 \cdot f = f$ and that $a \cdot f$ is an element of $\mathfrak{H}$;*

(3) *in $\mathfrak{H}$ there exists a null element, denoted by 0, with the properties*

$$f+0 = f, \qquad a \cdot 0 = 0, \qquad 0 \cdot f = 0.$$

POSTULATE B. *There exists a numerically-valued function (f, g) defined for every pair f, g of elements of $\mathfrak{H}$, with the properties:*

(1) $(af, g) = a(f, g)$,

(2) $(f_1 + f_2, g) = (f_1, g) + (f_2, g)$,

(3) $(g, f) = \overline{(f, g)}$,

(4) $(f, f) \geqq 0$,

(5) $(f, f) = 0$ *if and only if* $f = 0$.

The not-negative real number $(f, f)^{1/2}$ will be denoted for convenience by $\| f \|$.

POSTULATE C. *For every n, $n = 1, 2, 3, \cdots$, there exists a set of n linearly independent elements of $\mathfrak{H}$; that is, elements $f_1, \cdots, f_n$ such that the equation $a_1 f_1 + \cdots + a_n f_n = 0$ is true only when $a_1 = \cdots = a_n = 0$.*

POSTULATE D. *$\mathfrak{H}$ is separable; that is, there exists a denumerably infinite set of elements of $\mathfrak{H}$, $f_1, f_2, f_3, \cdots$, such that for every g in $\mathfrak{H}$ and every positive ε there exists an $n = n(g, \varepsilon)$ for which $\| f_n - g \| < \varepsilon$.*

POSTULATE E. *$\mathfrak{H}$ is complete; that is, if a sequence $\{f_n\}$ of elements of $\mathfrak{H}$ satisfies the condition*

$$\| f_m - f_n \| \to 0, \qquad m, n \to \infty,$$

then there exists an element f of $\mathfrak{H}$ such that

$$\|f - f_n\| \to 0, \qquad n \to \infty.$$

In order that our statement of the five postulates might be as simple and direct as possible, we have left certain notations unexplained, with the intention of adding the necessary comments here. On the basis of the operations $+$ and $\cdot$ described in Postulate A we can introduce a third operation—by means of the definitions $-f = (-1) \cdot f$, $f - g = f + (-g)$; it is easily verified that these operations have the formal properties of addition, scalar multiplication, and subtraction, respectively, in ordinary vector algebra. In the case of the operation $\cdot$ we find it convenient in most instances to write af in place of $a \cdot f$. For the null element introduced in Postulate A we have employed the familiar symbol 0. The danger of confusing this meaning of the symbol with its ordinary meaning as the notation for the zero of the real or complex number system is slight if not wholly illusory: for in any formula or equation it is evident which symbols are to represent numbers and which symbols are to represent elements of Hilbert space. Accordingly, we have avoided the real disadvantages of employing a special symbol for the null element. In Postulate B (3) we have used a dash to indicate the conjugate of a complex number: $\bar{a}$ is the conjugate of the complex number a. This practice will be continued. In this connection, we point out that we shall consistently use the standard notations $\mathfrak{R}\,a$ and $\mathfrak{I}\,a$ to indicate the real numbers which are respectively the real and imaginary parts of the complex number a. In Postulates D and E the notation — has the significance explained above.

We turn now to the derivation of certain elementary properties of the functions (f, g) and $\|f\|$ of Postulate B.

THEOREM 1.1. *The function (f, g) has the properties:*

(1) $(f, ag) = \bar{a}(f. g)$;

(2) $(f, g_1 + g_2) = (f, g_1) + (f, g_2)$;

(3) $\|af\| = |a| \cdot \|f\|$;

(4) $|(f, g)| \leqq \|f\| \cdot \|g\|$, *the equality sign holding if and only if f and g are linearly dependent (Schwarz's inequality)*;

(5) $\|f_1 + \cdots + f_n\| \leqq \|f_1\| + \cdots + \|f_n\|$, $n = 2, 3, 4, \cdots$.

Thus (f, g) may be described as a bilinear Hermitean symmetric form.

Properties (1) and (2) follow without difficulty from the first three postulated properties of (f, g); the third is easily established by means of the first property of Postulate B and the first of the present theorem.

In order to discuss the inequality (4), we define a complex number θ of unit modulus by means of the relations

$$\theta = (f, g)/|(f, g)| \text{ when } (f, g) \neq 0; \ \theta = 1 \text{ when } (f, g) = 0.$$

For an arbitrary pair of elements f, g and arbitrary real numbers a and b, we have

$$0 \leqq \|a\bar{\theta}f + bg\|^2 = a^2\|f\|^2 + 2ab|(f, g)| + b^2\|g\|^2.$$

Since the expression on the right is a not-negative definite quadratic form in a and b, its coefficients must satisfy the inequality $4|(f, g)|^2 - 4\|f\|^2\|g\|^2 \leqq 0$, which is equivalent to (4). In this inequality the equality sign can hold if and only if there exist values a and b, not both zero, such that the corresponding value of the quadratic form is zero; by Postulate B (5), the form vanishes if and only if $a\bar{\theta}f + bg = 0$.

For the case of two elements f and g, (5) takes the form $\|f+g\| \leqq \|f\| + \|g\|$, an inequality which follows immediately from the relations

$$\|f+g\|^2 = \|f\|^2 + 2|(f, g)| + \|g\|^2 \leqq \|f\|^2 + 2\|f\|\|g\| + \|g\|^2.$$

The general case of n elements is then treated by means of mathematical induction.

THEOREM 1.2. *The function $\|f - g\|$ has the properties of a distance between the elements f and g:*

(1) $\|f - g\| \geqq 0$;

(2) $\|f - g\| = 0$ *if and only if $f = g$*;

(3) $\|f-g\| = \|g-f\|$;

(4) $\|f-h\| \leqq \|f-g\| + \|g-h\|$ (*the triangle inequality*).

The space $\mathfrak{H}$ *can therefore be regarded as a metric space with the distance function* $\|f-g\|$. *This function has the further properties, characteristic of distance in a linear space*:

(5) $\|(f+h)-(g+h)\| = \|f-g\|$;

(6) $\|af-ag\| = |a|\,\|f-g\|$.

The assertions of the theorem are immediate consequences of the previously established properties of the function (f, g).

THEOREM 1.3. *The function* (f, g) *is a continuous function of its arguments in the sense that when* f_0, g_0, *and a positive number* ε *are given there exists a positive* $\delta = \delta(f_0, g_0, \varepsilon)$ *such that* $|(f, g)-(f_0, g_0)| < \varepsilon$ *for all* f *and* g *such that* $\|f-f_0\| < \delta$, $\|g-g_0\| < \delta$.

From the relations

$$(f, g)-(f_0, g_0) = (f-f_0, g)+(f_0, g-g_0),$$
$$\|g\| \leqq \|g_0\| + \|g-g_0\|$$

we see that

$$\begin{aligned}|(f, g)-(f_0, g_0)| &\leqq |(f-f_0, g)| + |(f_0, g-g_0)| \\ &\leqq \|f-f_0\|\,\|g\| + \|g-g_0\|\,\|f_0\| \\ &\leqq \|f-f_0\|\,\|g_0\| + \|g-g_0\|\,\|f_0\| + \|f-f_0\|\,\|g-g_0\|.\end{aligned}$$

Thus any positive number δ such that $\delta(\|f_0\| + \|g_0\|) + \delta^2 < \varepsilon$ has the requisite properties.

DEFINITION 1.2. *A sequence of elements* $\{f_n\}$ *in* $\mathfrak{H}$ *is said to converge if* $\|f_m - f_n\| \to 0$, *when* $m, n \to \infty$; *and is said to have a limit* f *in* $\mathfrak{H}$ *if* $\|f-f_n\| \to 0$ *when* $n \to \infty$.

Because of the fourth property ascribed to the distance function by Theorem 1.2, a sequence which has a limit is convergent. The function of Postulate E is to guarantee the converse of this statement: every convergent sequence has a limit in $\mathfrak{H}$. We shall write $f_n \to f$, $n \to \infty$, when we wish to express symbolically the convergence of the sequence $\{f_n\}$ to the limit f. We note that when $f_n \to f$ and $g_n \to g$, then $f_n + g_n \to f+g$, $af_n \to af$, and $(f_n, g_n) \to (f, g)$. With the meaning given to the term "limit" by this definition, it is possible to apply to sets of elements in $\mathfrak{H}$ the common language of point set theory, and refer to open sets, closed sets, derived sets, and so forth. For example, Postulate D asserts the existence of a enumerably infinite subset everywhere dense in $\mathfrak{H}$.

DEFINITION 1.3. *A subset $\mathfrak{M}$ of $\mathfrak{H}$ is said to be a linear manifold if, whenever $f_1, \cdots, f_n$ belong to $\mathfrak{M}$, $a_1 f_1 + \cdots + a_n f_n$ also belongs to $\mathfrak{M}$.*

DEFINITION 1.4. *The smallest linear manifold $\mathfrak{M}_1$ and the smallest closed linear manifold $\mathfrak{M}_2$ which contain a given subset $\mathfrak{S}$ of $\mathfrak{H}$ are called respectively the linear manifold and the closed linear manifold determined by $\mathfrak{S}$.*

$\mathfrak{M}_1$ is the class of all elements of $\mathfrak{H}$ which are expressible in the form $a_1 f_1 + \cdots + a_n f_n$ where $f_1, \cdots, f_n$ are elements of $\mathfrak{S}$; and $\mathfrak{M}_2$ is the derived set of $\mathfrak{M}_1$. It is apparent that $\mathfrak{M}_1$ is a subset of $\mathfrak{M}_2$.

DEFINITION 1.5. *Two elements f, g of $\mathfrak{H}$ are said to be orthogonal if (f, g) vanishes; two linear manifolds in $\mathfrak{H}$ are said to be orthogonal if every element of one is orthogonal to every element of the other.*

From a geometrical point of view, the concept of orthogonality is the analogue of the familiar concept of perpendicularity in Euclidean geometry.

DEFINITION 1.6. *A subset $\mathfrak{S}$ of $\mathfrak{H}$ is said to be an orthonormal set if, when f and g are elements of $\mathfrak{S}$,*

$$(f, g) = \begin{Bmatrix} 1, & f = g \\ 0, & f \neq g \end{Bmatrix}.$$

An orthonormal set is said to be complete if there exists no orthonormal set of which it is a proper subset. Two orthonormal sets in $\mathfrak{H}$ are said to be equivalent if they determine the same closed linear manifold.

THEOREM 1.4. *An orthonormal set $\mathfrak{S}$ is either finite or denumerably infinite.*

By Postulate D, there exists a denumerably infinite set of elements $\{f_n\}$ everywhere dense in $\mathfrak{H}$. Consequently it is possible to determine for each f in $\mathfrak{S}$ a number $k = k(f)$ such that $\|f - f_k\| \leq \frac{1}{4}\sqrt{2}$; in other words, there is a correspondence between $\mathfrak{S}$ and a subset of $\{f_n\}$. This correspondence must be a one-to-one correspondence; for, if f and g are elements of $\mathfrak{S}$ and f_i and f_k are their respective correspondents, we have, by Theorems 1.1 and 1.2,

$$\|f - g\| = (f - g, f - g)^{1/2} = \sqrt{2},$$
$$\|f - f_i\| + \|f_i - f_k\| + \|f_k - g\| \geqq \|f - g\|,$$
$$\|f_i - f_k\| \geqq \tfrac{1}{2}\sqrt{2},$$

so that f_i and f_k are distinct elements. Thus $\mathfrak{S}$ is either finite or denumerably infinite, since it is in one-to-one correspondence with a subset of $\{f_n$.

As a result of this theorem we see that every orthonormal set is a finite or infinite sequence such that, when its elements are denoted by $\varphi_1, \varphi_2, \varphi_3, \cdots,$

$$(\varphi_i, \varphi_k) = \delta_{ik} = \begin{Bmatrix} 0, & i \neq k \\ 1, & i = k \end{Bmatrix}.$$

In order to employ symbolism and language uniformly applicable to finite and to infinite orthonormal sets, we shall introduce the following convention: if we are dealing with an orthonormal set of n elements, sums such as $\sum_{\alpha=1}^{N} a_\alpha \varphi_\alpha$ and $\sum_{\alpha=1}^{N} a_\alpha \bar{b}_\alpha$ are to be interpreted as sums over the values $\alpha = 1, \cdots, n$ only, the terms for $\alpha > n$ being regarded as absent. It must be remembered that the sign $\sum_{\alpha=1}^{\infty}$, as used in connection with such sums, is an abbreviation for the symbol $\lim_{N \to \infty} \sum_{\alpha=1}^{N}$.

THEOREM 1.5. *If $\{\varphi_n\}$ is an orthonormal set, then the series $\sum_{\alpha=1}^{\infty} a_\alpha \bar{b}_\alpha$, in which $a_n = (f, \varphi_n)$ and $b_n = (g, \varphi_n)$ converges absolutely for every f and g in $\mathfrak{H}$. When $f = g$, we have Bessel's inequality, $\sum_{\alpha=1}^{\infty} |a_\alpha|^2 \leq \|f\|^2$.*

We have

$$\left\| f - \sum_{\alpha=1}^{n} a_\alpha \varphi_\alpha \right\|^2$$

$$= \left(f - \sum_{\alpha=1}^{n} a_\alpha \varphi_\alpha, f - \sum_{\beta=1}^{n} a_\beta \varphi_\beta \right)$$

$$= (f, f) - \sum_{\beta=1}^{n} (f, a_\beta \varphi_\beta) - \sum_{\alpha=1}^{n} (a_\alpha \varphi_\alpha, f) + \sum_{\alpha,\beta=1}^{n} (a_\alpha \varphi_\alpha, a_\beta \varphi_\beta)$$

$$= \|f\|^2 - \sum_{\beta=1}^{n} \bar{a}_\beta (f, \varphi_\beta) - \sum_{\alpha=1}^{n} a_\alpha (\varphi_\alpha, f) + \sum_{\alpha,\beta=1}^{n} a_\alpha \bar{a}_\beta (\varphi_\alpha, \varphi_\beta)$$

$$= \|f\|^2 - \sum_{\alpha=1}^{n} |a_\alpha|^2 \geqq 0.$$

Hence the series $\sum_{\alpha=1}^{\infty} |a_\alpha|^2$ converges and has a sum which cannot exceed $\|f\|^2$. From the familiar inequality

$$|a_n \bar{b}_n| \leqq \frac{1}{2}(|a_n|^2 + |b_n|^2)$$

we conclude that the series $\sum_{\alpha=1}^{\infty} a_\alpha \bar{b}_\alpha$ is absolutely convergent.

THEOREM 1.6. *When $\{\varphi_n\}$ is an orthonormal set, the sequence $\{f_n\} = \left\{ \sum_{\alpha=1}^{n} a_\alpha \varphi_\alpha \right\}$ converges if and only if the series $\sum_{\alpha=1}^{\infty} |a_\alpha|^2$ converges. If $f_n \to f$, then $a_n = (f, \varphi_n)$.*

For $m > n$, we have

$$\|f_m - f_n\|^2 = \left| \sum_{\alpha=n+1}^{m} a_\alpha \varphi_\alpha \right|^2 = \sum_{\alpha=n+1}^{m} |a_\alpha|^2.$$

Thus, in accordance with Definition 1.2, a necessary and sufficient condition for the convergence of the sequence $\{f_n\}$ is that the last expression have the limit zero when m and n become infinite; in other words, that the series $\sum_{\alpha=1}^{\infty} |a_\alpha|^2$ converge. If the sequence $\{f_n\}$ converges, it has the limit f in $\mathfrak{H}$; and by the continuity of (f, g) we have for $n > k$

$$a_k = (f_n, \varphi_k) \to (f, \varphi_k), \qquad n \to \infty.$$

THEOREM 1.7. *If $\{\varphi_n\}$ is an orthonormal set, then for every f in $\mathfrak{H}$ the series*

$$\sum_{\alpha=1}^{\infty} a_\alpha \varphi_\alpha, \qquad a_n = (f, \varphi_n),$$

converges to a limit f^* *and* $f - f^*$ *is orthogonal to every element of the set* $\{\varphi_n\}$.

The existence of the limit f^* follows at once from Theorems 1.5 and 1.6; furthermore

$$(f - f^*, \varphi_n) = (f, \varphi_n) - (f^*, \varphi_n) = 0$$

by Theorem 1.6.

THEOREM 1.8. *A necessary and sufficient condition that an element* f *of* $\mathfrak{H}$ *belong to the closed linear manifold* $\mathfrak{M}$ *determined by an orthonormal set* $\{\varphi_n\}$ *is that*

$$f = \sum_{\alpha=1}^{\infty} a_\alpha \varphi_\alpha, \qquad a_n = (f, \varphi_n).$$

The condition is obviously sufficient. In order to prove that it is also necessary we determine the element $f^* = \sum_{\alpha=1}^{\infty} a_\alpha \varphi_\alpha$, $a_n = (f, \varphi_n)$, in accordance with Theorem 1.7, f being an element of $\mathfrak{M}$. Evidently $g = f - f^*$ is an element of $\mathfrak{M}$; it is orthogonal to every element of $\{\varphi_n\}$. We can conclude, therefore, that g is orthogonal to every element of the linear manifold determined by $\{\varphi_n\}$ and to every element of the closed linear manifold $\mathfrak{M}$ determined by $\{\varphi_n\}$, since we know that the function (f, g) is linear and continuous. In particular, g is orthogonal to itself, $\|g\|^2 = 0$, so that $g = 0$ and $f = f^*$, as we wished to show.

THEOREM 1.9. *The five following assertions concerning the orthonormal set* $\{\varphi_n\}$ *are equivalent*:

(1) $\{\varphi_n\}$ *is complete*;

(2) $(f, \varphi_n) = 0$ *for every* n *implies* $f = 0$;

(3) *the closed linear manifold determined by* $\{\varphi_n\}$ *is* $\mathfrak{H}$;

(4) *for every* f *in* $\mathfrak{H}$, $f = \sum_{\alpha=1}^{\infty} a_\alpha \varphi_\alpha$, $a_n = (f, \varphi_n)$;

(5) *for every pair* f, g *in* $\mathfrak{H}$, *the Parseval identity*

$$(f, g) = \sum_{\alpha=1}^{\infty} a_\alpha \bar{b}_\alpha, \quad a_n = (f, \varphi_n), \quad b_n = (g, \varphi_n),$$

is true.

We shall show that the following inferences are possible:

$$1 \to 2 \to 3 \to 4 \to 5 \to 1,$$

each arrow being directed from hypothesis to conclusion. The equivalence of the five assertions is then obvious.

If $\{\varphi_n\}$ is complete, then the only element which is orthogonal to every φ_n is the null element; for otherwise there would exist an $f \neq 0$ orthogonal to every φ_n, and the element $\varphi_0 = f/\|f\|$ would form with $\{\varphi_n\}$ an orthonormal set including $\{\varphi_n\}$ as a proper subset.

If $(f, \varphi_n) = 0$ for every n implies $f = 0$, then the closed linear manifold determined by $\{\varphi_n\}$ is $\mathfrak{H}$. Let there exist an f not in $\mathfrak{M}$, the closed linear manifold in question. Then, by Theorems 1.7 and 1.8, the element $f^* = \sum_{\alpha=1}^{\infty} a_\alpha \varphi_\alpha$, $a_n = (f, \varphi_n)$ exists and belongs to $\mathfrak{M}$. Now $g = f - f^*$ is orthogonal to every φ_n and must therefore be the null element. Hence f coincides with f^* and belongs to $\mathfrak{M}$, contrary to hypothesis. To avoid this contradiction, we must have $\mathfrak{M} = \mathfrak{H}$.

If the closed linear manifold determined by $\{\varphi_n\}$ is $\mathfrak{H}$, then for every f in $\mathfrak{H}$

$$f = \sum_{\alpha=1}^{\infty} a_\alpha \varphi_\alpha, \qquad a_n = (f, \varphi_n),$$

by virtue of Theorem 1.8.

If the relations of (4) are satisfied, then

$$\sum_{\alpha=1}^{n} a_\alpha \varphi_\alpha = f_n \to f, \qquad \sum_{\alpha=1}^{n} b_\alpha \varphi_\alpha = g_n \to g$$

and, in consequence,

$$(f, g) = \lim_{n\to\infty} (f_n, g_n) = \lim_{n\to\infty} \sum_{\alpha=1}^{n} a_\alpha \bar{b}_\alpha = \sum_{\alpha=1}^{\infty} a_\alpha \bar{b}_\alpha,$$

as we wished to show.

If the relations of (5) are satisfied, then the set $\{\varphi_n\}$ must be complete. If $\{\varphi_n\}$ were not complete, there would exist an orthonormal set of which it is a proper subset; in particular, there would exist an element φ_0 orthogonal to every φ_n, with $\|\varphi_0\| = 1$. In the equations of (5) we put $f = g = \varphi_0$ and immediately obtain the contradictory result $\|\varphi_0\|^2 = 0$. Thus $\{\varphi_n\}$ must be a complete orthonormal set.

As an immediate consequence of this theorem and Postulate C, we have

THEOREM 1.10. *A complete orthonormal set is denumerably infinite.*

We know that the set is either finite or denumerably infinite. We can now exclude the possibility that it is finite. If it were finite, the various series encountered in the preceding theorem would reduce to finite sums, the theorems themselves still being true; in particular, every element of $\mathfrak{H}$ would be expressible in the form

$$f = a_1 \varphi_1 + \cdots + a_N \varphi_N,$$

by Theorem 1.9 (4); since N is fixed and independent of f this result is in contradiction with Postulate C.

THEOREM 1.11. *In order that two orthonormal sets $\{\varphi_n\}$ and $\{\psi_n\}$ in $\mathfrak{H}$ be equivalent, it is necessary and sufficient that*

$$\varphi_k = \sum_{\alpha=1}^{\infty} a_{k\alpha} \psi_\alpha, \qquad \psi_k = \sum_{\alpha=1}^{\infty} \bar{a}_{\alpha k} \varphi_\alpha$$

where $a_{kl} = (\varphi_k, \psi_l)$.

The conditions are easily seen to be necessary, in view of Theorem 1.8. If the conditions are satisfied, then $\{\varphi_n\}$ is a subset of the closed linear manifold determined by $\{\psi_n\}$, and *vice versa,* again by Theorem 1.8. By Definition 1.4, it is clear that $\{\varphi_n\}$ and $\{\psi_n\}$ determine the same closed linear manifold, and are therefore equivalent.

THEOREM 1.12. *Two complete orthonormal sets in $\mathfrak{H}$ are equivalent; two equivalent orthonormal sets in $\mathfrak{H}$ are both complete or both incomplete.*

This theorem is an obvious consequence of Theorem 1.9 (3).

THEOREM 1.13. *If $\{f_n\}$ is an arbitrary sequence of elements of $\mathfrak{H}$, then there exists an orthonormal set $\{\varphi_n\}$ which determines the same linear manifold and the same closed linear manifold as the set $\{f_n\}$, except in the case that $f_n = 0$ for every n.*

We first replace $\{f_n\}$ by a subsequence of linearly independent elements, excluding the case in which every f_n is the null element. We let f_1^* be the first element of the sequence which differs from the null element; if we have already determined $f_1^*, \cdots, f_k^*$ as elements of the sequence which are linearly independent, we select f_{k+1}^* as the first element f_n which is not a linear combination of $f_1^*, \cdots, f_k^*$. According to these rules we obtain a finite or denumerably infinite set $\{f_n^*\}$ which evidently determines the same linear manifold and, therefore, the same closed linear manifold as $\{f_n\}$.

It is now possible to apply to the set $\{f_n^*\}$ a process of orthogonalization: we define the sets $\{g_n\}$ and $\{\varphi_n\}$ step by step according to the equations

$$g_1 = f_1^*, \qquad \varphi_1 = g_1/|g_1|,$$

$$g_{k+1} = f_{k+1}^* - \sum_{\alpha=1}^{k} (f_{k+1}^*, \varphi_\alpha)\varphi_\alpha, \qquad \varphi_{k+1} = g_{k+1}/|g_{k+1}|.$$

The process indicated comes to an end if and only if the set $\{f_n^*\}$ is finite. When $g_1, \cdots, g_k, \varphi_1, \cdots, \varphi_k$ are defined, g_{k+1} is defined unless the sequence $\{f_n^*\}$ breaks off with f_k^*. When $g_1, \cdots, g_{k+1}, \varphi_1, \cdots, \varphi_k$ are defined, φ_{k+1} is defined unless $g_{k+1} = 0$. The latter circumstance can arise if and only if f_{k+1}^* is linearly dependent upon $\varphi_1, \cdots, \varphi_k$ and hence also upon $f_1^*, \cdots, f_k^*$; but our choice of the sequence $\{f_n^*\}$ forbids the occurrence of this situation. It is easily seen that $\{\varphi_n\}$ is an orthonormal set which determines the same linear manifold and, therefore, the same closed linear manifold as the given sequence $\{f_n\}$.†

THEOREM 1.14. *There exists a complete orthonormal set in* $\mathfrak{H}$.

By Postulate D, there exists a sequence $\{f_n\}$ which is everywhere dense in $\mathfrak{H}$ and which therefore determines the closed linear manifold $\mathfrak{H}$. We replace this sequence by an orthonormal set enjoying the second of these properties, by

† The process of orthogonalization is due to J. P. Gram, Journal für Mathematik, 94 (1883), pp. 41–73; and E. Schmidt, Mathematische Annalen, 63 (1907), p. 442.

the process described in the preceding theorem. This orthonormal set is complete by Theorem 1.9 (3).

The series of theorems which culminates in the present existence theorem enables us to describe $\mathfrak{H}$ as a space of infinitely many dimensions. Indeed, we can say with greater precision that $\mathfrak{H}$ has the dimension number $\aleph_0$: for there exists in $\mathfrak{H}$ a denumerably infinite (orthonormal) set such that every element in $\mathfrak{H}$ is uniquely expressible in the form $f = \sum_{\alpha=1}^{\infty} a_\alpha \varphi_\alpha$, $a_n = (f, \varphi_n)$.

THEOREM 1.15. *Let $\mathfrak{H}_0$ denote the class of all sequences of complex numbers $(x_1, x_2, x_3, \cdots)$ such that $\sum_{\alpha=1}^{\infty} |x_\alpha|^2$ converges, with the operations $+$ and $-$, the null element, and the function (f, g) defined as follows, for $f = (x_1, x_2, x_3, \cdots)$ and $g = (y_1, y_2, y_3; \cdots)$:*

$$\begin{aligned} f+g &= (x_1+y_1,\ x_2+y_2,\ x_3+y_3,\ \cdots), \\ a\cdot f &= (ax_1,\ ax_2,\ ax_3,\ \cdots), \\ 0 &= (0,\ 0,\ 0,\ \cdots), \\ (f, g) &= \sum_{\alpha=1}^{\infty} x_\alpha \bar{y}_\alpha. \end{aligned}$$

Then $\mathfrak{H}_0$ is a Hilbert space.

Because of the inequalities

$$\begin{aligned} |x_n + y_n|^2 &\leqq |x_n|^2 + 2|x_n y_n| + |y_n|^2, \\ |x_n \bar{y}_n| &\leqq \tfrac{1}{2}(|x_n|^2 + |y_n|^2), \end{aligned}$$

we see that, when f and g are elements of $\mathfrak{H}_0$, $f+g$ and $a\cdot f$ are also elements of $\mathfrak{H}_0$ and (f, g) is defined. The various properties attributed to the operations $+$ and $-$ and to the function (f, g) by Postulates A and B are easily verified.

The space $\mathfrak{H}_0$ satisfies Postulate C since it contains the linearly independent elements $(\delta_{1n}, \delta_{2n}, \delta_{3n}, \cdots)$, $n = 1, 2, 3, \cdots$. It also satisfies Postulate D, for the class of all finite sequences of complex numbers with rational real and imaginary parts is denumerably infinite and determines a set everywhere dense in $\mathfrak{H}_0$: if $f = (x_1, x_2, x_3, \cdots)$ and a positive ϵ are given, we can determine a positive integer N and complex numbers

$r_1, \cdots, r_N$ with rational real and imaginary parts such that if $g = (r_1, \cdots, r_N, 0, 0, 0, \cdots)$

$$\|f-g\|^2 = \sum_{\alpha=1}^{N} |x_\alpha - r_\alpha|^2 + \sum_{\alpha=N+1}^{\infty} |x_\alpha|^2 < \epsilon^2.$$

Finally, Postulate E is true for $\mathfrak{H}_0$. If $f_n = (x_1^{(n)}, x_2^{(n)}, x_3^{(n)}, \cdots)$ is a sequence such that $\|f_m - f_n\| \to 0$ when $m, n \to \infty$, then there exists for every positive ϵ an $N = N(\epsilon)$ such that

$$\sum_{\alpha=1}^{k} |x_\alpha^{(m)} - x_\alpha^{(n)}|^2 \leqq \sum_{\alpha=1}^{\infty} |x_\alpha^{(m)} - x_\alpha^{(n)}|^2 = \|f_m - f_n\|^2 < \epsilon^2, \quad m, n > N(\epsilon)$$

for every k. Consequently, $x_l^{(n)} \to x_l$ when n becomes infinite, for every l. In the inequality above we now allow m to become infinite obtaining

$$\sum_{\alpha=1}^{k} |x_\alpha - x_\alpha^{(n)}|^2 \leqq \epsilon^2, \quad n > N(\epsilon).$$

Thus the element $g_n = (x_1 - x_1^{(n)}, x_2 - x_2^{(n)}, x_3 - x_3^{(n)}, \cdots)$ is found to be an element of $\mathfrak{H}_0$, since in the inequality just obtained we can allow k to become infinite, with the result that $\sum_{\alpha=1}^{\infty} |x_\alpha - x_\alpha^{(n)}|^2$ is convergent and does not exceed ϵ^2 when $n > N(\epsilon)$. Since f_n and g_n both belong to $\mathfrak{H}_0$ their sum $f = (x_1, x_2, x_3, \cdots)$ also belongs to $\mathfrak{H}_0$. Lastly we see that when n becomes infinite $\|f - f_n\|^2 = \sum_{\alpha=1}^{\infty} |x_\alpha - x_\alpha^{(n)}|^2 \to 0$.

THEOREM 1.16. *If $\mathfrak{H}$ is a Hilbert space, it can be put into one-to-one correspondence with $\mathfrak{H}_0$ in such a manner that if the correspondence be indicated by*

$$f \sim (x_1, x_2, x_3, \cdots), \qquad g \sim (y_1, y_2, y_3, \cdots),$$

the relations

$$f + g \sim (x_1 + y_1, x_2 + y_2, x_3 + y_3, \cdots),$$
$$a \cdot f \sim (a x_1, a x_2, a x_3, \cdots),$$
$$0 \sim (0, 0, 0, \cdots),$$
$$(f, g) = \sum_{\alpha=1}^{\infty} x_\alpha \bar{y}_\alpha,$$

are true. This correspondence determines a system of coördinates in $\mathfrak{H}$.

We first determine a complete orthonormal set $\{\varphi_n\}$ in $\mathfrak{H}$; the correspondence between $\mathfrak{H}$ and $\mathfrak{H}_0$ is then defined to be

$$f \sim (x_1, x_2, x_3, \cdots), \qquad x_n = (f, \varphi_n).$$

Theorem 1.5 shows that to every f in $\mathfrak{H}$ corresponds a unique element $(x_1, x_2, x_3, \cdots)$ in $\mathfrak{H}_0$, Theorem 1.6 that every element $(x_1, x_2, x_3, \cdots)$ is the correspondent of at least one element in $\mathfrak{H}$. The correspondence is a one-to-one correspondence since from $(f, \varphi_n) = (g, \varphi_n)$ for every n we can conclude that

$$(f-g, \varphi_n) = 0, \qquad f-g = 0, \qquad f = g,$$

by the completeness of the set $\{\varphi_n\}$. Of the four properties of this correspondence between $\mathfrak{H}$ and $\mathfrak{H}_0$ enumerated in the theorem, the first three are evident and the fourth results from Theorem 1.9 (5).

We find in Theorem 1.15 a proof of the consistency of the system of postulates used to define abstract Hilbert space. Theorem 1.16 shows that all Hilbert spaces are formally identical, differing only in the significance which may be attributed to their elements.

§ 3. Abstract Unitary Spaces

In order to display more clearly the analogy between Hilbert space and ordinary Euclidean space, we describe briefly an intermediate type of space—the n-dimensional unitary space already mentioned at the end of § 1. Subsequently, we shall find it convenient to refer to the concepts and terminology which we introduce here.

Definition 1.7. *A class $\mathfrak{U}_n$ of elements $f, g, \cdots$ is called an n-dimensional unitary space, $n = 0, 1, 2, \cdots$ if it satisfies the following postulates:*

Postulate A *of Definition* 1.1;

Postulate B *of Definition* 1.1;

Postulate F. *The maximum number of linearly independent elements in $\mathfrak{U}_n$ is n; that is, there exist n elements $f_1, \cdots, f_n$ in $\mathfrak{U}_n$ such that the equation $a_1 f_1 + \cdots + a_n f_n = 0$ implies*

$a_1 = \cdots = a_n = 0$, *while if* $g_1, \cdots, g_{n+1}$ *are arbitrary elements of* $\mathfrak{U}_n$ *there exist complex numbers* $a_1, \cdots, a_{n+1}$ *not all zero such that* $a_1 g_1 + \cdots + a_{n+1} g_{n+1} = 0$.

The Postulates A, B, F are essentially those which have been given by Weyl.* Some unessential modifications have been made. The replacement of the three Postulates C, D, E of Definition 1.1 by the single Postulate F effects a considerable simplification which becomes apparent in proving the analogues of Theorems 1.15 and 1.16 of the preceding section. We give without proof the following theorem.

THEOREM 1.17. *Let* $\mathfrak{V}_n$ *denote the class of all ordered sets of* n *complex numbers* $f = (x_1, \cdots, x_n)$, $n = 1, 2, 3, \cdots$, *with the operations* $+$ *and* $\cdot$ *and the function* (f, g) *defined as follows, for* $f = (x_1, \cdots, x_n)$ *and* $g = (y_1, \cdots, y_n)$:

$$\begin{aligned} f+g &= (x_1+y_1, \cdots, x_n+y_n), \\ a\cdot f &= (ax_1, \cdots, ax_n), \\ 0 &= (0, \cdots, 0), \\ (f, g) &= \sum_{\alpha=1}^{n} x_\alpha \bar{y}_\alpha. \end{aligned}$$

Then $\mathfrak{V}_n$ *is an n-dimensional unitary space. If* $\mathfrak{U}_n$ *is an n-dimensional unitary space,* $n = 1, 2, 3, \cdots$, *it can be put in one-to-one correspondence with* $\mathfrak{V}_n$ *in such a manner that, if the correspondence be indicated by*

$$f \sim (x_1, \cdots, x_n), \qquad g \sim (y_1, \cdots, y_n),$$

the relations

$$\begin{aligned} f+g &\sim (x_1+y_1, \cdots, x_n+y_n), \\ a\cdot f &\sim (ax_1, \cdots, ax_n), \\ 0 &\sim (0, \cdots, 0), \\ (f, g) &= \sum_{\alpha=1}^{n} x_\alpha \bar{y}_\alpha \end{aligned}$$

are satisfied. By means of this correspondence there is determined a coördinate system in $\mathfrak{U}_n$.

---

* Weyl, *Raum, Zeit, Materie*, fifth edition, Berlin, 1923, pp. 10–30; *Gruppentheorie und Quantenmechanik*, first edition, Leipzig, 1928, pp. 4–5, 15–19.

The reader will have no difficulty in supplying a proof of this theorem along the lines followed in establishing Theorems 1.15 and 1.16. It will be observed that Theorem 1.17 does not cover the case $n = 0$; but this case is trivial, in that the space $\mathfrak{U}_0$ contains a single element, the null element. For $n = 1, 2, 3, \cdots$, we see that there exists at least one n-dimensional unitary space and that all n-dimensional unitary spaces are essentially equivalent, differing only in the significance attributed to their elements. More precisely, we can say that the properties of an n-dimensional unitary space $\mathfrak{U}_n$ constitute a geometry formally identical with the familiar coördinate geometry in $\mathfrak{B}_n$.* It is evident that Hilbert space is the natural generalization of the spaces treated here to a denumerable infinity of dimensions.

§ 4. Linear Manifolds in Hilbert Space

Before examining various realizations of the abstract Hilbert space $\mathfrak{H}$, we shall devote a few paragraphs to the study of linear manifolds in $\mathfrak{H}$. Such manifolds play a fundamental rôle in the developments of later chapters.

We first demonstrate a theorem dealing with an arbitrary set of elements in $\mathfrak{H}$.

Theorem 1.18. *An arbitrary not-empty subset $\mathfrak{S}$ of $\mathfrak{H}$ is separable in the sense that it contains a finite or a denumerably infinite subset everywhere dense in $\mathfrak{S}$.*

We know that $\mathfrak{H}$ itself is separable, so that there exists a sequence $\{f_n\}$ everywhere dense in $\mathfrak{H}$. We consider the class $\mathfrak{K}$ of all pairs (m, n) of integers such that the inequality $\|g - f_m\| < 1/n$ is true for at least one element g in $\mathfrak{S}$. Obviously, $\mathfrak{K}$ is not empty and is at most denumerably infinite. To each pair (m, n) of this class we order an element g_{mn} of $\mathfrak{S}$ within distance $1/n$ of f_m. Now if g is an arbitrary element of $\mathfrak{S}$ and ε is an arbitrary positive number, there exists an element g_{mn} within distance ε of g: for we

* Weyl, *Raum, Zeit, Materie,* fifth edition, Berlin, 1923, pp. 10–30; *Gruppentheorie und Quantenmechanik,* first edition, Leipzig, 1928, Chapter I.

can select integers m and n such that $n > 2/\varepsilon$, $\|g - f_m\| < 1/n$; we then see (m, n) belongs to $\mathfrak{K}$ and that the corresponding element g_{mn} satisfies the relations

$$\|g - g_{mn}\| \leqq \|g - f_m\| + \|f_m - g_{mn}\| < 2/n < \varepsilon.$$

Thus the set $\{g_{mn}\}$ is a finite or a denumerably infinite set everywhere dense in $\mathfrak{S}$, as we wished to show.*

THEOREM 1.19. *A closed linear manifold* $\mathfrak{M}$ *in* $\mathfrak{H}$ *is either an n-dimensional unitary space, for some* $n = 0, 1, 2, \cdots$, *or a Hilbert space, with the dimension number* $\aleph_0$.

Because $\mathfrak{M}$ is a linear manifold it automatically satisfies Postulate A of Definitions 1.1 and 1.7; it also satisfies Postulate B of these definitions, because the function (f, g) is defined throughout $\mathfrak{H}$ and serves *a fortiori* in $\mathfrak{M}$.

If there is an $n = 0, 1, 2, \cdots$, such that Postulate F of Definition 1.7 is true in $\mathfrak{M}$, then $\mathfrak{M}$ is an n-dimensional unitary space; if no such n exists, then Postulate C of Definition 1·1 holds.

In the latter case, we verify at once that the remaining Postulates D and E are true: $\mathfrak{M}$ is separable, by the preceding theorem, and $\mathfrak{M}$ is closed by hypothesis. Thus $\mathfrak{M}$ is seen to be a Hilbert space.

For future reference, we may comment on a few points related to this result. The only closed linear manifold in $\mathfrak{H}$ which has the dimension number 0 is the set $\mathfrak{O}$ which contains the null element 0 as its sole member. A linear manifold which contains only a finite number of linearly independent elements is at the same time a closed linear manifold with a finite dimension number. The detailed discussion of these cases may be left to the reader.

THEOREM 1.20. *The intersection of any collection of linear manifolds (closed linear manifolds)—that is, the set of elements common to all manifolds of the collection—is a linear manifold (closed linear manifold).*

---

* Compare Hausdorff, *Grundzüge der Mengenlehre*, first edition, Leipzig, 1914, Chapter VIII, § 3, Theorem VIII, p. 273, where the analogous theorem is stated and proved for an arbitrary separable space.

The proof of this theorem is immediate.

THEOREM 1.21. *If* $\mathfrak{M}_1$ *and* $\mathfrak{M}_2$ *are closed linear manifolds such that* $\mathfrak{M}_1 \subseteqq \mathfrak{M}_2$, *then the set* $\mathfrak{M}_3$ *of all elements of* $\mathfrak{M}_2$ *which are orthogonal to* $\mathfrak{M}_1$ *is a closed linear manifold. We shall denote the relation between the three manifolds by the equation*

$$\mathfrak{M}_3 = \mathfrak{M}_2 \ominus \mathfrak{M}_1.$$

The formal proof is simple and may be omitted.

DEFINITION 1.8. *If* $\mathfrak{M}$ *is a closed linear manifold, the closed linear manifold* $\mathfrak{H} \ominus \mathfrak{M}$ *is called the orthogonal complement of* $\mathfrak{M}$.

THEOREM 1.22. *If* $\mathfrak{M}_1, \cdots, \mathfrak{M}_l$ *are mutually orthogonal closed linear manifolds, then the linear manifold determined by their sum* $\mathfrak{M}_1 + \cdots + \mathfrak{M}_l$ *is closed.*

A necessary and sufficient condition that f belong to $\mathfrak{M}$, the linear manifold determined by $\mathfrak{M}_1 + \cdots + \mathfrak{M}_l$, is that it be expressible in the form

$$f_1 + \cdots + f_l, \qquad f_k \text{ in } \mathfrak{M}_k.$$

Hence, if $\{f^{(n)}\}$ is a convergent sequence of elements of $\mathfrak{M}$, we can write

$$f^{(n)} = f_1^{(n)} + \cdots + f_l^{(n)}, \qquad f_k^{(n)} \text{ in } \mathfrak{M}_k.$$

Since $\mathfrak{M}_1, \cdots, \mathfrak{M}_l$ are mutually orthogonal sets we have

$$\| f_1^{(m)} - f_1^{(n)} \|^2 + \cdots + \| f_l^{(m)} - f_l^{(n)} \|^2 = \| f^{(m)} - f^{(n)} \|^2 \to 0$$

when m and n become infinite. Thus each sequence $\{f_k^{(n)}\}$ is a convergent sequence; its limit element f_k belongs to $\mathfrak{M}_k$, which is closed. Consequently the limit of the sequence $\{f^{(n)}\}$ is $f_1 + \cdots + f_l$, which is an element of $\mathfrak{M}$, as we wished to show.

In many later discussions, special notations for sets of elements constructed from a finite or denumerably infinite collection $\{\mathfrak{M}_n\}$ of closed linear manifolds will be found very helpful. To indicate the logical sum of the sets $\mathfrak{M}_n$, we use one of the familiar notations

$$\sum_{\alpha=1}^{\infty} \mathfrak{M}_\alpha = \mathfrak{M}_1 + \mathfrak{M}_2 + \mathfrak{M}_3 + \cdots$$

for this purpose, as we have already done above. We denote the linear manifold and the closed linear manifold determined by this sum by means of the new symbols

$$\sum_{\alpha=1}^{\infty} (\mathfrak{M}_\alpha; \dot{+}) = \mathfrak{M}_1 \dot{+} \mathfrak{M}_2 \dot{+} \mathfrak{M}_3 + \cdots,$$

$$\sum_{\alpha=1}^{\infty} (\mathfrak{M}_\alpha; \oplus) = \mathfrak{M}_1 \oplus \mathfrak{M}_2 \oplus \mathfrak{M}_3 \oplus \cdots,$$

respectively. The three operations $+$, $\dot{+}$, $\oplus$ evidently obey the commutative and associative laws. In certain instances the operations $\oplus$ and $\ominus$ are inverse to one another as we see in the following important special relations:

$$\mathfrak{H} = \mathfrak{M} \oplus [\mathfrak{H} \ominus \mathfrak{M}], \qquad \mathfrak{H} \ominus [\mathfrak{H} \ominus \mathfrak{M}] = \mathfrak{M}.$$

The relations between the operations $\dot{+}$ and $\oplus$ are by no means simple. The theorem above shows that if $\mathfrak{M}_1, \cdots, \mathfrak{M}_l$ are mutually orthogonal closed linear manifolds, the two operations are equivalent, according to the equation

$$\mathfrak{M}_1 \dot{+} \cdots \dot{+} \mathfrak{M}_l = \mathfrak{M}_1 \oplus \cdots \oplus \mathfrak{M}_l;$$

but, in general, no such equivalence holds for infinite sums or for finite sums of manifolds not mutually orthogonal.

The following simple example illustrates the last remarks. Let $\{\varphi_n\}$ be a complete orthonormal set in $\mathfrak{H}$, $\{a_n\}$ a sequence such that $\sum_{\alpha=1}^{\infty} |a_\alpha|^2 = 1$, $a_n \neq 0$. We then determine a sequence $\{\theta_n\}$, $0 < \theta_n < \frac{\pi}{2}$, with the property that the series $\sum_{\alpha=1}^{\infty} \sec^2\theta_\alpha \; |a_{2\alpha-1}|^2$ diverges. If we put

$$\psi_n = \varphi_{2n}, \qquad \chi_n = \cos\theta_n \cdot \varphi_{2n-1} + \sin\theta_n \cdot \varphi_{2n},$$

the sets $\{\psi_n\}$ and $\{\chi_n\}$ are orthonormal sets determining closed linear manifolds $\mathfrak{M}$ and $\mathfrak{N}$ respectively. We shall show that the element f defined by the relation $f = \sum_{\alpha=1}^{\infty} a_\alpha \varphi_\alpha$ is in

$\mathfrak{M} \oplus \mathfrak{N}$ but not in $\mathfrak{M} \dot{+} \mathfrak{N}$. Since $\mathfrak{M} \dot{+} \mathfrak{N}$ contains the complete orthonormal set $\{\varphi_n\}$, it is evident that $\mathfrak{M} \oplus \mathfrak{N}$ coincides with $\mathfrak{H}$ and therefore contains f. If there exist elements g and h in $\mathfrak{M}$ and $\mathfrak{N}$ respectively such that $f = g + h$, we must have

$$\begin{aligned} a_{2n-1} &= (f, \varphi_{2n-1}) = (g, \varphi_{2n-1}) + (h, \varphi_{2n-1}) \\ &= \sum_{\alpha=1}^{\infty} (g, \psi_\alpha)(\psi_\alpha, \varphi_{2n-1}) + \sum_{\alpha=1}^{\infty} (h, \chi_\alpha)(\chi_\alpha, \varphi_{2n-1}) \\ &= \cos\theta_n \cdot (h, \chi_n). \end{aligned}$$

We obtain a contradiction at once from the relations

$$\|h\|^2 = \sum_{\alpha=1}^{\infty} |(h, \chi_\alpha)|^2 = \sum_{\alpha=1}^{\infty} \sec^2\theta_\alpha \cdot |a_{2\alpha-1}|^2.$$

Thus f does not belong to $\mathfrak{M} \dot{+} \mathfrak{N}$.

THEOREM 1.23. *If $\mathfrak{M}$ is a closed linear manifold and f an arbitrary element of $\mathfrak{H}$, then there exists a unique pair of elements g, h, such that $f = g + h$, g belongs to $\mathfrak{M}$, and h belongs to $\mathfrak{H} \ominus \mathfrak{M}$.*

By Theorem 1.19, $\mathfrak{M}$ is either an n-dimensional unitary space or a Hilbert space. In either case, we conclude from Theorems 1.13 and 1.14 that there exists an orthonormal set $\{\varphi_n\}$ which determines the closed linear manifold $\mathfrak{M}$. We set

$$g = \sum_{\alpha=1}^{\infty} a_\alpha \varphi_\alpha, \qquad a_n = (f, \varphi_n) \cdot$$

in accordance with Theorem 1.7. Now g is an element of $\mathfrak{M}$; and $h = f - g$ is orthogonal to every φ_n, is therefore orthogonal to $\mathfrak{M}$, and consequently belongs to $\mathfrak{H} \ominus \mathfrak{M}$. That the pair so constructed is uniquely determined by its properties may be seen as follows: if there are two such pairs g_1, h_1 and g_2, h_2 we have

$$g_1 + h_1 = g_2 + h_2, \qquad g_1 - g_2 = h_2 - h_1;$$

hence, $g_1 - g_2$ belongs to $\mathfrak{M}$ and to $\mathfrak{H} \ominus \mathfrak{M}$ at the same time; it is therefore orthogonal to itself, so that

$$g_1 - g_2 = 0, \qquad g_1 = g_2, \qquad h_1 = h_2.$$

DEFINITION 1.9. *The element g of Theorem* 1.23 *is called the projection of f on $\mathfrak{M}$.*

In Chapter II we shall investigate projections in an exhaustive manner, contenting ourselves for the present with their definition.

§ 5. REALIZATIONS OF ABSTRACT HILBERT SPACE

We have already exhibited the space $\mathfrak{H}_0$ as a concrete example of the abstract space $\mathfrak{H}$. This space $\mathfrak{H}_0$ was the first such example to be set up and studied; its importance was recognized by Hilbert, who employed it in formulating his general theory of integral equations.* For this reason, we shall refer to $\mathfrak{H}_0$ as ordinary Hilbert space.

Another realization of $\mathfrak{H}$, which has special interest for analysis, is the class $\mathfrak{L}_2$ of all complex-valued Lebesgue-measurable functions f for which the Lebesgue integral of $|f|^2$ exists. The real importance of the space $\mathfrak{H}_0$ is due to the fact that $\mathfrak{L}_2$ can be put into one-to-one correspondence with it, by Theorem 1.16, in such a manner that problems of analysis can be treated as quasi-algebraic problems in $\mathfrak{H}_0$. With due regard for generality, we may formulate the following theorem:

THEOREM 1.24. *Let E be a Lebesgue-measurable set of infinite or positive finite measure in Euclidean space of l dimensions, P a variable point in E. Let $\mathfrak{L}_2$ be the class of all complex-valued Lebesgue-measurable functions, $f(P)$, defined almost everywhere in E for which the Lebesgue integral $\int_E |f(P)|^2 \, dP$ exists, two functions f and g of this class being considered as identical if and only if $f(P) = g(P)$ almost everywhere in E. If the operations $+$ and $\cdot$ are defined as ordinary addition and multiplication, the null element is defined as the function which vanishes almost everywhere in E, and the function (f, g) is defined by the equation $(f, g) = \int_E f(P) \cdot \overline{g(P)} \, dP$, then $\mathfrak{L}_2$ is a Hilbert space.*

---

* Hilbert, *Grundzüge einer allgemeinen Theorie der linearen Integralgleichungen*, Leipzig, 1912.

We shall establish this result first in the particular case where E is the entire space, and then obtain the general case as a simple corollary.

We see at once from the familiar properties of the Lebesgue-measurable functions and of the Lebesgue integral that Postulates A and B of Definition 1.1 are satisfied.

Since for given n we can find n mutually exclusive sets $E_1, \cdots, E_n$ in E, each of positive finite measure, we are able to determine n linearly independent elements in $\mathfrak{L}_2$, namely, the characteristic functions of the sets $E_1, \cdots, E_n$; the characteristic function $f(P; E_0)$ of an arbitrary point-set E_0 is defined by the relations

$$f(P; E_0) = 1 \text{ when } P \text{ is in } E_0,$$
$$f(P; E_0) = 0 \text{ when } P \text{ is in } E - E_0.$$

Hence Postulate C is true for $\mathfrak{L}_2$.

We shall not discuss in full detail the demonstration of the fact that $\mathfrak{L}_2$ is separable with respect to the distance function $\| f - g \| = \left(\int_E |f - g|^2 \, dP \right)^{1/2}$. The process of constructing the entire class of Lebesgue-measurable functions from step-functions alone is sufficiently familiar that a repetition of the particulars is unnecessary. We first introduce a Cartesian coördinate system so that the point P is specified by the coördinates $(x_1, \cdots, x_l)$ and construct all the rational cells, defined by the inequalities $r_{1k} < x_k \leqq r_{2k}$, where r_{1k} and r_{2k} are rational numbers and $k = 1, \cdots, l$. We define a rational step-function as a function which assumes only a finite number of complex values (including zero), each with rational real and imaginary parts, the sets of points upon which it assumes any of its values except zero being composed of a finite number of non-overlapping rational cells. The class of all rational step-functions is a denumberably infinite subclass of $\mathfrak{L}_2$; we shall show that it is everywhere dense in $\mathfrak{L}_2$. Let an arbitrary function $f(P)$ in $\mathfrak{L}_2$ and an arbitrary positive number ε be given. By the definition of the Lebesgue integral, there exists a function $f_1(P)$ which satisfies the inequality

$\|f - f_1\| < \varepsilon/4$ and which assumes $N+1$ complex values (including zero), each with rational real and imaginary parts, the sets of points $E_1, \cdots, E_N$ upon which it assumes values other than zero being bounded Lebesgue-measurable sets. We next select closed sets F_k, $F_k \subseteqq E_k$, $k = 1, \cdots, N$, and define the function $f_2(P)$ by requiring that $f_2(P)$ take on in F_k the value assumed by $f_1(P)$ in E_k and that $f_2(P)$ vanish elsewhere; by the theory of Lebesgue measure we can select the sets $F_1, \cdots, F_N$ in such a manner that the set of points $(E_1 - F_1) + \cdots + (E_N - F_N)$, where $f_1(P)$ and $f_2(P)$ differ, has measure so small that the inequality $\|f_1 - f_2\| < \varepsilon/4$ is valid. By a similar step we replace the function $f_2(P)$ and the sets $F_1, \cdots, F_N$ by a function $f_3(P)$ and open sets $O_1, \cdots, O_N$ where $O_k \supseteqq F_k$. Since the closed sets $F_1, \cdots, F_N$ are bounded and disjoint—that is, no two of them have a point in common—the open sets $O_1, \cdots, O_N$ can be constructed as bounded and disjoint. We can select these open sets in such a manner that the set $(O_1 - F_1) + \cdots + (O_N - F_N)$, where $f_2(P)$ and $f_3(P)$ differ, has measure so small that the inequality $\|f_2 - f_3\| < \varepsilon/4$ is valid. Finally, we can replace the function $f_3(P)$ and the open sets $O_1, \cdots, O_N$ by a function $f_3(P)$ and sets $C_1, \cdots, C_N$, where C_k is a subset of O_k composed of a finite number of non-overlapping rational cells; and, as in the preceding cases, we can arrange that the inequality $\|f_3 - f_4\| < \varepsilon/4$ be satisfied. It is evident that the function $f_4(P)$ is a rational step-function satisfying the relation.

$$\|f - f_4\| \leqq \|f - f_1\| + \|f_1 - f_2\| + \|f_2 - f_3\| + \|f_3 - f_4\| < \varepsilon.$$

With this conclusion, the separability of $\mathfrak{L}_2$ is established.

We have still to show that $\mathfrak{L}_2$ is complete in the sense of Postulate E. The proof of this property is found in the celebrated Riesz-Fischer theorem* which asserts that if a

* F. Riesz, Göttinger Nachrichten (1907), pp. 116–122.—Fischer, Comptes Rendus de l'Académie des Sciences, Paris, 144 (1907), pp. 1022–1024.—W. H. and G. C. Young, Quarterly Journal of Mathematics, 44 (1912–13), pp. 49–88, give an excellent historical account and a great variety of proofs.

sequence of functions in the class $\mathfrak{L}_2$ "converges in the mean", that is, if

$$\int_E |f_m - f_n|^2 dP \to 0, \qquad m, n \to \infty,$$

then there exists a function f, likewise in $\mathfrak{L}_2$, to which the sequence "converges in the mean", that is, such that

$$\int_E |f - f_n|^2 dP \to 0, \qquad n \to \infty.$$

The term "convergence in the mean" was introduced to distinguish convergence in the space $\mathfrak{L}_2$, as defined in Definition 1.2, from ordinary point-wise convergence. The simplest and, at the same time, most penetrating proofs are those modelled on one given by Weyl.* We shall follow closely the demonstration of this type due to v. Neumann.† The fundamental motive of our reasoning is the expectation of replacing "convergence in the mean" by point-wise convergence. Let $\{f_n(P)\}$ be a sequence in $\mathfrak{L}_2$ which "converges in the mean" and let $N = N(\varepsilon)$ be a number such that for a given positive ε

$$\int_E |f_m - f_n|^2 dP < \varepsilon, \qquad m, n > N.$$

We then choose a sequence of positive integers $N_p > N(1/8^p)$ such that $N_{p+1} > N_p$. Since we have $\int_E |f_{N_{p+1}} - f_{N_p}|^2 dP < 1/8^p$, we see that the set of points on which $|f_{N_{p+1}} - f_{N_p}| \geqq 1/2^p$ cannot have measure greater than $1/2^p$. Consequently, the inequalities

$$|f_{N_{p+1}} - f_{N_p}| < 1/2^p, \qquad |f_{N_{p+2}} - f_{N_{p+1}}| < 1/2^{p+1}, \cdots$$

hold simultaneously on a set of points E_p such that $m(E - E_p) \leqq \sum_{\alpha=p}^{\infty} 1/2^\alpha = 1/2^{p-1}$. Evidently, E_p is a subset of E_{p+1} for every p. From the inequalities which define E_p, we conclude that the sequence $\{f_{N_n}\}$ converges uniformly on E_p, for every p; for we have on E_p

* Weyl, Mathematische Annalen, 67 (1909), pp. 225–245.

† J. v. Neumann, Mathematische Annalen, 102 (1929), pp. 109–111.

$$|f_{N_m} - f_{N_n}| \leqq \sum_{\alpha=n}^{m-1} |f_{N_{\alpha+1}} - f_{N_\alpha}| \leqq \sum_{\alpha=n}^{m-1} 1/2^\alpha < 1/2^{n-1}$$

when m and n are not less than p, $m > n$. Thus the sequence $\{f_{N_n}\}$ converges at every point of the set $E_0 = \lim_{p\to\infty} E_p$, which has the property that $E - E_0$ is of measure zero. We now define $f(P)$ by the relations

$$f(P) = \lim_{n\to\infty} f_{N_n}(P) \text{ when } P \text{ is in } E_0,$$
$$f(P) = 0 \text{ when } P \text{ is in } E - E_0.$$

In order to show that $f(P)$ belongs to $\mathfrak{L}_2$ we employ the inequality

$$\int_{E_k} |f_m - f_{N_p}|^2 \, dP \leqq \int_E |f_m - f_{N_p}|^2 \, dP < \varepsilon, \qquad m, N_p > N(\varepsilon)$$

Since the convergence of the sequence $\{f_{N_p}\}$ is uniform on the set E_k, we can allow p to become infinite in the first integral, with the result that

$$\int_{E_k} |f_m - f|^2 \, dP \leqq \varepsilon, \qquad m > N(\varepsilon),$$

for every k. Then, because $E_k \subseteq E_{k+1}$ and the integrand is positive in the last integral, it is permissible to pass to the limit, $k \to \infty$, with the result that

$$\int_E |f_m - f|^2 \, dP \leqq \varepsilon, \qquad m > N(\varepsilon),$$

the existence of the integral being established by the process. We now see that $f(P)$, as the sum of two functions in $\mathfrak{L}_2$, namely f_m and $f - f_m$, is itself in $\mathfrak{L}_2$; and we see also, from the last inequality, that the sequence $\{f_n\}$ "converges in the mean" to f. We have thus completed our proof of the Riesz-Fischer Theorem and have shown that $\mathfrak{L}_2$ satisfies Postulate E.

We shall now apply the facts so far established to the general case in which the functions in $\mathfrak{L}_2$ are defined over an arbitrary Lebesgue-measurable point-set E of infinite or positive finite measure in l-dimensional Euclidean space, which we denote by S. We denote by $\mathfrak{L}_2^*$ the particular class obtained by taking the domain of definition as S itself. The subset of $\mathfrak{L}_2^*$ comprising all those functions for which $\int_{S-E} |f|^2 \, dP$ vanishes is easily seen to be a closed linear manifold $\mathfrak{M}^*$ in $\mathfrak{L}_2^*$. Between $\mathfrak{L}_2$ and $\mathfrak{M}^*$ we determine a one-to-one correspondence by the relations

$$f^*(P) = f(P) \text{ when } P \text{ is in } E,$$
$$f^*(P) = 0 \qquad \text{when } P \text{ is in } S-E.$$

Evidently this correspondence has the property that

$$f^* + g^* \sim f + g,$$
$$a \cdot f^* \sim a \cdot f,$$
$$0 \sim 0$$
$$(f^*, g^*) = (f, g).$$

We know that $\mathfrak{M}^*$ is either an n-dimensional unitary space or a Hilbert space; the first possibility is very easily excluded, since, given an $n = 1, 2, 3, \cdots$, we can find n mutually exclusive subsets $E_1, \cdots, E_n$ of E, each of finite positive measure, and can then verify that the characteristic functions of these sets are linearly independent elements of $\mathfrak{M}^*$. Thus $\mathfrak{M}^*$ is a Hilbert space; and $\mathfrak{L}_2$, being abstractly identical with it, is also a Hilbert space. We could easily introduce slight modifications of the discussion relative to $\mathfrak{L}_2^*$ so as to make it cover the more general case; a certain interest attaches, however, to the proof we have just given, because it takes advantage of the known properties of $\mathfrak{H}$ instead of going back to the fundamental postulates.

Certain extensions of the preceding theorem promptly suggest themselves. We may substitute for Euclidean space any separable metric space in which the measure of properly restricted subsets can be suitably defined, thus replacing the Lebesgue integral by one of its generalizations. Since these generalizations consist essentially in abandoning the special properties of Euclidean space and of Lebesgue measure which have no influence on the fundamental structure of the theory of integration, it is evident that the demonstration given for Theorem 1.24 will apply to cases in which the generalized integral is involved. There is one significant exception to the statement made: the property by means of which the truth of Postulate C was established is not true for every generalization of Lebesgue measure and must be postulated of the particular type of measure considered, unless the space $\mathfrak{L}_2$ is to reduce to a unitary space. As an example, consider the following definition of the measure of a linear set of points: a set of points on a straight line (Euclidean space of one dimension) shall have the measure 1 or the measure 0 according as it does or does not include a fixed preassigned point on the line. In Chapter VI we shall find the extension of this theorem to the Lebesgue-Stieltjes or Radon-Stieltjes integral indispensable.* Many other types of integral of common application, such as integrals over curved surfaces and manifolds, are covered by the suggested extension.

An extension of the preceding theorem in a somewhat less obvious direction is of considerable utility in analysis; we phrase it in terms of ordinary Lebesgue integration, noting merely that a different concept of integration may be substituted by way of generalization.

THEOREM 1.25. *Let E be a Lebesgue-measurable set of infinite or finite positive measure in Euclidean space of l dimensions, and let $\mathfrak{L}_{2,m}$ be the class of all vector point-functions $\mathbf{f}$, with m components $(f_1, \cdots, f_m)$ defined over E and belonging to $\mathfrak{L}_2$. If the operations $+$ and $\cdot$ are defined as vector addition and scalar multiplication,*

$$\mathbf{f}+\mathbf{g} = (f_1+g_1, \cdots, f_m+g_m),$$
$$a\cdot\mathbf{f} = (af_1, \cdots, af_m);$$

* For the Lebesgue-Stieltjes or Radon-Stieltjes integral, see Radon, Sitzungsberichte, Akademie der Wissenschaften, Wien, Mathematische-Naturwissenschaftliche Klasse, 122^{2a^2} (1913), pp. 1295–1438, especially §§ I–IV; de la Vallée Poussin, *Intégrales de Lebesgue,* Paris, 1916, pp. 100–102; and Lebesgue, *Leçons sur l'intégration,* Paris, 1928, pp. 252–313.

the null element is defined to be $(0, \cdots, 0)$; *and the function* $(\mathbf{f}, \mathbf{g})$ *is determined by the equation* $(\mathbf{f}, \mathbf{g}) = \int_E (f_1 \bar{g}_1 + \cdots + f_m \bar{g}_m)\, dP$, *then* $\mathfrak{L}_{2,m}$ *is a Hilbert space.*

This theorem depends chiefly upon the properties of $\mathfrak{L}_2$. Postulates A and B are easily seen to be satisfied. Postulate C is true since it is true for the subclass of $\mathfrak{L}_{2,m}$ comprising those vectors all of whose components save the first are zero—that is, essentially for $\mathfrak{L}_2$. In view of the equation

$$|\mathbf{f} - \mathbf{g}| = \left(\int_E (|f_1 - g_1|^2 + \cdots + |f_m - g_m|^2)\, dP \right)^{1/2},$$

we see that Postulate D is satisfied; for if $\{f^{(n)}\}$ is a sequence everywhere dense in $\mathfrak{L}_2$ the denumerably infinite set of vectors

$$\mathbf{f}_{n_1, \cdots, n_m} = (f^{(n_1)}, \cdots, f^{(n_m)})$$

is readily shown to be everywhere dense in $\mathfrak{L}_{2,m}$. Similarly, if the vector sequence $\{\mathbf{f}_n\}$ converges in $\mathfrak{L}_{2,m}$ each of the component sequences converges in $\mathfrak{L}_2$ and therefore converges to a limiting element in $\mathfrak{L}_2$; hence $\mathbf{f}_n$ converges to a limiting vector $\mathbf{f}$ in $\mathfrak{L}_{2,m}$. Postulate E is thus seen to be true for $\mathfrak{L}_{2,m}$.

It is evident that this theorem is a particular case of the following theorem by which a new Hilbert space can be constructed from a finite set of known realizations. The proof may be left to the reader.

THEOREM 1.26. *If* $\mathfrak{H}_1, \cdots, \mathfrak{H}_m$ *are Hilbert spaces, then the class* $\mathfrak{H}$ *of all vectors* $\mathbf{f}$ *with components* $(f_1, \cdots, f_m)$, *where* f_k *is an element of* $\mathfrak{H}_k$, *is itself a Hilbert space if the operations* $+$ *and* $\cdot$, *the null-element, and the function* $(\mathbf{f}, \mathbf{g})$ *are defined thus:*

$$\begin{aligned} \mathbf{f} + \mathbf{g} &= (f_1 + g_1, \cdots, f_m + g_m), \\ a \cdot \mathbf{f} &= (af_1, \cdots, af_m), \\ \mathbf{0} &= (0, \cdots, 0), \\ (\mathbf{f}, \mathbf{g}) &= (f_1, g_1) + \cdots + (f_m, g_m). \end{aligned}$$

The relation between $\mathfrak{H}$ *and* $\mathfrak{H}_1, \cdots, \mathfrak{H}_m$ *may be indicated by the equation*

$$\mathfrak{H} = \mathfrak{H}_1 + \cdots + \mathfrak{H}_m = \mathfrak{H}_1 \oplus \cdots \oplus \mathfrak{H}_m.$$

This theorem may be extended as follows:

THEOREM 1.27. *If $\{\mathfrak{S}_m\}$ is a denumerably infinite set of n-dimensional unitary spaces, $n \geqq 1$, and Hilbert spaces, then the class $\mathfrak{H}$ of all ordered sets $\mathbf{f} = (f_1, f_2, f_3, \cdots)$, where f_k is an element of $\mathfrak{S}_k$, such that $\sum_{\alpha=1}^{\infty} \|f_\alpha\|^2$ converges, is a Hilbert space, when*

$$\begin{aligned} \mathbf{f}+\mathbf{g} &= (f_1+g_1, f_2+g_2, f_3+g_3, \cdots), \\ a\cdot\mathbf{f} &= (af_1, af_2, af_3, \cdots), \\ \mathbf{0} &= (0, 0, 0, \cdots), \\ (\mathbf{f}, \mathbf{g}) &= \sum_{\alpha=1}^{\infty} (f_\alpha, g_\alpha). \end{aligned}$$

We may indicate the relation between $\mathfrak{H}$ and the set $\{\mathfrak{S}_m\}$ by the equation

$$\mathfrak{H} = \mathfrak{S}_1 \oplus \mathfrak{S}_2 \oplus \mathfrak{S}_3 \oplus \cdots.$$

There is little difficulty in extending the proof of Theorem 1.15 to the more general theorem considered here. The relation between the two theorems becomes evident when we note that by taking each of the spaces $\mathfrak{S}_m$ as a one-dimensional unitary space we have the situation treated in Theorem 1.15. To generalize the proof of Theorem 1.15 we have to examine sequences of elements in the various spaces where previously we encountered sequences of complex numbers. It will be sufficient to illustrate these remarks by giving the proof that the space $\mathfrak{H}$ under consideration is complete in the sense of Postulate E. If $\mathbf{f}_n = (f_1^{(n)}, f_2^{(n)}, f_3^{(n)}, \cdots)$ is a sequence such that $\|\mathbf{f}_m - \mathbf{f}_n\| \to 0$ when $m, n \to \infty$, then there exists for every positive ε an $N = N(\varepsilon)$ such that

$$\begin{aligned} \sum_{\alpha=1}^{k} \|f_\alpha^{(m)} - f_\alpha^{(n)}\|^2 &\leqq \sum_{\alpha=1}^{\infty} \|f_\alpha^{(m)} - f_\alpha^{(n)}\|^2 \\ &= \|\mathbf{f}_m - \mathbf{f}_n\|^2 < \varepsilon^2, \ m, n > N(\varepsilon) \end{aligned}$$

for every k. Consequently $\|f_l^{(m)} - f_l^{(n)}\| \to 0$ when $m, n \to \infty$, for $l = 1, \cdots, k$, $k = 1, 2, 3, \cdots$; in other words, the sequence $\{f_l^{(n)}\}$ converges in $\mathfrak{S}_l$ and has a limiting element f_l

in $\mathfrak{S}_l$. Hence, on allowing m to become infinite in the inequality above, we obtain

$$\sum_{\alpha=1}^{k} |f_\alpha - f_\alpha^{(n)}|^2 \leqq \epsilon^2, \qquad n > N(\epsilon).$$

If we now let k become infinite, we find

$$\sum_{\alpha=1}^{\infty} |f_\alpha - f_\alpha^{(n)}|^2 \leqq \epsilon^2, \qquad n > N(\epsilon),$$

an inequality which shows that the element $\boldsymbol{g}_n = (f_1 - f_1^{(n)}, f_2 - f_2^{(n)}, f_3 - f_3^{(n)}, \cdots)$ belongs to $\mathfrak{H}$. Consequently $\boldsymbol{f}_n + \boldsymbol{g}_n = \boldsymbol{f} = (f_1, f_2, f_3, \cdots)$ is an element of $\mathfrak{H}$ such that $|\boldsymbol{f} - \boldsymbol{f}_n| \to 0$ when $n \to \infty$.

The realizations of Hilbert space which we have described are those which occur most frequently in analysis, and are sufficiently numerous to give an indication of the variety of possible concrete interpretations. We have not attempted to make an exhaustive catalogue of such realizations. We shall leave to the reader the interpretation in these concrete spaces of the various entities defined in the preceding sections—linear manifolds, orthonormal sets, and so forth. It has already been shown that for formal purposes all Hilbert spaces are alike, so that the study of concrete examples, while instructive, is not indispensable.

CHAPTER II

TRANSFORMATIONS IN HILBERT SPACE

§ 1. LINEAR TRANSFORMATIONS

A study of the geometry of Hilbert space involves the investigation of transformations in that space. From the geometrical point of view it would be sufficient to consider those transformations which are defined throughout $\mathfrak{H}$. The requirements of analysis, however, compel us to treat transformations which are not and cannot be defined over the entire space: in many analytical problems, some of the most important included, we have to study operations which take a variable element of a specified proper subset of the Hilbert space $\mathfrak{L}_2$ into a second element of that space. In this chapter we shall lay the foundations of a general theory of transformations, broad enough to include a variety of such analytical applications. The development will be effected in abstract terms, and the consideration of explicit examples, chosen largely from the field of analysis, will be postponed to the following chapter.

In this first section we shall discuss general concepts with particular emphasis upon the properties of linear transformations. The study of more special types of transformation will follow in the later sections.

We commence with the definition of a transformation in $\mathfrak{H}$:

DEFINITION 2.1. *If $\mathfrak{D}$ and $\mathfrak{R}$ are two subsets of $\mathfrak{H}$, each of which contains at least one element, a relation between them whereby to each element of $\mathfrak{D}$ is ordered one element of $\mathfrak{R}$ and each element of $\mathfrak{R}$ is ordered to at least one element of $\mathfrak{D}$ is called a transformation T. The sets $\mathfrak{D}$ and $\mathfrak{R}$ are called the domain and the range of T respectively; T is said* *to take* $\mathfrak{D}$ *into* $\mathfrak{R}$. *If* f *is an element of* $\mathfrak{D}$ *and* g *its correspondent in* $\mathfrak{R}$, *then the relation between them is written as an equation*

$$Tf = g.$$

We may think of the relation thus defined and symbolized by the letter T in three distinct ways: the form and nomenclature of the definition suggest the geometrical interpretation summed up in the word "transformation"; T may be treated as an operational symbol representing an operator which, when applied to an element of $\mathfrak{D}$, yields the corresponding element in $\mathfrak{R}$; and, finally, T may be treated as a functional symbol, the element g being expressed as a function of the variable element f according to the equation $g = Tf$. It will be convenient for us to employ the terms "transformation" and "operator" interchangeably in later developments. We must call attention to the fact that the definition just given does not require the domain of a transformation to contain more than a single element, though, of course, we shall usually ask more in practice. It is even more important to notice that a transformation is not determined unless its domain is known; this remark is particularly significant in connection with certain analytical problems discussed in the following chapter. We must therefore distinguish carefully between the identity of two transformations and their equality in a given set $\mathfrak{S}$: if T_1 and T_2 are transformations with domains $\mathfrak{D}_1$ and $\mathfrak{D}_2$ respectively, then they are identical if and only if $\mathfrak{D}_1 = \mathfrak{D}_2$ and $T_1 f = T_2 f$ for every element of their common domain; and T_1 and T_2 are said to be equal in the set $\mathfrak{S}$ if and only if $\mathfrak{S}$ is a subset of $\mathfrak{D}_1 \cdot \mathfrak{D}_2$ and $T_1 f = T_2 f$ for every element in $\mathfrak{S}$. We shall employ the usual symbols $\equiv$ and $=$ to denote identity and equality between transformations. It will be convenient to agree upon the notation for a transformation and the two related sets of elements once for all: we shall always denote a transformation by an ordinary capital, usually T, and its domain and range by the German capitals $\mathfrak{D}$ and $\mathfrak{R}$ respectively, except in cases where confusion would result; if several transformations denoted by the same letter and distinguished by subscripts or other suitable marks are to be considered simultaneously, the letters denoting the domains and the ranges of the various transformations will be distinguished by affixing corresponding subscripts or marks. By preserving these conventions, it will be possible to write a collection of symbols such as

$$T,\ \mathfrak{D},\ \mathfrak{R};\quad T^*,\ \mathfrak{D}^*,\ \mathfrak{R}^*;\quad T_1,\ \mathfrak{D}_1,\ \mathfrak{R}_1$$

without being under an obligation to explain their significance each time they occur.

We can now define in the customary manner the sum and the product of two transformations and the scalar product of a transformation by a complex number, according to the following rules:

(1) if T_1 and T_2 are transformations such that the set $\mathfrak{D}_1 \cdot \mathfrak{D}_2$ is not empty, then their sum is defined as the transformation T with domain $\mathfrak{D} = \mathfrak{D}_1 \cdot \mathfrak{D}_2$ such that $Tf = T_1 f + T_2 f$ throughout $\mathfrak{D}$, and is denoted by $T_1 + T_2$;

(2) if T_1 and T_2 are transformations such that the set $\mathfrak{D}_1 \cdot \mathfrak{R}_2$ is not empty, then their product, denoted by $T_1 T_2$, is defined as the transformation T whose domain is the set of all elements of $\mathfrak{D}_2$ transformed by T_2 into elements of $\mathfrak{D}_1$ and which transforms every element of its domain according to the equation $Tf = T_1(T_2 f)$;

(3) if a is a complex number and T_0 is a given transformation, then their product, denoted by $a \cdot T_0$, is defined as the transformation T such that $Tf = a \cdot (T_0 f)$ for every element in $\mathfrak{D} = \mathfrak{D}_0$.

In the class of all transformations with domains identical to $\mathfrak{H}$, these rules generate an algebra formally identical, so far as the operations of addition and scalar multiplication are concerned, with ordinary vector algebra. This algebra of transformations is complicated by the fact that the product of two transformations is in general non-commutative; that is, $T_1 T_2$ and $T_2 T_1$ are identical only in special circumstances. For transformations whose domains are proper subsets of $\mathfrak{H}$, it may happen that the sum or the product does not exist; but whenever the definitions just given are effective, it is clear that the operation $+$ is commutative and associative, the operation $\cdot$ associative and distributive in the sense indicated by the following equations:

$$T_1 + T_2 \equiv T_2 + T_1,$$

$$T_1 + (T_2 + T_3) \equiv (T_1 + T_2) + T_3 \equiv (T_1 + T_3) + T_2.$$

$$T_1(T_2 T_3) \equiv (T_1 T_2) T_3, \qquad (T_1 + T_2) T_3 \equiv T_1 T_3 + T_2 T_3.$$

When T_1 is a linear transformation as defined below, we have also

$$T_1(T_2 + T_3) \equiv T_1 T_2 + T_1 T_3.$$

Scalar multiplication is always defined and is seen to have the properties:

$$a \cdot (T_1 + T_2) \equiv a \cdot T_1 + a \cdot T_2, \qquad a \cdot (b \cdot T_0) \equiv (ab) \cdot T_0.$$

For linear transformations, as defined below, we have also:

$$(a \cdot T_1) \cdot (b \cdot T_2) \equiv (ab) \cdot T_1 T_2.$$

This algebra of transformations will be very useful in the following pages. We find it convenient to denote by O the transformation which takes every element of $\mathfrak{H}$ into the null element, and by I the transformation which takes every element of $\mathfrak{H}$ into itself; I is called the identity. Evidently, in the algebra of transformations, O plays the rôle of a null element, I that of an idempotent element or unity, as we see from the relations

$$T + O \equiv T, \qquad O \cdot T = O, \qquad I \cdot T \equiv T \cdot I \equiv T,$$

which hold in the domain of an arbitrary transformation T.

Two other types of relation involving transformations are introduced in the following definitions.

DEFINITION 2.2. *A transformation T_2 is said to be an extension of the transformation T_1—in symbols, $T_2 \supseteqq T_1$—when $\mathfrak{D}_2$ contains $\mathfrak{D}_1$ and T_1 and T_2 are equal in $\mathfrak{D}_1$. If T_2 is* *an extension of* T_1 *and* $\mathfrak{D}_2$ *contains at least one element not in* $\mathfrak{D}_1$, *then* T_2 *is said to be a proper extension of* T_1*—in symbols,* $T_2 \supset T_1$.

The use of the symbols $\supseteq$, $\supset$ to denote the relations involved here is easily justified by restating the definition in terms of the logic of classes: T_2 is an extension or a proper extension of T_1 according as the class of all significant pairs $(f, T_1 f)$ is a subclass or a proper subclass of the class of all significant pairs $(f, T_2 f)$. Consequently, these symbols obey the usual formal rules. In any general theory of transformations whose domains may be proper subsets of $\mathfrak{H}$, the concept of extension has an obvious and important part to play: whenever a given transformation T appears to be too narrowly defined, it is natural to seek to extend the definition of T by considering all possible extensions of T in an appropriately prescribed category.

DEFINITION 2.3. *A sequence of transformations* $\{T_n\}$ *with domains* $\{\mathfrak{D}_n\}$ *is said to converge in the set* $\mathfrak{S}$ *if to each element* f *in* $\mathfrak{S}$ *there corresponds an* $N = N(f)$ *such that*

(1) f *is an element of* $\mathfrak{D}_n$, $n > N$,

(2) *the sequence* $\{T_n f\}$, $n > N$, *converges.*

A sequence of transformations $\{T_n\}$ *is said to have the limit* T *in the set* $\mathfrak{S}$ *if to each element* f *in* $\mathfrak{S}$ *there corresponds an* $N = N(f)$ *such that*

(1) f *is an element of* $\mathfrak{D}_n$ *for* $n > N$ *and of* $\mathfrak{D}$,

(2) $T_n f \to Tf$;

we shall write $T_n \to T$, $n \to \infty$, *in* $\mathfrak{S}$.

By reference to the comments on Definition 1.2, we see immediately that the following theorem may be proved:

THEOREM 2.1. *If the sequence* $\{T_n\}$ *converges in* $\mathfrak{S}$, *then there exists a transformation* T *such that* $T_n \to T$ *in* $\mathfrak{S}$; *if* $T_n \to T$ *in* $\mathfrak{S}$, *then the sequence* $\{T_n\}$ *is convergent in* $\mathfrak{S}$.

We turn next to the introduction of the inverse of a given transformation or operator.

THEOREM 2.2. *If a transformation* T *takes its domain* $\mathfrak{D}$ *into its range* $\mathfrak{R}$ *in such a manner that to distinct elements of* $\mathfrak{D}$ *correspond distinct elements of* $\mathfrak{R}$, *then the correspondence between* $\mathfrak{D}$ *and* $\mathfrak{R}$ *is a one-to-one correspondence; there exists a transformation, denoted by* T^{-1}, *whose domain is* $\mathfrak{R}$ *and whose range is* $\mathfrak{D}$ *and which has the properties*

$$T^{-1} \cdot T = I \text{ in } \mathfrak{D}, \qquad T \cdot T^{-1} = I \text{ in } \mathfrak{R}.$$

The transformation T^{-1} *is called the inverse of* T. *The inverse of* T^{-1} *is* T.

In the preceding definitions and theorems, we have been concerned with relations between transformations with little or no reference to the internal properties of the transformations themselves. We shall now define and discuss a few internal characteristics of major importance.

DEFINITION 2.4. *A transformation* T *is said to be continuous at an element* f *in its domain if to each positive number* ε *there corresponds a positive number* $\delta = \delta(\varepsilon, f)$ *such that, whenever* g *is an element of* $\mathfrak{D}$ *satisfying the inequality* $\|f-g\| \leqq \delta$, *the element* Tg *satisfies the inequality* $\|Tf - Tg\| \leqq \varepsilon$. *A transformation* T *is said to be continuous if it is continuous at every element of its domain.*

DEFINITION 2.5. *A transformation* T *is said to be closed if, whenever the sequences* $\{f_n\}$ *and* $\{Tf_n\}$ *exist and converge to the limits* f *and* g *respectively, then* f *is in the domain of* T *and* $Tf = g$.

The definition may be phrased somewhat differently and more briefly as follows: a transformation T is said to be closed if the class of all significant pairs (f, Tf) is closed.

It is to be noted that a closed transformation need not possess a closed domain, need not be continuous: for example, the non-bounded self-adjoint transformations which we discuss below in § 2 and § 3 are closed, everywhere discontinuous transformations with domains which are not closed sets. The existence of transformations with such properties is possible because Hilbert space is neither compact nor locally compact:* in the neighborhood of an arbitrary element of $\mathfrak{H}$ there exists

---

* For definitions of the terms compact and locally compact, see, for instance, Fréchet, *Les espaces abstraits*, Paris, 1928, p. 69, p. 223.

a sequence which contains no convergent subsequence whatsoever. We can give an illustration of this property by means of the sequence $\{f + \epsilon \varphi_n\}$, where f is an arbitrary element, ϵ an arbitrary positive number, and $\{\varphi_n\}$ an arbitrary infinite orthonormal set; it is easily seen that every element of the sequence lies within distance ϵ of f and that the sequence has no convergent subsequence. It is instructive to compare the situation in Hilbert space with that which obtains in more familiar cases. For instance, let $y = f(x)$ define a transformation of a set $\mathfrak{D}$ on the interval $0 \leqq x \leqq 1$ into a set $\mathfrak{R}$ on the interval $0 \leqq y \leqq 1$, and let this transformation be closed in the sense that the relations $x_n \to \xi$, $y_n = f(x_n) \to \eta$ imply that $f(\xi)$ exists and is equal to η. The properties postulated of the function $f(x)$ enable us to show that $\mathfrak{D}$ is a closed set, and that $f(x)$ is continuous in $\mathfrak{D}$: for, if ξ is a limit point of $\mathfrak{D}$ and $\{x_n\}$ is any sequence in $\mathfrak{D}$ converging to ξ, the sequence $\{f(x_n)\}$ has at least one limit point η; and the closure property requires that $f(\xi)$ exist and have the value η. On the other hand, a continuous transformation in Hilbert space is not necessarily closed. In the case which is of most interest here, however, we can show that continuity does, in essence, imply closure: in Theorem 2.23 we prove that every continuous linear transformation has a certain uniquely determined closed continuous linear extension.

DEFINITION 2.6. *A transformation T is said to be linear if its domain $\mathfrak{D}$ is a linear manifold and $T(a_1 f_1 + \cdots + a_n f_n) = a_1 Tf_1 + \cdots + a_n Tf_n$ whenever $f_1, \cdots, f_n$ are elements of $\mathfrak{D}$.*

We must point out that the term "linear transformation" is here used in a narrower sense than usual, since it is applied to what is ordinarily described as a homogeneous linear transformation.

THEOREM 2.3. *The domain and range of a linear transformation T are both linear manifolds.*

The domain is a linear manifold, by definition. We see at once that $\mathfrak{R}$, the range of T, is also a linear manifold: for, if $g_1, \cdots, g_n$ are elements of $\mathfrak{R}$, then there exist elements

$f_1, \cdots, f_n$ of $\mathfrak{D}$ such that $Tf_1 = g_1, \cdots, Tf_n = g_n$; since T is a linear transformation we have

$$\begin{aligned} g = a_1 g_1 + \cdots + a_n g_n &= a_1 Tf_1 + \cdots + a_n Tf_n \\ &= T(a_1 f_1 + \cdots + a_n f_n); \end{aligned}$$

thus the element g belongs to $\mathfrak{R}$.

THEOREM 2.4. *If T_0, T_1, T_2 are linear transformations, then aT_0, $T_1 + T_2$, $T_1 T_2$ are linear transformations.*

This theorem is an almost immediate consequence of the definitions of the transformations $a \cdot T_0$, $T_1 + T_2$, $T_1 \cdot T_2$. We shall give the proof only in the case of the third transformation, the other two being established in an even simpler fashion. Since $\mathfrak{D}$, the domain of $T_1 T_2$, consists of all those elements of $\mathfrak{D}_2$ which are transformed by T_2 into elements of $\mathfrak{D}_1$, it is easily seen that $\mathfrak{D}$ is a linear manifold, containing the null element at least. Thus $T_1 T_2$ exists and, whenever $f_1, \cdots, f_n$ are elements of its domain, satisfies the relations

$$\begin{aligned} T_1 T_2 f &= T_1 T_2 (a_1 f_1 + \cdots + a_n f_n) \\ &= T_1 (a_1 T_2 f_1 + \cdots + a_n T_2 f_n) \\ &= a_1 T_1 T_2 f_1 + \cdots + a_n T_1 T_2 f_n, \end{aligned}$$

as we wished to show.

THEOREM 2.5. *If the linear transformation T possesses an inverse, then T^{-1} is linear; if the closed transformation T possesses an inverse then T^{-1} is closed.*

If T if a linear transformation such that T^{-1} exists, then T^{-1} has as its domain the linear manifold $\mathfrak{R}$ in accordance with Theorems 2.2 and 2.3. If $g_1 = Tf_1, \cdots, g_n = Tf_n$, are elements of $\mathfrak{R}$, then $a_1 g_1 + \cdots + a_n g_n = T(a_1 f_1 + \cdots + a_n f_n)$ is in the domain of T^{-1} and $T^{-1}(a_1 g_1 + \cdots + a_n g_n) = a_1 f_1 + \cdots + a_n f_n = a_1 T^{-1} g_1 + \cdots + a_n T_n^{-1} g_n$, so that T^{-1} is a linear transformation.

If T is a closed transformation such that T^{-1} exists, then the fact that the pair of limiting equations $f_n \to f$, $Tf_n \to f^*$ is equivalent to the pair $g_n \to f^*$, $T^{-1} g_n \to f$, where $g_n = Tf_n$, shows that the closure of T implies that of T^{-1}.

The remainder of the present section will be devoted to the study of a fundamental relation between transformations which we will term the relation of adjointness.

DEFINITION 2.7. *Two transformations T_1 and T_2 are said to be adjoint to each other, in symbols $T_1 \frown T_2$, if they satisfy the relation $(T_1 f, g) = (f, T_2 g)$ for every f in $\mathfrak{D}_1$, and every g in $\mathfrak{D}_2$.*

The relation $\frown$ is symmetrical, but neither reflexive nor transitive: $T_1 \frown T_2$ implies $T_2 \frown T_1$; the relation $T \frown T$ is not true in general; and the relations $T_1 \frown T_2$ and $T_2 \frown T_3$ do not imply the relation $T_1 \frown T_3$.

This concept of adjointness receives its name from its analogy with certain concepts of algebra and analysis. As we shall see in the next chapter, the adjoint operators of analysis satisfy the condition imposed by the definition, in a sense which will be made more precise at the appropriate moment. On the other hand, we could quite as well introduce a different terminology based upon a geometrical analogy, for the relation with which we are dealing is satisfactorily described as one of duality. We prefer the term with an analytical background because it will be more suggestive of the uses we shall find for it. The relation of adjointness has a very intimate connection with the property of linearity, as we shall show in detail below. It is for this reason that the relation is one of primary importance.

Definition 2.7 always has content: if T_1 and T_2 are given they are not necessarily adjoint to each other; but, if T_1 is given, T_2 can always be determined so that it is adjoint to T_1, since we may choose T_2 as the trivial transformation which takes the null element into itself. Under certain restrictions we are able to characterize the class of all transformations adjoint to a given transformation.

THEOREM 2.6. *A necessary and sufficient condition that every pair of transformations adjoint to a transformation T coincide in the set of elements common to their domains is that the domain of T determine the closed linear manifold $\mathfrak{H}$. If T is a transformation whose domain determines the closed* *linear manifold* $\mathfrak{H}$, *then there exists a uniquely determined transformation* T^* *adjoint to* T, *with the property that* $T_1 \frown T$ *implies* $T_1 \subseteqq T^*$: *its domain* $\mathfrak{D}^*$ *consists of those and only those elements* g *such that the relation* $(Tf, g) = (f, g^*)$ *holds for all elements* f *in* $\mathfrak{D}$ *and some element* g^* *in* $\mathfrak{H}$; *and, for such an element,* $T^*g = g^*$.

In order that the relations $T_1 \frown T$ and $T_2 \frown T$ imply $T_1 = T_2$ in $\mathfrak{D}_1 \mathfrak{D}_2$, no matter how T_1 and T_2 may be chosen, it is necessary that $\mathfrak{D}$ determine the closed linear manifold $\mathfrak{H}$. For, when T and T_1 are given and this condition is not fulfilled, we can construct a transformation T_2 which is adjoint to T, has the same domain as T_1, and is nowhere equal to T_1. We select an element $\varphi \neq 0$ which is orthogonal to every element in $\mathfrak{D}$ and then define T_2 throughout the domain of T_1 by the relation $T_2 f = T_1 f + \varphi$. The asserted properties of the transformation T_2 are easily verified. On the other hand, if T_1 and T_2 are adjoint to T and $\mathfrak{D}$ determines the closed linear manifold $\mathfrak{H}$, then $T_1 = T_2$ in $\mathfrak{D}_1 \mathfrak{D}_2$: for if g is an element of the latter set we have $(f, T_1 g - T_2 g) = (f, T_1 g) - (f, T_2 g) = (Tf, g) - (Tf, g) = 0$ for every f in $\mathfrak{D}$ and can therefore conclude that $T_1 g - T_2 g = 0$, $T_1 g = T_2 g$, as we wished to show.

If T is a transformation whose domain determines the closed linear manifold $\mathfrak{H}$, it is clear that to a given element g there can correspond at most one element g^* such $(Tf, g) = (f, g^*)$ for every f in $\mathfrak{D}$. Thus the transformation T^* described in the theorem exists and is unique, provided merely that $\mathfrak{D}^*$ is not empty; but $g = 0$ is evidently an element of $\mathfrak{D}^*$ with the correspondent $g^* = 0$. The relation $T^* \frown T$ follows at once from the definition of the transformation T^*; and the fact that every transformation T_1 adjoint to T satisfies the relation $T_1 \subseteqq T^*$ is obvious.

DEFINITION 2.8. *The transformation* T^* *of Theorem* 2.6 *is called the adjoint of the transformation* T.

We shall find this concept of the adjoint of a given transformation of the greatest importance. For this reason we shall note a few elementary properties of the operation of forming the adjoint. The following relations are true whenever they are significant:

$$T_1 \subseteq T_2 \text{ implies } T_1^* \supseteq T_2^*, \qquad T_1^* + T_2^* \subseteq (T_1 + T_2)^*,$$
$$T_2^* T_1^* \subseteq (T_1 T_2)^*, \qquad (a \cdot T)^* \supseteq \bar{a} \cdot T^*.$$

Formal proof is unnecessary. A somewhat deeper relation analogous to these concerns the connection between inverses and adjoints.

THEOREM 2.7. *If T is a transformation whose domain and range each determine the closed linear manifold $\mathfrak{H}$ and if T possesses an inverse T^{-1}, then T^* and $(T^{-1})^*$ both exist and are inverses of one another.*

The existence of the adjoints of T and T^{-1} follows at once from Theorem 2.6. From the defining properties of these adjoints, we see that the equations

$$(Tf, (T^{-1})^* g) = (T^{-1} Tf, g) = (f, g)$$

hold for every f in the domain of T and every g in the domain of $(T^{-1})^*$; consequently, $(T^{-1})^* g$ is in the domain of T^* and the relation $T^*(T^{-1})^* g = g$ is true for every g in the domain of $(T^{-1})^*$. In a similar manner, we conclude from the equations

$$(T^{-1}f, T^* g) = (T T^{-1} f, g) = (f, g)$$

that the relation $(T^{-1})^* T^* g = g$ is significant and true for every g in the domain of T^*. From the two relations connecting T^* and $(T^{-1})^*$ it is evident upon examination that these transformations are inverses of one another.

THEOREM 2.8. *When T^*, the adjoint of a transformation T, exists, it is a closed linear transformation.*

We prove that T^* is a linear transformation, as follows: if $g_1, \cdots, g_n$ are elements of $\mathfrak{D}^*$ and if g and g^* are the elements $a_1 g_1 + \cdots + a_n g_n$ and $a_1 T^* g_1 + \cdots + a_n T^* g_n$, respectively, we verify the relations.

$$(Tf, g) = \sum_{\alpha=1}^{n} \bar{a}_\alpha (Tf, g_\alpha) = \sum_{\alpha=1}^{n} \bar{a}_\alpha (f, T^* g_\alpha) = (f, g^*)$$

for every f in $\mathfrak{D}$; by definition, therefore, g is an element of $\mathfrak{D}^*$ and satisfies the relation $T^* g = g^*$; consequently, T^* is a linear transformation.

In order to show that T^* is a closed transformation, we must consider the properties of those sequences $\{g_n\}$ in $\mathfrak{D}^*$ such that

$$g_n \to g, \qquad T^* g_n \to g^*.$$

For such a sequence we have $(Tf, g_n) = (f, T^* g_n)$ for every f in the domain of T and for $n = 1, 2, 3, \cdots$. When n becomes infinite we obtain the equation

$$(Tf, g) = (f, g^*)$$

and conclude that g is in $\mathfrak{D}^*$ and that $T^* g = g^*$, as we wished to show.

THEOREM 2.9. *If T is a transformation such that T^* and $T^{**} \equiv (T^*)^*$ both exist, then T^{**} is a closed linear extension of T with the adjoint $(T^{**})^* \equiv T^*$. If in addition T^{-1} exists and the ranges of T and T^* both determine the closed linear manifold $\mathfrak{H}$, then $(T^{-1})^{**}$ exists and is the inverse of T^{**}.*

T^* exists if and only if $\mathfrak{D}$ determines the closed linear manifold $\mathfrak{H}$, T^{**} if and only if $\mathfrak{D}^*$ has the same property. When T^{**} exists we have $T^{**} \supseteqq T$ by virtue of the fact that $T \frown T^*$. By Theorem 2.8, T^{**} is a closed linear transformation. Since T^{**} is an extension of T its domain $\mathfrak{D}^{**}$ determines the closed linear manifold $\mathfrak{H}$ and $(T^{**})^*$ exists. Because of the relations $T^{**} \supseteqq T$ and $T^* \frown T^{**}$, this transformation must have the properties $(T^{**})^* \subseteqq T^*$ and $(T^{**})^* \supseteqq T^*$ respectively. Hence it coincides with T^*. The last assertion of the theorem is an obvious corollary of Theorem 2.7: the theorem can be used first to establish the existence of $(T^*)^{-1}$ and can then be applied to T^* and its adjoint T^{**} to obtain the result described.

As a result of Theorems 2.6–2.9 we see that the concept of adjointness is trivial unless we deal with transformations which enjoy linear properties. The relation between adjointness and linearity is made precise in the following theorem.

THEOREM 2.10. *If the transformation T has a linear extension, then there exists a unique linear transformation $\hat{T}$ with the properties.*

(1) *$\hat{T}$ is an extension of T;*

(2) *every linear extension of T is also an extension of $\hat{T}$. If the transformation T has a closed linear extension, then there exists a unique closed linear transformation $\tilde{T}$ with the properties:*

(1) *$\tilde{T}$ is an extension of T and of $\hat{T}$;*

(2) *every closed linear extension of T is also an extension of $\tilde{T}$. In particular, if T is linear, then $\hat{T} \equiv T$; if T is closed and linear, then $\tilde{T} \equiv \hat{T} \equiv T$; and, if the adjoints T^* and T^{**} exist, then $\hat{T}$ and $\tilde{T}$ exist and satisfy the relations $T \subseteqq \hat{T} \subseteqq \tilde{T} \subseteqq T^{**}$.*

When T has a linear extension T_0, we can define $\hat{T}$ by setting $\hat{T} = T_0$ in the linear manifold $\hat{\mathfrak{D}}$ determined by the domain $\mathfrak{D}$ of T. Since T_0 is a linear transformation with a linear manifold $\mathfrak{D}_0$ as its domain, we have $\mathfrak{D} \subseteqq \hat{\mathfrak{D}} \subseteqq \mathfrak{D}_0$ and infer that the definition of $\hat{T}$ is effective. The properties ascribed to $\hat{T}$ are evident.

When T has a closed linear extension T_0, the transformation $\hat{T}$ exists; and we define $\tilde{T}$ directly in terms of $\hat{T}$ as follows: the elements f and $\tilde{f}$ are assigned to the domain and to the range of $\tilde{T}$ respectively and are connected by the relation $\tilde{T}f = \tilde{f}$ if and only if to an arbitrary positive ϵ there corresponds an element g in $\hat{\mathfrak{D}}$ satisfying the inequalities.

$$\|f - g\| < \epsilon, \qquad \|\tilde{f} - \hat{T}g\| < \epsilon.$$

Special pairs of elements satisfying these requirements can evidently be obtained by selecting f arbitrarily in $\hat{\mathfrak{D}}$ and setting $\tilde{f} = \hat{T}f$; for the inequalities above are satisfied when f is used as the element g. On the other hand, an arbitrary pair of elements satisfying the indicated conditions has the property that f is in $\mathfrak{D}_0$ and that $T_0 f = \tilde{f}$. To show this, we use the conditions laid on f and $\tilde{f}$ to select a sequence $\{g_n\}$ in $\hat{\mathfrak{D}}$ such that

$$\|f - g_n\| < 1/n, \qquad \|\tilde{f} - \hat{T}g_n\| < 1/n.$$

We can write these relations in the form $g_n \to f$, $T_0 g_n = \hat{T} g_n \to \tilde{f}$, and can then conclude by virtue of the closure of T_0 that the equation $T_0 f = \tilde{f}$ is significant and true. Thus we see that $\tilde{T}$ is properly defined and that the relations $T \subseteqq \hat{T} \subseteqq \tilde{T} \subseteqq T_0$, $\mathfrak{D} \subseteqq \hat{\mathfrak{D}} \subseteqq \tilde{\mathfrak{D}} \subseteqq \mathfrak{D}_0$ are valid. We prove next that $\tilde{T}$ is a linear transformation: if $f_1, \cdots, f_n$ are arbitrary elements of $\tilde{\mathfrak{D}}$, if $a_1, \cdots, a_n$ are arbitrary complex constants, and if ε is a given positive number, we select a positive δ and elements $g_1, \cdots, g_n$ in $\hat{\mathfrak{D}}$ such that

$$\delta(|a_1| + \cdots + |a_n|) < \varepsilon, \quad \| f_k - g_k \| < \delta, \quad \| \tilde{T} f_k - \hat{T} g_k \| < \delta$$

for $k = 1, \cdots, n$; the elements $f = a_1 f_1 + \cdots + a_n f_n$, $\tilde{f} = a_1 \tilde{T} f_1 + \cdots + a_n \tilde{T} f_n$, and $g = a_1 g_1 + \cdots + a_n g_n$ then satisfy the relations

$$\| f - g \| \leqq |a_1| \| f_1 - g_1 \| + \cdots + |a_n| \| f_n - g_n \| < \varepsilon,$$
$$\| \tilde{f} - \hat{T} g \| \leqq |a_1| \| \tilde{T} f_1 - \hat{T} g_1 \| + \cdots + |a_n| \| \tilde{T} f_n - \hat{T} g_n \| < \varepsilon;$$

and we conclude that f is in $\tilde{\mathfrak{D}}$, $\tilde{T} f = \tilde{T}(a_1 f_1 + \cdots + a_n f_n) = a_1 \tilde{T} f_1 + \cdots + a_n \tilde{T} f_n$. The closure of $\tilde{T}$ is demonstrated in a similar manner. Let $\{f_n\}$ be a sequence in $\tilde{\mathfrak{D}}$ such that $f_n \to f$ and $\tilde{T} f_n \to \tilde{f}$. When a positive ε is assigned, we first choose m so large that

$$\| f - f_m \| < \varepsilon/2, \qquad \| \tilde{f} - \tilde{T} f_m \| < \varepsilon/2;$$

we can then select an element g in $\hat{\mathfrak{D}}$ such that

$$\| f_m - g \| < \varepsilon/2, \qquad \| \tilde{T} f_m - \hat{T} g \| < \varepsilon/2;$$

and we see finally that

$$\| f - g \| < \varepsilon, \qquad \| \tilde{f} - \hat{T} g \| < \varepsilon.$$

Hence $\tilde{T} f$ exists and is equal to $\tilde{f}$; in other words, $\tilde{T}$ is a closed transformation. By virtue of the fact that we have defined $\tilde{T}$ in terms of $\hat{T}$ and have used the existence of T_0 only to show that the definition is effective, we can vary the left-hand term in the relation $T_0 \supseteqq \tilde{T}$ without changing that on the right. Thus every closed linear extension of T is also an extension of $\tilde{T}$.

The closing remarks of the theorem require no particular comment, when it is recalled in connection with the last assertion that T^{**} is a closed linear extension of T by Theorem 2.9.

There exist transformations which have no linear extensions: the transformation with a closed linear manifold $\mathfrak{M}$ as its domain and the set comprising a single element $f \neq 0$ as its range is an obvious example. There exist linear transformations which have no closed linear extensions. As an instance we cite the transformation T defined as follows: the domain of T is the linear manifold determined by a complete orthonormal set $\{\varphi_n\}$; when f is an element of this manifold we have

$$f = \sum_{\alpha=1}^{n} a_\alpha \varphi_\alpha, \qquad Tf = \left(\sum_{\alpha=1}^{n} a_\alpha\right) \varphi_1 .$$

It is evident that T is a linear transformation. Let us suppose that T_0 is a closed linear extension of T. The element $f_n = \frac{1}{n^2} \sum_{\alpha=1}^{n} \alpha \varphi_\alpha$ is in the domain of T and has the properties

$$\|f_n\|^2 = \sum_{\alpha=1}^{n} \alpha^2/n^4 = (n+1)(2n+1)/6n^3 \to 0, \quad n \to \infty,$$

$$Tf_n = \left(\sum_{\alpha=1}^{n} \alpha/n^2\right) \varphi_1 = (n+1)/2n \cdot \varphi_1 \to \tfrac{1}{2} \varphi_1, \qquad n \to \infty.$$

Thus $f_n \to 0$, $T_0 f_n = Tf_n \to \frac{1}{2}\varphi_1$. Since T_0 is closed by hypothesis, we must have $T_0 0 = \frac{1}{2}\varphi_1$; on the other hand, since T_0 is linear by hypothesis, we must also have $T_0 0 = 0$. These two results are plainly incompatible, so that T can have no closed linear extension. From these examples we conclude that the assertions of Theorem 2.10 cannot be strengthened.

In the theorem just proved, we have not distinguished between proper and improper extensions. It is consequently of some interest to add a few remarks on this point. We can prove the assertion: a necessary and sufficient condition that a linear (closed linear) transformation T have a proper linear (proper closed linear) extension T_0 is that the domain of T be a proper subset of $\mathfrak{H}$. The necessity of the condition is trivial; and the sufficiency can be demonstrated in a few lines. Let $\mathfrak{D} \subset \mathfrak{H}$ be the linear manifold which is the domain of T, h an arbitrary element of the set $\mathfrak{H} - \mathfrak{D}$, and h_0 an arbitrary element in $\mathfrak{H}$. The domain of T_0 is to be the linear manifold comprising all the elements expressible in the form $f + ah$ where f is in $\mathfrak{D}$ and a is an arbitrary complex number; for such an element we set $T_0(f + ah) = Tf + ah_0$. It is clear that T_0 is a proper linear extension of T. Furthermore, when T is closed, T_0 has the same property, as we shall now show. Let $\{g_n\}$ be a sequence in $\mathfrak{D}_0$ such that $g_n \to g$, $T_0 g_n \to g_0$. We have $g_n = f_n + a_n h$ where f_n is in $\mathfrak{D}$. By retaining only an appropriately chosen subsequence of $\{g_n\}$ and renumbering the elements retained, we may suppose that we have

$$g_n = f_n + a_n h \to g, \qquad T_0 g_n = Tf_n + a_n h_0 \to g_0,$$

where the sequence $\{a_n\}$ has one of the two properties $|a_n| \to \infty$, $a_n \to a$. The first property, however, cannot be realized under our hypotheses concerning the transformation T. If $|a_n| \to \infty$ we have

$$(-1/a_n) f_n = (-1/a_n) g_n + h \to h,$$
$$T(-1/a_n) f_n = (-1/a_n) T_0 g_n + h_0 \to h_0,$$

and can conclude by virtue of the closure of T that h is in $\mathfrak{D}$, contrary to fact. Hence the sequence $\{a_n\}$ must enjoy the second property. Here we can conclude from the relations

$$f_n = g_n - a_n h \to g - ah, \qquad Tf_n = T_0 g_n - a_n h_0 \to g_0 - a h_0$$

that $g - ah$ is in $\mathfrak{D}$ and that $T(g - ah) = g_0 - ah_0$. It then follows that $g = (g - ah) + ah$ is in $\mathfrak{D}_0$ and that $T_0 g = (g_0 - ah_0) + ah_0 = g_0$. This result implies that T_0 is a closed transformation. It is instructive to compare the content of this paragraph with Theorems 2.13–2.16 of the following section.

§ 2. Symmetric Transformations

Those transformations H which satisfy the relation $H \frown H$ are of particular interest because of their frequent occurrence among the operators of analysis. The complete development of their properties is one of the chief aims of the following chapters. Here we investigate a few of their simpler and more elementary characteristics. In order to exclude trivial and relatively unimportant transformations H from our considerations, we shall confine our attention to transformations of the indicated type which have domains determining the closed linear manifold $\mathfrak{H}$. It has been customary to refer to such transformations as "Hermitian symmetric" but in the interests of brevity we call them merely "symmetric". Thus our definition reads:

Definition 2.9. *A transformation H is said to be a symmetric transformation if*

(1) *its domain determines the closed linear manifold $\mathfrak{H}$.*
(2) *it is adjoint to itself, $H \frown H$.*

Theorem 2.11. *If H is a symmetric transformation, then H^* and H^{**} exist. H^{**} is a closed linear symmetric transformation.* H symmetric $\Rightarrow H \subseteq H^{**} \subseteq H^{*}$

Since the domain of H determines the closed linear manifold $\mathfrak{H}$ we know that H^* exists. Now $H \frown H$ implies $H \subseteq H^*$, so that the condition for the existence of H^{**} is fulfilled. From the relations $H^{**} \frown H^*$ and $H^* \frown H$, we conclude that $H^{**} \supseteq H$; hence $H^{**} \subseteq H^*$ and $H^{**} \frown H^{**}$, as we wished to show.

Theorem 2.12. *If H is a symmetric transformation, then the linear transformation $\hat{H}$ and the closed linear transformation $\tilde{H}$ exist and are symmetric.*

The conditions for the existence of $\hat{H}$ and $\tilde{H}$, as set forth in Theorem 2.10, are satisfied; and since $H \subseteq \hat{H} \subseteq \tilde{H} \subseteq H^{**}$ we see that both $\hat{H}$ and $\tilde{H}$ are symmetric. In Theorem 9.5 we shall prove the more precise and more difficult result $\tilde{H} \equiv H^{**}$.

(See Naimark, p. 102)

Among the symmetric transformations, two types are of particular interest—the maximal symmetric transformations and the self-adjoint transformations.† In a sense which we shall explain in the next chapter, the self-adjoint transformations comprise the self-adjoint operators of analysis.

DEFINITION 2.10. *A symmetric transformation is said to be maximal if it has no proper symmetric extension.*

In symbols, H is maximal if the relations $H_0 \supseteq H$ and $H_0 \wedge H_0$ together imply $H_0 \equiv H$.

DEFINITION 2.11. *A symmetric transformation H is said to be self-adjoint if $H \equiv H^*$.*

THEOREM 2.13. *A self-adjoint transformation is a maximal symmetric transformation.*

Let H_0 be a symmetric extension of the self-adjoint transformation H. Then $H_0 \supseteq H$ implies that $H_0 \subseteq H^* \equiv H$ so that H_0 and H must be identical. In other words, H has no proper symmetric extension, and is therefore maximal.

It thus appears that the self-adjoint transformations form a subclass of the class of all maximal symmetric transformations. We shall see later that this subclass is a proper subclass; that is, that there exist maximal symmetric transformations which are not self-adjoint.

We are now in a position to formulate a series of fundamental problems concerning symmetric transformations:

(1) the determination of all the maximal symmetric extensions of a given symmetric transformation H;

(2) the determination of all maximal symmetric transformations;

(3) the determination of all self-adjoint transformations.

We shall obtain a solution of each of these problems in the course of this book. In Chapter IX we shall give an account of the beautiful methods by which J. v. Neumann has solved the first two problems. Since our subject matter is more intimately related to developments clustering around the third

† J. v. Neumann employs the term "hypermaximal" to denote the self-adjoint transformations, Mathematische Annalen, 102 (1929), pp. 49–131, Definition 9, p. 72.

problem, we shall investigate its solution and ramifications first.

A few elementary details relating to these problems will be given here.

THEOREM 2.14. *If H is a maximal symmetric transformation, then H and H^{**} are identical; H is therefore a closed linear transformation.*

Since H^{**} is a symmetric extension of H, we must have $H \equiv H^{**}$.

THEOREM 2.15. *Every maximal symmetric extension of a given symmetric transformation H is also an extension of H^{**}.*

If H_0 is a maximal symmetric extension of H, the relation $H_0 \supseteqq H$ implies the relations $H_0^* \subseteqq H^*$ and $H_0^{**} \supseteqq H^{**}$. Since H_0 and H_0^{**} are identical, our theorem is true.

As a consequence of this theorem we see that, in discussing our first problem above, we may limit our attention to closed linear symmetric transformations.

THEOREM 2.16. *If H is a symmetric transformation such that H^* and H^{**} are identical, then the self-adjoint transformation H^* is the only maximal symmetric extension of H.*

The symmetric transformation H^{**} is self-adjoint under our hypothesis, and is therefore maximal. If H_0 is a maximal symmetric extension of H, it is also an extension of H^{**}, by the preceding theorem. Hence H_0, H^{**}, and H^* must coincide.

As a consequence of this theorem we introduce

DEFINITION 2.12. *A symmetric transformation H such that H^* and H^{**} are identical is said to be essentially self-adjoint.*

THEOREM 2.17. *If H is a symmetric transformation whose domain is $\mathfrak{H}$, then H is self-adjoint.*

Since H^* and H^{**} are extensions of H, all three transformations must coincide.

THEOREM 2.18. *If H is a symmetric transformation whose range determines the closed linear manifold $\mathfrak{H}$, then H possesses a symmetric inverse H^{-1}. In order that H be maximal, self-adjoint, or essentially self-adjoint, it is necessary and sufficient that H^{-1} enjoy the same property.*

We denote by $\mathfrak{D}$, $\mathfrak{D}^*$, $\mathfrak{D}^{**}$ and $\mathfrak{R}$, $\mathfrak{R}^*$, $\mathfrak{R}^{**}$ the domains and the ranges of the respective transformations H, H^* and H^{**}, remarking the obvious relations

$$H \subseteqq H^{**} \subseteqq H^*, \qquad \mathfrak{D} \subseteqq \mathfrak{D}^{**} \subseteqq \mathfrak{D}^*, \qquad \mathfrak{R} \subseteqq \mathfrak{R}^{**} \subseteqq \mathfrak{R}^*.$$

If an element is orthogonal to any one of the sets $\mathfrak{R}$, $\mathfrak{R}^{**}$, $\mathfrak{R}^*$, then it must be the null element; for each of these sets determines the closed linear manifold $\mathfrak{H}$.

Evidently, H will possess an inverse with domain $\mathfrak{R}$ if and only if the relation $Hf_1 = Hf_2$ implies $f_1 = f_2$, whenever it is significant. If f_1 and f_2 are elements of $\mathfrak{D}$ satisfying this relation, we have

$$(f_1 - f_2, Hg) = (Hf_1, g) - (Hf_2, g) = 0$$

for every g in $\mathfrak{D}$; hence $f_1 - f_2$ must be the null element, and f_1 and f_2 are coincident. The condition that H have an inverse H^{-1} is therefore satisfied. From Theorems 2.7 and 2.9, we see that H^* and H^{**} also have inverses, identical with $(H^{-1})^*$ and $(H^{-1})^{**}$ respectively.

The assertions of the present theorem are obvious consequences of the relations $H^{-1} \subseteqq (H^{-1})^{**} \subseteqq (H^{-1})^*$, which are implied by the corresponding relations $H \subseteqq H^{**} \subseteqq H^*$. We see for instance that $(H^{-1})^{**} \curlywedge (H^{-1})^{**}$ follows from the fact that $(H^{-1})^*$ is an extension of $(H^{-1})^{**}$, and that $H^{-1} \curlywedge H^{-1}$ is then a consequence of the fact that $(H^{-1})^{**}$ is an extension of H^{-1}. Since $\mathfrak{R}$ and $\mathfrak{R}^{**}$ determine the closed linear manifold $\mathfrak{H}$, H^{-1} and $(H^{-1})^{**}$ are symmetric transformations in accord with Definition 2.8. From the evident equivalence of the two identities $H \equiv H^*$, $H^{-1} \equiv (H^{-1})^*$ and of the further identities $H^* \equiv H^{**}$, $(H^{-1})^* \equiv (H^{-1})^{**}$, we perceive that for H to be self-adjoint or essentially self-adjoint it is necessary and sufficient that H^{-1} enjoy the same property. If H is maximal, H^{-1} must be maximal, and conversely; for if H_0 is a proper symmetric extension of H, then H_0^{-1} exists and is a proper symmetric extension of H^{-1}, by the facts already established.

As an immediate result of the two preceding theorems, we have

THEOREM 2.19. *If H is a symmetric transformation whose range is $\mathfrak{H}$, then H is self-adjoint.*

§ 3. BOUNDED LINEAR TRANSFORMATIONS

From the geometrical point of view, those transformations which are continuous as well as linear are to be counted among the most interesting. As we shall see, the class of all continuous linear transformations is characterized by a certain property which we shall call boundedness and which serves as a very useful tool in many subsequent proofs: this property may be described by the assertion that no element within unit distance of the null element is taken by a continuous linear transformation into an element whose distance from the null element exceeds a certain positive constant determined by the transformation. The properties of continuous linear transformations are easily investigated by virtue of this circumstance.

We shall first prove the result:

THEOREM 2.20. *If a linear transformation T is continuous at one element of its domain, then it is uniformly continuous throughout its domain. A linear transformation is therefore either uniformly continuous or totally discontinuous in its domain.*

We denote by f_0 that element of $\mathfrak{D}$ at which T is known to be continuous, and by f an arbitrary element of $\mathfrak{D}$. When a positive ε is given, we must be able to determine a $\delta = \delta(\varepsilon)$ independent of f, so that, whenever g is an element of the domain of T for which $\|f-g\| \leqq \delta$, the truth of the inequality $\|Tf - Tg\| \leqq \varepsilon$ can be asserted. In order to determine δ we transfer our attention from f to f_0 by the introduction of the element $g_0 = g - f + f_0$, evidently in the linear manifold $\mathfrak{D}$, remarking that $Tg_0 = Tg - Tf_0 + Tf$. We have

$$\|f_0 - g_0\| = \|f - g\|, \qquad \|Tf_0 - Tg_0\| = \|Tf - Tg\|.$$

Since T is continuous at f_0, there exists a $\delta = \delta(\varepsilon)$, independent of g_0 such that $\|f_0 - g_0\| \leqq \delta$ implies $\|Tf_0 - Tg_0\| \leqq \varepsilon$. The equations just written down show that, for the same ε and δ, we have $\|Tf - Tg\| \leqq \varepsilon$ whenever $\|f - g\| \leqq \delta$, as we wished to prove.

THEOREM 2.21. *When T is a linear transformation and C is an appropriate constant, depending only on T, then the following assertions are equivalent:*

(1) *T is continuous;*

(2) $\|Tf\| \leqq C\|f\|$ *for every f in $\mathfrak{D}$;*

(3) $|(Tf, g)| \leqq C \cdot \|f\| \cdot \|g\|$ *for every f in $\mathfrak{D}$ and every g in $\mathfrak{H}$.*

When T is also symmetric, the assertion

(4) $|(Tf, f)| \leqq C\|f\|^2$ *for every f in $\mathfrak{D}$*

is equivalent to the preceding three.

Since T is linear its domain contains the null element and $T0 = 0$. The assertion that T is continuous is equivalent to the assertion that T is continuous at 0, as we showed in the previous theorem.

The equivalence of assertions (1) and (2) of the present theorem is established by showing directly that (2) is a necessary and sufficient condition for the continuity of T at 0. If the transformation T is continuous, then there exists a positive number $\delta = \delta(\varepsilon)$ such that $\|Tg\| \leqq \varepsilon$ whenever g is an element of $\mathfrak{D}$ for which $\|g\| \leqq \delta$. Consequently there must exist a positive constant C, not greater than $1/\delta(1)$, such that $\|Tg\| \leqq C$ whenever g is an element of $\mathfrak{D}$ for which $\|g\| \leqq 1$: for if there were an element f in $\mathfrak{D}$ such that $\|Tf\| > 1/\delta(1)$, $\|f\| \leqq 1$, we should find for $g = \delta(1) \cdot f$ the inequalities $\|Tg\| > 1$, $\|g\| \leqq \delta(1)$. The existence of such a constant C implies the truth of the inequality (2). For if f is the null element, the inequality of (2) is trivial; and if f is not the null element, we can set $g = f / \|f\|$, $Tg = Tf / \|f\|$ in the relations above, and thus obtain $\|g\| = 1$, $\|Tg\| = \|Tf\|/\|f\| \leqq C$, and $\|Tf\| \leqq C\|f\|$. On the other hand, if (2) is satisfied, T is continuous at 0: for upon setting $\varepsilon = C\delta$ we see that the inequality $\|Tf\| \leqq C\|f\|$ yields $\|Tf\| \leqq \varepsilon$ whenever $\|f\| \leqq \delta$.

We now perceive that (3) follows directly from (2), according to the inequalities $|(Tf, g)| \leqq \|Tf\| \cdot \|g\| \leqq C \cdot \|f\| \cdot \|g\|$. On the other hand, (2) is trivial when Tf is the null element, and can be derived from (3) when $Tf \neq 0$, as we find by the inequalities $\|Tf\|^2 = (Tf, Tf) \leqq C\|f\|\|Tf\|$, $\|Tf\| \leqq C\|f\|$, obtained by putting $g = Tf$ in (3).

When it is assumed that T is symmetric as well as linear, we can sharpen the condition (3), which obviously implies (4), by showing that (3) is in turn a consequence of (4). Before proceeding to a proof of this assertion, we note that if (3) holds for every f and g in the domain of the symmetric transformation T, it holds also for every f in the domain of T and every g in $\mathfrak{H}$, on account of the continuity of the function (f, g). We may restrict our attention, therefore, to the case in which both f and g are elements of $\mathfrak{D}$. Since the inequality (3) is trivial when (Tf, g) vanishes, we may suppose that this circumstance does not arise; in particular, neither f nor g is the null element. We now set $a = \|g\|/\|f\|$, and determine θ so that $|\theta| = 1$ and $\theta(Tf, g)$ is real and positive. We see that $|(Tf, g)| = \theta(Tf, g) = \left(Ta\theta f, \frac{1}{a} g\right)$. It is possible to obtain an appraisal of the magnitude of the real part of (Tf_1, g_1), where f_1 and g_1 are arbitrary elements of $\mathfrak{D}$ by applying (4) to the identity

$$\begin{aligned}\mathfrak{R}(Tf_1, g_1) &= \tfrac{1}{2}(Tf_1, g_1) + \tfrac{1}{2}(g_1, Tf_1) \\ &= \tfrac{1}{2}(Tf_1, g_1) + \tfrac{1}{2}(Tg_1, f_1) \\ &= \left(T\frac{f_1+g_1}{2}, \frac{f_1+g_1}{2}\right) - \left(T\frac{f_1-g_1}{2}, \frac{f_1-g_1}{2}\right).\end{aligned}$$

We have, therefore,

$$\begin{aligned}|\mathfrak{R}(Tf_1, g_1)| &\leqq \left|\left(T\frac{f_1+g_1}{2}, \frac{f_1+g_1}{2}\right)\right| + \left|\left(T\frac{f_1-g_1}{2}, \frac{f_1-g_1}{2}\right)\right| \\ &\leqq C\left[\left\|\frac{f_1+g_1}{2}\right\|^2 + \left\|\frac{f_1-g_1}{2}\right\|^2\right]\end{aligned}$$

and can so simplify the last expression as to obtain the desired inequality

$$|\mathfrak{R}(Tf_1, g_1)| \leqq C[\tfrac{1}{2}\|f_1\|^2 + \tfrac{1}{2}\|g_1\|^2].$$

In this equality we replace f_1 by $a\theta f$ and g_1 by g/a. Upon taking account of the definitions of a and θ, we are able to reduce the resulting inequality to (3):

$$|(Tf, g)| \leqq C\|f\|\|g\|.$$

Thus (3) follows from (4) under the conditions assumed.

In view of the theorem which has just been proved, it is desirable to introduce the following definitions.

DEFINITION 2.13. *A linear transformation T is said to be bounded if the equivalent inequalities $\|Tf\| \leqq C\|f\|$ and $|(Tf, g)| \leqq C\|f\|\|g\|$ hold for every f in $\mathfrak{D}$, every g in $\mathfrak{H}$, and some constant $C \geqq 0$. The least admissible value of C is called the bound of T.*

DEFINITION 2.14. *A linear symmetric transformation H is said to be bounded above if the inequality $(Hf, f) \leqq C_1\|f\|$ holds for every f in its domain and some constant C_1; and is said to be bounded below if the inequality $(Hf, f) \geqq C_2\|f\|$ holds for every f in its domain and some constant C_2. The least admissible value of C_1 is called the upper bound of H, the greatest admissible value of C_2 the lower bound of H. The transformation H is said to be positive definite or not-negative definite according as its lower bound is positive or zero; and is said to be negative definite or not-positive definite according as its upper bound is negative or zero.*

We can sum up the content of Theorem 2.21 and add a few obvious remarks concerning the bound of a linear transformation in the following statement.

THEOREM 2.22. *A linear transformation is bounded if and only if it is continuous; a linear symmetric transformation is bounded if and only if it is bounded above and below. If T, T_1, T_2 are bounded linear transformations with domain $\mathfrak{H}$ and the respective bounds C, C_1, C_2, then the transformations aT, $T_1 + T_2$, and $T_1 T_2$ are bounded linear transformations; and the bound of aT is $|a|C$, the bound of $T_1 + T_2$ does not exceed $C_1 + C_2$, and the bound of $T_1 T_2$ does not exceed $C_1 C_2$. The only linear transformation T with domain $\mathfrak{H}$ and bound 0 is the transformation O. If a linear symmetric* *transformation H has C, C_1, and C_2 as its bound, its upper bound, and its lower bound respectively, then C is the maximum of $|C_1|$ and $|C_2|$.*

The relation between continuity and boundedness which has just been discussed enables us to show that a continuous transformation fails to be closed only if its domain is too narrowly restricted.

THEOREM 2.23. *If T is a continuous linear transformation, then the transformation $\tilde{T}$ exists and is a closed continuous linear transformation whose domain is the closed linear manifold determined by the domain of T; the transformations T and $\tilde{T}$ have a common bound. If in particular the domain of T is everywhere dense in $\mathfrak{H}$, then $\tilde{T}$ is defined throughout $\mathfrak{H}$ and is the only closed linear extension of T.*

We first construct a closed linear extension T_0 of the transformation T, directly in terms of $T \equiv \hat{T}$: the elements f and f_0 are assigned to the domain and to the range of T_0 respectively and are connected by the relation $T_0 f = f_0$ if and only if to an arbitrary positive ε there corresponds an element g in $\mathfrak{D}$ satisfying the inequalities

$$\|f - g\| < \varepsilon, \qquad \|f_0 - Tg\| < \varepsilon.$$

By hypothesis, T is continuous and therefore bounded; we denote its bound by C. Consequently, if f is an arbitrary element of the closed linear manifold determined by $\mathfrak{D}$ and if $\{g_n\}$ is an arbitrary sequence in $\mathfrak{D}$ which converges to f, the sequence $\{Tg_n\}$ converges and has an element f_0 as its limit: the relation $\|Tg_m - Tg_n\| \leqq C\|g_m - g_n\| \to 0$ leads immediately to this result. Furthermore, it is clear that the limit f_0 is independent of the particular sequence $\{g_n\}$ converging to f. For if $\{g_n^{(1)}\}$ and $\{g_n^{(2)}\}$ are two sequences in $\mathfrak{D}$ such that $g_n^{(1)} \to f$, $g_n^{(2)} \to f$, $Tg_n^{(1)} \to f_0^{(1)}$, $Tg_n^{(2)} \to f_0^{(2)}$, we have

$$\|f_0^{(1)} - f_0^{(2)}\| = \lim_{n\to\infty} \|Tg_n^{(1)} - Tg_n^{(2)}\| \leqq C \lim_{n\to\infty} \|g_n^{(1)} - g_n^{(2)}\| = 0$$

and thus conclude that $f_0^{(1)} = f_0^{(2)}$. Finally, we note that the elements f and f_0 satisfy the conditions stated in the definition of T_0. Hence T is defined throughout the closed linear manifold determined by $\mathfrak{D}$. To show that the transformation T_0 is closed and linear, we can repeat the reasoning applied in the proof of Theorem 2.10 to the transformation $\tilde{T}$.

Since we now know that T has T_0 as a closed linear extension, we can infer the existence of the transformation $\tilde{T}$ described in Theorem 2.10. Evidently T_0 coincides with $\tilde{T}$, both transformations being constructed by the same process from the transformation $T \equiv \hat{T}$. We next show that $T_0 \equiv \tilde{T}$ has the same bound as T and is therefore continuous. If f is an arbitrary element in $\tilde{\mathfrak{D}}$ and ϵ is an arbitrary positive number, we can choose g in $\mathfrak{D}$ so that $\|f-g\| < \epsilon$ and $\|\tilde{T}f - Tg\| < \epsilon$ and thus obtain

$$\|\tilde{T}f\| = \|Tg + \tilde{T}(f-g\| \leqq \|Tg\| + \|\tilde{T}f - Tg\| \leqq C\|g\| + \epsilon$$
$$\leqq C\|f\| + C\|f-g\| + \epsilon \leqq C\|f\| + (C+1)\,\epsilon.$$

Since ϵ is arbitrary this inequality implies $\|\tilde{T}f\| \leqq C\|f\|$. Clearly the bound of $\tilde{T}$ cannot be less than that of $T \subseteqq \tilde{T}$; hence T and $\tilde{T}$ have the same bound C. The particular case in which $\mathfrak{D}$ is everywhere dense in $\mathfrak{H}$ so that $\tilde{\mathfrak{D}}$ must coincide with $\mathfrak{H}$ requires no detailed comment.

An interesting and useful application of these results can be made to the theory of continuous or bounded linear symmetric transformations.

THEOREM 2.24. *Every bounded linear symmetric transformation is essentially self-adjoint. A bounded linear symmetric transformation is self-adjoint if and only if its domain is $\mathfrak{H}$.*

Let H be a bounded linear symmetric transformation. Then $\tilde{H}$ can be constructed by the use of Theorem 2.10 or by the use of Theorem 2.23. According to the first theorem, as particularized in Theorem 2.12, $\tilde{H}$ is symmetric; and according to the second theorem $\tilde{H}$ has $\mathfrak{H}$ as its domain. By Theorem 2.17 $\tilde{H}$ is self-adjoint; and H must be essentially self-adjoint. H is itself a self-adjoint transformation if and only if it coincides with $\tilde{H}$; that is, if and only if its domain is $\mathfrak{H}$.

We shall next investigate the connection between the existence of the adjoint transformation T^* and boundedness, arriving at the important result that when T and T^* have $\mathfrak{H}$ as their common domain T is necessarily bounded and thus continuous. We encounter all the essential difficulties of the demonstration if we limit our attention to the special case of a symmetric transformation with $\mathfrak{H}$ as its domain. The general case is easily reduced to this apparently special one by a simple device, indicated in Theorem 2.26.

THEOREM 2.25. *If H is a symmetric transformation whose domain is $\mathfrak{H}$, then H is bounded and therefore continuous.*

(The fact that H is separable is not used in the proof.) Actually, the closed graph theorem is available.

We shall prove the theorem by a *reductio ad absurdum* based upon the following assertion: if $\mathfrak{M}_1$ is a closed linear manifold with dimension number n, $n = 1, 2, 3, \cdots$, and $\mathfrak{M}_2$ is the closed linear manifold $\mathfrak{H} \ominus \mathfrak{M}_1$, then a necessary and sufficient condition that H be bounded is that, for every f in $\mathfrak{M}_2$, $\|Hf\| \leq C_2 \|f\|$, where C_2 is a positive constant independent of f. The condition is obviously necessary. We show that it is also sufficient by proving that, whenever it is satisfied, there exists a positive constant C such that $|(Hf, f)| \leq C\|f\|$ for every f in $\mathfrak{H}$. Since $\mathfrak{M}_1$ is a closed linear manifold of n dimensions, it is determined by a finite orthonormal set $\varphi_1, \cdots, \varphi_n$. If f is in $\mathfrak{M}_1$, we have

$$f = a_1 \varphi_1 + \cdots + a_n \varphi_n, \qquad \|f\|^2 = |a_1|^2 + \cdots + |a_n|^2,$$

where $a_1, \cdots, a_n$ are constants. Hence, we see that

$$\|Hf\|^2 = \sum_{\alpha, \beta = 1}^{n} a_\alpha \bar{a}_\beta (H\varphi_\alpha, H\varphi_\beta) \leq M \sum_{\alpha, \beta = 1}^{n} |a_\alpha| \, |a_\beta|$$

$$\leq \tfrac{1}{2} M \sum_{\alpha, \beta = 1}^{n} [|a_\alpha|^2 + |a_\beta|^2] = Mn\|f\|^2 = C_1^2 \|f\|^2,$$

where $C_1 = \sqrt{Mn}$ and M is the greatest of the n^2 quantities $|(H\varphi_i, H\varphi_k)|$. Thus $\|Hf\|$ does not exceed $C_1\|f\|$ when f is in $\mathfrak{M}_1$. When f is an arbitrary element of $\mathfrak{H}$, we express it as the sum of its projections on $\mathfrak{M}_1$ and $\mathfrak{M}_2$ respectively, by means of Theorem 1.23. In view of the fact that these projections f_1 and f_2 are orthogonal, we have

$\|f_1\|^2+\|f_2\|^2=\|f\|^2$, $\|f_1\|\leqq\|f\|$, and $\|f_2\|\leqq\|f\|$. We have no difficulty, therefore, in deriving the inequalities

$$
\begin{aligned}
|(Hf,f)| &= \left|\sum_{\alpha,\beta=1}^{2}(Hf_\alpha,f_\beta)\right| \\
&\leqq \sum_{\alpha,\beta=1}^{2}|(Hf_\alpha,f_\beta)| \leqq \sum_{\alpha,\beta=1}^{2}\|Hf_\alpha\|\|f_\beta\| \\
&\leqq \sum_{\alpha,\beta=1}^{2}C_\alpha\|f_\alpha\|\|f_\beta\| \leqq 2(C_1+C_2)\|f\|^2.
\end{aligned}
$$

By setting $C=2(C_1+C_2)$, we find that $|(Hf,f)|$ does not exceed $C\|f\|^2$. Thus H is bounded, in accordance with Theorem 2.21.

We next show that, when $f_1,\cdots,f_m$ are m arbitrary elements of $\mathfrak{H}$, there exists an element φ such that φ and $H\varphi$ are orthogonal to each of the elements $f_1,\cdots,f_m$, $Hf_1,\cdots,Hf_m$; and such that, when H is assumed to be unbounded, φ has the properties $\|\varphi\|=1$ and $\|H\varphi\|\geqq C_m$, where C_m is an arbitrarily chosen positive constant. The proof depends upon the facts derived in the preceding paragraph. By hypothesis, H is defined throughout $\mathfrak{H}$, so that the transformation $H^2\equiv H\cdot H$ is defined and has $\mathfrak{H}$ for its domain. We choose $\mathfrak{M}_1$ as the n-dimensional linear manifold determined by $f_1,\cdots,f_m$, $Hf_1,\cdots,Hf_m$, $H^2f_1,\cdots,H^2f_m$; clearly, n does not exceed $3m$. We then set $\mathfrak{M}_2=\mathfrak{H}\ominus\mathfrak{M}_1$, as before. If φ is an arbitrary element of $\mathfrak{M}_2$, we have

$$(\varphi,f_i)=0,\qquad(\varphi,Hf_i)=0,\qquad(\varphi,H^2f_i)=0,$$

for $i=1,\cdots,m$; since H is symmetric, the last two equations become

$$(H\varphi,f_i)=0,\qquad(H\varphi,Hf_i)=0,\quad i=1,\cdots,m.$$

Thus every element of $\mathfrak{M}_2$ satisfies the first requirements laid on φ. If it is assumed that H is not bounded, then φ can be selected in $\mathfrak{M}_2$ so that $\|\varphi\|=1$ and $\|H\varphi\|\geqq C_m$, where C_m is an arbitrary positive constant; for, if $\|H\varphi\|\leqq C_m\|\varphi\|$ for every element in $\mathfrak{M}_2$, H is bounded, contrary to our assumption.

The final step in the proof consists in assuming that H is not bounded, and in constructing, as a result of this hypothesis, an orthonormal set $\{\varphi_n\}$ such that

$$(H\varphi_i, H\varphi_k) = 0, \quad i \neq k, \qquad \|H\varphi_k\| \geqq k.$$

The set $\{\varphi_n\}$ is easily determined by the method of the preceding paragraph: φ_1 is chosen arbitrarily save for the restrictions implied by the relations $\|\varphi_1\| = 1$ and $\|H\varphi_1\| \geqq 1$; when $\varphi_1, \cdots, \varphi_m$ have been chosen, we identify $f_1 = \varphi_1, \cdots, f_m = \varphi_m$, $C_m = m + 1$, in the preceding discussion, and thus select φ_{m+1} with the desired properties; this process is never halted, and thus determines an infinite orthonormal set with the required characteristics. If we set $g_n = \sum_{\alpha=1}^{n} \varphi_\alpha/\alpha$, then the sequence $\{g_n\}$ converges to an element g in $\mathfrak{H}$, and we have

$$\|Hg\|^2 = (g, H^2 g) = \lim_{n\to\infty} (g_n, H^2 g),$$

$$(g_n, H^2 g) = (H^2 g_n, g) = \sum_{\alpha=1}^{n} \sum_{\beta=1}^{\infty} (H^2 \varphi_\alpha, \varphi_\beta)/\alpha\beta$$

$$= \sum_{\alpha=1}^{n} \sum_{\beta=1}^{\infty} (H\varphi_\alpha, H\varphi_\beta)/\alpha\beta = \sum_{\alpha=1}^{n} \|H\varphi_\alpha\|^2/\alpha^2 \geqq n.$$

Thus the monotonely increasing sequence $(g_n, H^2 g)$ must satisfy the impossible inequalities $n \leqq (g_n, H^2 g) \leqq \|Hg\|^2$. This contradiction forces us to discard the hypothesis that H is not bounded. The theorem stands established.†

An immediate extension of this theorem to a broader class of transformations can be made.

THEOREM 2.26. *If T is a transformation whose domain is $\mathfrak{H}$ and if its adjoint T^* likewise has $\mathfrak{H}$ as its domain, then T and T^* are bounded linear transformations.*

Since T is defined throughout $\mathfrak{H}$, its adjoint T^* exists according to Theorem 2.6. The fact that T has $\mathfrak{H}$ as its

† This theorem is very closely related to the fundamental theorem concerning bilinear forms in infinitely many variables given by Hellinger and Toeplitz, Mathematische Annalen, 69 (1910), pp. 289–330, especially pp. 321–327.

Theorem. If $D_T = H$ and if $\overline{D_{T^*}} = H$ then in fact $D_{T^*} = H$ and both T and T^* are bounded.

Proof. Clearly $T^{**} = T$ and hence T is closed; according to the closed graph theorem, T is bounded. The remaining conclusions follow. ///

domain means that T and T^{**} are identical, so that the relation between T and T^* is symmetrical. It is sufficient, therefore, to prove that T, known to be linear by Theorems 2.8–2.10, is bounded and therefore continuous. The proof depends upon the fact that $H \equiv T^* T$ is a symmetric transformation with $\mathfrak{H}$ as its domain. By the preceding theorem, H is bounded; and we have $\|Tf\|^2 = (Tf, Tf) = (Hf, f) \leqq C^2 \|f\|^2$, for a suitably chosen positive constant C, independent of f. Consequently, T is bounded, as we wished to show.

It is now especially interesting to find that Theorem 2.26 has a valid converse, upon the proof of which we shall enter only after we have established a few preliminary results.

THEOREM 2.27. *If $L(f)$ is a numerically-valued function defined throughout a linear manifold $\mathfrak{M}$ everywhere dense in $\mathfrak{H}$; and if L is bounded and linear:*

$$|L(f)| \leqq C\|f\|, \qquad L(a_1 f_1 + a_2 f_2) = a_1 L(f_1) + a_2 L(f_2);$$

then there exists a unique element g in $\mathfrak{H}$ such that

$$L(f) = (f, g), \qquad \|g\| \leqq C,$$

for every f in $\mathfrak{M}$. Thus L can be defined in a uniquely determined way so as to be bounded and linear throughout $\mathfrak{H}$.

Since it is evident that two elements g_1 and g_2 satisfying the requirements of the theorem must be equal, because their difference is orthogonal to every element of $\mathfrak{M}$ and therefore of $\mathfrak{H}$, we have only to demonstrate the existence of one such element g. In order to do this, we determine a complete orthonormal set $\{\varphi_n\}$, each element of which belongs to $\mathfrak{M}$; we see by Theorems 1.9, 1.13 and 1.18 that such a set exists. We then put $L(\varphi_n) = a_n$ and show that the series $\sum_{\alpha=1}^{\infty} |\bar{a}_\alpha|^2$ is convergent, by the inequalities

$$\sum_{\alpha=1}^{n} |\bar{a}_\alpha|^2 = L\left(\sum_{\alpha=1}^{n} \bar{a}_\alpha \varphi_\alpha\right) \leqq C\left\|\sum_{\alpha=1}^{n} \bar{a}_\alpha \varphi_\alpha\right\| = C\left[\sum_{\alpha=1}^{n} |\bar{a}_\alpha|^2\right]^{1/2}$$

$$\sum_{\alpha=1}^{n} |\bar{a}_\alpha|^2 \leqq C^2.$$

Thus the element $g = \sum_{\alpha=1}^{\infty} \bar{a}_\alpha \varphi_\alpha$ exists and is such that $(\varphi_n, g) = L(\varphi_n)$ and $\|g\|^2 = \sum_{\alpha=1}^{\infty} |\bar{a}_\alpha|^2 \leqq C^2$. If f is an arbitrary element of $\mathfrak{M}$, we have $f = \sum_{\alpha=1}^{\infty} b_\alpha \varphi_\alpha$ where $b_n = (f, \varphi_n)$ and hence

$$|L(f) - L(f_n)| = |L(f - f_n)| \leqq C\|f - f_n\| \to 0,$$
$$L(f_n) \to L(f), \qquad L(f) = \lim_{n\to\infty} (f_n, g) = (f, g),$$

where $f_n = \sum_{\alpha=1}^{n} b_\alpha \varphi_\alpha$. This result shows that the element g has the properties asserted in the theorem. It is evident from the relation $L(f) = (f, g)$ that the function L is continuous in $\mathfrak{M}$ and can be extended by this equation so as to be defined throughout $\mathfrak{H}$. No other manner of defining L outside $\mathfrak{M}$ would lead to a continuous function.*

THEOREM 2.28. *Let $B(f, g)$ be a numerically-valued function for f in $\mathfrak{M}$ and g in $\mathfrak{N}$, where $\mathfrak{M}$ and $\mathfrak{N}$ are linear manifolds everywhere dense in $\mathfrak{H}$; and let $B(f, g)$ be bilinear in the sense that*

$$B(a_1 f_1 + a_1 f_1, g) = a_1 B(f_1, g) + a_2 B(f_2, g),$$
$$B(f, a_1 g_1 + a_2 g_2) = \bar{a}_1 B(f, g_1) + \bar{a}_2 B(f, g_2).$$

If $B(f, g)$ is bounded with respect to g for each f in accordance with the inequality $|B(f, g)| \leqq C(f)\,\|g\|$ where $C(f) \geqq 0$ is independent of g, then there exists a uniquely determined linear transformation T with domain $\mathfrak{M}$ such that $B(f, g) = (Tf, g)$ for every f in $\mathfrak{M}$ and every g in $\mathfrak{N}$; the function $B(f, g)$ can be extended in just one way so as to be bilinear and bounded with respect to g for every f in $\mathfrak{M}$ and every g in $\mathfrak{H}$, and the extended function coincides with (Tf, g). If $B(f, g)$ has the symmetry properties $\mathfrak{M} = \mathfrak{N}$ and $B(g, f) = \overline{B(f, g)}$, then T is a linear symmetric transformation. If $B(f, g)$ is bounded with respect to both f and g in the sense that $|B(f, g)| \leqq C\|f\|\|g\|$ for every f in $\mathfrak{M}$, every g

* This theorem is due essentially to Fréchet, Transactions of the American Mathematical Society, 8 (1907), pp. 439–441.

in $\mathfrak{N}$, *and some constant* C *independent of* f *and* g, *then the linear transformation* T *is bounded; the function* $B(f, g)$ *can be extended in just one way so as to be bilinear and bounded for every* f *and every* g *in* $\mathfrak{H}$, *and the extended function coincides with* $(\tilde{T}f, g)$.

We write $L(g) = \overline{B(f, g)}$ for fixed f in $\mathfrak{M}$ and apply Theorem 2.27 to the bounded linear function L. Since we have $|L(g)| \leqq C(f)\|g\|$, we see that there exists a unique element f^* such that $(g, f^*) = \overline{B(f, g)}$ and $\|f^*\| \leqq C(f)$, for every g in $\mathfrak{N}$; we see furthermore that for fixed f the function $L(g) = \overline{B(f, g)}$ can be extended in just one way so as to be linear and bounded for all g in $\mathfrak{H}$, the extension of $B(f, g)$ coinciding with (f^*, g). We define the transformation T with domain $\mathfrak{M}$ by setting $Tf = f^*$, so that we have $B(f, g) = (Tf, g)$, $\|Tf\| \leqq C(f)$. By virtue of the linear properties of $B(f, g)$ we obtain the equation

$$\begin{aligned}(T(a_1 f_1 + \cdots + a_n f_n) - a_1 Tf_1 - \cdots - a_n Tf_n, g)\\ = B(a_1 f_1 + \cdots + a_n f_n, g) - a_1 B(f_1, g) - \cdots - a_n B(f_n, g) = 0\end{aligned}$$

for every g in $\mathfrak{N}$. Since $\mathfrak{N}$ is everywhere dense in $\mathfrak{H}$, this equation implies that $T(a_1 f_1 + \cdots + a_n f_n) = a_1 Tf_1 + \cdots + a_n Tf_n$ and that T is linear. When $B(f, g)$ has the symmetry properties indicated above, we can write

$$(Tf, g) = B(f, g) = \overline{B(g, f)} = \overline{(Tg, f)} = (f, Tg)$$

for every f and every g in $\mathfrak{M} = \mathfrak{N}$, and can thus conclude that T is a symmetric transformation. Finally, if $B(f, g)$ is bounded with respect to f and g we can write $C(f) = C\|f\|$, $\|Tf\| \leqq C\|f\|$, so that the linear transformation T is bounded. It is evident that $B(f, g)$ can be extended in just one way so as to be bilinear and bounded for all f and g in $\mathfrak{H}$ and that the extended function must coincide with $(\tilde{T}f, g)$ in view of Theorem 2.23.

THEOREM 2.29. *If* T *is a bounded linear transformation whose domain determines the closed linear manifold* $\mathfrak{H}$, *then* T^* *exists and is a closed bounded linear transformation with bound equal to that of* T *and with domain* $\mathfrak{H}$.

Setting $B(f, g) = (f, \tilde{T}g)$ and denoting by C the common bound of T and $\tilde{T}$, we see that $B(f, g)$ is a bilinear function defined throughout $\mathfrak{H}$ and bounded in accordance with the inequality $|B(f, g)| \leqq C\|f\|\|g\|$. By the preceding theorem there exists a bounded linear transformation T_0 with domain $\mathfrak{H}$ such that $(T_0 f, g) = (f, \tilde{T}g)$ and $\|T_0 f\| \leqq C\|f\|$. Thus T_0 is adjoint to $\tilde{T}$ and to $T \subseteqq \tilde{T}$. Since T^* exists, and since T_0 is defined throughout $\mathfrak{H}$, these two transformations must coincide by virtue of the relation $T_0 \subseteqq T^*$. We have shown thereby that the bound of T^* is not greater than C. Now it is evident that T^{**} exists and is a bounded linear transformation with domain $\mathfrak{H}$ and bound not greater than that of T^*, since the preceding results can be applied to T^* as well as to T. T^{**}, however, must coincide with $\tilde{T}$ by virtue of the fact that both transformations are closed linear extensions of T with domain $\mathfrak{H}$. Hence we see that T, $\tilde{T}$, T^*, and T^{**} have C as their common bound.

The theorems which we have previously established lead to the following theorem concerning the class of all bounded continuous transformations.

THEOREM 2.30. *The class of all bounded linear transformations with domain* $\mathfrak{H}$ *is closed under the operations of addition, subtraction, multiplication, scalar multiplication, and formation of the adjoint; it contains the transformations* O *and* I. *In this class, the distance between two transformations* T_1, T_2 *can be defined as the bound of their difference* $T_1 - T_2$; *for this quantity is symmetric, is positive except when* $T_1 \equiv T_2$, *and obeys the triangle inequality. The class of all bounded linear transformations with domain* $\mathfrak{H}$ *which possess bounded inverses with domain* $\mathfrak{H}$ *is a transformation group.*

The proof will be left to the reader.

To bring the present section to a close, we shall consider briefly the definition of the norm of a bounded linear transformation and the properties of transformations of finite norm. With a bounded linear transformation T whose domain is $\mathfrak{H}$ and two complete orthonormal sets $\{\varphi_n\}$ and $\{\psi_n\}$, we can associate a number $N(T; \varphi, \psi)$, defined by the expression

$$\left[\sum_{\alpha,\beta=1}^{\infty}(T\varphi_\alpha, \psi_\beta)(\psi_\beta, T\varphi_\alpha)\right]^{1/2} = \left[\sum_{\alpha,\beta=1}^{\infty}|(T\varphi_\alpha, \psi_\beta)|^2\right]^{1/2};$$

if this series converges, its sum is not negative, and if it diverges, it can be assigned the "sum" $+\infty$. We show readily that $N(T; \varphi, \psi)$ is independent of the two orthonormal sets used to define it; for the obvious equations

$$N(T; \varphi, \psi) = N(T^*; \psi, \varphi),$$
$$N(T; \varphi, \psi) = \left[\sum_{\alpha=1}^{\infty}\|T\varphi_\alpha\|^2\right]^{1/2},$$
$$N(T^*; \psi, \varphi) = \left[\sum_{\alpha=1}^{\infty}\|T^*\psi_\alpha\|^2\right]^{1/2},$$

enable us to write

$$\begin{aligned} N(T; \varphi^{(1)}, \psi^{(1)}) &= N(T; \varphi^{(1)}, \psi^{(2)}) = N(T^*; \psi^{(2)}, \varphi^{(1)}) \\ &= N(T^*; \psi^{(2)}, \varphi^{(2)}) = N(T; \varphi^{(2)}, \psi^{(2)}). \end{aligned}$$

It is therefore permissible to employ the number N, which depends only on the transformation T, as a characteristic of the transformation.

DEFINITION 2.15. *A bounded linear transformation T whose domain is $\mathfrak{H}$ is said to have the norm $N(T) = N(T; \varphi, \psi)$, $0 \leqq N \leqq +\infty$; T is said to be of finite norm if $0 \leqq N < +\infty$.*†

THEOREM 2.31. *The class of all transformations of finite norm is closed under the operations of addition, scalar multiplication, and formation of the adjoint, in view of the relations*

$$N(T_1 + T_2) \leqq N(T_1) + N(T_2), \qquad N(aT) = |a|N(T),$$
$$N(T^*) = N(T).$$

Since T_1 and T_2 are assumed to be of finite norm, their sum $T_1 + T_2$ must be, like each of them, a bounded linear transformation with domain $\mathfrak{H}$. The norm of this sum can be computed, and, because of the familiar inequality

$$\left[\sum_{\alpha,\beta=1}^{n}|((T_1+T_2)\varphi_\alpha, \psi_\beta)|^2\right]^{1/2}$$
$$\leqq \left[\sum_{\alpha,\beta=1}^{n}|(T_1\varphi_\alpha, \psi_\beta)|^2\right]^{1/2} + \left[\sum_{\alpha,\beta=1}^{n}|(T_2\varphi_\alpha, \psi_\beta)|^2\right]^{1/2}$$

† The form of this definition is due to J. v. Neumann, Göttinger Nachrichten (1927), pp. 1–54, especially pp. 37–41.

for $n = 1, 2, 3, \cdots$, must be finite and satisfy the inequality stated in the theorem. The equations $N(aT) = |a| N(T)$, $N(T^*) = N(T)$ follow immediately from the definition of the norm.

THEOREM 2.32. *A transformation T of finite norm satisfies the relation $N(T) = 0$ if and only if $T \equiv O$.*

The proof will be left to the reader.

THEOREM 2.33. *If T_1 and T_2 are transformations of finite norm, then the function $Q(T_1 T_2; \varphi, \psi)$ defined by the equation.*

$$Q(T_1, T_2; \varphi, \psi) = \sum_{\alpha, \beta = 1}^{\infty} (T_1 \varphi_\alpha, \psi_\beta)(\psi_\beta, T_2 \varphi_\alpha)$$

exists and is independent of the orthonormal sets in terms of which it is defined.

We know that the series defining Q is absolutely convergent, since the series defining $N(T_1)$ and $N(T_2)$ are absolutely convergent. We see from the equations

$$\begin{aligned} Q(T_1, T_2; \varphi, \psi) &= Q(T_2^*, T_1^*; \psi, \varphi), \\ Q(T_1, T_2; \varphi, \psi) &= \sum_{\alpha=1}^{\infty} (T_1 \varphi_\alpha, T_2 \varphi_\alpha), \\ Q(T_2^*, T_1^*; \psi, \varphi) &= \sum_{\alpha=1}^{\infty} (T_2^* \psi_\alpha, T_1^* \psi_\alpha), \end{aligned}$$

that Q is independent of the particular sets employed to compute it. Evidently, $N(T)$ can be expressed in terms of Q as $\sqrt{Q(T, T)}$.

THEOREM 2.34. *Let $\mathfrak{T}$ be the class of all bounded linear transformations of finite norm; let the operations of addition and scalar multiplication have their usual significance in $\mathfrak{T}$; and let the function (T_1, T_2) be defined in $\mathfrak{T}$ by the equation $(T_1, T_2) = Q(T_1, T_2)$. Then $\mathfrak{T}$ is a Hilbert space.*

Of the characteristic properties A — E of a Hilbert space, the first two, A and B, evidently belong to $\mathfrak{T}$, as may be seen by elementary reckonings on the basis of Theorems 2.31–2.33.

We shall discuss first the properties C and D, starting with the construction of a doubly infinite sequence $\{T_{ik}\}$, $i, k = 1, 2, 3, \cdots$, in $\mathfrak{T}$. We select a complete orthonormal set $\{\varphi_n\}$

in $\mathfrak{H}$, to be held fast during the entire discussion. We define T_{ik} by the relations $T_{ik}f = (f, \varphi_i)\varphi_k$, and can verify easily that T_{ik} is a bounded linear transformation defined over $\mathfrak{H}$ with the norm 1. Since the transformations T_{ik} are linearly independent in the sense that the relation $\sum_{\alpha=1}^{m}\sum_{\beta=1}^{n} a_{\alpha\beta}T_{\alpha\beta} \equiv O$ is true if and only if the constant coefficients a_{ik} vanish, the property C must hold for $\mathfrak{T}$. In order to demonstrate the truth of D, we shall show that the denumerably infinite class of all transformations expressible in the form

$$\sum_{\alpha=1}^{m}\sum_{\beta=1}^{n} a'_{\alpha\beta}T_{\alpha\beta}, \qquad m, n = 1, 2, 3, \cdots,$$

where a'_{ik} is a complex number with rational real and imaginary parts, is everywhere dense in $\mathfrak{T}$. When T is a given transformation in $\mathfrak{T}$, its norm is given by

$$N^2 = \sum_{\alpha,\beta=1}^{\infty} |a_{\alpha\beta}|^2, \qquad a_{ik} = (T\varphi_i, \varphi_k).$$

Hence we can choose m and n so large that $R_{mn} = \sum_{\alpha=m+1}^{\infty}\sum_{\beta=n+1}^{\infty} |a_{\alpha\beta}|^2$ does not exceed $\epsilon^2/2$, ϵ being a given positive number. Then we can choose mn complex numbers a'_{ik} with rational real and imaginary parts so that $S_{mn} = \sum_{\alpha=1}^{m}\sum_{\beta=1}^{n} |a_{\alpha\beta} - a'_{\alpha\beta}|^2$ does not exceed $\epsilon^2/2$. It is now apparent that $N^2\left(T - \sum_{\alpha=1}^{m}\sum_{\beta=1}^{n} a'_{\alpha\beta}T_{\alpha\beta}\right) = S_{mn} + R_{mn}$ does not exceed ϵ^2, as we wished to show.

The preceding proof depends in essence upon the fact that to an arbitrary transformation in $\mathfrak{T}$ there can be ordered a doubly infinite sequence or matrix $\{a_{ik}\}$ such that $\sum_{\alpha,\beta=1}^{\infty} |a_{\alpha\beta}|^2$ is convergent. In order to show that $\mathfrak{T}$ is closed in the sense of the metric defined by $Q(T_1, T_2)$, we must prove that with each such sequence or matrix we can associate a transformation in $\mathfrak{T}$. When $\{a_{ik}\}$ is given, we employ the complete orthonormal set $\{\varphi_n\}$ to define a transformation T by means of the relation

$$Tf = \sum_{\beta=1}^{\infty} \left(\sum_{\alpha=1}^{\infty} a_{\alpha\beta} (f, \varphi_\alpha) \right) \varphi_\beta;$$

observing the inequalities

$$\left| \sum_{\alpha=1}^{\infty} a_{\alpha k} (f, \varphi_\alpha) \right|^2 \leqq \sum_{\alpha=1}^{\infty} | a_{\alpha k} |^2 \| f \|^2,$$

$$\sum_{\beta=1}^{\infty} \left| \sum_{\alpha=1}^{\infty} a_{\alpha\beta} (f, \varphi_\alpha) \right|^2 \leqq \sum_{\alpha, \beta=1}^{\infty} | a_{\alpha\beta} |^2 \| f \|^2,$$

we conclude that T is a bounded linear transformation defined throughout $\mathfrak{H}$ which satisfies the relation $\| Tf \|^2 \leqq \sum_{\alpha, \beta=1}^{\infty} | a_{\alpha\beta} |^2 \| f \|^2$. Evidently the transformation T has $\sum_{\alpha, \beta=1}^{\infty} | a_{\alpha\beta} |^2$ as its norm. The properties of orthonormal sets developed in the preceding chapter have been applied without special comment at each step in these considerations. Now let us suppose that in $\mathfrak{T}$ there is given a sequence $\{T_n\}$ such that $Q(T_m - T_n) = N^2(T_m - T_n) \to 0$. We wish to prove the existence of a transformation T in $\mathfrak{T}$ such that $Q(T - T_n) = N^2(T - T_n) \to 0$. We shall apply the connection between matrices and transformations to this end. If we set $a_{ik}^{(n)}$ equal to $(T_n \varphi_i, \varphi_k)$, we obtain

$$\sum_{\alpha, \beta=1}^{\infty} | a_{\alpha\beta}^{(m)} - a_{\alpha\beta}^{(n)} |^2 \to 0$$

when m and n become infinite. By the reasoning of the last paragraph in the proof of Theorem 1.15, it can be shown that there exists a doubly infinite sequence or matrix $\{a_{ik}\}$ such that

$$\sum_{\alpha, \beta=1}^{\infty} | a_{\alpha\beta} |^2 < +\infty, \qquad \sum_{\alpha, \beta=1}^{\infty} | a_{\alpha\beta} - a_{\alpha\beta}^{(n)} |^2 \to 0.$$

With $\{a_{ik}\}$ we associate a transformation T of finite norm by the process outlined above. The last property of $\{a_{ik}\}$ now becomes

$$Q(T - T_n) = N^2(T - T_n) \to 0,$$

as we wished to show.

The theorem which we have just proved enables us to apply to the class $\mathfrak{T}$ all the results of Chapter I concerning abstract Hilbert space and the bilinear function $(T_1, T_2) = Q(T_1, T_2)$. On this account we do not need to analyze further the properties of the class $\mathfrak{T}$. Before leaving the subject, however, it is well to remark that there are bounded linear transformations with domain $\mathfrak{H}$ which do not belong to $\mathfrak{T}$: the transformation I, for example, does not have finite norm.

§ 4. Projections

In the fourth section of Chapter I we defined the projection of an arbitrary element of $\mathfrak{H}$ on an arbitrary closed linear manifold $\mathfrak{M}$. We are thus led to introduce the following definition.

Definition 2.16. *The transformation E which takes each element of $\mathfrak{H}$ into its projection on the closed linear manifold $\mathfrak{M}$ is called the projection of $\mathfrak{H}$ on $\mathfrak{M}$.*

The projections are of such importance in later developments that we shall devote this section to an elaboration of their various properties. We first show how the projections are related to the types of transformation already discussed.

Theorem 2.35. *Let E be the projection of $\mathfrak{H}$ on the closed linear manifold $\mathfrak{M}$. Then E is a not-negative definite self-adjoint transformation with domain $\mathfrak{H}$ and range $\mathfrak{M}$, which has the properties*

$$0 \leqq (Ef, f) = |Ef|^2 \leqq |f|^2, \qquad E^2 \equiv E \cdot E \equiv E.$$

If f and g are arbitrary elements in $\mathfrak{H}$, each of them can be written in just one way as a sum, $f = f_1 + f_2$, $g = g_1 + g_2$, where f_1 and g_1 belong to $\mathfrak{M}$, f_2 and g_2 belong to $\mathfrak{H} \ominus \mathfrak{M}$. By definition, $Ef = f_1$ and $Eg = g_1$, so that

$$(Ef, g) = (f_1, g_1 + g_2) = (f_1, g_1) = (f_1 + f_2, g_1) = (f, Eg).$$

Thus E is a symmetric transformation defined throughout $\mathfrak{H}$; by Theorem 2.17 it must be self-adjoint. Since the projection of $\mathfrak{H}$ on $\mathfrak{M}$ takes every element of $\mathfrak{M}$ into itself we see that

$$E^2 f = E(Ef) = Ef_1 = f_1 = Ef, \qquad E^2 \equiv E.$$

Consequently,

$$\|Ef\| = \|f_1\| \leqq [\|f_1\|^2 + \|f_2\|^2]^{1/2} = \|f\|,$$
$$(Ef, f) = (E^2 f, f) = (Ef, Ef) \geqq 0.$$

These inequalities imply that E is bounded and not-negative definite.

THEOREM 2.36. *If E is a maximal symmetric transformation such that $E^2 \equiv E$ then E is the projection of $\mathfrak{H}$ on some closed linear manifold $\mathfrak{M}$.**

Let $\mathfrak{M}$ be the set of all elements in $\mathfrak{H}$ for which the equation $Ef = f$ is significant and true. Since E is a maximal symmetric transformation, it is closed and linear, in accordance with Theorem 2.14. We see at once that $\mathfrak{M}$ is a linear manifold; and, since $f_n \to f$ implies $Ef_n = f_n \to f$ and $Ef = f$, whenever $\{f_n\}$ is a convergent sequence in $\mathfrak{M}$, we see also that $\mathfrak{M}$ is a closed linear manifold. If f is an arbitrary element in the domain of E, we can write

$$f = f_1 + f_2, \qquad Ef = f_1, \qquad f - Ef = f_2.$$

Evidently, f_1 belongs to the closed linear manifold $\mathfrak{M}$, because $Ef_1 = E^2 f = Ef = f_1$; and f_2 belongs to $\mathfrak{H} \ominus \mathfrak{M}$, since for an arbitrary element g in $\mathfrak{M}$ we have the equations

$$(f_2, g) = (f_2, Eg) = (f - Ef, Eg) = (Ef - E^2 f, g) = 0.$$

Consequently, the projection of $\mathfrak{H}$ on $\mathfrak{M}$ is an extension of E; both E and the projection are maximal symmetric transformations, so that they must be identical.

THEOREM 2.37. *If E_1 and E_2 are projections of $\mathfrak{H}$ on the closed linear manifolds $\mathfrak{M}_1$ and $\mathfrak{M}_2$ respectively, then:*

(1) *$E_1 E_2$ is a projection if and only if $E_2 E_1 \equiv E_1 E_2$; its range is then $\mathfrak{M}_1 \mathfrak{M}_2$;*

---

* In the terminology of Hilbert and his followers such a transformation or operator and its associated matrix (see Chapter III, § 1) are called "Einzeltransformation" or "Einzeloperator" and "Einzelmatrix"; this terminology does not seem to admit a graceful or apt translation into English so that we shall use the term "projection", being justified by the present theorem.

(2) *$E_1 + E_2$ is a projection if and only if $E_1 E_2 \equiv O$ or $E_2 E_1 \equiv O$; then $\mathfrak{M}_1$ and $\mathfrak{M}_2$ are orthogonal, and $E_1 + E_2$ has $\mathfrak{M}_1 \oplus \mathfrak{M}_2$ as its range.*

(3) *$E_1 - E_2$ is a projection if and only if $E_1 E_2 \equiv E_2$ or $E_2 E_1 \equiv E_2$; $\mathfrak{M}_2$ is then a subset of $\mathfrak{M}_1$ and $E_1 - E_2$ has $\mathfrak{M}_1 \ominus \mathfrak{M}_2$ as its range.*

In order that $E_1 E_2$ be a projection it is necessary that it be symmetric. A necessary and sufficient condition for this is that the transformation $(E_1 E_2)^* \equiv E_2^* E_1^* \equiv E_2 E_1$ be identical with $E_1 E_2$. If $E_1 E_2$ is a projection, therefore, $E_2 E_1 \equiv E_1 E_2$. Conversely, if the latter condition is satisfied, $E_1 E_2$ is symmetric and has the property that $(E_1 E_2)^2 \equiv E_1 E_2 E_1 E_2 \equiv E_1 E_1 E_2 E_2 \equiv E_1 E_2$. Since $E_1 E_2$ has $\mathfrak{H}$ as its domain, Theorem 2.36 may be applied to show that $E_1 E_2$ is a projection. When $E_2 E_1 \equiv E_1 E_2$, the projection $E_1 E_2$ takes every element of $\mathfrak{H}$ into an element common to $\mathfrak{M}_1$ and $\mathfrak{M}_2$; and every element of $\mathfrak{M}_1 \mathfrak{M}_2$ into itself. Thus the range of $E_1 E_2$ is $\mathfrak{M}_1 \mathfrak{M}_2$.

In order that $E_1 + E_2$ be a projection, it is necessary and sufficient that the transformation $E_1 + E_2$, which is evidently a maximal symmetric transformation, should satisfy the condition $(E_1 + E_2)^2 \equiv E_1 + E_2$. For this it is necessary and sufficient that $E_1 E_2 + E_2 E_1 \equiv O$, in view of the equation

$$\begin{aligned}(E_1 + E_2)^2 &\equiv E_1^2 + E_1 E_2 + E_2 E_1 + E_2^2 \\ &\equiv E_1 + E_2 + E_1 E_2 + E_2 E_1.\end{aligned}$$

If $E_1 E_2 + E_2 E_1 \equiv O$, then both $E_1 E_2$ and $E_2 E_1$ coincide with O, because of the relations

$$\begin{aligned} O &\equiv E_1(E_1 E_2 + E_2 E_1) &&\equiv E_1 E_2 + E_1 E_2 E_1, \\ O &\equiv (E_1 E_2 + E_1 E_2 E_1) E_1 &&\equiv 2 E_1 E_2 E_1.\end{aligned}$$

Conversely, if $E_1 E_2 \equiv O$, we can show that $E_2 E_1 \equiv O$ and $E_1 E_2 + E_2 E_1 \equiv O$, thus proving that $E_1 + E_2$ is a projection: for the relation $E_1 E_2 \equiv O$ requires that the manifolds $\mathfrak{M}_1$ and $\mathfrak{M}_2$ be orthogonal by virtue of the equations

$$(f, g) = (E_1 f, E_2 g) = (f, E_1 E_2 g) = 0$$

which hold for every f in $\mathfrak{M}_1$, and every g in $\mathfrak{M}_2$. Thus, if E_1+E_2 is a projection it takes every element of $\mathfrak{H}$ into one in $\mathfrak{M}_1 \oplus \mathfrak{M}_2$ and every element of $\mathfrak{M}_1 \oplus \mathfrak{M}_2$ into itself; for $\mathfrak{M}_1$ and $\mathfrak{M}_2$ are orthogonal and every element f in $\mathfrak{H}$ can be written in a single way as a sum $f = f_1+f_2+f_3$, where f_1, f_2, f_3, are in $\mathfrak{M}_1, \mathfrak{M}_2, \mathfrak{H} \ominus [\mathfrak{M}_1 \oplus \mathfrak{M}_2]$, respectively, with the result that $E_1 f = f_1$, $E_2 f = f_2$, and $(E_1+E_2)f = f_1+f_2$. Thus the range of (E_1+E_2) is $\mathfrak{M}_1 \oplus \mathfrak{M}_2$.

If E is a symmetric transformation with $\mathfrak{H}$ as its domain, a necessary and sufficient condition that it be a projection is that $I-E$ be a projection, since the equations $(I-E)^2 \equiv (I-E)$ and $E^2 \equiv E$ are equivalent; the range of E is the closed linear manifold orthogonal to the range of $I-E$. Thus E_1-E_2, as a symmetric transformation defined throughout $\mathfrak{H}$, is a projection if and only if the transformation $I-(E_1-E_2) \equiv (I-E_1)+E_2$ is a projection. By (2), this transformation is a projection if and only if $(I-E_1)\cdot E_2 \equiv O$ or $E_2(I-E_1) \equiv O$; that is, if and only if $E_1 E_2 \equiv E_2$ or $E_2 E_1 \equiv E_2$. If E_1-E_2 is a projection, its range is therefore $\mathfrak{H} \ominus [(\mathfrak{H} \ominus \mathfrak{M}_1) \oplus \mathfrak{M}_2] = \mathfrak{M}_1 \ominus \mathfrak{M}_2$.

DEFINITION 2.17. *If the projections E_1 and E_2 are permutable, $E_1 E_2 \equiv E_2 E_1$, then their ranges are said to be permutable. The projections E_1 and E_2 are said to be orthogonal if their ranges are orthogonal. The projection E_2 is said to be part of E_1 if the range of E_2 is a subset of the range of E_1: in symbols, this relation will be denoted by $E_2 \leqq E_1$.*

THEOREM 2.38. *The projection E_2 is part of the projection E_1 if and only if, for every f in $\mathfrak{H}$, $\|E_2 f\| \leqq \|E_1 f\|$.*

The condition is necessary, because $E_2 \leqq E_1$ implies $E_2 E_1 \equiv E_2$ and hence

$$\|E_2 f\| = \|E_2 E_1 f\| \leqq \|E_1 f\|.$$

This condition is sufficient, since $f = E_2 f$ implies

$$\|f\| = \|E_2 f\| \leqq \|E_1 f\|, \quad (E_1 f, f) = \|E_1 f\|^2 \geqq \|f\|^2,$$
$$\|(I-E_1)f\|^2 = ((I-E_1)f, f) \leqq 0, \quad (I-E_1)f = 0;$$

in other words, $\mathfrak{M}_2$ is a subset of $\mathfrak{M}_1$ and $E_2 \leqq E_1$.

THEOREM 2.39. *The sum of the projections $E_1, \cdots, E_m$ is a projection if and only if $E_i E_k \equiv O$ whenever i and k are different, $i, k = 1, \cdots, m$.*

The condition is sufficient, since $E_1 + \cdots + E_m$ is a symmetric transformation with domain $\mathfrak{H}$ which satisfies the relation

$$(E_1 + \cdots + E_m)^2 \equiv \sum_{\alpha,\beta=1}^{m} E_\alpha E_\beta \equiv \sum_{\alpha=1}^{m} E_\alpha^2 \equiv E_1 + \cdots + E_m.$$

The condition is necessary since the inequalities

$$\begin{aligned} \|E_i f\|^2 + \|E_k f\|^2 \leqq \sum_{\alpha=1}^{m} \|E_\alpha f\|^2 &= \sum_{\alpha=1}^{m} (E_\alpha f, f) \\ &= \left(\left(\sum_{\alpha=1}^{m} E_\alpha\right) f, f\right) \leqq \|f\|^2. \end{aligned}$$

valid when i and k are different, yield by the replacement of f by $E_k g$ the result

$$\|E_i E_k g\| \leqq 0, \qquad E_i E_k g = 0, \qquad E_i E_k \equiv O,$$

as we wished to prove.

THEOREM 2.40. *If $\{E_n\}$ is a sequence of projections such that $E_m \leqq E_n$ ($E_m \geqq E_n$) whenever $m \leqq n$, then there exists a projection E such that $E_n \to E$ when $n \to \infty$ and $E_n \leqq E$ ($E_n \geqq E$), $n = 1, 2, 3, \cdots$.*

The case in which $E_m \geqq E_n$ whenever $m \leqq n$ can be reduced to the first case by the study of the sequence $\{I - E_n\}$. We shall on that account confine our attention to the case in which $E_m \leqq E_n$ whenever $m \leqq n$.

In view of the fact that $\|E_m f\| \leqq \|E_n f\|$ whenever $m \leqq n$, the sequence $\{\|E_n f\|\}$ is monotone increasing; and since $\|E_n f\|$ never exceeds $\|f\|$, the sequence is bounded. Consequently, this sequence converges; and, when a positive ε is assigned, an N can be found such that $|\|E_m f\|^2 - \|E_n f\|^2| < \varepsilon^2$ for all m and n greater than N. If in particular we take m greater than n, $E_m - E_n$ is a projection by Theorem 2.37 (3), so that the inequality above can be transformed into

$$\begin{aligned} \|(E_m - E_n) f\|^2 &= ((E_m - E_n) f, f) = (E_m f, f) - (E_n f, f) \\ &= \|E_m f\|^2 - \|E_n f\|^2 < \varepsilon^2, \qquad m, n > N. \end{aligned}$$

Thus, for every f in $\mathfrak{H}$, the sequence $\{E_n f\}$ converges to a limit Ef. We must show that the transformation E so defined is a projection. It is defined throughout $\mathfrak{H}$, and is symmetric by virtue of the equations.

$$(Ef, g) = \lim_{n\to\infty} (E_n f, g) = \lim_{n\to\infty} (f, E_n g) = (f, Eg).$$

Furthermore, it satisfies the identity $E^2 \equiv E$, since

$$\begin{aligned}(E^2 f, g) &= (Ef, Eg) = \lim_{n\to\infty} (E_n f, E_n g) \\ &= \lim_{n\to\infty} (E_n f, g) = (Ef, g).\end{aligned}$$

The conditions of Theorem 2.36 are satisfied, so that E is a projection. Because of the inequality $\|E_n f\| \leqq \|Ef\|$, which holds for every f in $\mathfrak{H}$, we see by Theorem 2.38 that $E_n \leqq E$.

THEOREM 2.41. *A projection has finite norm if and only if its range has the dimension number* n, $n = 0, 1, 2, \cdots$; *its norm is then equal to* $\sqrt{n}$.

We form an orthonormal set $\{\varphi_m^{(1)}\}$ which determines the range $\mathfrak{M}$ of the given projection E, and an orthonormal set $\{\varphi_m^{(2)}\}$ which determines the closed linear manifold $\mathfrak{H} \ominus \mathfrak{M}$; the orthonormal set obtained by combining these is complete, since it determines the closed linear manifold $\mathfrak{H}$. The details of the construction and properties of these sets are to be found in Chapter I. Using the equations $E\varphi_m^{(1)} = \varphi_m^{(1)}$ and $E\varphi_m^{(2)} = 0$, we find

$$N^2(E) = \sum_{\alpha,\beta=1}^{\infty} \sum_{\gamma,\delta=1}^{2} |(E\varphi_\alpha^{(\gamma)}, \varphi_\beta^{(\delta)})|^2 = \sum_{\alpha=1}^{\infty} \|\varphi_\alpha^{(1)}\|^2.$$

The sum in the last term is to be extended over the orthonormal set $\{\varphi_m^{(1)}\}$, which may be finite or infinite. If this set is infinite, the sum is a divergent infinite series and $N(E) = +\infty$. The sum is an ordinary sum with n terms, each equal to one, if and only if the orthonormal set $\{\varphi_m^{(1)}\}$ is finite and contains n elements; and this occurs if and only if $\mathfrak{M}$ has the dimension number n. When this is the case $N(E)$ evidently has the value $\sqrt{n}$.

§ 5. Isometric and Unitary Transformations

Klein's famous characterization of geometry in terms of the theory of groups finds application in the geometry of Hilbert space, in the following way: the geometrical properties of Hilbert space may be regarded as invariants of the transformations which take $\mathfrak{H}$ into itself without changing the fundamental bilinear form (f, g). While it is unnecessary to study this relation between geometry and group-theory *ab initio,* it is desirable to investigate the group of transformations involved and to show the invariance of the more important concepts previously introduced. As a natural generalization of these transformations, we shall consider transformations which leave invariant the function (f, g) but which no longer take $\mathfrak{H}$ into itself.

Definition 2.18. *A transformation U such that its domain and range coincide with $\mathfrak{H}$ and such that $(Uf, Ug) = (f, g)$ for every pair of elements in $\mathfrak{H}$ is called a unitary transformation.*

Definition 2.19. *A linear transformation U such that $(Uf, Ug) = (f, g)$ for every pair of elements in its domain is called an isometric transformation.*

Theorem 2.42. *A unitary transformation is continuous and isometric; its adjoint exists and is a unitary transformation; it has an inverse which coincides with its adjoint. The class of all unitary transformations constitutes a group.*

We shall first show that the unitary transformation U possesses an inverse whose domain is $\mathfrak{H}$. By the fundamental properties of U, we find that the equation $|Uf - Ug| = |f - g|$ holds for every pair of elements in $\mathfrak{H}$. Thus $Uf = Ug$ implies $f = g$, so that we can infer the existence of the inverse U^{-1} with domain $\mathfrak{H}$. In the fundamental equation $(Uf, Ug) = (f, g)$ we replace g by $U^{-1}h$, where h is an arbitrary element, obtaining the equation $(Uf, h) = (f, U^{-1}h)$ for every pair of elements in $\mathfrak{H}$. As a consequence U^* exists and coincides with U^{-1}. We may therefore apply Theorem 2.26 to the transformation U, with the result that it is seen to be a bounded continuous linear transformation.

In particular, U must satisfy the conditions of Definition 2.19 and is an isometric transformation. We already know that U^{-1}, as the inverse of U, has $\mathfrak{H}$ as its domain and range; in order to show that it is unitary, we have only to verify the equation $(U^{-1}f, U^{-1}g) = (f, g)$, which can be obtained by replacing f and g in the similar equation for U by $U^{-1}f$ and $U^{-1}g$ respectively.

The class of all unitary transformations is a group: it contains the identity I; it contains with U its inverse U^{-1}; it contains with U_1 and U_2 their product, by virtue of the fact that $U_2 U_1$ is a transformation which takes $\mathfrak{H}$ into itself and satisfies the relation $(U_2 U_1 f, U_2 U_1 g) = (U_1 f, U_1 g) = (f, g)$; and multiplication of transformations is associative.

THEOREM 2.43. *An isometric transformation U is bounded and possesses an isometric inverse. The transformation $\tilde{U}$ exists and is isometric; the domain and the range of $\tilde{U}$ are closed linear manifolds with the same dimension number n, $n = 0, 1, 2, \cdots, \aleph_0$.*

That U is bounded, and therefore continuous, appears immediately from the equation $|Uf| = |f|$, which holds for every element in $\mathfrak{D}$. The reasoning used in the preceding theorem can be applied without change to show that U has an inverse U^{-1}. By Theorem 2.5, U^{-1} is linear; and, by replacing f and g in the equation $(Uf, Ug) = (f, g)$ by $U^{-1}f$ and $U^{-1}g$ respectively, we find that U^{-1} satisfies a similar equation. It follows that U^{-1} is an isometric transformation. By Theorem 2.23, the closed linear transformation $\tilde{U}$ exists. If f and g are arbitrary elements of $\tilde{\mathfrak{D}}$, the domain of $\tilde{U}$, we can find sequences $\{f_n\}$ and $\{g_n\}$ in the domain of U such that $f_n \to f$, $g_n \to g$, $Uf_n \to \tilde{U}f$, and $Ug_n \to \tilde{U}g$. In consequence, we see that

$$(\tilde{U}f, \tilde{U}g) = \lim_{n\to\infty} (Uf_n, Ug_n) = \lim_{n\to\infty} (f_n, g_n) = (f, g).$$

$\tilde{U}$ is therefore an isometric transformation, and can be treated by the parts of the present theorem already established. In particular, the inverse of $\tilde{U}$ exists and must be closed and linear as well as isometric in accordance with Theorem 2.5, since $\tilde{U}$ itself has these properties. As a result, the domains of $\tilde{U}$ and its inverse are closed linear manifolds. If $n \geqq 1$ is the dimension number of the domain of $\tilde{U}$, we can find an orthonormal set $\{\varphi_m\}$ containing precisely n elements which determines the closed linear manifold $\tilde{\mathfrak{D}}$. In view of the equations $(\tilde{U}\varphi_i, \tilde{U}\varphi_k) = (\varphi_i, \varphi_k) = \delta_{ik}$, the set $\{\tilde{U}\varphi_m\}$ is an orthonormal set in the range of $\tilde{U}$. We see that the dimension number of the range is not less than n. By considering the inverse of $\tilde{U}$ we can show in the same way that n cannot be less than the dimension number of the range of $\tilde{U}$. Thus the domain and the range of U have the same dimension number n, when $n \geqq 1$. The case $n = 0$ is trivial.

THEOREM 2.44. *A necessary and sufficient condition that the transformation U be closed and isometric is that there exist two orthonormal sets $\{\varphi_n\}$ and $\{\psi_n\}$ with the same number of elements such that for every f expressible in the form $f = \sum_{\alpha=1}^{\infty} a_\alpha \varphi_\alpha$, $a_n = (f, \varphi_n)$, Uf is defined and is equal to $\sum_{\alpha=1}^{\infty} a_\alpha \psi_\alpha$, the trivial case where the domain of U is the set $\mathfrak{O}$ being excluded.*

The condition is necessary. For if U is a closed isometric transformation, we can select an orthonormal set $\{\varphi_n\}$ which determines the closed linear manifold in which U is defined and can put ψ_n equal to $U\varphi_n$. Since U is closed and bounded, we see that, whenever $\sum_{\alpha=1}^{n} a_\alpha \varphi_\alpha \to f$, then $Uf = \sum_{\alpha=1}^{\infty} a_\alpha \psi_\alpha$, in view of the relations

$$U\left(\sum_{\alpha=1}^{n} a_\alpha \varphi_\alpha\right) = \sum_{\alpha=1}^{n} a_\alpha U\varphi_\alpha = \sum_{\alpha=1}^{n} a_\alpha \psi_\alpha,$$

$$U\left(\sum_{\alpha=1}^{n} a_\alpha \varphi_\alpha\right) \to Uf, \qquad a_n = (f, \varphi_n).$$

The condition is sufficient. The transformation U defined by the relations stated in the theorem is evidently a linear transformation with closed linear manifolds for its domain and range. U is seen at once to be isometric, because, when $f = \sum_{\alpha=1}^{\infty} a_\alpha \varphi_\alpha$, $g = \sum_{\alpha=1}^{\infty} b_\alpha \varphi_\alpha$, we have $Uf = \sum_{\alpha=1}^{\infty} a_\alpha \psi_\alpha$,

$Ug = \sum_{\alpha=1}^{\infty} b_\alpha \psi_\alpha$, $(Uf, Ug) = \sum_{\alpha=1}^{\infty} a_\alpha \bar{b}_\alpha = (f, g)$. It is evident that U is closed, being identical with the related transformation $\tilde{U}$.

THEOREM 2.45. *If $\mathfrak{M}_1$ and $\mathfrak{M}_2$ are closed linear manifolds with the same dimension number, not* 0, *the class $\mathfrak{C}$ of all closed isometric transformations of $\mathfrak{M}_1$ into $\mathfrak{M}_2$ has the cardinal number $\mathfrak{c}$ of the continuum.*

From the preceding theorem, we know that $\mathfrak{C}$ contains at least one transformation U, for whose construction explicit directions have been given. Evidently, $\mathfrak{C}$ contains as a subclass the class of all transformations $a \cdot U$ with $|a| = 1$. This subclass has the property that $a_1 U$ and $a_2 U$ coincide only when $a_1 = a_2$; it therefore has the cardinal number $\mathfrak{c}$ of the class of all complex numbers a with modulus 1. If we denote the cardinal number of $\mathfrak{C}$ by the letter c, we can now write $c \geqq \mathfrak{c}$.

To obtain the reversed inequality $c \leqq \mathfrak{c}$, we commence by choosing an orthonormal set $\{\varphi_m\}$ which determines the closed linear manifold $\mathfrak{M}_1$. From the preceding theorem it is clear that the behavior of a transformation U belonging to $\mathfrak{C}$ is completely prescribed by its behavior in the set $\{\varphi_m\}$, in accordance with the equations

$$f = \sum_{\alpha=1}^{\infty} a_\alpha \varphi_\alpha, \qquad Uf = \sum_{\alpha=1}^{\infty} a_\alpha U\varphi_\alpha.$$

In other words, $\mathfrak{C}$ is in one-to-one correspondence with a certain subclass of the class of all transformations with domain $\{\varphi_m\}$; and c does not exceed the cardinal number of the latter class. We show that the cardinal number in question is $\mathfrak{c}$. As a first step we note that the cardinal number of an arbitrary Hilbert space is the same as that of the particular space $\mathfrak{H}_0$, with which it is in one-to-one correspondence; and that $\mathfrak{H}_0$ has the cardinal number $\mathfrak{c}$, as can be verified without difficulty from the fact that $\mathfrak{H}_0$ consists of all ordered sets $(x_1, x_2, x_3, \cdots)$ such that $\sum_{\alpha=1}^{\infty} |x_\alpha|^2$ is a convergent infinite series. Our next step is to observe than any transformation T with domain $\{\varphi_n\}$ is completely specified by means of a sequence $\{\psi_m\}$, with $T\varphi_m = \psi_m$. Thus the class of all such transformations has the same cardinal number as the class of all sequences with the same number of elements as the sequence $\{\varphi_m\}$. If the cardinal number of the set $\{\varphi_m\}$ is n, $1 \leqq n \leqq \aleph_0$, then this class of sequences has the cardinal number $\mathfrak{c}^n = \mathfrak{c}$. Consequently, we have established the truth of both relations $c \geqq \mathfrak{c}$, $c \leqq \mathfrak{c}$, and can conclude that $c = \mathfrak{c}$ as we wished to do.

THEOREM 2.46. *A necessary and sufficient condition that a linear transformation U be isometric is that it preserve distances in $\mathfrak{H}$, $\|Uf - Ug\| = \|f - g\|$.*

The condition is obviously necessary, since $\|Uf - Ug\| = \|U(f-g)\| = (U(f-g), U(f-g))^{1/2} = (f-g, f-g)^{1/2} = \|f-g\|$. On the other hand, the condition is sufficient; for if U is linear and preserves distances we have

$$\|Uf - Ug\|^2 = \|f-g\|^2, \quad \|Uf\|^2 = \|f\|^2, \quad \|Ug\|^2 = \|g\|^2,$$
$$\|Uf\|^2 - 2\mathfrak{R}(Uf, Ug) + \|Ug\|^2 = \|f\|^2 - 2\mathfrak{R}(f, g) + \|g\|^2,$$
$$\mathfrak{R}(Uf, Ug) = \mathfrak{R}(f, g),$$
$$\mathfrak{I}(Uf, Ug) = \mathfrak{R}(Uif, Ug) = \mathfrak{R}(if, g) = \mathfrak{I}(f, g),$$

so that $(Uf, Ug) = (f, g)$, as we wished to prove.

We shall now consider the problem of finding all the isometric extensions of a given isometric transformation. The results of Theorem 2.44 indicate that the tools for solving this problem with comparative ease are already at hand. In a later chapter we shall see that the analogous problem of determining all the symmetric extensions of a given symmetric transformation, already raised in § 2, can be solved by means of the facts which we shall establish here. We first introduce a useful definition.

DEFINITION 2.20. *An isometric transformation is said to be maximal if and only if it has no proper isometric extension.*

THEOREM 2.47. *If U_1 and U_2 are isometric transformations such that $U_2 \supseteqq U_1$, then $\tilde{U}_2 \supseteqq \tilde{U}_1$. Every maximal isometric extension of an isometric transformation U is also an extension* *of $\tilde{U}$; in particular every maximal isometric transformation is closed.*

The theorem follows at once from Theorem 2.23.

From the last theorem we see that we may confine our attention to closed isometric transformations without loss of generality.

DEFINITION 2.21. *If U is a closed isometric transformation with domain $\mathfrak{D}$ and range $\mathfrak{R}$, the pair of numbers (m, n) where m and n are the dimension numbers of $\mathfrak{H} \ominus \mathfrak{D}$ and $\mathfrak{H} \ominus \mathfrak{R}$ respectively, is called the deficiency-index of U.*†

THEOREM 2.48. *Let U_1 and U_2 be closed isometric transformations with the respective domains, ranges, and deficiency-indices $\mathfrak{D}_1$, $\mathfrak{D}_2$, $\mathfrak{R}_1$, $\mathfrak{R}_2$, (m_1, n_1), (m_2, n_2). If U_2 is a proper extension of U_1, then there exists a cardinal number p, $p = 1, 2, 3, \cdots, \aleph_0$, such that $m_2 + p = m_1$, $n_2 + p = n_1$. The transformation U with domain $\mathfrak{D} = \mathfrak{D}_2 \ominus \mathfrak{D}_1$ which coincides in its domain with U_2 is a closed isometric transformation with the range $\mathfrak{R} = \mathfrak{R}_2 \ominus \mathfrak{R}_1$. U_2 is completely determined by U_1 and U.*

Since $\mathfrak{D}_1$ is a proper subset of $\mathfrak{D}_2$, the closed linear manifold $\mathfrak{D} = \mathfrak{D}_2 \ominus \mathfrak{D}_1$ has the dimension number p, $p = 1, 2, 3, \cdots, \aleph_0$. Since the isometric transformation U_2 takes $\mathfrak{D}_1$ into $\mathfrak{R}_1$, $\mathfrak{D}_2$ into $\mathfrak{R}_2$, and preserves the relation of orthogonality, it takes $\mathfrak{D}$ into the closed linear manifold $\mathfrak{R} = \mathfrak{R}_2 \ominus \mathfrak{R}_1$. Hence $\mathfrak{R}$ has the dimension number p. We find that the equations $m_2 + p = m_1$, $n_2 + p = n_1$ express in terms of dimension numbers the relations

$$[\mathfrak{H} \ominus \mathfrak{D}_2] \oplus \mathfrak{D} = \mathfrak{H} \ominus \mathfrak{D}_1, \qquad [\mathfrak{H} \ominus \mathfrak{R}_2] \oplus \mathfrak{R} = \mathfrak{H} \ominus \mathfrak{R}_1.$$

It is now evident that the transformation U which coincides with U_2 in $\mathfrak{D}$ is isometric, and that its range is $\mathfrak{R}$. If f is an arbitrary element in $\mathfrak{D}_2 = \mathfrak{D}_1 \oplus \mathfrak{D}$, it can be expressed in just one way as a sum $g + h$ where g is in $\mathfrak{D}_1$ and h in $\mathfrak{D}$; and since U_2 is linear, the equations $U_2 f = U_2 g + U_2 h$

† This definition and the developments based upon it are due to J. v. Neumann, Mathematische Annalen, 102 (1929), pp. 49–131, especially pp. 71–72, 82–84, 87–91.

$= U_1 g + Uh$ serve to express U_2 in terms of its behavior in $\mathfrak{D}_1$ and $\mathfrak{D}$. Thus U_1 and U completely determine U_2.

THEOREM 2.49. *Let U_1 be a closed isometric transformation with domain $\mathfrak{D}_1$, range $\mathfrak{R}_1$, and deficiency-index (m_1, n_1) where $m_1 \geqq 1$, $n_1 \geqq 1$. Let p be an arbitrary cardinal number, $p = 1, 2, 3, \cdots, \min[m_1, n_1]$, where $\min[m_1, n_1]$ is the lesser of m_1 and n_1; and let $\mathfrak{D}$ and $\mathfrak{R}$ be arbitrary closed linear manifolds with dimension number p such that $\mathfrak{D} \subseteqq \mathfrak{H} \ominus \mathfrak{D}_1$, $\mathfrak{R} \subseteqq \mathfrak{H} \ominus \mathfrak{R}_1$. Then the class of all isometric transformations of $\mathfrak{D}$ into $\mathfrak{R}$ has the cardinal number $\mathfrak{c}$ of the continuum. If U is an arbitrary transformation of this class, then there exists a uniquely determined closed isometric transformation U_2 with domain $\mathfrak{D}_2 = \mathfrak{D}_1 \oplus \mathfrak{D}$, range $\mathfrak{R}_2 = \mathfrak{R}_1 \oplus \mathfrak{R}$, and deficiency-index $(m_2, n_2) = (m_1 - p, n_1 - p)$ which coincides with U_1 in $\mathfrak{D}_1$ and with U in $\mathfrak{D}$. U_2 is a proper isometric extension of U_1.*

Under the hypotheses concerning m_1, n_1, p, we can always choose $\mathfrak{D}$ and $\mathfrak{R}$ as subsets of $\mathfrak{H} \ominus \mathfrak{D}_1$ and $\mathfrak{H} \ominus \mathfrak{R}_1$ respectively. To ascertain the existence and cardinal number of the class of all closed isometric transformations of $\mathfrak{D}$ into $\mathfrak{R}$ we have only to refer to Theorems 2.44 and 2.45. If U is such a transformation we can use it in conjunction with U_1 to define a closed isometric transformation $U_2 \supseteqq U_1$ with domain $\mathfrak{D}_2$ and range $\mathfrak{R}_2$ identical respectively with $\mathfrak{D}_1 \oplus \mathfrak{D}$ and $\mathfrak{R}_1 \oplus \mathfrak{R}$. If f is an arbitrary element in $\mathfrak{D}_2$ it can be expressed in just one way as a sum $g + h$, where g is in $\mathfrak{D}_1$ and h is in $\mathfrak{D}$. Thus we can define $U_2 f = U_1 g + Uh$. It is clear that we obtain in this way a linear transformation with the desired domain and range which coincides with U_1 in $\mathfrak{D}_1$ and with U in $\mathfrak{D}$. In order to show that U_2 is isometric, we select arbitrary elements f_1 and f_2 in $\mathfrak{D}_2$ and set $f_1 = g_1 + h_1$, $f_2 = g_2 + h_2$, where g_1, g_2 are in $\mathfrak{D}_1$ and h_1, h_2 in $\mathfrak{D}$; since $\mathfrak{D}_1$, $\mathfrak{D}$ and $\mathfrak{R}_1$, $\mathfrak{R}$ are orthogonal we find

$$(U_2 f_1, U_2 f_2) = (U_1 g_1, U_1 g_2) + (Uh_1, Uh_2) = (g_1, g_2) + (h_1, h_2) \\ = (f_1, f_2).$$

Finally, the fact that U_2 has a closed linear manifold for its domain shows that U_2 and $\tilde{U}_2$ coincide; U_2 is closed in consequence. From the manner in which the domain and range of U_2 have been determined, the deficiency-index of the transformation is found to have the form indicated above.

THEOREM 2.50. *A closed isometric transformation is maximal if and only if its deficiency-index is of the form* $(m, 0)$ *or* $(0, n)$; *it is unitary if and only if its deficiency-index is* $(0, 0)$.

This theorem is an obvious consequence of the fundamental definitions and of Theorems 2.48 and 2.49.

THEOREM 2.51. *The maximal isometric extensions of a given closed isometric transformation which is not maximal constitute a class with the cardinal number* $\mathfrak{c}$. *A closed isometric transformation which is not maximal has unitary extensions if and only if its deficiency-index is of the form* (m, m): *if this condition is satisfied with* m *a finite cardinal number, then every maximal isometric extension is unitary*; *and if this condition is satisfied with* $m = \aleph_0$, *then the maximal isometric extensions with an assigned deficiency-index* $(p, 0)$ *or* $(0, p)$, $p = 0, 1, 2, \cdots, \aleph_0$, *constitute a class with the cardinal number* $\mathfrak{c}$.

This theorem also can be verified immediately from the preceding results.

§ 6. UNITARY INVARIANCE

We shall next discuss briefly some elementary facts in the invariant theory of the group of unitary transformation. Because of the unitary invariance of the underlying metric function as expressed in the equation $(Uf, Ug) = (f, g)$, all metric properties in $\mathfrak{H}$ must remain invariant under unitary transformations: thus, for example, a unitary transformation carries every closed set in $\mathfrak{H}$ into a closed set. Similarly, the fact that all unitary transformations are linear ensures the conservation of linearity properties by the group: for instance, a unitary transformation carries a linear manifold into a linear manifold. It is not necessary for us to discuss such rudimentary aspects of the invariant theory here.

The most important concept of group theory which we must mention and analyze at this point is the concept of the transform of a transformation by a member of the group.

If T is a given transformation and U a unitary transformation with the inverse U^{-1}, then the transform of T by U is the transformation UTU^{-1}. The significance of this definition is found in the fundamental logical character of the transformation T: since T may be regarded as the class of ordered pairs (f, Tf) where f is in the domain of T, the application of U to $\mathfrak{H}$ suggests the consideration of the new transformation defined by the class of ordered pairs (Uf, UTf); this new transformation is readily identified as UTU^{-1}. The domain and range of UTU^{-1} are obtained from the domain and range respectively of T by the application of the transformation U. Many simple properties of the transformation T are preserved in the passage to its transform by U; linearity, boundedness, the property of being closed, the property of possessing an inverse or an adjoint, and the property of being symmetric or isometric are such invariant properties. Some of these invariant properties will be stated in detail below. Furthermore, numerous relations between two transformations, such as adjointness, remain undisturbed when both transformations are replaced by their transforms by a given unitary transformation. Some relations of this character will be stated below.

For later reference it will be convenient to summarize certain of these facts of invariant theory in a series of formal theorems. The proof in each case is so elementary that we give no details.

THEOREM 2.52. *If T_0, T_1, T_2 are arbitrary transformations and U is a unitary transformation, then*

$$\begin{aligned} U(T_1 + T_2)U^{-1} &\equiv UT_1U^{-1} + UT_2U^{-1}, \\ U(T_1T_2)U^{-1} &\equiv (UT_1U^{-1})(UT_2U^{-1}), \\ U(aT_0)U^{-1} &\equiv a(UT_0U^{-1}). \end{aligned}$$

If $T_1 \frown T_2$, then $UT_1U^{-1} \frown UT_2U^{-1}$.

THEOREM 2.53. *If T has the adjoint T^*, then UTU^{-1} has the adjoint UT^*U^{-1}.*

THEOREM 2.54. *If T has the inverse T^{-1}, then UTU^{-1} has the inverse $UT^{-1}U^{-1}$.*

THEOREM 2.55. *If H is symmetric, maximal symmetric, essentially self-adjoint, or self-adjoint, then UHU^{-1} has the same property. If H is a symmetric transformation which is bounded, bounded above, bounded below, positive definite, or not-negative definite, then UHU^{-1} has the same property.*

THEOREM 2.56. *If T is isometric or unitary or closed and isometric with the deficiency index (m, n), then UTU^{-1} has the same character.*

THEOREM 2.57. *If T is a bounded linear transformation with bound C, then UTU^{-1} is a bounded linear transformation with bound C; if T has finite norm N, so has UTU^{-1}.*

While it would be possible to establish each of these theorems by a simple train of reasoning, a little reflection indicates that such a procedure may be regarded as unnecessary. Indeed, the truth of these theorems is a consequence of our abstract formulation of all definitions and concepts in Hilbert space. Since the result of applying a unitary transformation to Hilbert space is to determine a new Hilbert space abstractly identical with the old, we should expect every property and relation dependent only upon the abstract characteristics of Hilbert space to be preserved in such a passage from one fundamental space of reference to another.

CHAPTER III

EXAMPLES OF LINEAR TRANSFORMATIONS

§ 1. Infinite Matrices

We have already found that a system of coördinates can be introduced in an abstract Hilbert space $\mathfrak{H}$ by means of the one-to-one correspondence between $\mathfrak{H}$ and $\mathfrak{H}_0$ described in Theorems 1.15 and 1.16. The choice of a coördinate system made in this way is not unique; but the passage from one such set of coördinates to a second is effected by means of a unitary transformation in the coördinate space $\mathfrak{H}_0$. For if $\mathfrak{H}$ is put into correspondence with $\mathfrak{H}_0$ in each of two ways, the transformation of $\mathfrak{H}_0$ into itself defined by passing from $\mathfrak{H}_0$ to $\mathfrak{H}$ by the first correspondence and then back from $\mathfrak{H}$ to $\mathfrak{H}_0$ by the second is evidently a unitary transformation in $\mathfrak{H}_0$.

In terms of a system of coördinates $(x_1, x_2, x_3, \cdots)$ a transformation T in $\mathfrak{H}$ takes each element $(x_1, x_2, x_3, \cdots)$ of its domain into the corresponding element $(y_1, y_2, y_3, \cdots)$ of its range. T thus determines an infinite set of equations

$$y_m = f_m(x_1, x_2, x_3, \cdots), \qquad m = 1, 2, 3, \cdots$$

where the functions $f_1, f_2, f_3, \cdots$ are functions of infinitely many variables defined throughout $\mathfrak{D}$. Conversely, a set of equations of this type defines a transformation T in $\mathfrak{H}$. It is to be observed that the series $\sum_{\alpha=1}^{\infty} | y_\alpha |^2$ must be convergent, in order that $(y_1, y_2, y_3, \cdots)$ may represent an element of $\mathfrak{H}$.

We shall consider here the special case in which the functions $f_1, f_2, f_3, \cdots$ are linear and homogeneous: the equations above then take the special form

$$y_m = \sum_{\beta=1}^{\infty} a_{m\beta}\, x_\beta, \qquad m = 1, 2, 3, \cdots$$

characterized by the infinite matrix $A = \{a_{mn}\}$. In order that these equations have more than formal significance, we must specify the domain $\mathfrak{D}$ in which they are to be valid. The most extensive domain we can select is the set $\mathfrak{D}$ of all elements with coördinates $(x_1, x_2, x_3, \cdots)$ such that the infinite series

$$\sum_{\beta=1}^{\infty} a_{m\beta}\, x_\beta, \quad m = 1, 2, 3, \cdots, \qquad \sum_{\alpha=1}^{\infty} |y_\alpha|^2 = \sum_{\alpha=1}^{\infty} \left| \sum_{\beta=1}^{\infty} a_{\alpha\beta}\, x_\beta \right|^2$$

are convergent. The transformation T with domain $\mathfrak{D}$ which is thus defined by the equations is evidently linear. We see, therefore, that to an arbitrary infinite matrix A there corresponds a certain linear transformation T, defined in terms of the coördinate system introduced in $\mathfrak{H}$. This transformation may be only a trivial one if the matrix A is not suitably restricted. On the other hand, we see that certain linear transformations in $\mathfrak{H}$ determine such infinite matrices: if T is a linear transformation whose domain contains a complete orthonormal set $\{\varphi_n\}$, we assign to φ_n the coördinates $(\delta_{n1}, \delta_{n2}, \delta_{n3}, \cdots)$ and form the matrix $A = \{a_{mn}\} = \{(T\varphi_n, \varphi_m)\}$. If we now use this matrix A to determine the linear transformation described above, we cannot in general ascertain the precise relation between this transformation and the given transformation T, beyond the almost obvious fact that the two transformations are equal in the linear manifold determined by the set $\{\varphi_n\}$. A thoroughly satisfactory theory of the indicated connection between linear transformations and infinite matrices can be developed only in the case of bounded linear transformations.* In the more general cases, such as those considered below, we can establish some interesting results, which must be handled cautiously in practice.

In the following theorems, we shall require certain more or less familiar matrix operations and shall therefore enumerate

* J. v. Neumann, Journal für Mathematik, 161 (1929) pp. 208–236.

them here. We define matrix addition, matrix multiplication, and scalar multiplication as usual by the relations

$$A+B = \{a_{mn}\}+\{b_{mn}\} = \{a_{mn}+b_{mn}\},$$

$$A \cdot B = \{a_{mp}\}\{b_{pn}\} = \left\{\sum_{\alpha=1}^{\infty} a_{m\alpha} \cdot b_{\alpha n}\right\},$$

$$c \cdot A = c\{a_{mn}\} = \{c \cdot a_{mn}\}.$$

It must be noted that the associative law $A(BC) = (AB)C$ is not always true for infinite matrices. The matrix $A^* = \{a^*_{mn}\}$ where $a^*_{mn} = \bar{a}_{nm}$ is called the adjoint of the matrix A. A matrix A such that $A = A^*$ is ordinarily called Hermitian symmetric; in the following pages we shall refer to such a matrix as symmetric. The matrix with all its elements equal to zero will be denoted by O, the matrix $\{\delta_{mn}\}$ by I.

THEOREM 3.1. *If T is a linear transformation with the adjoint T^* such that the linear manifold $\mathfrak{M} = \mathfrak{D}\mathfrak{D}^*$ is everywhere dense in $\mathfrak{H}$, then the matrix*

$$A(\varphi) = \{a_{mn}\} = \{(T\varphi_n, \varphi_m)\}$$

exists when $\{\varphi_n\}$ is a complete orthonormal set in $\mathfrak{M}$, and has the property that the series $\sum_{\beta=1}^{\infty} |a_{m\beta}|^2$, $\sum_{\alpha=1}^{\infty} |a_{\alpha n}|^2$ converge for $m, n = 1, 2, 3, \cdots$. In terms of the matrices $A(\varphi)$ and $A^*(\varphi)$, the transformations T and T^* are calculated from the relations

$$y_m = (Tf, \varphi_m), \quad x_m = (f, \varphi_m), \quad f \text{ in } \mathfrak{D},$$

$$y_m = \sum_{\beta=1}^{\infty} a_{m\beta}\, x_\beta,$$

and

$$y_m = (T^*f, \varphi_m), \quad x_m = (f, \varphi_m), \quad f \text{ in } \mathfrak{D}^*,$$

$$y_m = \sum_{\alpha=1}^{\infty} \bar{a}_{\alpha m}\, x_\alpha.$$

Let $\{\psi_n\}$ be a second complete orthonormal set in $\mathfrak{M}$ and let U be the unitary transformation which takes ψ_n into φ_n, $n = 1, 2, 3, \cdots$. If $B(\psi)$ and $V(\psi)$ are the matrices associated with T and U respectively by the set $\{\psi_n\}$, then A, B, V are connected by the relations

$$B = V(AV^*) = (VA)V^*, \qquad VV^* = V^*V = I.$$

By means of the elementary properties of orthonormal sets and of the relations between T and T^*, every assertion of the theorem is the consequence of simple formal manipulations. Thus, for example, we have

$$\|T^*\varphi_m\|^2 = \sum_{\beta=1}^{\infty} |(\varphi_\beta, T^*\varphi_m)|^2 = \sum_{\beta=1}^{\infty} |(T\varphi_\beta, \varphi_m)|^2$$
$$= \sum_{\beta=1}^{\infty} |a_{m\beta}|^2,$$
$$\|T\varphi_n\|^2 = \sum_{\alpha=1}^{\infty} |(T\varphi_n, \varphi_\alpha)|^2 = \sum_{\alpha=1}^{\infty} |a_{\alpha n}|^2,$$

for $m, n = 1, 2, 3, \cdots$. The relations between T, T^* and $A(\varphi)$, $A^*(\varphi)$ are established by similar reasoning. The computation of $B(\psi)$ in terms of $A(\varphi)$ and $V(\psi)$ will be indicated in some detail. From Theorem 2.44 we know that the unitary transformation U exists; its matrix $V(\psi)$ then has the form $\{v_{mn}\} = \{(U\psi_n, \psi_m)\} = \{(\varphi_n, \psi_m)\}$. Now

$$b_{mn} = (T\psi_n, \psi_m) = \sum_{\alpha=1}^{\infty} (T\psi_n, \varphi_\alpha)(\varphi_\alpha, \psi_m)$$
$$= \sum_{\alpha=1}^{\infty} (\psi_n, T^*\varphi_\alpha)(\varphi_\alpha, \psi_m)$$
$$= \sum_{\alpha=1}^{\infty} \left[\sum_{\beta=1}^{\infty} (\psi_n, \varphi_\beta)(\varphi_\beta, T^*\varphi_\alpha)\right](\varphi_\alpha, \psi_m)$$
$$= \sum_{\alpha=1}^{\infty} \left[\sum_{\beta=1}^{\infty} (\psi_n, \varphi_\beta)(T\varphi_\beta, \varphi_\alpha)\right](\varphi_\alpha, \psi_m)$$
$$= \sum_{\alpha=1}^{\infty} \left(\sum_{\beta=1}^{\infty} v_{m\alpha}\, a_{\alpha\beta}\, v^*_{\beta n}\right),$$

$$b_{mn} = (T\psi_n, \psi_m) = (\psi_n, T^*\psi_m) = \sum_{\beta=1}^{\infty} (\psi_n, \varphi_\beta)(\varphi_\beta, T^*\psi_m)$$
$$= \sum_{\beta=1}^{\infty} (\psi_n, \varphi_\beta)(T\varphi_\beta, \psi_m)$$
$$= \sum_{\beta=1}^{\infty} (\psi_n, \varphi_\beta)\left[\sum_{\alpha=1}^{\infty} (T\varphi_\beta, \varphi_\alpha)(\varphi_\alpha, \psi_m)\right]$$
$$= \sum_{\beta=1}^{\infty} \left(\sum_{\alpha=1}^{\infty} v_{m\alpha}\, a_{\alpha\beta}\, v^*_{\beta n}\right),$$

so that the relation $B = V(AV^*) = (VA)V^*$ is established. The equations $VV^* = V^*V = I$ result almost immediately from the corresponding equations $UU^* \equiv U^*U \equiv I$, which refer to transformations.

THEOREM 3.2. *Let $A = \{a_{mn}\}$ be an infinite matrix such that the series $\sum_{\beta=1}^{\infty} |a_{m\beta}|^2$, $\sum_{\alpha=1}^{\infty} |a_{\alpha n}|^2$, $m, n = 1, 2, 3, \cdots$ are convergent. Let $\{\varphi_n\}$ be an arbitrary complete orthonormal set in $\mathfrak{H}$; let $\mathfrak{D}_1$ be the set $\{\varphi_n\}$; and let $\mathfrak{D}_2$ be the set of all elements f such that the series $\sum_{\alpha=1}^{\infty} \left| \sum_{\beta=1}^{\infty} a_{\alpha\beta} x_\beta \right|^2$, $x_n = (f, \varphi_n)$, converges. Then there exist two transformations $T_1(A)$ and $T_2(A)$ with domains $\mathfrak{D}_1$ and $\mathfrak{D}_2$ respectively defined in terms of A and $\{\varphi_n\}$ by the equations*

$$y_m = \sum_{\beta=1}^{\infty} a_{m\beta} x_\beta, \qquad x_n = (f, \varphi_n), \qquad y_m = (T_k f, \varphi_m)$$

where $k = 1, 2$ and $m, n = 1, 2, 3, \cdots$. The transformations associated in like manner with the adjoint matrix A^* exist and satisfy the relations

$$T_2(A^*) \equiv T_1^*(A), \qquad T_2(A) \equiv T_1^*(A^*).$$

Consequently $T_2(A)$ is a closed linear transformation. The transformations $\hat{T}_1(A)$ and $\tilde{T}_1(A)$ exist and satisfy the relations

$$T_1(A) \subseteq \hat{T}_1(A) \subseteq \tilde{T}_1(A) \subseteq T_2(A).$$

We first construct the transformation $T_1(A)$: since the series $\sum_{\beta=1}^{\infty} |a_{n\beta}|^2$ converges, the series $\sum_{\beta=1}^{\infty} a_{n\beta}\varphi_\beta$ converges in $\mathfrak{H}$ to a limiting element φ_n^*; on setting $T_1(A)\varphi_n = \varphi_n^*$, we have the desired transformation $T_1(A)$ with domain $\mathfrak{D}_1$ as we can verify without difficulty. Next, we consider the adjoint $T_1^*(A)$, which exists in accordance with Theorem 2.6. An element g is in the domain of $T_1^*(A)$, and $T_1^*(A)g = g^*$, if and only if there exists a corresponding element g^* such that

$$(T_1\varphi_n, g) = (\varphi_n, g^*) \text{ or } (g^*, \varphi_n) = (g, T_1\varphi_n),$$

for $n = 1, 2, 3, \cdots$. Evidently a necessary and sufficient condition for the existence of g^* is that the series

$$\sum_{\alpha=1}^{\infty}|(g, T_1\varphi_\alpha)|^2 = \sum_{\alpha=1}^{\infty}\left|\sum_{\beta=1}^{\infty}(g, \varphi_\beta)(\varphi_\beta, T_1\varphi_\alpha)\right|^2 = \sum_{\alpha=1}^{\infty}|\bar{a}_{\beta\alpha}x_\beta|^2$$

converge when we set $x_n = (g, \varphi_n)$. When g^* exists we have

$$y_m = (g^*, \varphi_m) = (g, T_1\varphi_m) = \sum_{\alpha=1}^{\infty}\bar{a}_{\alpha m}x_\alpha.$$

Hence $T_1^*(A)$ and $T_2(A^*)$ exist and are identical. By a similar argument $T_1(A^*)$ and $T_1^*(A^*) \equiv T_2(A)$ exist. The remaining assertions of the theorem are immediate consequences of Theorems 2.8 and 2.10.

THEOREM 3.3. *Let T be a linear transformation satisfying the hypotheses of Theorem 3.1 and let $A(\varphi)$ be the infinite matrix associated with T by a complete orthonormal set $\{\varphi_n\}$ in $\mathfrak{M} = \mathfrak{D}\mathfrak{D}^*$. Then the transformations $T_1(A)$ and $T_2(A)$ associated with the matrix $A = A(\varphi)$ by the same orthonormal set $\{\varphi_n\}$, as described in Theorem 3.2, satisfy the relations $T_1 \subseteqq \hat{T}_1 \subseteqq T \subseteqq T_2$. If T is closed $T_1 \subseteqq \hat{T}_1 \subseteqq \tilde{T}_1 \subseteqq T \subseteqq T_2$.*

It is evident that $T \supseteqq T_1$. From Theorem 2.10 it is thus apparent that $T \supseteqq \hat{T}_1$ and that, when T is closed, $T \supseteqq \tilde{T}_1$. By comparing the domains of T and T_2 as described in Theorems 3.1 and 3.2 we see that $\mathfrak{D} \subseteqq \mathfrak{D}_2$, $T \subseteqq T_2$.

The preceding theorems apply, in particular, to symmetric and bounded linear transformations. In these cases, we can sharpen our results to a considerable extent.

THEOREM 3.4. *If H is a symmetric transformation, the associated matrix $A(\varphi)$ of Theorem 3.1 exists and is symmetric, $A = A^*$. If A is a symmetric matrix satisfying the hypotheses of Theorem 3.2, then the associated transformations $T_1(A)$, $\hat{T}_1$, $\tilde{T}_1$ are symmetric. The associated transformation $T_2(A)$ is symmetric if and only if the matrix A has the property that whenever the series*

$$\sum_{\alpha=1}^{\infty}|x_\alpha|^2, \quad \sum_{\alpha=1}^{\infty}|y_\alpha|^2, \quad \sum_{\alpha=1}^{\infty}\left|\sum_{\beta=1}^{\infty}a_{\alpha\beta}x_\beta\right|^2, \quad \sum_{\alpha=1}^{\infty}\left|\sum_{\beta=1}^{\infty}a_{\alpha\beta}y_\beta\right|^2$$

are convergent, the necessarily convergent series

$$\sum_{\alpha=1}^{\infty}\left(\sum_{\beta=1}^{\infty}a_{\alpha\beta}\bar{x}_\alpha y_\beta\right), \qquad \sum_{\beta=1}^{\infty}\left(\sum_{\alpha=1}^{\infty}a_{\alpha\beta}\bar{x}_\alpha y_\beta\right)$$

have the same sum; and T_2 is then self-adjoint, T_1 essentially self-adjoint. If H determines A, and A determines an essentially self-adjoint transformation $T_1(A)$, then H is essentially self-adjoint; the converse is not necessarily true.

We see at once that when H is symmetric the associated matrix A is identical with its adjoint matrix A^*. When A is a symmetric matrix, the associated transformation $T_1(A)$ is symmetric, since $(T_1\varphi_n, \varphi_m) = a_{mn} = \bar{a}_{nm} = (\varphi_n, T_1\varphi_m)$. By Theorem 2.12, $\hat{T}_1$ and $\tilde{T}_1$ are also symmetric. In order that T_2 be symmetric it is necessary and sufficient that the equation $(T_2 f, g) = (f, T_2 g)$ be true for every pair of elements f, g in $\mathfrak{D}_2$. If we set $y_n = (f, \varphi_n)$ and $x_n = (g, \varphi_n)$, where $\{\varphi_n\}$ is the orthonormal set employed to construct $T_2(A)$ from A, the elements f and g belong to $\mathfrak{D}_2$ if and only if the series

$$\sum_{\alpha=1}^{\infty} |x_\alpha|^2, \quad \sum_{\alpha=1}^{\infty} |y_\alpha|^2, \quad \sum_{\alpha=1}^{\infty} \left| \sum_{\beta=1}^{\infty} a_{\alpha\beta} x_\beta \right|^2, \quad \sum_{\alpha=1}^{\infty} \left| \sum_{\beta=1}^{\infty} a_{\alpha\beta} y_\beta \right|^2$$

are convergent; thus if f and g are in $\mathfrak{D}_2$ the double series $\sum_{\alpha=1}^{\infty}\left(\sum_{\beta=1}^{\infty} a_{\alpha\beta}\bar{x}_\alpha y_\beta\right)$ and $\sum_{\beta=1}^{\infty}\left(\sum_{\alpha=1}^{\infty} a_{\alpha\beta}\bar{x}_\alpha y_\beta\right)$ converge to the sums $(T_2 f, g)$ and $(f, T_2 g)$ respectively. Consequently a necessary and sufficient condition that T_2 be symmetric is that these two series have the same sum. When T_2 is symmetric the relations $T_2 \frown T_2$ and $T_2 \supseteq T_1$ imply the relations

$$T_2 \subseteq T_2^*, \qquad T_2^* \subseteq T_1^*(A) \equiv T_2(A^*) \equiv T_2(A)$$

respectively; and these results require that $T_2 \equiv T_2^*$. Thus T_2 is self-adjoint. Further, when T_2 is self-adjoint, we have also $T_1^{**}(A) \equiv T_2^*(A) \equiv T_2(A) \equiv T_1^*(A)$ from the relations above, and conclude that T_1 is essentially self-adjoint. On the other hand, if T_1 is known to be essentially self-adjoint, the relations

$$T_2(A) \equiv T_2(A^*) \equiv T_1^*(A) \equiv T_1^{**}(A)$$

imply that T_2 and T_2^* are identical, whence T_2 is self-adjoint. Consequently, if we pass from H to A to T_1 and find that T_1 is essentially self-adjoint, we can apply the relations

$$T_1 \subseteq H \subseteq T_2, \qquad T_2 \equiv T_1^* \supseteq H^* \supseteq T_2^* \equiv T_2$$

to show that $H^{**} \equiv (H^*)^* \equiv T_2^* \equiv T_2 \equiv H^*$ and to conclude that H is essentially self-adjoint. On the other hand, if H is a self-adjoint transformation and A is the matrix associated with H by a complete orthonormal set in the domain of H, the transformation $T_1(A)$ is not necessarily essentially self-adjoint and the transformation $T_2(A)$ is not necessarily symmetric. We indicate examples at the end of § 3.

Lastly, we shall investigate the class of bounded linear transformations with domain $\mathfrak{H}$, for which the corresponding matrix theory is found to be complete and satisfactory. The fundamental definition which must be made here is the following:

DEFINITION 3.1. *A matrix $A = \{a_{mn}\}$ is said to be bounded if there exists a positive constant $C = C(A)$ such that*

$$\sum_{\alpha=1}^{\infty} \left| \sum_{\beta=1}^{n} a_{\alpha\beta}\, x_\beta \right|^2 \leqq C^2 \sum_{\beta=1}^{n} | x_\beta |^2$$

for all values $(x_1, \cdots, x_n)$, $n = 1, 2, 3, \cdots$.

THEOREM 3.5. *Let $\{\varphi_n\}$ be an arbitrary complete orthonormal set in $\mathfrak{H}$, $\mathfrak{B}$ the class of all bounded linear transformations with domain $\mathfrak{H}$, and $\mathfrak{M}$ the class of all bounded matrices. To each transformation T in $\mathfrak{B}$ there corresponds a matrix $A = A(\varphi) = \{(T\varphi_n, \varphi_m)\}$ in $\mathfrak{M}$; and each matrix A in $\mathfrak{M}$ satisfies the conditions of Theorem 3.2 and defines a transformation $T \equiv \tilde{T}_1(A) \equiv T_2(A)$ in $\mathfrak{B}$. This correspondence between $\mathfrak{B}$ and $\mathfrak{M}$ is a one-to-one correspondence such that if T_0, T_1, T_2 and A_0, A_1, A_2 are corresponding members of the two classes, then $T_1 + T_2 \sim A_1 + A_2$, $T_1 T_2 \sim A_1 A_2$, $a \cdot T_0 \sim a \cdot A_0$, and $T_0^* \sim A_0^*$. To the identity I corresponds the unit matrix $I = \{\delta_{mn}\}$, to the transformation O the matrix O all of whose elements are zero.*

If T is a bounded linear transformation with domain $\mathfrak{H}$, then T^* exists and is of the same nature according to Theorem 2.29. We can therefore apply Theorem 3.1 to construct the matrix $A(\varphi)$, using an arbitrary complete orthonormal set $\{\varphi_n\}$. Since T is bounded, there exists a positive constant C such that $\|Tf\|^2 \leqq C^2\|f\|^2$ for every f in $\mathfrak{H}$; when we apply this inequality to an arbitrary element of the form $f = \sum_{\beta=1}^{n} x_\beta \varphi_\beta$, we obtain the result

$$\|Tf\|^2 = \sum_{\alpha=1}^{\infty} |(Tf, \varphi_\alpha)|^2 = \sum_{\alpha=1}^{\infty} \left| \sum_{\beta=1}^{n} x_\beta (T\varphi_\beta, \varphi_\alpha) \right|^2$$
$$= \sum_{\alpha=1}^{\infty} \left| \sum_{\beta=1}^{n} a_{\alpha\beta} x_\beta \right|^2 \leqq C^2 \|f\|^2 = C^2 \sum_{\beta=1}^{n} |x_\beta|^2,$$

which signifies that $A(\varphi)$ is bounded. Further, we can apply Theorem 3.3 to obtain the relations $T_1(A) \subseteqq \hat{T}_1(A) \subseteqq \tilde{T}_1(A) \equiv T \equiv T_2(A)$.

If A is a bounded matrix, the inequality $\sum_{\alpha=1}^{\infty} |a_{\alpha n}|^2 \leqq C^2$ must be true as a special case of the inequality satisfied by A. Thus, as in Theorem 3.2, we can definite $T_1(A)$ with domain $\mathfrak{D}_1 \equiv \{\varphi_n\}$, where $\{\varphi_n\}$ is an arbitrary complete orthonormal set, by means of the equations

$$T_1(A)\varphi_n = \sum_{\alpha=1}^{\infty} a_{\alpha n} \varphi_\alpha.$$

Next, we can construct a linear extension $\hat{T}_1(A)$ of $T_1(A)$ in the following manner: if f is an arbitrary element of $\hat{\mathfrak{D}}_1$, the linear manifold determined by $\mathfrak{D}_1$, it can be expressed in just one way as a sum $f = \sum_{\beta=1}^{n} x_\beta \varphi_\beta$; and we can set $\hat{T}_1(A)f = \sum_{\beta=1}^{n} x_\beta T_1(A)\varphi_\beta = \sum_{\alpha=1}^{\infty} \left(\sum_{\beta=1}^{n} a_{\alpha\beta} x_\beta \right) \varphi_\alpha$. The transformation $\hat{T}_1(A)$ is obviously linear and is seen to be bounded, in view of the inequality

$$\|\hat{T}_1(A)f\|^2 = \sum_{\alpha=1}^{\infty} \left| \sum_{\beta=1}^{n} a_{\alpha\beta} x_\beta \right|^2 \leqq C^2 \sum_{\beta=1}^{n} |x_\beta|^2 = C^2\|f\|^2.$$

By Theorem 2.23, $\hat{T}_1(A)$ has a closed bounded linear extension $\tilde{T}_1(A)$ with domain $\mathfrak{H}$. It is evident now that A is the matrix associated with $T \equiv \tilde{T}_1(A)$ by the orthonormal set $\{\varphi_n\}$. To the matrix A and the transformation T we can apply the results of the preceding paragraph and thus establish the assertions of the theorem concerning the relations between them.

It is evident from the manner in which the correspondence between the classes $\mathfrak{B}$ and $\mathfrak{M}$ has been established that it is a one-to-one correspondence. Each complete orthonormal set determines one such correspondence; variation of the set will cause variation of the correspondence.

The isomorphism of the algebra of bounded linear transformations with domain $\mathfrak{H}$ and ordinary matrix algebra is a direct consequence of the relations

$$
\begin{aligned}
((T_1+T_2)\varphi_n, \varphi_m) &= (T_1\varphi_n, \varphi_m) + (T_2\varphi_n, \varphi_m),\\
(T_1T_2\varphi_n, \varphi_m) &= (T_2\varphi_n, T_1^*\varphi_m)\\
&= \sum_{\alpha=1}^{\infty}(T_2\varphi_n, \varphi_\alpha)(\varphi_\alpha, T_1^*\varphi_m)\\
&= \sum_{\alpha=1}^{\infty}(T_2\varphi_n, \varphi_\alpha)(T_1\varphi_\alpha, \varphi_m),\\
(aT_0\varphi_n, \varphi_m) &= a(T_0\varphi_n, \varphi_m),\\
(T_0^*\varphi_n, \varphi_m) &= (\varphi_n, T_0\varphi_m) = \overline{(T_0\varphi_m, \varphi_n)}.
\end{aligned}
$$

It is evident that the matrices corresponding to I and to O have the forms indicated.

THEOREM 3.6. *A necessary and sufficient condition that a bounded matrix A correspond to a unitary transformation is that the matrix equations $AA^* = A^*A = I$ hold. A matrix with these properties is called a unitary matrix.*

The condition is clearly necessary. It is sufficient since it implies that the transformation T corresponding to A by Theorem 3.5 satisfies the relations $TT^* \equiv T^*T \equiv I$, and hence that T is a transformation with domain and range identical to $\mathfrak{H}$ for which $(Tf, Tg) = (f, T^*Tg) = (f, g)$.

THEOREM 3.7. *A necessary and sufficient condition that two bounded matrices A and B correspond, in different correspondences, to the same bounded linear transformation T with domain $\mathfrak{H}$ is that there exist a unitary matrix V such that $B = VAV^*$; a necessary and sufficient condition that two* *bounded linear transformations S and T with domain $\mathfrak{H}$ correspond, in different correspondences, to the same matrix A is that there exist a unitary transformation U such that $T \equiv USU^* \equiv USU^{-1}$.*

If A and B correspond to T by means of the sets $\{\varphi_n\}$ and $\{\psi_n\}$ respectively, then Theorem 3.1 shows that there exists a matrix V such that $B = (VA)V^* = V(AV^*)$, $VV^* = V^*V = I$; the matrix V is unitary according to Theorem 3.7. We may note that the associative character of the product $(VA)V^* = V(AV^*)$ is in this case obvious, since matrix-multiplication is always associative for bounded matrices in consequence of Theorem 3.5. On the other hand, if $B = VAV^*$ where V is unitary, we choose an arbitrary complete orthonormal set $\{\psi_n\}$ and use it to construct from B and V respectively a bounded linear transformation T and a unitary transformation U, as indicated in Theorems 3.5 and 3.6. The matrix associated with T by the complete orthonormal set $\{\varphi_n\}$ where $\varphi_n = U\psi_n$ is readily identified with the matrix $A = V^*BV$.

If S and T correspond to the matrix A by means of the two complete orthonormal sets $\{\varphi_n\}$ and $\{\psi_n\}$ respectively, and if U is the unitary transformation determined by Theorem 2.44 which takes φ_n into ψ_n, we see at once that

$$(USU^*\psi_n, \psi_m) = (S\varphi_n, \varphi_m) = a_{mn} = (T\psi_n, \psi_m),$$

so that $USU^* = T$ in the set $\{\psi_n\}$ and thus $USU^* \equiv T$ in $\mathfrak{H}$. On the other hand, if there is a unitary transformation U such that $T \equiv USU^*$, we can take an arbitrary complete orthonormal set $\{\varphi_n\}$ and then determine a second $\{\psi_n\}$ so that $\psi_n = U\varphi_n$; we then find that

$$(T\psi_n, \psi_m) = (USU^*\psi_n, \psi_m) = (S\varphi_n, \varphi_m)$$

as we wished to show.

It is of some interest to evaluate the norm of a bounded linear transformation in terms of the associated matrix. It is found that, if A and T correspond by the complete orthonormal set $\{\varphi_n\}$, $N(T)$ can be computed by means of this set and has the value $\left(\sum_{\alpha,\beta=1}^{\infty} |a_{\alpha\beta}|^2\right)^{1/2}$. The latter quantity was introduced as the norm of the matrix A, at least in the case of a finite matrix, by Frobenius.†

The matrices which we have considered up to this point have been in the form $\{a_{mn}\}$ where the indices m, n run through the values $1, 2, 3, \cdots$. In many instances it is convenient to employ matrices $\{a_{mn}\}$ for which m, n assume the values $\cdots, -3, -2, -1, 0, +1, +2, +3, \cdots$. In order to develop a theory of infinite matrices of this type, it is sufficient for us to adopt a new numbering for a complete orthonormal set, writing it in the form

$$\cdots, \varphi_{-3}, \varphi_{-2}, \varphi_{-1}, \varphi_0, \varphi_{+1}, \varphi_{+2}, \varphi_{+3}, \cdots.$$

On this account, the theory of such matrices is essentially identical with that which we have developed above.

The connection between infinite matrices and linear transformations in $\mathfrak{H}$ can be utilized in two ways: we may employ the relation to suggest interesting theorems in the theory of transformations by generalizing known results concerning finite matrices to infinite matrices and hence to transformations; and we may use infinite matrices as mathematical tools in carrying out demonstrations. The part of the familiar theory of finite matrices which is of particular interest here has three distinct aspects, leading to the same fundamental problem: a matrix may be regarded abstractly, it may be regarded as the matrix of a linear transformation, and it may be regarded as the matrix of a bilinear form. In each case the aim of the theory is to determine a unitary matrix U such that a given matrix A is transformed by it into an appropriate normal form; the transformed matrix of A is the matrix $UAU^* = UAU^{-1}$. The generalization of this theory to infinite matrices and the corresponding transformations was begun systematically by Hilbert‡ and was developed by

† Frobenius, Sitzungsberichte der Preussischen Akademie zu Berlin, 1911, p. 242.

‡ Hilbert, *Grundzüge einer allgemeinen Theorie der linearen Integralgleichungen*, Leipzig, 1912.

him and by his followers, notably Hellinger and Toeplitz.* These writers laid particular emphasis upon the study of the forms $\sum_{\alpha,\beta=1}^{\infty} a_{\alpha\beta}\,\bar{x}_\alpha\, y_\beta$, $\sum_{\alpha,\beta=1}^{\infty} a_{\alpha\beta}\,\bar{x}_\alpha\, x_\beta$. Since they use infinite matrices as tools, the greater portion of their results is confined to the case in which the matrix $A = \{a_{mn}\}$ is bounded. A certain amount of progress along these lines has been made recently by Wintner.† Here we shall abandon the systematic use of infinite matrices as unsuited to the problems we wish to investigate, although we shall indicate briefly the algebraic analogues of our most important results in ordinary matrix theory. We shall find references to matrix theory valuable in giving us insight into the mathematical relations which interest us. By proceeding abstractly we simplify our notation and are able to deal directly with the mathematical elements of the problem, unhampered by extraneous details of manipulation. The great advantage of an abstract formulation was perceived by J. v. Neumann, who has also investigated the difficulties which arise from an attempt to use the matrix machinery.‡ In the following chapter we shall use the analogy of transformation theory with matrix theory to formulate in a precise manner the fundamental problems of the theory of linear transformations in Hilbert space.

§ 2. Integral Operators

We shall consider in this and the remaining sections of the present chapter the interpretation of transformations in the concrete Hilbert spaces $\mathfrak{L}_2$ and $\mathfrak{L}_{2,m}$ of Chapter I, § 5. Since $\mathfrak{L}_2$ is a space whose elements are functions of a point, a transformation in $\mathfrak{L}_2$ may be thought of as an operator which, applied to one function in $\mathfrak{L}_2$, yields a second. For

* See the general account in the article *Integralgleichungen und Gleichungen mit unendlichvielen Unbekannten* by Hellinger and Toeplitz, Encyklopädie der Mathematischen Wissenschaften 2^{3^2}, pp. 1335–1601.

† Wintner, *Spektraltheorie der unendlichen Matrizen,* Leipzig, 1929, especially Chapter VI.

‡ J. v. Neumann, Mathematische Annalen, 102 (1929), pp. 49–131; Journal für Mathematik, 161 (1929), pp. 208–236.

this reason we shall refer to transformations as operators throughout the remainder of this chapter; it must be remembered that a transformation or operator is not regarded as known unless its domain is precisely characterized. The operators of analysis are of many different kinds. It is our purpose to discuss a variety of typical operators, without any attempt at an exhaustive classification. We select our examples with a double end in view: in the first place, we wish to illustrate in an adequate way the concepts of the abstract theory which has already been developed; and, in the second place, we wish to introduce at this point a number of operators whose treatment by the general theory constructed in later chapters is of primary importance for analysis.

Here we shall consider integral operators, that is to say, operators defined in terms of integration. Let $\mathfrak{L}_2$ consist of complex-valued Lebesgue-measurable functions, defined over a Lebesgue-measurable set E in n-dimensional Euclidean space and restricted by the requirement that the Lebesgue-integral $\int_E |f(P)|^2 \, dP$ exist. If (P, Q) is a pair of points in n-dimensional Euclidean space, it can be represented as a single point in $2n$-dimensional Euclidean space. In this representation, pairs of points in E constitute a Lebesgue-measurable set E_2 in $2n$-dimensional Euclidean space. We shall denote by $K = K(P, Q)$ a complex-valued Lebesgue-measurable function defined almost everywhere in E_2. Concerning this function we recall the important fact that, for almost every P (or Q) in E, $K(P, Q)$ is a measurable function of the point Q (or P).* We can define an operator T in $\mathfrak{L}_2$ according to the formal equation

$$Tf(P) = g(P) = \int_E K(P, Q) f(Q) \, dQ.$$

This operator is called an integral operator, the function K being described as its kernel. The precise sense in which the formal equation defining T is to be interpreted may

* Carathéodory, *Vorlesungen über reelle Funktionen*, second edition, Leipzig, 1927, Chapter XI, pp. 621–641, especially p. 629. Here, as in the entire theory of multiple integrals, the theorem of Fubini is fundamental.

properly vary from instance to instance. For the present, we shall give it the following meaning: for a given function $f(P)$ in $\mathfrak{L}_2$, the number $g(P)$ corresponding to a fixed point P in E is defined if and only if the integral on the right exists as a Lebesgue integral and has the value $g(P)$; $f(P)$ belongs to the domain of T if and only if the function $g(P)$ determined by the equation is defined almost everywhere in E and has the property that the Lebesgue integral $\int_E |g(P)|^2 dP$ exists; and, if $f(P)$ is in the domain of T, $Tf(P)$ is the function $g(P)$. Under this convention, it is evident that the operator T is a linear operator or transformation, whose domain includes at least the null element in $\mathfrak{L}_2$.

An interesting type of integral operator is obtained by requiring the kernel to be (Hermitian) symmetric, $K(P, Q) = \overline{K(Q, P)}$ almost everywhere in E_2. The condition that the corresponding operator T be symmetric is seen to be that the iterated integrals

$$\int_E \left(\int_E K(P, Q) f(Q)\, dQ \right) \cdot \overline{g(P)}\, dP,$$

$$\int_E \left(\int_E K(P, Q)\, \overline{g(P)}\, dP \right) \cdot f(Q)\, dQ$$

have the same value for each pair (f, g) of elements in the domain of T: for the equality of these integrals expresses in the concrete space $\mathfrak{L}_2$ the abstract relation $(Tf, g) = (f, Tg)$.

The problems of analysis which lead us to study integral operators are concerned in the main with particular types of kernel, which can be discussed in greater detail than the entirely general kernel introduced above. The types which have received the most thorough investigation are those handled in various theories of integral equations due to Fredholm, Hilbert, Schmidt, and Carleman.* So far as essential properties are concerned, these kernels are of the following types:

---

* See the excellent general account of Hellinger and Toeplitz, *Integralgleichungen und Gleichungen mit unendlich vielen Unbekannten*, Encyklopädie der Mathematischen Wissenschaften, 2^{3^2}, pp. 1335–1601; this article does not cover Carleman's contributions adequately.

On pages 65, 66 the definition of N(T) appears. If N(T) < ∞, then T is bounded in the "usual" (see p. 56) sense, and ‖T‖ ≤ N(T). However, N(I) = ∞, where I is the identity operator in H.

(1) Kernels of Hilbert-Schmidt. The kernel $K(P, Q)$ has the property that the integral of $|K|^2$ over E_2 exists.

(2) Kernels of Carleman. The kernel $K(P, Q)$ is (Hermitian) symmetric and has the property that the integral $\int_E |K(P,Q)|^2 dQ$ exists for almost every P in E.

It is apparent that (Hermitian) symmetric kernels of Hilbert-Schmidt type are included among those of Carleman type. We shall discuss here the general properties of integral operators with kernels of Hilbert-Schmidt type, deferring the treatment of the Carleman kernel to a point where it can be taken up more advantageously. These fundamental properties can be stated in two theorems.

THEOREM 3.8. *If T is an integral operator whose kernel $K(P, Q)$ is of Hilbert-Schmidt type, then T is a bounded linear operator defined throughout $\mathfrak{L}_2$; T is of finite norm N, $N^2(T) = \int_E \int_E |K|^2 \, dP \, dQ$. The adjoint of T exists and is an integral operator whose kernel is $\overline{K(Q, P)}$. A necessary and sufficient condition that T be symmetric is that the kernel be (Hermitian) symmetric, $K(P, Q) = \overline{K(Q, P)}$ almost everywhere in E_2.*

Since, by hypothesis, $|K|^2$ is integrable over E_2, the integral $\int_E |K(P, Q)|^2 dQ$ exists for almost every P in E and defines a function such that $\int_E \left(\int_E |K(P, Q)|^2 dQ \right) dP$ exists.* Thus, if $f(P)$ is any function in $\mathfrak{L}_2$, the integral $\int_E K(P, Q) f(Q) \, dQ$ exists for almost every P in E and defines a function $g(P)$ such that

$$|g(P)|^2 \leqq \int_E |K(P, Q)|^2 \, dQ \cdot \int_E |f(Q)|^2 dQ.$$

Consequently, $g(P)$ is in $\mathfrak{L}_2$ and satisfies the relations

$$\int_E |g(P)|^2 dP \leqq \int_E \int_E |K(P, Q)|^2 \, dQ \, dP \cdot \int_E |f(Q)|^2 dQ.$$

* Carathéodory, *Vorlesungen über reelle Funktionen*, second edition, Leipzig, 1927, Chapter XI, especially p. 630. Here we neglect sets of measure zero, which are indicated in the more precise statements cited.

If the latter inequality is written in abstract terms, with $C^2 = \int_E \int_E |K(P, Q)|^2 \, dQ \, dP$, it becomes $\|Tf\|^2 \leqq C^2 \|f\|^2$, so that T is a bounded linear transformation with domain $\mathfrak{L}_2$.

We shall now compute the norm of the operator T. We first introduce a complete orthonormal set $\{\varphi_m(P)\}$ in $\mathfrak{L}_2$ and then form the set $\{\psi_{lm}(P, Q)\}$, where $\psi_{lm}(P, Q) = \varphi_l(P)\overline{\varphi}_m(Q)$. The latter set is a complete orthonormal set in the class $\mathfrak{L}_2^{(2)}$ of all Lebesgue-measurable functions $F(P, Q)$ in E_2 such that the Lebesgue integral of $|F|^2$ over E_2 exists, as we shall now show. The obvious equations

$$\int_E \int_E \psi_{lm} \overline{\psi}_{ik} \, dQ \, dP = \int_E \varphi_l(P) \cdot \overline{\varphi}_i(P) \, dP \int_E \varphi_k(Q) \overline{\varphi}_m(Q) \, dQ$$
$$= \delta_{li} \, \delta_{km}$$

show that the set $\{\psi_{lm}\}$ is an orthonormal set in $\mathfrak{L}_2^{(2)}$. If $F(P, Q)$ is a function in $\mathfrak{L}_2^{(2)}$ such that

$$\int_E \int_E F(P, Q) \overline{\psi}_{lm}(P, Q) \, dQ \, dP = 0$$

for $l, m = 1, 2, 3, \cdots$, we then have successively

$$\int_E \left(\int_E F(P, Q) \varphi_m(Q) \, dQ \right) \overline{\varphi}_l(P) \, dP = 0, \; l, m = 1, 2, 3, \cdots,$$

$$\int_E F(P, Q) \varphi_m(Q) \, dQ = 0 \text{ for almost every } P \text{ in } E,$$
$$m = 1, 2, 3, \cdots,$$

$$F(P, Q) = \overline{F(P, Q)} = 0 \text{ almost everywhere in } E_2,$$

and can therefore conclude that $\{\psi_{lm}\}$ is a complete orthonormal set in $\mathfrak{L}_2^{(2)}$. Since K is a function in $\mathfrak{L}_2^{(2)}$ we have

$$\int_E \int_E |K(P, Q)|^2 \, dQ \, dP = \sum_{\alpha, \beta = 1}^{\infty} |a_{\alpha\beta}|^2,$$
$$a_{lm} = \int_E \int_E K(P, Q) \overline{\psi}_{lm}(P, Q) \, dQ \, dP.$$

Now

$$a_{lm} = \int_E \left(\int_E K(P, Q) \varphi_m(Q) \, dQ \right) \overline{\varphi_l(P)} \, dP = (T\varphi_m, \varphi_l)$$

so that

$$N^2(T) = \sum_{\alpha,\beta=1}^{\infty} |(T\varphi_\alpha, \varphi_\beta)|^2 = \sum_{\alpha,\beta=1}^{\infty} |a_{\alpha\beta}|^2$$
$$= \int_E \int_E |K(P, Q)|^2 \, dQ \, dP,$$

as we wished to show.

Since T is a bounded linear transformation with domain $\mathfrak{L}_2$, we know by Theorem 2.29 that T^*, the adjoint of T, exists and is a bounded linear transformation with domain $\mathfrak{L}_2$. We may now compute it in the following way: if $f(P)$ and $g(P)$ are any two elements of $\mathfrak{L}_2$, the product $f(Q)\overline{g(P)} = F(P, Q)$ is such that $|F|^2$ is integrable over E_2; the product $F \cdot K$ is therefore integrable over E_2 and its integral may be expressed by iterated integration in either of the forms

$$\int_E \left(\int_E K(P, Q) \cdot f(Q) \, dQ \right) \overline{g(P)} \, dP,$$
$$\int_E f(Q) \cdot \overline{\left(\int_E \overline{K(P, Q)} \cdot g(P) \, dP \right)} \, dQ;$$

by the definition of T^*, it is evident that

$$T^* g(P) = \int_E \overline{K(Q, P)} \, g(Q) \, dQ$$

as we wished to show. It is now an easy matter to prove that T is symmetric if and only if the condition of the theorem is satisfied. Clearly, the condition is sufficient since it implies the identity of T and T^*. Whenever T and T^* are identical we must have, by the preceding results

$$\int_E [K(P, Q) - \overline{K(Q, P)}] \, f(Q) \, dQ = 0$$

for almost every P in E; and this equation implies that the function $[K(P, Q) - \overline{K(Q, P)}]$ must vanish almost everywhere in E_2, since $f(P)$ can vary over the entire class $\mathfrak{L}_2$.

THEOREM 3.9. *If T_0, T_1, and T_2 are integral operators with Hilbert-Schmidt kernels K_0, K_1, and K_2 respectively, then the operators $a \cdot T_0$, $T_1 + T_2$, $T_1 T_2$ are integral operators with Hilbert-Schmidt kernels $a \cdot K_0$, $K_1 + K_2$, and $\int_E K_1(P, R) K_2(R, Q) \, dR$ respectively.*

The only assertion of the theorem which is not immediately obvious is that which concerns the operator $T_1 T_2$. By definition, $T_1 T_2 f(P) = \int_E K_1(P, R)\left(\int_E K_2(R, Q) f(Q)\, dQ\right) dR$, so that the assertion of the theorem can be established by a change of order of integration in this iterated integral which will yield

$$T_1 T_2 f(P) = \int_E \left(\int_E K_1(P, R) K_2(R, Q)\, dR\right) f(Q)\, dQ.$$

The necessary change can be justified by the theorem of Fubini.* It is evident from the inequality

$$\left|\int_E K_1(P, R) K_2(R, Q)\, dR\right|^2$$
$$\leqq \int_E |K_1(P, R)|^2\, dR \cdot \int_E |K_2(R, Q)|^2\, dR,$$

where the function on the right is integrable over E_2, that the kernel of $T_1 T_2$ is of Hilbert-Schmidt type.

We shall now consider a number of integral operators whose kernels are altogether special but whose usefulness is particularly emphasized in various branches of analysis. In these examples, it is necessary to interpret in a new way the fundamental equations defining the operators in terms of their kernels.

We shall discuss in detail but one of these special integral operators, namely the Fourier transformation. There is no essential difference of principle encountered in the discussion of other cases.

THEOREM 3.10. (*Fourier-Plancherel Theorem*). *Let* $\mathfrak{L}_2$ *be the class of all Lebesgue-measurable functions defined over n-dimensional Euclidean space E, subject to the restriction that the integral* $\int_E |f(Q)|^2\, dQ$ *exist. Let* $K(P, Q)$ *be the function* $(2\pi)^{-n/2} e^{i(x_1 y_1 + \cdots + x_n y_n)}$ *where* $(x_1, \cdots, x_n)$ *and* $(y_1, \cdots, y_n)$ *are the coördinates of P and Q respectively in a fixed rectangular coördinate system. Let* $\{E_m\}$ *be a sequence of point-*

* Carathéodory, *Vorlesungen über reelle Funktionen*, second edition, Leipzig, 1927, Chapter XI.

sets determined by the inequalities $-A_{mp} < x_p \leqq A_{mp}$; $p = 1, 2, \cdots, n$, $m = 1, 2, 3, \cdots$ *where* A_{mp} *is positive and becomes infinite with* m. *Then the function* $g_m(P) = \int_{E_m} K(P, Q) f(Q)\, dQ$, *where* f *is an arbitrary element of* $\mathfrak{L}_2$, *converges in the mean (converges in* $\mathfrak{L}_2$*) to a function* $g(P)$ *independent of the sequence* $\{E_m\}$. *The transformation*

$$Tf(P) = g(P) = \int_E K(P, Q) f(Q)\, dQ$$

is a unitary transformation of $\mathfrak{L}_2$ *into itself. Similarly, the kernel* $\overline{K(Q, P)}$ *defines a transformation identical with* T^*. *As a consequence, the reciprocal relations*

$$g(P) = \int_E K(P, Q) f(Q)\, dQ, \qquad f(P) = \int_E \overline{K(Q, P)}\, g(Q)\, dQ$$

are true in $\mathfrak{L}_2$, *in the sense defined above.*†

We shall first note an elementary formula: if (a, b) and (c, d) are two intervals, then the absolutely convergent definite integral

$$L(a, b, c, d) = \int_{-\infty}^{+\infty} (e^{ib\xi} - e^{ia\xi})(e^{-id\xi} - e^{-ic\xi})/2\pi\xi^2\, d\xi$$

has as its value the length of the interval common to (a, b) and (c, d). This result may be obtained with the help of simple reductions leading to the familiar integral $\int_{-\infty}^{+\infty} (1 - \cos \xi)/\xi^2\, d\xi = \pi$. From this fact as our starting point, we propose to derive the assertions of the present theorem.

We denote by E_{ab} the point set specified by the inequalities $a_p < x_p \leqq b_p$, $p = 1, \cdots, n$, and define the function $\varphi(P; E_{ab})$, which is equal to 1 in E_{ab} and vanishes elsewhere, noting that φ is an element of $\mathfrak{L}_2$. We form the function

---

† Plancherel, Rendiconti di Palermo, 30 (1910), pp. 289–335. — Titchmarsh, Proceedings of the London Mathematical Society, (2) 23 (1924), pp. 279–289. — Berry, Journal of Mathematics and Physics of the Massachusetts Institute of Technology, 8 (1929), pp. 106–118.

$$\psi_m(P; E_{ab}) = \int_{E_m} K(P, Q)\,\varphi(Q; E_{ab})\,dQ$$
$$= \int_{E_m E_{ab}} K(P, Q)\,dQ,$$

which can be evaluated explicitly and is found to be an element of $\mathfrak{L}_2$. Since for all sufficiently large m the set $E_m E_{ab}$ reduces to E_{ab}, $\psi_m \to \psi = \int_{E_{ab}} K(P, Q)\,dQ$ in $\mathfrak{L}_2$. For ψ we find without difficulty the expression

$$\psi(P; E_{ab}) = (2\pi)^{-n/2} \prod_{\alpha=1}^{n} [(e^{ib_\alpha x_\alpha} - e^{ia_\alpha x_\alpha})/ix_\alpha].$$

It is now a simple matter to prove the equality

$$\int_E \varphi(Q; E_{ab})\,\overline{\varphi(Q; E_{cd})}\,dQ = \int_E \psi(Q; E_{ab})\,\overline{\psi(Q; E_{cd})}\,dQ,$$

by the use of iterated integration and the formula for $L(a, b, c, d)$ stated above.

Next we denote by $\mathfrak{D}_1$ the linear manifold determined in $\mathfrak{L}_2$ by the set of all functions $\varphi(P; E_{ab})$. Since $\mathfrak{D}_1$ contains as a subclass the class of all rational step-functions, defined and discussed in the proof of Theorem 1.24, it is evident that $\mathfrak{D}_1$ is everywhere dense in $\mathfrak{L}_2$. We define a linear transformation T_1 with domain $\mathfrak{D}_1$ by the following relations: if $f(P)$ is in $\mathfrak{D}_1$ it can be written in the form

$$f(P) = \sum_{\alpha=1}^{N} c_\alpha\,\varphi(P; E_{a_\alpha b_\alpha});$$

and $T_1 f(P)$ is then determined as

$$T_1 f(P) = \sum_{\alpha=1}^{N} c_\alpha\,\psi(P; E_{a_\alpha b_\alpha})$$
$$= \int_E K(P, Q) \sum_{\alpha=1}^{N} c_\alpha\,\varphi(Q; E_{a_\alpha b_\alpha})\,dQ$$
$$= \int_E K(P, Q)\,f(Q)\,dQ.$$

There is no difficulty in verifying the assertion that T_1 is a linear transformation which takes $\mathfrak{D}_1$ into a subset of $\mathfrak{L}_2$.

From the final result of the preceding paragraph, it is clear that T_1 has the property that the integrals

$$\int_E f(Q)\,\overline{g(Q)}\,dQ, \quad \int_E T_1 f(Q)\,\overline{T_1 g(Q)}\,dQ$$

are equal wherever f and g are in $\mathfrak{D}_1$.

The properties of the transformation T_1 may be resumed in our customary abstract terminology as follows: T_1 is a linear transformation whose domain is everywhere dense in $\mathfrak{L}_2$ and which has the property that $(T_1 f, T_1 g) = (f, g)$ for every pair of elements in $\mathfrak{D}_1$. In particular, we have $\| T_1 f \| = \| f \|$, so that T_1 is bounded. By Theorem 2.23, the transformation $\tilde{T}_1$ exists and is a bounded linear transformation with domain $\mathfrak{L}_2$. We shall set $T \equiv \tilde{T}_1$. Since T is bounded and continuous, we know that, whenever $\{f_m\}$ is a sequence converging to f in $\mathfrak{L}_2$, then Tf_m converges to Tf. If we choose f and g arbitrarily in $\mathfrak{L}_2$ and determine sequences $\{f_m\}$ and $\{g_m\}$ in $\mathfrak{D}_1$ so that $f_m \to f$, $g_m \to g$ we then have

$$(Tf, Tg) = \lim_{m\to\infty} (Tf_m, Tg_m) = \lim_{m\to\infty} (f_m, g_m) = (f, g).$$

By a similar argument, we can make use of certain elementary facts concerning sequences of functions in $\mathfrak{L}_2$ in order to show that T can be constructed from the kernel $K(P, Q)$ in the manner described in the theorem. We consider first an arbitrary element of $\mathfrak{L}_2$ which vanishes outside a properly chosen set E_{ab}, denoting this function by $f(P)$; we recall that it is possible to determine a sequence of elements $f_m(P)$, all in $\mathfrak{D}_1$ and all vanishing outside E_{ab}, such that $f_m \to f$. We know that

$$Tf_m = \int_{E_{ab}} K(P, Q) f_m(Q)\,dQ, \quad Tf_m \to Tf$$

and recall that, in terms of point-wise convergence,

$$\lim_{m\to\infty} \int_{E_{ab}} K(P, Q)\cdot f_m(Q)\,dQ = \int_{E_{ab}} K(P, Q)\cdot f(Q)\,dQ.$$

By reference to the proof of the Riesz-Fischer theorem, given in the discussion of Theorem 1.24, we see that we can identify the two limits assigned to the sequence Tf, one by convergence in $\mathfrak{L}_2$, the other by point-wise convergence; we have for any function $f(P)$ of the type considered the relation

$$Tf(P) = \int_{E_{ab}} K(P, Q) f(Q)\, dQ.$$

When $f(P)$ is an entirely unrestricted element in $\mathfrak{L}_2$, we form the sequence $\{f_m\}$, where f_m coincides with f in E_m and vanishes elsewhere, so that $f_m \to f$. We can then write

$$Tf_m = \int_{E_m} K(P, Q) f(Q)\, dQ, \quad Tf_m \to Tf.$$

We observe that this pair of relations expresses the desired connection between the transformation T and its kernel.

By an entirely analogous argument, we can construct the transformation $\bar{T}$ associated with the kernel $\overline{K(Q, P)}$: it is a bounded linear transformation whose domain is $\mathfrak{L}_2$ and which has the property that $(\bar{T}f, \bar{T}g) = (f, g)$ for every pair of elements in $\mathfrak{L}_2$. We shall show that $\bar{T}$ and T^* are identical. By the use of iterated integration we can obtain the expressions

$$I_1 = \int_E T\varphi(Q; E_{ab})\, \overline{\varphi(Q; E_{cd})}\, dQ = \int_{E_{cd}} \int_{E_{ab}} K(Q, R)\, dR\, dQ,$$

$$I_2 = \int_E \varphi(Q; E_{ab})\, \overline{\bar{T}\varphi(Q; E_{cd})}\, dQ = \int_{E_{ab}} \int_{E_{cd}} K(R, Q)\, dQ\, dR.$$

Since K is symmetrical in its arguments, and since the order of integration in the iterated integrals is immaterial, we conclude that I_1 and I_2 are equal. This result may be extended at once to the integrals

$$\int_E Tf(Q)\, \overline{g(Q)}\, dQ, \quad \int_E f(Q)\, \overline{\bar{T}g(Q)}\, dQ,$$

whenever f and g are elements of $\mathfrak{D}_1$. Thus, to revert to the abstract terminology once more, T and $\bar{T}$ satisfy the relation $(Tf, g) = (f, \bar{T}g)$ whenever f and g are in $\mathfrak{D}_1$. If f and g are arbitrary elements of $\mathfrak{L}_2$, we determine sequences $\{f_m\}$ and $\{g_m\}$ in $\mathfrak{D}_1$ so that $f_m \to f$, $g_m \to g$. We then have $Tf_m \to Tf$, $\bar{T}g_m \to \bar{T}g$, so that

$$(Tf, g) = \lim_{m\to\infty} (Tf_m, g_m) = \lim_{m\to\infty} (f_m, \bar{T}g_m) = (f, \bar{T}g),$$

whenever f, g are elements of $\mathfrak{L}_2$. This result shows that $\bar{T}$ and T^* are identical.

Finally, we shall prove that T is a unitary transformation. The properties which it is already known to possess characterize it as a closed isometric transformation with domain $\mathfrak{L}_2$. Its range is a closed linear manifold. The transformation T is unitary if and only if its range coincides with $\mathfrak{L}_2$; and the range coincides with $\mathfrak{L}_2$ if and only if the sole element orthogonal to it is the null element. In order to show that T is unitary, we have only to prove that $g = 0$, whenever (Tf, g) vanishes for every f in $\mathfrak{L}_2$. We can accomplish this easily as follows: $(f, T^*g) = (Tf, g) = 0$ implies $\bar{T}g = T^*g = 0$, when f ranges over $\mathfrak{L}_2$; and then $\bar{T}g = 0$ implies $|g| = |\bar{T}g| = 0$ and $g = 0$, as we wished to show. Hence T is a unitary transformation, and $\bar{T} \equiv T^*$ is its inverse.

There are many other "reciprocal theorems" which can be established by reasoning like that just employed for the Fourier-Plancherel theorem. We shall list a few of them here:

(1) if $\mathfrak{L}_2$ is the class of all Lebesgue-measurable functions on the interval $0 \leqq x < +\infty$ subject to the restriction that $\int_0^\infty |f(x)|^2 \, dx$ exist, then

$$g(x) = (2/\pi)^{1/2} \int_0^\infty \sin xy \, f(y) \, dy,$$

$$f(x) = (2/\pi)^{1/2} \int_0^\infty \sin xy \, g(y) \, dy;$$

the transformation T defined by the kernel $(2/\pi)^{1/2} \sin xy$ is both unitary and symmetric;

(2) if $\mathfrak{L}_2$ is the class discussed in (1), then

$$g(x) = (2/\pi)^{1/2} \int_0^\infty \cos xy \, f(y) \, dy,$$

$$f(x) = (2/\pi)^{1/2} \int_0^\infty \cos xy \, g(y) \, dy;$$

the transformation T defined by the kernel $(2/\pi)^{1/2} \cos xy$ is both unitary and symmetric;

(3) if $\mathfrak{L}_2$ is the class discussed in (1), then

$$g(x) = \int_0^\infty (xy)^{1/2} J_\nu(xy)\, f(y)\, dy,$$

$$f(x) = \int_0^\infty (xy)^{1/2} J_\nu(xy)\, g(y)\, dy,$$

where J_ν is the Bessel function of the first kind and order ν, $\nu \geqq -1/2$; the transformation T defined by the kernel $(xy)^{1/2} J_\nu(xy)$ is called the Hankel transformation and is both unitary and symmetric.

The formulas (1) and (2) may be proved directly, along the lines followed in establishing the Fourier-Plancherel theorem, or may be proved by specializing that theorem in an appropriate manner. Formula (3) may be proved by the method adopted for the Fourier transformation, though the specific formulas employed in the proof must be selected with some care in order to avoid complication. It turns out that, in discussing the Hankel transformation, we profit by defining $\varphi(P; E_{ab})$ as the function which is equal to $x^{\nu+1/2}$ in the interval $a < x \leqq b$ and vanishes elsewhere; for we can evaluate $T\varphi$ as follows:

$$\begin{aligned} T\varphi &= \int_a^b (xy)^{1/2} J_\nu(xy)\, y^{\nu+1/2}\, dy \\ &= \frac{b^{\nu+1} J_{\nu+1}(bx) - a^{\nu+1} J_{\nu+1}(ax)}{x^{1/2}}, \end{aligned}$$

using only elementary properties of Bessel functions.* In order to show that $(T\varphi_1, T\varphi_2) = (\varphi_1, \varphi_2)$, we must evaluate the absolutely convergent infinite integral

$$\int_0^\infty \frac{1}{x} [b^{\nu+1} J_{\nu+1}(bx) - a^{\nu+1} J_{\nu+1}(ax)] \times [c^{\nu+1} J_{\nu+1}(cx) - d^{\nu+1} J_{\nu+1}(dx)]\, dx.$$

It is evident that this integral can be evaluated if we can compute the absolutely convergent integral

$$\int_0^\infty \frac{J_{\nu+1}(\alpha x)\, J_{\nu+1}(\beta x)\, dx}{x}.$$

* Watson, *Theory of Bessel Functions,* Cambridge, 1922, § 5.1 (1).

The latter integral has been discussed by many writers and is shown by fairly simple methods to have the value $\frac{1}{2}(\beta/\alpha)^{\nu+1}/\nu+1$ when $\beta \leqq \alpha$, the value $\frac{1}{2}(\alpha/\beta)^{\nu+1}/\nu+1$ when $\alpha \leqq \beta$.† On substituting this result in the integral above it turns out to have the value $\int_{\Delta} y^{2\nu+1}\,dy$, where Δ is the interval common to (a, b) and (c, d). It thus appears that we can obtain, in the case of the Hankel transformation, simple analogues of the formulas used to discuss the Fourier transformation. Since the linear manifold $\mathfrak{D}_1$ determined by the set of all functions $\varphi(P; E_{ab})$ is everywhere dense in $\mathfrak{L}_2$ and can be used in the construction of sequences analogous to those set up in the proof of Theorem 3.10, we see that the theory of the Hankel transformation can be carried through successfully without the introduction of new processes.

Under certain general conditions, the theory of linear transformations in $\mathfrak{L}_{2,m}$ may be related in a simple manner to the theory of transformations in $\mathfrak{L}_2$, by taking advantage of the fact that each element of $\mathfrak{L}_{2,m}$ is a vector whose components are elements of $\mathfrak{L}_2$. After we have made the connection precise, we shall be able to define integral operators in $\mathfrak{L}_{2,m}$ in a natural manner.

THEOREM 3.11. *A necessary and sufficient condition that T be a linear transformation with domain $\mathfrak{L}_{2,m}$ is that there exist m^2 linear transformations T_{ik}, $i, k = 1, \cdots, m$, with $\mathfrak{L}_2$ as domain such that T takes the vector $\mathbf{f}$ with components $(f_1, \cdots, f_m)$ into the vector $\mathbf{g}$ with components $(g_1, \cdots, g_m)$, where*

$$g_i = \sum_{\beta=1}^{m} T_{i\beta} f_\beta, \qquad i = 1, \cdots, m.$$

If T is associated with the matrix $\{T_{ik}\}$, then T^* is associated with the matrix $\{T_{ki}^*\}$.

The condition enunciated in the theorem is obviously sufficient. We see immediately that it is necessary: for if T is a linear transformation which takes $\mathbf{f} = (f_1, \cdots, f_m)$ into $\mathbf{g} = (g_1, \cdots, g_m)$, we have only to define T_{ik} as the trans-

† Watson, *Theory of Bessel Functions*, Cambridge, 1922, § 13.42 (1).

formation which takes f_k into g_{ik}, the ith component of the vector into which T carries the vector $(0, \cdots, 0, f_k, 0, \cdots, 0)$, and can then verify without difficulty that the transformations T_{ik} enjoy all the properties ascribed to them by the theorem.

If T is associated with the matrix $\{T_{ik}\}$ and if T^*, assumed to have $\mathfrak{L}_{2,m}$ as its domain, is associated with the matrix $\{\overline{T}_{ik}\}$ we can show that $\overline{T}_{ik} \equiv T^*_{ki}$, by expressing the relation $(T\boldsymbol{f}, \boldsymbol{g}) = (\boldsymbol{f}, T^*\boldsymbol{g})$ explicitly in the form

$$\sum_{\alpha,\beta=1}^{m} \int_E T_{\alpha\beta} f_\beta \cdot \overline{g}_\alpha \, dQ = \sum_{\alpha,\beta=1}^{m} \int_E f_\alpha \cdot \overline{\overline{T}_{\alpha\beta} g_\beta} \, dQ.$$

For, on taking account of the fact that the components of $\boldsymbol{f}$ and $\boldsymbol{g}$ can be assigned arbitrarily, we can obtain from this single equation the m^2 relations

$$\int_E T_{ki} f_i \cdot \overline{g}_k \, dQ = \int_E f_i \overline{\overline{T}_{ik} g_k} \, dQ, \quad i, k = 1, \cdots, m,$$

which clearly imply that $\overline{T}_{ik}$ and T^*_{ki} are identical.

In view of the theorem just proved, we shall define an integral operator in $\mathfrak{L}_{2,m}$ as an operator T which is expressible in terms of m^2 integral operators T_{ik} defined in $\mathfrak{L}_2$ by the kernels $K_{ik}(P, Q)$; the matrix $\{K_{ik}(P, Q)\}$ may be called the kernel of T. More explicitly, T takes the vector $\boldsymbol{f}$ with components $(f_1, \cdots, f_m)$ into a vector $\boldsymbol{g}$ with components $(g_1, \cdots, g_m)$ where

$$g_i(P) = \sum_{\beta=1}^{m} \int_E K_{i\beta}(P, Q) f_\beta(Q) \, dQ,$$

We shall omit the consideration of detailed properties of such integral operators.

Finally we remark that the concept of the integral operator may be extended to admit generalizations of the Lebesgue integral to consideration.

§ 3. Differential Operators

Let $\mathfrak{L}_2$ be the Hilbert space introduced in Theorem 1.24 and let $\mathfrak{D}$ be an appropriately chosen linear manifold in $\mathfrak{L}_2$. If $f(P)$ is a function belonging to $\mathfrak{D}$, a new function can be formed from it by the application of an operator constructed from two particular types of operator by the use of addition, multiplication, and scalar multiplication as previously defined for operators and transformations; the two types of operator admitted are multiplication by an arbitrary complex-valued function of P and differentiation with respect to an arbitrary direction in space. Such an operator is called a linear differential operator, except in the special case where all differentiations can be eliminated by the usual manipulations. These operators are linear in the sense adopted throughout our previous considerations. The sum and product of any number of linear differential operators are operators of the same kind, an exception occurring in the case of the sum when it happens that all differentiations can be eliminated. The study of linear differential operators is complicated by an unavoidable difficulty: the domain $\mathfrak{D}$ must be chosen with regard to conditions of continuity and differentiability, but, at the outset, we have no satisfactory criterion at our disposal for determining whether or not a given function in $\mathfrak{L}_2$ is to be included in $\mathfrak{D}$. For this reason we shall confine our attention, in the sequel, to comparatively simple examples of recognized interest and importance.

We shall first discuss differential operators in the space $\mathfrak{L}_2$ defined on a bounded closed interval in one-dimensional Euclidean space. It will be convenient to distinguish between formal linear differential operators and operators in the sense of Definition 2.1. Denoting by x the coördinate of a variable point P in the interval of definition, $a \leqq x \leqq b$, we write down the symbolic expression

$$\Delta = p_0(x)\frac{d^n}{dx^n} + p_1(x)\frac{d^{n-1}}{dx^{n-1}} + \cdots$$

$$\cdots + p_{n-1}(x)\frac{d}{dx} + p_n(x)\cdot\,, \quad n \geqq 1,$$

where $p_i(x)$ is a complex-valued continuous function whose first $n-i$ derivatives exist and are continuous on (a, b) and where $|p_0(x)|$ does not vanish on that interval. This expression has in itself no precise meaning, but it directs the performance of certain definite operations on functions of a category as yet unspecified: we shall refer to Λ as a formal linear differential operator. When we provide a linear manifold $\mathfrak{D}$ as the domain of applicability of the formal operator Λ (observing the obvious condition that when f is in $\mathfrak{D}$ the calculated function Λf must be in $\mathfrak{L}_2$), we determine a linear differential operator or transformation T in the sense of Definition 2.1. Thus, a single formal operator may give rise to many operators or transformations, distinguished by means of their domains. For the sake of simplicity, we begin with the differential operator T whose domain is the class $\mathfrak{C}_n$ of all functions in $\mathfrak{L}_2$ which are continuous together with their first n derivatives; we note that $\mathfrak{C}_n$ is a linear manifold everywhere dense in $\mathfrak{L}_2$. One of our first concerns is to determine whether the adjoint T^* has any simple properties. The form of the expression $(Tf, g) = \int_a^b \Lambda f(x) \cdot \overline{g(x)}\, dx$ suggests that shifting the differential operator from f to g by means of integration by parts is the appropriate method for dealing with the problem. This procedure can be followed when f and g are both in $\mathfrak{C}_n$, and leads to the familiar formula of Lagrange†

$$\int_a^b (\Lambda f \cdot \overline{g} - f \cdot \overline{Mg})\, dx = B(f, g),$$

where M is the formal linear differential operator

$$M = (-1)^n \frac{d^n(\overline{p_0(x)} \cdot\,)}{dx^n} + (-1)^{n-1} \frac{d^{n-1}(\overline{p_1(x)} \cdot\,)}{dx^{n-1}} + \cdots + (-1) \frac{d(\overline{p_{n-1}(x)} \cdot\,)}{dx} + \overline{p_n(x)} \cdot\,,$$

and $B(f, g)$ is a numerically-valued bilinear function of f and g,

$$B(f, g) = \sum_{\alpha, \beta = 1}^{2n} a_{\alpha\beta}\, \xi_\alpha \overline{\eta_\beta}, \qquad \det\{a_{ik}\} \neq 0,$$
$$\xi_1 = f(a), \cdots, \xi_n = f^{(n-1)}(a),$$
$$\xi_{n+1} = f(b), \cdots, \xi_{2n} = f^{(n-1)}(b),$$

† See Bôcher, *Leçons sur les méthodes de Sturm*, Paris, 1917, Chapter II.

$$\eta_1 = g(a), \cdots, \eta_n = g^{(n-1)}(a),$$
$$\eta_{n+1} = g(b), \cdots, \eta_{2n} = g^{(n-1)}(b).$$

The formal linear differential operator M is called the formal or Lagrange adjoint of Λ. If we wish to associate with M an operator R which will be adjoint to T, we must choose its domain so that the term $B(f, g)$ in Lagrange's formula vanishes whenever f is in $\mathfrak{C}_n$ and g is in the domain of R: an obvious choice, for example, is that of the linear manifold comprising those functions g in $\mathfrak{C}_n$ such that

$$g(a) = \cdots = g^{(n-1)}(a) = g(b) = \cdots = g^{(n-1)}(b) = 0.$$

More generally, we can seek all possible pairs of linear differential operators (T, R) associated with (Λ, M) subject to the condition $T \frown R$. For the sake of simplicity, we shall select the domains of the desired linear operators as subsets of the linear manifold $\mathfrak{C}_n$. In order to choose linear manifolds for these domains, we are led to impose on each of the sets $\xi_1, \cdots, \xi_{2n}$ and $\eta_1, \cdots, \eta_{2n}$ a sufficient number of linear conditions to bring about the vanishing of $B(f, g)$.

At this point we find a certain freedom of choice granted us. We first assign arbitrarily $2n$ independent linear functions of $\xi_1, \cdots, \xi_{2n}$, which we shall denote by

$$W_i(f) = \sum_{\beta=1}^{2n} b_{i\beta}\, \xi_\beta, \qquad \det\{b_{ik}\} \neq 0.$$

By means of a well-known property of non-singular bilinear forms, we can determine $2n$ independent linear functions of $\eta_1, \cdots, \eta_{2n}$, which we shall write as

$$V_i(g) = \sum_{\beta=1}^{2n} c_{i\beta}\, \eta_\beta, \qquad \det\{c_{ik}\} \neq 0,$$

such that

$$B(f, g) = \sum_{\alpha,\beta=1}^{2n} a_{\alpha\beta}\, \xi_\alpha\, \bar{\eta}_\beta = \sum_{\alpha=1}^{2n} W_\alpha\, \bar{V}_{2n-\alpha+1}.$$

It thus appears that we can make $B(f, g)$ vanish by requiring that certain of the W's have the value zero and then setting the complementary V's equal to zero also. By numbering the functions W and V in an appropriate manner, we arrive at the following possibilities:

(1) the domain of T consists of the linear manifold in $\mathfrak{C}_n$ for which $W_1(f) = \cdots = W_{2n}(f) = 0$; the domain of R is $\mathfrak{C}_n$.

(2) the domain of T consists of the linear manifold in $\mathfrak{C}_n$ for which $W_1(f) = \cdots = W_k(f) = 0$; the domain of R consists of the linear manifold in $\mathfrak{C}_n$ for which $V_1(g) = \cdots \cdots = V_{2n-k}(g) = 0$; $k = 1, \cdots, 2n-1$.

(3) the domain of T is $\mathfrak{C}_n$; the domain of R consists of the linear manifold in $\mathfrak{C}_n$ for which $V_1(g) = \cdots = V_{2n}(g) = 0$.

In each case the domains involved are linear manifolds everywhere dense in $\mathfrak{L}_2$; it is easily verified that T^* and R^* exist and that the relations $T^* \supseteqq R$, $R^* \supseteqq T$, and $T \frown R$ are satisfied. In each case we have been forced to associate with the formal operators Λ and M appropriately chosen sets of linear boundary conditions characterizing the domains of the operators T and R respectively. In the ordinary language of the theory of differential equations, we have to deal with linear differential systems. In each of the various cases enumerated above, the two differential systems described are ordinarily called adjoint systems. Thus the terminology of the theory of transformations developed in the preceding chapter applies to linear differential systems rather than to formal linear differential operators without specified domains. Of primary interest and importance are the systems referred to as self-adjoint systems. They are characterized by the fact that the formal differential operator Λ coincides with its formal or Lagrange adjoint M and that the boundary conditions determining the domains of T and R respectively are equivalent.† The domain of T must therefore be specified

† This definition differs slightly from the standard one, where a distinction is made between the cases n even and n odd. In order to remain in the domain of real numbers, it is necessary to require that $\Lambda = M$ for n even and that $\Lambda = -M$ for n odd. Here we are working in the complex domain and can treat the case where n is odd as follows: since the

by n conditions $W_1(f) = \cdots = W_n(f) = 0$, the domain of R by n conditions $V_1(g) = \cdots = V_n(g) = 0$, where

$$V_i \equiv \sum_{\beta=1}^{n} A_{i\beta} W_\beta, \quad \det \{A_{ik}\} \neq 0.$$

Consequently, the operators T and R are allotted the same domain; they are identical and hence symmetric. We shall now prove the following theorem:

THEOREM 3.12. *A self-adjoint differential system defines a symmetric differential operator T. There exists a real number l such that the transformation $H \equiv T - l \cdot I$ has an inverse H^{-1} which is an integral operator with an Hermitian symmetric kernel $G(x, y)$ of Hilbert-Schmidt type. H and T are essentially self-adjoint operators.*

The differential system consists of a formal differential operator Λ and a set of boundary conditions. This system defines a transformation or operator, in our usual sense, by associating with each element f of $\mathfrak{C}_n$ which satisfies the boundary conditions the element $g = \Lambda f$. The operator determined in this domain $\mathfrak{D}$ we shall denote by T. By hypothesis, the adjoint system defines the same operator T in $\mathfrak{L}_2$. By previous remarks, T is a symmetric operator.

We must now rely upon certain facts established in the elementary theory of differential equations.* If we denote by H the transformation $T - l \cdot I$ where l is a real number, it is evident that H is symmetric. It is known that, if the equation $Hf = 0$ has no solution other than $f = 0$ in $\mathfrak{D}$, then H establishes a one-to-one correspondence between $\mathfrak{D}$

Lagrange adjoint of $i\Lambda$ is $-iM$ when Λ has M as its Lagrange adjoint, we have only to replace Λ by $i\Lambda$ in dealing with a differential system of odd order which is self-adjoint according to the prevailing definition. For example, the differential system defined by the formal operator $\frac{d}{dx}$ in the domain of all functions with continuous first derivative on the interval $0 \leqq x \leqq 1$, satisfying the boundary condition $W(f) = f(0) - f(1) = 0$, must be treated by substituting $i\frac{d}{dx}$ or $\frac{1}{i}\frac{d}{dx}$ for $\frac{d}{dx}$.

* Bôcher, *Leçons sur les méthodes de Sturm*, Paris 1917, Chapter V.

and $\mathfrak{C}_0$, the class of all continuous functions in $\mathfrak{L}_2$. Thus H^{-1} exists. It is known also that H^{-1} is an integral operator with domain $\mathfrak{C}_0$ and kernel $G(x, y)$, bounded, and continuous except possibly on the line $x = y$:

$$Hf = g, \qquad f = H^{-1}g = \int_a^b G(x, y)\, g(y)\, dy.$$

Since H is symmetric, H^{-1} is also symmetric, by Theorem 2.18. We can conclude, by Theorem 3.8, that $G(x, y)$ is an Hermitian symmetric kernel. Since H^{-1} is linear, symmetric, and bounded, it is essentially self-adjoint, as we have already proved in Theorem 2.24. Thus H itself is essentially self-adjoint, by Theorem 2.18. It is directly obvious that $T \equiv H + l \cdot I$ is also essentially self-adjoint.

We see, therefore, that the truth of the theorem depends upon the existence of a real number l such that $Hf = Tf - lf = 0$ is satisfied by no element of $\mathfrak{D}$ other than $f = 0$. Let L be the class of numbers l such that $Tf - lf = 0$ has a solution other than $f = 0$ in $\mathfrak{D}$. We shall show, by an argument of general character, that L is at most denumerably infinite. If l and m are distinct numbers in L, let f_l and f_m be elements of $\mathfrak{D}$ such that

$$Tf_l - lf_l = 0, \qquad Tf_m - mf_m = 0,$$
$$\|f_l\| = 1, \qquad \|f_m\| = 1.$$

Then $0 = (Tf_l, f_m) - (f_l, Tf_m) = (l - m)(f_l, f_m)$, so that f_l and f_m are orthogonal. Hence, there exists an orthonormal set in one-to-one correspondence with L; by Theorem 1.4, L is finite or denumerably infinite. Consequently, there exists a real number not in L; and the present theorem can therefore be established.

The extension of the foregoing considerations to differential operators in $\mathfrak{L}_{2,m}$ presents no unusual difficulties. We take $\mathfrak{L}_{2,m}$ as the class of all vectors with m components, each component being a function in $\mathfrak{L}_2$ defined on the interval (a, b). A formal differential operator Λ in $\mathfrak{L}_{2,m}$ is then defined by a matrix $\{\Lambda_{ik}\}$ of m^2 formal differential operators in $\mathfrak{L}_2$, as suggested by the result of Theorem 3.11. For vectors whose components are in $\mathfrak{C}_n$, the analogue of Lagrange's formula can be established without difficulty. We can therefore assign an appropriate domain by the introduction of boundary conditions, being led in this way to consider differential systems in $\mathfrak{L}_{2,m}$. In the case of a self-adjoint system, one which is equivalent to its adjoint system, the analogue of Theorem 3.12 is true.*

The precise generalization of the foregoing considerations to the case of an infinite interval $(a, +\infty)$, $(-\infty, b)$, or $(-\infty, +\infty)$ is obscured by the difficulty of choosing boundary conditions and stating them in appropriate form. We shall return to this question later, when the theory developed in Chapter IX is at our disposal.

We may also extend the concepts of the theory of differential operators to the classes $\mathfrak{L}_2$, $\mathfrak{L}_{2,m}$ defined over a suitably chosen set E in n-dimensional Euclidean space $n \geqq 2$. If we introduce rectangular coördinates $(x_1, \cdots, x_n)$, a typical formal differential operator in $\mathfrak{L}_2$ may be written

$$\Lambda \sim \sum_{\alpha=0}^{m} \sum_{\beta_1+\cdots+\beta_n=\alpha} p_{\alpha;\beta_1,\cdots,\beta_n}(x_1, \cdots, x_n) \frac{\partial^\alpha}{\partial x_1^{\beta_1} \cdots \partial x_n^{\beta_n}}.$$

The extension of Lagrange's formula to differential expressions of this type involves the use of familiar reduction formulas for multiple integrals, and is possible only if E is so restricted as to admit the application of such formulas. We are led to introduce boundary conditions, relative to the boundary points of E, and to define differential systems, adjoint differential systems, and so forth. We can then associate with Λ the operator T, defining its domain adequately. Since the theory of boundary value problems for partial differential equations is far from elementary, we shall content ourselves with the citation of the familiar Laplace operator.

THEOREM 3.13. *We introduce the following notations:*

(1) *E is a connected open set of finite measure in 3-dimensional Euclidean space for which the Dirichlet problem is solvable;*

---

* The matrix kernel $\{G_{ik}(x, y)\}$ corresponding to the function $G(x, y)$ has been discussed by Birkhoff and Langer, Proceedings of the American Academy of Arts and Sciences, 58 (1922–23), pp. 51–128.

(2) *E' is the derived set of E, $E'-E$ its boundary;*

(3) *$G(P, Q)$ is the Green's function for E;*

(4) *$\mathfrak{L}_2$ is the class of all Lebesgue-measurable functions defined in E such that $\int_E |f(Q)|^2 \, dQ$ exists;*

(5) *$\mathfrak{D}$ is the class of all continuous functions $f(P)$ defined in E with the properties*

(a) *$f(P)$ satisfies the boundary condition $\lim_{p \to P'} f(P) = 0$ whenever P' is in $E'-E$,*

(b) $\dfrac{\partial^2 f}{\partial x_1^2} + \dfrac{\partial^2 f}{\partial x_2^2} + \dfrac{\partial^2 f}{\partial x_3^2} = \nabla^2 f(P) = g(P)$ *exists at every point of E and defines a bounded continuous function $g(P)$ with continuous first partial derivatives;*

(6) *$\mathfrak{R}$ is the class of all bounded continuous functions in E with continuous first partial derivatives.*

Then the operator T defined by the relation $\nabla^2 f(P) = g(P)$ in the domain $\mathfrak{D}$ takes $\mathfrak{D}$ in a one-to-one manner into $\mathfrak{R}$; the inverse of T is an integral operator with Hermitian symmetric kernel of Hilbert-Schmidt type and domain $\mathfrak{R}$:

$$f(P) = T^{-1} g(P) = -\frac{1}{4\pi} \int_E G(P, Q)\, g(Q)\, dQ.$$

T and T^{-1} are therefore essentially self-adjoint operators.

The argument leading to these results is strictly analogous to that used in Theorem 3.12; it depends, of course, on facts concerning the Laplace operator ∇^2 which are far more difficult to establish than the corresponding facts for ordinary linear differential operators, such as those discussed in the preceding theorem.

By hypothesis the Green's function of E exists. It is expressible in the form

$$G(P, Q) = \frac{1}{r(P, Q)} + H(P, Q)$$

where $r(P, Q)$ is the distance between P and Q and $H(P, Q)$ is a real function harmonic in P and Q separately throughout E and assuming the boundary values $-\dfrac{1}{r(P, Q)}$ on $E'-E$.

Thus the Green's function enjoys the following familiar properties:*

(1) $0 < G(P, Q) < \dfrac{1}{r(P, Q)}$,

(2) $G(P, Q) = G(Q, P) = \overline{G(Q, P)}$,

(3) $\lim\limits_{P \to P'} G(P, Q) = 0, \quad P' \text{ in } E' - E, \quad Q \text{ in } E.$

From properties (1) and (2) we see that $G(P, Q)$ is an Hermitian symmetric kernel of Hilbert-Schmidt type such that

$$\int_E \int_E |G(P, Q)|^2 \, dQ \, dP \leqq \int_E \int_E \frac{1}{r^2(P, Q)} \, dP \, dQ,$$

the latter integral being convergent.

We now construct the function

$$f(P) = -\frac{1}{4\pi} \int_E G(P, Q) \, g(Q) \, dQ,$$

where $g(P)$ is in $\mathfrak{R}$ and, *a fortiori*, in $\mathfrak{L}_2$. It is easily seen that $f(P)$ is a continuous function such that $\lim_{P \to P'} f(P) = 0$ when P' is in $E' - E$. By a well-known theorem of potential theory $\nabla^2 f(P)$ exists and is equal to $g(P)$ for every point in E.† We see therefore that $f(P)$ is in $\mathfrak{D}$. If $f(P)$ is in $\mathfrak{D}$ we know, by definition, that $\nabla^2 f(P)$ is in $\mathfrak{R}$; and we can easily verify the relation

$$f(P) = -\frac{1}{4\pi} \int_E G(P, Q) \nabla^2 f(Q) \, dQ.$$

Thus the operator T defined in $\mathfrak{D}$ takes $\mathfrak{D}$ in an one-to-one manner into $\mathfrak{R}$.

Since $\mathfrak{D}$ and $\mathfrak{R}$ are both linear manifolds everywhere dense in $\mathfrak{L}_2$, the further properties of T and its inverse which have been enunciated can be derived without difficulty by an argument like that of Theorem 3.12.

The analogue of the theorem just proved for the Laplace operator in three-dimensional space holds in the plane; but in n-dimensional space, $n \geqq 4$, modifications must evidently be introduced because of the circumstance that the

* Kellogg, *Foundations of Potential Theory*, Berlin, 1929, pp. 236–240.

† See the proof in Kellogg, *Foundations of Potential Theory*, Berlin 1929, pp. 150–156.

fundamental function $[r(P, Q)]^{2-n}$ renders the integral $\int_E \int_E [r(P, Q)]^{-2n+4} \, dQ \, dP$ divergent. The extension of the theorem to more general elliptic partial differential operators with a variety of boundary conditions can be effected in the cases $n = 2$ and $n = 3$.

The consideration of partial differential operators in the class $\mathfrak{L}_{2,m}$ is particularly important in certain branches of physics. We shall indicate an example taken from the theory of elasticity. In a certain region E of three-dimensional space, occupied by an isotropic elastic medium, the vectors of $\mathfrak{L}_{2,3}$ represent the displacement vectors of the points of the medium. The fundamental partial differential operator Λ in $\mathfrak{L}_{2,3}$ which dominates the situation appears as the three-rowed matrix of operators in $\mathfrak{L}_2$

$$\begin{Bmatrix} \mu \nabla^2 + (\lambda+\mu) \dfrac{\partial^2}{\partial x^2} & (\lambda+\mu) \dfrac{\partial^2}{\partial x \, \partial y} & (\lambda+\mu) \dfrac{\partial^2}{\partial x \, \partial z} \\ (\lambda+\mu) \dfrac{\partial^2}{\partial x \, \partial y} & \mu \nabla^2 + (\lambda+\mu) \dfrac{\partial^2}{\partial y^2} & (\lambda+\mu) \dfrac{\partial^2}{\partial y \, \partial z} \\ (\lambda+\mu) \dfrac{\partial^2}{\partial x \, \partial z} & (\lambda+\mu) \dfrac{\partial^2}{\partial y \, \partial z} & \mu \nabla^2 + (\lambda+\mu) \dfrac{\partial^2}{\partial z^2} \end{Bmatrix}$$

where λ and μ are physical constants; this operator Λ may be expressed in vector notation as

$$\Lambda = (\lambda + \mu) \text{ grad div } - \mu \text{ curl curl.}$$

The various problems of elasticity give rise to certain types of boundary condition, holding at the boundary of E; as a typical example we may mention the case in which the displacement vector vanishes at the frontier points.*

Before leaving differential operators, let us take advantage of the insight gained in the preceding discussion to throw light on two important problems; the problem of finding symmetric extensions of a given symmetric operator, and the

* For a thorough mathematical discussion of the operator Λ, see Weyl, Rendiconti di Palermo, 39 (1915) pp. 1–49.

problem of associating an operator with a given infinite matrix. For this purpose we shall consider the formal differential operator $\frac{d^2}{dx^2}$ in the class $\mathfrak{L}_2$ defined on the interval (0,1). With this operator we associate the boundary conditions B_0:

$$f(0) = f(1) = f'(0) = f'(1) = 0,$$

and the boundary conditions $B(\theta, \varphi)$:

$$\cos\theta \cdot f'(0) + \sin\theta \cdot f(0) = 0,$$
$$\cos\varphi \cdot f'(1) + \sin\varphi \cdot f(1) = 0.$$

In the class of all functions in $\mathfrak{C}_2$ which satisfy the boundary conditions B_0, the formal differential operator $\frac{d^2}{dx^2}$ defines a symmetric operator T_0; in the class of all functions in $\mathfrak{C}_2$ which satisfy the boundary conditions $B(\theta, \varphi)$, it defines an essentially self-adjoint operator $T_{\theta,\varphi}$ which is an extension of T_0. It is easily shown that the operators T_{θ_1,φ_1}, T_{θ_2,φ_2} are not adjoint unless $\tan\theta_1 = \tan\theta_2$, $\tan\varphi_1 = \tan\varphi_2$; under the latter circumstances they coincide. The operator $T^*_{\theta,\varphi}$ is a self-adjoint extension of T_0. We see, therefore, that the symmetric operator T_0 possesses a continuum of self-adjoint extensions $T^*_{\theta,\varphi}$. When we come to Chapter IX, we shall recognize that this situation and certain analogous situations are of general occurrence. If we now undertake to describe these results in terms of matrices, we shall obtain a convincing example of the inadequacy of that language. In order to associate a matrix with the operator T_0, we must select a complete orthonormal set from the class of all functions in $\mathfrak{C}_2$ which satisfy the boundary conditions B_0. Since this class is a linear manifold everywhere dense in $\mathfrak{L}_2$, we may carry out the process detailed in § 1 and construct such a matrix. In the natural course of events, we should next use the same orthonormal set to find matrices for the operators $T^*_{\theta,\varphi}$. Unfortunately, we find that with the operators T_0, $T^*_{\theta,\varphi}$ we associate one and the same matrix A, which must do duty for them all.

Here are definite simple relations which elude the matrix theory completely. Even if we try to obtain a more favorable outcome by considering a single operator, say $T^{*}_{\theta,\varphi}$, we must admit failure, since it is only by a happy accident or a well-directed choice, springing from the special case considered, that we can associate an adequate matrix with the given operator. It seems, in view of such difficulties, that we must refrain from using the matrix theory as a tool, save in the case of bounded operators or in problems stated directly in matrix form. Differential operators are evidently non-bounded.

§ 4. Operators of Other Types

While we cannot profitably proceed to the enumeration of many further types of operator, we must mention one type which is important in the theory of groups.

Let the space $\mathfrak{L}_2$ be defined with reference to a set E in n-dimensional Euclidean space, and let the relation $P=F(Q)$ define a one-to-one transformation of E into itself. We can then define a linear operator or transformation by the equation $Tf(P)=f(F(P))$, the domain of the operator T consisting of all functions $f(P)$ such that $\int_E |f(F(P))|^2\,dP$ exists. An extremely important concrete example to illustrate the general concept refers to the class $\mathfrak{L}_2$ defined for the infinite interval $(-\infty, +\infty)$: the operator is that which carries $f(x)$ into $f(x+\tau)$. It is useful to note that we can consider also those more general cases in which the class $\mathfrak{L}_2$ is characterized in terms of a generalization of the Lebesgue integral. In practice, we must treat classes of functions defined over curved surfaces or manifolds, and must make use of surface and volume integrals in curved spaces.

It should further be observed that one can easily construct various mixed types of operator by setting up algebraic combinations of the types we have discussed explicitly. In this way can be obtained integro-differential operators, mixed differential and difference operators, integro-difference operators, and so on.

CHAPTER IV

RESOLVENTS, SPECTRA, REDUCIBILITY

§ 1. The Fundamental Problems

We have now reached a point where we can pose and illustrate two fundamental problems of the theory of linear transformations, the intimate connection of which will become clearer as we proceed. The first problem arises in a great variety of mathematical situations and assumes in abstract terms the following form:

Problem 1. *To study the equation $Tf - lf = g$ where l is a given complex number and g a given element in $\mathfrak{H}$.*

This problem obviously requires an investigation of the transformation $T_l \equiv T - lI$ and its inverse T_l^{-1}, when the latter exists, and leads to a classification of the points of the complex l-plane according to the behavior of these transformations. We shall analyze this problem in the following section. The second problem is inherent in the nature of linear transformations, and is seen to be more fundamental than the first. It may be stated as follows:

Problem 2. *To characterize the transformation T in terms of the linear manifolds which it leaves invariant.*

Of course, we must make the meaning of this problem more precise. We shall do so in the third section of the chapter.

Let us now examine the meaning of each of these problems for the specific transformations and operators discussed in Chapter III.

In terms of the matrix theory, the first problem is that of solving the system of infinitely many equations in infinitely many unknowns

$$\sum_{\beta=1}^{\infty}(a_{i\beta}-l\delta_{i\beta})x_\beta = y_i, \quad i = 1, 2, 3, \cdots,$$

where $(y_1, y_2, y_3, \cdots)$ is a given set of numbers such that $\sum_{\alpha=1}^{\infty}|y_\alpha|^2$ converges and the set $(x_1, x_2, x_3, \cdots)$ is required under the condition that $\sum_{\alpha=1}^{\infty}|x_\alpha|^2$ converge. One perceives immediately the suggestive analogy between this infinite system and the ordinary finite system of linear equations. When the matrix $\{a_{ik}\}$ is symmetric, this system of equations has an interesting heuristic connection with the Hermitian form $\sum_{\alpha,\beta=1}^{\infty} a_{\alpha\beta}\,\overline{x}_\alpha\, x_\beta$, which is best displayed by reference to the ordinary algebraic situation: if $\sum_{\alpha,\beta=1}^{n} a_{\alpha\beta}\,\overline{x}_\alpha\, x_\beta$ is an Hermitian form such that $\overline{a}_{ki} = a_{ik}$, then the n equations

$$\sum_{\beta=1}^{n}(a_{i\beta}-l\delta_{i\beta})x_\beta = 0, \quad i = 1, \cdots, n,$$

constitute a set of conditions that $(x_1, \cdots, x_n)$ yield an extremum of the Hermitian form, under the restriction $\sum_{\alpha=1}^{n}\overline{x}_\alpha\, x_\alpha = 1$. These equations lead also to a solution of the essentially equivalent problem of finding the "principal axes" of the manifold $\sum_{\alpha,\beta=1}^{n} a_{\alpha\beta}\,\overline{x}_\alpha\, x_\beta =$ constant, analogous to the axes of a quadric surface. By analogy, we may speak of the determination of the "principal axes" of a symmetric form in infinitely many variables by means of the infinite system of linear equations written out above. For "completely continuous" forms, Hilbert and his followers have converted what is here a mere intuitive connection into a sound theory; in general, we must be content with the analogy.*

In order to grasp the significance of the second problem for the matrix theory, we must recall that to a given trans-

* See the excellent outline of the theory of Hilbert, with special reference to the "completely continuous" or "vollstetig" forms, in Hellinger and Toeplitz, *Integralgleichungen und Gleichungen mit unendlichvielen Unbekannten*, Encyklopädie der Mathematischen Wissenschaften 2³, pp. 1553–1575.

formation T, which we shall assume to be bounded, we may order many different matrices through the use of different orthonormal sets. Two of these matrices A and B are always connected by a relation $B = UAU^*$ where U is a unitary matrix. Now, if T leaves a closed linear manifold invariant, we can so choose the representative matrix that it will exhibit this property of T through the appearance of zeros among its elements. It is manifest, therefore, that to our second problem corresponds the following matrix problem: when A is a matrix associated with the bounded linear transformation T, a unitary matrix U is to be found such that the matrix $UAU^* = B$, which also represents T, exhibits the character of T by the presence of zero elements. When U can be determined so that B has only a finite number of non-zero elements in each row and column, for example, the situation is more clearly revealed; and when B is in diagonal form (with all its elements not along the principal diagonal equal to zero) the problem of invariant linear manifolds is completely solved. In general, unfortunately, we cannot hope to accomplish so much. When A is a symmetric matrix which determines a "completely continuous" symmetric form $\sum_{\alpha,\beta=1}^{\infty} a_{\alpha\beta}\bar{x}_\alpha x_\beta$, U can be found so that B is in diagonal form. This means that we can introduce new variables $(y_1, y_2, y_3, \cdots)$ linearly related to $(x_1, x_2, x_3, \cdots)$ so that

$$\sum_{\alpha=1}^{\infty} \bar{x}_\alpha x_\alpha = \sum_{\alpha=1}^{\infty} \bar{y}_\alpha y_\alpha, \qquad \sum_{\alpha,\beta=1}^{\infty} a_{\alpha\beta}\bar{x}_\alpha x_\beta = \sum_{\alpha=1}^{\infty} l_\alpha \bar{y}_\alpha y_\alpha;$$

we have, therefore, a situation similar to that which arises in connection with the transformation theory of ordinary quadratic forms. The analogy is far from exact in more general cases.

When we turn to the discussion of integral and differential operators, we find that Problem 1 embraces some of the most important problems of analysis and of mathematical physics. For an integral operator, the equation of that problem is seen to be an integral equation

$$\int_E K(P, Q) f(Q)\, dQ - l f(P) = g(P),$$

which may be written in the more usual and familiar form

$$f(P) = \varphi(P) + \lambda \int_E K(P, Q) f(Q)\, dQ,$$

by means of the substitutions

$$\lambda = 1/l, \qquad \varphi(P) = -g(P)/l.$$

We shall retain the first form, which is better adapted to our purposes. Similarly, a differential operator leads to the study of a differential equation such as $\sum_{\alpha=0}^{n} p_\alpha(x) \frac{d^{n-\alpha}}{dx^{n-\alpha}} f(x) - l f(x) = g(x)$ or $\nabla^2 f(P) - l f(P) = g(P)$ with suitable boundary conditions. For the other types of operator mentioned in Chapter III, we do not obtain interpretations of Problem 1 which have so great an intrinsic interest. Our second problem, being primarily of geometrical and algebraic nature, is only implicit in most analytical considerations and so need not be considered until a later stage.

§ 2. Resolvents and Spectra

In the present section we shall undertake a preliminary study of the first problem stated in § 1, introducing the requisite definitions and theorems, and considering in particular bounded linear transformations and symmetric transformations.

We shall first consider the classification of the points of the finite l-plane according to the character of the transformation $T_l \equiv T - lI$, where T is a given linear transformation; such a classification must reflect the essential peculiarities of the equation $Tf - lf = g$ for different values of l. For this purpose we have to study the relation between the domain $\mathfrak{D}$ and the range $\mathfrak{R}_l$ of T_l. All the necessary information concerning this relation is at hand when we can determine whether T_l has an inverse and can characterize the inverse T_l^{-1} if it exists.

Definition 4.1. *Let T denote a linear transformation, T_l the linear transformation $T - lI$, and T_l^{-1} the linear trans-*

formation inverse to T_l, when it exists. The set of points $A(T)$ in the complex l-plane such that T_l has no inverse is called the point spectrum of T. The set of points $B(T)$ such that T_l^{-1} exists and is an unbounded linear transformation with domain $\mathfrak{R}_l$ everywhere dense in $\mathfrak{H}$ is called the continuous spectrum of T. The set of points $C(T)$ such that T_l^{-1} exists and is a linear transformation with domain $\mathfrak{R}_l$ not everywhere dense in $\mathfrak{H}$ is called the residual spectrum of T. The set of points $D(T)$ such that T_l^{-1} exists and is a bounded linear transformation with domain $\mathfrak{R}_l$ everywhere dense in $\mathfrak{H}$ is called the resolvent set of T. The set of points $S(T) = A(T) + B(T) + C(T)$ is called the spectrum of T.†

THEOREM 4.1. *The four sets of points A, B, C, D of Definition 4.1 are mutually exclusive and their sum is the finite l-plane.*

This assertion is an obvious consequence of the definition.

THEOREM 4.2. *The complex number l is in the point-spectrum of the linear transformation T if and only if the equation $T_l f = 0$ has a solution in $\mathfrak{H}$ distinct from the null element.*

The transformation T_l will fail to have an inverse if and only if there exist distinct elements f_1 and f_2 in $\mathfrak{D}$ such that $T_l f_1 = T_l f_2$, $T_l(f_1 - f_2) = 0$.

DEFINITION 4.2. *A complex number l in the point spectrum of T is called a characteristic value of T; an element $f \neq 0$ such that $T_l f = 0$, $Tf = lf$, is called a characteristic element of T and is said to be normalized when $\|f\| = 1$. The closed linear manifold determined by the set of all characteristic elements of T corresponding to a characteristic value l is called a characteristic manifold of T; its dimension number n is called the multiplicity of the characteristic value l.*

Occasionally we shall refer to a complex number l as a characteristic value of multiplicity zero for a given transformation: the statement is to be interpreted as a conventional circumlocution meaning that l is not a characteristic value in the sense of the definition above.

† The terminology of this definition is adopted primarily for historical reasons.

9

THEOREM 4.3. *The spectral classification of the points of the l-plane, inclusive of the multiplicities of characteristic values, is invariant under unitary transformations; in other words, if U is an arbitrary unitary transformation, the transformations T and UTU^{-1} have the same point spectra (inclusive of the multiplicities of characteristic values), continuous spectra, residual spectra, and resolvent sets.*

Let $U\mathfrak{D}$ and $U\mathfrak{R}_l$ denote the sets into which U carries the domain and the range of the transformation T_l. Since $UTU^{-1}-lI\equiv UT_lU^{-1}$, the classification of the complex number l with respect to UTU^{-1} depends upon the correspondence which the transformation UT_lU^{-1} sets up between its domain $U\mathfrak{D}$ and its range $U\mathfrak{R}_l$. This correspondence is an image of the correspondence determined between $\mathfrak{D}$ and $\mathfrak{R}_l$ by the transformation T. Thus the complex number l must be classified in exactly the same way with respect to the transformations T and UTU^{-1}.

THEOREM 4.4. *The spectra of the transformation $aT+bI$, where a and b are arbitrary complex numbers with $a\neq 0$, are obtained from the corresponding spectra of the transformation T by the application of the transformation $l'=al+b$ to the l-plane.*

This theorem is an obvious consequence of the fact, that the equations

$$(aT+b)f-l'f=ag,\qquad Tf-lf=g$$

are equivalent when $l'=al+b$, $a\neq 0$.

DEFINITION 4.3. *The set of all points l such that for some element f in the domain of the transformation T the equations $(Tf,f)=l$ and $\|f\|=1$ are satisfied, is called the numerical range of T and is denoted by $W(T)$.*

THEOREM 4.5. *The numerical range is invariant under unitary transformations; in other words, if U is an arbitrary unitary transformation, then $W(T)=W(UTU^{-1})$.*

Evidently, a necessary and sufficient condition that f satisfy the equations $(Tf,f)=l$, $\|f\|=1$, is that $g=Uf$ satisfy the equations

$$(UTU^{-1}g, g) = (UTf, Uf) = (Tf, f) = l,$$
$$\|g\| = \|Uf\| = \|f\| = 1.$$

THEOREM 4.6. *If $a \neq 0$ and b are arbitrary complex numbers, then $W(aT+b)$ is obtained from $W(T)$ by applying the transformation $l' = al+b$ to the complex l-plane.*

This theorem is an obvious consequence of the fact that the equations

$$((aT+b)f, f) = al+b = l', \quad (Tf, f) = l$$

are equivalent when $a \neq 0$ and $\|f\| = 1$.

THEOREM 4.7. *If T is a linear transformation, then $W(T)$ is a convex point set. When $W(T)$ contains more than one point, its derived set $W'(T)$ is a perfect convex set including $W(T)$.**

By a convex point set we mean a point set which contains the entire line segment joining each pair of points in the set.

We exclude the trivial case where $W(T)$ contains just one point. If l_1 and l_2 are distinct points in $W(T)$, there exist corresponding elements f_1 and f_2 in the domain of T satisfying the relations $(Tf_1, f_1) = l_1$, $(Tf_2, f_2) = l_2$ $\|f_1\| = 1$, $\|f_2\| = 1$. We must show that whenever l is a point of the line segment joining l_1 and l_2, there exists an element f for which $(Tf, f) = l$, $\|f\| = 1$. We find f as a linear combination of f_1 and f_2.

We consider for this purpose the binary Hermitian forms

$$\begin{aligned} A_1(x_1, x_2) &= (T(x_1 f_1 + x_2 f_2), x_1 f_1 + x_2 f_2) \\ &= l_1 \bar{x}_1 x_1 + (Tf_2, f_1)\bar{x}_1 x_2 + (Tf_1, f_2) x_1 \bar{x}_2 + l_2 \bar{x}_2 x_2, \\ A_2(x_1, x_2) &= \|x_1 f_1 + x_2 f_2\|^2 \\ &= \bar{x}_1 x_1 + (f_2, f_1)\bar{x}_1 x_2 + (f_1, f_2) x_1 \bar{x}_2 + \bar{x}_2 x_2. \end{aligned}$$

* A similar theorem was suggested originally by Toeplitz, Mathematische Zeitschrift, 2 (1918), pp. 195–6 (actually for transformations and Hermitian forms in n-dimensional unitary space); and a proof of Toeplitz's surmise was given by Hausdorff, Mathematische Zeitschrift, 3 (1919), pp. 314–316; Wintner, *Spektraltheorie der unendlichen Matrizen,* Leipzig, 1930, § 18, considered special cases of the theorem as stated here. An entirely general theorem, which is not restricted to the case of Hilbert space, was given by the writer, Bulletin of the American Mathematical Society, 36 (1930), pp. 259–261.

We shall show that, while $A_2 = 1$, the form A_1 takes on every value l represented by a point of the line segment joining l_1 and l_2. We first introduce a new binary form

$$A(x_1, x_2) = \frac{A_1 - l_2 A_2}{l_1 - l_2} = \bar{x}_1 x + a_{12} \bar{x}_1 x_2 + a_{21} x_1 \bar{x}_2.$$

In order that A_1 should have the requisite behavior, it is necessary and sufficient that, while $A_2 = 1$, the form A should take on every real value from 0 to 1 inclusive. We shall now exhibit values of x_1 and x_2 which bring about the desired result. We first define a complex number γ as follows:

$$\gamma = \pm 1 \text{ when } \bar{a}_{12} = a_{21},$$

$$\gamma = \pm \frac{\bar{a}_{12} - a_{21}}{|\bar{a}_{12} - a_{21}|} \text{ when } \bar{a}_{12} \neq a_{21},$$

the plus or the minus sign being chosen in each case so as to render the real number $\beta = \Re(\gamma(f_2, f_1))$ not negative. We set $x_1 = x$, $x_2 = \gamma y$, where x and y are real variables, and substitute these values in the expressions for A and A_2, obtaining

$$A = x^2 + \alpha x y, \qquad \alpha = \bar{\alpha} = \pm(a_{12} + a_{21}) \text{ or }$$

$$\pm \frac{a_{12}\bar{a}_{12} - a_{21}\bar{a}_{21}}{|\bar{a}_{12} - a_{21}|},$$

$$A_2 = x^2 + 2\beta x y + y^2, \qquad \beta = \bar{\beta} \geqq 0.$$

We observe that $0 \leqq \beta \leqq 1$, because of the inequalities

$$|\beta| = |\Re\gamma(f_2, f_1)| \leqq |\gamma|\,|(f_2, f_1)|$$
$$= |(f_2, f_1)| \leqq \|f_1\|\,\|f_2\| = 1.$$

Thus the equation $A_2 = 1$ can be satisfied by taking

$$y = -\beta x + (1 - (1 - \beta^2) x)^{1/2}, \qquad -1 \leqq x \leqq 1,$$

and, with this value for y, A becomes

$$A = x^2(1 - \alpha\beta) + \alpha x (1 - (1 - \beta^2) x)^{1/2}.$$

For $x = 0$, A takes on the value 0; for $x = 1$, the value 1. Since A is now a continuous real-valued function of the real variable x, it assumes at least once every real value between

0 and 1 when x varies between these same values. This result shows that $W(T)$ is a convex set. The statement concerning the derived set $W'(T)$ is obvious.

THEOREM 4.8. *If H is a linear symmetric transformation, closed or not, the set $W(H)$ lies on the real axis in the l-plane: it may be a set consisting of a single point; a finite interval containing both, one, or neither of its endpoints; a semi-infinite interval with or without its endpoint; or the entire axis of reals.*

Since $(Hf, f) = (f, Hf) = \overline{(Hf, f)}$ we see that $W(H)$ must lie on the real axis. In order to construct examples illustrating the various possibilities indicated in the theorem, we select an infinite sequence of real numbers $\{l_n\}$ with greatest lower bound a and least upper bound b, $-\infty \leqq a \leqq b \leqq +\infty$. We define a symmetric transformation H whose domain is a complete orthonormal set $\{\varphi_n\}$ by the equations $H\varphi_n = l_n \varphi_n$, $n = 1, 2, 3, \cdots$. The transformations $\hat{H}$ and $\tilde{H}$ exist according to Theorem 2.12 and can be described as follows: f is in $\hat{\mathfrak{D}}$ if and only if it can be expressed in the form $f = \sum_{\alpha=1}^{N} x_\alpha \varphi_\alpha$ and for such an element $\hat{H}f = \sum_{\alpha=1}^{N} l_\alpha x_\alpha \varphi_\alpha$; f is in $\tilde{\mathfrak{D}}$ if and only if it is expressible in the form $f = \sum_{\alpha=1}^{\infty} x_\alpha \varphi_\alpha$ where the series $\sum_{\alpha=1}^{\infty} |x_\alpha|^2$ and $\sum_{\alpha=1}^{\infty} l_\alpha^2 |x_\alpha|^2$ converge, and for such an element $\tilde{H}f = \sum_{\alpha=1}^{\infty} l_\alpha x_\alpha \varphi_\alpha$. Evidently we have to consider the relations

$$l = (\hat{H}f, f) = \sum_{\alpha=1}^{N} l_\alpha |x_\alpha|^2, \qquad |f|^2 = \sum_{\alpha=1}^{N} |x_\alpha|^2 = 1,$$

$$l = (\tilde{H}f, f) = \sum_{\alpha=1}^{\infty} l_\alpha |x_\alpha|^2, \qquad |f|^2 = \sum_{\alpha=1}^{\infty} |x_\alpha|^2 = 1.$$

Clearly, the closed interval $a \leqq l \leqq b$ on the real axis contains $W(\tilde{H})$, $W(\tilde{H})$ contains $W(\hat{H})$, and $W(\hat{H})$ contains the open interval $a < l < b$. It is easily seen that $W(\tilde{H})$ includes the point $l = a$ (the point $l = b$) or fails to include it according as the sequence $\{l_n\}$ attains or fails to attain the bound a (the bound b). In this statement we can replace $W(\tilde{H})$ by $W(\hat{H})$. Thus we can select the sequence $\{l_n\}$ so as to realize each of the possibilities described in the theorem. We have shown in particular that $W(\tilde{H})$ and $W(\hat{H})$ are coincident and must therefore infer that closure has little to do with the character of the numerical range.

In many cases, the most powerful tool we possess for the study of a transformation T is the family of inverses T_l^{-1}, where l lies in the resolvent set. For this reason it is convenient to introduce an appropriate terminology and notation:

DEFINITION 4.4. *If T is a linear transformation whose resolvent set is not empty, the family of transformations T_l^{-1}, where l is in the resolvent set, is called the resolvent of T and is denoted by R_l.*

We shall now apply the various concepts introduced above to a detailed study of the spectrum of a linear transformation with special reference to symmetric and bounded transformations.

THEOREM 4.9. *Let T be a linear transformation which has a closed linear extension, so that the transformation $\tilde{T}$ exists. Then the spectra of T and $\tilde{T}$ are related as follows:*

$$S(\tilde{T}) \equiv S(T), \quad D(\tilde{T}) \equiv D(T), \quad A(\tilde{T}) \supseteqq A(T),$$
$$B(\tilde{T}) \subseteqq B(T), \quad C(\tilde{T}) \subseteqq C(T).$$

In particular, every characteristic manifold of T is a subset of a characteristic manifold of $\tilde{T}$ corresponding to the same characteristic value; and every characteristic value of T is a characteristic value of $\tilde{T}$ of the same or greater multiplicity. Every characteristic manifold of $\tilde{T}$ lies in the domain of T. If $\tilde{T}$ has a resolvent, then the resolvent of $\tilde{T}$ has $\mathfrak{H}$ as its domain.

We apply Theorem 2.10 to T in order to show that $\tilde{T}$ exists and is an extension of $T \equiv \hat{T}$. Since T_l has $(\tilde{T})_l \equiv \tilde{T} - lI$ as a closed linear extension, we can apply the theorem to T_l also, showing that $\tilde{T}_l$ exists and is an extension of $T_l = \hat{T}_l$. It is easily verified that $\tilde{T}_l$ and $(\tilde{T})_l$ are identical. Following our usual convention we employ the letters $\mathfrak{D}$, $\mathfrak{R}$, $\tilde{\mathfrak{D}}$, $\tilde{\mathfrak{R}}$, $\mathfrak{D}_l$, $\mathfrak{R}_l$, $\tilde{\mathfrak{D}}_l$, $\tilde{\mathfrak{R}}_l$ to denote the linear manifolds which are the domains and ranges respectively of the linear transformations

T, $\tilde{T}$, T_l, $\tilde{T}_l$. The relations $\mathfrak{D}_l \equiv \mathfrak{D}$, $\tilde{\mathfrak{D}}_l \equiv \tilde{\mathfrak{D}}$, $\mathfrak{D} \subseteqq \tilde{\mathfrak{D}}$, $\mathfrak{R} \subseteqq \tilde{\mathfrak{R}}$, $\mathfrak{R}_l \subseteqq \tilde{\mathfrak{R}}_l$ are evident; and the definition of $\tilde{T}_l$ in terms of $T_l \equiv \hat{T}_l$ requires that $\mathfrak{D}$ and $\mathfrak{R}_l$ be everywhere dense in $\tilde{\mathfrak{D}}$ and $\tilde{\mathfrak{R}}_l$ respectively. Hence $\mathfrak{R}_l$ and $\tilde{\mathfrak{R}}_l$ determine the same closed linear manifold $\mathfrak{M}_l$.

We show that $A(\tilde{T}) \supseteqq A(T)$ and give other particulars concerning the point spectra of T and $\tilde{T}$. The situation depends upon the theory of the equations

$$Tf - lf = T_l f = 0, \qquad \tilde{T}f - lf = \tilde{T}_l f = 0.$$

By virtue of the relation $\tilde{T} \supseteqq T$, it is obvious that the solutions of the first equation for fixed l are included among those of the second for the same l. In other words, every characteristic manifold for T is a subset of a characteristic manifold for $\tilde{T}$ corresponding to the same characteristic value l; and every characteristic value for T is a characteristic value for $\tilde{T}$ of the same or greater multiplicity. The relation $A(\tilde{T}) \supseteqq A(T)$ follows directly. If f is an element of the characteristic manifold of $\tilde{T}$ corresponding to the characteristic value l, then there exists a sequence of characteristic elements $\{f_n\}$ of $\tilde{T}$ which converges to f. Thus we have $f_n \to f$, $\tilde{T}f_n = lf_n \to lf$, and conclude, by virtue of the fact that $\tilde{T}$ is closed, that f is in the domain of $\tilde{T}$ and is a characteristic element (except in the case $f = 0$). Hence every characteristic manifold of $\tilde{T}$ lies in $\tilde{\mathfrak{D}}$.

By referring to Theorem 4.2, we can pass from the results of the preceding paragraph to the consideration of the inverses of T_l and $\tilde{T}_l$. If $\tilde{T}_l^{-1}$ exists, it is a closed linear transformation in accordance with Theorem 2.5; and the inverse T_l^{-1} must also exist and be a linear transformation with $\tilde{T}_l^{-1}$ as an extension. On the other hand, the existence of T_l^{-1} need not involve the existence of $\tilde{T}_l^{-1}$.

We prove next that $B(\tilde{T}) \subseteqq B(T)$. If l is in $B(\tilde{T})$, then $\tilde{T}_l^{-1}$ exists and is a non-bounded closed linear transformation with domain $\tilde{\mathfrak{R}}_l$ everywhere dense in $\mathfrak{H} \equiv \mathfrak{M}_l$. Thus T_l^{-1} exists and is a linear transformation with domain $\mathfrak{R}_l$ everywhere dense in $\mathfrak{H}$. Furthermore T_l^{-1} is non-bounded:

for, if it were bounded, its closed linear extension $\tilde{T}_l^{-1}$ would be bounded by Theorem 2.23. Hence l is in $B(T)$, as we wished to prove.

By a similar argument we find that $C(\tilde{T}) \subseteqq C(T)$. If l is in $C(\tilde{T})$, then $\tilde{T}_l^{-1}$ exists and is defined in $\tilde{\mathfrak{R}}_l \subseteqq \mathfrak{M}_l \subset \mathfrak{H}$. Hence T_l^{-1} exists and is defined in $\mathfrak{R}_l \subseteqq \tilde{\mathfrak{R}}_l \subseteqq \mathfrak{M}_l \subset \mathfrak{H}$. It follows immediately that l is in $C(T)$.

We show next that $D(\tilde{T}) \equiv D(T)$. The relation $D(\tilde{T}) \subseteqq D(T)$ is easily established by reasoning of the same type as that used in the two paragraphs which precede. If l is in $D(\tilde{T})$, then $\tilde{T}_l^{-1}$ is a bounded linear transformation with domain $\tilde{\mathfrak{R}}_l$ everywhere dense in $\mathfrak{H} \equiv \mathfrak{M}_l$. Hence T_l^{-1} exists and is a linear transformation with domain $\mathfrak{R}_l$ everywhere dense in $\mathfrak{H}$. Since $T_l^{-1} \subseteqq \tilde{T}_l^{-1}$, it is clear that T_l^{-1} is bounded and that l is in $D(T)$. We note that $\tilde{T}_l^{-1}$, being closed and bounded, must have $\mathfrak{H}$ as its domain in accordance with Theorem 2.23; in other words, the resolvent of $\tilde{T}$, if it exists, has $\mathfrak{H}$ as its domain. In order to establish the desired identity, we must also prove that $D(T) \subseteqq D(\tilde{T})$. If l is in $D(T)$, then T_l^{-1} is a bounded linear transformation with domain $\mathfrak{R}_l$ everywhere dense in $\mathfrak{H}$; in particular, there exists a positive constant C such that $\| T_l^{-1} f \| \leqq C \| f \|$ for every element f in $\mathfrak{R}_l$. We show that $\tilde{T}_l$ has an inverse, by proving that $\tilde{T}_l f = 0$ implies $f = 0$. If f is an element such that $\tilde{T}_l f = 0$ and if ϵ is an arbitrary positive number, then there exists an element g in the domain of $T_l \equiv \hat{T}_l$ such that $\| f - g \| < \epsilon$, $\| T_l g \| = \| \tilde{T}_l f - \hat{T}_l g \| < \epsilon$. Consequently we can write

$$\begin{aligned} \| f \| \leqq \| g \| + \| f - g \| < \| g \| + \epsilon &= \| T_l^{-1}(T_l g) \| + \epsilon \\ \leqq C \| T_l g \| + \epsilon < (C+1)\epsilon. \end{aligned}$$

Since ϵ is arbitrary, we must have $\| f \| = 0$ and $f = 0$, as we wished to prove. Thus $\tilde{T}_l^{-1}$ exists and is a closed linear extension of the bounded linear transformation T_l^{-1}. By Theorem 2.23, $\tilde{T}_l^{-1}$ must be a bounded linear transformation with domain $\mathfrak{H}$. It follows that l is in $D(\tilde{T})$. With this result, our demonstration is complete.

The identity $D(\tilde{T}) \equiv D(T)$ implies the identity $S(\tilde{T}) \equiv S(T)$ since the sets D and S are complementary sets in the l-plane.

It is of some interest to note the manner in which a number l is transferred from the residual or the continuous spectrum of T to the point spectrum of $\tilde{T}$. A number l can become a characteristic value if and only if there exists an element $f \neq 0$ which is the limit of a sequence $\{f_n\}$ in the domain of T with the property that $Tf_n \to lf$; for it is only under these conditions that we can assert the existence of an element $f \neq 0$ such that $\tilde{T}f = lf$. It is also of some interest to exhibit examples in which such a transfer actually takes place. We leave to the reader the detailed study of the following simple instance: we define T and $\tilde{T}$ with reference to a complete orthonormal set $\{\varphi_n\}$; the domain $\mathfrak{D}$ of T is to be the linear manifold determined by this set, while the domain of $\tilde{T}$ is to be $\mathfrak{H}$ itself; if $f = \sum_{\alpha=1}^{N} a_\alpha \varphi_\alpha$ we define $Tf = \sum_{\alpha=1}^{N-1} a_{\alpha+1} \varphi_\alpha$, while if $f = \sum_{\alpha=1}^{\infty} a_\alpha \varphi_\alpha$ we define $\tilde{T}f = \sum_{\alpha=1}^{\infty} a_{\alpha+1} \varphi_\alpha$, so that $T \subseteqq \tilde{T}$ and both transformations are bounded and linear.

THEOREM 4.10. *If T is a closed linear transformation with resolvent R_l then*

(1) $(l - m) R_l R_m \equiv R_l - R_m$ *throughout $\mathfrak{H}$, for every pair of points (l, m) in the resolvent set of T;*

(2) *if $R_l f = 0$, then $f = 0$.*

Conversely, if X_l is a family of bounded linear transformations with domain $\mathfrak{H}$, defined for every point l in a set Δ of the complex l-plane, such that

(1) $(l - m) X_l X_m \equiv X_l - X_m$ *throughout $\mathfrak{H}$, for every pair of points (l, m) in Δ;*

(2) $X_l f = 0$ *implies $f = 0$ for at least one l in Δ;*

then there exists a unique closed linear transformation T whose resolvent exists and coincides with X_l for every point in the set Δ.

If T is a closed linear transformation with resolvent R_l, then the preceding theorem shows that $R_l \equiv T_l^{-1} \equiv \tilde{T}_l^{-1}$ is a bounded linear transformation with domain $\mathfrak{H}$. From the obvious relations $R_l T_m R_m \equiv R_l$, $R_l T_l R_m \equiv R_m$, $T_m - T_l = (l-m) I$ in $\mathfrak{D}$, we find immediately that $R_l - R_m \equiv R_l(T_m - T_l) R_m \equiv (l-m) R_l R_m$. As to the second property of R_l, we observe that it expresses the fact that R_l determines a one-to-one correspondence between $\mathfrak{H}$ and the domain of T_l.

We turn now to the converse. If X_l is a family of transformations possessing the properties enumerated above, then we can show that X_l establishes a one-to-one correspondence between $\mathfrak{H}$ and its range. By hypothesis, there is at least one value of l, say $l = m$, such that $X_l f = 0$ implies $f = 0$; if l is an arbitrary point of Λ, and f an element for which $X_l f = 0$, we obtain from (1) the relations

$$X_m f = X_l f + (m-l) X_m X_l f = 0, \qquad f = 0,$$

so that $X_l f = 0$ always implies $f = 0$. The one-to-one relation between the domain and range of the linear transformation X_l is thus apparent. By inverting this correspondence and using the inverse in an appropriate manner we can define the transformation T whose existence we are to establish. If $f = X_l g$, we assign as correspondent to f the element $g + l X_l g$. Since there is at most one element g such that $X_l g = f$, we assign at most one correspondent to a given element f. For a given value of l, therefore, we have defined a transformation whose domain consists of the range of X_l. We wish to show that this transformation is independent of l. This fact results if we can prove that, when l, m, and g are given, there exists a unique element g' satisfying the equation $X_l g = X_m g'$ and that this element g' satisfies the further equation $g + l X_l g = g' + m X_m g'$. It is evident that, if one element g' exists such that $X_m g' = X_l g$, it must be unique; now we can construct g' explicitly in terms of l, m, and g, by the defining equation $g' = g + (l-m) X_l g$, and can verify directly by the use of (1) that

$$X_m g' = X_m g + (l-m) X_m X_l g = X_l g.$$

Furthermore, if we write the defining equation for g' in the form $g' + m X_l g = g + l X_l g$ and then replace $X_l g$ on the left by $X_m g'$, we see that g and g' satisfy the second required relation. We now let $\mathfrak{D}$ be the linear manifold which is the range of X_l and $\mathfrak{R}$ the corresponding set obtained by the transformation T, which takes the element $f = X_l g$ of $\mathfrak{D}$ into the element $g + lX_l g$; the sets $\mathfrak{D}$ and $\mathfrak{R}$ and the transformation T are independent of l as we have seen. If l is a point of the set Δ, then we have

$$Tf - lf = g, \quad f = X_l g$$

so that $T - lI$ and X_l are inverse to each other. Since X_l is bounded and linear with domain $\mathfrak{H}$, it is a closed linear transformation. By Theorem 2.5, $T - lI$ is also closed and linear, and thus T has the same properties. It is now apparent that the resolvent set of the closed linear transformation T includes the points of Δ and that R_l, the resolvent of T, coincides with X_l throughout $\mathfrak{H}$ for every point in Δ. It is evident also that T is uniquely determined by X_l.

The functional equation satisfied by the resolvent, when it exists, suggests strongly that the resolvent must depend "analytically" on the complex variable l, in a sense yet to be explained. That this is indeed the case, we shall demonstrate immediately.

THEOREM 4.11. *If the resolvent set of a closed linear transformation T contains the point m, then there exists a positive constant C, dependent on m, such that the resolvent R_l satisfies the inequality $\| R_m g \| \leqq C \| g \|$ for every element g in $\mathfrak{H}$. Every point l within distance $1/C$ of the point m belongs to the resolvent set of T; and, furthermore,*

$$\sum_{\alpha=0}^{n} (l-m)^\alpha R_m^{\alpha+1} \to R_l, \quad n \to \infty, \quad |l-m| < 1/C,$$

$$(R_l f, g) = \sum_{\alpha=0}^{\infty} (l-m)^\alpha (R_m^{\alpha+1} f, g), \quad |l-m| < 1/C.$$

By hypothesis, R_m is a bounded linear transformation with domain $\mathfrak{H}$, so that the constant C exists as stated. If we form from an arbitrary element f in $\mathfrak{H}$ the element $f_n = \sum_{\alpha=0}^{n} (l-m)^\alpha R_m^{\alpha+1} f$, we can show without difficulty that the sequence $\{f_n\}$ is convergent when $|l-m|<1/C$. For $p>q$ we have

$$\begin{aligned}\|f_p-f_q\|^2 &= \left(\sum_{\alpha=q+1}^{p}(l-m)^\alpha R_m^{\alpha+1}f,\ \sum_{\beta=q+1}^{p}(l-m)^\beta R_m^{\beta+1}f\right)\\ &= \sum_{\alpha,\beta=q+1}^{p}(l-m)^\alpha\overline{(l-m)}^\beta(R_m^{\alpha+1}f, R_m^{\beta+1}f)\\ &\leqq \sum_{\alpha,\beta=q+1}^{p}|l-m|^{\alpha+\beta}\|R_m^{\alpha+1}f\|\,\|R_m^{\beta+1}f\|\\ &\leqq \sum_{\alpha,\beta=q+1}^{p}|l-m|^{\alpha+\beta}C^{\alpha+\beta+2}\|f\|^2\\ &\leqq \left[\sum_{\gamma=2q+2}^{2p}(\gamma+1)(C|l-m|)^\gamma\right]C^2\|f\|^2\end{aligned}$$

Since $\lim_{p,q\to\infty}\sum_{\gamma=2q+2}^{2p}(\gamma+1)x^\gamma=0$ when $|x|<1$, we see that $\lim_{p,q\to\infty}\|f_p-f_q\|=0$ when $|l-m|$ is less than $1/C$. Consequently the sequence $\{f_n\}$ converges and has a limiting element f^*. The transformation X_l which takes f into f^* is evidently a bounded linear transformation with domain $\mathfrak{H}$, defined when $|l-m|$ is less than $1/C$; its boundedness follows upon the easily verified inequality

$$\|X_lf\|^2=\lim_{n\to\infty}\|f_n\|^2\leqq\left[\sum_{\gamma=0}^{\infty}(\gamma+1)(C|l-m|)^\gamma\right]C^2\|f\|^2.$$

With the notation of Definition 2.3, we can write

$$\sum_{\alpha=0}^{n}(l-m)^\alpha R_m^{\alpha+1}\to X_l,\quad |l-m|<1/C.$$

We must now show that X_l is the inverse of T_l. When f is in the domain of T and of T_l, we have

$$\begin{aligned}X_l(T_lf) &= \lim_{n\to\infty}\sum_{\alpha=0}^{n}(l-m)^\alpha R_m^{\alpha+1}(T_lf)\\ &= \lim_{n\to\infty}\sum_{\alpha=0}^{n}(l-m)^\alpha(R_m^\alpha f-(l-m)R_m^{\alpha+1}f)\\ &= \lim_{n\to\infty}[f-(l-m)^{n+1}R_m^{n+1}f]\\ &= f,\quad |l-m|<1/C.\end{aligned}$$

On the other hand, if f is an arbitrary element of $\mathfrak{H}$ we have, in the notation used above, $f^* = X_l f = \lim_{n \to \infty} f_n$; and, in addition

$$T_l f_n = \sum_{\alpha=0}^{n} (l-m)^\alpha T_l R_m^{\alpha+1} f$$
$$= \sum_{\alpha=0}^{n} (l-m)^\alpha (R_m^\alpha f - (l-m) R_m^{\alpha+1} f)$$
$$= f - (l-m)^{n+1} R_m^{n+1} f \to f, \quad |l-m| < 1/C.$$

Since $f_n \to X_l f$, $T_l f_n \to f$, and T_l is a closed linear transformation, we conclude that $X_l f$ is in the domain of T_l and that $T_l X_l f = f$. In view of the relations between the transformations T_l and X_l, we see that X_l is the inverse of T_l and that l is a point of the resolvent set of the given transformation T. Consequently we can identify X_l with the resolvent R_l, when $|l-m| < 1/C$. Since, therefore, $\sum_{\alpha=0}^{n} (l-m)^\alpha R_m^{\alpha+1} \to R_l$ for such values of l, we have

$$(R_l f, g) = \lim_{n \to \infty} \sum_{\alpha=0}^{n} (l-m)^\alpha (R_m^{\alpha+1} f, g)$$

as we wished to show.

THEOREM 4.12. *The resolvent set of a closed linear transformation T is either empty or open; in the latter case the resolvent exists and the function of l defined as $(R_l f, g)$ is an analytic function of l in the resolvent set. The spectrum of T is the entire plane, a closed set, or an empty set.*

This theorem is an obvious consequence of the results just established in the preceding one.

To one who has some familiarity with the theory of integral equations, the theorems which we have just proved, introducing the functional equation and the "analytic" character of the resolvent, will appear as well-known results, extended to the general cases here under consideration. The functional equation is precisely that satisfied by Fredholm's resolvent kernel, or, more accurately, by the integral operator defined by the kernel; and the series developments for R_l and $(R_l f, g)$ were introduced at an early stage by C. Neumann*.

We shall now investigate in some detail the spectrum of a symmetric transformation. We can obtain a fairly complete characterization of the spectrum with the general material already developed.

THEOREM 4.13. *The point spectrum of a linear symmetric transformation H is an empty, finite, or denumerably infinite set of points on the real axis. Characteristic elements corresponding to distinct characteristic values of H are orthogonal.*

First, let l be a characteristic value of H and f a corresponding normalized characteristic element. Since H is symmetric, we have $(Hf, f) = (f, Hf)$, an equation which reduces to $l = \bar{l}$ when Hf is replaced by lf and (f, f) by 1. Consequently l is a point on the real axis.

Next, let l_1 and l_2 be distinct characteristic values of H, f_1 and f_2 corresponding normalized characteristic elements. The equation $(Hf_1, f_2) = (f_1, Hf_2)$ becomes $l_1(f_1, f_2) = l_2(f_1, f_2)$, so that $(f_1, f_2) = 0$ and f_1 and f_2 are orthogonal. If to each characteristic value of H we order a corresponding normalized characteristic element, then these elements must constitute an orthonormal set; consequently, the set of characteristic values, the point spectrum, can be at most denumerably infinite, by Theorem 1.4.

THEOREM 4.14. *If H is a linear symmetric transformation, and l a point such that $\mathfrak{J}(l) \neq 0$, then H_l^{-1} exists and is a bounded linear transformation satisfying the inequality*

$$\|H_l^{-1} g\| \leq \|g\| / |\mathfrak{J}(l)|$$

for every g in its domain. The point l is therefore in the resolvent set or in the residual spectrum of H.

By the preceding theorem, l cannot be a characteristic value of H since it is not real. Hence H_l^{-1} exists. If g

* For a discussion of these subjects, see Hellinger and Toeplitz, *Integralgleichungen und Gleichungen mit unendlichvielen Unbekannten*, Encyklopädie der Mathematischen Wissenschaften 2^{3}, pp. 1347–1350, 1370–1376.

is in the domain of H_l^{-1}, then $H_l^{-1}g$ is in the domain of H, and we have

$$(HH_l^{-1}g, H_l^{-1}g) = (H_l^{-1}g, HH_l^{-1}g),$$
$$(g+lH_l^{-1}g, H_l^{-1}g) = (H_l^{-1}g, g+lH_l^{-1}g),$$
$$(l-\bar{l})\|H_l^{-1}g\|^2 = (H_l^{-1}g, g)-(g, H_l^{-1}g).$$

As a result, we find the inequalities

$$2|\mathfrak{J}(l)|\|H_l^{-1}g\|^2 \leqq |(H_l^{-1}g, g)|+|(g, H_l^{-1}g)| \leqq 2\|H_l^{-1}g\|\|g\|$$

and

$$\|H_l^{-1}g\| \leqq \|g\|/|\mathfrak{J}(l)|,$$

as we wished to prove. H_l^{-1} is therefore bounded, so that the statement of the theorem is justified.

THEOREM 4.15. *If T is an arbitrary linear transformation whose adjoint T^* exists, then each point l of the residual spectrum of T determines a characteristic value $\bar{l}$ of T^* and each characteristic value $\bar{l}$ of T^* determines a point l either in the residual spectrum or in the point spectrum of T.*

If l is in the residual spectrum of T, then T_l^{-1} exists and has a domain which is a linear manifold not everywhere dense in $\mathfrak{H}$. Thus there exists an element $g \neq 0$ such that $(T_l f, g) = 0$ for every f in the domain of T. This equation may be written $(Tf, g) = (f, \bar{l}g)$ and thus implies that T^*g exists and is equal to $\bar{l}g$. It follows therefore that $\bar{l}$ is a characteristic value of T^*.

Now let $\bar{l}$ be a characteristic value of T^* and g a corresponding characteristic element. We have, for every f in the domain of T,

$$\begin{aligned}(T_l f, g) &= (Tf, g)-l(f, g)\\ &= (f, T^*g)-l(f, g)\\ &= (f, \bar{l}g)-l(f, g) = 0.\end{aligned}$$

Hence g is orthogonal to every element of the range of T_l. If T_l^{-1} exists its domain is not everywhere dense in $\mathfrak{H}$, so that l is a point of the residual spectrum of T; and if T_l^{-1} does not exist, then l is a characteristic value of T.

We may now apply this general theorem to the special case of symmetric transformations.

THEOREM 4.16. *If H is a linear symmetric transformation and l an arbitrary point such that $\mathfrak{J}(l) \neq 0$, then either l is in the resolvent set of H or $\bar{l}$ is in the point spectrum of H^*; if $\bar{l}$ is a characteristic value of H^*, then l is in the residual spectrum of H. The half plane $\mathfrak{J}(l) > 0$ $[\mathfrak{J}(l) < 0]$ consists entirely of points of the resolvent set or entirely of points of the residual spectrum of H.*

We have already shown that l, $\mathfrak{J}(l) \neq 0$, is either in the resolvent set or in the residual spectrum of H; by the preceding theorem, if l is in the residual spectrum, then $\bar{l}$ is a characteristic value of H^*. If $\bar{l}$ is a characteristic value of H^*, then l is either a point of the residual spectrum of H or a characteristic value of H. Since H has only real characteristic values, the latter possibility is excluded.

If we consider $\tilde{H}$ in place of H, we do not change the spectral classification of those points in the l-plane which do not lie on the real axis, as we see by Theorem 4.9 and the fact that H has only real characteristic values. If l, $\mathfrak{J}(l) > 0$, is a point of the resolvent set of H, it is also in the resolvent set of $\tilde{H}$. If we apply to the resolvent of $\tilde{H}$, the results of Theorems 4.11 and 4.14, we see that every point of the half-plane $\mathfrak{J}(l) > 0$ is in the resolvent set; for we can construct circles $C_0, C_1, \cdots, C_n$, none of which intersects the axis of reals, C_0 having its center at l, C_{k+1} having its center in C_k, and C_n containing a preassigned point of the half-plane $\mathfrak{J}(l) > 0$, and then, by the theorems cited, ascertain the existence of the resolvent in each of the circles in order and thus at the point in C_n in particular. As a result, we find that the points of the half-plane $\mathfrak{J}(l) > 0$, shared between the resolvent set and the residual spectrum, must belong all to one set or all to the other. Similar arguments apply to the lower half-plane, $\mathfrak{J}(l) < 0$.

THEOREM 4.17. *A necessary and sufficient condition that the linear symmetric transformation H be essentially self-adjoint is that its resolvent set contain a point in each half-plane*

$\mathfrak{J}(l) > 0$, $\mathfrak{J}(l) < 0$. *If H is essentially self-adjoint, then* $\tilde{H} \equiv H^{**} \equiv H^*$.

If H is essentially self-adjoint, then H^* is symmetric and so has only real characteristic values. By the preceding theorem, the half-planes $\mathfrak{J}(l) > 0$, $\mathfrak{J}(l) < 0$ belong to the resolvent set of H.

If H has a resolvent set with one point in each of the half-planes $\mathfrak{J}(l) > 0$, $\mathfrak{J}(l) < 0$, then the resolvent set includes both half-planes by the preceding theorem. We consider in place of H, the closed linear symmetric transformation $\tilde{H}$. We consider also the transformation H^*, the adjoint of H and $\tilde{H}$. We know that the relations $\tilde{H} \subseteqq H^{**} \subseteqq H^*$ hold, so that if we can show that $\tilde{H} \subset H^*$ is impossible we can conclude that $\tilde{H} \equiv H^{**} \equiv H^*$ and that H is essentially self-adjoint. Now $\tilde{H}$ is a closed symmetric transformation whose resolvent set includes the half-planes $\mathfrak{J}(l) > 0$, $\mathfrak{J}(l) < 0$; and the inverse $\tilde{H}_l^{-1}$ has $\mathfrak{H}$ as its domain when $\mathfrak{J}(l) \neq 0$, by Theorem 4.9. Consequently, the range of $\tilde{H}_l \equiv \tilde{H} - lI$ is $\mathfrak{H}$ when $\mathfrak{J}(l) \neq 0$. If we had $H^* \supset \tilde{H}$, we should be able to find an f not in the domain of $\tilde{H}$ such that $H^* f$ exists. If we take $\mathfrak{J}(l) \neq 0$, we can find a g in the domain of $\tilde{H}$ so that $\tilde{H}_l g = H_l^* f$. Hence $H_l^* (f - g) = 0$, $f - g \neq 0$. We conclude that every point of the half-planes $\mathfrak{J}(l) > 0$, $\mathfrak{J}(l) < 0$ is in the residual spectrum of H, contrary to fact. This establishes the theorem.

THEOREM 4.18. *If H is a self-adjoint transformation, its spectrum is empty, or is a closed subset of the real axis, or is the entire real axis. Its point spectrum is empty, finite, or denumerably infinite. Its residual spectrum is empty. Its continuous spectrum may be empty or not. If l is real, the transformation H_l has an inverse H_l^{-1} except when l is a characteristic value of H; when l is in the continuous spectrum the inverse is an unbounded self-adjoint transformation whose domain is a proper subset of $\mathfrak{H}$; when l is in the resolvent set of H the inverse is a bounded self-adjoint transformation with domain $\mathfrak{H}$.*

The assertions of the theorem are easy consequences of theorems already established. The characterization of the spectrum is simple once it has been shown that the residual spectrum is absent; and this is proved by the following *reductio ad absurdum*: if l were a point of the residual spectrum, it would be real in view of the preceding theorem, and thus $l = \bar{l}$ would be a characteristic value of $H^* \equiv H$ and not a point of the residual spectrum, as we had supposed. When l is real and the inverse H_l^{-1} exists, the properties of the inverse can be obtained at once from our fundamental classification. It is to be noted that we cannot yet exclude the possibility that the spectrum of H is empty.

THEOREM 4.19. *A family X_l of bounded linear transformations with domain $\mathfrak{H}$, defined for every not-real l, is the resolvent of some self-adjoint transformation H if and only if the following conditions are satisfied:*

(1) $(l - m) X_l X_m \equiv X_l - X_m$ in $\mathfrak{H}$;
(2) $X_l f = 0$ implies $f = 0$ for at least one not-real l;
(3) $(X_l)^* \equiv X_{\bar{l}}$ in $\mathfrak{H}$.

When H exists, it is unique.

If H is a self-adjoint transformation, it is linear and closed. Its resolvent R_l is defined for all not-real l and satisfies (1) and (2) as we have already proved in Theorems 4.10 and 4.18. We see from the relations

$$(R_l f, g) = (R_l f, H_{\bar{l}} R_{\bar{l}} g) = (H_l R_l f, R_{\bar{l}} g) = (f, R_{\bar{l}} g),$$

holding for every f and g in $\mathfrak{H}$, that $(R_l)^*$ and $R_{\bar{l}}$ are identical. Thus the three conditions (1), (2), (3) are necessary.

If, on the other hand, the family X_l is given, with the properties (1), (2), (3), we know that there exists a unique closed linear transformation T whose resolvent exists and coincides with X_l for all not-real l; for, by Theorem 4.10, conditions (1) and (2) suffice to prove the existence of T. We must now apply (3) to show that T is self-adjoint. First we must prove that its domain $\mathfrak{D}$ is everywhere dense in $\mathfrak{H}$: since $\mathfrak{D}$ consists of all elements f expressible in the form

$f = X_l g$ for some not-real l, every element h orthogonal to $\mathfrak{D}$ must vanish in view of the equations

$$0 = (X_l g, h) = (g, X_{\bar{l}} h), \quad X_{\bar{l}} h = 0, \quad h = 0,$$

which follow from conditions (2) and (3); thus $\mathfrak{D}$ has the desired property. Next we must show that, if h and h^* are elements of $\mathfrak{H}$ satisfying the equation $(Tf, h) = (f, h^*)$ for every f in $\mathfrak{D}$, then Th exists and is equal to h^*. On putting $f = X_l g$ and $Tf = g + l X_l g$, we can write the equation for h and h^* in the form

$$(g + l X_l g, h) = (X_l g, h^*), \quad (g, h + \bar{l} X_{\bar{l}} h - X_{\bar{l}} h^*) = 0.$$

Hence $h + \bar{l} X_{\bar{l}} h - X_{\bar{l}} h^* = 0$, $h = X_{\bar{l}} h^* - \bar{l} X_{\bar{l}} h$. From the second equation, we see that Th exists, since both elements on the right are in the domain of T. On computing Th we find that it is equal to h^*, as it should be. Thus T and T^* are identical. By definition, therefore, T is self-adjoint.

We shall find this theorem exceedingly useful in numerous situations which arise subsequently. In particular, it is fundamental in the further analysis of self-adjoint transformations carried out in the following chapter.

In the case of a bounded linear transformation with domain $\mathfrak{H}$, we find an interesting connection between the spectrum and the set $W(T)$ discussed in Theorems 4.5–4.8.

THEOREM 4.20. *The resolvent set of a bounded linear transformation T with domain $\mathfrak{H}$ includes every point at positive distance from the set $W(T)$. $W(T) + W'(T)$ is a bounded closed set which includes every point of the spectrum of T. If l is a point at positive distance d from $W(T)$ then R_l exists and satisfies the inequality $\|R_l f\| \leq \|f\| / d$ for every element f in $\mathfrak{H}$.*

Since T is a bounded linear transformation with domain $\mathfrak{H}$ it possesses an adjoint T^* with similar properties, as asserted in Theorem 2.29. If l is a point at positive distance d from the set $W(T)$, the inequality

$$|(T_l f, f)| = |(Tf, f) - l| \geqq d$$

holds whenever $\|f\| = 1$. In consequence l cannot be a characteristic value of T, since a normalized characteristic element g would satisfy the relations

$$\|T_l g\| = 0, \quad |(T_l g, g)| \geqq d > 0,$$

which are evidently incompatible. We can show also that l cannot belong to the residual spectrum of T. By Theorem 4.15, the assumption that l is in the residual spectrum of T implies that $\bar{l}$ is a characteristic value of T^*. We must therefore consider the relation between $\bar{l}$ and $W(T^*)$. Since $(T^*f, f) = (f, Tf) = \overline{(Tf, f)}$ it is evident that $W(T^*)$ is obtained from $W(T)$ by a reflection in the real axis, just as $\bar{l}$ is obtained from l. Consequently, $\bar{l}$ is at distance d from $W(T^*)$ and, by an argument logically identical with that just used, cannot be a characteristic value of T^*. This contradiction shows that l is not in the residual spectrum of T.

We now recognize that T_l^{-1} exists and that l belongs to the resolvent set or to the continuous spectrum of T according as T_l^{-1} is bounded or not. We shall show that T_l^{-1} is bounded. If $f \neq 0$ is an element in the domain of T_l^{-1}, then $T_l^{-1} f \neq 0$ and the element $f / \|T_l^{-1} f\|$ is also an element in the domain of the linear transformation T_l^{-1}. We put $g = T_l^{-1} f / \|T_l^{-1} f\|$ and determine the significance of the relations $\|g\| = 1$, $|(T_l g, g)| \geqq d$, for the element f. We find that the inequality becomes $|(f, T_l^{-1} f) / \|T_l^{-1} f\|^2| \geqq d$ so that we can write

$$\|f\| / \|T_l^{-1} f\| \geqq |(f, T_l^{-1} f) / \|T_l^{-1} f\|^2| \geqq d,$$
$$\|T_l^{-1} f\| \leqq \|f\| / d.$$

The final inequality, trivial in case $f = 0$, is now seen to hold throughout the domain of the linear transformation T_l^{-1} and implies that T_l^{-1} is bounded. Thus l belongs to the resolvent set of T.

The remaining parts of the theorem follow easily. Any point outside the closed set $W(T) + W'(T)$ is at positive distance from $W(T)+W'(T)$ and, *a fortiori*, from $W(T)$; such a point must belong to the resolvent set of T, the spectrum of T being contained as a result in $W(T)+W'(T)$. From Theorem 4.9, the resolvent of the closed linear transformation T has $\mathfrak{H}$ as its domain. Thus, if l is at positive distance d from $W(T)$, R_l exists, has $\mathfrak{H}$ as its domain, and satisfies the inequality given in the theorem and proved above.

It is worthy of comment that this theorem cannot be extended to unbounded transformations in general. In the case of a symmetric transformation which is not essentially self-adjoint, the set W is confined to the real axis, but the spectrum contains at least one of the two half-planes $\mathfrak{I}(l)>0$, $\mathfrak{I}(l)<0$. This situation is not modified even if the transformation is required to be maximal; in other words, the appearance of an extended residual spectrum is not a consequence of an avoidable incompleteness in the definition of the transformation considered.

THEOREM 4.21. *If T is a bounded linear transformation with domain $\mathfrak{H}$ such that $|Tf|\leq C|f|$ for every element f in $\mathfrak{H}$, then whenever $|l|>C$*

$$-\sum_{\alpha=1}^{n} l^{-\alpha} T^{\alpha-1}\to R_l, \qquad n\to\infty,$$

$$(R_l f, g) = -\sum_{\alpha=1}^{\infty} l^{-\alpha}(T^{\alpha-1}f, g).$$

The proof is much the same as that of Theorem 4.11, with occasional simplifications due to the fact that the transformation T is bounded and continuous, with domain $\mathfrak{H}$. We shall not go into details.

THEOREM 4.22. *The spectrum of a bounded linear transformation T with domain $\mathfrak{H}$ contains at least one point.*

We consider the function of the complex variable l defined by the expression $(R_l f, g)$. If T has no spectrum, then this function is a single-valued analytic function defined over the entire finite l-plane. By Theorems 4.20 and 4.21 it is bounded over the entire finite plane and can be defined so as to be analytic at $l=\infty$. By applying Liouville's Theorem, we find that $(R_l f, g) = 0$. Since this equation holds for every f and g in $\mathfrak{H}$, we have $R_l f = 0$ for every f, contrary to fact. We see therefore that the spectrum of T must contain at least one point.

§ 3. Reducibility

We shall now discuss briefly the second problem raised in § 1—the study of linear manifolds left invariant by a linear transformation T. In formulating the problem in precise terms, it is convenient to restrict attention to closed manifolds and closed transformations. Since a closed linear manifold is conveniently characterized by means of the projection of $\mathfrak{H}$ upon it, it is natural to express the relation between a given transformation T and a given manifold as a relation between T and the corresponding projection. This is the course which we shall pursue. Here it will be sufficient to introduce a few fundamental definitions and theorems.

Definition 4.5. *Let T be a closed linear transformation, $\mathfrak{M}$ a closed linear manifold, and E the projection of $\mathfrak{H}$ on $\mathfrak{M}$. $\mathfrak{M}$ is said to reduce T if T and E are permutable in the following precise sense: whenever f is in the domain of T, Ef is also in the domain of T and $TEf = ETf$. If $T(a)$ is a family of closed linear transformations defined over a class $\mathfrak{A}$ whose general element is a, then $\mathfrak{M}$ is said to reduce the family $T(a)$ if it reduces every transformation of the family.*

In succeeding definitions and theorems of general bearing we shall state the facts for a single transformation, leaving aside the obvious extensions to the case of a family of transformations.

Our next theorem indicates the relation between reducibility and invariance in a more precise form than we have yet given.

Theorem 4.23. *If the closed linear manifold $\mathfrak{M}$ reduces the closed linear transformation T, then T leaves $\mathfrak{M}$ invariant in the sense that it carries every element common to its domain and the manifold $\mathfrak{M}$ into an element of $\mathfrak{M}$. The manifolds $\mathfrak{H}$ and $\mathfrak{O}$ both reduce T. If $\mathfrak{M}_1$ and $\mathfrak{M}_2$ are orthogonal mani-*

folds which both reduce T, then $\mathfrak{M}_1 \oplus \mathfrak{M}_2$ *reduces T; and, if* $\mathfrak{M}_1$ *and* $\mathfrak{M}_2 \subseteqq \mathfrak{M}_1$ *both reduce T, then* $\mathfrak{M}_1 \ominus \mathfrak{M}_2$ *reduces T.*

If f is an element common to the domain of T and the manifold $\mathfrak{M}$, we have $f = Ef$ and thus obtain $Tf = TEf = ETf$; the form of the last expression shows that it is an element of $\mathfrak{M}$. Thus T leaves $\mathfrak{M}$ invariant in the sense described. If $\mathfrak{M}_1$ and $\mathfrak{M}_2$ are orthogonal manifolds which reduce T and if E_1 and E_2 are the respective projections of $\mathfrak{H}$ upon them, we have $TE_1f = E_1Tf$, $TE_2f = E_2Tf$ for every element f in the domain of T. According to Theorem 2.37, $E_1 + E_2$ is the projection of $\mathfrak{H}$ on $\mathfrak{M}_1 \oplus \mathfrak{M}_2$. By adding the equations noted above, we find that $(E_1+E_2)f$ is in the domain of T and that $T(E_1+E_2)f = (E_1+E_2)Tf$. Thus $\mathfrak{M}_1 \oplus \mathfrak{M}_2$ reduces T. Similarly, when $\mathfrak{M}_1$ and $\mathfrak{M}_2 \subseteqq \mathfrak{M}_1$ reduce T, we find by subtraction that $(E_1 - E_2)f$ is an element in the domain of T and that $T(E_1-E_2)f = (E_1-E_2)Tf$. Since $E_1 - E_2$ is the projection of $\mathfrak{H}$ on $\mathfrak{M}_1 \ominus \mathfrak{M}_2$, it follows that the latter manifold reduces T. We call attention to the special case $\mathfrak{M}_1 \equiv \mathfrak{H}$. The fact that the manifolds $\mathfrak{H}$ and $\mathfrak{O}$ reduce T is evident.

DEFINITION 4.6. *A closed linear transformation T is said to be irreducible if it is reduced by no closed linear manifold other than $\mathfrak{H}$ and $\mathfrak{O}$; and is said to be reducible in the contrary case.*

THEOREM 4.24. *If H is a closed linear symmetric transformation and if $\mathfrak{M}$ is a closed linear manifold such that*

(1) *when f is in the domain of H, Ef is in the domain of H,*

(2) *when f is an element of $\mathfrak{M}$ in the domain of H, Hf is in $\mathfrak{M}$, — then $\mathfrak{M}$ reduces H.*

We have to show that, when f is in the domain of H, $HEf = EHf$. Since $E(HEf) = HEf$, by (2), we have for every g in the domain of H

$$(HEf - EHf, g) = (EHEf - EHf, g) = (HEf - Hf, Eg)$$
$$= (Ef - f, HEg).$$

Now $Ef - f$ is in $\mathfrak{H} \ominus \mathfrak{M}$, and HEg is in $\mathfrak{M}$, so that the last expression vanishes. Remembering that the elements g

are everywhere dense in $\mathfrak{H}$, we see that $HEf - EHf = 0$, as we wished to prove.

THEOREM 4.25. *If T is a bounded linear transformation with domain $\mathfrak{H}$, T^* its adjoint, and $\mathfrak{M}$ a closed linear manifold which contains Tf and T^*f whenever it contains f, then $\mathfrak{M}$ reduces T. This result includes the case where $T \equiv U$ is a unitary transformation and $T^* \equiv U^* \equiv U^{-1}$ is its inverse.*

Since T and T^* both have $\mathfrak{H}$ as their domain according to Theorem 2.29, TEf and T^*Ef always exist. As in the proof of the preceding theorem, we find

$$\begin{aligned}(TEf - ETf, g) = (ETEf - ETf, g) &= (TEf - Tf, Eg) \\ = (Ef - f, T^*Eg) &= 0\end{aligned}$$

so that $TEf - ETf = 0$ as we wished to prove. We observe that the part of the hypothesis affecting T^* cannot be eliminated, even in the case that T is a unitary transformation. As an example we indicate the following situation: let $\{\varphi_n\}$ be a complete orthonormal set with the index n running through the values $0, \pm 1, \pm 2, \cdots$; let U be the unitary transformation which carries $f = \sum_{\alpha=-\infty}^{+\infty} a_\alpha \varphi_\alpha$ into $Uf = \sum_{\alpha=-\infty}^{+\infty} a_\alpha \varphi_{\alpha+1}$; and let $\mathfrak{M}$ be the closed linear manifold which comprises all elements g expressible in the form $g = \sum_{\alpha=0}^{+\infty} a_\alpha \varphi_\alpha$. It is easily verified that Ug is in $\mathfrak{M}$ whenever g is an element of $\mathfrak{M}$; but that $EU\varphi_{-1} = E\varphi_0 = \varphi_0$, $UE\varphi_{-1} = U0 = 0$, and $\mathfrak{M}$ does not reduce U.

THEOREM 4.26. *Let $\{\mathfrak{M}_n\}$ be a finite or infinite sequence of mutually orthogonal closed linear manifolds; let $\{E_n\}$ be the corresponding sequence of projections; let $\mathfrak{M}$ be the closed linear manifold $\mathfrak{M}_1 \oplus \mathfrak{M}_2 \oplus \mathfrak{M}_3 \oplus \cdots$, E the projection of $\mathfrak{H}$ on $\mathfrak{M}$; and let T be a closed linear transformation which is reduced by every manifold of the sequence $\{\mathfrak{M}_n\}$. Then $\mathfrak{M}$ reduces T. An element f in $\mathfrak{M}$ belongs to the domain of T if and only if the sequence $\{T(E_1 + \cdots + E_n)f\}$ exists and is convergent; if f is an element in $\mathfrak{M}$ which belongs to the* *domain of T, then $T(E_1 + \cdots + E_n)f \to Tf$. The case $\mathfrak{M} \equiv \mathfrak{H}$ is to be noted.*

If $m \geqq n$ we have

$$\begin{aligned}&(E_1 + \cdots + E_m)(E_1 + \cdots + E_n)\\ &\equiv (E_1 + \cdots + E_n)(E_1 + \cdots + E_m)\\ &\equiv (E_1^2 + \cdots + E_n^2) \equiv (E_1 + \cdots + E_n)\end{aligned}$$

so that

$$(E_1 + \cdots + E_n) \leqq (E_1 + \cdots + E_m).$$

By Theorem 2.40, the projection $E_1 + \cdots + E_n$ tends to a limit when $n \to \infty$; since $E_1 + \cdots + E_n$ is the projection of $\mathfrak{H}$ on $\mathfrak{M}_1 \oplus \cdots \oplus \mathfrak{M}_n$, we see that the limit is the projection E. Thus, if f is an element of $\mathfrak{M}$ such that $T(E_1 + \cdots + E_n)f$ exists and is convergent to a limit f^* we conclude that Tf exists and is equal to f^*; for we have

$$(E_1 + \cdots + E_n)f \to Ef = f, \quad T(E_1 + \cdots + E_n)f \to f^*,$$

and T is a closed linear transformation. If f is an element of $\mathfrak{M}$ such that Tf exists, then

$$T(E_1 + \cdots + E_n)f = (E_1 + \cdots + E_n)Tf \to ETf;$$

according to the result just established this means that Tf is equal to the limit element ETf. More generally, if f is an arbitrary element in the domain of T, we have

$$T(E_1 + \cdots + E_n)f = (E_1 + \cdots + E_n)Tf;$$

since

$$\begin{aligned}(E_1 + \cdots + E_n)f \to Ef, \quad &T(E_1 + \cdots + E_n)f\\ &= (E_1 + \cdots + E_n)Tf \to ETf,\end{aligned}$$

we see that TEf exists and is equal to ETf. Thus $\mathfrak{M}$ reduces T.

In the following theorem we indicate a connection between the spectral theory developed in the preceding section and the concept of reducibility now under consideration. It will be applied to the study of self-adjoint transformations subsequently.

THEOREM 4.27. *Let T be a closed linear transformation with the resolvent R_l. A necessary and sufficient condition that a closed linear manifold $\mathfrak{M}$ reduce T is that it reduce R_l.*

If $\mathfrak{M}$ reduces T it reduces T_l, since T and T_l have the same domain and $ETf = TEf$ implies $ET_lf = T_lEf$. If l is a value such that $R_l \equiv T_l^{-1}$ exists we know that R_lEg and ER_lg exist for every g in $\mathfrak{H}$. Since $T_l(R_lg)$ exists, we see that $T_l(ER_lg)$ exists and is equal to Eg. Hence

$$R_l(T_lER_lg) = R_lEg, \quad ER_lg = R_lEg$$

for every g in $\mathfrak{H}$; and $\mathfrak{M}$ reduces R_l.

Conversely, if $\mathfrak{M}$ reduces R_l, it must reduce T. If f is in the domain of T, we can write $f = R_lg$ for some g in $\mathfrak{H}$, l being fixed. Hence $Ef = ER_lg = R_lEg$ is also in the domain of T, and

$$\begin{aligned} TEf &= T(ER_lg) = T(R_lEg) = Eg + lR_lEg \\ &= E(g + lR_lg) = E(TR_lg) = ETf, \end{aligned}$$

as we were to prove.

CHAPTER V

SELF-ADJOINT TRANSFORMATIONS

§ 1. Analytical Methods

The analysis developed in the present chapter depends upon methods originated by Stieltjes and extended by Carleman. In a celebrated memoir on continued fractions, Stieltjes introduced the integral to which his name is now attached and discussed the character of certain analytic functions defined in terms of it.* Since there is an intimate connection between the theory of continued fractions and that of symmetric transformations in Hilbert space,† it is not surprising that the methods devised by Stieltjes should be suitable for treating the problem considered here. Carleman, in fact, has extended the methods of Stieltjes and has applied them successfully to a general theory of integral operators and integral equations. His monograph on the subject marks the first substantial advance in the general theory of non-bounded symmetric transformations.‡ More recent developments along the same lines have been published by Wintner, who uses the terminology of matrix theory.§ The theorems which are proved below

* Stieltjes, Annales de la Faculté des Sciences de Toulouse, (1) 8 (1894), pp. J1–J122.

† We shall describe the connection in Chapter X and apply our general theory to the study of continued fractions. For a brief account, the reader is referred to Hellinger and Toeplitz, *Integralgleichungen und Gleichungen mit unendlichvielen Unbekannten*, Encyclopädie der Mathematischen Wissenschaften 2^3, pp. 1586–1590.

‡ Carleman, *Sur les équations intégrales singulières à noyau réel et symétrique*, Uppsala, 1923.

§ Wintner, *Spektraltheorie der unendlichen Matrizen*, Leipzig, 1929, Chapter VI.

carry the concepts introduced by Stieltjes and elaborated by his followers to a fitting culmination in a complete characterization of the class of all self-adjoint transformations. It is possible to obtain these results by methods conceived in quite a different spirit. Two schemes of proof distinct from the one given here have been devised by J. v. Neumann* and F. Riesz† respectively.

In the present section we shall give a descriptive survey of the analytical methods in question. We begin with a discussion of the Stieltjes integral and then pass to the examination of a certain Stieltjes integral depending analytically upon a parameter.

We denote by Δ the interval $\alpha \leqq \lambda \leqq \beta$, the improper values $\alpha = -\infty$ and $\beta = +\infty$ being admitted for convenience. We use the symbol $\mathfrak{D}$ to denote a finite collection of intervals $\Delta_1, \cdots, \Delta_n$, no two of which have a common interior point. Such a collection is said to be a subdivision of Δ if $\Delta = \Delta_1 + \cdots + \Delta_n$.

If $\varrho(\lambda)$ is an arbitrary complex-valued function of the real variable λ defined on the open interval $-\infty < \lambda < +\infty$, we write $\varrho(\Delta) = \varrho(\beta) - \varrho(\alpha)$ when Δ is a finite interval. When $\mathfrak{D}$ is a collection of intervals $\Delta_1, \cdots, \Delta_n$ contained in Δ, we write $V_{\mathfrak{D}}(\varrho; \Delta) = \sum_{\gamma=1}^{n} |\varrho(\Delta_\gamma)|$. The set of real numbers $V_{\mathfrak{D}}(\varrho; \Delta)$ formed for all such collections $\mathfrak{D}$ has a finite or infinite least upper bound, called the variation of the function $\varrho(\lambda)$ on the interval Δ and denoted by $V(\varrho; \Delta)$. We have $0 \leqq V(\varrho; \Delta) \leqq +\infty$. The variation of the function $\varrho(\lambda)$ over a semi-infinite interval or an infinite interval is defined in a similar manner; we shall give details in the case of the interval $-\infty < \lambda < +\infty$. In the latter case, we consider the least upper bound of the set of numbers $V_{\mathfrak{D}}(\varrho) = \sum_{\gamma=1}^{n} |\varrho(\Delta_\gamma)|$ formed for all collections $\mathfrak{D}$ containing only finite intervals; this quantity is called the

* J. v. Neumann, Mathematische Annalen, 102 (1929), pp. 49–131.

† F. Riesz, Acta Litterarum ac Scientiarum, Sectio Scientiarum Mathematicarum, Szeged, 5 (1930), pp. 19–54.

variation of the function $\varrho(\lambda)$ and is denoted by $V(\varrho)$. If $\varrho, \varrho_1, \varrho_2$ are arbitrary functions and a is an arbitrary complex number, the relations

$$V(\mathfrak{R}\varrho) \leqq V(\varrho), \quad V(\mathfrak{J}\varrho) \leqq V(\varrho), \quad V(a\varrho) = |a| V(\varrho),$$
$$V(\varrho_1 + \varrho_2) \leqq V(\varrho_1) + V(\varrho_2),$$

follow directly from the definitions. We note also that the variation of $\varrho(\lambda)$ over a finite or semi-infinite interval cannot exceed $V(\varrho)$. A function is said to be of bounded variation if $V(\varrho) < +\infty$.

By virtue of the relations just indicated, the class $\mathfrak{B}$ of all functions of bounded variation is seen to be a linear class, containing $a\varrho$ and $\varrho_1 + \varrho_2$ whenever it contains $\varrho, \varrho_1, \varrho_2$; this class is seen also to contain $\mathfrak{R}\varrho$ and $\mathfrak{J}\varrho$ whenever it contains ϱ. We observe that $\mathfrak{B}$ becomes a metric space when the quantity $V(\varrho_1 - \varrho_2)$ is introduced as the distance between ϱ_1 and ϱ_2, although we shall make no further use of this fact. If ϱ is an arbitrary element of $\mathfrak{B}$, then it can be represented in the form $\varrho = (\varrho_1 - \varrho_2) + i(\varrho_3 - \varrho_4)$ where $\varrho_1, \varrho_2, \varrho_3, \varrho_4$ are real monotone functions in $\mathfrak{B}$ which do not decrease as λ increases. In the case of real functions this representation assumes the well-known form $\varrho = \varrho_1 - \varrho_2$; the more general case of complex-valued functions is deduced from this by considering the functions $\mathfrak{R}\varrho$ and $\mathfrak{J}\varrho$ separately. It is now evident that when $\varrho(\lambda)$ is a function in $\mathfrak{B}$ the limits

$$\varrho(+\infty) = \lim_{\lambda \to +\infty} \varrho(\lambda), \quad \varrho(-\infty) = \lim_{\lambda \to -\infty} \varrho(\lambda),$$
$$\varrho(\lambda + 0) = \lim_{\varepsilon \to 0} \varrho(\lambda + \varepsilon), \quad \varrho(\lambda - 0) = \lim_{\varepsilon \to 0} \varrho(\lambda - \varepsilon), \quad \varepsilon > 0,$$

exist. The set of points at which the equations $\varrho(\lambda - 0) = \varrho(\lambda) = \varrho(\lambda + 0)$ are not satisfied is empty, finite, or denumerably infinite. With a given function $\varrho(\lambda)$ in $\mathfrak{B}$ we can therefore associate the function $\varrho^*(\lambda) = \varrho(\lambda + 0) - \varrho(-\infty)$. Since $\varrho^*(\lambda)$ is continuous on the right, we see that $V(\varrho^*)$ is the least upper bound of the set of numbers $V_{\mathfrak{D}}(\varrho^*)$ formed for all collections of finite intervals whose end-points are distinct from the points of discontinuity of ϱ and of ϱ^*: for if $\mathfrak{D}$ is a collection which does not satisfy this condition we can shift each interval contained in it slightly to the right and thus obtain a new collection $\mathfrak{D}'$ of the type desired with the property that $V_{\mathfrak{D}'}(\varrho^*)$ differs only slightly from $V_{\mathfrak{D}}(\varrho^*)$. For every collection of the type now admitted the equation $V_{\mathfrak{D}}(\varrho^*) = V_{\mathfrak{D}}(\varrho)$ is satisfied. It follows immediately that $V(\varrho^*) \leqq V(\varrho)$ and that ϱ^* belongs to $\mathfrak{B}$. We remark that in certain cases the inequality sign must be used; the function $\varrho(\lambda)$ which assumes the value zero or one according as λ is different from or equal to zero is an example, since we have here $\varrho^*(\lambda) \equiv 0$, $V(\varrho) = 2$, $V(\varrho^*) = 0$. The properties of the operation $*$ indicated in the equations

$$(\varrho^*)^* = \varrho^*, \quad (a\varrho)^* = a\varrho^*, \quad (\varrho_1 + \varrho_2)^* = \varrho_1^* + \varrho_2^*$$

are readily established, as are the relations

$$\varrho^*(-\infty) = 0, \qquad \varrho^*(\lambda + 0) = \varrho^*(\lambda)$$

A function in $\mathfrak{B}$ such that $\varrho(\lambda) \equiv \varrho^*(\lambda)$ is said to be in normal form: a necessary and sufficient condition that a function in $\mathfrak{B}$ be in normal form is that it vanish at $\lambda = -\infty$ and be continuous on the right. The class $\mathfrak{B}^*$ of all functions which belong to $\mathfrak{B}$ and are in normal form is a linear subclass of $\mathfrak{B}$. We note that when $\varrho(\lambda)$ is in $\mathfrak{B}^*$ the inequality $|\varrho(\lambda)| = |\varrho(\lambda) - \varrho(-\infty)| \leqq V(\varrho)$ is satisfied.†

We are now prepared to define the Stieltjes integral of a bounded continuous complex-valued function $F(\lambda)$, $-\infty < \lambda < +\infty$, with respect to a function $\varrho(\lambda)$ of bounded variation. Let $\{\mathfrak{D}_n\}$ be a sequence of subdivisions of an arbitrary interval Δ, and let Δ_{nk}, $k = 1, \cdots, n$, be the intervals belonging to $\mathfrak{D}_n$. The sequence $\{\mathfrak{D}_n\}$ will be called admissible if the longest interval Δ_{nk} with end-points between

† The facts summarized in this paragraph and the preceding one are, for the most part, fundamental in the theory of functions of a real variable; cf. Hobson, *The Theory of Functions of a Real Variable*, vol. I, third edition, Cambridge, 1927, pp. 325–330, 337–338, 341–342, or Carathéodory, *Vorlesungen über reelle Funktionen*, second edition, Leipzig, 1927, pp. 180–191.

$-A$ and A has length tending to zero with $1/n$, for every positive A. If $\{\mathfrak{D}_n\}$ is an admissible sequence and if λ_{nk} is a point of the interval Δ_{nk}, then the sum $\sum_{\gamma=1}^{n} F(\lambda_{n\gamma})\,\varrho(\Delta_{n\gamma})$ tends with increasing n to a limit which depends only upon Δ, $\varrho(\lambda)$, and the values assumed by $F(\lambda)$ on the interval Δ; this limit is called the Stieltjes integral of $F(\lambda)$ with respect to $\varrho(\lambda)$ over the interval Δ and is denoted by the symbol $\int_{\Delta} F(\lambda)\,d\varrho(\lambda)$ or the symbol $\int_{\alpha}^{\beta} F(\lambda)\,d\varrho(\lambda)$. We define $\int_{\beta}^{\alpha} F(\lambda)\,d\varrho(\lambda)$ as $-\int_{\alpha}^{\beta} F(\lambda)\,d\varrho(\lambda)$, in the usual way. We observe that the case of an infinite or semi-infinite interval Δ is included under the preceding statements, provided that we restrict the numbers λ_{nk} to proper values and extend the notation $\varrho(\Delta) = \varrho(\beta) - \varrho(\alpha)$ to cover the cases $\alpha = -\infty$ and $\beta = +\infty$. Thus the integral $\int_{-\infty}^{+\infty} F(\lambda)\,d\varrho(\lambda)$ is not regarded as an improper integral, and the equation

$$\int_{-\infty}^{+\infty} F(\lambda)\,d\varrho(\lambda) = \lim_{\substack{\alpha\to-\infty\\ \beta\to+\infty}} \int_{\alpha}^{\beta} F(\lambda)\,d\varrho(\lambda)$$

is a simple theorem and not a definition of the term on the left. On the other hand, we shall treat the relations

$$\int_{-\infty}^{+\infty} \lambda^k\,d\varrho(\lambda) = \lim_{\substack{\alpha\to-\infty\\ \beta\to+\infty}} \int_{\alpha}^{\beta} \lambda^k\,d\varrho(\lambda), \qquad k = 1, 2, 3, \cdots,$$

$$\int_{-\infty}^{+\infty} \frac{1}{\lambda-\mu}\,d\varrho(\lambda) = \lim_{\varepsilon\to 0}\left[\int_{-\infty}^{\mu-\varepsilon} + \int_{\mu+\varepsilon}^{+\infty}\right] \frac{1}{\lambda-\mu}\,d\varrho(\lambda),\ \varepsilon > 0,$$

as definitions of the improper integrals on the left whenever the respective right-hand members are significant. Since the integral $\int_{\Delta} F(\lambda)\,d\varrho(\lambda)$ is independent of the values assumed by $F(\lambda)$ outside the interval Δ, we may relinquish the requirement that $F(\lambda)$ be defined for all values of λ, $-\infty < \lambda < +\infty$, and demand merely that $F(\lambda)$ be defined on the finite part of the interval Δ and be bounded and continuous there; conversely, if $F(\lambda)$ is given only on the interval Δ, we can define the function arbitrarily outside Δ so as to be bounded and continuous for all values of λ, $-\infty<\lambda<+\infty$, without affecting the value of the integral of $F(\lambda)$ with respect to $\varrho(\lambda)$ over Δ. For the purposes of the present chapter the form of statement given above is the most convenient. The principal properties of the Stieltjes integral may now be formulated.

LEMMA 5.1. *If $F(\lambda)$ and $G(\lambda)$ are bounded continuous complex-valued functions on the range $-\infty<\lambda<+\infty$, if $|F(\lambda)|$ and $|G(\lambda)|$ have on the interval Δ the respective least upper bounds $M(\Delta)$ and $N(\Delta)$, and if $\varrho(\lambda)$, $\varrho_1(\lambda)$, $\varrho_2(\lambda)$ are complex-valued functions of bounded variation, then we have:*

(1) $\left|\int_\Delta F(\lambda)\,d\varrho(\lambda)\right| \leqq M(\Delta)\,V(\varrho;\Delta) \leqq M(\Delta)\,V(\varrho)$;

(2) *for arbitrary values of α, β, γ,*

$$\int_\alpha^\beta F(\lambda)\,d\varrho(\lambda)+\int_\beta^\gamma F(\lambda)\,d\varrho(\lambda) = \int_\alpha^\gamma F(\lambda)\,d\varrho(\lambda);$$

(3) $\int_\Delta aF(\lambda)\,d\varrho(\lambda) = a\int_\Delta F(\lambda)\,d\varrho(\lambda)$;

(4) $\int_\Delta [F(\lambda)+G(\lambda)]\,d\varrho(\lambda) = \int_\Delta F(\lambda)\,d\varrho(\lambda)+\int_\Delta G(\lambda)\,d\varrho(\lambda)$;

(5) *if $\sigma(\lambda) = \int_\alpha^\lambda F(\mu)\,d\varrho(\mu)$, then $V(\sigma;\Delta) \leqq M(\Delta)\,V(\varrho;\Delta)$ and $\sigma(\lambda)$ belongs to $\mathfrak{B}$;*

(6) *if $\sigma(\lambda)$ is the function introduced in* (5), *then*

$$\int_\Delta G(\lambda)\,d\sigma(\lambda) = \int_\Delta F(\lambda)\,G(\lambda)\,d\varrho(\lambda);$$

(7) *for all values of α and β, the difference $\int_\alpha^\beta F(\lambda)\,d\varrho^*(\lambda) - \int_\alpha^\beta F(\lambda)\,d\varrho(\lambda)$ is equal to $F(\beta)\,[\varrho(\beta+0)-\varrho(\beta)] - F(\alpha)\,[\varrho(\alpha+0)-\varrho(\alpha)]$, provided that when $\alpha = \pm\infty$ or $\beta = \pm\infty$ the corresponding term in the last expression be equated to zero;*

(8) *if $\sigma(\lambda)$ is the function introduced in* (5), *then*

$$\sigma^*(\lambda) = \int_{-\infty}^\lambda F(\mu)\,d\varrho^*(\mu);$$

(9) *if* $F(\lambda)$ *is continuous at* $\lambda = -\infty$ *and* $\lambda = +\infty$ *and if* $F'(\lambda)$ *is bounded and continuous,* $-\infty < \lambda < +\infty$, *then*

$$\int_\alpha^\beta F(\lambda)\, d\varrho(\lambda) = [F(\beta)\,\varrho(\beta) - F(\alpha)\,\varrho(\alpha)] - \int_\alpha^\beta F'(\lambda)\,\varrho(\lambda)\, d\lambda;$$

(10)
$$\int_\Delta F(\lambda)\, d(a_1\varrho_1(\lambda) + a_2\varrho_2(\lambda)) = a_1\int_\Delta F(\lambda)\, d\varrho_1(\lambda) + a_2\int_\Delta F(\lambda)\, d\varrho_2(\lambda).$$

We shall not give detailed proofs of any of these properties. Evidently those numbered (1)-(4), (10) are elementary consequences of the fundamental definitions introduced above. The inequality stated in (5) follows almost immediately from (1) and (2). The relation given in (6) is not quite so evident, so we outline the demonstration. If $\{\mathfrak{D}_n\}$ is an admissible sequence of subdivisions of Δ, we have

$$\begin{aligned}\left|\int_\Delta G\, d\sigma - \int_\Delta FG\, d\varrho\right| &= \left|\lim_{n\to\infty} \sum_{\gamma=1}^{n} G(\lambda_{n\gamma})[\sigma(\Delta_{n\gamma}) - F(\lambda_{n\gamma})\varrho(\Delta_{n\gamma})]\right| \\ &\leqq N(\Delta) \limsup_{n\to\infty} \sum_{\gamma=1}^{n} \left|\int_{\Delta_{n\gamma}} [F(\mu) - F(\lambda_{n\gamma})]\, d\varrho(\mu)\right| \\ &\leqq N(\Delta) \limsup_{n\to\infty} \sum_{\gamma=1}^{n} M_{n\gamma}\, V(\varrho;\, \Delta_{n\gamma}),\end{aligned}$$

where M_{nk} is the least upper bound of $|F(\mu) - F(\lambda_{nk})|$ for values of μ in Δ_{nk}. When the postulated properties of the functions $F(\lambda)$ and $\varrho(\lambda)$ are brought to bear, the last expression is found to have the value zero. The equation given in (7) can be established by writing the difference to be evaluated as the limit of the difference between the corresponding sums formed for an admissible sequence of subdivisions $\{\mathfrak{D}_n\}$ such that the end-points of Δ_{nk} interior to Δ are points of continuity for ϱ and ϱ^*; in the latter difference all terms save those arising from the intervals which contain end-points of Δ are found to cancel, and the evaluation follows at once. The statement in (8) is an immediate consequence of the result in (7). The formula for integration by parts given in (9) is standard in the theory of the Stieltjes integral. Since the proof cannot be outlined in a few words, we refer the reader to the literature.†

The proofs of the fundamental theorems concerning self-adjoint transformations center about the properties of the Stieltjes integral $I(l; \varrho) = \int_{-\infty}^{+\infty} \frac{1}{\lambda - l}\, d\varrho(\lambda)$, where $\varrho(\lambda)$ is a complex-valued function of bounded variation and l is a complex parameter. By applying the formula for integration by parts given in Lemma 5.1 (9) we can express this integral as an improper Riemann integral $I(l; \varrho) = \int_{-\infty}^{+\infty} \frac{\varrho(\lambda)}{(\lambda - l)^2}\, d\lambda$. In the sequel we shall distinguish between these two representations by calling them the Stieltjes representation and the Riemann representation respectively. By reference to Lemma 5.1 (7) we see that $I(l; \varrho^*) = I(l; \varrho)$. Thus we may suppose without loss of generality that the function $\varrho(\lambda)$ is in normal form. By doing so we are able to impart

† Most texts ignore the Stieltjes integral or treat it in a summary fashion, as we have been compelled to do here. The various definitions and generalizations of the Stieltjes integral introduced by different writers are surveyed by Hildebrandt, Bulletin of the American Mathematical Society, 14 (1917-18), pp. 177–194. Papers of similar character are listed by Montel and Rosenthal, *Integration und Differentiation*, Enzyklopädie der mathematischen Wissenschaften, 2^{3^2}, pp. 1071–73. The definition and properties of the Stieltjes integral, as described above, can be found in Hobson, *The Theory of Functions of a Real Variable*, vol. I, third edition, Cambridge, 1927, pp. 538–546, and Bray, Annals of Mathematics (2) 20 (1918), pp. 177–186. We may also mention the treatment of Lebesgue, *Leçons sur l'intégration*, Paris, 1928, pp. 252–313. These writers discuss many properties of the Stieltjes integral which we have not indicated in our descriptive statement; on the other hand, they do not refer specifically to the properties (5)–(8). Property (5) is stated without proof by Carleman, *Sur les équations intégrales singulières à noyau réel et symètrique*, Uppsala, 1923, pp. 11–12; it is proved under more general conditions by Francis, Proceedings of the Cambridge Philosophical Society, 22 (1923-25), pp. 935–950. Wintner, *Spektraltheorie der unendlichen Matrizen*, Leipzig, 1929, §§ 34–48, presents a full account of the Stieltjes integral with the same purpose as that which motivates the brief exposition above; his treatment is perhaps the one best suited to give the reader detailed information on those points which are of primary importance later in the chapter.

a particularly simple form to the statements of subsequent lemmas and theorems. At the same time we do not sacrifice facility of manipulation for the sake of this advantage, since the class $\mathfrak{B}^*$ of all functions in normal form is a linear class. In all statements which follow, it is therefore assumed that $\varrho(\lambda)$ is in $\mathfrak{B}^*$, unless a specific exception be made.

LEMMA 5.2. *The integral* $I(l;\varrho)$ *is a single-valued analytic function of* l *in each of the half-planes* $\mathfrak{J}(l)>0$, $\mathfrak{J}(l)<0$; *it satisfies the inequality*

$$|I(l;\varrho)| \leqq V(\varrho)/|\mathfrak{J}(l)|.$$

The function $\varrho(\lambda)$ *is expressible in terms of* $I(l;\varrho)$ *by means of a contour integral according to the formula*

$$\frac{1}{2}[\{\varrho(\mu)+\varrho(\mu-0)\}-\{\varrho(\nu)+\varrho(\nu-0)\}] = \lim_{\varepsilon\to 0} -\frac{1}{2\pi i}\int_{C(\mu,\nu,\alpha,\varepsilon)} I(l;\varrho)\,dl,$$

where the contour $C(\mu,\nu,\alpha,\varepsilon)$ *consists of two oriented polygonal lines whose vertices, in order, are* $\mu+i\varepsilon$, $\mu+i\alpha$, $\nu+i\alpha$, $\nu+i\varepsilon$, *and* $\nu-i\varepsilon$, $\nu-i\alpha$, $\mu-i\alpha$, $\mu-i\varepsilon$, *respectively, the real numbers* $\mu,\nu,\alpha,\varepsilon$ *being subject to the inequalities* $\nu<\mu$, $0<\varepsilon<\alpha$. *In consequence, a necessary and sufficient condition that* $I(l;\varrho_1)$ *and* $I(l;\varrho_2)$ *coincide is that* $\varrho_1(\lambda)$ *and* $\varrho_2(\lambda)$ *coincide.*†

The analytic character of $I(l;\varrho)$ is most easily deduced from the Riemann representation: the integral $\int_{-\infty}^{+\infty}\frac{\varrho(\lambda)}{(\lambda-l)^2}\,d\lambda$ converges absolutely and uniformly on any bounded closed set in the l-plane which is not intersected by the real axis, and therefore represents a function analytic in each of the half-planes $\mathfrak{J}(l)>0$, $\mathfrak{J}(l)<0$. The inequality satisfied by $I(l;\varrho)$ is established by applying Lemma 5.1 (1) to the Stieltjes representation. To determine $\varrho(\lambda)$ in terms of $I(l;\varrho)$ we substitute the Riemann representation for $I(l;\varrho)$ in the contour integral given in the statement of the lemma and then invert the order of integration. After a few elementary reductions, we obtain

† This lemma is due to Stieltjes, Annales de la Faculté des Sciences de Toulouse, (1) 8 (1894), pp. J72–J75.

$$-\frac{1}{2\pi i}\int_{C(\mu,\nu,\alpha,\varepsilon)} I(l;\varrho)\,dl$$
$$=\frac{1}{\pi}\int_{-\infty}^{+\infty}\left[\frac{\varepsilon}{(\lambda-\mu)^2+\varepsilon^2}-\frac{\varepsilon}{(\lambda-\nu)^2+\varepsilon^2}\right]\varrho(\lambda)\,d\lambda.$$

The expression on the right is the difference of two integrals analogous to the Poisson integral and its limit when ε tends to zero is readily computed.* The result is the one stated above. Thus when $\varrho(\lambda)$ is in normal form it is uniquely determined by $I(l;\varrho)$; and the relations $I(l;\varrho_1)\equiv I(l;\varrho_2)$ and $\varrho_1(\lambda)\equiv\varrho_2(\lambda)$ are equivalent.

LEMMA 5.3. *Let $F(l)$ be a single-valued function defined for all not-real l, and let $\{\varrho_n\}$ be a sequence of functions of the real variable λ, $-\infty<\lambda<+\infty$, such that*

$$V(\varrho_n)\leqq K; \qquad \lim_{n\to\infty} I(l;\varrho_n)=F(l), \qquad \mathfrak{I}(l)\neq 0.$$

Then the sequence $\{I(l;\varrho_n)\}$ converges uniformly on any bounded closed point-set in the l-plane which is not intersected by the real axis; its limit function $F(l)$ is expressible in the form $F(l)=I(l;\varrho)$ where $\varrho(\lambda)$ is a uniquely determined function in normal form satisfying the inequality $V(\varrho)\leqq K$.

Since, by the preceding lemma, we have $|I(l;\varrho_n)|\leqq V(\varrho_n)/|\mathfrak{I}(l)|\leqq K/|\mathfrak{I}(l)|$ we see that the sequence $\{I(l;\varrho_n)\}$ converges boundedly on any point-set in the l-plane which is at positive distance from the real axis, such as a set of the type described above. A well-known theorem of Vitali now ensures the uniform convergence of the sequence on any bounded closed set which is not intersected by the real axis.† We may remark that we shall not need to make any use of the fact that the sequence is uniformly convergent. In order to express $F(l)$ as a Stieltjes integral, we employ

* When the unit circle is replaced by a half-plane the ordinary Poisson integral is transformed into an integral of the type under consideration here. All the relevant calculations are given by Osgood, *Lehrbuch der Funktionentheorie*, second edition, Leipzig, 1912, pp. 635–641.

† The theorem in question is stated and proved by Montel, *Leçons sur les familles normales*, Paris, 1927, pp. 28–30.

a theorem due to Helly which provides an explicit construction of the function $\varrho(\lambda)$ in terms of a properly chosen subsequence of the sequence $\{\varrho_n\}$.† Since $\varrho_n(\lambda)$ is in normal form, we have $|\varrho_n| \leqq V(\varrho_n) \leqq K$. Thus $\{\varrho_n\}$ is a sequence of uniformly bounded functions of uniformly bounded variation. According to the theorem of Helly, there exists a sequence of integers $\{n_k\}$ such that

$$n_k \to \infty, \qquad \varrho_{n_k}(\lambda) \to \varrho(\lambda), \qquad k \to \infty.$$

The limit-function $\varrho(\lambda)$ is defined for every value of λ, $-\infty < \lambda < +\infty$, and is a function of bounded variation satisfying the inequalities $|\varrho| \leqq K$ and $V(\varrho) \leqq K$. Obviously, there is no reason why $\varrho(\lambda)$ should be in normal form. We can now show that $\lim\limits_{k\to\infty} I(l; \varrho_{n_k}) = I(l; \varrho)$, $\mathfrak{I}(l) \neq 0$. If the term on the left is expressed by means of the Riemann representation, we see that our task becomes the one of showing that the sequence of integrands $\dfrac{\varrho_{n_k}(\lambda)}{(\lambda - l)^2}$ can be integrated term-by-term. Clearly, this sequence converges to the limit $\dfrac{\varrho(\lambda)}{(\lambda - l)^2}$ and is dominated by the absolutely integrable function $\dfrac{K}{|\lambda - l|^2}$. According to a theorem of Lebesgue, the sequence can be integrated term-wise, as we wished to prove.‡ Thus $F(l) = I(l; \varrho)$. We must now replace ϱ by ϱ^* in the latter expression so as to obtain the representation specified in the statement of the lemma. By Lemma 5.2 this representation is clearly unique.

§ 2. Analytical Representation of the Resolvent

In order to apply the tools provided in the preceding section to the study of a given self-adjoint transformation H,

---

† Helly, Sitzungsberichte, Akademie der Wissenschaften zu Wien, Mathematische-Naturwissenschaftliche Klasse, 121[2a1] (1912), p. 283–290.

‡ This theorem is stated and proved by de la Vallée Poussin, *Intégrales de Lebesgue*, Paris, 1916, p. 49.

we construct a sequence $\{H^{(n)}\}$ of self-adjoint transformations which approximate to H in a sense to be specified presently.

First, we determine a linear manifold $\mathfrak{M}$ which has in relation to the given transformation H the following properties:

(1) $\mathfrak{M}$ is a subset of the domain of H;

(2) $\mathfrak{M}$ is everywhere dense in $\mathfrak{H}$;

(3) the linear manifolds into which $\mathfrak{M}$ is carried by the transformations $H_i \equiv H - iI$ and $H_{-i} \equiv H + iI$ respectively are both everywhere dense in $\mathfrak{H}$.

On the basis of Theorem 1.18 we can select a sequence $\{f_n\}$ which is everywhere dense in the domain of H and in consequence everywhere dense in $\mathfrak{H}$ as well. When $l = \pm i$ the resolvent of H exists so that the sequences $\{R_l f_n\}$ are significant and consist entirely of elements in the domain of H; the requisite facts concerning the resolvent have already been established in Theorem 4.18. We arrange the three sequences $\{f_n\}$, $\{R_l f_n\}$, $l = \pm i$, as a single sequence $\{g_p\}$ and define $\mathfrak{M}$ as the linear manifold determined by this sequence. The manifold $\mathfrak{M}$ has the required properties: it is a subset of the domain of the linear transformation H because the sequence $\{g_p\}$ consists of elements of that domain; it is everywhere dense in $\mathfrak{H}$ because it contains the sequence $\{f_n\}$; the linear manifolds into which it is carried by the transformations H_l, $l = \pm i$, are everywhere dense in $\mathfrak{H}$, since each contains the sequence $\{f_n\}$, $f_n = H_l R_l f_n$, where $R_l f_n$ is in $\mathfrak{M}$, $l = \pm i$.

Next, we select an orthonormal set $\{\varphi_n\}$ which determines the linear manifold $\mathfrak{M}$; we can do so conveniently by applying to the sequence $\{g_p\}$ the process of orthogonalization outlined in Theorem 1.13. We denote by $\mathfrak{M}_n$ the closed linear manifold determined by the set $(\varphi_1, \cdots, \varphi_n)$ and by E_n the projection of $\mathfrak{H}$ on $\mathfrak{M}_n$. When $m \geqq n$ we have $\mathfrak{M}_n \subseteqq \mathfrak{M}_m$, $E_n \leqq E_m$. In consequence, we can apply Theorem 2.40 to establish the existence of a projection E such that $E_n \to E$. In view of the fact that the orthonormal set $\{\varphi_n\}$ is complete, it is clear that E is the identity I.

On the basis of these preliminary considerations we can prove the key theorems of the present chapter.

THEOREM 5.1. *If H is an arbitrary self-adjoint transformation and $\mathfrak{M}$, $\mathfrak{M}_n$, E_n have the meanings described above, then the transformation $H^{(n)} \equiv E_n H E_n$ is a bounded self-adjoint transformation with domain $\mathfrak{H}$, such that $H^{(n)} \to H$ in the linear manifold $\mathfrak{M}$.*

Since the range of E_n is the closed linear manifold $\mathfrak{M}_n$, lying in the domain of H, it is evident that the transformation $H^{(n)}$ is a linear transformation with $\mathfrak{H}$ as its domain. In view of the relations $(E_n H E_n f, g) = (H E_n f, E_n g) = (E_n f, H E_n g) = (f, E_n H E_n g)$ it is obvious that $H^{(n)}$ is symmetric. Theorem 2.25 now assures us that $H^{(n)}$ is a bounded self-adjoint transformation, as we can verify independently by direct computation.

If f is an element of the linear manifold $\mathfrak{M}$, it is expressible in the form $f = \sum_{\alpha=1}^{N} a_\alpha \varphi_\alpha$, which shows that f belongs to $\mathfrak{M}_n$ when $n \geqq N$. Thus $H^{(n)} f = E_n H E_n f = E_n H f$, $n \geqq N$; and, since $E_n \to I$, $H^{(n)} f \to H f$ when $n \to \infty$. The limiting relation $H^{(n)} \to H$ is thus seen to hold throughout $\mathfrak{M}$.

It is of some interest to comment on another relation between $H^{(n)}$ and H, which we do not intend to use directly. If $A(\varphi) = \{a_{pq}\}$ and $A^{(n)}(\varphi) = \{a_{pq}^{(n)}\}$ are the matrices associated with H and $H^{(n)}$ respectively by the orthonormal set $\{\varphi_n\}$, it is evident upon inspection that

$$a_{pq}^{(n)} = a_{pq}, \qquad p, q = 1, \cdots, n;$$
$$a_{pq}^{(n)} = 0, \qquad p > n \quad \text{or} \quad q > n.$$

Thus the matrix $A^{(n)}$ is obtained from A by replacing all rows and columns after the first n by rows and columns of zeros. Our choice of the set $\{\varphi_n\}$ enables us to determine H completely from the matrix A.

THEOREM 5.2. *There exists a complex-valued function $\varrho_n(\lambda)$, dependent upon f and g, such that*

$$V(\varrho_n) \leqq |f|\,|g|, \qquad (R_l^{(n)} f, g) = I(l; \varrho_n),$$

where $R_l^{(n)}$ is the resolvent of the self-adjoint transformation $H^{(n)}$, l is not real, and f and g are arbitrary.

The proof of this theorem, which is equivalent to the purely algebraic theorem that the matrix $A^{(n)}$ described at the close of the preceding theorem can be reduced to diagonal form, will be made to depend upon an analysis of the reducibility of the transformation $H^{(n)}$. By reference to Definition 4.5 and to Theorem 4.23, we see that the closed linear manifolds $\mathfrak{M}_n$ and $\mathfrak{H} \ominus \mathfrak{M}_n$ reduce $H^{(n)}$. Furthermore, if f is in $\mathfrak{H} \ominus \mathfrak{M}_n$, we have $(I - E_n)f = f$ and, in consequence,

$$H^{(n)} f = E_n H E_n (I - E_n) f = E_n H O f = 0;$$

in other words, $\mathfrak{H} \ominus \mathfrak{M}_n$ is contained in a characteristic manifold of $H^{(n)}$ corresponding to the characteristic value $l = 0$.

If l is a characteristic value of $H^{(n)}$ different from zero and f a corresponding characteristic element, we see at once that f is orthogonal to every element of $\mathfrak{H} \ominus \mathfrak{M}_n$ by Theorem 4.13 and must therefore belong to the manifold $\mathfrak{M}_n$. Now $\mathfrak{M}_n$, being an n-dimensional linear manifold, contains no orthonormal set with more than n elements; by referring once more to Theorem 4.13 we see that $H^{(n)}$ can have no more than n characteristic values different from zero and thus at most $n+1$ in all. Since the $N \leqq n+1$ characteristic values of $H^{(n)}$ are all real, we can denote them in order of magnitude as $l_1 < \cdots < l_N$. We introduce the characteristic manifold $\mathfrak{N}_k$ of $H^{(n)}$ corresponding to $l = l_k$, and the projection F_k of $\mathfrak{H}$ on $\mathfrak{N}_k$, $k = 1, \cdots, N$. In view of the closure of the transformation $H^{(n)}$, it is evident by Theorem 4.9 that every element of $\mathfrak{N}_k$ save the null element is a characteristic element of $H^{(n)}$ corresponding to the characteristic value $l = l_k$. We see by Theorem 4.24 that $\mathfrak{N}_k$ reduces $H^{(n)}$. Since the closed linear manifolds $\mathfrak{N}_1, \cdots, \mathfrak{N}_N$ are mutually orthogonal, the closed linear manifold $\mathfrak{N}_1 \oplus \cdots \oplus \mathfrak{N}_N$ and its orthogonal complement $\mathfrak{N}_{N+1} = \mathfrak{H} \ominus [\mathfrak{N}_1 \oplus \cdots \oplus \mathfrak{N}_N]$ both reduce $H^{(n)}$. We shall show that $\mathfrak{N}_{N+1}$ is the manifold $\mathfrak{O}$.

Since $\mathfrak{H} \ominus \mathfrak{M}_n$ is a subset of some one of $\mathfrak{N}_1, \cdots, \mathfrak{N}_N$ we see that $\mathfrak{N}_{N+1}$ is a subset of $\mathfrak{M}_n$ and that accordingly its dimension number m must satisfy the inequality $0 \leqq m \leqq n$. If $m = 0$, then $\mathfrak{N}_{N+1} = \mathfrak{O}$, so that we can gain our end by showing that the relation $1 \leqq m \leqq n$ is impossible. If we assume tentatively that the latter inequality is true, we can select an orthonormal set $(\psi_1, \cdots, \psi_m)$ which determines the closed linear manifold $\mathfrak{N}_{N+1}$; an arbitrary element f in $\mathfrak{N}_{N+1}$ can then be expressed uniquely as a sum $f = \sum_{\alpha=1}^{m} a_\alpha \psi_\alpha$. Since $\mathfrak{N}_{N+1}$ reduces $H^{(n)}$, the element $H^{(n)} \psi_k$ is in $\mathfrak{N}_{N+1}$ and can be written in the form $H^{(n)} \psi_k = \sum_{\alpha=1}^{m} c_{k\alpha} \psi_\alpha$, $k = 1, \cdots, m$. For a general element in $\mathfrak{N}_{N+1}$ we then have $H^{(n)} f = \sum_{\alpha, \beta=1}^{m} c_{\alpha\beta}\, a_\alpha \psi_\beta$. We can determine the element f so as to be a characteristic element of $H^{(n)}$ corresponding to the characteristic value l if and only if we can select the numbers $l, a_1, \cdots, a_m$ so that

$$\sum_{\alpha=1}^{m} c_{\alpha k}\, a_\alpha = l a_k, \qquad \sum_{\alpha=1}^{m} | a_\alpha |^2 \neq 0,$$

since these relations are equivalent to the relations $H^{(n)} f = lf$, $\| f \| \neq 0$. Such numbers can be found if and only if l is selected as a root of the algebraic equation $\det (c_{ik} - l \delta_{ik}) = 0$. Since the determinant is a polynomial of degree $m \geqq 1$, it has at least one root. Thus the manifold $\mathfrak{N}_{N+1}$ must contain at least one characteristic value of $H^{(n)}$, contrary to the property by which it was defined. Hence our assumption that $1 \leqq m \leqq n$ is impossible and must be abandoned.

We can now obtain the desired analytic representation of the resolvent $R_l^{(n)}$ of the transformation $H^{(n)}$. Since $\mathfrak{N}_k$ reduces $H^{(n)}$ it reduces $R_l^{(n)}$ as well, in accordance with Theorem 4.27; and, since $\mathfrak{N}_1 \oplus \cdots \oplus \mathfrak{N}_N = \mathfrak{H}$ from the results just obtained, the projections $F_1, \cdots, F_N$ satisfy the relations

$$F_i F_k \equiv O, \qquad i \neq k; \qquad F_1 + \cdots + F_N \equiv I.$$

Consequently we can write

$$R_l^{(n)} f = \sum_{\alpha=1}^{N} F_\alpha R_l^{(n)} f = \sum_{\alpha=1}^{N} R_l^{(n)} F_\alpha f.$$

Now $R_l^{(n)} F_k f$ is evidently an element of $\mathfrak{N}_k$. Thus the two equations

$$H^{(n)} R_l^{(n)} F_k f = l_k R_l^{(n)} F_k f,$$
$$H^{(n)} R_l^{(n)} F_k f = F_k f + l R_l^{(n)} F_k f,$$

are valid and together imply that $R_l^{(n)} F_k f = \dfrac{1}{l_k - l} F_k f$. The resolvent can therefore be represented as a finite sum

$$R_l^{(n)} f = \sum_{\alpha=1}^{N} \frac{1}{l_\alpha - l} F_\alpha f, \qquad l \neq l_1, \cdots, l_N.$$

To complete our proof of the present theorem we must represent the sum $(R_l^{(n)} f, g) = \sum_{\alpha=1}^{N} \dfrac{1}{l_\alpha - l} (F_\alpha f, g)$ as a Stieltjes integral. For this purpose we have only to introduce the function $\varrho_n = \varrho_n(\lambda; f, g)$ through the defining equations

$$\varrho_n(\lambda) = 0, \qquad -\infty < \lambda < l_1;$$
$$\varrho_n(\lambda) = \sum_{\alpha=1}^{k} (F_\alpha f, g), \qquad l_k \leqq \lambda < l_{k+1}, \qquad k = 1, \cdots, N-1;$$
$$\varrho_n(\lambda) = \sum_{\alpha=1}^{N} (F_\alpha f, g) = (f, g), \qquad l_N \leqq \lambda < +\infty;$$

in terms of this function we have $(R_l^{(n)} f, g) = I(l; \varrho_n)$. As λ increases from $-\infty$ to $+\infty$, the function $\varrho_n(\lambda)$ remains constant save for an increment $(F_k f, g) = (F_k f, F_k g)$ acquired at the point $\lambda = l_k$, $k = 1, \cdots, N$. Thus its variation must satisfy the relations

$$V(\varrho_n) = \sum_{\alpha=1}^{N} |(F_\alpha f, F_\alpha g)| \leqq \sum_{\alpha=1}^{N} \| F_\alpha f \| \cdot \| F_\alpha g \|$$
$$\leqq \left[\sum_{\alpha=1}^{N} \| F_\alpha f \|^2 \cdot \sum_{\alpha=1}^{N} \| F_\alpha g \|^2 \right]^{1/2} = \| f \| \cdot \| g \|.$$

This completes the proof of the theorem.

THEOREM 5.3. *If H is a self-adjoint transformation, $H^{(n)}$ the related transformation described above, and R_l and $R_l^{(n)}$ are the respective resolvents of these transformations, then*

$$(R_l^{(n)} f, g) \to (R_l f, g), \qquad \mathfrak{J}(l) \neq 0$$

for arbitrary elements f and g.

Let l be a fixed not-real point and let g be an element of the linear manifold $\mathfrak{M}(\bar{l})$ into which $\mathfrak{M}$ is carried by the transformation $H_{\bar{l}} \equiv H - \bar{l} \cdot I$. If we put $g = H_{\bar{l}} h$ where h is in $\mathfrak{M}$, we have

$$\begin{aligned}(R_l^{(n)} f, g) - (R_l f, g) &= (R_l^{(n)} f, g) - (f, h) \\ &= (R_l^{(n)} f, H_{\bar{l}} h) - (R_l^{(n)} f, H_{\bar{l}}^{(n)} h);\end{aligned}$$

and with the aid of the inequality satisfied by the resolvent $R_l^{(n)}$, we can obtain the inequalities

$$\begin{aligned}|(R_l^{(n)} f, g) - (R_l f, g)| &= |(R_l^{(n)} f, (H_{\bar{l}} - H_{\bar{l}}^{(n)}) h)| \\ &\leqq \|R_l^{(n)} f\| \|(H - H^{(n)}) h\| \\ &\leqq \|f\| \|(H - H^{(n)}) h\| / |\mathfrak{J}(l)|.\end{aligned}$$

Since $H^{(n)} \to H$ in $\mathfrak{M}$ according to Theorem 5.1 and since h is an element of $\mathfrak{M}$, we see that the right-hand member of this inequality tends to zero with $1/n$ and that $(R_l^{(n)} f, g)$ converges to the limit $(R_l f, g)$.

In order to remove the restrictions imposed upon the element g we shall now prove that the linear manifold $\mathfrak{M}(\bar{l})$ is everywhere dense in $\mathfrak{H}$. Let T be the transformation with domain $\mathfrak{M}$ defined by the equation $T = H$ (in $\mathfrak{M}$); T is a linear symmetric transformation since $T \subseteqq H$ and since $\mathfrak{M}$ is everywhere dense in $\mathfrak{H}$. Our construction of $\mathfrak{M}$ shows that the manifolds $\mathfrak{M}(\pm i)$ into which it is carried by $T_{\pm i} = H_{\pm i}$ (in $\mathfrak{M}$) are everywhere dense in $\mathfrak{H}$; in other words, according to Theorem 4.16, the points $l = \pm i$ belong to the resolvent set of T. From Theorem 4.16 it results immediately that the resolvent set of T includes every point l not on the real axis. This fact implies that the manifold

$\mathfrak{M}(\bar{l})$, $\mathfrak{J}(l) \neq 0$, into which $\mathfrak{M}$ is carried by the transformation $T_{\bar{l}} = H_{\bar{l}}$ (in $\mathfrak{M}$) is everywhere dense in $\mathfrak{H}$.

It is now a simple matter to complete the proof of the theorem. We allow l, f, g to be arbitrary save for the restriction that l is not real. If ε is an arbitrary positive number, we can select an element g^* belonging to $\mathfrak{M}(\bar{l})$ and yet lying so close to g that the inequality $\|f\|\|g-g^*\|/|\mathfrak{J}(l)| < \varepsilon/2$ is true. We have immediately

$$|(R_l^{(n)} f, g)-(R_l f, g)| \leqq |(R_l^{(n)} f, g^*)-(R_l f, g^*)| \\ +|(R_l^{(n)} f, g-g^*)|+|(R_l f, g-g^*)|;$$

and we observe that the first term on the right tends to zero with $1/n$ while neither of the remaining terms can exceed $\|f\|\|g-g^*\|/|\mathfrak{J}(l)|$. Combining these facts, we see that as $1/n$ tends to zero the inferior and superior limits of the expression $|(R_l^{(n)} f, g)-(R_l f, g)|$ lie between 0 and ε. Since ε can be taken small at pleasure, the expression must in fact converge to zero. In other words,

$$(R_l^{(n)} f, g) \to (R_l f, g), \qquad \mathfrak{J}(l) \neq 0,$$

without restriction on f and g, as we wished to prove.

By combining the results of Theorems 5.2 and 5.3 we obtain the fundamental theorem of the present section. We have only to apply Lemmas 5.2 and 5.3 in order to establish it.

THEOREM 5.4. *If H is an arbitrary self-adjoint transformation and R_l is its resolvent, then there exists a unique function $\varrho(\lambda)$, dependent upon f and g, such that*

$$V(\varrho) \leqq \|f\|\cdot\|g\|, \qquad (R_l f, g) = I(l; \varrho)$$

for every f and g in $\mathfrak{H}$ and for every not-real l. The function $(R_l f, g)$ can be determined explicitly as the limit of the sequence $(R_l^{(n)} f, g)$, converging uniformly in any bounded closed point set at positive distance from the real axis.

§ 3. THE REDUCIBILITY OF THE RESOLVENT

The further study of the self-adjoint transformation H and its resolvent R_l is to be effected by considering the nature of the function $\varrho(\lambda)$ introduced in Theorem 5.4.

THEOREM 5.5. *There exists a family of bounded self-adjoint transformations* $E(\lambda)$ *with domain* $\mathfrak{H}$ *such that* $\varrho(\lambda; f, g) = (E(\lambda)f, g)$, $-\infty < \lambda < +\infty$, *for every* f *and* g *in* $\mathfrak{H}$.

We first note that in its dependence on f and g the function ϱ enjoys the linear properties expressed by the relations

$$\begin{aligned} \varrho(\lambda; af, g) &= a\varrho(\lambda; f, g), \\ \varrho(\lambda; f, ag) &= \bar{a}\varrho(\lambda; f, g), \\ \varrho(\lambda; f_1 + f_2, g) &= \varrho(\lambda; f_1, g) + \varrho(\lambda; f_2, g), \\ \varrho(\lambda; f, g_1 + g_2) &= \varrho(\lambda; f, g_1) + \varrho(\lambda; f, g_2). \end{aligned}$$

Each of these identities in λ, f, and g depends upon three facts: the linear nature of the resolvent, the uniqueness of the relation between the resolvent and the function $\varrho(\lambda)$, and the appearance of $\varrho(\lambda)$ in normal form. We shall indicate the proof of the third identity as typical. From the equation

$$(R_l(f_1 + f_2), g) = (R_l f_1, g) + (R_l f_2, g)$$

we see that

$$I(l; \varrho(\lambda; f_1 + f_2, g)) = I(l; \varrho(\lambda; f_1, g)) + I(l; \varrho(\lambda; f_2, g))$$

for all not-real l. Since the class $\mathfrak{B}^*$ of all functions in normal form is a linear class, the function $\varrho(\lambda; f_1, g) + \varrho(\lambda; f_2, g)$ belongs to $\mathfrak{B}^*$ and, by Lemma 5.2, is identical with $\varrho(\lambda; f_1 + f_2, g)$, as we wished to prove.

Next we observe that $\varrho(\lambda; f, g)$, being in normal form, satisfies the inequality $|\varrho(\lambda; f, g)| \leq V(\varrho) \leq \|f\| \cdot \|g\|$. Thus we can apply Theorem 2.28 and can write $\varrho(\lambda; f, g) = (E(\lambda)f, g)$, where $E(\lambda)$ is a bounded linear transformation with domain $\mathfrak{H}$.

Finally we show that the relation $R_l^* \equiv R_{\bar{l}}$, proved in Theorem 4.19, implies that $\varrho(\lambda; g, f) = \overline{\varrho(\lambda; f, g)}$. We have

$$\begin{aligned} I(l; \varrho(\lambda; g, f)) &= (R_l g, f) = (g, R_{\bar{l}} f) = \overline{(R_{\bar{l}} f, g)} \\ &= \overline{I(\bar{l}; \varrho(\lambda; f, g))} = I(l; \overline{\varrho(\lambda; f, g)}) \end{aligned}$$

and deduce the desired result on the basis of Lemma 5.2. Theorem 2.28 then shows that the transformation $E(\lambda)$ is symmetric and hence self-adjoint.

Before proceeding to the further investigation of the family of transformations $E(\lambda)$, it will be well to recall a few facts about projections. If $E(\lambda)$ be used for a moment to indicate a family of projections defined for $-\infty<\lambda<+\infty$ with the property

$$E(\lambda)\,E(\mu)\equiv E(\lambda),\quad \lambda<\mu,$$

we see by Theorem 2.40 that the limits

$$\lim_{\varepsilon\to 0} E(\lambda+\varepsilon),\quad \varepsilon>0;\qquad \lim_{\lambda\to+\infty} E(\lambda);\qquad \lim_{\lambda\to-\infty} E(\lambda)$$

exist and are projections. It is therefore possible to introduce the following definition.

DEFINITION 5.1. *If $E(\lambda)$ is a family of projections defined for $-\infty<\lambda<+\infty$ with the properties*

(1) $E(\lambda)\,E(\mu)\equiv E(\lambda),\ \lambda\leq\mu;\quad E(\lambda)\,E(\mu)\equiv E(\mu),\ \lambda>\mu;$
(2) $E(\lambda+\varepsilon)\to E(\lambda),\ \varepsilon>0,\ \varepsilon\to 0;$
(3) $E(\lambda)\to I,\ \lambda\to+\infty;$
(4) $E(\lambda)\to O,\ \lambda\to-\infty;$ —

then $E(\lambda)$ is called a resolution of the identity.

THEOREM 5.6. *The family of transformations $E(\lambda)$ introduced in Theorem 5.5 is a resolution of the identity.*

The theorem rests almost exclusively upon the functional equation for the resolvent, as stated in Theorems 4.10 and 4.19. The relation

$$(l-m)\,(R_l R_m f,\, g) = (R_l f,\, g)-(R_m f,\, g)$$

is expressed in terms of Stieltjes integrals by the theorems of this section and the last, the two integrals on the right then being combined and a common factor, $l-m$, being removed from the equation. As a result we obtain

$$\int_{-\infty}^{+\infty}\frac{1}{\lambda-l}\,d(E(\lambda)R_m f, g)=\int_{-\infty}^{+\infty}\frac{1}{(\lambda-l)(\lambda-m)}\,d(E(\lambda)f, g).$$

From Lemma 5.1(8) it follows that the integral $\int_{-\infty}^{\lambda}\frac{1}{\mu-m}$ $d(E(\mu)f, g)$ is a function of bounded variation in normal form. On introducing this function and applying the result of Lemma 5.1 (6), we find that the equation above can be thrown into the more convenient form

$$\int_{-\infty}^{+\infty} \frac{1}{\lambda - l}\, d\,(E(\lambda)\, R_m f,\, g)$$
$$= \int_{-\infty}^{+\infty} \frac{1}{\lambda - l}\, d\left[\int_{-\infty}^{\lambda} \frac{1}{\mu - m}\, d\,(E(\mu) f,\, g)\right].$$

This equation is evidently an identity for every f and g in $\mathfrak{H}$ and every not-real l and m. We now apply the uniqueness argument based on Lemma 5.2, finding

$$(E(\lambda)\, R_m f,\, g) = \int_{-\infty}^{\lambda} \frac{1}{\mu - m}\, d\,(E(\mu) f,\, g).$$

The term on the left is next expressed as a Stieltjes integral by the equations

$$\begin{aligned}(E(\lambda)\, R_m f,\, g) &= (R_m f,\, E(\lambda)\, g)\\ &= \int_{-\infty}^{+\infty} \frac{1}{\mu - m}\, d\,(E(\mu) f,\, E(\lambda)\, g).\end{aligned}$$

We are therefore in a position to make a second application of the uniqueness argument, this time to the identity

$$\int_{-\infty}^{+\infty} \frac{1}{\mu - m}\, d\,(E(\mu) f,\, E(\lambda)\, g) = \int_{-\infty}^{\lambda} \frac{1}{\mu - m}\, d\,(E(\mu) f,\, g).$$

We note that the function $\varrho(\mu)$ equal to $(E(\mu) f, g)$, $\mu < \lambda$, and equal to $(E(\lambda) f, g)$, $\mu \geqq \lambda$, is a function of bounded variation in normal form and that the integral on the right of the identity is $I(m; \varrho(\mu))$. We see therefore that

$$(E(\lambda)\, E(\mu) f,\, g) = (E(\mu) f,\, E(\lambda)\, g) = \varrho(\mu)$$

for every f and g in $\mathfrak{H}$. On expressing this equation without reference to elements of $\mathfrak{H}$, we find

$$\begin{aligned} E(\lambda)\, E(\mu) &\equiv E(\mu), \quad \mu < \lambda,\\ E(\lambda)\, E(\mu) &\equiv E(\lambda), \quad \mu = \lambda,\\ E(\lambda)\, E(\mu) &\equiv E(\lambda), \quad \mu > \lambda.\end{aligned}$$

From the second equation, written $E^2(\lambda) \equiv E(\lambda)$, the self-adjoint transformation $E(\lambda)$ must be a projection, in accordance with Theorem 2.36.

In order to show that the family of projections $E(\lambda)$ is a resolution of the identity, we must prove that

$$E(\lambda) \equiv E(\lambda+0) \equiv \lim_{\varepsilon \to 0} E(\lambda+\varepsilon), \qquad \varepsilon > 0,$$

$$E(-\infty) \equiv \lim_{\lambda \to -\infty} E(\lambda) \equiv O, \qquad E(+\infty) \equiv \lim_{\lambda \to +\infty} E(\lambda) \equiv I.$$

The first relation follows at once from the equations

$$(E(\lambda+0)f, g) = \lim_{\varepsilon \to 0} (E(\lambda+\varepsilon)f, g) = \lim_{\varepsilon \to 0} \varrho(\lambda+\varepsilon; f, g)$$
$$= \varrho(\lambda+0; f, g) = \varrho(\lambda; f, g) = (E(\lambda)f, g).$$

The second follows in a similar manner, since

$$(E(-\infty)f, g) = \lim_{\lambda \to -\infty} \varrho(\lambda; f, g) = 0.$$

The third relation depends upon a fundamental property of the resolvent. It is evident that the projection $F \equiv I - E(+\infty)$ satisfies the relation $E(\lambda)F \equiv FE(\lambda) \equiv O$, $-\infty < \lambda < +\infty$; hence, if f is an element of the closed linear manifold into which F takes $\mathfrak{H}$, $\varrho(\lambda; f, g) = (E(\lambda)f, g) = 0$ and $(R_l f, g) = I(l; \varrho) = 0$. The latter equation implies $R_l f = 0$, which in turn implies $f = 0$, as was pointed out in Theorems 4.10 and 4.19. Consequently F takes every element of $\mathfrak{H}$ into the null element, and we have

$$F \equiv O, \qquad E(+\infty) \equiv I,$$

as we wished to show. This completes the proof of the theorem.

THEOREM 5.7. *Between the class $\mathfrak{H}$ of all self-adjoint transformations H and the class $\mathfrak{E}$ of all resolutions of the identity $E(\lambda)$ there is a one-to-one correspondence such that if H and $E(\lambda)$ are corresponding members of these classes the resolvent of H is expressed by the formula*

$$(R_l f, g) = \int_{-\infty}^{+\infty} \frac{1}{\lambda - l}\, d(E(\lambda)f, g).$$

In the preceding theorem, we have proved that if H is an arbitrary self-adjoint transformation its resolvent R_l determines a unique resolution of the identity $E(\lambda)$ such that the formula $(R_l f, g) = I(l; (E(\lambda)f, g))$ is true. It is thus evident that the class of all self-adjoint transformations is in one-to-one correspondence with a certain subset of the class of all resolutions of the identity. We must show that this subset is identical with the entire class.

Let $E(\lambda)$ be a given resolution of the identity. We first show that the function $\varrho(\lambda; f, g) = (E(\lambda)f, g)$ is a function of bounded variation in normal form. From the fundamental properties of the family $E(\lambda)$ we derive the relations

$$\varrho(\lambda+0; f, g) = \varrho(\lambda; f, g), \qquad \lim_{\lambda\to-\infty} \varrho(\lambda; f, g) = 0,$$

so that it remains for us to prove that $V(\varrho)$ is finite. We let $\mathfrak{D}$ be a collection of intervals $\Delta_1, \cdots, \Delta_n$, no two of which have interior points in common, and introduce the notation $E(\Delta)$ for the difference $E(\beta) - E(\alpha)$ associated with the interval Δ, $\alpha \leqq \lambda \leqq \beta$. By reference to Theorem 2.37 we see that $E(\Delta_1), \cdots, E(\Delta_n)$ are projections such that $E(\Delta_i)\,E(\Delta_k) \equiv O$, $i \neq k$; thus, by Theorem 2.39, $E \equiv E(\Delta_1) + \cdots + E(\Delta_n)$ is a projection. It is now a simple matter to appraise the sum $V_{\mathfrak{D}}(\varrho)$ and the related quantity $V(\varrho)$. We have

$$V_{\mathfrak{D}}(\varrho) = \sum_{\alpha=1}^{n} |\varrho(\Delta_\alpha)| = \sum_{\alpha=1}^{n} |(E(\Delta_\alpha)f, g)|$$

$$= \sum_{\alpha=1}^{n} |(E(\Delta_\alpha)f, E(\Delta_\alpha)g)|$$

$$\leqq \sum_{\alpha=1}^{n} \|E(\Delta_\alpha)f\|\,\|E(\Delta_\alpha)g\| \leqq \left(\sum_{\alpha=1}^{n} \|E(\Delta_\alpha)f\|^2 \cdot \sum_{\alpha=1}^{n} \|E(\Delta_\alpha)g\|^2\right)^{1/2}$$

Since the projections $E(\Delta_i)$ and $E(\Delta_k)$ are orthogonal when $i \neq k$, we have

$$\sum_{\alpha=1}^{n} \|E(\Delta_\alpha)f\|^2 = \left\|\left(\sum_{\alpha=1}^{n} E(\Delta_\alpha)\right) f\right\|^2 = \|Ef\|^2 \leqq \|f\|^2,$$

$$\sum_{\alpha=1}^{n} \|E(\Delta_\alpha)g\|^2 = \left\|\left(\sum_{\alpha=1}^{n} E(\Delta_\alpha)\right) g\right\|^2 = \|Eg\|^2 \leqq \|g\|^2.$$

Thus $V_{\mathfrak{D}}(\varrho) \leqq \|f\| \cdot \|g\|$, independent of the collection $\mathfrak{D}$. We conclude that $V(\varrho)$ does not exceed $\|f\| \cdot \|g\|$, as we wished to prove.

We now form the Stieltjes integral $I(l; \varrho(\lambda; f, g))$; by Lemma 5.2 it is a single-valued analytic function of l when $\mathfrak{J}(l) \neq 0$. By the properties of the Stieltjes integral and those of the function $\varrho(\lambda; f, g)$ we see that

$$
\begin{aligned}
&I(l; \varrho(\lambda; a_1 f_1 + a_2 f_2, g)) \\
&\qquad = a_1 I(l; \varrho(\lambda; f_1, g)) + a_2 I(l; \varrho(\lambda; f_2, g)) \\
&I(l; \varrho(\lambda; f, a_1 g_1 + a_2 g_2)) \\
&\qquad = \bar{a}_1 I(l; \varrho(\lambda; f, g_1)) + \bar{a}_2 I(l; \varrho(\lambda; f, g_2)) \\
&|I(l; \varrho(\lambda; f, g))| \leqq V(\varrho) / |\mathfrak{J}(l)| \leqq \|f\| \cdot \|g\| / |\mathfrak{J}(l)|.
\end{aligned}
$$

Hence we can apply Theorem 2.28 and write $I(l; \varrho(\lambda; f, g)) = (X_l f, g)$ where X_l is a bounded linear transformation with domain $\mathfrak{H}$ and $\mathfrak{J}(l) \neq 0$. We note that the bound of X_l does not exceed $1/|\mathfrak{J}(l)|$.

The remainder of the proof consists in showing that the family of transformations X_l satisfies the conditions of Theorem 4.19 and can therefore be regarded as the resolvent of a uniquely determined self-adjoint transformation H. The first condition is met if we can show that the relation

$$(l - m)(X_l X_m f, g) = (X_l f, g) - (X_m f, g)$$

holds for every f and g in $\mathfrak{H}$ as a consequence of the fundamental equation $E(\lambda) E(\mu) \equiv E(\lambda)$, $\lambda < \mu$. In the proof of Theorem 5.6 we passed from the functional equation for the resolvent to the fundamental equation for the family $E(\lambda)$. It is here a question of reversing the procedure. Upon examination it turns out that we can accomplish our aim simply by retracing the steps of the proof given in that theorem. As a consequence, we do not need to write out the details of a proof here. Having verified in this manner that the family X_l has the first property required by Theorem 4.19, we turn to the second. We let m be a fixed value and show that $X_m f = 0$ implies $f = 0$. By means of the relation $(l-m) X_l X_m f = X_l f - X_m f$ we see that $X_m f = 0$ implies $X_l f = 0$ for arbitrary l. Thus $(X_l f, g) = I(l; \varrho) = 0$ for arbitrary l and g. By applying the uniqueness argument drawn from Lemma 5.2, we conclude that $(E(\lambda) f, g) = \varrho(\lambda; f, g) = 0$. On allowing λ to become positively infinite we have

$$E(\lambda) \to I, \ (f, g) = 0.$$

Hence we have shown that $f = 0$. Finally we can establish the truth of the third condition without difficulty: we have

$$\begin{aligned}(X_l f, g) &= I(l; \varrho(\lambda; f, g)) = \overline{I(\bar{l}; (g, E(\lambda) f))} \\ &= \overline{I(\bar{l}; (E(\lambda) g, f))} = \overline{(X_{\bar{l}} g, f)} = (f, X_{\bar{l}} g).\end{aligned}$$

Since all the conditions of Theorem 4.19 are met, we see that there is a unique self-adjoint transformation H whose resolvent R_l satisfies the relation

$$(R_l f, g) = (X_l f, g) = \int_{-\infty}^{+\infty} \frac{1}{\lambda - l} \, d(E(\lambda) f, g)$$

for every f and g in $\mathfrak{H}$. The proof of the theorem is thus complete.

The expression for the resolvent of a self-adjoint transformation given in the preceding theorems yields an immediate knowledge of properties of reducibility. If we denote by $\mathfrak{M}(\Delta)$ the range of the projection $E(\Delta) \equiv E(\beta) - E(\alpha)$ where Δ is a finite or infinite interval (α, β) and $E(-\infty) \equiv O$, $E(+\infty) \equiv I$, we can state the facts as follows:

THEOREM 5.8. *If H is a self-adjoint transformation and R_l is its resolvent, then H and R_l are both reduced by $\mathfrak{M}(\Delta)$.*

According to Theorem 4.27, it is sufficient to prove that R_l is reduced by $\mathfrak{M}(\Delta)$; and, by the definition of reducibility, it is in turn sufficient to prove that $E(\Delta) R_l$ and $R_l E(\Delta)$ are identical. This follows at once from the equations

$$\begin{aligned}
(E(\Delta)\,R_l f,\, g) \\
&= (R_l f,\, E(\Delta)\,g) &&= I(l;\,(E(\lambda)\,f,\, E(\Delta)\,g)) \\
&= I(l;\,(E(\Delta)\,E(\lambda)\,f,\, g)) &&= I(l;\,E(\lambda)\,E(\Delta)\,f,\, g)) \\
&= (R_l\,E(\Delta)\,f,\, g).
\end{aligned}$$

It is to be noted that Theorem 5.8 does not give complete information concerning the reducibility of H or of its resolvent. The precise characterization of these transformations in terms of reducibility can be effected only at a somewhat later stage, which we shall reach in Chapter VII.

§ 4. The Analytical Representation of a Self-Adjoint Transformation

We must now obtain a method for representing a given self-adjoint transformation H directly in terms of the corresponding resolution of the identity $E(\lambda)$. The correspondence between the classes $\mathfrak{H}$ and $\mathfrak{E}$ is rendered more transparent by the direct characterization provided in the following theorem.

THEOREM 5.9. *The correspondence between the classes $\mathfrak{H}$ and $\mathfrak{E}$ defined in Theorem 5.7 is determined directly as follows: if H and $E(\lambda)$ are corresponding elements of the two classes, then the domain of H comprises those and only those elements f such that the Stieltjes integral $\int_{-\infty}^{+\infty} \lambda^2 d\,\|E(\lambda)f\|^2$ is convergent; when f is in the domain of H, the element Hf is determined by the relations*

$$(Hf,\, g) = \int_{-\infty}^{+\infty} \lambda\, d(E(\lambda)\,f,\, g), \quad \|Hf\|^2 = \int_{-\infty}^{+\infty} \lambda^2 d\|E(\lambda)\,f\|^2,$$

where g is an arbitrary element in $\mathfrak{H}$.

Let Δ be a finite interval $\alpha \leqq \lambda \leqq \beta$, $E(\Delta)$ the projection $E(\beta) - E(\alpha)$, $\mathfrak{M}(\Delta)$ the range of this projection, and f an arbitrary element of $\mathfrak{H}$. Our first step is to show that $E(\Delta)\,f$ is in the domain of H and that

$$(H\,E(\Delta)\,f,\, g) = \int_{\alpha}^{\beta} \lambda\, d\,(E(\lambda)\,f,\, g),$$

$$\|H\,E(\Delta)\,f\|^2 = \int_{\alpha}^{\beta} \lambda^2 d\|E(\lambda)\,f\|^2.$$

Since $E(\lambda)$ is a resolution of the identity, we see that the transformation $E(\lambda)\,E(\Delta)$ coincides with O, $E(\lambda)-E(\alpha)$, or $E(\Delta)$ according as $\lambda<\alpha$, $\alpha\leqq\lambda<\beta$, or $\beta\leqq\lambda$. It follows immediately that

$$\int_\alpha^\beta \lambda\, d(E(\lambda)\,E(\Delta)f,g) = \int_\alpha^\beta \lambda\, d(E(\lambda)f,g).$$

We denote the conjugate of this integral by $L(g)$, observing that L is a complex-valued linear function of g which satisfies the inequality

$$|L(g)| \leqq M V(\overline{(E(\lambda)f,g)}) \leqq M|f|\cdot|g|$$

where M is the maximum of the two numbers $|\alpha|$ and $|\beta|$. Theorem 2.27 shows that there exists an element f^* in $\mathfrak{H}$ such that

$$L(g) = (g,f^*), \qquad \int_\alpha^\beta \lambda\, d(E(\lambda)\,E(\Delta)f,g) = (f^*,g).$$

We next compute $R_l(f^*-lE(\Delta)f)$ for an arbitrary not-real value of l. From the proofs of Theorems 5.8 and 5.6 we recall the equations

$$\begin{aligned}(R_l\,E(\lambda)\,E(\Delta)f,g) &= (E(\lambda)\,R_l\,E(\Delta)f,g)\\ &= \int_{-\infty}^{\lambda}\frac{1}{\mu-l}\,d(E(\mu)\,E(\Delta)f,g).\end{aligned}$$

With this expression at our disposal, we obtain

$$\begin{aligned}&(R_l(f^*-lE(\Delta)f),g)\\ &\quad= (f^*-lE(\Delta)f,R_{\bar l}\,g)\\ &\quad= \int_\alpha^\beta(\lambda-l)\,d(E(\lambda)\,E(\Delta)f,R_{\bar l}\,g)\\ &\quad= \int_\alpha^\beta(\lambda-l)\,d(R_l\,E(\lambda)\,E(\Delta)f,g) = \int_\alpha^\beta d(E(\lambda)\,E(\Delta)f,g)\\ &\quad= (E(\Delta)f,g),\end{aligned}$$

where g is an arbitrary element in $\mathfrak{H}$. Thus we see that $R_l(f^*-lE(\Delta)f) = E(\Delta)f$. This equation implies that $E(\Delta)f$ is in the domain of H and that

$$HE(\Delta)f = H_l E(\Delta)f + lE(\Delta)f$$
$$= (f^* - lE(\Delta)f) + lE(\Delta)f = f^*,$$
$$(HE(\Delta)f, g) = (f^*, g) = \int_\alpha^\beta \lambda\, d(E(\lambda)f, g).$$

We calculate $\|HE(\Delta)f\|^2$ as follows:

$$\|HE(\Delta)f\|^2 = (HE(\Delta)f, HE(\Delta)f) = \int_\alpha^\beta \lambda\, d(E(\lambda)f, HE(\Delta)f)$$
$$= \int_\alpha^\beta \lambda\, d\left(\int_\alpha^\lambda \mu\, d(E(\lambda)f, E(\mu)f)\right)$$
$$= \int_\alpha^\beta \lambda\, d\left(\int_\alpha^\lambda \mu\, d\|E(\mu)f\|^2\right) = \int_\alpha^\beta \lambda^2\, d\|E(\lambda)f\|^2.$$

Thus the relations indicated above are valid.

Let us suppose now that f belongs to the domain of H. From Theorem 5.8 we know that $HE(\Delta)f = E(\Delta)Hf$. When α and β tend to $-\infty$ and $+\infty$ respectively, we see therefore that $E(\Delta) \to I$, $HE(\Delta)f \to Hf$. Hence we can pass directly from the results obtained in the preceding paragraph to the corresponding integral representations for (Hf, g) and $\|Hf\|^2$ stated in the theorem.

On the other hand, let f be an element such that the integral $\int_{-\infty}^{+\infty} \lambda^2\, d\|E(\lambda)f\|^2$ is convergent. We determine a sequence of intervals $\{\Delta_n\}$ expanding so as to cover the interval $-\infty < \lambda < +\infty$; we suppose that Δ_n is contained in Δ_{n+1}, $n = 1, 2, 3, \cdots$. The properties of $E(\lambda)$ show that

$$E(\Delta_m)E(\Delta_n) \equiv E(\Delta_n), \quad (E(\Delta_m) - E(\Delta_n))E(\Delta_n) \equiv O,$$
$$m \geqq n.$$

By Theorem 2.37, therefore, $E(\Delta_m) - E(\Delta_n)$ and $E(\Delta_n)$ are orthogonal projections with the respective ranges $\mathfrak{M}(\Delta_m) \ominus \mathfrak{M}(\Delta_n)$ and $\mathfrak{M}(\Delta_n.)$ From Theorems 4.23 and 5.8 we see that the elements $H(E(\Delta_m) - E(\Delta_n))f$ and $HE(\Delta_n)f$ belong to $\mathfrak{M}(\Delta_m) \ominus \mathfrak{M}(\Delta_n)$ and to $\mathfrak{M}(\Delta_n)$ respectively, and are therefore orthogonal. Hence their sum $HE(\Delta_m)f$ satisfies the equation

$$\|HE(\Delta_m)f\|^2 = \|H(E(\Delta_m) - E(\Delta_n))f\|^2 + \|HE(\Delta_n)f\|^2.$$

By an obvious rearrangement of terms we obtain

$$\| HE(\Delta_m)f - HE(\Delta_n)f \|^2 = \| HE(\Delta_m)f \|^2 - \| HE(\Delta_n)f \|^2$$
$$= \int_{\Delta_m - \Delta_n} \lambda^2 \, d\| E(\lambda)f \|^2.$$

Since the last expression tends to zero when m and n become infinite, we conclude that $HE(\Delta_n)f \to f^*$, $E(\Delta_n)f \to f$. By virtue of the fact that H is a closed transformation these relations imply that f is in the domain of H and that $Hf = f^*$. This result completes the proof of the theorem.

This theorem gives us a specific method for passing from a given resolution of the identity $E(\lambda)$ to the corresponding transformation H. By reference to Lemma 5.2 and Theorem 5.7, we may formulate a somewhat indirect but none the less specific method for computing $E(\lambda)$ in terms of H. We have

THEOREM 5.10. *If H is a given self-adjoint transformation and R_l is its resolvent, then the corresponding resolution of the identity $E(\lambda)$ can be determined from the relation*

$$\tfrac{1}{2}[\{(E(\mu)f,g)+(E(\mu-0)f,g)\}-\{(E(\nu)f,g)+(E(\nu-0)f,g)\}]$$
$$= \lim_{\varepsilon\to 0} -\frac{1}{2\pi i}\int_{C(\mu,\nu,\alpha,\varepsilon)} (R_l f, g)\, dl,$$

where $C(\mu, \nu, \alpha, \varepsilon)$ is the contour described in Lemma 5.2.

We make particular mention of this theorem because of its great practical and historical significance. It is the key to many important researches concerning boundary value problems for differential equations, including those discussed in Chapter III, § 3; we shall indicate the connection in greater detail in the following section. It is also the fundamental tool in Hellinger's discussion of the spectrum of bounded symmetric transformations.† More precisely, Hellinger uses the contour integral to define the family of projections $E(\lambda)$ and is thus able to obtain important information concerning the transformation H. It should be noted that his method cannot be extended directly to the general self-adjoint trans-

† Hellinger, Journal für Mathematik, 136 (1909), pp. 210–271.

formation; we are able, however, to verify its validity *a posteriori*. Under these circumstances, it is natural to inquire whether or not the procedure of Hellinger can be so modified as to give a direct treatment of unbounded as well as bounded transformations; we have not been able to do so.

§ 5. The Spectrum of a Self-Adjoint Transformation

The relation between the spectrum of a self-adjoint transformation and the corresponding resolution of the identity remains to be considered. A few paragraphs will be sufficient to extablish the salient features of the relationship.

First of all it is necessary to classify the points of the real axis in the l-plane with reference to a given resolution of the identity.

Definition 5.2. *If $E(\lambda)$ is a resolution of the identity, the point $\lambda = \mu$ is called*

(1) *a point of constancy of $E(\lambda)$ when μ is interior to some interval Δ such that $E(\Delta) \equiv O$;*

(2) *a point of continuity of $E(\lambda)$ when μ is interior to no such interval but has the property that $E(\mu) \equiv E(\mu - 0)$;*

(3) *a point of discontinuity of $E(\lambda)$ when $E(\mu) \not\equiv E(\mu - 0)$. The set of all points of constancy will be denoted by D_E, that of all points of continuity by B_E, and that of all points of discontinuity by A_E.*

It is evident that the three sets A_E, B_E, D_E are mutually exclusive sets whose sum is the entire range $-\infty < \lambda < +\infty$; some of them may be empty, but not all. At every point $\lambda = \mu$ of the set $B_E + D_E$, $E(\lambda)$ is continuous in the sense that $E(\mu - 0) \equiv E(\mu) \equiv E(\mu + 0)$; the terminology used in the definition to distinguish between points in B_E and those in D_E has been adopted for reasons of convenience.

Theorem 5.11. *If H is a self-adjoint transformation, $E(\lambda)$ is the corresponding resolution of the identity, and A_E, B_E, and D_E are the sets introduced in Definition 5.2, then*

(1) *the point-spectrum $A(H)$ coincides with A_E;*

(2) *the continuous spectrum $B(H)$ coincides with B_E;*

(3) *the residual spectrum* $C(H)$ *is empty*;

(4) *the real points in the resolvent set* $D(H)$ *coincide with* D_E.

The spectrum $S(H) \equiv A_E + B_E$ *contains at least one point. If* l *is in the set* $B(H) + D(H)$, *then* H_l^{-1} *exists*; *its domain comprises those and only those elements* f *such that the integral* $\int_{-\infty}^{+\infty} \frac{1}{|\lambda - l|^2}\, d\|E(\lambda)f\|^2$ *is convergent, and it is determined by the equation* $(H_l^{-1}f, g) = \int_{-\infty}^{+\infty} \frac{1}{\lambda - l}\, d(E(\lambda)f, g)$.

It is convenient to begin with the study of the integral representation of H_l^{-1}. When l is not real, we know that $H_l^{-1} \equiv R_l$ so that this case needs no further discussion. We shall therefore suppose that $l = \mu$ is a point of the set $B_E + D_E$ on the real axis. We then define a transformation T whose domain consists of those and only those elements f such that the integral $\int_{-\infty}^{+\infty} \frac{1}{(\lambda - \mu)^2}\, d\|E(\lambda)f\|^2$ exists and whose behavior is described by the equation

$$(Tf, g) = \int_{-\infty}^{+\infty} \frac{1}{\lambda - \mu}\, d(E(\lambda)f, g).$$

If we make the transformation $\nu = \frac{1}{\lambda - \mu}$, we can draw on the results of Theorem 5.9 to show that this definition is effective and provides us with a transformation T which is self-adjoint. In order to introduce the new variable ν we define a family of transformations $F(\nu)$ according to the identities

$$F(\nu) \equiv E(\mu - 0) - E(1/\nu + \mu - 0), \quad \nu < 0,$$

$$F(0) \equiv E(\mu),$$

$$F(\nu) \equiv I + E(\mu) - E(1/\nu + \mu - 0), \quad \nu > 0;$$

it is easily verified by direct computation that $F(\nu)$ is a resolution of the identity. The introduction of the variable ν leads to the equations

$$\left[\int_{-\infty}^{\mu-\varepsilon} + \int_{\mu+\varepsilon}^{+\infty}\right] \frac{1}{(\lambda-\mu)^2}\, d\,|E(\lambda)f|^2$$

$$= \left[\int_{-1/\varepsilon}^{0} + \int_{0}^{1/\varepsilon}\right] \nu^2\, d\,|F(\nu)f|^2, \quad \varepsilon > 0,$$

$$\left[\int_{-\infty}^{\mu-\varepsilon} + \int_{\mu+\varepsilon}^{+\infty}\right] \frac{1}{\lambda-\mu}\, d(E(\lambda)f, g)$$

$$= \left[\int_{-1/\varepsilon}^{0} + \int_{0}^{1/\varepsilon}\right] \nu\, d(F(\nu)f, g).$$

When ε tends to zero we find

$$\int_{-\infty}^{+\infty} \frac{1}{(\lambda-\mu)^2}\, d\,|E(\lambda)f|^2 = \int_{-\infty}^{+\infty} \nu^2\, d\,|F(\nu)f|^2,$$

$$\int_{-\infty}^{+\infty} \frac{1}{\lambda-\mu}\, d(E(\lambda)f, g) = \int_{-\infty}^{+\infty} \nu\, d(F(\nu)f, g),$$

where the existence of either of the two integrals in the first equation entails the existence of the remaining three. Thus the transformation T may be defined as indicated and is a self-adjoint transformation corresponding to the resolution of the identity $F(\nu)$. It is now a matter of simple manipulations to show that in the domain of H the transformation TH_μ exists and reduces to the identity. By reference to Theorems 5.8 and 5.9 we can write

$$|E(\lambda)H_\mu f|^2 = |H_\mu E(\lambda)f|^2$$

$$= \int_{-\infty}^{+\infty} (\nu-\mu)^2\, d\,|E(\nu)E(\lambda)f|^2 = \int_{-\infty}^{\lambda} (\nu-\mu)^2\, d\,|E(\nu)f|^2,$$

$$(E(\lambda)H_\mu f, g) = \int_{-\infty}^{\lambda} (\nu-\mu)\, d(E(\nu)f, g),$$

whenever $H_\mu f$ exists. Thus, for $\varepsilon > 0$,

$$\int_{-\infty}^{\mu-\varepsilon} \frac{1}{(\lambda-\mu)^2}\, d\,|E(\lambda)H_\mu f|^2 + \int_{\mu+\varepsilon}^{+\infty} \frac{1}{(\lambda-\mu)^2}\, d\,|E(\lambda)H_\mu f|^2$$

$$= \int_{-\infty}^{\mu-\varepsilon} d\,|E(\lambda)f|^2 + \int_{\mu+\varepsilon}^{+\infty} d\,|E(\lambda)f|^2 \to \int_{-\infty}^{+\infty} d\,|E(\lambda)f|^2 = |f|^2,$$

so that $TH_\mu f$ exists. Furthermore, we can evaluate $TH_\mu f$ by means of the relations

$$\int_{-\infty}^{\mu-\varepsilon} \frac{1}{\lambda-\mu}\, d(E(\lambda) H_\mu f, g) + \int_{\mu+\varepsilon}^{+\infty} \frac{1}{\lambda-\mu}\, d(E(\lambda) H_\mu f, g)$$
$$= \int_{-\infty}^{\mu-\varepsilon} d(E(\lambda) f, g) + \int_{\mu+\varepsilon}^{+\infty} d(E(\lambda) f, g) \to \int_{-\infty}^{+\infty} d(E(\lambda) f, g)$$
$$= (f, g),$$

which imply that $(TH_\mu f, g) = (f, g)$, $TH_\mu f = f$. From these facts it appears immediately that H_μ takes its domain in a one-to-one manner into its range, and thus possesses an inverse H_μ^{-1} of which T is a self-adjoint extension. Since, as we have already proved in Theorem 4.18, the inverse H_μ^{-1} is self-adjoint, it must coincide with T. The desired integral representation for H_μ^{-1} is thus established, and is seen to be valid whenever $\lambda = \mu$ is not a point of discontinuity of $E(\lambda)$.

There is no further difficulty in proving the assertions of the theorem. In Theorem 4.18 we have already proved (3). From what has just been demonstrated, the point spectrum $A(H)$ must be a subset of A_E. In order to show that $A(H)$ and A_E are identical, we have to prove that every point of the second set belongs to the first. That the latter assertion is true we see as follows: $\lambda = \mu$ is a point of discontinuity of $E(\lambda)$ if and only if there exists an element f such that $(E(\mu) - E(\mu - 0))f = f$, $\|f\| = 1$; such an element f is obviously a characteristic element of H for $l = \mu$, in view of the relations

$$E(\lambda) f = 0,\ \lambda < \mu; \qquad E(\lambda) f = f,\ \lambda \geqq \mu;$$
$$\int_{-\infty}^{+\infty} \lambda^2\, d\|E(\lambda) f\|^2 = \mu^2;$$
$$(Hf, g) = \int_{-\infty}^{+\infty} \lambda\, d(E(\lambda) f, g) = \mu(f, g).$$

In order to complete the proof of the theorem, we must still show that every point of constancy of $E(\lambda)$ is a real point of the resolvent set $D(H)$, and conversely. If $\lambda = \mu$ is a point of constancy of $E(\lambda)$, then ε can be chosen so small that $E(\lambda) \equiv E(\mu)$ for $\mu - \varepsilon \leqq \lambda \leqq \mu + \varepsilon$. Thus H_μ^{-1} has $\mathfrak{H}$ as its domain since the integral

$$\int_{-\infty}^{+\infty} \frac{1}{(\lambda-\mu)^2}\, d\| E(\lambda) f\|^2$$

$$= \int_{-\infty}^{\mu-\varepsilon} \frac{1}{(\lambda-\mu)^2}\, d\| E(\lambda) f\|^2 + \int_{\mu+\varepsilon}^{+\infty} \frac{1}{(\lambda-\mu)^2}\, d\| E(\lambda) f\|^2$$

is always convergent; by Theorem 4.18, $l = \mu$ is in $D(H)$. On the other hand, if $l = \mu$ is in $D(H)$, $(R_l f, g)$ is defined and analytic in a sufficiently small circle $|l-\mu| < \varepsilon$; if $E(\lambda)$ be computed by means of the contour integral of Theorem 5.10, it is evident from Cauchy's theorem that $E(\lambda) \equiv E(\mu)$ for $\mu-\varepsilon < \lambda < \mu+\varepsilon$, so that $\lambda = \mu$ is a point of constancy of $E(\lambda)$. Thus the real part of the set $D(H)$ coincides with D_E. Since we have already shown that $A(H) = A_E$, it must now follow that $B(H) = B_E$.

From the fact that the spectrum of H can be empty only when $E(\lambda)$ is everywhere constant, we conclude that the spectrum of H must contain at least one point; for we know that $E(-\infty) \equiv O$, $E(+\infty) \equiv I$. In the case of the identity I, the spectrum consists of just one point $l = 1$.

THEOREM 5.12. *If H is a self-adjoint transformation satisfying one of the inequalities*

$$(Hf, f) \leqq C\|f\|^2, \quad (Hf, f) \geqq C\|f\|^2, \quad |(Hf, f)| \leqq C\|f\|^2$$

where C is a real number, then the points $l = \lambda$ of the spectrum of H satisfy the corresponding inequality

$$\lambda \leqq C, \quad \lambda \geqq C, \quad |\lambda| \leqq C;$$

and conversely.

We shall discuss the first pair of inequalities alone, the second pair being subject to analogous treatment, and the third pair being a combination of the first two. If $(Hf, f) \leqq C\|f\|^2$ we can conclude that every point $l = \lambda$ in the spectrum of H satisfies the inequality $\lambda \leqq C$; that is, that every point $\lambda > C$ is a point of constancy of $E(\lambda)$, the resolution of the identity for H. For, if $\lambda = \mu > C$ is a point of the spectrum of H, we can find a positive ε such that $\mu - \varepsilon > C$ and an element f such that $E(\mu+\varepsilon) - E(\mu-\varepsilon)) f = f$, $\|f\| \neq 0$; and we can then obtain the inequality

$$(Hf, f) = \int_{-\infty}^{+\infty} \lambda\, d(E(\lambda)f, f) = \int_{\mu-\varepsilon}^{\mu+\varepsilon} \lambda\, d(E(\lambda)f, f)$$

$$(Hf, f) \geqq (\mu-\varepsilon) \int_{\mu-\varepsilon}^{\mu+\varepsilon} d(E(\lambda)f, f) = (\mu-\varepsilon)\|f\|^2 > C\|f\|^2,$$

which is contrary to our hypothesis. On the other hand, if every point of the spectrum satisfies the inequality $\lambda \leqq C$, we have $E(\lambda) \equiv I$, $\lambda \geqq C$, so that

$$(Hf, f) = \int_{-\infty}^{+\infty} \lambda\, d(E(\lambda)f, f) = \int_{-\infty}^{C} \lambda\, d(E(\lambda)f, f)$$

$$\leqq C \int_{-\infty}^{C} d(E(\lambda)f, f) = C\|f\|^2.$$

This completes our discussion.

By reference to Theorem 2.38 we see that the expression $\|E(\lambda)f\|^2$ is a real monotone-increasing function of λ. It is of considerable interest to determine how the properties of this function are connected with the mutual relations between the element f and the self-adjoint transformation H associated with $E(\lambda)$.

THEOREM 5.13. *Let H be a self-adjoint transformation, $E(\lambda)$ the corresponding resolution of the identity, and $\mathfrak{C}$ the set of all characteristic elements of H; let $\mathfrak{M}$ be the closed linear manifold determined by $\mathfrak{C} + \mathfrak{O}$, m its dimension number; and let $\mathfrak{N}$ be the orthogonal complement of $\mathfrak{M}$, n its dimension number. Then one of the three following cases must occur:*

(1) *$\mathfrak{C}$ is empty; $\mathfrak{M} = \mathfrak{O}$, $\mathfrak{N} = \mathfrak{H}$; $m = 0$, $n = \aleph_0$;*

(2) *$\mathfrak{C}$ contains an incomplete orthonormal set $\{\varphi_k\}$ which determines the closed linear manifold $\mathfrak{M}$; the manifolds $\mathfrak{M}$ and $\mathfrak{N}$ are both proper subsets of $\mathfrak{H}$; and $m = 1, \cdots, \aleph_0$, $n = \aleph_0$;*

(3) *$\mathfrak{C}$ contains a complete orthonormal set $\{\varphi_k\}$; $\mathfrak{M} = \mathfrak{H}$, $\mathfrak{N} = \mathfrak{O}$; and $m = \aleph_0$, $n = 0$.*

In each of the three cases, the closed linear manifolds $\mathfrak{M}$ and $\mathfrak{N}$ reduce H. A necessary and sufficient condition that the element f be a characteristic element of H corresponding to the characteristic value l is that

$$\|E(\lambda)f\|^2 = 0, \ \lambda < l; \quad \|E(\lambda)f\|^2 = \|f\|^2 \neq 0, \ \lambda \geqq l.$$

A necessary and sufficient condition that $f \neq 0$ be an element of $\mathfrak{M}$ (in cases (2) and (3)) is that

$$f = \sum_{\alpha=1}^{\infty} a_\alpha \varphi_\alpha, \quad a_k = (f, \varphi_k), \quad \|f\|^2 = \sum_{\alpha=1}^{\infty} |a_\alpha|^2 \neq 0;$$

for such an element f we have

$$\|E(\lambda)f\|^2 = \sum_{(\alpha)}^{l_\alpha \leqq \lambda} |a_\alpha|^2, \quad H\varphi_k = l_k \varphi_k.$$

A necessary and sufficient condition that $f \neq 0$ be an element of $\mathfrak{N}$ is that $\|E(\lambda)f\|^2$ be a continuous function not identically zero. If f is an arbitrary element of $\mathfrak{H}$ and if g and h are its projections on $\mathfrak{M}$ and $\mathfrak{N}$ respectively, then the equation

$$\|E(\lambda)f\|^2 = \|E(\lambda)g\|^2 + \|E(\lambda)h\|^2$$

is valid and provides the standard resolution of the monotone function on the left into its discontinuous and continuous monotone components.

We first consider under what circumstances an element $f \neq 0$ is a characteristic element of H for the characteristic value l. We see at once that a necessary and sufficient condition is the vanishing of the expression $\|Hf - lf\|^2$. If we expand this quantity in the form $\|Hf\|^2 - 2l(Hf, f) + l^2\|f\|^2$ and apply Theorem 5.9, we obtain the equivalent condition $\int_{-\infty}^{+\infty} (\lambda - l)^2 \, d\|E(\lambda)f\|^2 = 0$. This integral vanishes if and only if the function $\|E(\lambda)f\|^2$ is constant on each of the open intervals $-\infty < \lambda < l$ and $l < \lambda < +\infty$, a condition which is evidently equivalent to the one stated in the theorem. We can show also that, for $f \neq 0$ to be a characteristic element, it is necessary and sufficient that f belong to the range of the projection $E(l) - E(l-0) \neq O$ or, in other words, that $f = (E(l) - E(l-0))f$. The condition given in the theorem requires

$$\|E(\lambda)f\|^2 = 0, \quad E(\lambda)f = 0, \quad \lambda < l,$$
$$\|f - E(\lambda)f\|^2 = \|f\|^2 - \|E(\lambda)f\|^2 = 0, \quad E(\lambda)f = f, \quad \lambda \geqq l,$$

and hence $E(l)f = f$, $E(l-0)f = 0$, $(E(l) - E(l-0))f = f$. On the other hand, the last relation enables us to show that

$$E(\lambda)f = E(\lambda)(E(l) - E(l-0))f = 0, \quad \lambda < l,$$
$$E(\lambda)f = E(\lambda)(E(l) - E(l-0))f = f, \quad \lambda \geqq l;$$

thus $\|E(\lambda)f\|^2$ has the properties described in the theorem and f is a characteristic element. The range of the projection $E(l) - E(l-0) \neq O$ is a closed linear manifold which reduces H, since it can be shown that the projection is permutable with the resolvent and hence with H, by the argument used in Theorem 5.8. Obviously, this manifold is in the domain of H, and is a characteristic manifold of H corresponding to the characteristic value l.

When $\mathfrak{C}$ is empty, it is evident that the various conditions stated under case (1) are fulfilled; the fact that $\mathfrak{M} = \mathfrak{O}$ and $\mathfrak{N} = \mathfrak{H}$ reduce H is trivial. When $\mathfrak{C}$ is not empty we construct all the characteristic manifolds of H and arrange them in a finite or denumerably infinite sequence $\{\mathfrak{M}_i\}$. Clearly we have

$$\mathfrak{C} + \mathfrak{O} = \sum_{\alpha=1}^{\infty} \mathfrak{M}_\alpha, \qquad \mathfrak{M} = \mathfrak{M}_1 \oplus \mathfrak{M}_2 \oplus \mathfrak{M}_3 \oplus \cdots.$$

Since each of the manifolds $\mathfrak{M}_i$ reduces H we can apply Theorem 4.26 to show that $\mathfrak{M}$ reduces H; hence, by Theorem 4.23, $\mathfrak{N}$ reduces H. The construction of the orthonormal set $\{\varphi_k\}$ now offers little difficulty. In $\mathfrak{M}_i$ we can select an everywhere dense sequence, by Theorem 1.18, and can then replace this sequence, according to Theorem 1.13, by an orthonormal set which determines the same closed linear manifold as the sequence itself. The orthonormal sets in the different manifolds $\mathfrak{M}_i$ are then arranged in a simple sequence. Since the manifolds $\mathfrak{M}_i$ are mutually orthogonal, $\{\varphi_k\}$ is an orthonormal set. It is evident that $\{\varphi_k\}$ is contained in $\mathfrak{C}$ and that it determines the closed linear manifold $\mathfrak{M}$. In the case where $\mathfrak{M} = \mathfrak{H}$, $\mathfrak{N} = \mathfrak{O}$, the set $\{\varphi_k\}$ is complete; but there is no reason in general why it should be complete.

According to Theorem 1.8 a necessary and sufficient condition that $f \neq 0$ belong to $\mathfrak{M}$ (in case (2) and (3)) is the one stated in the theorem. To compute $\|E(\lambda)f\|^2$ when $f \neq 0$ is in $\mathfrak{M}$, we make use of the continuity of the projection $E(\lambda)$. We have

$$\|E(\lambda)f\|^2 = \lim_{n\to\infty} \left\| E(\lambda)\left(\sum_{\alpha=1}^{n} a_\alpha \varphi_\alpha\right)\right\|^2$$
$$= \lim_{n\to\infty} \left\| \sum_{\alpha=1}^{n} a_\alpha E(\lambda)\varphi_\alpha \right\|^2 = \sum_{\alpha=1}^{\infty} |a_\alpha|^2 \|E(\lambda)\varphi_\alpha\|^2.$$

Since φ_k is a normalized characteristic element for the characteristic value l_k, we know that $\|E(\lambda)\varphi_k\|^2$ has the value 0 or 1 according as $\lambda < l_k$ or $\lambda \geqq l_k$; the expression for $\|E(\lambda)f\|^2$ thus reduces to the form given in the theorem.

If $f \neq 0$ is an element of $\mathfrak{N}$ (in cases (1) and (2)) the function $\|E(\lambda)f\|^2$ is continuous and not identically zero. If the function were discontinuous at $\lambda = l$ we should put $\varphi = (E(l) - E(l-0))f$ and then find that

$$(E(l) - E(l-0))\varphi = \varphi,$$
$$(f, \varphi) = \|\varphi\|^2 = \|E(l)f\|^2 - \|E(l-0)f\|^2 \neq 0.$$

These relations show that φ is a characteristic element of H to which f is not orthogonal. Since φ and f lie in the orthogonal manifolds $\mathfrak{M}$ and $\mathfrak{N}$ respectively, this result is in contradiction with the facts, and our hypothesis must be rejected. A sufficient condition that $f \neq 0$ belong to $\mathfrak{N}$ is of interest only in case (2). Then f has the desired properties if $\|E(\lambda)f\|^2$ is continuous and not identically zero. If φ_k is an arbitrary element of the set $\{\varphi_k\}$ we have

$$\begin{aligned} |(f, \varphi_k)|^2 &= |(f, (E(l_k) - E(l_k-0))\varphi_k)|^2 \\ &= |((E(l_k) - E(l_k-0))f, \varphi_k)|^2 \\ &\leqq \|(E(l_k) - E(l_k-0))f\|^2 \|\varphi_k\|^2 \\ &= \|E(l_k)f\|^2 - \|E(l_k-0)f\|^2 = 0 \end{aligned}$$

so that $(f, \varphi_k) = 0$. Hence f is orthogonal to every element of the set $\{\varphi_k\}$ and must belong to $\mathfrak{N}$. Since $\|f\|^2 = \|E(+\infty)f\|^2 \neq 0$, we see that $f \neq 0$. On the basis of these conditions it is easy to show that in case (2) $\mathfrak{N}$ has the dimension number $\aleph_0$. We select an arbitrary element $f \neq 0$ in $\mathfrak{N}$ and an arbitrary integer N. We can then choose N closed intervals $\Delta_1, \cdots, \Delta_N$, no two of which have a point in common, on each of which the function $\|E(\lambda)f\|^2$ increases and does not remain constant. Since the projections

$E(\Delta_1), \cdots, E(\Delta_N)$ are mutually orthogonal, we find that the set $\{\psi_k\}$ where $\psi_k = E(\Delta_k)f/\|E(\Delta_k)f\|$ is an orthonormal set. It is easily verified that this set is contained in $\mathfrak{N}$. Hence we must have $n \geqq N$ and conclude that $n = \aleph_0$.

THEOREM 5.14. *A set of necessary and sufficient conditions that the self-adjoint transformation H have finite norm N is the following:*

(1) *the points of the spectrum of H other than $l=0$ are characteristic values of finite multiplicity and can accordingly be denoted as a sequence $l_1, l_2, l_3, \cdots$, with corresponding multiplicities $n_1, n_2, n_3, \cdots$; they all lie on a finite interval of the real axis;*

(2) *the point $l=0$ is a characteristic value of finite or infinite multiplicity, or is a point of the continuous spectrum of H;*

(3) *the set of all characteristic elements of H contains a complete orthonormal set;*

(4) *the equation $N^2 = \sum_{\alpha=1}^{\infty} n_\alpha l_\alpha^2$ is valid and implies that the sequence $\{l_n\}$ is finite or is convergent to zero.*

The necessity of these conditions is less readily proved than the sufficiency and will be taken up first. The key to the behavior of H is found in the following assertion: if H is a self-adjoint transformation of finite norm, $E(\lambda)$ the corresponding resolution of the identity, and Δ an arbitrary interval on the real axis at positive distance δ from the origin, then the range of $E(\Delta)$ must have a finite dimension number. To prove this result, we assume that the range of $E(\Delta)$ has the dimension number $\aleph_0$ and derive a contradiction. Under this assumption, the range of $E(\Delta)$ contains an infinite orthonormal set $\{\varphi_n\}$. Making use of the equation $E(\Delta)\varphi_n = \varphi_n$, we can derive the inequalities

$$|(H\varphi_n, \varphi_n)| = \left|\int_{-\infty}^{+\infty} \lambda\, d(E(\lambda)\varphi_n, \varphi_n)\right| = \left|\int_{\Delta} \lambda\, d(E(\lambda)\varphi_n, \varphi_n)\right|$$
$$\geqq \delta \int_{\Delta} d(E(\lambda)\varphi_n, \varphi_n) = \delta\|\varphi_n\|^2 = \delta,$$
$$N^2 \geqq \sum_{\alpha=1}^{m} |(H\varphi_\alpha, \varphi_\alpha)|^2 \geqq m\delta^2, \; m = 1, 2, 3, \cdots,$$

the second of which is evidently false when m is large.

If $l = \mu$ is a point of the continuous spectrum of H and Δ is any interval containing μ as an interior point, then the range of the projection $E(\Delta)$ must have the dimension number $\aleph_0$. To prove this assertion we must assume that the range has a finite dimension number $n(\Delta)$ and establish a contradiction. If the interval Δ be contracted about the point $l = \mu$ so that its length tends to zero, then $E(\Delta) \to O$ and $n(\Delta)$ decreases and tends to zero. Since $n(\Delta)$ can assume only a finite number of integral values, it must vanish for a sufficiently small interval Δ and the identity $E(\Delta) \equiv O$ must then be satisfied. According to Theorem 5.11, the point $l = \mu$ belongs to the resolvent set of H, contrary to hypothesis. Thus the range of $E(\Delta)$ cannot have a finite dimension number.

By combining the results of the preceding paragraphs we see that the continuous spectrum of a self-adjoint transformation of finite norm can contain no point other than $l = 0$; all points of the spectrum other than $l = 0$ are characteristic values. An analogous but somewhat simpler argument shows that each characteristic value different from zero must have finite multiplicity. As to the point $l = 0$ it may belong either to the point spectrum or to the continuous spectrum; we shall distinguish various cases below. Since H has finite norm, it is bounded, and its entire spectrum is confined to a finite interval of the real axis, by Theorem 5.12.

The first two conditions having been derived, we turn to an investigation of the third. By a slight further refinement of the argument used above, we find that the number of characteristic values of H at positive distance ε from $l = 0$ must be finite, so that the range of the projection $I - E(\varepsilon - 0) + E(-\varepsilon + 0)$ has a finite dimension number and is actually the closed linear manifold determined by the characteristic elements of H corresponding to such characteristic values. In a similar way, it is verified that $l = 0$ is a characteristic value if and only if the range of the projection $E(0+0) - E(0-0)$ is different from $\mathfrak{O}$, and that when $l = 0$ is a characteristic value the range of the projection is the corresponding characteristic manifold. The two projections so determined are evidently mutually orthogonal, so that their sum $F(\varepsilon) \equiv I - E(\varepsilon - 0) - E(-\varepsilon + 0) + E(0+0) - E(0-0)$ is also a projection in accordance with Theorem 2.37. When ε tends to zero, $F(\varepsilon) \to I$. We conclude therefore that the closed linear manifold determined by the set of all characteristic elements of H is the entire space $\mathfrak{H}$. By Theorem 5.13, we can select a complete orthonormal set $\{\varphi_n\}$ of characteristic elements.

The fourth condition of the theorem follows at once. We have merely to compute the norm of H by means of the set $\{\varphi_n\}$ and obtain at once the equation

$$N^2 = \sum_{\alpha, \beta = 1}^{\infty} |(H\varphi_\alpha, \varphi_\beta)|^2 = \sum_{\alpha=1}^{\infty} n_\alpha \, l_\alpha^2.$$

We can employ these facts to analyze in somewhat greater detail the nature of the point $l = 0$. When the characteristic values of H other than $l = 0$ are finite in number, $l = 0$ must be a characteristic value of infinite multiplicity, in order that the set $\{\varphi_n\}$ be complete. When the characteristic values of H other than $l = 0$ are infinite in number and every element of the set $\{\varphi_n\}$ corresponds to one of them, then $l = 0$ cannot be a characteristic value; since, by the fourth condition, it is a limit point of the point spectrum it belongs to the spectrum and, more precisely, lies in the continuous spectrum. Finally there are intermediate cases where the spectrum of H reduces to the point spectrum alone and consists of infinitely many characteristic values, among which is the value $l = 0$ with finite or infinite multiplicity.

The sufficiency of our conditions is almost immediately clear. With the given transformation H we associate a matrix $A(\varphi)$ by the use of the orthonormal set $\{\varphi_n\}$ of condition (3). The only elements of $A(\varphi)$ different from zero are evidently along the principal diagonal; in the diagonal each characteristic value l_k will appear n_k times. The norm of H is thus precisely the number determined by the equation

$$N^2 = \sum_{\alpha=1}^{\infty} n_\alpha l_\alpha^2.$$

Since the series converges by condition (4), H has finite norm.

We shall apply the theorem just proved to the various operators or transformations of finite norm which were discussed in Chapter III. In the case of a bounded symmetric matrix A of finite norm, we infer the existence of a unitary matrix U such that $B = UAU^*$ is a diagonal matrix: we begin by introducing the self-adjoint transformation H associated with A by an arbitrary complete orthonormal set $\{\psi_n\}$; then we compute B as the matrix associated with H by the complete orthonormal set $\{\varphi_n\}$ of condition (4). For a symmetric kernel of Hilbert-Schmidt type and the related integral operator, the theorem informs us that

(1) the integral equation $\int_E K(P, Q) f(Q)\, dQ - lf(P) = g(P)$ has a unique solution $f(P)$, when f and g are to be in $\mathfrak{L}_2$, unless l takes on one of a finite or denumerably infinite set of values;

(2) the integral equation $\int_E K(P, Q) f(Q)\, dQ = lf(P)$ has a solution different from 0 if and only if l is one of the exceptional values in (1) and for each such value other than $l = 0$ possesses only a finite number of linearly independent solutions;

(3) an arbitrary function $f(P)$ in $\mathfrak{L}_2$ can be represented in the sense of the Riesz-Fischer theorem (convergence in the mean) by a series whose terms are solutions of the homogeneous equation of (2).

In connection with the self-adjoint differential systems discussed in Chapter III, we must proceed indirectly. If T is the transformation or operator defined by such a differential system, we can select a real number μ such that T_μ^{-1} exists and is an integral operator of Hilbert-Schmidt type. Since $T_\mu^{-1} f = 0$ implies that $f = 0$, it is clear that $l = 0$ is not a characteristic value for T_μ^{-1}. Now a necessary and sufficient condition that an element f be a characteristic element of T_μ^{-1} for a characteristic value $l \neq 0$ is that it be a characteristic value of T for the characteristic value $\mu + \frac{1}{l}$: we prove this statement by observing that the three equations

$$T_\mu^{-1} f = lf, \quad T_\mu f = \frac{1}{l} f, \quad Tf = \left(\mu + \frac{1}{l}\right) f$$

are equivalent when $l \neq 0$. It is at once evident that the demonstrated properties of the spectrum of T_μ^{-1} require that the spectrum of T consist of infinitely many characteristic values with no limit-point on the finite real axis. We can state the following facts for the operator or transformation T, the first of which we had to use in the construction of T and had already recognized as true:

(1) the differential system $Tf - lf = g$ has at most one solution $f(P)$ in $\mathfrak{L}_2$ unless l takes on one of a denumerably infinite set of real values;

(2) the differential system $Tf = lf$ has a solution different from zero if and only if l is one of the exceptional values of (1) and for each such value has only a finite number of linearly independent solutions;

(3) an arbitrary function $f(P)$ in $\mathfrak{L}_2$ can be represented in the sense of the Riesz-Fischer theorem by a series whose terms are solutions of the homogeneous equation in (2).

If $E(\lambda)$ is the resolution of the identity corresponding to the transformation T and f is an arbitrary element of $\mathfrak{L}_2$, then $(E(\varrho) - E(-\varrho))f$, $\varrho > 0$, is a finite sum of characteristic elements of T which converges "in the mean" to f when ϱ tends to infinity. For the purpose of examining the behavior of this sum in greater detail the representation of $E(\varrho) - E(-\varrho)$ in terms of the contour integral of Theorem 5.10 has proved itself of great value.

CHAPTER VI

THE OPERATIONAL CALCULUS

§ 1. The Radon-Stieltjes Integral

The formulas of the preceding chapter, especially those concerning the resolvent, suggest the construction of a general operational calculus applicable to an arbitrary self-adjoint transformation or operator H. We base our development of such a calculus upon the Radon-Stieltjes integral: if $E(\lambda)$ is the resolution of the identity corresponding to H and if $F(\lambda)$ is a complex-valued function, we define a transformation $T(F)$ by means of the equation

$$(T(F)f, g) = \int_{-\infty}^{+\infty} F(\lambda)\, d(E(\lambda)f, g)$$

where the integral is a Radon-Stieltjes integral. In this section we shall discuss the theory of integration which is to be used, in the next we shall develop the operational calculus. An important application is immediately made in Chapter VII, and an explicit illustration of the general theory by means of the Heaviside operational calculus is pointed out in Chapter X, § 2.

It will be convenient to summarize at the outset a few general facts of algebraic nature concerning the theory of point sets. We denote by I an arbitrary fixed set and by E an arbitrary subset of I. We define the sum, intersection (or product), difference, complement, inner and outer limit sets, and limit sets in the usual manner, and assume the rules for reckoning with the operations involved. The symbols $+$ and $\sum$ will be used to indicate the formation of sums, with our usual convention of applying the symbol $\sum_{\alpha=1}^{\infty}$ both to infinite and to finite sums; the symbols $\cdot$ and $\prod$ will be used to indicate the formation of intersections; and the symbol $-$ will be used to denote the formation of the difference. The complement of a set E with respect to I will be denoted by $\bar{E}$. We can now proceed to describe some of the properties which a family $\mathfrak{F}$ of sets E may enjoy with respect to the various operations introduced. The family $\bar{\mathfrak{F}}$ which consists of the complements of the sets belonging to $\mathfrak{F}$ is called the complementary family of $\mathfrak{F}$; a family $\mathfrak{F}$ is said to be self-complementary if $\mathfrak{F} = \bar{\mathfrak{F}}$. It should be observed that the dash plays quite different rôles in the symbols $\bar{E}$ and $\bar{\mathfrak{F}}$. A family $\mathfrak{F}$ is said to be additive (multiplicative) if it contains the sum (intersection) of every pair of its members; and is said to be completely additive (completely multiplicative) if it contains the sum (intersection) of the sets of every denumerable subfamily of $\mathfrak{F}$. By virtue of the relation $\overline{\sum_{\alpha=1}^{\infty} E_\alpha} = \prod_{\alpha=1}^{\infty} \bar{E}_\alpha$, a family $\mathfrak{F}$ is additive (completely additive) if and only if the complementary family $\bar{\mathfrak{F}}$ is multiplicative (completely multiplicative). In particular, a necessary and sufficient condition that a self-complementary family be additive (completely additive) is that it be multiplicative (completely multiplicative). A family which is completely additive and completely multiplicative contains the outer and inner limit sets of any sequence of its members, the limit set of any convergent sequence of its members. A complex-valued function $\mathfrak{f}(E)$ defined for every set E in an additive family is said to be additive if $\mathfrak{f}(E_1 + E_2) = \mathfrak{f}(E_1) + \mathfrak{f}(E_2)$ whenever E_1 and E_2 are disjoint sets belonging to the family; and a complex-valued function $\mathfrak{f}(E)$ defined for every set E in a completely additive family is said to be completely additive if $\mathfrak{f}\left(\sum_{\alpha=1}^{\infty} E_\alpha\right) = \sum_{\alpha=1}^{\infty} \mathfrak{f}(E_\alpha)$ whenever $\{E_n\}$ is a sequence of mutually disjoint sets belonging to the family. Finally, a complex-valued function $\mathfrak{f}(E)$ is said to be continuous whenever $\mathfrak{f}(\lim_{n\to\infty} E_n) = \lim_{n\to\infty} \mathfrak{f}(E_n)$ for every convergent sequence $\{E_n\}$ such that $\mathfrak{f}(E_n)$ and $\mathfrak{f}(\lim_{n\to\infty} E_n)$ exist.

Here we shall take I as the set of all real numbers λ, $-\infty < \lambda \leqq +\infty$, the improper value $+\infty$ being included for formal reasons. The interval $\alpha < \lambda \leqq \beta$ will be denoted by the symbol $^*\Delta$, where the asterisk indicates by its position that the interval is open at the left; we retain the earlier notation Δ for the closed interval $\alpha \leqq \lambda \leqq \beta$. The improper values $\alpha = -\infty$, $\beta = +\infty$ will be admitted; and the case where $\alpha = \beta$ will also be admitted so that we may consider the empty set as included among the intervals $^*\Delta$. The family consisting of every set representable as the sum of a finite number of mutually disjoint intervals $^*\Delta$ will be denoted by $\mathfrak{A}$, any set belonging to $\mathfrak{A}$ by A. Similarly, the family consisting of every set representable as the sum of a denumerable collection of mutually disjoint intervals $^*\Delta$ will be denoted by $\mathfrak{B}$, any set belonging to $\mathfrak{B}$ by B. The properties of the family $\mathfrak{A}$ are used only in establishing the properties of $\mathfrak{B}$; but the families $\mathfrak{B}$ and $\overline{\mathfrak{B}}$ are basic in subsequent developments. Every interval $^*\Delta$ belongs to $\mathfrak{A}$; the family $\mathfrak{A}$ is self-complementary, multiplicative, and, in consequence, additive; and, furthermore, $\mathfrak{A}$ contains $A_1 - A_2$ whenever it contains A_1 and A_2. $\mathfrak{A}$ is contained in $\mathfrak{B}$ and $\overline{\mathfrak{B}}$. The family $\mathfrak{B}$ is completely additive and multiplicative, the complementary family $\overline{\mathfrak{B}}$ is therefore additive and completely multiplicative. If B_1 and B_2 are sets belonging to $\mathfrak{B}$ such that $\overline{B}_2 \subseteq B_1$, then $B_1 - \overline{B}_2$ belongs to $\mathfrak{B}$. Every open set belongs to $\mathfrak{B}$, every closed set to $\overline{\mathfrak{B}}$.

If $\varrho(\lambda)$ is a function in the class $\mathfrak{V}^*$ defined in Chapter V, § 1, we introduce the complex number $\mathfrak{m}(^*\Delta) = \varrho(\beta) - \varrho(\alpha)$ as the weight or measure of the interval $\alpha < \lambda \leqq \beta$. The measure of a more general set E is then defined by a constructive process which starts from the interval as a foundation. In outline the theory is parallel to the Lebesgue theory of measure, where the weight assigned to an interval is its length. The fact that $\mathfrak{m}(^*\Delta)$ is a complex number necessitates certain deviations from the procedure of Lebesgue. A further variation is introduced for reasons of convenience by using the families $\mathfrak{B}$ and $\overline{\mathfrak{B}}$ in place of the families of open and of closed sets which play dominating rôles in the Lebesgue theory. Our development is thus a slightly modified form of that due to Radon.† It should be noted that Radon uses a different normal form for the function $\varrho(\lambda)$ in his treatment, preferring to have functions continuous on the left; the difference is unessential.

For an arbitrary set B we define $\mathfrak{m}(B)$ as follows: we determine a representation of B in the form $\sum_{\alpha=1}^{\infty} {}^*\Delta_\alpha$ and set $\mathfrak{m}(B) = \sum_{\alpha=1}^{\infty} \mathfrak{m}({}^*\Delta_\alpha)$. This infinite series is absolutely convergent by virtue of the fact that $\varrho(\lambda)$ is a function of bounded variation; and it can be shown that the sum of the series depends only upon the set B and not upon the particular representation of B which has been chosen. A few remarks about the proof of the latter fact are in order. First, we may consider the special case where $\varrho(\lambda)$ is real and monotone-increasing, so that $\mathfrak{m}({}^*\Delta)$ is real and not-negative. In the discussion of this case it is found that the normal form of $\varrho(\lambda)$ and the covering theorem of Heine-Borel are essential to a proof. The general case is easily reduced to this by writing

$$\varrho(\lambda) = [\varrho_1(\lambda) - \varrho_2(\lambda)] + i[\varrho_3(\lambda) - \varrho_4(\lambda)],$$

where $\varrho_k(\lambda)$ is a real monotone-increasing function in $\mathfrak{B}^*$ for $k = 1, 2, 3, 4$. We also define $\mathfrak{v}(B)$ as the least upper bound of the set of numbers $\sum_{\alpha=1}^{\infty} |\mathfrak{m}({}^*\Delta_\alpha)|$ formed for all possible representations $B = \sum_{\alpha=1}^{\infty} {}^*\Delta_\alpha$. The numbers $\mathfrak{m}(B)$ and $\mathfrak{v}(B)$ will be called respectively the ϱ-measure of B and the variation of $\varrho(\lambda)$ over the set B. The functions $\mathfrak{m}$ and $\mathfrak{v}$ are both completely additive in the completely additive

† For the Lebesgue theory, see de la Vallée Poussin, *Intégrales de Lebesgue,* Paris, 1916, pp. 1–27. Radon's discussion is found in Sitzungsberichte der Akademie der Wissenschaften zu Wien, 122^{2a^2} (1913), pp. 1298–1322. The reader will be able to supply the proofs which we omit by consulting these references. Further important aspects of the theory will be found in articles by Fréchet, Fundamenta Mathematicae, 4 (1923), pp. 329–365; 5 (1924), pp. 206–251.

family $\mathfrak{B}$; and $\mathfrak{m}$ and $\mathfrak{v}$ are both continuous in $\mathfrak{B}$. The relations

$$|\mathfrak{m}(B)| \leqq \mathfrak{v}(B) \leqq \mathfrak{v}(I) = V(\varrho), \quad \mathfrak{v}(B_2) \leqq \mathfrak{v}(B_1) \text{ when } B_2 \subseteqq B_1,$$

$$\mathfrak{v}\left(\sum_{\alpha=1}^{\infty} B_\alpha\right) \leqq \sum_{\alpha=1}^{\infty} \mathfrak{v}(B_\alpha)$$

are easily established. An important property of $\mathfrak{v}$ is expressed by the equation $\mathfrak{v}(^*\Delta) = V(\varrho; \Delta)$, where the right-hand term is the variation of $\varrho(\lambda)$ over Δ according to the definition given in Chapter V, § 1. Thus we are here using the word "variation" in a slightly modified sense. If we define $\sigma(\lambda)$ as the variation of $\varrho(\lambda)$ over the interval $^*\Delta$ with extremities $-\infty$ and λ, then $\sigma(\lambda)$ is a real monotone-increasing function in $\mathfrak{B}^*$ and $\mathfrak{v}(B)$ is the σ-measure of the set B. When $\varrho(\lambda)$ is a real monotone-increasing function, we have $\mathfrak{m}(B) = \mathfrak{v}(B)$.

If $\varrho(\lambda)$ is in $\mathfrak{B}^*$, the functions $\mathfrak{R}\varrho(\lambda)$ and $\mathfrak{I}\varrho(\lambda)$ are both in $\mathfrak{B}^*$. Hence we can define $\varrho_1(\lambda)$ and $\varrho_3(\lambda)$ as the respective variations of $\mathfrak{R}\varrho(\lambda)$ and $\mathfrak{I}\varrho(\lambda)$ over the interval $^*\Delta$ with extremities $-\infty$ and λ; we know that $\varrho_1(\lambda)$ and $\varrho_3(\lambda)$ are real monotone-increasing functions in $\mathfrak{B}^*$. It is now easily verified that the functions $\varrho_2(\lambda) = \varrho_1(\lambda) - \mathfrak{R}\varrho(\lambda)$ and $\varrho_4(\lambda) = \varrho_3(\lambda) - \mathfrak{I}\varrho(\lambda)$ are also real monotone-increasing functions in $\mathfrak{B}^*$. The equation $\varrho = (\varrho_1 - \varrho_2) + i(\varrho_3 - \varrho_4)$ is evidently true. The resolution of $\varrho(\lambda)$ into its real monotone-increasing components obtained by this particular method will be referred to as the canonical resolution. Using subscripts to distinguish between measures and variations associated with different functions, we have

$$\begin{aligned} \mathfrak{m}_k &= \mathfrak{v}_k, \qquad k = 1, 2, 3, 4, \\ \mathfrak{m} &= (\mathfrak{v}_1 - \mathfrak{v}_2) + i(\mathfrak{v}_3 - \mathfrak{v}_4), \\ \mathfrak{v} &\leqq \mathfrak{v}_1 + \mathfrak{v}_2 + \mathfrak{v}_3 + \mathfrak{v}_4, \end{aligned}$$

for every set B. On the other hand, the relations

$$\begin{aligned} \mathfrak{v}_1 &\leqq \mathfrak{v}, \qquad \mathfrak{v}_2 = \mathfrak{v}_1 - \mathfrak{R}\,\mathfrak{m} \leqq 2\mathfrak{v}, \\ \mathfrak{v}_3 &\leqq \mathfrak{v}, \qquad \mathfrak{v}_4 = \mathfrak{v}_3 - \mathfrak{I}\,\mathfrak{m} \leqq 2\mathfrak{v} \end{aligned}$$

hold for every set B, by virtue of the definition of $\varrho_k(\lambda)$, $k = 1, 2, 3, 4$. Thus we find the inequality $\frac{1}{6}(\mathfrak{v}_1 + \mathfrak{v}_2 + \mathfrak{v}_3 + \mathfrak{v}_4) \leqq \mathfrak{v} \leqq (\mathfrak{v}_1 + \mathfrak{v}_2 + \mathfrak{v}_3 + \mathfrak{v}_4)$, holding for every set B. This inequality is important in the sequel.

The final steps in constructing the theory of measure can now be taken. If E is an arbitrary subset of I, we define $\mathfrak{v}^*(E)$ as the greatest lower bound of the set of numbers $\mathfrak{v}(B)$ where $B \supseteqq E$, and we call this number the outer variation of $\varrho(\lambda)$ over the set E. The properties

$$\mathfrak{v}^*(B) = \mathfrak{v}(B), \qquad \mathfrak{v}^*(E) \leqq \mathfrak{v}^*(I) = \mathfrak{v}(I),$$
$$\mathfrak{v}^*(E_2) \leqq \mathfrak{v}^*(E_1) \quad \text{when} \quad E_2 \subseteqq E_1,$$
$$\mathfrak{v}^*\left(\sum_{\alpha=1}^{\infty} E_\alpha\right) \leqq \sum_{\alpha=1}^{\infty} \mathfrak{v}^*(E_\alpha),$$

are readily established. We define a second number $\mathfrak{v}_*(E)$ as the least upper bound of the set of numbers $\mathfrak{v}^*(\bar{B})$ where $\bar{B} \subseteqq E$; and we call this number the inner variation of $\varrho(\lambda)$ over the set E. We can then verify the relations

$\mathfrak{v}_*(\bar{B}) = \mathfrak{v}^*(\bar{B}), \quad \mathfrak{v}_*(E) \leqq \mathfrak{v}^*(E),$

$\mathfrak{v}_*(E_2) \leqq \mathfrak{v}_*(E_1)$ when $E_2 \subseteqq E_1$,

$\mathfrak{v}_*\left(\sum_{\alpha=1}^{\infty} E_\alpha\right) \geqq \sum_{\alpha=1}^{\infty} \mathfrak{v}_*(E_\alpha)$ when the sets E_k are mutually disjoint, $k = 1, 2, 3, \cdots$.

We shall now restrict our attention to the family $\mathfrak{M}$ defined by the equation $\mathfrak{v}_*(E) = \mathfrak{v}^*(E)$. A set E belonging to $\mathfrak{M}$ will be called ϱ-measurable or measurable with respect to $\varrho(\lambda)$; and the common value of $\mathfrak{v}_*(E)$ and $\mathfrak{v}^*(E)$ for such a set will be denoted by $\mathfrak{v}(E)$ and will be called the variation of $\varrho(\lambda)$ over the set E. Since $\mathfrak{v}^*(B) = \mathfrak{v}(B)$, this terminology cannot conflict with that introduced above. It can be shown that $\mathfrak{M}$ is self-complementary, completely additive and completely multiplicative and that $\mathfrak{M}$ contains $\mathfrak{B}$ and $\bar{\mathfrak{B}}$. Consequently $\mathfrak{M}$ contains the inner and outer limit sets of any sequence of its members, the limit set of any convergent sequence of its members. In particular, every Borel set—every set which can be constructed by applying addition, subtraction, multiplication, and limiting operations in any order to a denumerable collection of intervals $^*\Delta$—belongs to $\mathfrak{M}$. It is important to state various criteria which will enable us to test whether or not a given set E belongs to $\mathfrak{M}$. We have:

(1) E belongs to $\mathfrak{M}$ if and only if to arbitrary positive ε there corresponds a pair of sets B_1 and B_2 such that $\bar{B}_2 \subseteqq E \subseteqq B_1$, $\mathfrak{v}(B_1 - \bar{B}_2) < \varepsilon$;

(2) E belongs to $\mathfrak{M}$ if and only if to arbitrary positive ε there corresponds a set B such that $\bar{B} \subseteqq E$, $\mathfrak{v}^*(E - \bar{B}) < \varepsilon$;

(3) E belongs to $\mathfrak{M}$ if and only if to arbitrary positive ε there correponds a pair of sets E_1 and E_2, respectively open and closed, such that $E_2 \subseteqq E \subseteqq E_1$, $\mathfrak{v}(E_1 - E_2) < \varepsilon$;

(4) E belongs to $\mathfrak{M}$ if and only if to arbitrary positive ε there corresponds a closed set E_1 such that $E_1 \subseteqq E$, $\mathfrak{v}^*(E - E_1) < \varepsilon$;

(5) E belongs to $\mathfrak{M}$ if and only if there exist Borel sets E_1 and E_2 such that $E_1 \subseteqq E \subseteqq E_2$, $\mathfrak{v}(E_1) = \mathfrak{v}(E_2)$.

It can be shown that $\mathfrak{v}(E)$ is a completely additive continuous function over the family $\mathfrak{M}$. A ϱ-measurable set E such that $\mathfrak{v}(E) = 0$ will be called a null set with respect to $\varrho(\lambda)$; and a property which holds for every λ, $-\infty < \lambda \leqq +\infty$, with the exception of values belonging to a null set with respect to $\varrho(\lambda)$ will be said to hold almost everywhere with respect to $\varrho(\lambda)$.

It is now possible to define the ϱ-measure of an arbitrary set E belonging to the family $\mathfrak{M}$. We make use of the canonical resolution of $\varrho(\lambda)$ into its real monotone-increasing components given by the equation $\varrho = (\varrho_1 - \varrho_2) + i(\varrho_3 - \varrho_4)$, employing subscripts to distinguish between the symbols associated with the various functions $\varrho(\lambda)$, $\varrho_k(\lambda)$, $k = 1, 2, 3, 4$. We first show that the family $\mathfrak{M}$ contains those and only those sets E which belong simultaneously to the families $\mathfrak{M}_1$, $\mathfrak{M}_2$, $\mathfrak{M}_3$, and $\mathfrak{M}_4$. If E is an arbitrary set, and if B_1 and B_2 are sets such that $\bar{B}_2 \subseteqq E \subseteqq B_1$, we have for $B = B_1 - \bar{B}_2$ the inequality

$$\tfrac{1}{6}(\mathfrak{v}_1(B) + \mathfrak{v}_2(B) + \mathfrak{v}_3(B) + \mathfrak{v}_4(B)) \leqq \mathfrak{v}(B) \leqq (\mathfrak{v}_1(B) + \mathfrak{v}_2(B) + \mathfrak{v}_3(B) + \mathfrak{v}_4(B)).$$

Thus $\mathfrak{v}(B)$ is small if and only if $\mathfrak{v}_1(B)$, $\mathfrak{v}_2(B)$, $\mathfrak{v}_3(B)$, $\mathfrak{v}_4(B)$ are simultaneously small. Our first criterion for measurability shows us that E belongs to $\mathfrak{M}$ if and only if it belongs simultaneously to $\mathfrak{M}_1$, $\mathfrak{M}_2$, $\mathfrak{M}_3$, $\mathfrak{M}_4$, as we wished to prove. We define $\mathfrak{m}(E)$, the ϱ-measure of a set E belonging to $\mathfrak{M}$, by the equation

$$\begin{aligned}\mathfrak{m}(E) &= [\mathfrak{m}_1(E) - \mathfrak{m}_2(E)] + i[\mathfrak{m}_3(E) - \mathfrak{m}_4(E)] \\ &= [\mathfrak{v}_1(E) - \mathfrak{v}_2(E)] + i[\mathfrak{v}_3(E) - \mathfrak{v}_4(E)].\end{aligned}$$

It is evident that this definition includes the earlier definition of $\mathfrak{m}(B)$; and it is clear that $\mathfrak{m}$ is a completely additive continuous function in the family $\mathfrak{M}$. The various relations which hold between $\mathfrak{m}$, $\mathfrak{v}$, $\mathfrak{v}_1$, $\mathfrak{v}_2$, $\mathfrak{v}_3$, $\mathfrak{v}_4$ in $\mathfrak{B}$ can be extended directly to the family $\mathfrak{M}$.

The theory of integration founded upon the theory of measure which has been described is essentially the same as the Lebesgue theory, even in detail. A complex-valued function $F(\lambda)$ is said to be ϱ-measurable if the set of numbers λ specified by the inequalities $a_1 < \mathfrak{R}F(\lambda) \leqq a_2$, $a_3 < \mathfrak{I}F(\lambda) \leqq a_4$ is always a ϱ-measurable set. The class of ϱ-measurable functions is closed under addition, subtraction, multiplication by a complex constant, multiplication, and limiting operations; it contains all Borel measurable functions. We shall consider ϱ-measurable functions which are defined almost everywhere with respect to $\varrho(\lambda)$. Two such functions are said to be ϱ-equivalent if they are equal almost everywhere with respect to $\varrho(\lambda)$. The integral of a bounded function $F(\lambda)$ of this type may be defined as follows: we choose a positive number M such that $-M < \mathfrak{R}F(\lambda) \leqq M$, $-M < \mathfrak{I}F(\lambda) \leqq M$ almost everywhere with respect to $\varrho(\lambda)$ and set

$$c_k^{(n)} = \frac{2k-n}{n} M, \qquad c_{jk}^{(n)} = c_j^{(n)} + i c_k^{(n)}, \quad j, k = 0, \cdots, n;$$

we then construct the set $E_{jk}^{(n)}$ specified by the inequalities

$$c_j^{(n)} < \mathfrak{R}F(\lambda) \leqq c_{j+1}^{(n)}, \qquad c_k^{(n)} < \mathfrak{I}F(\lambda) \leqq c_{k+1}^{(n)},$$

for $j, k = 0, \cdots, n-1$; we find that $\lim_{n\to\infty} \sum_{\alpha,\beta=0}^{n-1} c_{\alpha\beta}^{(n)} \mathfrak{m}(E_{\alpha\beta}^{(n)})$ exists and depends only upon $F(\lambda)$ and $\varrho(\lambda)$; we therefore denote this limit by $\int_I F(\lambda)\, d\varrho(\lambda)$ and call it the Radon-Stieltjes integral of $F(\lambda)$ with respect to $\varrho(\lambda)$ over the set I. We note that the limit $\lim_{n\to\infty} \sum_{\alpha,\beta=0}^{n-1} c_{\alpha\beta}^{(n)}\, \mathfrak{v}(E_{\alpha\beta}^{(n)})$ also exists and is equal to $\int_I F(\lambda)\, d\sigma(\lambda)$, where $\sigma(\lambda)$ is the function described above, equal to the variation of $\varrho(\lambda)$ over the interval $*\Delta$ with extremities $-\infty$ and λ; it is customary to denote this limit by $\int_I F(\lambda)\, d\,|\,\varrho(\lambda)\,|$. When $F(\lambda)$ is an arbitrary ϱ-measurable function defined almost everywhere with respect to $\varrho(\lambda)$, we form the truncated function $F_M(\lambda)$, equal to $F(\lambda)$ or to zero according as the inequalities $-M < \Re F(\lambda) \leqq M$, $-M < \Im F(\lambda) \leqq M$ are satisfied or not. The integral $\int_I |\, F_M(\lambda)\,|\, d\,|\,\varrho(\lambda)\,|$ then exists and depends upon M; if it is bounded with respect to M, the function $F(\lambda)$ is said to be ϱ-integrable. The Radon-Stieltjes integral of $F(\lambda)$ with respect to $\varrho(\lambda)$ is then defined as $\lim_{M\to\infty} \int_I F_M(\lambda)\, d\varrho(\lambda)$ and is denoted by $\int_I F(\lambda)\, d\varrho(\lambda)$; we also set

$$\int_I F(\lambda)\, d\,|\,\varrho(\lambda)\,| = \lim_{M\to\infty} \int_I F_M(\lambda)\, d\,|\,\varrho(\lambda)\,|.$$

These definitions are found to be effective, and are found to be in accord with the earlier definitions when $F(\lambda)$ is bounded. The integrals $\int_E F(\lambda)\, d\varrho(\lambda)$ and $\int_E F(\lambda)\, d\,|\,\varrho(\lambda)\,|$ where $F(\lambda)$ is ϱ-integrable and E is ϱ-measurable are defined in a similar manner. Two ϱ-integrable functions which are ϱ-equivalent have equal Radon-Stieltjes integrals with respect to $\varrho(\lambda)$ over every ϱ-measurable set. It is convenient to write $\int_\alpha^\beta$ in place of $\int_{*\Delta}$ for an arbitrary interval $*\Delta: \alpha < \lambda \leqq \beta$. With this notation we find that whenever $F(\lambda)$ is a bounded continuous function the Radon-Stieltjes integral $\int_\alpha^\beta F(\lambda)\, d\varrho(\lambda)$ exists and is equal to the Stieltjes integral $\int_\alpha^\beta F(\lambda)\, d\varrho(\lambda)$ defined in Chapter V, § 1. The main properties of the Radon-Stieltjes integral which will be required later will be formulated in a series of lemmas.

LEMMA 6.1. *In its dependence on* $F(\lambda)$, $\varrho(\lambda)$, *and* E, *the Radon-Stieltjes integral has the following properties:*

(1) $\left|\int_E F(\lambda)\,d\varrho(\lambda)\right| \leqq \int_E |F(\lambda)|\,d\,|\varrho(\lambda)|$;

(2) *if* $|F(\lambda)| \leqq M$, *then* $\int_E |F(\lambda)|\,d\,|\varrho(\lambda)| \leqq M\mathfrak{v}(E)$;

(3) *if* $E = \sum_{\alpha=1}^{\infty} E_\alpha$ *where the sets* E_n *are mutually disjoint, then* $\int_E F(\lambda)\,d\varrho(\lambda) = \sum_{\alpha=1}^{\infty} \int_{E_\alpha} F(\lambda)\,d\varrho(\lambda)$;

(4) $\int_E F(\lambda)\,d\varrho(\lambda) + \int_E G(\lambda)\,d\varrho(\lambda) = \int_E [F(\lambda)+G(\lambda)]\,d\varrho(\lambda)$, *where the existence of the integrals on the left implies the existence of the integral on the right;*

(5) $a\int_E F(\lambda)\,d\varrho(\lambda) = \int_E a\,F(\lambda)\,d\varrho(\lambda)$ *where the existence of the integral on the left implies the existence of the integral on the right;*

(6) *if* $\{F_n(\lambda)\}$ *is a sequence such that* $\lim_{n\to\infty} F_n(\lambda) = F(\lambda)$ *and* $|F_n(\lambda)| \leqq \Phi(\lambda)$, *where* $\Phi(\lambda)$ *is* ϱ*-integrable and both relations hold almost everywhere with respect to* $\varrho(\lambda)$, *then*

$$\lim_{n\to\infty} \int_E F_n(\lambda)\,d\varrho(\lambda) = \int_E F(\lambda)\,d\varrho(\lambda),$$

where the existence of the integral on the left for $n = 1, 2, 3, \cdots$ *implies the existence of that on the right;*

(7) *if* $\varrho(\lambda) = \varrho_1(\lambda) + \varrho_2(\lambda)$, *then*

$$\int_E F(\lambda)\,d\varrho_1(\lambda) + \int_E F(\lambda)\,d\varrho_2(\lambda) = \int_E F(\lambda)\,d\varrho(\lambda),$$

where the existence of the integrals on the left implies the existence of that on the right;

(8) *if* $\varrho_2(\lambda) = a\,\varrho_1(\lambda)$, *then*

$$a\int_E F(\lambda)\,d\varrho_1(\lambda) = \int_E F(\lambda)\,d\varrho_2(\lambda),$$

where the existence of the integral on the left implies the existence of that on the right;

(9) *if* $\varrho = (\varrho_1 - \varrho_2) + i(\varrho_3 - \varrho_4)$ *is the canonical resolution of* $\varrho(\lambda)$ *into its real monotone-increasing components, then*

$$\int_E F(\lambda)\,d\varrho(\lambda) = \left[\int_E F(\lambda)\,d\varrho_1(\lambda) - \int_E F(\lambda)\,d\varrho_2(\lambda)\right] + i\left[\int_E F(\lambda)\,d\varrho_3(\lambda) - \int_E F(\lambda)\,d\varrho_4(\lambda)\right],$$

where the existence of the integral on the left implies and is implied by the existence of those on the right.

The properties (1)-(6) follow without difficulty from the definition of the Radon-Stieltjes integral; with minor exceptions the methods for proving them are identical with the methods used in establishing the corresponding properties of the Lebesgue integral. Property (7) is also an immediate consequence of the definition of the integral, once we have shown that every set or function measurable with respect to both ϱ_1 and ϱ_2 is also measurable with respect to $\varrho = \varrho_1 + \varrho_2$. The inequality $\mathfrak{v}(B) \leqq \mathfrak{v}_1(B) + \mathfrak{v}_2(B)$ is readily proved; in conjunction with the first criterion for measurability, it yields the desired result. We treat (8) in a similar manner, starting from the inequality $\mathfrak{v}_2(B) \leqq |a|\,\mathfrak{v}_1(B)$. Finally, property (9) follows directly from (7), (8) and the known properties of the canonical resolution.†

We shall require the following generalization of Theorem 1.24 to the case of the Radon-Stieltjes integral:

LEMMA 6.2. *Let* $\varrho(\lambda)$ *be a real monotone-increasing function in the class* $\mathfrak{B}^*$ *and let* $\mathfrak{L}_2(\varrho)$ *be the class of all complex-valued* ϱ*-measurable functions* $F(\lambda)$ *such that* $|F(\lambda)|^2$ *is* ϱ*-integrable, two functions* $F(\lambda)$ *and* $G(\lambda)$ *being regarded as identical if and only if they are* ϱ*-equivalent. If the operations* $+$ *and* $\cdot$ *are defined as ordinary addition and multiplication, the null*

† For the Lebesgue theory, see de la Vallée Poussin, *Intégrales de Lebesgue,* Paris 1916, pp. 27–50; Radon's treatment appears in Sitzungsberichte der Akademie der Wissenschaften, Wien, 122[2a^2] (1913), pp. 1322–1342. A recent discussion is due to O. Nikodym, Fundamenta Mathematicae 15 (1930), pp. 131–179.

element is defined as the function $F(\lambda) \equiv 0$, *and the function* (F, G) *is defined by the equation*

$$(F, G) = \int_{-\infty}^{+\infty} F(\lambda) \overline{G(\lambda)} \, d\varrho(\lambda),$$

then $\mathfrak{L}_2(\varrho)$ *is either a Hilbert space or an* n*-dimensional unitary space for some* n, $n = 0, 1, 2, \cdots$. *The two cases are distinguished as follows:*

(1) $\mathfrak{L}_2(\varrho)$ *is a Hilbert space if and only if* $\varrho(\lambda)$ *assumes infinitely many different values;*

(2) $\mathfrak{L}_2(\varrho)$ *is an* n*-dimensional unitary space if and only if* $\varrho(\lambda)$ *assumes exactly* $n+1$ *distinct values.*

We leave the discussion to the reader, since the proof given for Theorem 1.24 applies with little change.

LEMMA 6.3. *If* $\varrho_1(\lambda)$ *is a real monotone-increasing function in the class* $\mathfrak{B}^*$ *and* $F(\lambda)$ *is a real not-negative* ϱ_1*-integrable function, then*

(1) $\varrho_2(\lambda) = \int_{-\infty}^{\lambda} F(\mu) \, d\varrho_1(\mu)$ *is a real monotone-increasing function in* $\mathfrak{B}^*$;

(2) *every* ϱ_1*-measurable set* E *is* ϱ_2*-measurable; and for such a set* E *we have* $\mathfrak{m}_2(E) = \mathfrak{v}_2(E) = \int_E F(\mu) \, d\varrho_1(\mu)$;

(3) *a necessary and sufficient condition that a set* E *be* ϱ_2*-measurable is that* $E = E_1 + E_2$ *where* E_1 *and* E_2 *are disjoint,* E_1 *is* ϱ_1*-measurable, and* E_2 *is contained in the set specified by the equation* $F(\lambda) = 0$; *when this condition is satisfied,* $\mathfrak{m}_2(E_2) = 0$ *and* $\mathfrak{m}_2(E) = \mathfrak{m}_2(E_1)$;

(4) *if* $G(\lambda)$ *is* ϱ_2*-measurable and if* $F(\lambda) \circ G(\lambda)$ *is equal to the product* $F(\lambda) \cdot G(\lambda)$ *whenever both factors are defined and equal to zero elsewhere, then*

$$\int_{-\infty}^{+\infty} G(\lambda) \, d\varrho_2(\lambda) = \int_{-\infty}^{+\infty} F(\mu) \circ G(\mu) \, d\varrho_1(\mu)$$

where the existence of the integral on the left implies and is implied by the existence of that on the right.

It is obvious that $\varrho_2(\lambda)$ is a real monotone-increasing function such that $V(\varrho_2) = \int_{-\infty}^{+\infty} F(\mu) \, d\varrho_1(\mu)$. If we set $F_\lambda(\mu)$ equal to $F(\mu)$ or to zero according as $\mu \leqq \lambda$ or $\mu > \lambda$, we see that $\varrho_2(\lambda) = \int_{-\infty}^{+\infty} F_\lambda(\mu)\, d\varrho_1(\mu)$; and if $\{\lambda_n\}$ is a sequence such that $\lambda_n \geqq \lambda_{n+1}$, $\lambda_n \to \lambda$ where $-\infty \leqq \lambda$, we can apply Lemma 6.1 (6) to show that $\varrho_2(\lambda_n) \to \varrho_2(\lambda)$, $\varrho_2(-\infty) = 0$. Thus $\varrho_2(\lambda)$ belongs to $\mathfrak{B}^*$.

To prove (2) we first note that

$$\mathfrak{v}_2(B) = \int_B F(\mu)\, d\varrho_1(\mu), \qquad \mathfrak{v}_2(\bar{B}) = \int_{\bar{B}} F(\mu)\, d\varrho_1(\mu):$$

for if $B = \sum_{\alpha=1}^{\infty} {}^*\Delta_\alpha$, we have

$$\begin{aligned} \mathfrak{v}_2(B) &= \mathfrak{m}_2(B) = \sum_{\alpha=1}^{\infty} \mathfrak{m}_2({}^*\Delta_\alpha) \\ &= \sum_{\alpha=1}^{\infty} \int_{{}^*\Delta_\alpha} F(\mu)\, d\varrho_1(\mu) = \int_B F(\mu)\, d\varrho_1(\mu) \end{aligned}$$

and, therefore,

$$\begin{aligned} \mathfrak{v}_2(\bar{B}) &= \mathfrak{v}_2(I - B) = \mathfrak{v}_2(I) - \mathfrak{v}_2(B) \\ &= \int_I F(\mu)\, d\varrho_1(\mu) - \int_B F(\mu)\, d\varrho_1(\mu) = \int_{\bar{B}} F(\mu)\, d\varrho_1(\mu). \end{aligned}$$

It will be noted that we have here made use of Lemma 6.1, (3)–(5). Now if the set E is ϱ_1-measurable and ε is an arbitrary positive number, we choose $M > 0$ and sets B_1 and B_2 so that

$$0 \leqq \int_{-\infty}^{+\infty} F(\mu)\, d\varrho_1(\mu) - \int_{-\infty}^{+\infty} F_M(\mu)\, d\varrho_1(\mu) < \varepsilon/2,$$
$$\bar{B}_2 \subseteqq E \subseteqq B_1, \qquad \mathfrak{v}_1(B_1 - \bar{B}_2) < \varepsilon/2M,$$

where $F_M(\lambda)$ is the truncated function equal to $F(\lambda)$ or to zero according as $F(\lambda) \leqq M$ or not. Since $B_1 - \bar{B}_2$ belongs to $\mathfrak{B}$, we have

$$\begin{aligned} \mathfrak{v}_2(B_1 - \bar{B}_2) &= \int_{B_1 - \bar{B}_2} [F(\mu) - F_M(\mu)]\, d\varrho_1(\mu) + \int_{B_1 - \bar{B}_2} F_M(\mu)\, d\varrho_1(\mu) \\ &< \varepsilon/2 + M \cdot \varepsilon/2M = \varepsilon. \end{aligned}$$

Out first criterion for measurability shows that E is ϱ_2-measurable. By virtue of the further inequalities

$$\mathfrak{v}_2(\bar{B}_2) \leqq \mathfrak{v}_2(E) \leqq \mathfrak{v}_2(B_1), \quad \mathfrak{v}_2(B_1) - \mathfrak{v}_2(\bar{B}_2) = \mathfrak{v}_2(B_1 - \bar{B}_2) < \varepsilon,$$

$$\int_{\bar{B}_2} F(\mu)\, d\varrho_1(\mu) \leqq \int_E F(\mu)\, d\varrho_1(\mu) \leqq \int_{B_1} F(\mu)\, d\varrho_1(\mu),$$

we see that $\mathfrak{v}_2(E)$ and $\int_E F(\mu)\, d\varrho_1(\mu)$ differ by an amount less than ε and must therefore be equal.

To prove (3), let E be ϱ_2-measurable; and let E_0, $E^{(0)}$, $E^{(n)}$, and E_u denote respectively the ϱ_1-measurable sets where $F(\lambda) = 0$, $F(\lambda) > 1$, $\frac{1}{n+1} < F(\lambda) \leqq \frac{1}{n}$, and $F(\lambda)$ is undefined, $n = 1, 2, 3, \cdots$. We then set

$$E_1 = EE_u + \sum_{\alpha=0}^{\infty} EE^{(\alpha)}, \quad E_2 = EE_0,$$

so that E_1 and E_2 are disjoint sets with sum E and so that $F(\lambda) = 0$ on $E_2 \subseteqq E_0$. We shall prove that E_1 is ϱ_1-measurable by showing that each term of the sum which defines it is ϱ_1-measurable. First, we have $\mathfrak{v}_1^*(EE_u) \leqq \mathfrak{v}_1^*(E_u) = \mathfrak{v}_1(E_u) = 0$ so that EE_u is ϱ_1-measurable and is a null set with respect to $\varrho_1(\lambda)$. Next, we observe that $E^{(n)}$, being ϱ_1-measurable, is ϱ_2-measurable, and that $EE^{(n)}$, as the intersection of two ϱ_2-measurable sets, is itself ϱ_2-measurable. If ε is an arbitrary positive number, we can therefore choose sets $B_1^{(n)}$ and $B_2^{(n)}$ such that $\bar{B}_2^{(n)} \subseteqq EE^{(n)} \subseteqq B_1^{(n)}$, $\mathfrak{v}_2(B_1^{(n)} - \bar{B}_2^{(n)}) < \frac{\varepsilon}{n+1}$, for $n = 0, 1, 2, \cdots$. The set $D^{(n)} = E^{(n)}(B_1^{(n)} - \bar{B}_2^{(n)})$ is ϱ_1-measurable and satisfies the relations

$$EE^{(n)} - \bar{B}_2^{(n)} \subseteqq D^{(n)}, \quad D^{(n)} \subseteqq B_1^{(n)} - \bar{B}_2^{(n)}, \quad D^{(n)} \subseteqq E^{(n)}.$$

Hence the inequalities

$$\mathfrak{v}_1^*(EE^{(n)} - \bar{B}_2^{(n)}) \leqq \mathfrak{v}_1(D^{(n)}),$$

$$\frac{\varepsilon}{n+1} > \mathfrak{v}_2(B_1^{(n)} - \bar{B}_2^{(n)}) \geqq \mathfrak{v}_2(D^{(n)})$$

$$= \int_{F^{(n)}} F(\mu)\, d\varrho_1(\mu) \geqq \frac{\mathfrak{v}_1(D^{(n)})}{n+1}$$

are valid and in conjunction yield the inequality

$$\mathfrak{v}_1^*(EE^{(n)} - \bar{B}_2^{(n)}) < \varepsilon.$$

The second criterion for measurability shows that $EE^{(n)}$ is ϱ_1-measurable for $n = 0, 1, 2, \cdots$. Thus E_1 is ϱ_1-measurable, as we were to prove. By virtue of the relations $E_1 \subseteqq E_1 + E_2 \subseteqq E_1 + E_0$ we have $\mathfrak{v}_2(E_1) \leqq \mathfrak{v}_2(E) \leqq \mathfrak{v}_2(E_1 + E_0)$ and hence $\mathfrak{v}_2(E) = \int_{E_1} F(\mu)\, d\varrho_1(\mu) = \int_{E_1+E_0} F(\mu)\, d\varrho_1(\mu)$. On the other hand, the sufficiency of the condition given in (3) is evident.

To prove (4) we commence by showing that the function $F(\lambda) \circ G(\lambda)$ is ϱ_1-measurable. If we put $F_0(\lambda)$ equal to $F(\lambda)$ wherever the latter function is defined and equal to zero elsewhere, then $F_0(\lambda)$ is ϱ_1-measurable. If we put $G_0(\lambda)$ equal to $G(\lambda)$ wherever the latter function is defined and $F(\lambda) \neq 0$, equal to zero elsewhere, then $G_0(\lambda)$ is ϱ_1-measurable: for if we consider the ϱ_2-measurable set E where $a_1 < \mathfrak{R}\, G_0(\lambda) \leqq a_2$, $a_3 < \mathfrak{I}\, G_0(\lambda) \leqq a_4$ and represent it in the form $E = E_1 + E_2$ by (3), we see that E_2 is empty or is the ϱ_1-measurable set defined by the equation $F(\lambda) = 0$ according as the inequalities $a_1 < 0 \leqq a_2$, $a_3 < 0 \leqq a_4$ are false or simultaneously true; and in either case, E is ϱ_1-measurable. Since $F(\lambda) \circ G(\lambda) \equiv F_0(\lambda) \cdot G_0(\lambda)$, we see that $F(\lambda) \circ G(\lambda)$ is ϱ_1-measurable. To establish the integral relation in (4), we first treat the case where $G(\lambda)$ is real, not-negative, and bounded. If $G(\lambda) \leqq M$, we form the set $E_k^{(n)}$ where $\dfrac{kM}{n} < G(\lambda) \leqq \dfrac{(k+1)M}{n}$ for $k = 0, \cdots, n-1$, and put $G_n(\lambda)$ equal to $\dfrac{kM}{n}$ on $E_k^{(n)}$ and equal to zero elsewhere. It is clear that $\int_{-\infty}^{+\infty} G_n(\lambda)\, d\varrho_2(\lambda) = \sum_{\alpha=0}^{n-1} \dfrac{\alpha M}{n} \mathfrak{v}_2(E_\alpha^{(n)})$. Applying (3) to the ϱ_2-measurable set $E_k^{(n)}$, we have

$$\sum_{\alpha=0}^{n-1} \frac{\alpha M}{n} \mathfrak{v}_2(E_\alpha^{(n)}) = \sum_{\alpha=0}^{n-1} \frac{\alpha M}{n} \int_{E_{\alpha,1}^{(n)}} F(\mu)\, d\varrho_1(\mu),$$

an expression readily identified with the integral

$$\int_{-\infty}^{+\infty} F(\mu) \circ G_n(\mu)\, d\varrho_1(\mu).$$

We let n become infinite in the equation

$$\int_{-\infty}^{+\infty} G_n(\lambda)\, d\varrho_2(\lambda) = \int_{-\infty}^{+\infty} F(\mu) \circ G_n(\mu)\, d\varrho_1(\mu),$$

observing that the left-hand side tends to the integral of $G(\lambda)$ as its limit, by definition. On the right, we apply Lemma 6.1 (6), noting that

$$\lim_{n\to\infty} F(\mu) \circ G_n(\mu) = F(\mu) \circ G(\mu),$$
$$F(\mu) \circ G_n(\mu) \leqq F(\mu) \circ G(\mu) \leqq M \circ F(\mu).$$

Hence, we find that the equation

$$\int_{-\infty}^{+\infty} G(\lambda)\, d\varrho_2(\lambda) = \int_{-\infty}^{+\infty} F(\mu) \circ G(\mu)\, d\varrho_1(\mu)$$

is significant and true in this case. Next we consider the case where $G(\lambda)$ is a real not-negative ϱ_2-measurable function. We form the truncated function $G_M(\lambda)$ equal to $G(\lambda)$ or to zero according as $G(\lambda) \leqq M$ or $G(\lambda) > M$. In the equation

$$\int_{-\infty}^{+\infty} G_M(\lambda)\, d\varrho_2(\lambda) = \int_{-\infty}^{+\infty} F(\mu) \circ G_M(\mu)\, d\varrho_1(\mu)$$

we allow M to become infinite. The integral on the left has a finite limit if and only if $G(\lambda)$ is ϱ_2-integrable, that on the right if and only if $F(\mu) \circ G(\mu)$ is ϱ_1-integrable; and, when these limits exist, they are equal to $\int_{-\infty}^{+\infty} G(\lambda)\, d\varrho_2(\lambda)$ and $\int_{-\infty}^{+\infty} F(\mu) \circ G(\mu)\, d\varrho_1(\mu)$ respectively. Thus (4) is established for this case. When $G(\lambda)$ is an unrestricted ϱ_2-measurable function defined almost everywhere with respect to $\varrho_2(\lambda)$, we study the functions $G_1 = |\Re G|$, $G_2 = |\Re G| - \Re G \geqq 0$, $G_3 = |\Im G|$, $G_4 = |\Im G| - \Im G \geqq 0$ separately. By the aid of the relations $G = (G_1 - G_2) + i(G_3 - G_4)$ and $F \circ G = (F \circ G_1 - F \circ G_2) + i(F \circ G_3 - F \circ G_4)$ and of the facts stated in Lemma 6.1, (4) and (5), we bring the proof of (4) to a close.

In order to express symbolically the existence of a relation $\varrho_2(\lambda) = \int_{-\infty}^{\lambda} F(\mu)\, d\varrho_1(\mu)$ connecting two real monotone-increasing functions in $\mathfrak{B}^*$, where $F(\lambda)$ is real and not-negative, we shall write $\varrho_1 \succ \varrho_2$ or $\varrho_2 \prec \varrho_1$. Since the function $F(\lambda)$ in this relation is uniquely determined by ϱ_1 and ϱ_2 in the sense that two such functions must be ϱ_1-equivalent, we write $F(\lambda) = \dfrac{d\varrho_2}{d\varrho_1}$. The use of the ordinary notation of the differential calculus for this purpose is partially justified below and can be further supported by facts which we do not need to discuss in the present connection.† To indicate that the relations $\varrho_1 \succ \varrho_2$, $\varrho_2 \succ \varrho_1$ both hold, we write $\varrho_1 \sim \varrho_2$ or $\varrho_2 \sim \varrho_1$.

LEMMA 6.4. *If $\varrho_1(\lambda)$, $\varrho_2(\lambda)$, and $\varrho_3(\lambda)$ are real monotone-increasing functions in $\mathfrak{B}^*$ and if the symbols $\succ$ and $\sim$ have the meanings described above, then*

(1) *when $\varrho_1 \succ \varrho_2$, every ϱ_1-measurable set is ϱ_2-measurable; and if E is ϱ_1-measurable,*

$$\mathfrak{m}_2(E) = \mathfrak{v}_2(E) = \int_E \frac{d\varrho_2}{d\varrho_1}\, d\varrho_1(\lambda);$$

(2) *when $\varrho_1 \succ \varrho_2$, a necessary and sufficient condition that a set E be ϱ_2-measurable is that $E = E_1 + E_2$ where E_1 and E_2 are disjoint, E_1 is ϱ_1-measurable, and E_2 is contained in the set specified by the equation $\dfrac{d\varrho_2}{d\varrho_1} = 0$; when this condition is satisfied $\mathfrak{m}_2(E) = \mathfrak{m}_2(E_1)$;*

(3) *if $\varrho_1 \succ \varrho_2$ and $\varrho_2 \succ \varrho_3$, then $\varrho_1 \succ \varrho_3$ and*

$$\frac{d\varrho_3}{d\varrho_1} = \frac{d\varrho_3}{d\varrho_2} \circ \frac{d\varrho_2}{d\varrho_1}$$

almost everywhere with respect to $\varrho_1(\lambda)$;

(4) *if $\varrho_1 \sim \varrho_2$ then $\dfrac{d\varrho_1}{d\varrho_2} \circ \dfrac{d\varrho_2}{d\varrho_1} = 1$ almost everywhere with respect to ϱ_1 and with respect to ϱ_2;*

(5) *if $\varrho_1 \sim \varrho_2$ and $\varrho_2 \sim \varrho_3$, then $\varrho_1 \sim \varrho_3$.*

Statements (1) and (2) are mere paraphrases of Lemma 6.3 (2) and (3). The property (3) follows immediately from

† Compare Banach, Fundamenta Mathematicae, 6 (1924), pp. 170–188.

Lemma 6.3 (4). We put $F(\lambda) = \dfrac{d\varrho_2}{d\varrho_1}$ and define $G_\lambda(\mu)$ to be equal to $\dfrac{d\varrho_3}{d\varrho_2}$ or to zero according as $\mu \leqq \lambda$ or $\mu > \lambda$. We then have

$$\varrho_3(\lambda) = \int_{-\infty}^{\lambda} \frac{d\varrho_3}{d\varrho_2}\, d\varrho_2(\mu) = \int_{-\infty}^{+\infty} G_\lambda(\mu)\, d\varrho_2(\mu),$$

$$\varrho_2(\mu) = \int_{-\infty}^{\mu} F(\nu)\, d\varrho_1(\nu),$$

so that

$$\varrho_3(\lambda) = \int_{-\infty}^{+\infty} F(\nu) \circ G_\lambda(\nu)\, d\varrho_1(\nu) = \int_{-\infty}^{\lambda} \frac{d\varrho_3}{d\varrho_2} \circ \frac{d\varrho_2}{d\varrho_1}\, d\varrho_1(\nu).$$

Hence $\varrho_1 \succ \varrho_3$ and $\dfrac{d\varrho_3}{d\varrho_1}$ is ϱ_1-equivalent to $\dfrac{d\varrho_3}{d\varrho_2} \circ \dfrac{d\varrho_2}{d\varrho_1}$. Property (4) is a special case of (3); and (5) follows directly from the first part of (3).

In order to test whether the relation $\varrho_1 \succ \varrho_2$ is satisfied, we may appeal to the following criteria:

LEMMA 6.5. *If $\varrho_1(\lambda)$ and $\varrho_2(\lambda)$ are real monotone-increasing functions in $\mathfrak{B}^*$, then*

(1) *a necessary and sufficient condition that $\varrho_1 \succ \varrho_2$ is that every set of ϱ_1-measure zero have ϱ_2-measure zero;*

(2) *a necessary and sufficient condition that $\varrho_1 \succ \varrho_2$ is that every Borel set of ϱ_1-measure zero have ϱ_2-measure zero;*

(3) *a sufficient condition that $\varrho_1 \succ \varrho_2$ is that $\mathfrak{v}_1(E) \geqq \mathfrak{v}_2(E)$ for every ϱ_1-measurable set E;*

(4) *a sufficient condition that $\varrho_1 \succ \varrho_2$ is that $\mathfrak{v}_1(B) \geqq \mathfrak{v}_2(B)$ for every set in $\mathfrak{B}$;*

(5) *a sufficient condition that $\varrho_1 \succ \varrho_2$ is that $\mathfrak{v}_1(^*\Delta) \geqq \mathfrak{v}_2(^*\Delta)$ for every interval $^*\Delta$.*

If $\varrho(\lambda) = \sum_{\alpha=1}^{\infty} a_\alpha \varrho_\alpha(\lambda)$ where a_n is a positive constant and ϱ and ϱ_n are real monotone-increasing functions in $\mathfrak{B}^*$, $n = 1, 2, 3, \cdots$, then $\varrho \succ \varrho_n$ for every n.

The condition in (5) implies that in (4), the condition in (4) implies that in (3), by arguments already suggested above. We may note that the condition in (5) can be stated in the form: $\varrho_1(\beta)-\varrho_1(\alpha) \geqq \varrho_2(\beta)-\varrho_2(\alpha)$ for every pair of numbers α, β such that $\beta \geqq \alpha$. The condition in (3) implies the condition in (1). By virtue of our fifth criterion for measurability the conditions in (1) and (2) are equivalent. The assertion (1) is a well-known analogue of one of the fundamental criteria for determining whether or not a given function is an indefinite integral in the sense of Lebesgue.† Since, in the final statement of the lemma, $a_n \varrho_n(\lambda)$ is a real monotone-increasing function in $\mathfrak{B}^*$ which obviously satisfies the sufficient condition of (5), relative to $\varrho(\lambda)$, we have $\varrho \succ a_n \varrho_n$. The fact that $a_n > 0$ enables us to conclude that $\varrho \succ \varrho_n$.

With the aid of Lemmas 6.4 and 6.5 we proceed to remove some of the hypotheses used in Lemma 6.3.

LEMMA 6.6. *If $\varrho_1(\lambda)$ is a function in $\mathfrak{B}^*$ and $F(\lambda)$ is a complex-valued ϱ_1-integrable function, then*

(1) $\varrho_2(\lambda) = \int_{-\infty}^{\lambda} F(\mu)\, d\varrho_1(\mu)$ is in $\mathfrak{B}^*$;

(2) if $G(\lambda)$ is a ϱ_2-measurable function defined almost everywhere with respect to $\varrho_2(\lambda)$, then

$$\int_{-\infty}^{+\infty} G(\lambda)\, d\varrho_2(\lambda) = \int_{-\infty}^{+\infty} F(\mu) \circ G(\mu)\, d\varrho_1(\mu),$$

where the existence of the integral on the left implies and is implied by the existence of that on the right.

We commence with the special case where $\varrho_1(\lambda)$ is still assumed to be real and monotone-increasing. If we put

$$F_1 = |\Re F|, \qquad F_2 = |\Re F| - \Re F \geqq 0,$$
$$F_3 = |\Im F|, \qquad F_4 = |\Im F| - \Im F \geqq 0,$$
$$\varrho_{2k}(\lambda) = \int_{-\infty}^{\lambda} F_k(\mu)\, d\varrho_1(\mu), \qquad k = 1, 2, 3, 4,$$

then each of the four functions $\varrho_{2k}(\lambda)$ is a real monotone-increasing function in $\mathfrak{B}^*$ by Lemma 6.3. Thus

$$\varrho_2 = (\varrho_{21} - \varrho_{22}) + i(\varrho_{23} - \varrho_{24})$$

† The proof is given by Radon, Sitzungsberichte der Akademie der Wissenschaften, Wien, 122[2a] (1913), pp. 1342–1351; O. Nikodym, Fundamenta Mathematicae, 15 (1930), pp. 166–179; Wileński, Fundamenta Mathematicae, 16 (1930), pp. 399–400.

is a function in $\mathfrak{B}^*$. Furthermore, this resolution of $\varrho_2(\lambda)$ into real monotone-increasing components is the canonical resolution: for it can be shown† that the variations of $\Re\varrho_2(\lambda) = \int_{-\infty}^{\lambda} \Re F(\mu)\, d\varrho_1(\mu)$ and $\Im\varrho_2(\lambda) = \int_{-\infty}^{\lambda} \Im F(\mu)\, d\varrho_1(\mu)$ over the interval $*\Delta$ with extremities $-\infty$ and λ are equal respectively to $\varrho_{21}(\lambda)$ and $\varrho_{23}(\lambda)$. By making appropriate use of Lemma 6.1, (4), (5), and (9), and Lemma 6.3, (4), we obtain the result asserted in (2) for the case in hand. We shall give the details when $G(\lambda)$ is assumed to be ϱ_2-integrable, leaving the case where $F(\lambda)\circ G(\lambda)$ is ϱ_1-integrable to the reader. By Lemma 6.1 (9) we can express the integral $\int_{-\infty}^{+\infty} G(\lambda)\, d\varrho_2(\lambda)$ as a linear combination of the integrals $\int_{-\infty}^{+\infty} G(\lambda)\, d\varrho_{2k}(\lambda)$, $k = 1, 2, 3, 4$; by Lemma 6.3 (4) we can write each of the latter integrals in the form

$$\int_{-\infty}^{+\infty} F_k(\mu)\circ G(\mu)\, d\varrho_1(\mu), \quad k = 1, 2, 3, 4;$$

and finally we can write the appropriate linear combination of these four integrals in the form $\int_{-\infty}^{+\infty} F(\mu)\circ G(\mu)\, d\varrho_1(\mu)$ by Lemma 6.1, (4) and (5).

We can now extend our result to the case of an arbitrary function $\varrho_1(\lambda)$ in $\mathfrak{B}^*$. We define $\varrho_3(\lambda)$ as the variation of $\varrho_1(\lambda)$ over the interval $*\Delta$ with extremities $-\infty$ and λ; and we form the canonical resolution $\varrho_1(\lambda) = (\varrho_{11}-\varrho_{12})+i(\varrho_{13}-\varrho_{14})$. We know that $\varrho_3(\lambda)$ is a real monotone-increasing function in $\mathfrak{B}^*$. Thus, by noting the relations $\mathfrak{v}_{1k}(B) \leqq \mathfrak{v}_1(B) = \mathfrak{v}_3(B)$ and making use of Lemma 6.5 (4), we see that $\varrho_3 \succ \varrho_{1k}$ for $k = 1, 2, 3, 4$. Consequently there exist functions $H_k(\lambda)$, $H(\lambda)$ such that

$$H = (H_1 - H_2) + i(H_3 - H_4),$$

$$\varrho_{1k}(\lambda) = \int_{-\infty}^{\lambda} H_k(\mu)\, d\varrho_3(\mu), \qquad \varrho_1(\lambda) = \int_{-\infty}^{\lambda} H(\mu)\, d\varrho_3(\mu).$$

† The proof of the corresponding theorem for the Lebesgue integral is typical; see, for example, Hobson, *The Theory of Functions of a Real Variable*, third edition, Cambridge, 1927, vol. I, pp. 605-606.

Since $\varrho_3(\lambda)$ is a real monotone-increasing function in $\mathfrak{B}^*$ we can apply the results of the preceding paragraph. We have first

$$\varrho_2(\lambda) = \int_{-\infty}^{\lambda} F(\mu)\, d\varrho_1(\mu) = \int_{-\infty}^{\lambda} F(\nu) \circ H(\nu)\, d\varrho_3(\nu)$$

and, in consequence,

$$\int_{-\infty}^{+\infty} G(\lambda)\, d\varrho_2(\lambda) = \int_{-\infty}^{+\infty} F(\nu) \circ G(\nu) \circ H(\nu)\, d\varrho_3(\nu).$$

On the other hand we have also

$$\int_{-\infty}^{+\infty} F(\mu) \circ G(\mu)\, d\varrho_1(\mu) = \int_{-\infty}^{+\infty} F(\nu) \circ G(\nu) \circ H(\nu)\, d\varrho_3(\nu).$$

In each of these three equations, the existence of the integral on the left implies and is implied by the existence of that on the right. Thus the last two equations imply the truth of (2) and our proof is complete.

In applying the preceding lemmas to the construction of the operational calculus, we shall have to deal exclusively with functions in $\mathfrak{B}^*$ of the form $\varrho(\lambda) = (E(\lambda)f, g)$ where $E(\lambda)$ is a resolution of the identity; the case where $f = g$, $\varrho(\lambda) = \|E(\lambda)f\|^2$ is particularly important. We find the following extension of the inequality of Schwarz indispensable in treating such functions.

LEMMA 6.7. *If $E(\lambda)$ is a resolution of the identity, if*

$$\varrho_1(\lambda) = \|E(\lambda)f\|^2, \quad \varrho_2(\lambda) = \|E(\lambda)g\|^2, \quad \varrho_3(\lambda) = (E(\lambda)f, g),$$

and if $F(\lambda)$ and $G(\lambda)$ are functions in $\mathfrak{L}_2(\varrho_1)$ and $\mathfrak{L}_2(\varrho_2)$ respectively, then $F(\lambda)\, G(\lambda)$ is ϱ_3-integrable and

$$\left| \int_{-\infty}^{+\infty} F(\lambda)\, G(\lambda)\, d\varrho_3(\lambda) \right|^2 \leqq \int_{-\infty}^{+\infty} |F(\lambda)|\, |G(\lambda)|\, d\, |\varrho_3(\lambda)|$$
$$\leqq \int_{-\infty}^{+\infty} |F(\lambda)|^2\, d\varrho_1(\lambda) \cdot \int_{-\infty}^{+\infty} |G(\lambda)|^2\, d\varrho_2(\lambda).$$

If B is an arbitrary set in $\mathfrak{B}$ and ε is an arbitrary positive number, then there exists a representation of B as a sum of intervals, $\sum_{\alpha=1}^{\infty} {}^*\Delta_\alpha$, such that $\mathfrak{v}_3(B) - \varepsilon < \sum_{\alpha=1}^{\infty} |\mathfrak{m}_3({}^*\Delta_\alpha)|$. Since we have also

$$\sum_{\alpha=1}^{\infty} |\mathfrak{m}_3(^*\Delta_\alpha)| = \sum_{\alpha=1}^{\infty} |(E(^*\Delta_\alpha)f, g)|$$
$$\leqq \sum_{\alpha=1}^{\infty} \| E(^*\Delta_\alpha f) \| \cdot \| E(^*\Delta_\alpha) g \|$$
$$\leqq \left[\sum_{\alpha=1}^{\infty} \| E(^*\Delta_\alpha) f \|^2 \sum_{\alpha=1}^{\infty} \| E(^*\Delta_\alpha) g \|^2 \right]^{1/2} \leqq [\mathfrak{v}_1(B) \cdot \mathfrak{v}_2(B)]^{1/2},$$

where the transformation $E(^*\Delta)$ associated with the interval $^*\Delta : \alpha < \lambda \leqq \beta$ is the projection $E(\beta) - E(\alpha)$, we see that $\mathfrak{v}_3^2(B) \leqq \mathfrak{v}_1(B) \cdot \mathfrak{v}_2(B)$. Let E be a ϱ_1-measurable set, ε an arbitrary positive number, and B_1 and B_2 sets such that $\bar{B}_2 \subseteqq E \subseteqq B_1$, $\mathfrak{v}_1(B) < \varepsilon^2$, where $B = B_1 - \bar{B}_2$. Then the inequality just proved shows that $\mathfrak{v}_3(B) \leqq \varepsilon[\mathfrak{v}_2(B)]^{1/2} \leqq \varepsilon[\mathfrak{v}_2(I)]^{1/2} = \varepsilon \| g \|$ and, by virtue of our first criterion for measurability, enables us to assert that E is ϱ_3-measurable. A similar discussion shows that every ϱ_2-measurable set is ϱ_3-measurable. Obviously a set which is a null set with respect to $\varrho_1(\lambda)$ or $\varrho_2(\lambda)$ is also a null set with respect to $\varrho_3(\lambda)$. Furthermore the inequality $\mathfrak{v}_3^2(E) \leqq \mathfrak{v}_1(E) \cdot \mathfrak{v}_2(E)$ holds for every set E which is both ϱ_1-measurable and ϱ_2-measurable. For an arbitrary set E, we have similarly $(\mathfrak{v}_3^*(E))^2 \leqq \mathfrak{v}_1^*(E) \cdot \mathfrak{v}_2^*(E)$ and $(\mathfrak{v}_{3*}(E))^2 \leqq \mathfrak{v}_{1*}(E) \cdot \mathfrak{v}_{2*}(E)$.

If $F(\lambda)$ is in $\mathfrak{L}_2(\varrho_1)$, then it is evidently a ϱ_3-measurable function defined almost everywhere with respect to $\varrho_3(\lambda)$, by the facts established in the preceding paragraph; and similar remarks apply to a function $G(\lambda)$ in $\mathfrak{L}_2(\varrho_2)$. Thus $F(\lambda)\,G(\lambda)$ is a ϱ_3-measurable function defined almost everywhere with respect to $\varrho_3(\lambda)$ under the hypotheses admitted. We see that the integral $\int_{-\infty}^{+\infty} F(\lambda)\,G(\lambda)\,d\varrho_3(\lambda)$ exists if and only if the integral $\int_{-\infty}^{+\infty} |F(\lambda)\,G(\lambda)|\,d|\varrho_3(\lambda)|$ exists. By reference to Lemma 6.1 (1), we see that the present lemma is established as soon as we can prove that

$$\left[\int_{-\infty}^{+\infty} |F(\lambda)|\,|G(\lambda)|\,d|\varrho_3(\lambda)|\right]^2$$
$$\leqq \int_{-\infty}^{+\infty} |F(\lambda)|^2\,d\varrho_1(\lambda) \int_{-\infty}^{+\infty} |G(\lambda)|^2\,d\varrho_2(\lambda).$$

Hence it is sufficient to consider the case of real not-negative functions $F(\lambda)$ and $G(\lambda)$.

First, let us suppose that $F(\lambda)$ and $G(\lambda)$ are, in addition, bounded. Since $F(\lambda)\,G(\lambda)$ is also bounded, the integral $\int_{-\infty}^{+\infty} F(\lambda)\,G(\lambda)\,d\,|\varrho_3(\lambda)|$ exists. If we choose M so that $F(\lambda) \leqq M$, $G(\lambda) \leqq M$ and define the sets $E_{nk}^{(1)}$ and $E_{nk}^{(2)}$ specified by the inequalities $\frac{kM}{n} < F(\lambda) \leqq \frac{(k+1)M}{n}$ and $\frac{kM}{n} < G(\lambda) \leqq \frac{(k+1)M}{n}$ respectively, for $k = 0, \cdots, n-1$ and $n = 1, 2, 3, \cdots$, then the functions $F_n(\lambda)$ and $G_n(\lambda)$ equal to $\frac{kM}{n}$ on the sets $E_{nk}^{(1)}$ and $E_{nk}^{(2)}$ respectively and equal to zero elsewhere are both ϱ_3-measurable functions. Clearly we have

$$\int_{-\infty}^{+\infty} F_n(\lambda)\,G_n(\lambda)\,d\,|\varrho_3(\lambda)| = \sum_{\alpha,\,\beta=0}^{n-1} \frac{\alpha\beta M^2}{n^2}\,\mathfrak{v}_3(E_{n\alpha}^{(1)}E_{n\beta}^{(2)})$$

$$\leqq \sum_{\alpha,\,\beta=0}^{n-1} \frac{\alpha\beta M^2}{n^2}\,[\mathfrak{v}_{1*}(E_{n\alpha}^{(1)}E_{n\beta}^{(2)})\,\mathfrak{v}_{2*}(E_{n\alpha}^{(1)}E_{n\beta}^{(2)})]^{1/2}$$

$$\leqq \Big[\sum_{\alpha,\,\beta=0}^{n-1} \frac{\alpha^2 M^2}{n^2}\,\mathfrak{v}_{1*}(E_{n\alpha}^{(1)}E_{n\beta}^{(2)})\Big]^{1/2} \times\Big[\sum_{\alpha,\,\beta=0}^{n-1} \frac{\beta^2 M^2}{n^2}\,\mathfrak{v}_{2*}(E_{n\alpha}^{(1)}E_{n\beta}^{(2)})\Big]^{1/2}$$

$$\leqq \Big[\sum_{\alpha=0}^{n-1} \frac{\alpha^2 M^2}{n^2}\,\mathfrak{v}_{1*}(E_{n\alpha}^{(1)})\Big]^{1/2}\cdot\Big[\sum_{\beta=0}^{n-1} \frac{\beta^2 M^2}{n^2}\,\mathfrak{v}_{2*}(E_{n\beta}^{(2)})\Big]^{1/2}$$

The last expression reduces to

$$\Big[\sum_{\alpha=0}^{n-1} \frac{\alpha^2 M^2}{n^2}\,\mathfrak{v}_{1}(E_{n\alpha}^{(1)})\Big]^{1/2}\cdot\Big[\sum_{\beta=0}^{n-1} \frac{\beta^2 M^2}{n^2}\,\mathfrak{v}_{2}(E_{n\beta}^{(2)})\Big]^{1/2}$$

$$= \Big[\int_{-\infty}^{+\infty} F_n^2(\lambda)\,d\varrho_1(\lambda)\Big]^{1/2}\cdot\Big[\int_{-\infty}^{+\infty} G_n^2(\lambda)\,d\varrho_2(\lambda)\Big]^{1/2}.$$

When we allow n to become infinite we obtain the desired inequality

$$\int_{-\infty}^{+\infty} F(\lambda)\,G(\lambda)\,d\,|\varrho_3(\lambda)| \leqq \Big[\int_{-\infty}^{+\infty} F^2(\lambda)\,d\varrho_1(\lambda)\Big]^{1/2} \times\Big[\int_{-\infty}^{+\infty} G^2(\lambda)\,d\varrho_2(\lambda)\Big]^{1/2}$$

We can now treat the case where $F(\lambda)$ and $G(\lambda)$ are not required to be bounded. We form the truncated functions $F_M(\lambda)$ and $G_M(\lambda)$ in the usual manner and then allow M to become infinite in the inequality which, as bounded functions, they satisfy. The inequality

$$\int_{-\infty}^{+\infty} F(\lambda)\,G(\lambda)\,d\,|\varrho_3(\lambda)|$$
$$\leqq \left[\int_{-\infty}^{+\infty} F^2(\lambda)\,d\varrho_1(\lambda)\right]^{1/2}\cdot\left[\int_{-\infty}^{+\infty} G^2(\lambda)\,d\varrho_2(\lambda)\right]^{1/2}$$

is obtained without difficulty, the existence of the integral on the left being established at the same time. With this result the proof of the lemma is complete.

It is convenient to apply some of the notations introduced above directly to elements in $\mathfrak{H}$, when we are discussing the properties of a fixed self-adjoint transformation H with the corresponding resolution of the identity $E(\lambda)$. We make the following formal definitions:

DEFINITION 6.1. *If* $\varrho(\lambda) = \|E(\lambda)f\|^2$, *then* $\mathfrak{L}_2(f)$ *is the space* $\mathfrak{L}_2(\varrho)$.

DEFINITION 6.2. *If* $\varrho_1(\lambda) = \|E(\lambda)f_1\|^2$ *and* $\varrho_2(\lambda) = \|E(\lambda)f_2\|^2$, *we write* $f_1 \succ f_2$ *in place of* $\varrho_1 \succ \varrho_2$, $f_1 \sim f_2$ *in place of* $\varrho_1 \sim \varrho_2$.

The reader will have no difficulty in establishing such relations as the following: $\mathfrak{L}_2(af) = \mathfrak{L}_2(f)$ when $a \neq 0$, $f_1 \succ f_2$ implies $a_1 f_1 \succ a_2 f_2$ when $a_1 \neq 0$. We shall use such simple properties without further comment and without proof.

§ 2. THE OPERATIONAL CALCULUS

In all the theorems of this section, save the last, we shall consider a fixed self-adjoint transformation H with the corresponding resolution of the identity $E(\lambda)$.

THEOREM 6.1. *If* $F(\lambda)$ *is an arbitrary complex-valued function of* λ *and if* $\mathfrak{D}(F)$ *is the class of all elements* f *in* $\mathfrak{H}$ *such that* $F(\lambda)$ *belongs to* $\mathfrak{L}_2(f)$, *then* $\mathfrak{D}(F)$ *is a linear manifold and there exists a linear transformation* $T(F)$ *with domain* $\mathfrak{D}(F)$ *defined in terms of the Radon-Stieltjes integral by the relations*

$$(T(F)f, g) = \int_{-\infty}^{+\infty} F(\lambda)\, d(E(\lambda)f, g),$$

$$\|T(F)f\|^2 = \int_{-\infty}^{+\infty} |F(\lambda)|^2\, d\|E(\lambda)f\|^2,$$

holding for every f in $\mathfrak{D}(F)$ and every g in $\mathfrak{H}$. In its dependence on $F(\lambda)$, the transformation $T(F)$ has the properties

(1) $T(aF) = aT(F)$ *in* $\mathfrak{D}(F) \subseteqq \mathfrak{D}(aF)$;

(2) $T(F+G) = T(F)+T(G)$ *in* $\mathfrak{D}(F)\cdot\mathfrak{D}(G) \subseteqq \mathfrak{D}(F+G)$;

(3) $T(F)$ *and* $T(\bar{F})$ *are adjoint to each other;*

(4) *if f is an element in $\mathfrak{D}(F)$, then $E(\lambda)f$ is also in $\mathfrak{D}(F)$ and $T(F)E(\lambda)f = E(\lambda)T(F)f$; in case the linear transformation $T(F)$ is closed it is reduced by the range of the projection $E(\lambda)$;*

(5) *if f and g are elements of $\mathfrak{D}(F)$ and $\mathfrak{D}(G)$ respectively, then*

$$(T(F)f, T(G)g) = \int_{-\infty}^{+\infty} F(\lambda)\overline{G(\lambda)}\, d(E(\lambda)f, g);$$

(6) *if f is an element of $\mathfrak{D}(F)$, the element $T(F)f$ belongs to $\mathfrak{D}(G)$ if and only if f is an element of $\mathfrak{D}(F\circ G)$; when this condition is satisfied*

$$T(G)T(F)f = T(F\circ G)f;$$

(7) *if f is an element of $\mathfrak{H}$ such that $T(G)T(F)f$ and $T(G)f$ exist, then $T(F)T(G)f$ exists and*

$$T(F)T(G)f = T(G)T(F)f = T(F\circ G)f;$$

similarly, if $T(F)f$ exists and $G(\lambda)$ is a bounded function in $\mathfrak{L}_2(f)$, then $T(F)T(G)f$ and $T(G)T(F)f$ exist and are equal.

First we shall prove that $\mathfrak{D}(F)$ is a linear manifold. Obviously, $\mathfrak{D}(F)$ contains the null element. If f belongs to $\mathfrak{D}(F)$ and a is complex number different from zero, then af belongs to $\mathfrak{D}(F)$ since $\mathfrak{L}_2(af)$ is equal to $\mathfrak{L}_2(f)$ and therefore contains $F(\lambda)$. If f and g both belong to $\mathfrak{D}(F)$, we write

$$\|E(\lambda)(f+g)\|^2 = \|E(\lambda)f\|^2 + (E(\lambda)f, g) + (E(\lambda)g, f) + \|E(\lambda)g\|^2$$

and refer to Lemma 6.7 and Lemma 6.1 (7) to show that whenever $F(\lambda)$ belongs to $\mathfrak{L}_2(f)$ and $\mathfrak{L}_2(g)$ it also belongs $\mathfrak{L}_2(f+g)$; we conclude that $f+g$ is in $\mathfrak{D}(F)$. Thus $\mathfrak{D}(F)$ is a linear manifold, as we wished to prove.

Next we shall define the transformation $T(F)$. By Lemma 6.7, we have

$$\left|\int_{-\infty}^{+\infty} F(\lambda)\, d(E(\lambda)f, g)\right|^2$$
$$\leqq \int_{-\infty}^{+\infty} |F(\lambda)|^2 d\|E(\lambda)f\|^2 \cdot \int_{-\infty}^{+\infty} d\|E(\lambda)g\|^2$$

for every f in $\mathfrak{D}(F)$ and every g in $\mathfrak{H}$. Hence the expression $B(f, g) = \int_{-\infty}^{+\infty} F(\lambda)\, d(E(\lambda)f, g)$ is a numerically-valued function of the pair f, g defined for f in $\mathfrak{D}(F)$ and g in $\mathfrak{H}$, bounded for each f according to the inequality

$$|B(f, g)| \leqq C(f) \cdot \|g\|, \qquad C^2(f) = \int_{-\infty}^{+\infty} |F(\lambda)|^2\, d\|E(\lambda)f\|^2.$$

Furthermore, by virtue of Lemma 6.1, (7), (8), we find that

$$B(a_1 f_1 + a_2 f_2, g) = a_1 B(f_1, g) + a_2 B(f_2, g)$$

whenever f_1 and f_2 are in $\mathfrak{D}(F)$, and that

$$B(f, a_1 g_1 + a_2 g_2) = \bar{a}_1 B(f, g_1) + \bar{a}_2 B(f, g_2)$$

whenever g_1 and g_2 are arbitrary elements in $\mathfrak{H}$. By Theorem 2.28 there exists a linear transformation $T(F)$ with domain $\mathfrak{D}(F)$ such that $B(f, g) = (T(F)f, g)$, $\|T(F)f\| \leqq C(f)$. In the latter relation the equality always holds, as we prove below in connection with (5).

If $a \neq 0$, then $\mathfrak{D}(aF)$ and $\mathfrak{D}(F)$ are identical since $\mathfrak{L}_2(f)$ contains both functions $aF(\lambda)$ and $F(\lambda)$ or neither; and if $a = 0$, then $\mathfrak{D}(aF) \supseteqq \mathfrak{D}(F)$. When f is in $\mathfrak{D}(F)$ we see by Lemma 6.1 (5) that

$$\int_{-\infty}^{+\infty} aF(\lambda)\, d(E(\lambda)f, g) = a\int_{-\infty}^{+\infty} F(\lambda)\, d(E(\lambda)f, g)$$

for every g in $\mathfrak{H}$. Thus $T(aF)f = aT(F)f$, and (1) is established. A similar argument based on Lemma 6.1 (4) yields a proof of (2). In order to show that $T(F) \frown T(\bar{F})$, we select arbitrary elements f and g in $\mathfrak{D}(F)$ and $\mathfrak{D}(\bar{F})$ respectively and then write

$$\begin{aligned}(T(F)f, g) &= \int_{-\infty}^{+\infty} F(\lambda)\, d(E(\lambda)f, g) = \int_{-\infty}^{+\infty} F(\lambda)\, d(f, E(\lambda)g) \\ &= \overline{\int_{-\infty}^{+\infty} \bar{F}(\lambda)\, d(E(\lambda)g, f)} \\ &= \overline{(T(\bar{F})g, f)} = (f, T(\bar{F})g).\end{aligned}$$

We may note that $\mathfrak{D}(\bar{F})$ is identical with $\mathfrak{D}(F)$. To prove (4), we first observe that $\| E(\mu)E(\lambda)f\|^2$ is equal to $\|E(\mu)f\|^2$ or to $\|E(\lambda)f\|^2$ according as $\mu \leqq \lambda$ or $\mu > \lambda$. If $F(\lambda)$ is in $\mathfrak{L}_2(f)$, the integral $\int_{-\infty}^{+\infty} |F(\mu)|^2 d\|E(\mu)E(\lambda)f\|^2$ therefore exists and is equal to $\int_{-\infty}^{\lambda} |F(\mu)|^2 d\|E(\mu)f\|^2$; thus $F(\lambda)$ belongs to $\mathfrak{L}_2(E(\lambda)f)$, $E(\lambda)f$ is in $\mathfrak{D}(F)$ together with f. We now verify directly that

$$\begin{aligned}(T(F)E(\lambda)f, g) &= \int_{-\infty}^{+\infty} F(\mu)\, d(E(\mu)E(\lambda)f, g) \\ &= \int_{-\infty}^{+\infty} F(\mu)\, d(E(\mu)f, E(\lambda)g) \\ &= (T(F)f, E(\lambda)g) = (E(\lambda)T(F)f, g)\end{aligned}$$

for every f in $\mathfrak{D}(F)$ and every g in $\mathfrak{H}$. This suffices to establish (4). The proof of (5) is founded on (4). By Lemma 6.7 we know that whenever f is in $\mathfrak{D}(F)$ and g is in $\mathfrak{D}(G)$ the function $F(\lambda)\overline{G(\lambda)}$ is integrable with respect to $(E(\lambda)f, g)$. We can therefore proceed as follows:

$$\begin{aligned}(T(F)f, T(G)g) &= \int_{-\infty}^{+\infty} F(\lambda)\, d(E(\lambda)f, T(G)g) \\ &= \int_{-\infty}^{+\infty} F(\lambda)\, d(f, E(\lambda)T(G)g) \\ &= \int_{-\infty}^{+\infty} F(\lambda)\, d\overline{(T(G)E(\lambda)g, f)} \\ &= \int_{-\infty}^{+\infty} F(\lambda)\, d\left(\int_{-\infty}^{+\infty} \overline{G(\mu)}\, d\overline{(E(\mu)E(\lambda)g, f)}\right)\end{aligned}$$

$$= \int_{-\infty}^{+\infty} F(\lambda)\, d\left(\int_{+\infty}^{\lambda} \overline{G(\mu)\, d(g, E(\mu) f)}\right)$$
$$= \int_{-\infty}^{+\infty} F(\lambda)\, d\left(\int_{-\infty}^{\lambda} \overline{G(\mu)}\, d(E(\mu) f, g)\right)$$
$$= \int_{-\infty}^{+\infty} F(\lambda) \overline{G(\lambda)}\, d(E(\lambda) f, g),$$

where the last equation follows from Lemma 6.6. We shall also use (4) in the proof of (6). We must show that, when $T(F)f$ exists, $T(G) \cdot T(F) f$ exists if and only if $T(F \circ G) f$ exists — in other words, that, when the integral

$$\int_{-\infty}^{+\infty} | F(\lambda) |^2 d \| E(\lambda) f \|^2$$

exists, the integral

$$\int_{-\infty}^{+\infty} | G(\lambda) |^2 d \| E(\lambda) T(F) f \|^2$$

exists if and only if the integral

$$\int_{-\infty}^{+\infty} | F(\lambda) \circ G(\lambda) |^2 d \| E(\lambda) f \|^2$$

exists. If we interchange $E(\lambda)$ and $T(F)$ by means of (4) and then make use of (5) we have

$$\int_{-\infty}^{+\infty} | G(\lambda) |^2 d \| T(F) E(\lambda) f \|^2$$
$$= \int_{-\infty}^{+\infty} | G(\lambda) |^2 d \left(\int_{-\infty}^{+\infty} | F(\mu) |^2 d \| E(\mu) E(\lambda) f \|^2 \right)$$
$$= \int_{-\infty}^{+\infty} | G(\lambda) |^2 d \left(\int_{-\infty}^{\lambda} | F(\mu) |^2 d \| E(\mu) f \|^2 \right)$$
$$= \int_{-\infty}^{+\infty} | F(\mu) |^2 \circ | G(\mu) |^2 d \| E(\mu) f \|^2$$
$$= \int_{-\infty}^{+\infty} | F(\lambda) \circ G(\lambda) |^2 d \| E(\lambda) f \|^2,$$

where the next to the last equation depends on Lemma 6.3; and the existence of the first integral implies and is implied by the existence of the last. To complete the proof of (6) we note the equations

$$(T(G)\, T(F) f, g) = \int_{-\infty}^{+\infty} G(\lambda)\, d(E(\lambda)\, T(F) f, g)$$
$$= \int_{-\infty}^{+\infty} G(\lambda)\, d\left(\int_{-\infty}^{\lambda} F(\mu)\, d(E(\mu) f, g)\right)$$

$$= \int_{-\infty}^{+\infty} F(\mu) \circ G(\mu)\, d(E(\mu) f, g)$$
$$= (T(F \circ G) f, g);$$

in deriving them we make use of Lemma 6.6. Finally, (7) is merely a specialization of (6) and does not require discussion.

THEOREM 6.2. *If $\mathfrak{M}(f)$ is the set of all elements $f^* = T(F)f$ where $F(\lambda)$ belongs to $\mathfrak{L}_2(f)$, then the correspondence $f^* \sim F(\lambda)$ defines an isomorphism between $\mathfrak{M}(f)$ and $\mathfrak{L}_2(f)$ in the sense that*

(1) *$f^* = g^*$ if and only if the corresponding functions $F(\lambda)$ and $G(\lambda)$ are identical in $\mathfrak{L}_2(f)$—that is, are equivalent with respect to $\|E(\lambda)f\|^2$;*

(2) *$af^* \sim aF(\lambda)$ if $f^* \sim F(\lambda)$;*

(3) *$f^* + g^* \sim F(\lambda) + G(\lambda)$ if $f^* \sim F(\lambda)$ and $g^* \sim G(\lambda)$;*

(4) $(f^*, g^*) = (F, G) = \int_{-\infty}^{+\infty} F(\lambda)\, \overline{G(\lambda)}\, d\|E(\lambda)f\|^2$.

$\mathfrak{M}(f)$ is therefore a closed linear manifold. A sequence of elements $f_n^* = T(F_n)f$ converges and has the limit $f^* = T(F)f$ in $\mathfrak{M}(f)$ if and only if the sequence of functions $F_n(\lambda)$ is convergent in $\mathfrak{L}_2(f)$ and has the limit $F(\lambda)$ in $\mathfrak{L}_2(f)$.

The symbol $\sim$, which is used to express the fact that an element in $\mathfrak{M}(f)$ and an element in $\mathfrak{L}_2(f)$ are in correspondence, here plays a different rôle from that assigned to it in Definition 6.2. The various properties of the correspondence between $\mathfrak{M}(f)$ and $\mathfrak{L}_2(f)$ are merely properties (1), (2), and (5) of the preceding theorem phrased in different form: this is obvious save in the case of property (1) of the correspondence; and in that case we have only to remark that $f^* \sim F(\lambda)$, $g^* \sim G(\lambda)$ imply

$$\|f^* - g^*\|^2 = \int_{-\infty}^{+\infty} |F(\lambda) - G(\lambda)|^2\, d\|E(\lambda)f\|^2$$

by virtue of Theorem 6.1, (1), (2), and (5). The existence of such a correspondence between $\mathfrak{M}(f)$ and $\mathfrak{L}_2(f)$ shows at once that $\mathfrak{M}(f)$ must be a closed linear manifold. The facts about corresponding sequences in $\mathfrak{M}(f)$ and $\mathfrak{L}_2(f)$ are obvious implications of the isomorphism.

In order to make further progress in the study of the transformation $T(F)$ defined above we shall lay certain restrictions on the function $F(\lambda)$ which so far has been entirely arbitrary.

DEFINITION 6.3. *If H is a self-adjoint transformation and $E(\lambda)$ is the corresponding resolution of the identity, the terms "H-measurable", "null set with respect to H", "almost everywhere with respect to H", and "H-equivalent" shall mean respectively "ϱ-measurable", "null set with respect to $\varrho(\lambda)$", "almost everywhere with respect to $\varrho(\lambda)$", and "ϱ-equivalent", for $\varrho(\lambda) = \| E(\lambda) f \|^2$ and every element f in $\mathfrak{H}$.*

We find the following criterion useful in testing for measurability with respect to H.

THEOREM 6.3. *Let $\{\varphi_n\}$ be a complete orthonormal set and write $\varrho_n(\lambda) = \| E(\lambda) \varphi_n \|^2$. A necessary and sufficient condition that a set (or, a function) be H-measurable is that it be ϱ_n-measurable for every n. A necessary and sufficient condition that a set be a null set with respect to H is that it be a null set with respect to $\varrho_n(\lambda)$ for every n.*

Since the necessity of the conditions stated is trivial, we turn at once to a proof of their sufficiency.

Let E be ϱ_n-measurable. It is at once obvious that E is measurable with respect to $\| E(\lambda) a \varphi_n \|^2 = |a|^2 \| E(\lambda) \varphi_n \|^2$. If E is measurable with respect to $\| E(\lambda) f \|^2$ and $\| E(\lambda) g \|^2$, it is measurable also with respect to $\| E(\lambda)(f+g) \|^2 = \| E(\lambda) f \|^2 + (E(\lambda) f, g) + (E(\lambda) g, f) + \| E(\lambda) g \|^2$. In fact, by the aid of results obtained in the course of the proof of Lemma 6.7, we see that on setting

$$\varrho(\lambda) = \| E(\lambda)(f+g) \|^2, \ \varrho_1(\lambda) = \| E(\lambda) f \|^2, \ \varrho_2(\lambda) = \| E(\lambda) g \|^2,$$
$$\varrho_3(\lambda) = (E(\lambda) f, g)$$

we have

$$\varrho = \varrho_1 + \varrho_3 + \bar{\varrho}_3 + \varrho_2, \quad \mathfrak{v} \leqq \mathfrak{v}_1 + 2\mathfrak{v}_3 + \mathfrak{v}_2 \leqq (\mathfrak{v}_1^{1/2} + \mathfrak{v}_2^{1/2})^2.$$

The final inequality here holds for every set B in $\mathfrak{B}$ and leads at once to the result we wish. The inequality can then be extended to hold for every set E measurable with respect to both $\varrho_1(\lambda)$ and $\varrho_2(\lambda)$. Combining the results so far obtained, we see that E is measurable with respect to $\|E(\lambda)f\|^2$ for every element f in the linear manifold determined by the set $\{\varphi_n\}$. We now make use of the fact that this manifold is everywhere dense in $\mathfrak{H}$. If f is an arbitrary element in $\mathfrak{H}$ and ε is an arbitrary positive number, we select an element g in this manifold such that $\|f-g\|<\varepsilon$. We set $\varrho_1(\lambda)=\|E(\lambda)g\|^2$, $\varrho_2(\lambda)=\|E(\lambda)f\|^2$. Since E is ϱ_1-measurable, we can select sets B_1 and B_2 such that $\bar{B}_2 \subseteqq E \subseteqq B_1$, $\mathfrak{v}_1(B_1-\bar{B}_2)<\varepsilon$. Writing $B = B_1-\bar{B}_2$ and applying Lemma 6.1 (7), we have

$$\mathfrak{v}_2(B) = \int_B d\|E(\lambda)f\|^2$$

$$= \int_B d(E(\lambda)f, f-g) + \int_B d(E(\lambda)(f-g), g) + \int_B d\|E(\lambda)g\|^2.$$

By Lemma 6.1 (1) and Lemma 6.7, we obtain the inequalities

$$\left|\int_B d(E(\lambda)f, f-g)\right| \leqq \int_{-\infty}^{+\infty} d\,|(E(\lambda)f, f-g)|$$
$$\leqq \|f\|\cdot\|f-g\| \leqq \varepsilon\|f\|,$$

$$\left|\int_B d(E(\lambda)(f-g), g)\right| \leqq \int_{-\infty}^{+\infty} d\,|(E(\lambda)(f-g), g)|$$
$$\leqq \|f-g\|\cdot\|g\|$$
$$\leqq \|f-g\|(\|f-g\|+\|f\|) < \varepsilon\|f\|+\varepsilon^2.$$

We have also $\int_B d\|E(\lambda)g\|^2 = \mathfrak{v}_1(B) < \varepsilon$. Combining these inequalities, we find that $\mathfrak{v}_2(B) < \varepsilon(1+2\|f\|)+\varepsilon^2$. Our first criterion for measurability shows that E is measurable with respect to $\varrho_2(\lambda) = \|E(\lambda)f\|^2$. Since f is arbitrary, E is H-measurable, as we wished to prove.

If E is a null set with respect to $\varrho_n(\lambda)$ for every n, we have only to follow the argument of the preceding paragraph to show that E is a null set with respect to $\|E(\lambda)f\|^2$ for every f and, therefore, with respect to H. By specializing the earlier steps of the proof given above, we see that E is a null set with respect to $\|E(\lambda)f\|^2$ for every element f in the linear manifold determined by the set $\{\varphi_n\}$. It is only in the final step, where f is allowed to be arbitrary, that we need to make any modification. We select g as before. We can then take $B_2 = I$, so that $B = B_1 \supseteqq E$. We find that $\mathfrak{v}_2^*(E) \leqq \mathfrak{v}_2(B_1) = \mathfrak{v}_2(B)$ can be made as small as we please by an appropriate initial choice of ϵ. Hence $\mathfrak{v}_2^*(E) = 0$ and $\mathfrak{v}_2(E) = 0$, as we wished to prove.

THEOREM 6.4. *If $F(\lambda)$ is an H-measurable function defined almost everywhere with respect to H, then $T(F)$ is a closed linear transformation with domain $\mathfrak{D}(F)$ everywhere dense in $\mathfrak{H}$. The adjoint transformation $T^*(F)$ exists and is identical with $T(\bar{F})$.*

Since $F(\lambda)$ is an H-measurable function defined almost everywhere with respect to H, the functions $|F| + a$ and $F(|F| + a)^{-1}$, where a is a positive real constant, have the same properties. Evidently, $\mathfrak{D}(|F| + a) = \mathfrak{D}(F)$; and, since $F(|F| + a)^{-1}$ is bounded, $\mathfrak{D}(F(|F| + a)^{-1}) = \mathfrak{H}$. By Theorem 6.1 (7) we see that

$$\begin{aligned} T(F) &= T(|F| + a)\, T(F(|F| + a)^{-1}) \\ &= T(F(|F| + a)^{-1})\, T(|F| + a) \end{aligned}$$

throughout $\mathfrak{D}(F)$. We can show without difficulty that $T(F(|F| + a)^{-1})$ and $T(\bar{F}(|F| + a)^{-1})$ are two bounded linear transformations with domain $\mathfrak{H}$; and, by reference to Theorem 6.1 (3), we see that each is the adjoint of the other. Next we shall show that $T(|F| + a)$ is a self-adjoint transformation; we can then conclude that its domain $\mathfrak{D}(|F| + a) = \mathfrak{D}(F)$ is everywhere dense in $\mathfrak{H}$. By Theorem 6.1 (6) we have

$$f = T(|F| + a)\, T((|F| + a)^{-1}) f \text{ for every } f \text{ in } \mathfrak{H},$$
$$f = T((|F| + a)^{-1})\, T(|F| + a) f \text{ for every } f \text{ in } \mathfrak{D}(F).$$

Hence $T(|F| + a)$ and $T((|F| + a)^{-1})$ are inverse to each other. The latter transformation is defined throughout $\mathfrak{H}$ and, by Theorem 6.1 (3), is symmetric; it is therefore a self-adjoint transformation. According to Theorem 2.18 its inverse $T(|F| + a)$ is also self-adjoint.

Since the linear transformation $T(F)$ has domain $\mathfrak{D}(F)$ everywhere dense in $\mathfrak{H}$, we can construct the adjoint transformation $T^*(F)$. Whenever g and g^* are elements of $\mathfrak{H}$ such that $(T(F)f, g) = (f, g^*)$ for every f in $\mathfrak{D}(F)$, we have $T^*(F)g = g^*$; we wish to prove that g is in $\mathfrak{D}(\bar{F}) = \mathfrak{D}(F)$ and that $T(\bar{F})g = g^*$. Using some of the transformations discussed in the preceding paragraph, we find that

$$\begin{aligned}(T(F)f, g) &= (T(F(|F|+a)^{-1})\, T(|F|+a)f, g) \\ &= (T(|F|+a)f,\ T(\bar{F}(|F|+a)^{-1})g)\end{aligned}$$

for every f in $\mathfrak{D}(F)$. Since the equation

$$(T(|F|+a)f,\ T(\bar{F}(|F|+a)^{-1})g) = (f, g^*)$$

holds for every f in the domain $\mathfrak{D}(F)$ of the self-adjoint transformation $T(|F|+a)$, we conclude that $T(\bar{F}(|F|+a)^{-1})g$ is in the domain of this transformation and that

$$T(|F|+a)\cdot T(\bar{F}(|F|+a)^{-1})g = g^*.$$

By Theorem 6.1 (6) we infer that $T(\bar{F})g$ exists and is equal to g^*. Thus $T^*(F) \equiv T(\bar{F})$ as we wished to prove. In view of the fact that $T(F) \equiv T^*(\bar{F})$, we see that $T(F)$ is a closed linear transformation in accordance with Theorem 2.8. This completes the proof of the present theorem.

THEOREM 6.5. *If $F(\lambda)$ is an H-measurable function defined almost everywhere with respect to H, then a necessary and sufficient condition that $T(F)$ have $\mathfrak{H}$ as its domain is that it be a bounded linear transformation, satisfying the inequality $\|T(F)f\| \leqq C\|f\|$ for some constant C and every f in $\mathfrak{H}$; a necessary and sufficient condition that the transformation $T(F)$ satisfy such an inequality for a given value of C is that $F(\lambda)$ satisfy the inequality $|F(\lambda)| \leqq C$ almost everywhere with respect to H.*

If $T(F)$ is bounded, then its domain $\mathfrak{D}(F)$ must coincide with $\mathfrak{H}$ since, by the preceding theorem, $T(F)$ is a closed linear transformation and $\mathfrak{D}(F)$ is everywhere dense in $\mathfrak{H}$. On the other hand, if $\mathfrak{D}(F) = \mathfrak{H}$, then $T(F)$ and its adjoint $T^*(F) \equiv T(\bar{F})$ have $\mathfrak{H}$ as their common domain; hence $T(F)$ is bounded in accordance with Theorem 2.26.

If $|F(\lambda)| \leqq C$ almost everywhere with respect to H, it is evident that $T(F)$ is defined throughout $\mathfrak{H}$ and is therefore bounded; we have furthermore the inequality

$$\|T(F)f\|^2 = \int_{-\infty}^{+\infty} |F(\lambda)|^2 \, d\|E(\lambda)f\|^2$$
$$\leqq C^2 \int_{-\infty}^{+\infty} d\|E(\lambda)f\|^2 = C^2\|f\|^2$$

or

$$\|T(F)f\| \leqq C\|f\|.$$

On the other hand, if the latter inequality is satisfied for a given value of C, we introduce the H-measurable set E_ε defined by the inequality $|F(\lambda)| \geqq C + \varepsilon$. We wish to show that E_ε is a null set with respect to H for every $\varepsilon > 0$. We let $G_\varepsilon(\lambda)$ denote the bounded H-measurable function equal to 1 or to 0 according as λ is in E_ε or $\bar{E}_\varepsilon$. The transformation $T(G_\varepsilon)$ is a bounded linear transformation with domain $\mathfrak{H}$; and the identities $T(F)\,T(G_\varepsilon) \equiv T(G_\varepsilon)\,T(F) \equiv T(F \circ G_\varepsilon)$ hold throughout $\mathfrak{H}$ by Theorem 6.1 (7). If f is an arbitrary element of $\mathfrak{H}$ and $\varrho(\lambda) = \|E(\lambda)f\|^2$ then

$$\mathfrak{v}(E_\varepsilon) = \int_{E_\varepsilon} d\varrho(\lambda) = \int_{-\infty}^{+\infty} G_\varepsilon^2(\lambda)\, d\|E(\lambda)f\|^2 = \|T(G_\varepsilon)f\|^2.$$

In the inequality satisfied by $T(F)$ we replace f by $T(G_\varepsilon)f$, obtaining $\|T(F \circ G_\varepsilon)f\| \leqq C\|T(G_\varepsilon)f\| = C\mathfrak{v}^{1/2}(E_\varepsilon)$; but we also have

$$\|T(F \circ G_\varepsilon)f\|^2 = \int_{-\infty}^{+\infty} |F(\lambda) \circ G_\varepsilon(\lambda)|^2 \, d\|E(\lambda)f\|^2$$
$$= \int_{E_\varepsilon} |F(\lambda)|^2 \, d\|E(\lambda)f\|^2 \geqq (C+\varepsilon)^2 \int_{E_\varepsilon} d\varrho(\lambda)$$
$$= (C+\varepsilon)^2\, \mathfrak{v}(E_\varepsilon).$$

Hence the inequality $C^2\mathfrak{v}(E_\varepsilon) \geqq (C+\varepsilon)^2\mathfrak{v}(E_\varepsilon)$ is true and implies that $\mathfrak{v}(E_\varepsilon) = 0$ whenever $\varepsilon > 0$. Since f is arbitrary, E_ε is a null set with respect to H and the inequality $|F(\lambda)| \leqq C$ holds almost everywhere with respect to H.

We can now characterize the transformation $T(F)$ by means of the properties of the function $F(\lambda)$ in some simple and important cases.

THEOREM 6.6. *If $F(\lambda)$ and $G(\lambda)$ are H-measurable functions defined almost everywhere with respect to H, then $T(F) = T(G)$ in $\mathfrak{D}(F) \cdot \mathfrak{D}(G)$ if and only if $F(\lambda)$ and $G(\lambda)$ are H-equivalent. If $F(\lambda)$ has the properties described above, then the transformation $T(F)$ is*

(1) *self-adjoint if and only if $F(\lambda)$ and $\bar{F}(\lambda)$ are H-equivalent;*

(2) *unitary if and only if $|F(\lambda)|^2 = F(\lambda)\bar{F}(\lambda)$ is equal to 1 almost everywhere with respect to H;*

(3) *a projection if and only if $F(\lambda)$ assumes the values 0 and 1 except on a null set with respect to H.*

Since the sufficiency of these conditions is obvious, we do not consider it here, but pass directly to the discussion of the necessity. When $F(\lambda)$ and $G(\lambda)$ have the properties described, we see that $F-G$, $|F|$, $|G|$, $|F|+|G|$ are all H-measurable functions defined almost everywhere with respect to H. We note the relations

$$\mathfrak{D}(F) = \mathfrak{D}(|F|), \qquad \mathfrak{D}(G) = \mathfrak{D}(|G|),$$
$$\mathfrak{D}(F) \cdot \mathfrak{D}(G) = \mathfrak{D}(|F|+|G|) \subseteq \mathfrak{D}(F-G),$$

which show that the linear manifold $\mathfrak{D}(F) \cdot \mathfrak{D}(G)$ is everywhere dense in $\mathfrak{H}$. If $T(F) = T(G)$ in this manifold, we have $T(F-G) = T(F) - T(G) = O$ in $\mathfrak{D}(F) \cdot \mathfrak{D}(G)$. Since $T(F-G)$ is a closed linear transformation, the identity $T(F-G) \equiv O$ must hold throughout $\mathfrak{H}$. The relation $\|T(F-G)f\| \leqq C\|f\|$ holds therefore for $C = 0$ and every f in $\mathfrak{H}$. The preceding theorem shows that $F(\lambda) = G(\lambda)$ almost everywhere with respect to H. When $F(\lambda)$ has the properties described in the theorem, the transformation $T(F)$ is self-adjoint only if $\mathfrak{D}(F) = \mathfrak{D}(\bar{F})$ and $T(F) \equiv T(\bar{F})$; that is, only if $F(\lambda) = \bar{F}(\lambda)$ almost everywhere with respect to H. Similarly, $T(F)$ is unitary only if $T(F)\,T^*(F) \equiv T(F)\,T(\bar{F}) \equiv I$ throughout $\mathfrak{H}$: that is, only if $F(\lambda)\bar{F}(\lambda)$ is equal to 1 almost everywhere with respect to H. Finally $T(F)$ is a projection only if it is self-adjoint and $T^2(F)$

$\equiv T(F^2) \equiv T(F)$ throughout $\mathfrak{H}$: that is, only if $F(\lambda) = \bar{F}(\lambda)$ and $F^2(\lambda) = F(\lambda)$ almost everywhere with respect to H, a pair of conditions equivalent to the one stated in (3).

We are now in a position to characterize the spectrum of the transformation $T(F)$ when $F(\lambda)$ is an H-measurable function defined almost everywhere with respect to H. Roughly speaking, we can assert that the spectrum of $T(F)$ is the smallest closed set which contains the map of the spectrum of H on the l-plane by means of the relation $l = F(\lambda)$. More precisely, we have:

THEOREM 6.7. *Let $F(\lambda)$ be an H-measurable function defined almost everywhere with respect to H. Then l belongs to the resolvent set of $T(F)$ if and only if there exists a positive real number C such that $|F(\lambda) - l| \geqq \frac{1}{C}$ almost everywhere with respect to H: and l is a characteristic value of $T(F)$ if and only if there exists an element f in $\mathfrak{H}$ such that $\|E(\lambda)f\|^2$ has positive variation over the set of points where $F(\lambda) = l$. The transformation $T((F-l)^{-1})$ is defined except when l is a characteristic value of $T(F)$ and is the inverse of $T(F-l) \equiv T(F) - lI$; it is bounded if and only if l is a point of the resolvent set of $T(F)$.*

We begin by considering the function $(F(\lambda) - l)^{-1}$; this function is H-measurable, and is defined almost everywhere with respect to H whenever the set specified by the equation $F(\lambda) = l$ is a null set with respect to H. When the latter condition is satisfied, the transformation $T((F-l)^{-1})$ exists and has a domain everywhere dense in $\mathfrak{H}$; by reference to Theorem 6.1 (6) we verify easily that the transformations $T(F-l) \equiv T(F) - lI$ and $T((F-l)^{-1})$ are inverse to each other. By Theorem 6.5 we know that $T((F-l)^{-1})$ is defined throughout $\mathfrak{H}$ if and only if there exists a positive constant C such that $|F(\lambda) - l|^{-1} \leqq C$ almost everywhere with respect to H; in other words, l is a point of the resolvent set of $T(F)$ if and only if there is a positive constant C such that $|F(\lambda) - l| \geqq \frac{1}{C}$ almost everywhere with respect to H.

We have thus disposed of all values l except those for which there is some element f in $\mathfrak{H}$ with the property that $\|E(\lambda)f\|^2$ has positive variation over the set where $F(\lambda) = l$. We denote this set by E and define $G(\lambda)$ as the function which is equal to 1 or to 0 according as λ is in E or in $\bar{E}$. Since $F(\lambda)\,G(\lambda) \equiv F(\lambda)\circ G(\lambda) \equiv l\,G(\lambda)$, we conclude that $T(G)\,g$ is an element belonging to $\mathfrak{D}(F)$ for every g in $\mathfrak{H}$ and that

$$T(F)\,T(G)\,g = T(F\circ G)\,g = T(lG)\,g = l\,T(G)\,g.$$

If in particular we choose g as the element f described above, we have also

$$\|T(G)f\|^2 = \int_{-\infty}^{+\infty} |G|^2\, d\|E(\lambda)f\|^2 = \int_E d\|E(\lambda)f\|^2 = \mathfrak{v}(E) > 0.$$

Thus $T(G)f$ is a characteristic element, l a characteristic value of $T(F)$. On the other hand, if l is a characteristic value of $T(F)$, the inverse of $T(F-l)$ cannot exist and l must be a point of the type described at the opening of the paragraph.

The theorems which we have established show that we can perform algebraic manipulations with the transformations $T(F)$ just as we do with the functions $F(\lambda)$. That the isomorphism between the calculus of transformations or operators and the calculus of functions is not confined to the algebraic operations alone, has been pointed out by J. v. Neumann*. We shall extend his results to the case of unbounded functions. In order to carry out this extension we shall employ the following theorem:

THEOREM 6.8. *If $F(\lambda)$ is an H-measurable function defined almost everywhere with respect to H, then there exists an H-equivalent function which is defined everywhere and which is in the kth class of Baire, for $k = 0, 1, 2$ or 3. In particular, if $F(\lambda)$ is real, then there exist two real functions $F_1(\lambda)$ and $F_2(\lambda)$ which, in addition to the properties described, have the property $F_1(\lambda) \leqq F(\lambda) \leqq F_2(\lambda)$; and if $F(\lambda)$ is also bounded, then $F_1(\lambda)$ and $F_2(\lambda)$ may be taken as bounded.*

---

* J. v. Neumann, Annals of Mathematics, (2) 32 (1931), pp. 191–226, especially pp. 205–207.

We commence with the case of a real bounded function $F(\lambda)$. If $F(\lambda)$ fails to be defined on a null set with respect to H, we shall, for the sake of convenience, supply the missing values by setting $F(\lambda) = 0$ on this set. We determine a number $M > 0$ such that $-M < F(\lambda) \leqq M$ and form the H-measurable set E_{km} specified by the inequality

$$\frac{2k-m}{m} M < F(\lambda) \leqq \frac{2(k+1)-m}{m} M$$

for $k = 0, \cdots, m-1$ and $m = 1, 2, 3, \cdots$. Next we choose a complete orthonormal set $\{\varphi_n\}$ and introduce the functions $\varrho_n(\lambda) = \| E(\lambda)\varphi_n \|^2$. Since E_{km} is ϱ_n-measurable for every n, we can apply our fourth criterion for measurability and select a closed set $E_{km}^{(n)} \subseteqq E_{km}$ such that $\mathfrak{v}_n(E_{km} - E_{km}^{(n)}) \leqq \mathfrak{v}_n(E_{km})/2^m$. We now define a function $F_m^{(n)}(\lambda)$ equal to $\frac{2k-m}{m} M$ on the set $E_{km}^{(n)}$, $k = 0, \cdots, m-1$, and equal to $-M$ on the open set $\overline{\sum_{\alpha=0}^{m-1} E_{\alpha m}^{(n)}}$. It is immediately obvious that $F_m^{(n)}(\lambda)$ is in the first class of Baire, unless it happens to be a constant and belong to the 0th class.* We see furthermore that $F_m^{(n)}(\lambda) \leqq F(\lambda)$.

Our next step is to define the function $F^{(n)}(\lambda)$ equal to the least upper bound of the set of numbers $F_m^{(n)}(\lambda)$, $m = 1, 2, 3, \cdots$. The function $F^{(n)}(\lambda)$ belongs to a class of Baire not higher than the second.† It is evident that $F^{(n)}(\lambda) \leqq F(\lambda)$. Furthermore we can show that $F^{(n)}$ and F are ϱ_n-equivalent. If E_n is the ϱ_n-measurable set defined by the inequality $F^{(n)}(\lambda) < F(\lambda)$, we see that a point λ in E_n can belong to only a finite number of the sets $E_{km}^{(n)}$ for fixed n, and must therefore belong to infinitely many sets of the sequence $\left\{\sum_{\alpha=0}^{m-1} (E_{\alpha m} - E_{\alpha m}^{(n)})\right\}$; in other words, $E_n \subseteqq E^{(n)}$, where $E^{(n)}$ is the outer limit set of this sequence. By virtue of the relation

---

* de la Vallée Poussin, *Intégrales de Lebesgue,* Paris, 1916, pp. 117–118.

† Carathéodory, *Vorlesungen über reelle Funktionen,* second edition, Leipzig 1927, Chapter VII, Theorem 10, p. 398.

$$E^{(n)} = \lim_{l\to\infty} \sum_{\beta=l+1}^{\infty} \sum_{\alpha=0}^{\beta-1} (E_{\alpha\beta} - E_{\alpha\beta}^{(n)})$$

we have

$$\mathfrak{v}_n(E_n) \leqq \mathfrak{v}_n(E^{(n)}) = \lim_{l\to\infty} \mathfrak{v}_n\left(\sum_{\beta=l+1}^{\infty} \sum_{\alpha=0}^{\beta-1} (E_{\alpha\beta} - E_{\alpha\beta}^{(n)})\right)$$

$$\leqq \lim_{l\to\infty}\left(\sum_{\beta=l+1}^{\infty} \sum_{\alpha=0}^{\beta-1} \mathfrak{v}_n(E_{\alpha\beta} - E_{\alpha\beta}^{(n)})\right)$$

$$\leqq \lim_{l\to\infty}\left(\sum_{\beta=l+1}^{\infty} \sum_{\alpha=0}^{\beta-1} \frac{\mathfrak{v}_n(E_{\alpha\beta})}{2^\beta}\right)$$

$$= \lim_{l\to\infty} \sum_{\beta=l+1}^{\infty} \frac{\mathfrak{v}_n(I)}{2^\beta} = 0.$$

Our contention is thus established.

The last step consists in defining $F_1(\lambda)$ as the least upper bound of the numbers $F^{(n)}(\lambda)$ for $n = 1, 2, 3, \cdots$. Since $F^{(n)}(\lambda) \leqq F_1(\lambda) \leqq F(\lambda)$ for every n, we see that F_1 and F are ϱ_n-equivalent for every n; by Theorem 6.3 it follows that F_1 and F are H-equivalent. Finally $F_1(\lambda)$ belongs to a class of Baire not higher than the third.

We define $F_2(\lambda)$ indirectly. Setting $G(\lambda) = -F(\lambda)$ we construct the corresponding function $G_1(\lambda) \leqq G(\lambda)$ by the method just described. We then put $F_2(\lambda) = -G_1(\lambda)$ and find that this function has the various properties desired.

Next let us consider a real function $F(\lambda)$, not necessarily bounded, which is defined everywhere and satisfies the inequality $F(\lambda) \geqq 1$. We set $F^{(n)}(\lambda)$ equal to $F(\lambda)$ or to n according as $F(\lambda) \leqq n$ or $F(\lambda) > n$, $n = 1, 2, 3, \cdots$. We then apply our preceding results to the function $F^{(n)}(\lambda)$, obtaining the two functions $F_1^{(n)}(\lambda)$ and $F_2^{(n)}(\lambda)$. We put $F_1(\lambda) = \lim_{n\to\infty} F_1^{(n)}(\lambda)$, $F_2(\lambda) = \lim_{n\to\infty} F_2^{(n)}(\lambda)$. It is evident that $F_1(\lambda) \leqq F(\lambda) \leqq F_2(\lambda)$ and that these three functions are H-equivalent. If we consider the function $\dfrac{1}{F_1(\lambda)} = \lim_{n\to\infty} \dfrac{1}{F_1^{(n)}(\lambda)}$ we see that the defining sequence converges uniformly by virtue of the inequality $0 \leqq \dfrac{1}{F_1^{(n)}(\lambda)} - \dfrac{1}{F_1(\lambda)} \leqq \dfrac{1}{n}$. Hence

$\frac{1}{F_1(\lambda)}$ and $F_1(\lambda)$ belong to a class of Baire no higher than the third.* Similar arguments apply to $F_2(\lambda)$.

We can now take up the case of a real function $F(\lambda)$, not necessarily bounded. We begin by setting $F(\lambda) = 0$ wherever the function happens to be undefined. We then form the functions $F^{(1)} = |F| + 1$ and $F^{(2)} = |F| - F + 1$, noting the relations

$$F = F^{(1)} - F^{(2)}, \qquad F^{(1)} \geqq 1, \qquad F^{(2)} \geqq 1.$$

By the preceding paragraph, we construct functions $F_1^{(1)}$, $F_2^{(1)}$, $F_1^{(2)}$, $F_2^{(2)}$. The functions $F_1(\lambda)$ and $F_2(\lambda)$ are then given by the equations

$$F_1 = F_1^{(1)} - F_2^{(2)}, \qquad F_2 = F_2^{(1)} - F_1^{(2)}$$

and are easily shown to have all the desired properties.

Finally, we may consider the general case of a complex-valued function $F(\lambda)$ by treating its real and imaginary parts separately. We conclude by the preceding results that the given function $F(\lambda)$ is H-equivalent to a function which belongs to a class of Baire no higher than the third.

THEOREM 6.9. *Let H be a self-adjoint transformation, $F(\lambda)$ a real H-measurable function defined almost everywhere with respect to H, and $T(F)$ the transformation associated with $F(\lambda)$ and H. Then $T(F)$ is a self-adjoint transformation, which will be denoted by $H^{(1)}$. Now let $G^{(1)}(\lambda)$ be a complex-valued $H^{(1)}$-measurable function defined almost everywhere with respect to $H^{(1)}$, and let $T^{(1)}(G^{(1)})$ be the transformation associated with $G^{(1)}(\lambda)$ and $H^{(1)}$. If $\Phi(\lambda) = G^{(1)}(F(\lambda))$, then $\Phi(\lambda)$ is an H-measurable function defined almost everywhere with respect to H; and the transformation $T(\Phi)$ associated with $\Phi(\lambda)$ and H is identical with $T^{(1)}(G^{(1)})$.*

By Theorem 6.6 (1) the transformation $H^{(1)} \equiv T(F)$ is self-adjoint; we can therefore construct the transformation $T^{(1)}(G^{(1)})$ introduced in the statement of the theorem.

---

* Carathéodory, *Vorlesungen über reelle Funktionen*, second edition, Leipzig 1927, Chapter VII, Theorems 8 and 10, pp. 398–399.

When $G^{(1)}(\lambda) \equiv \dfrac{1}{\lambda - l}$ for arbitrary not-real l, the assertion of the theorem reduces, by reference to Theorem 5.11, to the following statement: the resolvent of $H^{(1)}$ is identical with the transformation $T\left(\dfrac{1}{F(\lambda) - l}\right)$, for arbitrary not-real l. The latter statement has already been established in Theorem 6.7.

Let us consider what class of functions we can build by the use of linear combinations and of limiting processes if we start with the class of all functions $\dfrac{1}{\lambda - l}$, where l is not real. We observe first that the infinite series

$$\frac{1}{2c}\sum_{\alpha=0}^{\infty}\left[\frac{1}{\lambda-(\mu+i(\alpha+\frac{1}{2})\pi/c)}+\frac{1}{\lambda-(\mu-i(\alpha+\frac{1}{2})\pi/c)}\right]$$
$$=\sum_{\alpha=0}^{\infty}\frac{c(\lambda-\mu)}{c^2(\lambda-\mu)^2+(\alpha+\frac{1}{2})^2\pi^2}$$

converges boundedly to the function $\frac{1}{2}\tanh c(\lambda-\mu)$. Next we let c become positively infinite, finding that $\frac{1}{2}\tanh c(\lambda-\mu)$ converges boundedly to the function $\varphi_\mu(\lambda)$ which assumes the value $\frac{1}{2}$, 0, or $-\frac{1}{2}$ according as $\lambda > \mu$, $\lambda = \mu$, or $\lambda < \mu$. If α, β, and ε are arbitrary real numbers such that $\alpha < \beta$, $\varepsilon > 0$, we form the function $\varphi_{\alpha+\varepsilon}(\lambda) - \varphi_{\beta+\varepsilon}(\lambda)$ and allow ε to tend to zero; the function converges boundedly to the limit function $\psi_{\alpha\beta}(\lambda)$ which is equal to 1 or to 0 according as λ belongs to the interval $^*\Delta$ with extremities α and β, or not. With the aid of the functions $\psi_{\alpha\beta}(\lambda)$ it is possible to construct every bounded complex-valued continuous function $F(\lambda)$, $-\infty < \lambda < +\infty$; and we may suppose that in any limiting process employed we have bounded convergence. It is now clear that we can obtain every bounded function $F(\lambda)$ in every class of Baire, by repeated applications of limiting processes; in fact, we may restrict ourselves to limiting processes in which we have bounded convergence.†

In order to apply the results of the preceding paragraph, we establish a few preliminary facts. First, let $G_k^{(1)}(\lambda)$,

† Carathéodory, *Vorlesungen über reelle Funktionen*, second edition, Leipzig 1927, Chapter VII, Theorem 5, p. 396.

$k = 1, \cdots, n$, be bounded $H^{(1)}$-measurable functions defined everywhere, and let us suppose that the theorem has been proved for each of them. We prove that the theorem is true for the function $G^{(1)}(\lambda) = \sum_{\alpha=1}^{n} a_\alpha G_\alpha^{(1)}(\lambda)$. We introduce the functions $\Phi_k(\lambda) = G_k^{(1)}(F(\lambda))$, $k = 1, \cdots, n$, and $\Phi(\lambda) = G^{(1)}(F(\lambda)) = \sum_{\alpha=1}^{n} a_\alpha \Phi_\alpha(\lambda)$. By Theorem 6.1, (1) and (2), we have

$$T^{(1)}(G^{(1)}) \equiv \sum_{\alpha=1}^{n} a_\alpha T^{(1)}(G_\alpha^{(1)}) \equiv \sum_{\alpha=1}^{n} a_\alpha T(\Phi_\alpha) \equiv T(\Phi)$$

throughout $\mathfrak{H}$. Next, let the theorem be true for each of the $H^{(1)}$-measurable functions $G_n^{(1)}(\lambda)$ defined everywhere, $n = 1, 2, 3, \cdots$; and let the relations

$$|G_n^{(1)}(\lambda)| \leqq M, \quad \lim_{n\to\infty} G_n^{(1)}(\lambda) = G^{(1)}(\lambda)$$

hold for every λ. We show that the theorem is true for $G^{(1)}(\lambda)$. Obviously $G^{(1)}(\lambda)$ is a bounded $H^{(1)}$-measurable function defined everywhere. Furthermore, we see that $T^{(1)}(G_n^{(1)})f \to T^{(1)}(G^{(1)})f$ for every f in $\mathfrak{H}$: to prove this assertion, we note that $G_n^{(1)} \to G^{(1)}$ in $\mathfrak{L}_2(f)$ and then apply Theorem 6.2. By hypothesis $\Phi_n(\lambda) = G_n^{(1)}(F(\lambda))$ is an H-measurable function defined wherever $F(\lambda)$ is defined. Obviously, $|\Phi_n(\lambda)| \leqq M$ and $\lim_{n\to\infty} \Phi_n(\lambda) = G^{(1)}(F(\lambda)) = \Phi(\lambda)$ wherever $F(\lambda)$ is defined. Hence $\Phi(\lambda)$ is a bounded H-measurable function defined almost everywhere with respect to H. Clearly we must have $T(\Phi_n)f \to T(\Phi)f$ for every f in $\mathfrak{H}$, by the reasoning used above. Our hypothesis that $T^{(1)}(G_n^{(1)})$ and $T(\Phi_n)$ are identical for every n now shows that $T^{(1)}(G^{(1)}) \equiv T(\Phi)$, as we wished to prove.

Combining the facts deduced in the two preceding paragraphs, we see that the theorem is true for every bounded function $G^{(1)}(\lambda)$ defined everywhere which belongs to one of Baire's classes. Thus we can appeal for further progress to Theorem 6.8.

Let $G^{(1)}(\lambda)$ be a bounded real $H^{(1)}$-measurable function defined almost everywhere with respect to $H^{(1)}$; and let $G_1^{(1)}(\lambda)$ and $G_2^{(1)}(\lambda)$ be the two equivalent functions of Baire described in Theorem 6.8. By Theorem 6.6 we have $T^{(1)}(G_1^{(1)}) \equiv T^{(1)}(G^{(1)}) \equiv T^{(1)}(G_2^{(1)})$. If we write $\Phi_1(\lambda) = G_1^{(1)}(F(\lambda))$, $\Phi(\lambda) = G^{(1)}(F(\lambda))$, $\Phi_2(\lambda) = G_2^{(1)}(F(\lambda))$, we have

$$\Phi_1(\lambda) \leqq \Phi(\lambda) \leqq \Phi_2(\lambda), \quad T(\Phi_1) \equiv T(\Phi_2)$$

by virtue of the identities $T^{(1)}(G_1^{(1)}) \equiv T(\Phi_1)$, $T^{(1)}(G_2^{(1)}) \equiv T(\Phi_2)$. Theorem 6.6 shows that Φ_1 and Φ_2 are H-equivalent; hence Φ must be H-equivalent to them both. Thus we must have $T(\Phi) \equiv T(\Phi_1) \equiv T(\Phi_2) \equiv T^{(1)}(G^{(1)})$, as we wished to prove.

If $G^{(1)}(\lambda)$ is not real, we consider its real and imaginary parts separately and combine them to obtain the desired result. We leave the details to the reader, since they differ only slightly from those given in the discussion of linear combinations above.

Finally we take up the general case of a complex-valued function $G^{(1)}(\lambda)$ which may not be bounded. We introduce the bounded function $G_M^{(1)}(\lambda)$ equal to $G^{(1)}(\lambda)$ or to zero according as the inequalities $-M < \Re G^{(1)}(\lambda) \leqq M$, $-M < \Im G^{(1)}(\lambda) \leqq M$ are true or not; and we put $\Phi_M(\lambda) = G_M^{(1)}(F(\lambda))$, $\Phi(\lambda) = G^{(1)}(F(\lambda))$. Clearly $\lim_{M\to\infty} G_M^{(1)}(\lambda) = G^{(1)}(\lambda)$, $\lim_{M\to\infty} \Phi_M(\lambda) = G^{(1)}(F(\lambda)) = \Phi(\lambda)$. We know that $T^{(1)}(G_M^{(1)}) \equiv T(\Phi_M)$. Now it is easily verified that $G_M^{(1)}$ converges in $\mathfrak{L}_2(f)$ if and only if $G^{(1)}$ is in $\mathfrak{L}_2(f)$; and then $G_M^{(1)} \to G^{(1)}$. Hence $T^{(1)}(G_M^{(1)})f$ converges in $\mathfrak{H}$ if and only if f is in the domain of $T^{(1)}(G^{(1)})$, in accordance with Theorem 6.2; and when this condition is satisfied, $T^{(1)}(G_M^{(1)})f \to T^{(1)}(G^{(1)})f$. This means that $T(\Phi_M)f$ converges in $\mathfrak{H}$ if and only if f is in the domain of $T^{(1)}(G^{(1)})$. By Theorem 6.2, Φ_M converges in $\mathfrak{L}_2(f)$ if and only if f is in the domain of $T^{(1)}(G^{(1)})$; and hence Φ is in $\mathfrak{L}_2(f)$ if and only if the same condition is satisfied. Thus $\Phi(\lambda)$ is a ϱ-measurable function defined almost everywhere, with respect to $\varrho(\lambda)$, for $\varrho(\lambda) = \|E(\lambda)f\|^2$ and every f in the domain of $T^{(1)}(G^{(1)})$.

Since the admissible elements f constitute a linear manifold everywhere dense in $\mathfrak{H}$ by Theorem 6.4, we see that Φ is an H-measurable function defined almost everywhere with respect to H, by virtue of Theorem 6.3. Thus $T(\Phi)$ exists and has the same domain as $T^{(1)}(G^{(1)})$. Finally, the relation $T(\Phi_M) \to T(\Phi)$ holds throughout this domain, and we conclude that $T^{(1)}(G^{(1)}) \equiv T(\Phi)$. This completes the proof of the theorem.

It is now convenient to revise our notation so that it reflects more fully the established isomorphism between the operational calculus with the transformations or operators $T(F)$ and the functional calculus with the functions $F(\lambda)$. We have used a less colorful notation in stating our theorems in order to avoid giving them a specious appearance of simplicity.

DEFINITION 6.4. *If H is a self-adjoint transformation and $F(\lambda)$ is an arbitrary complex-valued function, the transformation $T(F)$ will be denoted by the new symbol $F(H)$.*

We leave to the reader the task of interpreting the preceding theorems by means of the new notation and that of calculating the transformation $F(H)$ for special choices of $F(\lambda)$ such as $F(\lambda) \equiv 0, 1, \lambda, \lambda^n$. We shall use this notation without further comment; such symbols as e^{iH}, $\sqrt{H}$, $|H|$ are to be interpreted in terms of it whenever they occur.

CHAPTER VII

THE UNITARY EQUIVALENCE OF SELF-ADJOINT TRANSFORMATIONS

§ 1. Preparatory Theorems

On the basis of the operational calculus to which the preceding chapter was devoted, it is possible to solve the problem of the unitary equivalence of self-adjoint transformations, formulated in the last section of Chapter II: the problem of determining what conditions enable us to connect two self-adjoint transformations H_1 and H_2 by the relation $UH_1U^{-1} \equiv H_2$, where U is a unitary transformation. Our solution differs in no essential respect from that which Hellinger and Hahn have given in the case of bounded self-adjoint transformations.* Because of our geometrical approach, the solution assumes a form more transparent than theirs.

We first show that the problem of unitary equivalence for two self-adjoint transformations reduces to the corresponding problem for the associated resolutions of the identity.

Theorem 7.1. *If H_1 and H_2 are self-adjoint transformations and if $E_1(\lambda)$ and $E_2(\lambda)$ are the corresponding resolutions of the identity, then the relation $UH_1U^{-1} \equiv H_2$, where U is a unitary transformation, implies and is implied by the relation $UE_1(\lambda)U^{-1} \equiv E_2(\lambda)$, $-\infty < \lambda < +\infty$.*

It is easily verified that $UE_1(\lambda)U^{-1}$ is a resolution of the identity whenever U is a unitary transformation independent of λ. The corresponding self-adjoint transformation can be calculated by means of Theorem 5.9 and is found to be the

---

* Hellinger, Journal für Mathematik, 136 (1909), pp. 210–271; Hahn, Monatshefte für Mathematik und Physik, 23 (1912), pp. 161–224.

transformation UH_1U^{-1}. Evidently, an element f is in the domain of the self-adjoint transformation determined by $UE_1(\lambda)U^{-1}$ if and only if

$$\int_{-\infty}^{+\infty} \lambda^2 d\|UE_1(\lambda)U^{-1}f\|^2$$

$$= \int_{-\infty}^{+\infty} \lambda^2 d\|E_1(\lambda)U^{-1}f\|^2 = \|H_1U^{-1}f\|^2 = \|UH_1U^{-1}f\|^2;$$

and the transformation in question is determined from the equation

$$\int_{-\infty}^{+\infty} \lambda\, d(UE_1(\lambda)U^{-1}f, g) = \int_{-\infty}^{+\infty} \lambda\, d(E_1(\lambda)U^{-1}f, U^{-1}g)$$
$$= (H_1U^{-1}f, U^{-1}g) = (UH_1U^{-1}f, g),$$

holding for every f in its domain and every g in $\mathfrak{H}$. The present theorem now follows directly.

THEOREM 7.2. *Let H be a self-adjoint transformation, $E(\lambda)$ the corresponding resolution of the identity, f an arbitrary element of $\mathfrak{H}$, $\mathfrak{M}(f)$ the closed linear manifold of Theorem 6.2, and E the projection of $\mathfrak{H}$ on $\mathfrak{M}(f)$. Then $\mathfrak{M}(f)$ is the closed linear manifold determined by the set of elements $E(\lambda)f$, $-\infty < \lambda < +\infty$; it reduces both $E(\lambda)$ and H. The projection E is permutable with $E(\lambda)$ and with H.*

Let us denote by $\mathfrak{N}$ the closed linear manifold determined by the set of elements $E(\lambda)f$. By virtue of the fact that $E(\lambda) \equiv F_\lambda(H)$ where $F_\lambda(\mu)$ is equal to 1 or to 0 according as $\mu \leqq \lambda$ or $\mu > \lambda$, it is evident that $\mathfrak{N} \subseteq \mathfrak{M}(f)$. In order to prove that $\mathfrak{N} = \mathfrak{M}(f)$ it is therefore sufficient to prove that the orthogonal complement of $\mathfrak{M}(f)$ contains that of $\mathfrak{N}$. We can do so, as follows: if $(E(\lambda)f, g) = 0$ for every value of λ, then we have

$$(F(H)f, g) = \int_{-\infty}^{+\infty} F(\lambda)\, d(E(\lambda)f, g) = 0$$

for every function $F(\lambda)$ in $\mathfrak{L}_2(f)$, and conclude that g is orthogonal to $\mathfrak{M}(f)$ whenever it is orthogonal to $\mathfrak{N}$.

We can show that the projection $E(\lambda)$ transforms $\mathfrak{M}(f)$ into a subset of itself: for if $F(H)f$ is a given element of $\mathfrak{M}(f)$ and $F_\lambda(\mu)$ is the function defined in the preceding paragraph, we set $G_\lambda(\mu) = F(\mu) \circ F_\lambda(\mu)$ and obtain the equation

$E(\lambda)F(H)f = F(H)E(\lambda)f = F(H)\cdot F_\lambda(H)f = G_\lambda(H)f$, by the help of Theorem 6.1, (4) and (7). Since $E(\lambda)$ is a self-adjoint transformation, we can apply Theorem 4.24 and thus show that $\mathfrak{M}(f)$ reduces $E(\lambda)$. This means that E is permutable with $E(\lambda)$.

Finally, since E is permutable with $E(\lambda)$, it is also permutable with R_l, the resolvent of H, as can be verified with the aid of Theorem 5.7; this means that $\mathfrak{M}(f)$ reduces R_l. Theorem 4.27 shows that $\mathfrak{M}(f)$ reduces H and that H and E are permutable.

This theorem suggests the fundamental rôle which will be played by manifolds of the type of $\mathfrak{M}(f)$ in the further study of self-adjoint transformations and their reducibility. If we have two self-adjoint transformations H_1 and H_2 with the corresponding resolutions of the identity $E_1(\lambda)$ and $E_2(\lambda)$, we construct the closed linear manifolds $\mathfrak{M}_1(f_1)$ and $\mathfrak{M}_2(f_2)$ associated with H_1, f_1 and H_2, f_2 respectively. The question of the unitary equivalence of H_1 and H_2 then hinges upon the following problem: to determine whether or not there exists an isometric transformation V with domain $\mathfrak{M}_1(f_1)$ and range $\mathfrak{M}_2(f_2)$ such that $VE_1(\lambda)V^{-1} = E_2(\lambda)$ in $\mathfrak{M}_2(f_2)$. If we take $H_1 \equiv H_2 \equiv H$, a solution of the specialized problem so obtained affords us a criterion for determining when $\mathfrak{M}(f_1) = \mathfrak{M}(f_2)$ for two elements f_1 and f_2 in $\mathfrak{H}$. We shall answer the questions raised, in the following theorem.

THEOREM 7.3. *Let $H_1, H_2, E_1(\lambda), E_2(\lambda), \mathfrak{M}_1(f_1)$, and $\mathfrak{M}_2(f_2)$ have the meanings stated above; and let $\varrho_1(\lambda) = |E_1(\lambda)f_1|^2$, $\varrho_2(\lambda) = |E_2(\lambda)f_2|^2$. There exists an isometric transformation V with domain $\mathfrak{M}_1(f_1)$ and range $\mathfrak{M}_2(f_2)$, such that $VE_1(\lambda)V^{-1} = E_2(\lambda)$ in $\mathfrak{M}_2(f_2)$, if and only if $\varrho_1 \sim \varrho_2$. When $H_1 \equiv H_2 \equiv H$, a necessary and sufficient condition that $\mathfrak{M}(f_1) = \mathfrak{M}(f_2)$ is that f_2 be an element of $\mathfrak{M}(f_1)$ with the property $f_2 \sim f_1$.*

First, let us consider the necessity of the condition for the existence of V. If we put $g_1 = V^{-1}f_2$, we have

$$|E_1(\lambda)g_1|^2 = |E_1(\lambda)V^{-1}f_2|^2 = |VE_1(\lambda)V^{-1}f_2|^2$$
$$= |E_2(\lambda)f_2|^2$$

so that $\varrho_2(\lambda) = \|E_1(\lambda)\, g_1\|^2$. Now g_1 is an element of $\mathfrak{M}_1(f_1)$ and can therefore be expressed in the form $F_1(H_1)f_1$ where $F_1(\lambda)$ is in $\mathfrak{L}_2^{(1)}(f_1)$. By reference to Theorem 6.1, we find

$$\begin{aligned}\varrho_2(\lambda) = \|E_1(\lambda)\, g_1\|^2 &= \|E_1(\lambda)\, F_1\,(H_1)\, f_1\|^2 \\ &= \int_{-\infty}^{\lambda} |F_1(\mu)|^2\, d\|E_1(\mu)\, f_1\|^2\end{aligned}$$

and conclude that $\varrho_1 \succ \varrho_2$. In the same manner we show that $\varrho_1 \prec \varrho_2$. Thus $\varrho_1 \sim \varrho_2$, as we wished to prove.

Next we consider the sufficiency of the condition for the existence of V. We suppose that $\varrho_1 \sim \varrho_2$. Then the functions $\dfrac{d\varrho_2}{d\varrho_1}$ and $\dfrac{d\varrho_1}{d\varrho_2}$ exist and are respectively ϱ_1-integrable and ϱ_2-integrable functions which assume real not-negative values. Hence $\left(\dfrac{d\varrho_2}{d\varrho_1}\right)^{1/2}$ is a real not-negative function in $\mathfrak{L}_2(\varrho_1) = \mathfrak{L}_2^{(1)}(f_1)$ which we shall denote by $F_{12}(\lambda)$. The element $F_{12}(H_1)\, f_1 = g_1$ thus exists and belongs to $\mathfrak{M}_1(f_1)$; clearly, it has the property that

$$\|E_1(\lambda)\, g_1\|^2 = \int_{-\infty}^{\lambda} |F_{12}(\mu)|^2\, d\|E_1(\mu)\, f_1\|^2 = \varrho_2(\lambda).$$

We introduce the closed linear manifold $\mathfrak{M}_1(g_1)$. By reference to Theorem 6.2, we see that the space $\mathfrak{L}_2^{(1)}(g_1) = \mathfrak{L}_2(\varrho_2) = \mathfrak{L}_2^{(2)}(f_2)$ is in one-to-one isometric correspondence with $\mathfrak{M}_1(g_1)$ on one hand and with $\mathfrak{M}_2(f_2)$ on the other. By eliminating $\mathfrak{L}_2(\varrho_2)$ we obtain a direct isometric correspondence between $\mathfrak{M}_1(g_1)$ and $\mathfrak{M}_2(f_2)$ and thus define an isometric transformation V with domain $\mathfrak{M}_1(g_1)$ and range $\mathfrak{M}_2(f_2)$. If $h_2 = F(H_2)\, f_2$, where $F(\lambda)$ is in $\mathfrak{L}_2(\varrho_2)$, is an arbitrary element of $\mathfrak{M}_2(f_2)$ its correspondent $V^{-1}h_2$ in $\mathfrak{M}_1(g_1)$ is $h_1 = F(H_1)\, g_1$. If $F_\lambda(\mu)$ is the function equal to 1 or to 0 according as $\mu \leqq \lambda$ or $\mu > \lambda$, we have

$$E_2(\lambda)\, h_2 = E_2(\lambda)\, F(H_2)\, f_2 = F_\lambda(H_2)\, F(H_2)\, f_2 = G_\lambda(H_2)\, f_2$$

where $G_\lambda(\mu) = F_\lambda(\mu) \circ F(\mu)$, and

$$E_1(\lambda)\, h_1 = E_1(\lambda)\, F(H_1)\, g_1 = F_\lambda(H_1)\, F(H_1)\, g_1 = G_\lambda(H_1)\, g_1.$$

These relations can be written in the form

$$E_1(\lambda)\,V^{-1}h_2 = V^{-1}E_2(\lambda)\,h_2.$$

Since h_2 is an arbitrary element in $\mathfrak{M}_2(f_2)$, this means that $VE_1(\lambda)\,V^{-1} = E_2(\lambda)$ in $\mathfrak{M}_2(f_2)$. To complete the discussion, we show that $\mathfrak{M}_1(g_1) = \mathfrak{M}_1(f_1)$. Since g_1 is an element in $\mathfrak{M}_1(f_1)$, it is evident by reference to Theorem 7.2 that $\mathfrak{M}_1(g_1) \subseteqq \mathfrak{M}_1(f_1)$. On the other hand, f_1 is an element of $\mathfrak{M}_1(g_1)$ so that $\mathfrak{M}_1(f_1) \subseteqq \mathfrak{M}_1(g_1)$ and $\mathfrak{M}_1(g_1) = \mathfrak{M}_1(f_1)$. In fact, if we put $F_{21}(\lambda) = \left(\dfrac{d\varrho_1}{d\varrho_2}\right)^{1/2}$, we have $f_1 = F_{21}(H_1)g_1$: for, by Theorem 6.1 (6), $F_{21}(H_1)\,g_1 = F_{21}(H_1)\,F_{12}(H_1)\,f_1 = G(H_1)f_1$ where $G(\lambda) = F_{21}(\lambda)\circ F_{12}(\lambda) = \left(\dfrac{d\varrho_1}{d\varrho_2}\circ\dfrac{d\varrho_2}{d\varrho_1}\right)^{1/2}$; and by Lemma 6.4 (5) the last expression has the value 1 almost everywhere with respect to $\varrho_1(\lambda)$.

We may now consider the special case $H_1 \equiv H_2 \equiv H$. If $\mathfrak{M}(f_1) = \mathfrak{M}(f_2)$, we may take $V = I$ in $\mathfrak{M}(f_1)$; it follows that $\varrho_1 \sim \varrho_2$. Furthermore f_2 must evidently be an element of $\mathfrak{M}(f_1)$. If we write $f_1 \sim f_2$ in place of $\varrho_1 \sim \varrho_2$, the necessary conditions assume the form given in the statement of the theorem. When these conditions are satisfied, we can write $f_2 = F(H)f_1$ for some function $F(\lambda)$ in $\mathfrak{L}_2(f_1) = \mathfrak{L}_2(\varrho_1)$. By virtue of the relations

$$\varrho_2(\lambda) = \|E(\lambda)\,f_2\|^2 = \int_{-\infty}^{\lambda} |F(\mu)|^2\, d\|E(\mu)\,f_1\|^2$$
$$= \int_{-\infty}^{\lambda} |F(\mu)|^2\, d\varrho_1(\mu)$$

we see that $\left(\dfrac{d\varrho_2}{d\varrho_1}\right)^{1/2} = |F(\lambda)|$ almost everywhere with respect to $\varrho_1(\lambda)$. The function $|F(\lambda)|/F(\lambda)$ is a bounded ϱ_1-measurable function defined almost everywhere with respect to $\varrho_1(\lambda)$; by virtue of the relation $\varrho_1 \succ \varrho_2$, it is ϱ_2-measurable and is defined almost everywhere with respect to $\varrho_2(\lambda)$. The function $G(\lambda) = \left(\dfrac{d\varrho_1}{d\varrho_2}\right)^{1/2}\circ|F(\lambda)|/F(\lambda)$, as the product of a bounded function and the function $\left(\dfrac{d\varrho_1}{d\varrho_2}\right)^{1/2}$ in $\mathfrak{L}_2(\varrho_2)$, is itself a function in $\mathfrak{L}_2(\varrho_2) = \mathfrak{L}_2(f_2)$. Furthermore

$$\Phi(\lambda) = F(\lambda) \circ G(\lambda) = \left(\frac{d\varrho_2}{d\varrho_1} \circ \frac{d\varrho_1}{d\varrho_2}\right)^{1/2} = 1$$

almost everywhere with respect to $\varrho_1(\lambda)$. We find therefore that $G(H) f_2 = G(H) F(H) f_1 = \Phi(H) f_1 = f_1$; thus f_1 belongs to $\mathfrak{M}(f_2)$ and $\mathfrak{M}(f_1) \subseteq \mathfrak{M}(f_2)$. On the other hand the fact that f_2 is in $\mathfrak{M}(f_1)$ implies $\mathfrak{M}(f_2) \subseteq \mathfrak{M}(f_1)$ and hence $\mathfrak{M}(f_1) = \mathfrak{M}(f_2)$, as we wished to prove.

§ 2. Unitary Equivalence

From the three theorems just established, it is plain that we can solve the problem of unitary equivalence by constructing a canonical set of manifolds of the type of $\mathfrak{M}(f)$ for a given self-adjoint transformation H, and by comparing the canonical sets corresponding to two given transformations H_1 and H_2. Hahn introduced such canonical sets and called them "ordered sets" because the relation $\succ$, which we have defined in Chapter VI, § 1, serves to arrange them in a definite sequence.* We find it convenient to carry through our discussion in two parts, embodied in the two theorems which follow. The first theorem is equivalent to one due to Hellinger, who stated and proved it in terms of "orthogonal differential forms".†

Theorem 7.4. *Let H be a self-adjoint transformation, $E(\lambda)$ the corresponding resolution of the identity, $\mathfrak{M}$ and $\mathfrak{N}$ the two closed linear manifolds defined in Theorem* 5.13; *and, when $\mathfrak{M} \neq \mathfrak{O}$, let $\{\varphi_k\}$ be the orthonormal set introduced in that theorem. Then the following three cases arise:*

(1) *when $\mathfrak{M} = \mathfrak{O}$, $\mathfrak{N} = \mathfrak{H}$, there exists an orthonormal set $\{g_k\}$ in $\mathfrak{N}$ such that the closed linear manifolds $\mathfrak{M}(g_k)$ are mutually orthogonal and together determine $\mathfrak{N} = \mathfrak{H}$ according to the relation*

$$\mathfrak{N} = \mathfrak{M}(g_1) \oplus \mathfrak{M}(g_2) \oplus \mathfrak{M}(g_3) \oplus \cdots;$$

(2) *when $\mathfrak{M}$ and $\mathfrak{N}$ are both proper subsets of $\mathfrak{H}$, then there exists an orthonormal set $\{g_k\}$ in $\mathfrak{N}$; the closed linear manifolds*

* Hahn, Monatshefte für Mathematik und Physik, 23 (1912), pp. 161–206.
† Hellinger, Journal für Mathematik, 136 (1909), pp. 210–271.

$\mathfrak{M}(\varphi_k)$, $\mathfrak{M}(g_k)$ *are mutually orthogonal and together determine* $\mathfrak{H}$ *according to the relations*

$$\mathfrak{M} = \mathfrak{M}(\varphi_1) \oplus \mathfrak{M}(\varphi_2) \oplus \mathfrak{M}(\varphi_3) \oplus \cdots,$$
$$\mathfrak{N} = \mathfrak{M}(g_1) \oplus \mathfrak{M}(g_2) \oplus \mathfrak{M}(g_3) \cdots,$$
$$\mathfrak{H} = \mathfrak{M} \oplus \mathfrak{N};$$

(3) *when* $\mathfrak{M} = \mathfrak{H}$, $\mathfrak{N} = \mathfrak{O}$, *the closed linear manifolds* $\mathfrak{M}(\varphi_k)$ *are mutually orthogonal and together determine* $\mathfrak{M} = \mathfrak{H}$ *according to the relation*

$$\mathfrak{M} = \mathfrak{M}(\varphi_1) \oplus \mathfrak{M}(\varphi_2) \oplus \mathfrak{M}(\varphi_3) \cdots.$$

In cases (1) *and* (2) *the functions* $\|E(\lambda) g_k\|^2$ *are continuous.*

In cases (2) and (3), where $\mathfrak{M} \neq \mathfrak{O}$, we show that $\mathfrak{M}(\varphi_k)$ is the one-dimensional linear manifold determined by the characteristic element φ_k. According to Theorem 5.13 the function $\varrho_k(\lambda) = \|E(\lambda) \varphi_k\|^2$ takes on just two distinct values, zero and one. By Lemma 6.2, the space $\mathfrak{L}_2(\varrho_k)$ is a one-dimensional unitary space. Theorem 6.2 shows therefore that $\mathfrak{M}(\varphi_k)$ is a one-dimensional linear manifold. Since $\mathfrak{M}(\varphi_k)$ contains φ_k the assertion that $\mathfrak{M}(\varphi_k)$ is the linear manifold determined by φ_k is established. In consequence the closed linear manifolds $\mathfrak{M}(\varphi_k)$ are mutually orthogonal and together determine $\mathfrak{M}$ according to the relation stated under cases (2) and (3).

In cases (1) and (2), we commence by selecting an orthonormal set $\{f_{1n}\}$ which determines the closed linear manifold $\mathfrak{N}$. By a process described below, we then construct successive sequences $\{f_{mn}\}$ in $\mathfrak{N}$, $n = m, m+1, m+2, \cdots$, $m = 1, 2, 3, \cdots$, such that the closed linear manifolds $\mathfrak{M}(f_{mm})$, $m = 1, 2, 3, \cdots$, are mutually orthogonal and together determine $\mathfrak{N}$ according to the relation $\mathfrak{N} = \mathfrak{M}(f_{11}) \oplus \mathfrak{M}(f_{22}) \oplus \mathfrak{M}(f_{33}) \oplus \cdots$. In effecting the details of this construction, we must use the following assertion: if f is an arbitrary element of $\mathfrak{N}$, then $\mathfrak{M}(f) \subseteqq \mathfrak{N}$. To prove this assertion, it is sufficient to show that $E(\lambda) f$ belongs to $\mathfrak{N}$ for every λ, as we see by reference to Theorem 7.2; and for this it is in turn sufficient to show that $\|E(\mu) E(\lambda) f\|^2$ is a continuous function of μ for each value of λ, as we see by reference to Theorem 5.13. Since $|E(\mu)E(\lambda)f|^2$ is equal to $|E(\mu)f|^2$ when $\mu \leqq \lambda$ and equal to $|E(\lambda)f|^2$ when $\mu > \lambda$ and since f is in $\mathfrak{N}$, this function has the desired properties. We now suppose that the sequences $\{f_{1n}\}, \cdots, \{f_{mn}\}$ have been constructed so that $\mathfrak{M}(f_{11}), \cdots, \mathfrak{M}(f_{mm})$ are mutually orthogonal closed linear manifolds and so that every element f_{kn}, $k = 1, \cdots, m$, $n = k, k+1, k+2, \cdots$ is in $\mathfrak{N}$. We then define $f_{m+1,n}$ for $n = m+1, m+2, m+3, \cdots$ as the projection of f_{mn} on the closed linear manifold

$$\mathfrak{N}_{m+1} = \mathfrak{N} \ominus [\mathfrak{M}(f_{11}) \oplus \cdots \oplus \mathfrak{M}(f_{mm})].$$

It is clear from the remarks made above that the new sequence $\{f_{m+1,n}\}$ exists and is contained in $\mathfrak{N}$. We must show that $\mathfrak{M}(f_{m+1,m+1})$ is orthogonal to $\mathfrak{M}(f_{11}), \cdots, \mathfrak{M}(f_{mm})$. By construction $f_{m+1,m+1}$ is orthogonal to each of the latter manifolds. Hence if we let ν be the lesser of λ and μ, we have

$$(E(\lambda)f_{m+1,m+1}, E(\mu)f_{nn}) = (f_{m+1,m+1}, E(\nu)f_{nn}) = 0$$

for all values of λ and μ and for $n = 1, \cdots, m$. By Theorem 7.2, this equation implies that $\mathfrak{M}(f_{m+1,m+1})$ is orthogonal to each of the closed linear manifolds $\mathfrak{M}(f_{11}), \cdots, \mathfrak{M}(f_{mm})$. The principle of mathematical induction now permits us to carry out the construction for $m = 1, 2, 3, \cdots$ and to obtain in this manner the succession of sequences $\{f_{mn}\}$ desired. We have still to show that $\mathfrak{N} = \mathfrak{M}(f_{11}) \oplus \mathfrak{M}(f_{22}) \oplus \mathfrak{M}(f_{33}) \oplus \cdots$. Let f be an element of $\mathfrak{N}$ orthogonal to $\mathfrak{M}(f_{mm})$, $m = 1, 2, 3, \cdots$; we have to prove that $f = 0$. Now the element f_{1n} is contained in the closed linear manifold $\mathfrak{M}(f_{11}) \oplus \cdots \oplus \mathfrak{M}(f_{nn})$ by the manner in which the sequences $\{f_{mn}\}$ were constructed. Hence f is orthogonal to f_{1n} for $n = 1, 2, 3, \cdots$ and must reduce to the null element by virtue of our choice of the orthonormal set $\{f_{1n}\}$. Our construction therefore yields all that we required of it. The desired orthonormal set $\{g_k\}$ is formed by casting out of the sequence $\{f_{mm}\}$ every element such that $f_{mm} = 0$, normalizing each of the elements which remain, and numbering the new sequence in a suitable manner. Of the manifolds $\mathfrak{M}(f_{mm})$ we have discarded only those of the form $\mathfrak{M}(0) = \mathfrak{O}$ and the rest are unaltered, since

$$\mathfrak{M}(f_{mm}) = \mathfrak{M}(f_{mm}/\|f_{mm}\|) = \mathfrak{M}(g_k)$$

whenever $f_{mm} \neq 0$, and $g_k = f_{mm}/\|f_{mm}\|$. In case (2) the relations $\mathfrak{M}(\varphi_k) \subseteqq \mathfrak{M}$, $\mathfrak{M}(g_k) \subseteqq \mathfrak{N}$ show that every manifold $\mathfrak{M}(\varphi_k)$ is orthogonal to every manifold $\mathfrak{M}(g_k)$. In both cases (1) and (2), $\|E(\lambda) g_k\|^2$ is continuous by virtue of Theorem 5.13. It should be noted that the orthonormal set $\{g_k\}$ does not necessarily determine the closed linear manifold $\mathfrak{N}$; in fact, it may contain a single element.

THEOREM 7.5. *If H is a self-adjoint transformation for which case* (1) *or case* (2) *of Theorem* 7.4 *occurs, then the set $\{g_k\}$ can be replaced by an orthonormal set $\{\psi_k\}$ with the following properties:*

(1) $\mathfrak{N} = \mathfrak{M}(\psi_1) \oplus \mathfrak{M}(\psi_2) \oplus \mathfrak{M}(\psi_3) \oplus \cdots$;

(2) $\|E(\lambda) \psi_k\|^2$ *is a continuous function of* λ;

(3) $\psi_1 \succ \psi_2 \succ \psi_3 \succ \cdots$.

The symbol $\succ$ has the meaning stated in Definition 6.2. The construction of the set $\{\psi_k\}$ from the set $\{g_k\}$ of the preceding theorem is accomplished by repeated internal adjustments which we shall first isolate and analyze independently.

We start with an arbitrary sequence $\{f_n\}$ in $\mathfrak{N}$ such that the manifolds $\mathfrak{M}(f_1), \mathfrak{M}(f_2), \mathfrak{M}(f_3), \cdots$ are mutually orthogonal. We may suppose without loss of generality that those elements of the sequence which do not coincide with the null element have been normalized; thus $\|f_n\| = 0$ or $\|f_n\| = 1$ for every n. By a process which we call a regularizing transposition, we replace the given sequence by a new sequence $\{f_{1n}\}$ in $\mathfrak{N}$ which has the following properties:

(1) $\mathfrak{M}(f_{11})$ contains $\mathfrak{M}(f_1)$ and certain subsets of $\mathfrak{M}(f_2)$, $\mathfrak{M}(f_3), \mathfrak{M}(f_4), \cdots$;

(2) $\mathfrak{M}(f_{1n}) \subseteqq \mathfrak{M}(f_n)$, $n = 2, 3, 4, \cdots$;

(3) $\mathfrak{M}(f_1) \oplus \mathfrak{M}(f_2) \oplus \mathfrak{M}(f_3) \oplus \cdots = \mathfrak{M}(f_{11}) \oplus \mathfrak{M}(f_{12}) \oplus \mathfrak{M}(f_{13}) \oplus \cdots$;

(4) $f_{11} \succ f_{1n}$, $n = 2, 3, 4, \cdots$.

The validity of this process depends primarily upon the last part of Lemma 6.5. In detail, the regularizing transposition can be described as follows. Each element of the sequence $\{f_n\}$ is to be resolved into two components g_{1n} and g_{2n} such that $f_n = g_{1n} + g_{2n}$, the closed linear manifolds $\mathfrak{M}(g_{1n})$ and $\mathfrak{M}(g_{2n})$ are orthogonal to each other, and

$$\mathfrak{M}(f_n) = \mathfrak{M}(g_{1n}) \oplus \mathfrak{M}(g_{2n}).$$

This resolution is accomplished by induction. We first set $g_{11} = f_1$, $g_{12} = 0$. When the mutually orthogonal elements $g_{11}, g_{21}, \cdots, g_{1,n-1}, g_{2,n-1}$ have been defined, we put

$$\varrho_n(\lambda) = \left| E(\lambda) \sum_{\alpha=1}^{n-1} g_{1\alpha}/\alpha \right|^2 = \sum_{\alpha=1}^{n-1} |E(\lambda) g_{1\alpha}|^2/\alpha^2,$$

$$\tau_n(\lambda) = |E(\lambda) f_n|^2/n^2, \qquad \sigma_n(\lambda) = \varrho_n(\lambda) + \tau_n(\lambda).$$

Since $\sigma_n \succ \varrho_n$, $\sigma_n \succ \tau_n$, the set of points specified by the equation $\dfrac{d\varrho_n}{d\sigma_n} = 0$ exists and is σ_n-measurable; by Lemma 6.4 (1) this set is also τ_n-measurable. We introduce the function $G_{1n}(\lambda)$ which is equal to 1 or to 0 according as $\dfrac{d\varrho_n}{d\sigma_n} = 0$ or not; since $G_{1n}(\lambda)$ is bounded and τ_n-measurable, it belongs to $\mathfrak{L}_2(\tau_n) = \mathfrak{L}_2(f_n)$. We define $g_{1n} = G_{1n}(H) f_n$, $g_{2n} = f_n - g_{1n} = [1 - G_{1n}(H)] f_n$. Thus g_{1n} and g_{2n} are elements of $\mathfrak{M}(f_n)$, and $\mathfrak{M}(g_{1n})$, $\mathfrak{M}(g_{2n})$ are subsets of $\mathfrak{M}(f_n)$. It is easily shown that $\mathfrak{M}(g_{1n})$ and $\mathfrak{M}(g_{2n})$ are orthogonal: for if $F_1(H) g_{1n}$ and $F_2(H) g_{2n}$ are arbitrary elements of these two linear manifolds respectively, we have

$$\begin{aligned} &(F_1(H) g_{1n}, F_2(H) g_{2n}) \\ &= (F_1(H) G_{1n}(H) f_n, F_2(H)(1 - G_{1n}(H)) f_n) \\ &= \int_{-\infty}^{+\infty} F_1(\lambda) \circ G_{1n}(\lambda) \cdot \overline{F_2(\lambda)} \circ \overline{(1 - G_{1n}(\lambda))}\, d|E(\lambda) f_n|^2 = 0. \end{aligned}$$

Finally, if $F(H) f_n$ is an arbitrary element of $\mathfrak{M}(f_n)$ we can express it as the sum of elements in $\mathfrak{M}(g_{1n})$ and $\mathfrak{M}(g_{2n})$ respectively, by virtue of the equation

$$F(H)f_n = F(H)G_{1n}(H)f_n + F(H)(1 - G_{1n}(H))f_n$$
$$= F(H)g_{1n} + F(H)g_{2n}.$$

It follows that $\mathfrak{M}(f_n) = \mathfrak{M}(g_{1n}) \oplus \mathfrak{M}(g_{2n})$. Thus we can suppose that every element of the sequence $\{f_n\}$ has been resolved into components of the type described. The desired sequence $\{f_{1n}\}$ is defined by the equations

$$f_{11} = \sum_{\alpha=1}^{\infty} g_{1\alpha}/\alpha; \qquad f_{1n} = g_{2n}, \qquad n = 2, 3, 4, \cdots.$$

Since the elements g_{1n} are mutually orthogonal and satisfy the inequality $|g_{1n}|^2 \leqq |f_n|^2$ where $|f_n|$ is equal to zero or to one, there is no difficulty in showing that the series defining f_{11} converges in $\mathfrak{H}$. It is evident that the sequence $\{f_{1n}\}$ is a subset of $\mathfrak{N}$.

In order to verify that the sequence $\{f_{1n}\}$ has the properties demanded above, we need a few preliminary facts. In addition to the functions $\varrho_n(\lambda)$, $\sigma_n(\lambda)$, $\tau_n(\lambda)$ of the preceding paragraph, we introduce the functions

$$\varrho(\lambda) = |E(\lambda)f_{11}|^2 = \sum_{\alpha=1}^{\infty} |E(\lambda)g_{1\alpha}|^2/\alpha^2,$$
$$\tau(\lambda) = \left|E(\lambda)\sum_{\alpha=1}^{\infty} f_\alpha/\alpha\right|^2 = \sum_{\alpha=1}^{\infty} |E(\lambda)f_\alpha|^2/\alpha^2 = \sum_{\alpha=1}^{\infty} \tau_\alpha(\lambda),$$
$$\omega_n(\lambda) = |E(\lambda)g_{1n}|^2, \qquad v_n(\lambda) = |E(\lambda)g_{2n}|^2.$$

By means of Lemma 6.5 we show that

$$\tau \succ \begin{cases} \varrho \succ \varrho_n \succ \omega_1, \cdots, \omega_{n-1}. \\ \sigma_n \succ \begin{cases} \varrho_n. \\ \tau_n \succ \omega_n, v_n. \end{cases} \end{cases}$$

The relations

$$\tau(\lambda) = \varrho(\lambda) + \sum_{\alpha=1}^{\infty} v_\alpha(\lambda)/\alpha^2,$$
$$\tau(\lambda) = \sigma_n(\lambda) + \sum_{\alpha=1}^{n-1} v_\alpha(\lambda)/\alpha^2 + \sum_{\alpha=n+1}^{\infty} \tau_\alpha(\lambda)$$

are easily verified, and show that $\tau \succ \varrho, \sigma_n$. The relations

$$\varrho(\lambda) = \varrho_n(\lambda) + \sum_{\alpha=n}^{\infty} \omega_\alpha(\lambda)/\alpha^2, \qquad \varrho_n(\lambda) = \sum_{\alpha=1}^{n-1} \omega_\alpha(\lambda)/\alpha^2$$

show similarly that $\varrho \succ \varrho_n \succ \omega_1, \cdots, \omega_{n-1}$. We have already noted that $\sigma_n \succ \varrho_n, \tau_n$ by virtue of the defining equation $\sigma_n(\lambda) = \varrho_n(\lambda) + \tau_n(\lambda)$. Finally, we have

$$\tau_n(\lambda) = \omega_n(\lambda)/n^2 + v_n(\lambda)/n^2$$

and conclude that $\tau_n \succ \omega_n, v_n$. By virtue of the equations

$$\tau_n(\lambda) = \|E(\lambda) f_n\|^2/n^2,$$
$$\omega_n(\lambda) = \|E(\lambda) g_{1n}\|^2 = \int_{-\infty}^{\lambda} |G_{1n}(\lambda)|^2 \, d\|E(\lambda) f_n\|^2,$$

we find that $\dfrac{d\omega_n}{d\tau_n} = n^2 |G_{1n}(\lambda)|^2 = n^2 G_{1n}(\lambda)$ almost everywhere with respect to $\tau_n(\lambda)$. With the aid of this result we show that $\varrho_{n+1} \sim \sigma_n$. The relation $\sigma_n(\lambda) = \varrho_{n+1}(\lambda) + v_n(\lambda)/n^2$ implies $\sigma_n \succ \varrho_{n+1}$ so that we can shift our attention immediately to the proof of the relation $\sigma_n \prec \varrho_{n+1}$. We appeal for this purpose to Lemma 6.5 (2). Let E be a Borel set over which the variation of $\varrho_{n+1}(\lambda)$ is zero. Then E is σ_n-measurable, and we have

$$\int_E \frac{d\varrho_{n+1}}{d\sigma_n} \, d\sigma_n(\lambda) = \int_E d\varrho_{n+1}(\lambda) = 0.$$

This equation requires that the not-negative σ_n-integrable function $\dfrac{d\varrho_{n+1}}{d\sigma_n}$ vanish almost everywhere in E, with respect to $\sigma_n(\lambda)$. It is possible, however, to prove that $\dfrac{d\varrho_{n-1}}{d\sigma_n}$ can vanish only on a null set with respect to $\sigma_n(\lambda)$. Using the equations $\varrho_{n+1}(\lambda) = \varrho_n(\lambda) + \omega_n(\lambda)/n^2$, $\sigma_n(\lambda) = \varrho_n(\lambda) + \tau_n(\lambda)$, $G_{1n}(\lambda) \circ \dfrac{d\varrho_n}{d\sigma_n} = 0$, we obtain

$$\begin{aligned}
\frac{d\varrho_{n+1}}{d\sigma_n} &= \frac{d\varrho_n}{d\sigma_n} + \frac{1}{n^2} \frac{d\omega_n}{d\tau_n} \circ \frac{d\tau_n}{d\sigma_n} = \frac{d\varrho_n}{d\sigma_n} + G_{1n}(\lambda) \circ \frac{d\tau_n}{d\sigma_n} \\
&= \frac{d\varrho_n}{d\sigma_n} + G_{1n}(\lambda) \circ \left[\frac{d\sigma_n}{d\sigma_n} - \frac{d\varrho_n}{d\sigma_n} \right] \\
&= \frac{d\varrho_n}{d\sigma_n} + G_{1n}(\lambda),
\end{aligned}$$

an expression which, by the definition of $G_{1n}(\lambda)$, is positive wherever it is defined. Consequently, E must be a null set with respect to $\sigma_n(\lambda)$. This result shows that $\varrho_{n+1} \succ \sigma_n$. The relations

$$\varrho \succ \varrho_{n+1} \sim \sigma_n \succ v_n, \qquad \varrho \succ v_n$$

are therefore satisfied.

We shall now take up the proof that the regularizing transposition supplies us with a sequence $\{f_{1n}\}$ enjoying the four properties enumerated above. Since $f_{1n} = g_{2n}$ the relation $\mathfrak{M}(f_{1n}) = \mathfrak{M}(g_{2n}) \subseteqq \mathfrak{M}(f_n)$ of (2) holds when $n = 2, 3, 4, \cdots$. The properties (1) and (3) are consequences of a relation proved below, namely,

$$\mathfrak{M}(f_{11}) = \mathfrak{M}(g_{11}) \oplus \mathfrak{M}(g_{12}) \oplus \mathfrak{M}(g_{13}) \oplus \cdots,$$

as we see at once: for (1) is satisfied by virtue of the fact that $f_1 = g_{11}$, (3) by virtue of the relations

$$\begin{aligned}\sum_{\alpha=1}^{\infty}[\mathfrak{M}(f_{1\alpha}); \oplus] &= \sum_{\alpha=1}^{\infty}[\mathfrak{M}(g_{1\alpha}); \oplus] \oplus \sum_{\alpha=1}^{\infty}[\mathfrak{M}(g_{2\alpha}); \oplus] \\ &= \sum_{\alpha=1}^{\infty}[\mathfrak{M}(g_{1\alpha}) \oplus \mathfrak{M}(g_{2\alpha}); \oplus] \\ &= \sum_{\alpha=1}^{\infty}[\mathfrak{M}(f_{\alpha}); \oplus].\end{aligned}$$

The element f_{11} has the property that $E(\lambda) f_{11} = \sum_{\alpha=1}^{\infty} E(\lambda) g_{1\alpha}/\alpha$ is an element of $\sum_{\alpha=1}^{\infty}[\mathfrak{M}(g_{1\alpha}); \oplus]$ for every λ; and in consequence, $\mathfrak{M}(f_{11}) \subseteqq \sum_{\alpha=1}^{\infty}[\mathfrak{M}(g_{1\alpha}); \oplus]$. We shall now show that the only element in the closed linear manifold $\sum_{\alpha=1}^{\infty}[\mathfrak{M}(g_{1\alpha}); \oplus]$ orthogonal to every element of $\mathfrak{M}(f_{11})$ is the null element, and thus conclude that the two manifolds are identical. Let $h = \sum_{\alpha=1}^{\infty} h_\alpha$ where $h_n = F_n(H) g_{1n}$ is an element of $\mathfrak{M}(g_{1n})$ and let h be orthogonal to every element of $\mathfrak{M}(f_{11})$. We then have

$$\|h\|^2 = \sum_{\alpha=1}^{\infty} \|h_\alpha\|^2 = \sum_{\alpha=1}^{\infty} \int_{-\infty}^{+\infty} |F_\alpha(\lambda)|^2 \, d\|E(\lambda)\, g_{1\alpha}\|^2$$
$$= \sum_{\alpha=1}^{\infty} \int_{-\infty}^{+\infty} |F_\alpha(\lambda)|^2 \circ \frac{d\,\omega_\alpha}{d\,\tau} \, d\tau(\lambda),$$

with the aid of the relations established in the preceding paragraph. Let $\Phi_n(\lambda)$ be the τ-measurable function which is equal to one or to zero according as $\frac{d\,\omega_n}{d\,\tau}$ is positive or zero, $n = 1, 2, 3, \cdots$. Then $\Phi_n(\lambda)$ is also ϱ-measurable and ω_k-measurable, $k = 1, 2, 3, \cdots$; being a bounded function as well, it belongs to $\mathfrak{L}_2(\varrho) = \mathfrak{L}_2(f_{11})$ and to $\mathfrak{L}_2(\omega_k) = \mathfrak{L}_2(g_{1k})$ for $k = 1, 2, 3, \cdots$. Hence we have

$$\begin{aligned}(h, \Phi_n(H)\, E(\lambda) f_{11}) &= \sum_{\alpha=1}^{\infty} (h_\alpha, \Phi_n(H)\, E(\lambda) f_{11}) \\ &= \sum_{\alpha=1}^{\infty} (E(\lambda)\, \Phi_n(H)\, h_\alpha, f_{11}) \\ &= \sum_{\alpha=1}^{\infty} (E(\lambda)\, \Phi_n(H)\, h_\alpha, g_{1\alpha})/\alpha \\ &= \sum_{\alpha=1}^{\infty} \frac{1}{\alpha} \int_{-\infty}^{\lambda} \Phi_n(\lambda) \circ F_\alpha(\lambda) \, d\|E(\lambda)\, g_{1\alpha}\|^2 \\ &= \sum_{\alpha=1}^{\infty} \frac{1}{\alpha} \int_{-\infty}^{\lambda} \Phi_n(\lambda) \circ F_\alpha(\lambda) \circ \frac{d\,\omega_\alpha}{d\,\tau} \, d\tau(\lambda).\end{aligned}$$

We show next that $\Phi_n(\lambda) \circ \frac{d\,\omega_m}{d\,\tau} = 0$ whenever $m \neq n$. This equation is implied by the relation $\frac{d\,\omega_m}{d\,\tau} \circ \frac{d\,\omega_n}{d\,\tau} = 0$ which holds almost everywhere with respect to $\tau(\lambda)$, as we can prove directly. Without loss of generality, we may confine ourselves to the case $m < n$. We have

$$\frac{d\,\omega_m}{d\,\tau} \leqq m^2 \sum_{\alpha=1}^{n-1} \frac{1}{\alpha^2} \frac{d\,\omega_\alpha}{d\,\tau} = m^2 \frac{d\,\varrho_n}{d\,\tau} = m^2 \frac{d\,\varrho_n}{d\,\sigma_n} \circ \frac{d\,\sigma_n}{d\,\tau},$$
$$\frac{d\,\omega_n}{d\,\tau} = \frac{d\,\omega_n}{d\,\tau_n} \circ \frac{d\,\tau_n}{d\,\tau} = n^2\, G_{1n}(\lambda) \circ \frac{d\,\tau_n}{d\,\tau},$$

and therefore

$$0 \leqq \frac{d\omega_m}{d\tau} \circ \frac{d\omega_n}{d\tau} \leqq m^2 n^2 G_{1n}(\lambda) \circ \frac{d\varrho_n}{d\sigma_n} \circ \frac{d\sigma_n}{d\tau} \circ \frac{d\tau_n}{d\tau} = 0$$

by the manner in which $G_{1n}(\lambda)$ was defined. Hence the equation given above assumes the simplified form

$$\begin{aligned}(h, \Phi_n(H) E(\lambda) f_{11}) &= \frac{1}{n} \int_{-\infty}^{\lambda} \Phi_n(\lambda) \circ F_n(\lambda) \circ \frac{d\omega_n}{d\tau} d\tau(\lambda) \\ &= \frac{1}{n} \int_{-\infty}^{\lambda} F_n(\lambda) \circ \frac{d\omega_n}{d\tau} d\tau(\lambda).\end{aligned}$$

Since h is orthogonal to every element of $\mathfrak{M}(f_{11})$, this expression must vanish for every value of λ. It follows that $F_n(\lambda) \circ \frac{d\omega_n}{d\tau} = 0$ for all values of λ except for a null set with respect to $\tau(\lambda)$, and also that $|F_n(\lambda)|^2 \circ \frac{d\omega_n}{d\tau} = 0$ almost everywhere with respect to $\tau(\lambda)$. If we now calculate the integral $\int_{-\infty}^{+\infty} |F_n(\lambda)|^2 \circ \frac{d\omega_n}{d\tau} d\tau(\lambda) = 0$ and substitute its value in the expression for $\|h\|^2$, we find $\|h\|^2 = 0$, $h = 0$, as we wished to prove. With this conclusion, we have established that the sequence $\{f_{1n}\}$ enjoys properties (1) and (3). To show that (4) is also satisfied, we use the relation $\varrho \succ v_n$ derived in the preceding paragraph. We have merely to write it in the form

$$\|E(\lambda) f_{11}\|^2 \succ \|E(\lambda) g_{2n}\|^2 = \|E(\lambda) f_{1n}\|^2, \; n \geqq 2,$$

to conclude that $f_{11} \succ f_{1n}$, $n \geqq 2$. The regularizing transposition therefore has all the properties predicted.

Before applying the regularizing transposition to the proof of the theorem under consideration, let us examine the effect of two successive applications to a sequence $\{f_n\}$, carried out as follows: from $\{f_n\}$ we construct the sequence $\{f_{1n}\}$ exactly as described above; from the latter sequence we drop the first term and then apply the regularizing transformation to the sequence which remains, thus obtaining a new sequence $\{f_{2n}\}$ where the index n assumes the values

2, 3, 4, $\cdots$. We wish to show that $f_{11} \succ f_{22}$. In the second application of the regularizing transformation we resolve f_{1n} into components g_{11n} and g_{21n} and put $f_{22} = \sum_{\alpha=2}^{\infty} g_{11\alpha}/(\alpha - 1)$. We consider the functions

$$\varrho^{(1)}(\lambda) = |E(\lambda) f_{11}|^2, \qquad \varrho^{(2)}(\lambda) = |E(\lambda) f_{22}|^2,$$
$$\omega_n^{(2)}(\lambda) = |E(\lambda) g_{11n}|^2,$$
$$v_n^{(1)}(\lambda) = |E(\lambda) f_{1n}|^2, \qquad v_n^{(2)}(\lambda) = |E(\lambda) g_{21n}|^2,$$

for $n = 2, 3, 4, \cdots$. We have $v_n^{(1)}(\lambda) = \omega_n^{(2)}(\lambda) + v_n^{(2)}(\lambda)$ so that $v_n^{(1)} \succ \omega_n^{(2)}$. As we proved above, $\varrho^{(1)} \succ v_n^{(1)}$ so that we obtain $\varrho^{(1)} \succ \omega_n^{(2)}$. We now write

$$\begin{aligned} \varrho^{(2)}(\lambda) = |E(\lambda) f_{22}|^2 &= \sum_{\alpha=2}^{\infty} |E(\lambda) g_{11\alpha}|^2/(\alpha - 1)^2 \\ &= \sum_{\alpha=2}^{\infty} \omega_\alpha^{(2)}(\lambda)/(\alpha - 1)^2 \\ &= \sum_{\alpha=2}^{\infty} \int_{-\infty}^{\lambda} \frac{1}{(\alpha - 1)^2} \cdot \frac{d\omega_\alpha^{(2)}}{d\varrho^{(1)}} \, d\varrho^{(1)}(\lambda). \end{aligned}$$

This means that the positive-term series $\sum_{\alpha=2}^{\infty} \frac{1}{(\alpha - 1)^2} \frac{d\omega_\alpha^{(2)}}{d\varrho^{(1)}}$ yields, by integration with respect to $\varrho^{(1)}(\lambda)$ over the interval $^*\Delta$ with extremities $-\infty$ and λ, a convergent positive term series. This can occur if and only if the series itself converges almost everywhere with respect to $\varrho^{(1)}(\lambda)$ and has as its sum a $\varrho^{(1)}$-integrable function $F(\lambda)$ such that the equation $\varrho^{(2)}(\lambda) = \int_{-\infty}^{\lambda} F(\lambda)\, d\varrho^{(1)}(\lambda)$ is valid.* Thus we have $\varrho^{(1)} \succ \varrho^{(2)}$, $f_{11} \succ f_{22}$ as we were to prove.

There is now no difficulty in constructing the set $\{\psi_k\}$. We start from the sequence $\{g_k\}$ already introduced in Theorem 7.4 and apply the regularizing transposition successively to form the sequences $\{g_{mn}\}$, $n = m, m+1, m+2, \cdots$, $m = 1, 2, 3, \cdots$: the sequence $\{g_{1n}\}$ is obtained by a direct

* The proof is like that of a theorem of B. Levi for Lebesgue integrals; compare de la Vallée Poussin, *Intégrales de Lebesgue*, Paris 1916, pp. 49–50.

application from the sequence $\{g_n\}$; when the sequence $\{g_{mn}\}$ has been obtained, its first term is dropped and the regularizing transposition is then applied to yield the sequence $\{g_{m+1,n}\}$. By the properties of the regularizing transposition, we see that the closed linear manifolds $\mathfrak{M}(g_{mm})$ are mutually orthogonal, that they satisfy the relations

$$\sum_{\alpha=1}^{n} [\mathfrak{M}(g_\alpha);\ \oplus] \subseteq \sum_{\alpha=1}^{n} [\mathfrak{M}(g_{\alpha\alpha});\ \oplus] \subseteq \sum_{\alpha=1}^{\infty} [\mathfrak{M}(g_\alpha);\ \oplus],$$

$$\sum_{\alpha=1}^{\infty} [\mathfrak{M}(g_{\alpha\alpha});\ \oplus] = \sum_{\alpha=1}^{\infty} [\mathfrak{M}(g_\alpha);\ \oplus] = \mathfrak{N},$$

and that the elements g_{mm} satisfy the relations

$$g_{11} \succ g_{22} \succ g_{33} \succ \cdots.$$

If for some value of m we find that $g_{mm} = 0$, then the relation $g_{mm} \succ g_{nn}$, $n \geqq m$, implies that $g_{nn} = 0$, $n \geqq m$. We now cast from the sequence $\{g_{mm}\}$ every term which reduces to the null element and then normalize the remaining terms to obtain the sequence $\{\psi_k\}$, $\psi_k = g_{kk}/\|g_{kk}\|$, with the properties stated in the theorem.

THEOREM 7.6. *If H is a self-adjoint transformation for which case* (1) *or case* (2) *of Theorem* 7.4 *occurs, and if $\{\psi_k\}$ and $\{\chi_k\}$ are associated orthonormal sets of the type described in Theorem* 7.5, *then*

(1) *the sets $\{\psi_k\}$ and $\{\chi_k\}$ have the same number of elements;*

(2) $\psi_k \sim \chi_k$ *for* $k = 1, 2, 3, \cdots$.

At the outset we must envisage several possibilities: the two sets $\{\psi_k\}$ and $\{\chi_k\}$ may contain the same number of elements or different numbers; each set may be either finite or infinite. By an artifice we modify each set so that we do not need to classify and examine special cases. If the set $\{\psi_k\}$ is finite, we augment it by the introduction of a denumerable infinity of new terms, each equal to the null element; and if the set $\{\chi_k\}$ is finite, we treat it in the same way. We thus obtain two infinite sets $\{\psi_k\}$ and $\{\chi_k\}$ with the properties:

(1) $|\psi_k| = 1$ or $|\psi_k| = 0$, $|\chi_k| = 1$ or $|\chi_k| = 0$;

(2) $\mathfrak{N} = \sum_{\alpha=1}^{\infty}[\mathfrak{M}(\psi_\alpha); \oplus] = \sum_{\alpha=1}^{\infty}[\mathfrak{M}(\chi_\alpha); \oplus]$;

(3) $\psi_1 > \psi_2 > \psi_3 > \cdots$, $\chi_1 > \chi_2 > \chi_3 > \cdots$.

We shall prove immediately that the augmented sets satisfy the relations $\psi_k \sim \chi_k$, $k = 1, 2, 3, \cdots$. If the original set $\{\psi_k\}$ contains exactly n elements, then in the augmented set $\psi_k = 0$ for $k \geqq n+1$; it follows that in the augmented set $\{\chi_k\}$ we have $\chi_k \sim 0$ and hence $\chi_k = 0$ for $k \geqq n+1$. Similarly, if the original set $\{\chi_k\}$ contains exactly m elements we find that in the augmented set $\{\psi_k\}$ we have $\psi_k = 0$ for $k \geqq m+1$. Thus we conclude that the original sets are both infinite or both finite, containing the same number of elements. The entire theorem depends therefore on the truth of the indicated relation between the augmented sets.

First we shall show that $\psi_1 \sim \chi_1$. We introduce the functions $\varrho_k(\lambda) = |E(\lambda)\psi_k|^2$, $\sigma_k(\lambda) = |E(\lambda)\chi_k|^2$ for $k = 1, 2, 3, \cdots$, and note the relations $\varrho_1 > \varrho_2 > \varrho_3 > \cdots$ and $\sigma_1 > \sigma_2 > \sigma_3 > \cdots$. We let $\psi_{ik} = F_{ik}(H)\psi_k$ be the projection of χ_i on $\mathfrak{M}(\psi_k)$, $i, k = 1, 2, 3, \cdots$. We then have

$$\sigma_1(\lambda) = |E(\lambda)\chi_1|^2 = \sum_{\alpha=1}^{\infty}|E(\lambda)\psi_{1\alpha}|^2 = \sum_{\alpha=1}^{\infty}|E(\lambda)F_{1\alpha}(H)\psi_\alpha|^2$$

$$= \sum_{\alpha=1}^{\infty}\int_{-\infty}^{\lambda}|F_{1\alpha}(\lambda)|^2\, d|E(\lambda)\psi_\alpha|^2$$

$$= \sum_{\alpha=1}^{\infty}\int_{-\infty}^{\lambda}|F_{1\alpha}(\lambda)|^2 \circ \frac{d\varrho_\alpha}{d\varrho_1}\, d\varrho_1(\lambda).$$

By the argument used in the next to the last paragraph of the proof of Theorem 7.5, we see that the positive term series $\sum_{\alpha=1}^{\infty}|F_{1\alpha}(\lambda)|^2 \circ \frac{d\varrho_\alpha}{d\varrho_1}$ converges almost everywhere with respect to $\varrho_1(\lambda)$ and has as its sum a ϱ_1-integrable function $F(\lambda)$ such that $\sigma_1(\lambda) = \int_{-\infty}^{\lambda} F(\lambda)\, d\varrho_1(\lambda)$. Consequently, we have $\varrho_1 > \sigma_1$, $\frac{d\sigma_1}{d\varrho_1} = F(\lambda)$. In a similar manner we conclude that $\sigma_1 > \varrho_1$ and hence obtain the desired relations $\varrho_1 \sim \sigma_1$, $\psi_1 \sim \chi_1$.

Next we shall show that if $\psi_1 \sim \chi_1, \cdots, \psi_n \sim \chi_n$, the relation $\psi_{n+1} \sim \chi_{n+1}$ is also true. Our proof hinges upon the equations

$$\frac{d\sigma_i}{d\varrho_1} = \sum_{\alpha=1}^{\infty} |F_{i\alpha}(\lambda)|^2 \circ \frac{d\varrho_\alpha}{d\varrho_1}, \qquad i = 1, 2, 3, \cdots,$$

$$0 = \sum_{\alpha=1}^{\infty} F_{i\alpha}(\lambda)\, \overline{F_{k\alpha}(\lambda)} \circ \frac{d\varrho_\alpha}{d\varrho_1}, \quad i \neq k, \quad i, k = 1, 2, 3, \cdots.$$

The relation $\varrho_1 \succ \sigma_1 \succ \sigma_i$ shows that $\varrho_1 \succ \sigma_i$ and that $\dfrac{d\sigma_i}{d\varrho_1}$ exists. We calculate this function exactly as we calculated $\dfrac{d\sigma_1}{d\varrho_1}$ in the preceding paragraph: we have only to replace the subscript 1 by the subscript i. By virtue of the inequality

$$|F_{im}(\lambda)\, \overline{F_{km}(\lambda)}| \circ \frac{d\varrho_m}{d\varrho_1} \leqq \frac{1}{2}[|F_{im}(\lambda)|^2 + |F_{km}(\lambda)|]^2 \circ \frac{d\varrho_m}{d\varrho_1}$$

and of the facts stated in Lemma 6.1 (6), we see that the series $\sum_{\alpha=1}^{\infty} F_{i\alpha}(\lambda)\, \overline{F_{k\alpha}(\lambda)} \circ \dfrac{d\varrho_\alpha}{d\varrho_1}$ converges absolutely, has a ϱ_1-integrable sum, and can be integrated term by term. When i and k are different we calculate its sum by the use of the equation $(E(\lambda)\chi_i, \chi_k) = 0$. We have

$$\begin{aligned}
(E(\lambda)\chi_i, \chi_k) &= \sum_{\alpha=1}^{\infty} (E(\lambda)\psi_{i\alpha}, \psi_{k\alpha}) \\
&= \sum_{\alpha=1}^{\infty} (E(\lambda) F_{i\alpha}(H)\psi_\alpha, F_{k\alpha}(H)\psi_\alpha) \\
&= \sum_{\alpha=1}^{\infty} \int_{-\infty}^{\lambda} F_{i\alpha}(\lambda)\, \overline{F_{k\alpha}(\lambda)}\, d\| E(\lambda)\psi_\alpha \|^2 \\
&= \sum_{\alpha=1}^{\infty} \int_{-\infty}^{\lambda} F_{i\alpha}(\lambda)\, \overline{F_{k\alpha}(\lambda)} \circ \frac{d\varrho_\alpha}{d\varrho_1}\, d\varrho_1(\lambda) \\
&= \int_{-\infty}^{\lambda} \left[\sum_{\alpha=1}^{\infty} F_{i\alpha}(\lambda)\, \overline{F_{k\alpha}(\lambda)} \circ \frac{d\varrho_\alpha}{d\varrho_1}\right] d\varrho_1(\lambda).
\end{aligned}$$

Hence the sum of this series vanishes almost everywhere with respect to $\varrho_1(\lambda)$. After these preliminaries, we turn to the problem in hand. By using the equations just established and Lemma 6.5 (2), we show that $\varrho_{n+1} \succ \sigma_{n+1}$. Let E be a Borel set which is a null set with respect to $\varrho_{n+1}(\lambda)$; we must show that E is also a null set with respect to $\sigma_{n+1}(\lambda)$. In other words, we must prove that the relations

$$\int_E \frac{d\varrho_{n+1}}{d\varrho_1} d\varrho_1(\lambda) = \int_E d\varrho_{n+1}(\lambda) = 0,$$

$$\int_E \frac{d\sigma_{n+1}}{d\varrho_1} d\varrho_1(\lambda) = \int_E d\sigma_{n+1}(\lambda) > 0,$$

are incompatible. We suppose that E can be chosen so that these relations are satisfied, with a view to establishing a contradiction. The equation $\int_E \frac{d\varrho_{n+1}}{d\varrho_1} d\varrho_1(\lambda) = 0$ implies that $\frac{d\varrho_{n+1}}{d\varrho_1} = 0$ everywhere in E except possibly for a null set with respect to $\varrho_1(\lambda)$. In view of the relations $\varrho_1 \succ \varrho_{n+1} \succ \varrho_k$, $\frac{d\varrho_k}{d\varrho_1} = \frac{d\varrho_k}{d\varrho_{n+1}} \circ \frac{d\varrho_{n+1}}{d\varrho_1}$, holding for $k \geqq n+1$, we see that $\frac{d\varrho_k}{d\varrho_1} = 0$, $k \geqq n+1$, everywhere on E except possibly for a null set with respect to $\varrho_1(\lambda)$. On the other hand, the inequality $\int_E \frac{d\sigma_{n+1}}{d\varrho_1} d\varrho_1(\lambda) > 0$ implies that E contains a subset E_0 over which $\varrho_1(\lambda)$ has positive variation and on which $\frac{d\sigma_{n+1}}{d\varrho_1} > 0$. In view of the relations $\varrho_1 \succ \sigma_k \succ \sigma_{n+1}$, $\frac{d\sigma_{n+1}}{d\varrho_1} = \frac{d\sigma_{n+1}}{d\sigma_k} \circ \frac{d\sigma_k}{d\varrho_1}$, holding for $k \leqq n+1$, we see that $\frac{d\sigma_k}{d\varrho_1} > 0$, $k \leqq n+1$, everywhere on E_0 except possibly for a null set with respect to $\varrho_1(\lambda)$. In consequence of these facts, we can choose a point $\lambda = \mu$ in E_0 such that

$$\frac{d\varrho_k}{d\varrho_1} = 0, \quad k \geqq n+1; \qquad \frac{d\sigma_k}{d\varrho_1} > 0, \quad k \leqq n+1,$$

at $\lambda = \mu$, while the equations discussed in our preliminary remarks are valid at $\lambda = \mu$. Thus for $i = 1, \cdots, n+1$, $k = 1, \cdots, n$ and $\lambda = \mu$, we have

$$\sum_{\alpha=1}^{n} | F_{i\alpha}(\mu) |^2 \circ \frac{d\varrho_\alpha}{d\varrho_1} = \frac{d\sigma_i}{d\varrho_1} > 0,$$

$$\sum_{\alpha=1}^{n} F_{i\alpha}(\mu) \overline{F_{k\alpha}(\mu)} \circ \frac{d\varrho_\alpha}{d\varrho_1} = 0, \qquad i \neq k.$$

By means of the abbreviations $A_{ik} = F_{ik}(\mu) \circ \left(\frac{d\varrho_k}{d\varrho_1} \Big/ \frac{d\sigma_i}{d\varrho_1}\right)^{1/2}$, $x_k = \overline{F_{n+1,k}(\mu)} \circ \left(\frac{d\varrho_k}{d\varrho_1} \Big/ \frac{d\sigma_{n+1}}{d\varrho_1}\right)^{1/2}$ where $i, k = 1, \cdots, n$, this system of equations reduces to the system

$$\sum_{\alpha=1}^{n} A_{i\alpha} \bar{A}_{k\alpha} = \delta_{ik}, \quad \sum_{\alpha=1}^{n} A_{i\alpha} x_\alpha = 0, \quad \sum_{\alpha=1}^{n} |x_\alpha|^2 = 1,$$

where $i, k = 1, \cdots, n$, after suitable manipulations. If we interpret $(x_1, \cdots, x_n)$ as a point in n-dimensional unitary space and $\{A_{ik}\}$ as the matrix of a linear transformation T in this space, then the system of equations above signifies that the point $(x_1, \cdots, x_n)$ is at unit distance from $(0, \cdots, 0)$, that the transformation T is isometric, and that T carries $(x_1, \cdots, x_n)$ into $(0, \cdots, 0)$. Obviously, this result is the contradiction which we sought. Thus the relation $\varrho_{n+1} \succ \sigma_{n+1}$ is valid. We prove in the same way that $\sigma_{n+1} \succ \varrho_{n+1}$, and therefore conclude that $\varrho_{n+1} \sim \sigma_{n+1}$, $\psi_{n+1} \sim \chi_{n+1}$.

By mathematical induction the relation $\psi_k \sim \chi_k$ must now hold for every k; and the proof of the theorem is complete.

We are now prepared to prove the fundamental theorem of the present section, giving the solution of the problem of unitary equivalence for self-adjoint transformations.

THEOREM 7.7. *If H_1 and H_2 are self-adjoint transformations with the corresponding resolutions of the identity $E_1(\lambda)$ and $E_2(\lambda)$ respectively and if the symbols $\{\varphi_k^{(1)}\}$, $\{\varphi_k^{(2)}\}$ and $\{\psi_k^{(1)}\}$, $\{\psi_k^{(2)}\}$ denote associated sets of the types described in Theorems 7.4 and 7.5 respectively, then for the unitary equivalence of H_1 and H_2 it is necessary and sufficient that*

(1) *H_1 and H_2 fall under the same case in Theorems 5.13 and 7.4;*

(2) *in cases* (2) *and* (3), *the sets* $\{\varphi_k^{(1)}\}$ *and* $\{\varphi_k^{(2)}\}$ *can be put in one-to-one correspondence in such a way that corresponding elements are characteristic elements for* H_1 *and* H_2 *respectively, corresponding to the same characteristic value;*

(3) *in cases* (1) *and* (2), *the sets* $\{\psi_k^{(1)}\}$ *and* $\{\psi_k^{(2)}\}$ *have the same number of elements and the functions* $\varrho_k^{(1)}(\lambda) = \| E_1(\lambda)\psi_k^{(1)} \|^2$ *and* $\varrho_k^{(2)}(\lambda) = \| E_2(\lambda)\psi_k^{(2)} \|^2$ *satisfy the relation* $\varrho_k^{(1)} \sim \varrho_k^{(2)}$, $k = 1, 2, 3, \cdots$.

We shall denote by $\mathfrak{M}_1$, $\mathfrak{M}_2$, $\mathfrak{N}_1$, $\mathfrak{N}_2$, the manifolds associated with H_1 and H_2 in the manner indicated in Theorem 5.13.

Let H_1 and H_2 be connected by the relation $H_2 \equiv UH_1U^{-1}$ where U is a unitary transformation. If f_1 is a characteristic element of H_1 corresponding to the characteristic value l, then $f_2 = Uf_1$ is a characteristic element of H_2 corresponding to the same characteristic value, since

$$H_2 f_2 = UH_1U^{-1} \cdot Uf_1 = UH_1f_1 = lUf_1 = lf_2.$$

Hence U takes the characteristic manifold of H_1 corresponding to the characteristic value l into a subset of the characteristic manifold of H_2 corresponding to the same characteristic value. By considering the relation $U^{-1}H_2U \equiv H_1$, we see that U^{-1} takes the characteristic manifold of H_2 corresponding to the characteristic value l into a subset of the characteristic manifold of H_1 corresponding to the same characteristic value. We see therefore that H_1 and H_2 have the same characteristic values with the same multiplicities, and that characteristic manifolds of H_1 and H_2 corresponding to the same characteristic value are in isometric correspondence with each other by the unitary transformation U. It follows that $\mathfrak{M}_1$ and $\mathfrak{M}_2$ are in isometric correspondence with each other by the unitary transformation U, even in the special cases $\mathfrak{M}_1 = \mathfrak{O}$, $\mathfrak{M}_2 = \mathfrak{O}$. Since $\mathfrak{N}_1$ and $\mathfrak{N}_2$ are the orthogonal complements of $\mathfrak{M}_1$ and $\mathfrak{M}_2$ respectively, the unitary transformation U must take $\mathfrak{N}_1$ into $\mathfrak{N}_2$. We conclude therefore that H_1 and H_2 fall under the same one of the three cases distinguished in Theorems 5.13 and 7.4. If

H_1 and H_2 both fall under case (2) or case (3), then the remarks made above show immediately that the sets $\{\varphi_k^{(1)}\}$ and $\{\varphi_k^{(2)}\}$ can be put in one-to-one correspondence in the manner described in the present theorem. Indeed, we can number the two sets in such a way that $\varphi_k^{(1)}$ and $\varphi_k^{(2)}$ correspond to each other and to the same characteristic value l_n. By Theorem 5.13 we then have

$$|E_1(\lambda)\varphi_k^{(1)}|^2 = |E_2(\lambda)\varphi_k^{(2)}|^2 = \begin{Bmatrix} 1, & \lambda \geqq l_n \\ 0, & \lambda < l_n \end{Bmatrix},$$

where $k = 1, 2, 3, \cdots$, and n depends upon k. If H_1 and H_2 both fall under case (1) or case (2), we consider the elements $\chi_k^{(2)} = U\psi_k^{(1)}$, $k = 1, 2, 3, \cdots$. Evidently $\{\chi_k^{(2)}\}$ is an orthonormal set in $\mathfrak{N}_2$. By Theorem 7.1 we have

$$E_2(\lambda)\chi_k^{(2)} = UE_1(\lambda)U^{-1}\cdot U\psi_k^{(1)} = UE_1(\lambda)\psi_k^{(1)},$$
$$|E_2(\lambda)\chi_k^{(2)}|^2 = |UE_1(\lambda)\psi_k^{(1)}|^2 = |E_1(\lambda)\psi_k^{(1)}|^2.$$

The first line here shows that U carries the set of elements $E_1(\lambda)\psi_k^{(1)}$ into the set of elements $E_2(\lambda)\chi_k^{(2)}$ for each value of k; hence U takes $\mathfrak{M}_1(\psi_k^{(1)})$ into $\mathfrak{M}_2(\chi_k^{(2)})$, by Theorem 7.2 and the unitary character of the transformation U. Consequently $\mathfrak{N}_2 = \mathfrak{M}_2(\chi_1^{(2)}) \oplus \mathfrak{M}_2(\chi_2^{(2)}) \oplus \mathfrak{M}(\chi_3^{(2)}) \oplus \cdots$. The second line shows that the relations $\psi_1^{(1)} \succ \psi_2^{(1)} \succ \psi_3^{(1)} \succ \cdots$ (with respect to H_1) go over into the relations $\chi_1^{(2)} \succ \chi_2^{(2)} \succ \chi_3^{(2)} \succ \cdots$ (with respect to H_2). Thus we can apply Theorem 7.6 to the sets $\{\psi_k^{(2)}\}$ and $\{\chi_k^{(2)}\}$, finding that they have the same number of elements and that $\psi_k^{(2)} \sim \chi_k^{(2)}$ for every k. Writing the last result in the form

$$|E_2(\lambda)\psi_k^{(2)}|^2 \sim |E_2(\lambda)\chi_k^{(2)}|^2 = |E_1(\lambda)\psi_k^{(1)}|^2$$

we see that we must have $\varrho_k^{(1)} \sim \varrho_k^{(2)}$ as asserted in the theorem. Hence the three conditions are necessary.

To prove their sufficiency, we use Theorem 7.3 to construct isometric transformations V_1 and V_2 taking $\mathfrak{M}_1$ into $\mathfrak{M}_2$ and $\mathfrak{N}_1$ into $\mathfrak{N}_2$ respectively, and enjoying the properties

$$E_2(\lambda) = V_1 E_1(\lambda) V_1^{-1} \text{ in } \mathfrak{M}_1, \quad E_2(\lambda) = V_2 E_1(\lambda) V_2^{-1} \text{ in } \mathfrak{N}_1.$$

If $\mathfrak{M}_1 = \mathfrak{O}$, then $\mathfrak{M}_2 = \mathfrak{O}$ by condition (1), and we let V_1 be the transformation which takes the null element into itself. If $\mathfrak{M}_1 \neq \mathfrak{O}$, we can apply (1) and (2), supposing that the two sets $\{\varphi_k^{(1)}\}$ and $\{\varphi_k^{(2)}\}$ are put in correspondence with each other in the specific manner described in the preceding paragraph. Since $\|E_1(\lambda)\varphi_k^{(1)}\|^2 = \|E_2(\lambda)\varphi_k^{(2)}\|^2$, we can apply Theorem 7.3 and determine an isometric transformation V_{1k} with domain $\mathfrak{M}_1(\varphi_k^{(1)})$ and range $\mathfrak{M}_2(\varphi_k^{(2)})$ such that $E_2(\lambda) = V_{1k}E_1(\lambda)V_{1k}^{-1}$ in $\mathfrak{M}_2(\varphi_k^{(2)})$, for $k = 1, 2, 3, \cdots$. If f is an arbitrary element of $\mathfrak{M}_1$ and f_k is its projection on $\mathfrak{M}_1(\varphi_k^{(1)})$, we define V_1 by means of the equations

$$f = f_1 + f_2 + f_3 + \cdots, \qquad V_1 f = V_{11}f_1 + V_{12}f_2 + V_{13}f_3 + \cdots.$$

From the definition it is evident that V_1 is a linear transformation with the closed linear manifold $\mathfrak{M}_1$ as its domain and that $(V_1 f, V_1 g) = (f, g)$. Hence V_1 is a closed isometric transformation. Its range is a closed linear manifold containing $\mathfrak{M}_2(\varphi_k^{(2)})$ for every k, contained in $\mathfrak{M}_2$; and must therefore coincide with $\mathfrak{M}_2 = \mathfrak{M}_2(\varphi_1^{(2)}) \oplus \mathfrak{M}_2(\varphi_2^{(2)}) \oplus \mathfrak{M}_2(\varphi_3^{(2)}) \oplus \cdots$. We must show also that $E_2(\lambda) = V_1 E_1(\lambda) V_1^{-1}$ in $\mathfrak{M}_2$. Now if f is an arbitrary element in $\mathfrak{M}_1$, we have, in the notation used above,

$$E_2(\lambda)V_1 f = E_2(\lambda)V_{11}f_1 + E_2(\lambda)V_{12}f_2 + E_2(\lambda)V_{13}f_3 + \cdots.$$

Recalling that, according to Theorem 7.2, $\mathfrak{M}_1(\varphi_k^{(1)})$ reduces $E_1(\lambda)$, we see that the projection of $E_1(\lambda)f$ on $\mathfrak{M}_1(\varphi_k^{(1)})$ is equal to $E_1(\lambda)f_k$; and we have therefore

$$V_1 E_1(\lambda) f = V_{11}E_1(\lambda)f_1 + V_{12}E_1(\lambda)f_2 + V_{13}E_1(\lambda)f_3 + \cdots.$$

Since the relation $E_2(\lambda) = V_{1k}E_1(\lambda)V_{1k}^{-1}$ holding in $\mathfrak{M}_2(\varphi_k^{(2)})$ can be written in the form $E_2(\lambda)V_{1k} = V_{1k}E_1(\lambda)$ in $\mathfrak{M}_1(\varphi_k^{(1)})$, we conclude that $E_2(\lambda)V_1 f = V_1 E_1(\lambda) f$ for every element f in $\mathfrak{M}_1$. This result is equivalent to the equation $E_2(\lambda) = V_1 E_1(\lambda) V_1^{-1}$ in $\mathfrak{M}_2$. We construct V_2 is an entirely similar manner. If $\mathfrak{N}_1 = \mathfrak{O}$, then $\mathfrak{N}_2 = \mathfrak{O}$ by (1), and we let V_2 be the transformation which takes the null element into itself.

If $\mathfrak{N}_1 \neq \mathfrak{O}$ we can apply (1) and (3). Since the functions $\varrho_k^{(1)}(\lambda) = \| E_1(\lambda) \psi_k^{(1)} \|^2$, $\varrho_k^{(2)}(\lambda) = \| E_2(\lambda) \psi_k^{(2)} \|^2$ satisfy the relation $\varrho_k^{(1)} \sim \varrho_k^{(2)}$, we can apply Theorem 7.3 and determine an isometric transformation V_{2k} with domain $\mathfrak{M}_1(\psi_k^{(1)})$ and and range $\mathfrak{M}_2(\psi_k^{(2)})$ such that $E_2(\lambda) = V_{2k} E_1(\lambda) V_{2k}^{-1}$ in $\mathfrak{M}_2(\psi_k^{(2)})$, $k = 1, 2, 3, \cdots$. If f is an arbitrary element of $\mathfrak{N}_1$ and f_k is its projection on $\mathfrak{M}_1(\psi_k^{(1)})$, we define V_2 by means of the equations

$$f = f_1 + f_2 + f_3 + \cdots, \qquad V_2 f = V_{21} f_1 + V_{22} f_2 + V_{23} f_3 + \cdots.$$

By reasoning exactly like that applied to V_1, we show that V_2 has all the properties predicted. Finally, we can combine the two transformations V_1 and V_2 so as to obtain the desired unitary transformation U. If f is an arbitrary element in $\mathfrak{H}$ and f_1 and f_2 are its projections on $\mathfrak{M}_1$ and $\mathfrak{N}_1$ respectively, we define U by means of the equations

$$f = f_1 + f_2, \qquad Uf = V_1 f_1 + V_2 f_2 .$$

From the definition it is evident that U is a linear transformation with domain $\mathfrak{H}$ and that $(Uf, Ug) = (f, g)$. Hence U is a closed isometric transformation. Its range is a closed linear manifold containing $\mathfrak{M}_2$ and $\mathfrak{N}_2$, and must therefore coincide with $\mathfrak{H} = \mathfrak{M}_2 \oplus \mathfrak{N}_2$. Thus U is a unitary transformation. We must show that $E_2(\lambda) \equiv U E_1(\lambda) U^{-1}$. If f is an arbitrary element in $\mathfrak{H}$, we have, in the notation used above,

$$E_2(\lambda) Uf = E_2(\lambda) V_1 f_1 + E_2(\lambda) V_2 f_2 .$$

Recalling that, according to Theorems 4.26, 5.13, and 7.2, $\mathfrak{M}_1$ and $\mathfrak{N}_1$ reduce H_1 and $E_1(\lambda)$, we see that the projections of $E_1(\lambda) f$ on $\mathfrak{M}_1$ and $\mathfrak{N}_1$ are $E_1(\lambda) f_1$ and $E_1(\lambda) f_2$ respectively; and we have in consequence

$$U E_1(\lambda) f = V_1 E_1(\lambda) f_1 + V_2 E_1(\lambda) f_2 .$$

We conclude directly by the properties of V_1 and V_2 that

$$E_2(\lambda) U \equiv U E_1(\lambda), \qquad E_2(\lambda) \equiv U E_1(\lambda) U^{-1}$$

Theorem 7.1 shows finally that H_1 and H_2 are connected by the relation $H_2 \equiv U H_1 U^{-1}$, as we wished to prove.

It is of considerable interest to restate this theorem in a different form, after introducing a suitable terminology and establishing some hitherto unnecessary facts about the Radon-Stieltjes integral. If H, $E(\lambda)$, $\{\varphi_k\}$, $\{\psi_k\}$ have the usual meanings, we introduce

DEFINITION 7.1. *The point $\lambda = \mu$ is said to have multiplicity zero with respect to the continuous spectrum of the self-adjoint transformation H if the closed linear manifold $\mathfrak{N}$ of Theorem* 5.13 *is equal to $\mathfrak{O}$ or if $\mathfrak{N} \neq \mathfrak{O}$ and $\lambda = \mu$ is interior to an interval of constancy of the function $\varrho_1(\lambda) = \|E(\lambda)\psi_1\|^2$; the point $\lambda = \mu$ is said to have finite multiplicity n, $n = 1, 2, 3, \cdots$, with respect to the continuous spectrum of H, if for $k = 1, \cdots, n$ the functions $\varrho_k(\lambda) = \|E(\lambda)\psi_k\|^2$ exist and are constant on no interval containing $\lambda = \mu$ as an interior point while the set $\{\psi_k\}$ contains exactly n elements or the function $\varrho_{n+1}(\lambda) = \|E(\lambda)\psi_{n+1}\|^2$ exists and is constant on an interval containing $\lambda = \mu$ as an interior point; the point $\lambda = \mu$ is said to have infinite multiplicity $\aleph_0$ with respect to the continuous spectrum of H if the set $\{\psi_k\}$ is infinite and if none of the functions $\varrho_k(\lambda)$, $k = 1, 2, 3, \cdots$, is constant on an interval to which $\lambda = \mu$ is interior.*

It is easily verified that this definition assigns a definite multiplicity to each point $\lambda = \mu$. The relations $\varrho_1 \succ \varrho_2 \succ \varrho_3 \succ \cdots$ play an important rôle since they enable us to assert that, whenever $\varrho_n(\lambda)$ is constant on an interval Δ, then $\varrho_k(\lambda) \succ \varrho_n(\lambda)$, $k \geqq n$, is also constant on Δ, by virtue of the fact that Δ is a null set with respect to $\varrho_n(\lambda)$. It should be observed that a point $\lambda = \mu$ now has two multiplicities, one with respect to the point spectrum, the other with respect to the continuous spectrum. In terms of these numbers we can characterize the relation of an arbitrary point $\lambda = \mu$ to the spectrum of H. The set of all points with multiplicity greater than or equal to one with respect to the point spectrum, coincides with the point spectrum. The set of all points with multiplicity zero with respect to the point spectrum and multiplicity greater than zero with respect to the continuous spectrum, is a subset of the continuous spectrum. The set of all points with both multiplicities equal to zero is divided into two subsets: the first, consisting of those points which are limit points of characteristic values, is a subset of the continuous spectrum; and the second, consisting of those points which are not limit points of characteristic values, coincides with the set of all real points in the resolvent set of H.

We shall now state the requisite facts concerning the Radon-Stieltjes integral in the form of a lemma.

LEMMA 7.1. *If $\varrho(\lambda)$ is a real monotone-increasing continuous function in the class $\mathfrak{B}^*$ and if $r = \varrho(+\infty) > 0$, then the equation $x = \varrho(\lambda)$ defines a correspondence between the ranges $-\infty \leqq \lambda \leqq +\infty$ and $0 \leqq x \leqq r$ which is one-to-one except that each interval of constancy of $\varrho(\lambda)$ corresponds to a single value x. If E_λ and E_x are corresponding sets of points λ and x respectively, then*

(1) *the outer and inner variations of $\varrho(\lambda)$ over E_λ are equal respectively to the outer and inner Lebesgue measures of the set E_x;*

(2) *E_λ is ϱ-measurable if and only if E_x is Lebesgue-measurable; and the ϱ-measure of E_λ is equal to the Lebesgue measure of E_x.*

Let C_λ be the sum of all intervals $\Delta: \alpha \leqq \lambda \leqq \beta$ on which $\varrho(\lambda)$ is constant and let C_x be the corresponding set of points x. Let two functions of λ be regarded as identical if they differ only at points of C_λ; and let two functions of x be regarded as identical if they differ only at points of C_x. Then the relations $F(\lambda) = f(x)$, $x = \varrho(\lambda)$ define a one-to-one correspondence between the class of all complex-valued functions of λ and the class of all complex-valued functions of x. This correspondence has the properties:

(1) *a necessary and sufficient condition that $F(\lambda)$ be ϱ-measurable is that its correspondent $f(x)$ be Lebesgue-measurable; ϱ-equivalent functions of λ are in correspondence with Lebesgue-equivalent functions of x, and conversely;*

(2) *a necessary and sufficient condition that $F(\lambda)$ be ϱ-integrable is that its correspondent $f(x)$ be Lebesgue-integrable; when $F(\lambda)$ and $f(x)$ satisfy these conditions, $\int_{-\infty}^{+\infty} F(\lambda)\, d\varrho(\lambda) = \int_0^r f(x)\, dx$;*

(3) *the correspondence sets up a one-to-one isometric correspondence between the Hilbert space* $\mathfrak{L}_2(\varrho)$ *and the Hilbert space* $\mathfrak{L}_2$ *consisting of all Lebesgue-measurable functions* $f(x)$ *such that* $\int_0^r |f(x)|^2 dx$ *exists.*

In essence, this lemma is merely a ponderous tautology. It means that, if we describe the development of ϱ-measure and of the ϱ-integral, sketched in Chapter VI, § 1, in terms of the variable x instead of the variable λ, we find ourselves retracing the classical theory of Lebesgue measure and of the Lebesgue integral. To verify the details of the lemma, we merely translate from the λ-language into the x-language and *vice versa*, displaying sufficient sensitiveness to the idiom of each language to guard against clumsy translations. In discussing the various points made in the lemma, we shall use the notations of Chapter VI, § 1, and shall denote Lebesgue measure, outer Lebesgue measure, and inner Lebesgue measure by the symbols m, m^*, m_* respectively.

If the set E_λ is an interval, the set E_x is an interval and conversely; but corresponding intervals may be of different types with regard to the inclusion or exclusion of their extremities. We always have $\mathfrak{m}(E_\lambda) = \mathfrak{v}(E_\lambda) = m(E_x)$ when E_λ and E_x are corresponding intervals. If E_λ is an interval $^*\Delta : \alpha < \lambda \leqq \beta$, this is obvious from the relation $\mathfrak{m}(^*\Delta) = \varrho(\beta) - \varrho(\alpha)$. If E_λ is an interval of different type we can modify it by proper treatment of its end points. The omission or inclusion of an end point does not change the ϱ-measure of the interval, since the ϱ-measure of a single point λ is zero by virtue of the fact that $\varrho(\lambda)$ is continuous. If the corresponding interval E_x undergoes the corresponding modification, its Lebesgue measure is unchanged since the Lebesgue measure of a single point is zero. Hence the equation above is valid for arbitrary corresponding intervals. If E_λ is the sum of a denumerable family of intervals such that the intersection of any two intervals of the family (unless it is empty) is a point or an interval on which $\varrho(\lambda)$ is constant, then the corresponding set E_x is the sum of a denumerable family of intervals such that the intersection of any two intervals of the family (unless it is empty) is a point; and conversely. By virtue of the fact that a point λ and an interval on which $\varrho(\lambda)$ is constant are both sets of ϱ-measure zero, we see that the additive properties of ϱ-measure and of Lebesgue measure lead at once to the equation $\mathfrak{m}(E_\lambda) = \mathfrak{v}(E_\lambda) = m(E_x)$ for corresponding sets of this type. If E_λ and E_x are arbitrary corresponding sets and ε is an arbitrary positive number, we can find a set B_λ in the family $\mathfrak{B}$ and an open set O_x with the properties

$$E_\lambda \subseteqq B_\lambda, \qquad \mathfrak{v}^*(E_\lambda) > \mathfrak{v}(B_\lambda) - \varepsilon, \qquad E_x \subseteqq B_x,$$
$$E_x \subseteqq O_x, \qquad m^*(E_x) > m(O_x) - \varepsilon, \qquad E_\lambda \subseteqq O_\lambda.$$

Hence we have

$$\mathfrak{v}^*(E_\lambda) > \mathfrak{v}(B_\lambda) - \varepsilon = m(B_x) - \varepsilon \geqq m^*(E_x) - \varepsilon,$$
$$m^*(E_x) > m(O_x) - \varepsilon = \mathfrak{v}(O_\lambda) - \varepsilon \geqq \mathfrak{v}^*(E_\lambda) - \varepsilon,$$

and conclude that $\mathfrak{v}^*(E_\lambda) = m^*(E_x)$. Similarly, we can find a set $\bar{B}_\lambda$ and a closed set F_x with the properties

$$\bar{B}_\lambda \subseteqq E_\lambda, \qquad \mathfrak{v}_*(E_\lambda) < \mathfrak{v}^*(\bar{B}_\lambda) + \varepsilon, \qquad \bar{B}_x \subseteqq E_x,$$
$$F_x \subseteqq E_x, \qquad m_*(E_x) < m^*(F_x) + \varepsilon, \qquad F_\lambda \subseteqq E_\lambda.$$

Hence we have

$$\mathfrak{v}_*(E_\lambda) < \mathfrak{v}^*(\bar{B}_\lambda) + \varepsilon = m^*(\bar{B}_x) + \varepsilon = m(\bar{B}_x) + \varepsilon \leqq m_*(E_x) + \varepsilon,$$
$$m_*(E_x) < m^*(F_x) + \varepsilon = \mathfrak{v}^*(F_x) + \varepsilon = \mathfrak{v}(F_x) + \varepsilon \leqq \mathfrak{v}_*(E_x) + \varepsilon,$$

and conclude that $\mathfrak{v}_*(E_\lambda) = m_*(E_x)$. It follows immediately that $\mathfrak{m}(E_\lambda) = \mathfrak{v}(E_\lambda) = m(E_x)$, where the existence of any of the three terms implies the existence of the two others.

The facts concerning the correspondence between functions $F(\lambda)$ and functions $f(x)$ by the relations $F(\lambda) = f(x)$, $x = \varrho(\lambda)$, are easily ascertained. Our agreement to neglect what occurs on the sets C_λ and C_x simply allows us to avoid all values of λ and of x for which the correspondence is not a one-to-one correspondence. Since $\mathfrak{m}(C_\lambda) = \mathfrak{v}(C_\lambda) = m(C_x) = 0$, we encounter no difficulties when we restrict our attention to ϱ-measurable functions $F(\lambda)$ and Lebesgue-measurable functions $f(x)$. Thus we can proceed without need of discussion until we reach the statement concerning the simultaneous integrability of $F(\lambda)$ and $f(x)$. First, if $F(\lambda)$ and $f(x)$ are corresponding functions, it is evident that $F(\lambda)$ is a bounded ϱ-measurable function defined almost everywhere with respect to $\varrho(\lambda)$ if and only if $f(x)$ is a bounded Lebesgue-measurable function defined almost everywhere. In this case both the integrals $\int_{-\infty}^{+\infty} F(\lambda)\, d\varrho(\lambda)$, $\int_0^r f(x)\, d(x)$ exist. To show that they are equal, we let $M > 0$ be a number such that

$$-M < \Re F(\lambda) \leqq M, \qquad -M < \Im F(\lambda) \leqq M,$$
$$-M < \Re f(x) \leqq M, \qquad -M < \Im f(x) \leqq M,$$

and set

$$c_k^{(n)} = \frac{2k-n}{n} M, \qquad c_{jk}^{(n)} = c_j^{(n)} + i\, c_k^{(n)}, \qquad j, k = 0, \cdots, n.$$

We then form the sets $E_{jk,\lambda}^{(n)}$ specified by the inequalities

$$c_j^{(n)} < \Re F(\lambda) \leqq c_{j+1}^{(n)}, \qquad c_k^{(n)} < \Im F(\lambda) \leqq c_{k+1}^{(n)},$$
$$j, k = 0, \cdots, n-1,$$

and the sets $F_{jk,x}^{(n)}$ specified by the inequalities

$$c_j^{(n)} < \Re f(x) \leqq c_{j+1}^{(n)}, \qquad c_k^{(n)} < \Im f(x) \leqq c_{k+1}^{(n)},$$

for the same values of j and k. Since $E_{jk,\lambda}^{(n)}$ and $F_{jk,\lambda}^{(n)}$ differ only with regard to points in C_λ, $E_{jk,x}^{(n)}$ and $F_{jk,x}^{(n)}$ only with regard to points in C_x, we see that

$$\mathfrak{m}(E_{jk,\lambda}^{(n)}) = \mathfrak{v}(E_{jk,\lambda}^{(n)}) = m(F_{jk,x}^{(n)}).$$

Thus the integrals

$$\int_{-\infty}^{+\infty} F(\lambda)\, d\varrho(\lambda) = \lim_{n\to\infty} \sum_{\alpha,\beta=0}^{n-1} c_{\alpha\beta}^{(n)}\, \mathfrak{m}(E_{\alpha\beta,\lambda}^{(n)}),$$

$$\int_0^r f(x)\, dx = \lim_{n\to\infty} \sum_{\alpha,\beta=0}^{n-1} c_{\alpha\beta}^{(n)}\, m(F_{\alpha\beta,x}^{(n)})$$

are equal. When $F(\lambda)$ is a ϱ-measurable function defined almost everywhere with respect to $\varrho(\lambda)$ and $f(x)$ is a Lebesgue-measurable function defined almost everywhere, the requirement of boundedness being relinquished, we form the truncated functions $F_M(\lambda)$ and $f_M(x)$ in the usual manner. When $F(\lambda)$ and $f(x)$ correspond, the truncated functions $F_M(\lambda)$ and $f_M(x)$ also correspond. Thus we have

$$\int_{-\infty}^{+\infty} F_M(\lambda)\, d\varrho(\lambda) = \int_0^r f_M(x)\, dx,$$
$$\int_{-\infty}^{+\infty} |F_M(\lambda)|\, d\varrho(\lambda) = \int_0^r |f_M(x)|\, dx.$$

When M becomes infinite we find that the integrals $\int_{-\infty}^{+\infty} F(\lambda)\, d\varrho(\lambda)$ and $\int_{-\infty}^{+\infty} |F(\lambda)|\, d\varrho(\lambda)$ exist if and only if the integrals $\int_0^r f(x)\, d(x)$ and $\int_0^r |f(x)|\, dx$ exist and are respectively equal to them. The closing remarks concerning the relation between the spaces $\mathfrak{L}_2(\varrho)$ and $\mathfrak{L}_2$ are then evident.

THEOREM 7.8. *Let $m_1^{(1)}(\lambda)$ and $m_1^{(2)}(\lambda)$ be the multiplicities of λ with respect to the point spectra of the self-adjoint transformations H_1 and H_2 respectively; and let $m_2^{(1)}(\lambda)$ and $m_2^{(2)}(\lambda)$ be the multiplicities of λ with respect to the continuous spectra of H_1 and H_2 respectively. Let $\{\psi_k^{(1)}\}$ and $\{\psi_k^{(2)}\}$ be orthonormal sets of the type described in Theorem* 7.5,

$$\varrho_k^{(1)}(\lambda) = \|E_1(\lambda)\,\psi_k^{(1)}\|^2, \qquad \varrho_k^{(2)}(\lambda) = \|E_2(\lambda)\,\psi_k^{(2)}\|^2$$

the corresponding real monotone-increasing continuous functions in $\mathfrak{B}^*$; and let $f_k^{(1)}(x)$ and $f_k^{(2)}(x)$ be the real monotone-increasing functions of x defined on the interval $0 \leqq x \leqq 1$ by the relations

$$f_k^{(1)}(x) = \varrho_k^{(1)}(\lambda), \qquad x = \varrho_k^{(2)}(\lambda),$$
$$f_k^{(2)}(x) = \varrho_k^{(2)}(\lambda), \qquad x = \varrho_k^{(1)}(\lambda),$$

respectively, whenever k is a value for which they are significant. Then for the unitary equivalence of H_1 and H_2 it is necessary and sufficient that

(1) $m_1^{(1)}(\lambda) = m_1^{(2)}(\lambda)$, $m_2^{(1)}(\lambda) = m_2^{(2)}(\lambda)$, $-\infty < \lambda < +\infty$;

(2) *$f_k^{(1)}(x)$ and $f_k^{(2)}(x)$ be continuous functions such that*

$$\int_0^1 \frac{df_k^{(1)}(x)}{dx}\, dx = 1, \qquad \int_0^1 \frac{df_k^{(2)}(x)}{dx}\, dx = 1.$$

If H_1 and H_2 are unitary-equivalent, the preceding theorem shows that they have the same characteristic values with the same multiplicities and that $m_1^{(1)}(\lambda) = m_1^{(2)}(\lambda)$ in consequence. Furthermore H_1 and H_2 fall under the same one of the three cases described in Theorems 5.13 and 7.4; in cases (1) and (2) the sets $\{\psi_k^{(1)}\}$ and $\{\psi_k^{(2)}\}$ have the same number of elements and the functions $\varrho_k^{(1)}(\lambda)$ and $\varrho_k^{(2)}(\lambda)$ satisfy the relation $\varrho_k^{(1)} \sim \varrho_k^{(2)}$. This relation implies that $\varrho_k^{(1)}(\lambda)$ is constant on every interval of constancy of $\varrho_k^{(2)}(\lambda)$, and conversely; and this fact in turn implies that $m_2^{(1)}(\lambda) = m_2^{(2)}(\lambda)$, as we can verify directly. The relation $\varrho_k^{(1)} \sim \varrho_k^{(2)}$ can be put in more explicit form if we set $F_k^{(1)}(\lambda) = \dfrac{d\varrho_k^{(1)}}{d\varrho_k^{(2)}}$, $F_k^{(2)}(\lambda) = \dfrac{d\varrho_k^{(2)}}{d\varrho_k^{(1)}}$ and write

$$\varrho_k^{(1)}(\lambda) = \int_{-\infty}^{\lambda} F_k^{(1)}(\lambda)\, d\varrho_k^{(2)}(\lambda), \qquad \varrho_k^{(2)}(\lambda) = \int_{-\infty}^{\lambda} F_k^{(2)}(\lambda)\, d\varrho_k^{(1)}(\lambda).$$

We now apply Lemma 7.1 to throw these equations into a new form. Setting $\varphi_k^{(1)}(x) = F_k^{(1)}(\lambda)$ for $x = \varrho_k^{(2)}(\lambda)$ and $\varphi_k^{(2)}(x) = F_k^{(2)}(\lambda)$ for $x = \varrho_k^{(1)}(\lambda)$, we have

$$f_k^{(1)}(x) = \int_0^x \varphi_k^{(1)}(x)\, dx, \qquad f_k^{(2)}(x) = \int_0^x \varphi_k^{(2)}(x)\, dx$$

where the integrals are Lebesgue integrals. It is to be noted that x ranges over the interval $0 \leqq x \leqq 1$ by virtue of the relations $\varrho_k^{(1)}(+\infty) = \|\psi_k^{(1)}\|^2 = 1$, $\varrho_k^{(2)}(+\infty) = \|\psi_k^{(2)}\|^2 = 1$. These equations for $f_k^{(1)}(x)$ and $f_k^{(2)}(x)$ show us that the functions in question are continuous and have the derivatives

$$\frac{df_k^{(1)}(x)}{dx} = \varphi_k^{(1)}(x), \qquad \frac{df_k^{(2)}(x)}{dx} = \varphi_k^{(2)}(x).$$

If we substitute these expressions for the integrands and take $x = 1$ for the upper limit in each integral, we obtain the equations given in (2). Hence the conditions (1) and (2) are necessary.

If the conditions (1) and (2) are satisfied, then we can show that H_1 and H_2 are unitary-equivalent by reverting to Theorem 7.7. Condition (1) of the present theorem immediately implies condition (1) of Theorem 7.7; and if H_1 and H_2 fall under case (2) or case (3) of Theorems 5.13 and 7.4, then condition (7) of the present theorem also implies condition (2) of Theorem 7.7. If H_1 and H_2 fall under case (1) or case (2) of Theorems 5.13 or 7.4, then the equation $m_2^{(1)}(\lambda) = m_2^{(2)}(\lambda)$ shows that the sets $\{\psi_k^{(1)}\}$ and $\{\psi_k^{(2)}\}$ have the same number of elements, and that $\varrho_k^{(1)}(\lambda)$ is constant on every interval of constancy of $\varrho_k^{(2)}(\lambda)$ and conversely. Thus the functions $f_k^{(1)}(x)$ and $f_k^{(2)}(x)$ are defined for the same range of the index k and are real monotone-increasing continuous functions assuming values from 0 to 1 inclusive on the interval $0 \leqq x \leqq 1$. These functions therefore have Lebesgue-integrable derivatives such that

$$\int_0^1 \frac{d f_k^{(1)}(x)}{dx}\, dx \leqq f_k^{(1)}(1) = 1,$$

$$\int_0^1 \frac{d f_k^{(2)}(x)}{dx}\, dx \leqq f_k^{(2)}(1) = 1;$$

and in each of these inequalities, the equality sign holds* if and only if the corresponding function $f(x)$ is the indefinite integral of its derivative $\dfrac{d f(x)}{dx}$. Hence we can write

$$f_k^{(1)}(x) = \int_0^x \frac{d f_k^{(1)}(x)}{dx}\, dx, \quad f_k^{(2)}(x) = \int_0^x \frac{d f_k^{(2)}(x)}{dx}\, dx;$$

and we can then apply Lemma 7.1, setting

$$F_k^{(1)}(\lambda) = \frac{d f_k^{(1)}(x)}{dx}, \quad x = \varrho_k^{(2)}(\lambda),$$

$$F_k^{(2)}(\lambda) = \frac{d f_k^{(2)}(x)}{dx}, \quad x = \varrho_k^{(1)}(\lambda),$$

and obtaining

$$\varrho_k^{(1)}(\lambda) = \int_{-\infty}^{\lambda} F_k^{(1)}(\lambda)\, d\varrho_k^{(2)}(\lambda),$$

$$\varrho_k^{(2)}(\lambda) = \int_{-\infty}^{\lambda} F_k^{(2)}(\lambda)\, d\varrho_k^{(1)}(\lambda),$$

---

* Carathéodory, *Vorlesungen über reelle Funktionen*, second edition, Leipzig, 1927, pp. 563–573.

These equations imply $\varrho_k^{(1)} \sim \varrho_k^{(2)}$ and hence lead to condition (3) of Theorem 7.7. Thus the two conditions of the present theorem have been shown to imply the three conditions of Theorem 7.7, so that H_1 and H_2 are unitary-equivalent, as we wished to prove.

§ 3. Self-adjoint Transformations with Simple Spectra.

In view of Definitions 4.2 and 7.1, it is evidently of some interest to consider self-adjoint transformations of the following type:

Definition 7.2. *A self-adjoint transformation H is said to have a simple spectrum if the point $\lambda = \mu$ has multiplicity 0 or 1 with respect to the point spectrum of H and also has multiplicity 0 or 1 with respect to the continuous spectrum of H, for $-\infty < \mu < +\infty$.*

These transformations are satisfactorily characterized by the theorem which we shall now prove.

Theorem 7.9. *A necessary and sufficient condition that the self-adjoint transformation H have a simple spectrum is that there exist an element f such that $\mathfrak{M}(f) = \mathfrak{H}$.*

We suppose first that H has a simple spectrum. If H falls under case (2) or case (3) of Theorems 5.13 and 7.4, we see that each of its characteristic values has multiplicity one and that the elements of the set $\{\varphi_k\}$ are characteristic elements, no two of which correspond to the same characteristic value. If H falls under case (1) or case (2), the set $\{\psi_k\}$ of Theorem 7.5 has just one element ψ_1. In case (1) we set $f = \psi_1$, in case (2) we set $f = \psi_1 + \sum_{\alpha=1}^{\infty} \varphi_\alpha/\alpha$, and in case (3) we set $f = \sum_{\alpha=1}^{\infty} \varphi_\alpha/\alpha$. We shall now prove that $\mathfrak{M}(f) = \mathfrak{H}$; that is, that the equation $(E(\lambda)f, g) = 0$, $-\infty < \lambda < +\infty$, implies $g = 0$. In case (1), we have immediately $\mathfrak{M}(f) = \mathfrak{M}(\psi_1) = \mathfrak{N} = \mathfrak{H}$. In cases (2) and (3) we select a characteristic value $\lambda = l_n$ and let φ_n be the corresponding element in the set $\{\varphi_k\}$. The projection $E(l_n) - E(l_n - 0)$ will be denoted by E_n. We know that $E_n \psi_1 = 0$, $E_n \varphi_n$

$= \varphi_n$, $E_n \varphi_k = 0$ for $k \neq n$, by the facts established in Theorem 5.13. Hence we have $E_n f = \varphi_n/n$, $(E_n f, g) = (\varphi_n, g)/n = 0$. Thus g is an element of the closed linear manifold $\mathfrak{N}$, orthogonal to all the characteristic elements of H. In case (2), we have $\mathfrak{N} = \mathfrak{M}(\psi_1)$; and we know that the relation $(E(\lambda)f, g) = 0$ assumes the form $(E(\lambda)\psi_1, g) = 0$ for g in $\mathfrak{N} = \mathfrak{M}(\psi_1)$ and therefore implies $g = 0$. In case (3) we have $\mathfrak{N} = \mathfrak{O}$, $g = 0$. We have thus shown that the condition stated in the theorem is necessary.

Now let there exist an element f such that $\mathfrak{M}(f) = \mathfrak{H}$. If H falls under case (2) or case (3) of Theorems 5.13 and 7.4, we can show that it has no characteristic value of multiplicity greater than one. Let $\lambda = \mu$ be a characteristic value of multiplicity two or greater and let $\mathfrak{M}_\mu$ be the corresponding characteristic manifold. Since $\mathfrak{M}_\mu$ has dimension number greater than one, it contains a normalized characteristic element φ which is orthogonal to the projection of f on $\mathfrak{M}_\mu$. We now obtain a contradiction by showing that $(E(\lambda)f, \varphi) = 0$ while $\|\varphi\| = 1$. By reference to Theorem 5.13 we have

$$\begin{aligned}(E(\lambda)f, \varphi) &= (f, E(\lambda)\varphi) = (f, \varphi), \ \lambda \geqq \mu, \\ &= (f, E(\lambda)\varphi) = 0, \ \lambda < \mu;\end{aligned}$$

we have chosen φ so that $(f, \varphi) = 0$ and therefore conclude that $(E(\lambda)f, \varphi) = 0$. This result is incompatible with the relations $\|\varphi\| = 1$, $\mathfrak{M}(f) = \mathfrak{H}$. If H falls under case (1) or case (2) of Theorems 5.13 and 7.4, then we can show that the set $\{\psi_k\}$ of Theorem 7.5 contains exactly one element. In case (1), we may set $\psi_1 = f$ and have $\mathfrak{M}(\psi_1) = \mathfrak{M}(f) = \mathfrak{H}$. In case (2) we may set $\psi_1 = f - \sum_{\alpha=1}^{\infty} a_\alpha \varphi_\alpha$ where $a_n = (f, \varphi_n)$, $n = 1, 2, 3, \cdots$ and $\{\varphi_n\}$ is the orthonormal set described in Theorems 5.13 and 7.4. Clearly ψ_1 is the projection of f on the closed linear manifold $\mathfrak{N}$. Since $\mathfrak{N}$ reduces H and $E(\lambda)$ by Theorems 4.26, 5.13, and 7.2, we can prove that when g is in $\mathfrak{N}$ the equation $(E(\lambda)\psi_1, g) = 0$ implies $g = 0$: for $E(\lambda)g$ is also an element of $\mathfrak{N}$ and we have

$$(E(\lambda)\psi_1, g) = (\psi_1, E(\lambda)g) = (f, E(\lambda)g) - \sum_{\alpha=1}^{\infty} a_\alpha(\varphi_\alpha, E(\lambda)g)$$
$$= (f, E(\lambda)g) = (E(\lambda)f, g) = 0;$$

it follows by virtue of the relation $\mathfrak{M}(f) = \mathfrak{H}$ that $g = 0$. Hence we have $\mathfrak{M}(\psi_1) = \mathfrak{N}$ and infer that the set $\{\psi_k\}$ contains just one element. The facts established about the sets $\{\varphi_k\}$ and $\{\psi_k\}$ yield the result that H has a simple spectrum in accord with Definition 7.2. The proof of the theorem is thus complete.

With the aid of this characterization of the self-adjoint transformations with simple spectra, we are able to establish a number of interesting theorems concerning representations and realizations of such transformations.

THEOREM 7.10. *If H is a self-adjoint transformation with simple spectrum and f is an element such that $\mathfrak{M}(f) = \mathfrak{H}$, then the isometric correspondence between $\mathfrak{M}(f)$ and $\mathfrak{L}_2(f)$ described in Theorems 6.2 and 7.2 carries H into a self-adjoint transformation H_1 in $\mathfrak{L}_2(f)$. The domain of H_1 comprises those and only those elements $F(\lambda)$ in $\mathfrak{L}_2(f)$ for which the integral $\int_{-\infty}^{+\infty} \lambda^2 |F(\lambda)|^2 d\|E(\lambda)f\|^2$ exists and H_1 takes an element $F(\lambda)$ of its domain into the element $\lambda F(\lambda)$.*

The only point in this theorem which requires investigation is the characterization of the transformation H_1. By combining Theorem 5.9 and Theorem 6.1 (5) and (6), we obtain the results stated.

When H has no point spectrum, we can apply Lemma 7.1 and throw this theorem into a different form.

THEOREM 7.11. *If H is a self-adjoint transformation with simple spectrum which falls under case* (1) *of Theorem* 5.13, *and if f is an element such that $\mathfrak{M}(f) = \mathfrak{H}$, then the correspondence between $\mathfrak{L}_2(f)$ and the space $\mathfrak{L}_2$ described in Lemma* 7.1 *carries the transformation H_1 of the preceding theorem into a self-adjoint transformation H_2. If we set $\varrho(\lambda) = \|E(\lambda)f\|^2$, $r = \|f\|^2$, then $\varrho(\lambda)$ is a real monotone-increasing continuous function in $\mathfrak{B}^*$; and the function $H(x)$ defined by the relations $\lambda = H(x)$, $x = \varrho(\lambda)$ is a real monotone-*

increasing function with no interval of constancy. The transformation H_2 *has as its domain the set of all functions* $f(x)$ *in* $\mathfrak{L}_2$ *such that the Lebesgue integral* $\int_0^r H^2(x)\,|f(x)|^2\,dx$ *exists; it takes an element* $f(x)$ *of its domain into the function* $H(x)\,f(x)$.

This theorem leads at once to information concerning matrices associated with the transformation H in the manner described in Chapter III, § 1. We have:

THEOREM 7.12. *If* H *is a self-adjoint transformation with a simple spectrum and* f *an element for which* $\mathfrak{M}(f) = \mathfrak{H}$, *then a complete orthonormal set* $\{g_n\}$ *can be selected such that the matrix* $A(g)$ *associated with* H *by* $\{g_n\}$ *has the following properties:*

(1) *its elements are expressible in the form*

$$a_{mn} = (Hg_n, g_m) = \int_{-\infty}^{+\infty} \lambda\, G_n(\lambda)\, \bar{G}_m(\lambda)\, d\varrho(\lambda),$$
$$\varrho(\lambda) = \|E(\lambda) f\|^2,$$

where $\{G_n\}$ *is a complete orthonormal set in* $\mathfrak{L}_2(f)$ *such that the integrals* $\int_{-\infty}^{+\infty} \lambda^2\,|G_n(\lambda)|^2\,d\varrho(\lambda)$ *exist;*

(2) *the transformation* $T_1(A)$ *associated with the matrix* A *by the set* $\{g_n\}$ *is essentially self-adjoint.*

Conversely, if the matrix A *has the properties*

(1) *each element of* A *is expressible in the form*

$$a_{mn} = \int_{-\infty}^{+\infty} \lambda\, G_n(\lambda)\, \bar{G}_m(\lambda)\, d\varrho(\lambda),$$

where $\varrho(\lambda)$ *is a real monotone-increasing function in* $\mathfrak{B}^*$ *which assumes infinitely many distinct values, where* $\{G_n\}$ *is a complete orthonormal set in the Hilbert space consisting of all those functions* $F(\lambda)$ *such that the Radon-Stieltjes integral* $\int_{-\infty}^{+\infty} |F(\lambda)|^2\,d\varrho(\lambda)$ *exists, and where the integrals* $\int_{-\infty}^{+\infty} \lambda^2\,|G_n(\lambda)|^2\,d\varrho(\lambda)$, $n = 1, 2, 3, \cdots$, *exist;*

(2) *the transformation* $T_1(A)$ *associated with* A *by a complete orthonormal set* $\{g_n\}$ *is essentially self-adjoint;—then there exists*

a unique self-adjoint transformation H and an element f such that

$$a_{mn} = (Hg_n, g_m), \quad \varrho(\lambda) = \|E(\lambda)f\|^2, \quad \mathfrak{M}(f) = \mathfrak{H}.$$

The transformation H has a simple spectrum and is related to the matrix A in the manner described in the first part of the theorem.

When H is given we select a complete orthonormal set $\{g_n\}$ in its domain by the process described in the introductory paragraphs of Chapter V, § 2. We then define a transformation T with the set $\{g_n\}$ as its domain by means of the equation $Tg_n = Hg_n$. As we proved in demonstrating Theorem 5.3, the transformation T is essentially self-adjoint, so that $H \equiv T^* \equiv T^{**} \equiv \tilde{T}$ by Theorem 4.17. By reference to Theorem 6.2 we now express g_n in the form $g_n = G_n(H)f$; since $\{g_n\}$ is a complete orthonormal set in $\mathfrak{H} = \mathfrak{M}(f)$, the corresponding set $\{G_n\}$ is a complete orthonormal set in $\mathfrak{L}_2(f)$. The matrix A associated with the transformation H by the set $\{g_n\}$ is now readily computed by the relations

$$a_{mn} = (Hg_n, g_m) = \int_{-\infty}^{+\infty} \lambda\, G_n(\lambda)\, \overline{G_m(\lambda)}\, d\varrho(\lambda)$$

where $\varrho(\lambda) = \|E(\lambda)f\|^2$, as we see by Theorem 6.1. Since g_n is in the domain of H, we have

$$\|Hg_n\|^2 = \int_{-\infty}^{+\infty} \lambda^2 |G_n(\lambda)|^2\, d\varrho(\lambda), \qquad n = 1, 2, 3, \cdots.$$

Lastly, we can reconstruct the transformation H from the matrix A and the set $\{g_n\}$: for the transformation $T_1(A)$ associated with A by the set $\{g_n\}$ is identical with the essentially self-adjoint transformation T discussed above.

When the matrix A with the various properties enumerated in the theorem is given, we first study the concrete Hilbert space $\mathfrak{L}_2(\varrho)$ of all functions $F(\lambda)$ such that the Radon-Stieltjes integral $\int_{-\infty}^{+\infty} |F(\lambda)|^2\, d\varrho(\lambda)$ exists. We have assumed that the function $\varrho(\lambda)$ takes on infinitely many distinct values: thus, when we introduce in $\mathfrak{L}_2(\varrho)$ the metric defined by the bilinear function $(F, G) = \int_{-\infty}^{+\infty} F(\lambda)\,\overline{G(\lambda)}\,d\varrho(\lambda)$ we are certain of obtaining a Hilbert space rather than a unitary space. In $\mathfrak{L}_2(\varrho)$ we define a transformation T with the given set $\{G_n\}$ as its domain by means of the relations $TG_n(\lambda) = \lambda G_n(\lambda)$; this definition is significant since our hypotheses require that $\lambda G_n(\lambda)$ be an element of $\mathfrak{L}_2(\varrho)$. It is plain that the transformation T is identical with the transformation $T_1(A)$ associated in the concrete Hilbert space $\mathfrak{L}_2(\varrho)$ with the matrix A by the set $\{G_n\}$. Thus T is essentially self-adjoint and has a unique self-adjoint extension $H \equiv \tilde{T} \equiv T^* \equiv T^{**}$ by Theorem 4.17. We must now characterize the transformation H in such a manner that the properties of its spectrum can be ascertained. We define a family of transformations R_l with domain $\mathfrak{L}_2(\varrho)$ by the equation $R_l F(\lambda) \equiv \frac{1}{\lambda - l} F(\lambda)$, where l is not real, and can verify immediately that this family satisfies the three conditions of Theorem 4.19. The family R_l is therefore the resolvent of a self-adjoint transformation H_1. It is obvious that R_l is an extension of T_l^{-1} so that H_1 is an extension of T. We see therefore that H and H_1 are identical. By the use of the known relations between H and R_l, it is easily shown that the domain of H comprises those and only those functions $F(\lambda)$ such that $\lambda F(\lambda)$ is in $\mathfrak{L}_2(\varrho)$ and that when $F(\lambda)$ is such a function $HF(\lambda)$ is equal to $\lambda F(\lambda)$. For our purposes it is more important to compute the resolution of the identity corresponding to H. While this computation can be effected in various ways, we shall find the method indicated in Theorem 5.10 suitable in the present case. The contour integral of that theorem takes the form

$$-\frac{1}{2\pi i}\int_{C(\mu,\nu,\alpha,\varepsilon)}\left[\int_{-\infty}^{+\infty}\frac{1}{\lambda - l}F(\lambda)\,\overline{G(\lambda)}\,d\varrho(\lambda)\right]dl.$$

Because of the simple way in which the variable l enters into the inner integral, we can justify changing the order of integration and can thus quickly reduce this expression to the new form

$$\frac{1}{\pi}\int_{-\infty}^{+\infty}\left[\arctan\frac{\nu-\lambda}{\varepsilon}-\arctan\frac{\mu-\lambda}{\varepsilon}\right]F(\lambda)\,\overline{G(\lambda)}\,d\varrho(\lambda).$$

When ε tends to zero through positive values, the expression in square brackets multiplied by the factor $\frac{1}{\pi}$ tends boundedly to the limiting function $\varphi_{\mu,\nu}(\lambda)$ defined by the equations

$$\begin{aligned} &\varphi_{\mu,\nu}(\lambda)=0,\quad \lambda<\nu, &&\varphi_{\mu,\nu}(\lambda)=0,\quad \lambda>\mu,\\ &\varphi_{\mu,\nu}(\nu)=\tfrac{1}{2}, &&\varphi_{\mu,\nu}(\mu)=\tfrac{1}{2},\end{aligned}$$
$$\varphi_{\mu,\nu}(\lambda)=1,\quad \nu<\lambda<\mu.$$

Since the convergence to this limiting function is bounded, it is easily shown that the result of allowing ε to tend to zero in the integral above is the integral $\int_{-\infty}^{+\infty}\varphi_{\mu,\nu}(\lambda)\,F(\lambda)\,\overline{G(\lambda)}\,d\varrho(\lambda)$. Finally, we allow ν to tend to $-\infty$, μ to a value α under the restriction $\mu>\alpha$. By reference to Theorem 5.10 we see that we obtain in this manner the equation

$$(E(\alpha)\,F,\,G)=\int_{-\infty}^{+\infty}\varphi_{\alpha}(\lambda)\,F(\lambda)\,\overline{G(\lambda)}\,d\varrho(\lambda)$$

where $\varphi_{\alpha}(\lambda)=1$, $\lambda\leqq\alpha$, and $\varphi_{\alpha}(\lambda)=0$, $\lambda>\alpha$. Thus $E(\alpha)$ takes $F(\lambda)$ into $\varphi_{\alpha}(\lambda)\,F(\lambda)$. With this explicit expression for the resolution of the identity corresponding to H, we can show immediately that H has a simple spectrum. Indeed, the function $F(\lambda)\equiv 1$ is evidently an element of $\mathfrak{L}_2(\varrho)$ with the properties

$$\varrho(\alpha)=\|E(\alpha)\,F\|^2=\int_{-\infty}^{\alpha}d\varrho(\lambda),\qquad \mathfrak{M}(F)\equiv\mathfrak{L}_2(\varrho).$$

The first property is obvious, while the second can be proved by the customary argument. For if $G(\lambda)$ is an element orthogonal to $\mathfrak{M}(F)$ it is orthogonal to $E(\alpha)\,F$ for every value of α; but the equation

$$0=(G,\,E(\alpha)\,F)=\int_{-\infty}^{\alpha}G(\lambda)\,d\varrho(\lambda)$$

implies that $G(\lambda)$ is equivalent to the null element in $\mathfrak{L}_2(\varrho)$. Thus it is evident that H has a simple spectrum.

We have now proved all the assertions of the theorem concerning the situation that arises when the matrix A is given. We have carried on all our operations in the concrete Hilbert space $\mathfrak{L}_2(\varrho)$, but we can parallel these operations in an arbitrary abstract Hilbert space $\mathfrak{H}$. In order to do so, it is sufficient to put $\mathfrak{H}$ in one-to-one isometric correspondence with $\mathfrak{L}_2(\varrho)$ by selecting a complete orthonormal set $\{g_n\}$ in $\mathfrak{H}$ and pairing off the element g_n with the given function $G_n(\lambda)$ in $\mathfrak{L}_2(\varrho)$; then every operation in $\mathfrak{L}_2(\varrho)$ is mirrored in $\mathfrak{H}$ and our results hold in the case of the abstract space.

We can refine the results of the preceding theorem in important respects by being more judicious in our choice of the elements f, $\{g_n\}$: in fact, if we select these elements properly we can cause most of the elements of the matrix A to vanish.

THEOREM 7.13. *If H is a self-adjoint transformation with a simple spectrum, the element f and the complete orthonormal set $\{g_n\}$ of Theorem 7.12 can be so chosen that the matrix $A(g)$ associated with H by $\{g_n\}$ enjoys, in addition to the properties enumerated in Theorem 7.12, the special property that $G_n(\lambda)$ is a polynomial of degree $n-1$ for $n = 1, 2, 3, \cdots$. In consequence, the matrix $A = \{a_{mn}\}$ assumes a particular form noted by Jacobi: the only elements of the matrix different from zero are found along the principal diagonal or adjacent to it, while no element $a_{n+1,n}$, $a_{n,n+1}$ vanishes. Such a Jacobi matrix can be written*

$$\begin{array}{cccccccccc} a_1 & b_1 & 0 & \cdot & \cdot & \cdot & \cdot & \cdot & \cdot & \cdot \\ \bar{b}_1 & a_2 & b_2 & \cdot & \cdot & \cdot & \cdot & \cdot & \cdot & \cdot \\ 0 & \bar{b}_2 & a_3 & \cdot & \cdot & \cdot & \cdot & \cdot & \cdot & \cdot \\ \cdot & \cdot & \cdot & \cdot & \cdot & \cdot & \cdot & \cdot & \cdot & \cdot \\ \cdot & \cdot & \cdot & \cdot & \cdot & \cdot & \cdot & \cdot & \cdot & \cdot \\ \cdot & \cdot & \cdot & \cdot & \cdot & \cdot & \cdot & \cdot & \cdot & \cdot \\ \cdot & \cdot & \cdot & \cdot & \cdot & \bar{b}_{n-1} & a_n & b_n & \cdot & \\ \cdot & \cdot & \cdot & \cdot & \cdot & \cdot & \cdot & \cdot & \cdot & \cdot \end{array}$$

where a_n is real and $b_n \neq 0$, $n = 1, 2, 3, \cdots$.

The proof of this theorem is not difficult once we have stated certain theorems concerning the approximation of bounded continuous functions on an infinite interval. We shall denote by f, for the time being, an arbitrary element in $\mathfrak{H}$ and by $\mathfrak{L}_2(f)$ the corresponding space of all functions $F(\lambda)$ such that $\int_{-\infty}^{+\infty} |F(\lambda)|^2 d\|E(\lambda)f\|^2$ exists. We shall denote by $\mathfrak{C}$, $\mathfrak{C}_0$, $\mathfrak{M}_\alpha$ the following classes of functions in $\mathfrak{L}_2(f)$:

(1) $\mathfrak{C}$ is the class of all bounded continuous functions;

(2) $\mathfrak{C}_0$ is the class of all bounded continuous functions $F(\lambda)$ with continuous first derivative such that $F(\lambda) = 0$ for $|\lambda| \geqq L = L(F)$;

(3) $\mathfrak{M}_\alpha$ is the class of all functions of the form $e^{-\alpha\lambda^2} P(\lambda)$ where α is a fixed positive number and $P(\lambda)$ is a polynomial.

These classes are linear manifolds in $\mathfrak{L}_2(f)$; $\mathfrak{C}$ is everywhere dense* in $\mathfrak{L}_2(f)$ and contains $\mathfrak{C}_0$ and $\mathfrak{M}_\alpha$ as subsets. We wish to show that $\mathfrak{M}_\alpha$ is everywhere dense in $\mathfrak{C}$ and hence also in $\mathfrak{L}_2(f)$.

If $F(\lambda)$ is an arbitrary bounded continuous function and ε is an arbitrary positive number we can determine a value $L \geqq 0$ so large that

$$\int_{-\infty}^{+\infty} |F(\lambda)|^2 d\|E(\lambda)f\|^2 - \int_{-L}^{+L} |F(\lambda)|^2 d\|E(\lambda)f\|^2$$

does not exceed $\varepsilon/2$; and we can suppose that the points $\lambda = \pm L$ are points of continuity of $\|E(\lambda)f\|^2$. Next we can choose a function $F_0(\lambda)$ in $\mathfrak{C}_0$ which vanishes outside the interval $-L \leqq \lambda \leqq +L$ and which satisfies the inequality

$$\int_{-L}^{+L} |F(\lambda) - F_0(\lambda)|^2 d\|E(\lambda)f\|^2 < \frac{\varepsilon}{2}.$$

In order to do so we choose $F_0(\lambda)$ so that it approximates closely to $F(\lambda)$ throughout the interval $-L \leqq \lambda \leqq +L$ save at the two end-points where $F_0(\lambda)$ must vanish. Due to the fact that these points are points of continuity of $\|E(\lambda)f\|^2$, the unsatisfactory nature of the approximation does not prevent us from rendering the contribution of the neighborhoods of these points to the integral as small as we please by taking the range of poor approximation sufficiently small.

* Compare the discussion under Theorem 6.9.

We shall not write down the obvious elementary inequalities concerned. When $F_0(\lambda)$ has been selected, we see that the value of the integral

$$\int_{-\infty}^{+\infty} |F(\lambda) - F_0(\lambda)|^2 \, d\|E(\lambda)f\|^2$$

is less than ε. Thus $\mathfrak{C}_0$ is everywhere dense in $\mathfrak{C}$.

We prove next that when $F_0(\lambda)$ is an arbitrary function in $\mathfrak{C}_0$ and ε is an arbitrary positive number there exists a function $F_\alpha(\lambda)$ in $\mathfrak{M}_\alpha$ such that

$$\int_{-\infty}^{+\infty} |F_0(\lambda) - F_\alpha(\lambda)|^2 \, d\|E(\lambda)f\|^2 < \varepsilon.$$

In order to do so it is sufficient to show that when $F_0(\lambda)$ and $\eta > 0$ are given the function $F_\alpha(\lambda)$ can be formed so that $|F_0(\lambda) - F_\alpha(\lambda)| < \eta$, $-\infty < \lambda < +\infty$: for if we choose η so that $\eta^2\|f\|^2 < \varepsilon$, the integral inequality can then be obtained immediately. Furthermore, we can restrict our attention to the case $\alpha = 1$ insofar as the last inequality alone is concerned: when $F_0(\lambda)$ is in $\mathfrak{C}_0$, $F_0(\lambda'/\sqrt{\alpha})$ is also in $\mathfrak{C}_0$; and the inequality $|F_0(\lambda'/\sqrt{\alpha}) - F_1(\lambda')| < \eta$ is transformed by the substitution $\lambda' = \sqrt{\alpha}\,\lambda$ into the inequality $|F_0(\lambda) - F_1(\sqrt{\alpha}\,\lambda)| < \eta$ where $F_1(\sqrt{\alpha}\,\lambda)$ is evidently a function in $\mathfrak{M}_\alpha$. In the case $\alpha = 1$ the desired approximation is a consequence of the elementary theory of developments in Hermite polynomials. If we take a sufficiently large number of terms in the development of $e^{\lambda^2} F_0(\lambda)$ according to the Hermite polynomials and denote their sum by $P(\lambda)$, then the inequality $|F_0(\lambda) - e^{-\lambda^2} P(\lambda)| < \eta$ is true for $-\infty < \lambda < +\infty$. Our choice of the class $\mathfrak{C}_0$ was made with this application of the theory of Hermite polynomials in view.*

On combining the results of the two preceding paragraphs we find that $\mathfrak{M}_\alpha$ is everywhere dense in $\mathfrak{C}$ and hence also in $\mathfrak{L}_2(f)$, as we wished to show.

We are now prepared for the proof of the present theorem. We select an element f in the domain of H^n for $n = 1, 2, 3, \cdots$

---

* See Stone, Annals of Mathematics, (2) 29 (1927), p. 7.

such that $\mathfrak{M}(f) = \mathfrak{H}$: since H has a simple spectrum there exists an element g such that $\mathfrak{M}(g) = \mathfrak{H}$, and in terms of g we take $f = e^{-H^2} g$. It is evident that $H^n f = H^n e^{-H^2} g$ exists; and the relations $f = e^{-H^2} g$, $g = e^{H^2} f$ imply that $\mathfrak{M}(f) = \mathfrak{M}(g) = \mathfrak{H}$. The equation

$$\| H^n f \|^2 = \int_{-\infty}^{+\infty} \lambda^{2n} \, d \, \| E(\lambda) f \|^2 = \int_{-\infty}^{+\infty} \lambda^{2n} e^{-\lambda^2} \, d \, \| E(\lambda) g \|^2$$

shows that $\| H^n f \| \neq 0$, $H^n f \neq 0$. We must show further that the sequence $\{H^n f\}$ determines the closed linear manifold $\mathfrak{H}$. Using the relation $\mathfrak{M}(g) = \mathfrak{H}$, we introduce the correspondence between $\mathfrak{H}$ and $\mathfrak{L}_2(g)$ established by Theorem 6.2. The correspondent of $H^n f = H^n e^{-H^2} g$ is the function $\lambda^n e^{-\lambda^2}$. The closed linear manifold determined in $\mathfrak{L}_2(g)$ by the sequence $\{\lambda^n e^{-\lambda^2}\}$ is $\mathfrak{L}_2(g)$ itself, as we proved above; and in consequence the sequence $\{H^n f\}$ must determine the closed linear manifold $\mathfrak{H}$. From the sequence $\{H^n f\}$ we now form an orthonormal set $\{g_n\}$, $g_n = G_n(H) f$, by means of the process described in Theorem 1.13. When the details of that process are scrutinized with regard to this application, it is found that $G_n(\lambda)$ is a polynomial of degree $n - 1$. The orthonormal set $\{g_n\}$ is evidently complete.

Next we define a transformation T, with the set $\{g_n\}$ as its domain, by means of the equations $T g_n = H g_n$; we wish to show that T is essentially self-adjoint. For this purpose, it is sufficient to show that the range of the transformation T_l, l not real, determines the closed linear manifold $\mathfrak{H}$, as we have already proved in Theorem 4.17. Thus we have to prove that the equations $(T_l g_n, h) = 0$, $n = 1, 2, 3, \cdots$ imply $h = 0$ when l is not real. If we put $h = G(H) g$, these equations take the form

$$\int_{-\infty}^{+\infty} (\lambda - l) \, G_n(\lambda) \, e^{-\lambda^2} \, \overline{G(\lambda)} \, d \| E(\lambda) g \|^2 = 0, \quad n = 1, 2, 3, \cdots.$$

Since the sequence $\{G_n(\lambda) e^{-\lambda^2/2}\}$ determines the closed linear manifold $\mathfrak{L}_2(g)$ in accordance with our preparatory results, the function $(\lambda - l) e^{-\lambda^2/2} \overline{G(\lambda)}$ must be equivalent to the null element in $\mathfrak{L}_2(g)$. In this function, the factor $(\lambda - l) e^{-\lambda^2/2}$

never vanishes when l is not real, so that $G(\lambda)$ must itself be equivalent to the null element in $\mathfrak{L}_2(g)$. It follows at once that $h = 0$ and that T is essentially self-adjoint.

The matrix $A = \{a_{mn}\}$, where

$$a_{mn} = (Hg_n, g_m) = \int_{-\infty}^{+\infty} \lambda\, G_n(\lambda)\, \overline{G_m(\lambda)}\, d\|E(\lambda)f\|^2$$

is therefore a matrix with the general properties described in the preceding theorem. The fact that $G_n(\lambda)$ is a polynomial of degree $n-1$ requires that the matrix A have the Jacobi form indicated above. If $m < n-1$ we must have $a_{mn} = \bar{a}_{nm} = 0$: for $\lambda\, G_m(\lambda)$ is a polynomial of degree m and can therefore be expressed as a linear combination of $G_1(\lambda), \cdots, G_{m+1}(\lambda)$; since $G_n(\lambda)$ is by construction orthogonal in $\mathfrak{L}_2(f)$ to each of the latter polynomials we have

$$a_{mn} = \int_{-\infty}^{+\infty} \lambda\, G_n(\lambda)\, \bar{G}_m(\lambda)\, d\|E(\lambda)f\|^2 = 0.$$

Thus the only elements of the matrix A which differ from zero must lie on the principal diagonal or adjacent to it. We can show furthermore that $a_{n-1,n} = \bar{a}_{n,n-1}$ cannot vanish. We have only to set

$$\lambda\, G_{n-1}(\lambda) = c_1 G_1(\lambda) + \cdots + c_n G_n(\lambda), \qquad c_n \neq 0,$$

and then to compute

$$a_{n-1,n} = \int_{-\infty}^{+\infty} \lambda\, G_n(\lambda)\, \overline{G_{n-1}(\lambda)}\, d\|E(\lambda)f\|^2 = \bar{c}_n \neq 0.$$

Thus none of the elements adjacent to the principal diagonal can vanish. In short, the matrix A can be written in the form displayed in the theorem, when its (Hermitian) symmetric character is taken into account.

It is interesting to establish the converse of the theorem just proved.

THEOREM 7.14. *Let A be a symmetric matrix in the Jacobi form*

$$a_{mn} = \bar{a}_{nm}; \qquad a_{mn} = 0, \qquad m < n-1;$$
$$a_{nn} = a_n; \qquad a_{n,n+1} = b_n \neq 0;$$

and let the transformation $T_1(A)$ *associated with* A *by the complete orthonormal set* $\{g_n\}$ *be essentially self-adjoint. Then the self-adjoint transformation* $H \equiv \tilde{T}_1(A) \equiv T_1^*(A) \equiv T_1^{**}(A)$ *has a simple spectrum and is represented by the matrix* A *in terms of the set* $\{g_n\}$. *The matrix* A *is expressible according to the relations*

$$\varrho(\lambda) = \|E(\lambda) g_1\|^2, \qquad \mathfrak{M}(g_1) \equiv \mathfrak{H},$$

$$\int_{-\infty}^{+\infty} G_m(\lambda) \overline{G_n(\lambda)}\, d\varrho(\lambda) = \delta_{mn},$$

$$\int_{-\infty}^{+\infty} \lambda\, G_n(\lambda) \overline{G_m(\lambda)}\, d\varrho(\lambda) = a_{mn}.$$

The functions $G_n(\lambda)$ *are polynomials determined by the recurrence relations*

$$G_1(\lambda) = 1, \qquad G_2(\lambda) = \frac{\lambda - a_1}{\bar{b}_1},$$

$$G_n(\lambda) = \frac{(\lambda - a_{n-1}) G_{n-1}(\lambda) - b_{n-2} G_{n-2}(\lambda)}{\bar{b}_{n-1}};$$

the degree of $G_n(\lambda)$ *is therefore* $n-1$.

Since the self-adjoint transformation H is an extension of the symmetric transformation $T_1(A)$, we can write down expressions for the elements Hg_n directly from our knowledge of the matrix A, finding

$$Hg_1 = \sum_{\alpha=1}^{\infty} a_{\alpha 1} g_\alpha = a_1 g_1 + \bar{b}_1 g_2,$$

$$Hg_{n-1} = \sum_{\alpha=1}^{\infty} a_{\alpha, n-1} g_\alpha = b_{n-2} g_{n-2} + a_{n-1} g_{n-1} + \bar{b}_{n-1} g_n,$$

where $n = 3, 4, 5, \cdots$. It is evident that $g_1 = G_1(H) g_1$, $g_2 = \dfrac{Hg_1 - a_1 g_1}{\bar{b}_1} = G_2(H) g_1$. If we assume that the equation $g_k = G_k(H) g_1$ holds for $k = 1, \cdots, n-1$, we can deduce the relation

$$\begin{aligned} g_n &= \frac{Hg_{n-1} - a_{n-1} g_{n-1} - b_{n-2} g_{n-2}}{\bar{b}_{n-1}} \\ &= \frac{(H - a_{n-1}) G_{n-1}(H) - b_{n-2} G_{n-2}(H) g_1}{\bar{b}_{n-1}} \\ &= G_n(H) g_1. \end{aligned}$$

By the principle of mathematical induction, the equation $g_n = G_n(H)g_1$ is true for every integer n. Since the closed linear manifold $\mathfrak{M}(g_1)$ contains the elements $G_n(H)g_1 = g_n$, the identity $\mathfrak{M}(g_1) = \mathfrak{H}$ is satisfied and the transformation H has a simple spectrum. If $E(\lambda)$ is the resolution of the identity for H and $\varrho(\lambda)$ is the function $\|E(\lambda)g_1\|^2$, we have

$$\int_{-\infty}^{+\infty} G_m(\lambda)\overline{G_n(\lambda)}\,d\varrho(\lambda) = (G_m(H)g_1, G_n(H)g_1) = (g_m, g_n) = \delta_{mn},$$

$$\int_{-\infty}^{+\infty} \lambda G_n(\lambda)\overline{G_m(\lambda)}\,d\varrho(\lambda) = (HG_n(H)g_1, G_m(H)g_1) = (Hg_n, g_m) = a_{mn}.$$

The proof of the theorem is thus completed.

In Chapter X we shall examine in detail the general theory of symmetric transformations defined by matrices in Jacobi form. It will be possible then to relinquish the condition that these transformations be essentially self-adjoint.

§ 4. The Reducibility of Self-Adjoint Transformations

The investigations of the preceding sections serve as a basis for a number of interesting and important results concerning self-adjoint transformations in general and those with simple spectra in particular. In this section we shall characterize the closed linear manifolds which reduce a given self-adjoint transformation H; and in the next we shall combine the results of § 3 and § 4 to study the reduction to "principal axes" of a self-adjoint transformation.

We find in the case of a general self-adjoint transformation that we can describe its behavior with regard to reducibility in the following way:

THEOREM 7.15. *Let H be a self-adjoint transformation with domain $\mathfrak{D}$, $E(\lambda)$ the corresponding resolution of the identity, $\mathfrak{M}$ a closed linear manifold, and E the projection of $\mathfrak{H}$ on $\mathfrak{M}$. The four ensuing statements are equivalent:*

(1) *$\mathfrak{M}$ reduces H;*

(2) *$\mathfrak{M}$ reduces $E(\lambda)$;*

(3) *there exists a finite or denumerably infinite set* $\{f_n\}$ *such that the closed linear manifolds* $\{\mathfrak{M}(f_n)\}$ *are mutually orthogonal and* $\mathfrak{M} = \mathfrak{M}(f_1) \oplus \mathfrak{M}(f_2) \oplus \mathfrak{M}(f_3) \oplus \cdots$;

(4) *there exist a self-adjoint transformation* H_0 *and functions* $F(\lambda)$, $G(\lambda)$ *measurable with respect to* H_0 *such that* $H \equiv F(H_0)$, $E \equiv G(H_0)$.

If $\mathfrak{M}$ *reduces* H *and if* T *and* $F(\lambda)$ *are the transformations defined in* $\mathfrak{M}$ *by the equations*

$$T = H \text{ in } \mathfrak{M} \cdot \mathfrak{D}, \qquad F(\lambda) = E(\lambda) \text{ in } \mathfrak{M},$$

respectively, then these transformations have the properties:

(1) *when* $\mathfrak{M}$ *has the dimension-number* $\aleph_0$, *it can be regarded as an abstract Hilbert space; in this space* T *is a self-adjoint transformation with the corresponding resolution of the identity* $F(\lambda)$;

(2) *when* $\mathfrak{M}$ *has the finite dimension-number* n, *it can be regarded as a unitary space of* n *dimensions; in this space* T *behaves like the familiar (Hermitian) symmetric transformation of* n*-dimensional unitary geometry.*

We first examine the equivalence of the four statements by showing that we can perform the deductions indicated by the scheme

$$1 \longrightarrow 2 \begin{array}{c} \nearrow\, 3 \,\searrow \\ \searrow\, 4 \,\nearrow \end{array} 1.$$

It is then evident that the statements are equivalent in pairs.

If $\mathfrak{M}$ reduces H then $\mathfrak{M}$ reduces the resolvent R_l of H according to Theorem 4.27; in other words $ER_l \equiv R_l E$ for $\mathfrak{J}(l) \neq 0$. This identity implies the truth of the equation

$$(R_l f, Eg) = (ER_l f, g) = (R_l Ef, g),$$

which by Theorem 5.7 can be written in the form

$$\int_{-\infty}^{+\infty} \frac{1}{\lambda - l}\, d(E(\lambda)f, Eg) = \int_{-\infty}^{+\infty} \frac{1}{\lambda - l}\, d(E(\lambda)Ef, g).$$

The uniqueness argument based on Lemma 5.2 enables us to assert that

$$(EE(\lambda)f, g) = (E(\lambda)f, Eg) = (E(\lambda)Ef, g).$$

Thus $EE(\lambda)$ and $E(\lambda)E$ are identical; in other words, $\mathfrak{M}$ reduces $E(\lambda)$, $-\infty<\lambda<+\infty$.

If $\mathfrak{M}$ reduces $E(\lambda)$, then we can conclude that whenever f is an element of $\mathfrak{M}$

$$f = Ef, \qquad E(\lambda)f = E(\lambda)Ef = EE(\lambda)f$$

and that the entire family of elements $E(\lambda)f$, $-\infty<\lambda<+\infty$ lies in $\mathfrak{M}$. Thus, whenever f is in $\mathfrak{M}$ we have $\mathfrak{M}(f) \subseteqq \mathfrak{M}$. By the argument employed in proving Theorem 7.4 we construct a sequence $\{f_n\}$ with the properties described in our third statement: we select an orthonormal set $\{\varphi_n\}$ which determines the closed linear manifold $\mathfrak{M}$ and by the successive steps described in detail in the earlier discussion obtain the set $\{f_n\}$.

If $\mathfrak{M}$ has the property indicated in (3), it reduces H by Theorem 4.26.

If $\mathfrak{M}$ reduces $E(\lambda)$, then there is no difficulty in constructing explicitly the transformation H_0 and the functions F and G of our fourth statement. The resolution of the identity determining H_0 is defined by means of the equations

$$\begin{aligned} E_0(\mu) &\equiv EE(-\log(-\mu)), & -\infty<\mu<0, \\ E_0(0) &\equiv E, \\ E_0(\mu) &\equiv E+(I-E)\,E(\log\mu), & 0<\mu<+\infty. \end{aligned}$$

The systematic checking of the properties of $E_0(\mu)$ is rendered quite simple by the use of the relation $EE(\lambda) \equiv E(\lambda)E$. We define the functions F and G by the equations

$$\begin{aligned} F(\mu) &= -\log(-\mu), & -\infty<\mu<0, \\ F(0) &= 0, \\ F(\mu) &= \log\mu, & 0<\mu<+\infty, \\ G(\mu) &= 1, & -\infty<\mu\leqq 0, \\ G(\mu) &= 0, & 0<\mu<+\infty; \end{aligned}$$

since $\mu=0$ is evidently not a point of discontinuity of $E_0(\mu)$ the definition of F and of G at that point is actually a matter of indifference. The proof of the identity $E \equiv G(H_0)$ is exceedingly simple so that we shall omit it. It is worth while, however, to examine more carefully the proof of the relation $H \equiv F(H_0)$. An element f is in the domain of $F(H_0)$ if and only if the integral $\int_{-\infty}^{+\infty} |F(\mu)|^2 \, d\|E_0(\mu)f\|^2$ exists; and when f is in the domain of $F(H_0)$, the equation

$$(F(H_0)f, g) = \int_{-\infty}^{+\infty} F(\mu) \, d(E_0(\mu)f, g)$$

is valid. We translate these statements into new forms by making the change of variable $\lambda = -\log(-\mu)$ when $-\infty < \mu < 0$ and the change $\lambda = \log \mu$ when $0 < \mu < +\infty$; the omission of the point $\mu = 0$ is unessential since it is not a point of discontinuity of $E_0(\mu)$. The two integrals above are then replaced by

$$\int_{-\infty}^{+\infty} \lambda^2 d\|E(\lambda)Ef\|^2 + \int_{-\infty}^{+\infty} \lambda^2 d\|E(\lambda)(I-E)f\|^2$$
$$= \int_{-\infty}^{+\infty} \lambda^2 d\|E(\lambda)f\|^2,$$
$$\int_{-\infty}^{+\infty} \lambda \, d(E(\lambda)Ef, g) + \int_{-\infty}^{+\infty} \lambda \, d(E(\lambda)(I-E)f, g)$$
$$= \int_{-\infty}^{+\infty} \lambda \, d(E(\lambda)f, g).$$

By Theorem 5.9 these relations imply that $F(H_0) \equiv H$.

Finally, when the conditions of (4) are satisfied we can apply Theorem 6.1 (7) to assert that HE exists whenever EH is significant and that $HE = EH$ whenever both transformations are significant. Since E is defined throughout $\mathfrak{H}$, the conditions that $\mathfrak{M}$ reduce H are fulfilled.

In order to study the behavior of H in a closed linear manifold $\mathfrak{M}$ which reduces it, we shall define transformations T and $F(\lambda)$ as follows: the domain of T consists of all the elements of $\mathfrak{M} \cdot \mathfrak{D}$ and T coincides there with H; the domain of $F(\lambda)$ is the manifold $\mathfrak{M}$ and $F(\lambda)$ coincides there with $E(\lambda)$. From the knowledge that $\mathfrak{M}$ reduces H and $E(\lambda)$ we conclude that the ranges of T and of $F(\lambda)$ are subsets of $\mathfrak{M}$ in accordance with Theorem 4.23. If $\mathfrak{M}$ has the dimension number $\aleph_0$ it can be regarded as an abstract Hilbert space, as we pointed out in Theorem 1.19. We can verify immediately the assertion that the family of transformations $F(\lambda)$ is a resolution of the identity in this Hilbert space. Thus it is possible to take the assertion of Theorem 5.9 in regard to H and $E(\lambda)$ and specialize it by considering its effect in the manifold $\mathfrak{M}$ alone. Since we are permitted to replace H and $E(\lambda)$ by T and $F(\lambda)$ respectively for this purpose, we obtain the following statement: an element f in $\mathfrak{M}$ belongs to the domain of T if and only if the integral $\int_{-\infty}^{+\infty} \lambda^2 d \| F(\lambda) f \|^2$ exists; if f is in the domain of T then

$$(Tf, g) = \int_{-\infty}^{+\infty} \lambda \, d\, (F(\lambda) f, g)$$

for every element g in $\mathfrak{M}$. This result signifies that with reference to the abstract Hilbert space $\mathfrak{M}$ the transformation T is a self-adjoint transformation with the corresponding resolution of the identity $F(\lambda)$. When $\mathfrak{M}$ has finite dimension number n, we can apply Theorem 5.9 in a somewhat similar manner. $\mathfrak{M}$ is to be regarded now as a unitary space of n dimensions. As before we find that $F(\lambda)$ has the usual formal properties of a resolution of the identity, holding in $\mathfrak{M}$:

$$\begin{aligned} &F(\lambda) F(\mu) = F(\mu) F(\lambda) = F(\lambda), \quad \lambda \leqq \mu, \\ &F(\lambda) \to O, \qquad \lambda \to -\infty, \\ &F(\lambda) \to I, \qquad \lambda \to +\infty. \end{aligned}$$

Since the range of $F(\lambda)$ is the set of elements common to $\mathfrak{M}$ and the range of $E(\lambda)$, it is a closed linear manifold with dimension number $n(\lambda)$. In view of the fact that $n(\lambda)$ is a monotone-increasing function which can assume no values other that $0, 1, \cdots, n$, we see that $F(\lambda)$ must have the following properties: there exist points $\lambda = l_1, \cdots, l_m$, $m \leqq n$, numbered in order of magnitude, such that

$$\begin{aligned} &F(\lambda) = O, \qquad \lambda < l_1, \\ &F(\lambda) = F(l_k), \qquad l_k \leqq \lambda < l_{k+1}, \qquad k = 1, \cdots, m-1, \\ &F(\lambda) = I, \qquad l_m \leqq \lambda. \end{aligned}$$

Theorem 5.9 now shows us that $Tf = Hf$ exists when f is in $\mathfrak{M}$ if and only if the integral $\int_{-\infty}^{+\infty} \lambda^2 d\|F(\lambda)f\|^2$ converges; but in view of the properties of $F(\lambda)$ this integral is convergent for every element f in $\mathfrak{M}$. Thus T coincides with H throughout $\mathfrak{M}$ and the equation $(Tf, g) = (f, Tg)$ is true for every pair of elements f and g in $\mathfrak{M}$. It is to be noted of course that the numbers $l_1, \cdots, l_m$ are characteristic values of T and of H, and that the corresponding characteristic elements determine the closed linear manifold $\mathfrak{M}$. The behavior of T is completely characterized by this description and is seen to be precisely that indicated in the statement of the theorem.

In the case of a self-adjoint transformation with simple spectrum we can sharpen the results of the preceding theorem to a remarkable degree. We obtain the following result:

THEOREM 7.16. *If the symbols H, $E(\lambda)$, $\mathfrak{M}$, and E have the same meanings as in the preceding theorem, then the statements*

(1) *$\mathfrak{M}$ reduces H,*

(2) *$E \equiv G(H)$, where $G(\lambda)$ is a function measurable with respect to H which assumes the values zero and one almost everywhere with respect to H,*

are equivalent whenever H has a simple spectrum.

The results of the preceding theorem show that the first statement is a consequence of the second; for (2) is a special form of Theorem 7.15 (4). If we once show that $E \equiv G(H)$ where $G(\lambda)$ is measurable with respect to H, we can then apply Theorem 6.6 to ascertain the range of values assumed by $G(\lambda)$.

The crux of the proof is the deduction of the identity $E \equiv G(H)$ from the assertion that $\mathfrak{M}$ reduces H. It is convenient to replace the latter assertion by the equivalent statement that $\mathfrak{M}$ reduces $E(\lambda)$; in other words, by the statement that $EE(\lambda)$ and $E(\lambda)E$ are identical. We can generalize the last statement by replacing E by an arbitrary bounded linear transformation T with $\mathfrak{H}$ as its domain. The present theorem follows at once, therefore, if the identity $TE(\lambda)$

$\equiv E(\lambda)T$ implies $T \equiv G(H)$ when H has a simple spectrum. By hypothesis there exists an element f such that $\mathfrak{M}(f) = \mathfrak{H}$; in other words, the set $\mathfrak{D}_0$ of all elements $E(\lambda)f$, $-\infty < \lambda < +\infty$, determines the closed linear manifold $\mathfrak{H}$. We define a transformation T_0 with $\mathfrak{D}_0$ as its domain by the equation $T_0 = T$ in $\mathfrak{D}_0$. By reference to Theorems 2.10 and 2.23 it is evident that the only closed linear extension of T_0 is the bounded linear transformation T. Now in accordance with Theorem 6.2 we can find a function $G(\lambda)$ in $\mathfrak{L}_2(f)$ such that $Tf = G(H)f$; indeed, we can suppose that $G(\lambda)$ is measurable with respect to H or even Borel measurable because every function in $\mathfrak{L}_2(f)$ is equivalent to a corresponding Borel measurable function. The transformation $G(H)$ is then a closed linear transformation with its domain everywhere dense in $\mathfrak{H}$. We consider the behavior of $G(H)$ in the set $\mathfrak{D}_0$. We have

$$T_0 E(\mu) f = T E(\mu) f = E(\mu) T f = E(\mu) G(H) f = G(H) E(\mu) f$$

since

$$E(\mu) G(H) \equiv G_\mu(H) \equiv G(H) E(\mu),$$

where $G_\mu(\lambda) = G(\lambda)$ for $\lambda \leqq \mu$ and $G_\mu(\lambda) = 0$ for $\lambda > \mu$. In other words, $G(H)$ is a closed linear extension of T_0; and the remark made above enables us to identify T and $G(H)$, as we wished to do. Since $G(H)$ is defined throughout $\mathfrak{H}$, the function $G(\lambda)$ must satisfy the restrictions enumerated in Theorem 6.5; that is, there must exist a constant C such that the inequality $|G| \leqq C$ is satisfied except possibly on a set of points of zero measure with respect to H.

§ 5. Reduction to Principal Axes

We have already indicated in Chapter III, § 1, that our investigation of self-adjoint transformations presents analogies with the algebraic theory of the reduction of (Hermitian) symmetric matrices to diagonal form and with the theory of the reduction of (Hermitian) symmetric forms to principal axes. Due to the occurrence of the continuous spectrum in the case of the self-adjoint transformation, the resemblance between the transcendental and algebraic theories is distorted and imperfect. We are able, however, to rehabilitate the analogy in a satisfactory and perspicuous manner. From the detailed study of the continuous spectrum carried out in the preceding paragraphs we can select and combine results so as to obtain the appropriate generalization of the algebraic theories we have mentioned. We find that we can introduce coördinates in Hilbert space which reduce a given self-adjoint transformation to the simplest possible canonical form. If we insist on the use of ordinary infinite matrices, we choose coördinates so that the matrix representing the given self-adjoint transformation can be built up out of matrices in diagonal form or in Jacobi form. If we admit the use of generalized matrices with a continuum of rows and columns we can represent the transformation by such a generalized matrix in diagonal form.

THEOREM 7.17. *If H is a self-adjoint transformation, there exists a complete orthonormal set $\{f_{mn}\}$, $m, n = 1, 2, 3, \cdots$ with the following properties:*

(1) *the closed linear manifold $\mathfrak{M}_m$ determined by the orthonormal set $f_{m1}, f_{m2}, f_{m3}, \cdots$ reduces H;*

(2) *the behavior of H in $\mathfrak{M}_m$ is completely determined by the matrix $A^{(m)}$ whose general term is $a_{ik}^{(m)} = (Hf_{mk}, f_{mi})$;*

(3) *the matrix $A^{(m)}$ is either a finite or infinite (Hermitian) symmetric matrix in diagonal form or an infinite (Hermitian) symmetric matrix in Jacobi form.*

The matrix A associated with H by the complete orthonormal set $\{f_{mn}\}$ has the properties:

(1) *each row and each column of A contains only a finite number of non-zero elements;*

(2) $\tilde{T}_1(A) \equiv T_1^*(A) \equiv T_1^{**}(A) \equiv H$.

The proof of this theorem is immediate. There are three cases to consider, corresponding to the three cases of Theorem 7.4. In case (3) we set $\mathfrak{M}_1 = \mathfrak{M} = \mathfrak{H}$, $f_{1n} = \varphi_n$. Since $f_{1n} = \varphi_n$ is a characteristic element of H, the matrix $A^{(1)} = A$ associated with H by the set $\{f_{1n}\}$ is in diagonal form. By the usual analysis based on Theorem 4.17, the transformation $T_1(A)$

associated with A by the set $\{f_{1n}\}$ is found to be essentially self-adjoint. Since H is an extension of $T_1(A)$, we have $\tilde{T}_1(A) \equiv T_1^*(A) \equiv T_1^{**}(A) \equiv H$.

In case (2), we set $\mathfrak{M}_1 = \mathfrak{M}$, $\mathfrak{M}_{m+1} = \mathfrak{M}(g_m)$, $m = 1, 2, 3, \cdots$. Instead of the set $\{g_k\}$ of Theorem 7.4 we might also use the set $\{\psi_k\}$ of Theorem 7.5. We set $f_{1n} = \varphi_n$, as in case (3). We observe that $\mathfrak{M}_1$ and $\mathfrak{M}_m$, $m \geqq 2$, reduce H by Theorems 5.13 and 7.2, respectively, that these manifolds are mutually orthogonal, and that they determine $\mathfrak{H}$ according to the equation $\mathfrak{H} = \mathfrak{M}_1 \oplus \mathfrak{M}_2 \oplus \mathfrak{M}_3 \oplus \cdots$. We define $T^{(m)}$ by the equation $T^{(m)} = H$ in $\mathfrak{M}_m \cdot \mathfrak{D}$, where $\mathfrak{D}$ is the domain of H and $m = 1, 2, 3, \cdots$, and then apply Theorem 7.15 to ascertain the properties of these transformations. When $\mathfrak{M}_1$ has finite dimension number, the matrix $A^{(1)}$ associated with $T^{(1)}$ by the orthonormal set $\{f_{1n}\}$ is a finite square matrix in diagonal form; and when $\mathfrak{M}_1$ has the dimension number $\aleph_0$, $T^{(1)}$ is self-adjoint in $\mathfrak{M}_1$ and the matrix $A^{(1)}$ associated with it by the orthonormal set $\{f_{1n}\}$ is in diagonal form. For $m \geqq 1$, the manifold $\mathfrak{M}_{m+1}$ has the dimension number $\aleph_0$ because $\| E(\lambda) g_m \|^2$ is continuous and not identically zero. The transformation $T^{(m+1)}$ is self-adjoint in $\mathfrak{M}_{m+1}$ and has a simple spectrum, by virtue of the relation $\mathfrak{M}_{m+1} = \mathfrak{M}(g_m)$. We therefore select the orthonormal set $\{f_{m+1,n}\}$ in $\mathfrak{M}_{m+1}$ by means of Theorem 7.13 so that the associated matrix $A^{(m+1)}$ is in Jacobi form. The set $\{f_{mn}\}$, $m = 1, 2, 3, \cdots$, $n = 1, 2, 3, \cdots$, is obviously a complete orthonormal set. Theorem 4.26 shows that H is completely determined by its behavior in the manifolds $\mathfrak{M}_m$, $m = 1, 2, 3, \cdots$; that is, by the transformations $T^{(m)}$ or by the corresponding matrices $A^{(m)}$. The matrix A is obtained by appropriate arrangement of the elements of the matrices $A^{(m)}$ together with the introduction of zero elements. It is easily seen that no row or column of A contains an infinite number of non-zero elements. From what was said above, it is clear that the matrix A determines the transformation H. In fact, we can show that the transformation $T_1(A)$ associated with A by the complete orthonormal set $\{f_{mn}\}$ satisfies the relation

$T_1^*(A) \equiv H$; it is thus an essentially self-adjoint transformation. If f and f^* are elements such that $(T_1(A)f_{mn}, f) = (f_{mn}, f^*)$ for $m, n = 1, 2, 3, \cdots$, we consider their respective projections f_m and f_m^* on the closed linear manifold $\mathfrak{M}_m$. Since $T_1(A)f_{mn}$ is an element in $\mathfrak{M}_m$ for $n = 1, 2, 3, \cdots$, we see that the equations satisfied by f and f^* can be written in the form

$$(T^{(m)}f_{mn}, f_m) = (f_{mn}, f_m^*) \qquad m, n = 1, 2, 3, \cdots.$$

By virtue of the fact that the transformation $T^{(m)}$ is completely determined in $\mathfrak{M}_m$ by the matrix A^m or by its behavior in the set $\{f_{mn}\}$, we conclude that f_m is in the domain of $T^{(m)}$ and that $T^{(m)}f_m = f_m^*$. We now have

$$f = f_1 + f_2 + f_3 + \cdots,$$
$$f^* = T^{(1)}f_1 + T^{(2)}f_2 + T^{(3)}f_3 + \cdots = Hf_1 + Hf_2 + Hf_3 + \cdots.$$

According to Theorem 4.26, these equations imply that Hf exists and is equal to f^*. This result means that $T_1^*(A) \equiv H$, as we wished to prove.

In case (1) we set $\mathfrak{M}_m = \mathfrak{M}(g_m)$ and proceed in the same way as in case (2), except for the difference due to the absence of the point spectrum.

THEOREM 7.18. *If H is a self-adjoint transformation with domain $\mathfrak{D}$ and $\{\mathfrak{M}_m\}$ is the set of closed linear manifolds described in Theorem 7.17, then one of the three following cases occurs:*

(1) *H has no point spectrum; $\mathfrak{M}_m$ can be put in one-to-one correspondence with the space $\mathfrak{L}_2$ of all Lebesgue-measurable functions $f(x)$, $0 \leq x \leq 1$, such that the Lebesgue integral $\int_0^1 |f(x)|^2\,dx$ exists; the image of the transformation $T^{(m)}$, where $T^{(m)} = H$ in $\mathfrak{M}_m \cdot \mathfrak{D}$, is a self-adjoint transformation which takes every function $f(x)$ in $\mathfrak{L}_2$ such that the Lebesgue integral $\int_0^1 H_m^2(x)\,|f(x)|^2\,dx$ exists into the function $H_m(x)f(x)$, where $H_m(x)$ is a real monotone-increasing function of x with no interval of constancy;*

(2) *H has an incomplete orthonormal set of characteristic elements; H is represented in $\mathfrak{M}_1 = \mathfrak{M} \dotplus \mathfrak{H}$ by a finite or infinite matrix in diagonal form; and H is represented in each of the manifolds $\mathfrak{M}_m$, $m \geqq 2$, in the manner described in* (1);

(3) *H has a complete orthonormal set of characteristic elements and is represented in $\mathfrak{M}_1 = \mathfrak{M} = \mathfrak{H}$ by an infinite matrix in diagonal form.*

The three cases correspond to the three cases of Theorems 5.13 and 7.4. The proof of the theorem is analogous to that of the preceding theorem, the only difference being that in each of the manifolds $\mathfrak{M}(g_m)$ we apply Theorem 7.11 instead of Theorem 7.13. It is to be noted that each representation in $\mathfrak{L}_2$ may be regarded as a representation in terms of a diagonal matrix with a continuum of rows and of columns, where the variable x, $0 \leqq x \leqq 1$, specifies position along the principal diagonal.

Theorems 7.17 and 7.18 enable us to survey at a glance all possible types of self-adjoint transformation; and at the same time they provide methods for constructing actual examples of each conceivable type. We shall not enter upon a detailed consideration of this phase of the theory, though we shall suggest some of the questions which it raises. In cases (1) and (2) of Theorem 7.18, it is of considerable interest to determine whether the functions $H_m(x)$ can be chosen arbitrarily, subject only to the restrictions stated there. The reader should have no difficulty in verifying that they can. A more involved question concerning these cases is the following: what conditions are necessary and sufficient in order that the functions $H_m(x)$ be associated with the set $\{\psi_k\}$ of Theorem 7.5? All the materials for answering this question are at hand. If we consider the function $H_m(x)$ as a monotone function, we think at once of its resolution into discontinuous, absolutely continuous, and non-absolutely continuous components; and we are thus led to consider the significance of this resolution for the theory of self-adjoint transformations.

CHAPTER VIII

GENERAL TYPES OF LINEAR TRANSFORMATIONS

§ 1. PERMUTABILITY

In this chapter we shall consider certain general properties of linear transformations with special attention to unitary and normal transformations, the latter type being defined below in § 3. All our results are based upon the theory of self-adjoint transformations which has been developed in the preceding chapters. An important tool in applying those developments is the concept of permutability, to which the present section is devoted.

In the case of two bounded linear transformations T_1 and T_2 defined throughout $\mathfrak{H}$ we define permutability in the most natural and obvious manner, saying that T_1 and T_2 are permutable if and only if $T_1 T_2 \equiv T_2 T_1$ throughout $\mathfrak{H}$. In the more complicated case of a bounded linear transformation T_1 and a closed linear transformation T_2 which may be non-bounded, the formulation of an appropriate definition of permutability is attended by some difficulties. If we examine the preliminary discussion of reducibility given in Chapter IV, § 3, which centers about the requirements for the permutability of a projection and a self-adjoint transformation laid down in Definition 4.5, we are led to adopt the following definition:

DEFINITION 8.1. *Two closed linear transformations T_1 and T_2, the first of which is bounded and has $\mathfrak{H}$ as its domain, are said to be permutable if $T_1 T_2 \subseteq T_2 T_1$: that is, if, whenever f is in the domain of T_2, $T_1 f$ is also in the domain of T_2 and $T_1 T_2 f = T_2 T_1 f$.*

It is evident that this definition includes the more restricted definition indicated in the case of two bounded transformations with domain $\mathfrak{H}$.

The characterization of permutability for two non-bounded transformations T_1 and T_2 is beset with difficulties of such a serious nature that we avoid them by a subterfuge in those cases where a precise definition is needed. The source of these difficulties is located in the fact that domains of non-bounded transformations are proper subsets of $\mathfrak{H}$ in the cases which are of most interest. For the investigations of this chapter we shall have occasion to define permutability only for self-adjoint transformations. In order to justify our choice of a definition, we prove the following generalization of particular results which we have already noted in the preceding chapters:

THEOREM 8.1. *A necessary and sufficient condition that a closed bounded linear transformation T with domain $\mathfrak{H}$ be permutable with a self-adjoint transformation H is that T be permutable with $E(\lambda)$, the resolution of the identity corresponding to H, for all values of λ, $-\infty<\lambda<+\infty$. If H has a simple spectrum, a necessary and sufficient condition that T be permutable with H is that T be expressible in the form $T\equiv G(H)$, where $G(\lambda)$ is an H-measurable function satisfying the inequality $|G(\lambda)|\leqq C$ for some constant C almost everywhere with respect to H.*

We first remark that T and H are permutable if and only if T and R_l, the resolvent of H, are permutable for every l in the resolvent set of H. The domain of H consists of all the elements expressible in the form $R_l f$, where l is fixed and f is an arbitrary element in $\mathfrak{H}$. If T and H are permutable, $TR_l f$ is in the domain of H and we can write $(H-lI)\,TR_l f = T(H-lI)\,R_l f = Tf$. Applying the transformation R_l to both members of this equation, we obtain $TR_l f = R_l\,Tf$, $TR_l\equiv R_l\,T$, as we wished to prove. On the other hand, if T and R_l are permutable, we consider an arbitrary element f in the domain of H and write $(H-lI)f = g$, $f = R_l\,g$. We then have $Tf = TR_l\,g$

$= R_l Tg$, so that Tf is in the domain of H. Applying the transformation $H - lI$ to both members of this equation, we obtain $(H-lI)Tf = Tg = T(H-lI)f$, $HTf = THf$. Thus T and H are permutable.

Since T is a bounded linear transformation with domain $\mathfrak{H}$ its adjoint T^* exists and has the same properties. By Theorem 5.7, therefore, we have

$$(R_l Tf, g) = \int_{-\infty}^{+\infty} \frac{1}{\lambda - l} d(E(\lambda)Tf, g),$$

$$(TR_l f, g) = (R_l f, T^* g) = \int_{-\infty}^{+\infty} \frac{1}{\lambda - l} d(E(\lambda)f, T^* g)$$

$$= \int_{-\infty}^{+\infty} \frac{1}{\lambda - l} d(TE(\lambda)f, g).$$

Applying the uniqueness argument drawn from Lemma 5.2, we see that T and R_l are permutable if and only if $(TE(\lambda)f, g) = (E(\lambda)Tf, g)$ for arbitrary f, g and λ; that is, if and only if T and $E(\lambda)$ are permutable, $-\infty < \lambda < +\infty$.

The case where H has a simple spectrum was discussed in full under Theorem 7.16, with the result formulated in the present theorem.

The criterion for permutability given in this theorem shows that we may consistently define permutability for general self-adjoint transformations as follows:

DEFINITION 8.2. *Two self-adjoint transformations H_1 and H_2 with the corresponding resolutions of the identity $E_1(\lambda)$ and $E_2(\lambda)$ respectively are said to be permutable if $E_1(\lambda_1)$ and $E_2(\lambda_2)$ are permutable for $-\infty < \lambda_1 < +\infty$, $-\infty < \lambda_2 < +\infty$.*

In § 3 and § 4, we shall obtain theorems which show that the transformations H_1 and H_2 behave in a appropriate manner under multiplication if they are permutable according to this definition.

In recent investigations, J. v. Neumann has studied the algebra of bounded and of normal transformations in considerable detail, obtaining many interesting results which bear on the question of permutability.† He formulates no

† J. v. Neumann, Mathematische Annalen, 102 (1929), pp. 370–427.

general definition of permutability and does not succeed in finding a definition for the permutability of self-adjoint transformations any more direct than that which we have given. In this chapter, we have no need to inquire more deeply into the situation and shall not attempt to do so.

§ 2. Unitary Transformations

A few facts concerning the resolvent set and the spectrum of an arbitrary unitary transformation U will serve to motivate our subsequent considerations. We obtain the following theorem:

Theorem 8.2. *If U is a unitary transformation, its resolvent set and its spectrum have the following properties:*

(1) *the resolvent set comprises all points of the finite l-plane such that $|l| < 1$, $|l| > 1$;*

(2) *the spectrum lies on the unit circle $|l| = 1$;*

(3) *the residual spectrum is empty;*

(4) *the point spectrum is empty, finite, or denumerably infinite, characteristic elements corresponding to distinct characteristic values being orthogonal.*

The specific properties of the unitary transformation U on which this theorem is based are expressed by the relations

$$U^{-1} \equiv U^*, \quad (Uf, Ug) = (f, g), \quad (U^{-1}f, U^{-1}g) = (f, g).$$

We see at once that the origin in the l-plane is a point of the resolvent set since U^{-1} exists and is unitary. From Theorem 4.11 it follows, in view of the equation $\| U^{-1}f \| = \| f \|$, that every point l for which $|l| < 1$ is also a point of the resolvent set. By applying Theorem 4.21 and the equation $\| Uf \| = \| f \|$, we show similarly that every point for which $|l| > 1$ is in the resolvent set. Consequently (1) and (2) are established.

By definition the point l belongs to the residual spectrum of U if and only if the range of $(U - l \cdot I)$ is not everywhere dense in $\mathfrak{H}$. As pointed out in Theorem 4.15, this implies that $\bar{l}$ is a characteristic value of $U^* \equiv U^{-1}$, so that there exists an element $f \neq 0$ for which $U^{-1}f = \bar{l}f$. If we apply to both sides of this equation the transformation $l \cdot U$, noting that $l\bar{l} = 1$ by virtue of (2), we obtain $lf = Uf$ so that we must admit, contrary to hypothesis, that l is a characteristic value of U. We infer therefore that the residual spectrum is empty.

In discussing (4) we may suppose that the point spectrum contains at least two points. If, then, l_1 and l_2 are two distinct characteristic values with correspondig characteristic elements f_1 and f_2, we can write

$$(f_1, f_2) = (Uf_1, Uf_2) = (l_1 f, l_2 f) = l_1 \bar{l}_2 (f_1, f_2).$$

Since $l_1 \neq l_2$, $|l_1| = |l_2| = 1$, we see that $l_1 \bar{l}_2$ is different from 1 and that (f_1, f_2) vanishes: the characteristic elements for distinct characteristic values are orthogonal. To each characteristic value of U we order a corresponding characteristic element, which we may choose as already normalized. The set thus obtained is an orthonormal set as we have just proved, and is at most denumerably infinite. Thus when the point spectrum contains at least two points it is either finite or denumerably infinite.

In view of the theorem just proved it is natural to attempt to build a bridge between the class of unitary transformations and the class of self-adjoint transformations by a transformation of the unit circle into the real axis in the l-plane, the object being to map the spectrum of a transformation in one class on the spectrum of a transformation in the other. Of the available maps, we shall consider only the linear fractional transformations $w = \dfrac{az+b}{cz+d}$ which map the extended z-plane on the extended w-plane and which take the unit circle in the z-plane into the real axis in the w-plane. This device leads to an important relationship, first perceived by Cayley and Frobenius in the algebraic problem of finite unitary and (Hermitian) symmetric matrices and by v. Neumann in the problem which confronts us here.* These writers consider

* J. v. Neumann, Mathematische Annalen, 102 (1929), pp. 62–63, 80–84.

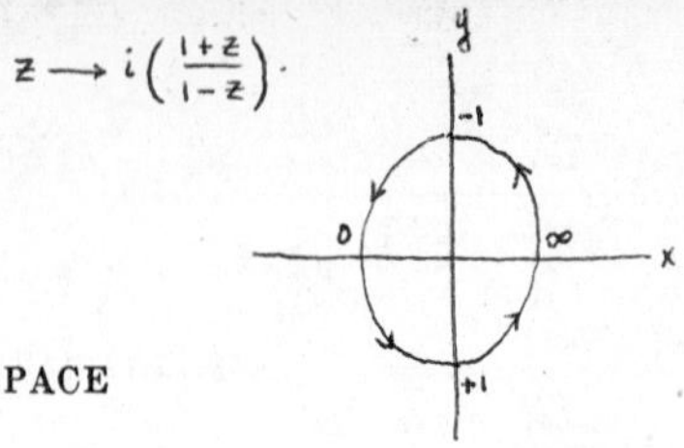

the special case $w = i\,\dfrac{1+z}{1-z}$, which differs in no essential respect from that which we shall study. The relationship in question may be stated as follows:

THEOREM 8.3. *Let* $w = \dfrac{az+b}{cz+d}$, $-d/c = \zeta$, *be a linear fractional transformation such that the loci* $|z| = 1$, $\mathfrak{J}(w) = 0$, *and the points* $z = \zeta$, $w = \infty$ *correspond. Let* $\mathfrak{U}_\zeta$ *be the class of all unitary transformations whose point spectra do not contain the point* $l = \zeta$; *and let* $\mathfrak{H}$ *be the class of all self-adjoint transformations. Then there exists a one-to-one correspondence between* $\mathfrak{U}_\zeta$ *and* $\mathfrak{H}$ *such that corresponding transformations* U *and* H *are related in the following way:*

(1) *H is expressible symbolically by the relation* $H \equiv \dfrac{aU+bI}{cU+dI}$, *which, interpreted, means that the domain of H comprises those and only those elements f which are expressible in the form* $f = (cU+dI)g$ *and that in its domain H is defined by the equation*

$$Hf = H(cU+dI)g = (aU+bI)g;$$

(2) $U \equiv \dfrac{-dH+b}{cH-a}$.

We first note a few specific properties of the transformation $w = \dfrac{az+b}{cz+d}$. Since it is not degenerate, the determinant $ad-bc$ does not vanish. Its inverse is given by the equation $z = \dfrac{-dw+b}{cw-a}$. The fact that the circle $z = e^{i\theta}$, θ real, is mapped on the real axis in the w-plane requires that the equations

$$\bar{a}c+\bar{b}d = a\bar{c}+b\bar{d}, \qquad \bar{b}c = a\bar{d}, \qquad b\bar{c} = \bar{a}d$$

be true; this property of the transformation also requires that for all real values of λ the fraction $\dfrac{-d\lambda+b}{c\lambda-a}$ have the absolute value 1.

When U is a given transformation in $\mathfrak{U}_\zeta$, we know from the preceding theorem that $\zeta = -d/c$, $|\zeta| = 1$, is a point either of the resolvent set or of the continuous spectrum. The range of the transformation $cU+dI$ is accordingly a linear manifold everywhere dense in $\mathfrak{H}$. In this manifold we define T as that transformation which takes $f=(cU+dI)g$ into $(aU+bI)g$, noting that this definition can lead to no inconsistency because $(cU+dI)$ defines a one-to-one correspondence between its domain and its range. The transformation T is symmetric: for if $f_1=(cU+dI)g_1$ and $f_2=(cU+dI)g_2$ are arbitrary elements of its domain we have

$$\begin{aligned}(Tf_1, f_2) &= ((aU+bI)g_1, (cU+dI)g_2)\\ &= (g_1, [(\bar{a}c+\bar{b}d)I+\bar{b}cU+\bar{a}dU^*]g_2),\\ (f_1, Tf_2) &= ((cU+dI)g_1, (aU+bI)g_2)\\ &= (g_1, [(a\bar{c}+b\bar{d})I+a\bar{d}U+b\bar{c}U^*]g_2),\\ (Tf_1, f_2) &= (f_1, Tf_2).\end{aligned}$$

In order to prove that T is self-adjoint we consider the character of the range of the transformation $T_l \equiv T-l\cdot I$ when l is not real. Since, by definition,

$$T_l f=(aU+bI)g-l(cU+dI)g=[(a-cl)U+(b-dl)I]g,$$

the range of T_l coincides with the range of $[(a-cl)U+(b-dl)I]$. If $(a-cl)$ vanishes the latter transformation has the range $\mathfrak{H}$ since it reduces to $\dfrac{bc-ad}{c}I$, $(bc-ad)\neq 0$, $c\neq 0$. If $(a-cl)$ does not vanish its range is the same as that of the transformation $U-l'I$ where $l'=\dfrac{-dl+b}{cl-a}$. If l is not real, l' is a point either interior or exterior to the unit circle and belongs to the resolvent set of U. The range in question is therefore the entire space $\mathfrak{H}$. According to Theorems 4.16 and 4.17, T is a self-adjoint transformation. As a member of the class $\mathfrak{H}$ we denote it by the letter H. To each transformation in $\mathfrak{U}_\xi$ there corresponds a uniquely determined transformation in $\mathfrak{H}$ constructed explicitly in the manner described. We shall now show that to distinct transformations U_1 and U_2 there correspond distinct transformations H_1 and H_2

20

respectively; in other words, that $H_1 \equiv H_2$ implies $U_1 \equiv U_2$. If H_1 and H_2 are identical, then an arbitrary element f in their common domain can be expressed in the form

$$f = (cU_1 + dI)g_1 = (cU_2 + dI)g_2;$$

when f ranges over the domain in question g_1 and g_2 range over $\mathfrak{H}$. By hypothesis $H_1 f = H_2 f$, $(aU_1 + bI)g_1 = (aU_2 + bI)g_2$. From the equations

$$c(U_1 g_1 - U_2 g_2) + d(g_1 - g_2) = 0,$$
$$a(U_1 g_1 - U_2 g_2) + b(g_1 - g_2) = 0$$

we conclude that $g_1 = g_2$, $U_1 g_1 = U_2 g_2$ since the determinant $ad - bc$ does not vanish. Thus U_1 and U_2 are identical and our proof is completed.

Next, if H is an arbitrary transformation in $\mathfrak{H}$ we can form the transformation $U \equiv \dfrac{-dH + b}{cH - a}$ and can use Theorem 6.6, coupled with the fact that $\left| \dfrac{-d\lambda + b}{c\lambda - a} \right| = 1$ when λ is real, to verify its unitary character. The transformation $cU + dI = \dfrac{bc - ad}{cH - a}$, where $bc - ad$ is different from zero, has an inverse $\dfrac{cH - a}{bc - ad}$ whose domain is the same as that of H and is therefore a linear manifold everywhere dense in $\mathfrak{H}$. Thus the point $l = \zeta$ is not a characteristic value of U and the transformation itself belongs to the class $\mathfrak{U}_\zeta$. If, starting with H, we construct $U \equiv \dfrac{-dH + b}{cH - a}$ as indicated and then form the transformation T of the preceding paragraph, it is clear that T and H are identical. In order to prove this relation we apply the operational calculus of Chapter VI to show that we have

$$H(cU + dI) = \frac{(bc - ad)H}{cH - a}, \qquad aU + bI = \frac{(bc - ad)H}{cH - a}$$

throughout $\mathfrak{H}$. From the relation $H(cU + dI) = aU + bI$ thus established, we see that H is an extension of the self-adjoint transformation T. Since H is itself self-adjoint the identity $H \equiv T$ must hold.

On combining the information concerning the correspondence between $\mathfrak{U}_\zeta$ and $\mathfrak{H}$ now at our disposal we see that the theorem is established.

As an immediate consequence of this result we can prove a theorem somewhat simpler in statement and somewhat more complete in its assertions. This theorem is the analogue of our fundamental theorems concerning self-adjoint transformations.

THEOREM 8.4. *Let $\mathfrak{U}$ be the class of all unitary transformations, $\mathfrak{H}_0$ the class of all bounded self-adjoint transformations whose spectra are confined to the interval $0 \leqq \lambda \leqq 1$ and whose point spectra do not include the point $\lambda = 0$. Then there exists a one-to-one correspondence between $\mathfrak{U}$ and $\mathfrak{H}_0$ such that corresponding transformations U and H_0 satisfy the identity $U \equiv e^{2\pi i H_0}$.*

The principal point of the proof is to construct at least one correspondent in $\mathfrak{H}_0$ for each member of $\mathfrak{U}$. When U is a given unitary transformation we select a point ζ on the unit circle which is not a characteristic value, recalling from Theorem 8.2 (4) that this is possible. The transformation $w = \frac{1}{i}\frac{z+\zeta}{z-\zeta}$ with the inverse $z = \zeta\frac{w-i}{w+i}$, takes the circle $|z| = 1$ into the line $\mathfrak{I}(w) = 0$, the point $z = \zeta$ into $w = \infty$. By the preceding theorem there exists a unique self-adjoint transformation H such that $U \equiv \zeta\,\frac{H-i}{H+i}$. If the resolution of the identity for H is $E(\lambda)$, we have

$$(Uf, g) = \zeta \int_{-\infty}^{+\infty} \frac{\lambda - i}{\lambda + i}\, d(E(\lambda)f, g)$$

for every pair of elements f, g in $\mathfrak{H}$. We express this integral in terms of a new variable μ, where $\lambda = -\cot \pi\mu$, $\mu = -\frac{1}{\pi}$ arccot λ. We introduce a new resolution of the identity $F(\mu)$, defined for all real values of μ according to the relations

$$F(\mu) \equiv E(-\cot \pi\mu), \qquad 0 < \mu < 1;$$
$$F(\mu) \equiv O, \quad \mu \leqq 0; \qquad F(\mu) \equiv I, \quad 1 \leqq \mu.$$

The integral then takes the form

$$(Uf, g) = \int_0^1 \zeta e^{2\pi i\mu}\, d(F(\mu)f, g) = \int_{-\infty}^{+\infty} \zeta e^{2\pi i\mu}\, d(F(\mu)f, g).$$

In order to eliminate the factor ζ from this expression we change the variables once more. We set $\zeta = e^{-2\pi i\theta}$, where $0 \leqq \theta < 1$, and then define a correspondence between μ and ν as follows:

$$\nu = \mu, \quad \mu \leqq 0; \quad \nu = \mu + 1 - \theta, \quad 0 < \mu \leqq \theta;$$
$$\nu = \mu - \theta, \quad \theta < \mu \leqq 1; \quad \nu = \mu, \quad \mu > 1.$$

We introduce at the same time a family of projections $E_0(\nu)$, which can be verified from the defining relations to be a resolution of the identity. This family is constructed by means of the equations

$$E_0(\nu) \equiv F(\mu) \text{ when } \nu = \mu, \; \nu \leqq 0, \; \mu \leqq 0;$$
$$E_0(\nu) \equiv F(\mu) - F(\theta) \text{ when } \nu = \mu - \theta, \; 0 < \nu \leqq 1 - \theta, \; \theta < \mu \leqq 1;$$
$$E_0(\nu) \equiv F(\mu) + F(1) - F(\theta) \text{ when } \nu = \mu + 1 - \theta, \; 1 - \theta < \nu \leqq 1, \; 0 < \mu \leqq \theta;$$
$$E_0(\nu) \equiv F(\mu) \text{ when } \nu = \mu, \; \nu > 1, \; \mu > 1.$$

It is found by substitution that the integral at last takes the desired form

$$(Uf, g) = \int_{-\infty}^{+\infty} e^{2\pi i\nu}\, d(E_0(\nu)f, g).$$

The self-adjoint transformation H_0 determined by $E_0(\nu)$ evidently belongs to the class $\mathfrak{H}_0$ and is connected with U by the identity $U \equiv e^{2\pi i H_0}$. Thus each transformation in $\mathfrak{U}$ has at least one correspondent in $\mathfrak{H}_0$.

When H_0 is a given member of $\mathfrak{H}_0$, the transformation $e^{2\pi i H_0}$ exists and is unitary, by Theorem 6.6.

We must still show that two distinct members of $\mathfrak{H}_0$ cannot yield the same unitary transformation. In order to establish this result it is sufficient to prove that from a knowledge of $U \equiv e^{2\pi i H_0}$ we can determine the resolution of the identity $E_0(\lambda)$ corresponding to H_0. The series

$$\sum_{\alpha=-\infty}^{+\infty} a_\alpha e^{2\pi i \alpha \nu}, \qquad a_n = \int_b^c e^{2\pi i n \nu} d\nu, \qquad 0 < b < c < 1,$$

is known to be a Fourier series which converges boundedly to the limit function $F_{bc}(\nu)$, where

$$\begin{aligned} F_{bc}(\nu) &= 0, \qquad 0 \leqq \nu < b, \qquad c < \nu \leqq 1, \\ F_{bc}(\nu) &= \tfrac{1}{2}, \qquad \nu = b, \qquad \nu = c, \\ F_{bc}(\nu) &= 1, \qquad b < \nu < c, \end{aligned}$$

insofar as the closed interval $0 \leqq \nu \leqq 1$ is concerned. Recalling that the spectrum of H_0 is confined to this interval, we see that the transformation $F_{bc}(H_0)$ is determined by the transformation $U \equiv e^{2\pi i H_0}$ from the operational series

$$F_{bc}(H_0) \equiv \sum_{\alpha=-\infty}^{+\infty} a_\alpha e^{2\pi i \alpha H_0} \equiv \sum_{\alpha=-\infty}^{+\infty} a_\alpha U^\alpha.$$

We now allow b and c to tend to the limits 0 and λ respectively in such manner that $0 < b$, $0 < \lambda < c < 1$. The function $F_{bc}(\nu)$ then tends to the limiting function $G_\lambda(\nu)$, where

$$\begin{aligned} G_\lambda(\nu) &= 0, \qquad \nu = 0, \qquad \lambda < \nu \leqq 1, \\ G_\lambda(\nu) &= 1, \qquad 0 < \nu \leqq \lambda. \end{aligned}$$

Since $\nu = 0$ is a point of continuity or a point of constancy of $E_0(\nu)$ we conclude that

$$E_0(\lambda) \equiv G_\lambda(H_0), \qquad 0 < \lambda < 1.$$

Since $E_0(\lambda) \equiv O$, $\lambda \leqq 0$, and $E_0(\lambda) \equiv I$, $\lambda \geqq 1$, we see that $E_0(\lambda)$ is completely determined by U, as we wished to prove. This completes the proof of the theorem.

This connection between unitary and self-adjoint transformations was noted by J. v. Neumann who used it to prove the main theorems concerning self-adjoint transformations, the general course of the discussion running from a treatment of unitary transformations through the facts stated here to the desired fundamental theorems.*

For subsequent applications we note an interesting theorem concerning the correspondence between $\mathfrak{U}_\zeta$ and $\mathfrak{H}$ discussed in Theorem 8.3 above. It gives a new criterion for permutability.

THEOREM 8.5. *A necessary and sufficient condition that a bounded linear transformation T with domain $\mathfrak{H}$ and a self-adjoint transformation H be permutable is that the transformation T and the unitary transformation $U \equiv \dfrac{-dH+b}{cH-a}$ be permutable.*

If T and H are permutable, then T and R_l are permutable whenever l is in the resolvent set of H, as we have already shown in the proof of Theorem 8.1. The fact that U is unitary requires that $\left|\dfrac{-d\lambda+b}{c\lambda-a}\right| = 1$; hence $c \neq 0$, $\mathfrak{I}(a/c) \neq 0$. If we put $l = a/c$, we see therefore that U can be written in the form

$$U \equiv \left(\frac{-dH+b}{c}\right) R_l.$$

Thus

$$TU \equiv T\left(\frac{-dH+b}{c}\right) R_l \equiv \left(\frac{-dH+b}{c}\right) TR_l$$
$$\equiv \left(\frac{-dH+b}{c}\right) R_l T \equiv UT$$

throughout $\mathfrak{H}$.

On the other hand, if T and U are permutable we can show that T and H are permutable. An element f is in the domain of H if and only if it is expressible in the form $f = (cU+dI)g$. Thus, when f is in the domain of H,

$$Tf = T(cU+dI)g = (cU+dI)Tg$$

is also in the domain of H. Evidently

$$Hf = (aU+bI)g, \quad THf = T(aU+bI)g,$$
$$HTf = (aU+bI)Tg = T(aU+bI)g = THf,$$

so that T and H are permutable.

* J. v. Neumann, Mathematische Annalen, 102 (1929), pp. 91–96, 111–122.

We note also the following immediate application of this theorem:

THEOREM 8.6. *Two self-adjoint transformations H_1 and H_2 are permutable if and only if the unitary transformation* $U_1 \equiv \dfrac{-d_1 H_1 + b_1}{c_1 H_1 - a_1}$ *is permutable with H_2 or with the unitary transformation* $U_2 \equiv \dfrac{-d_2 H_2 + b_2}{c_2 H_2 - a_2}$.

By Definition 8.2, the transformations H_1 and H_2 are permutable if and only if the corresponding resolutions of the identity $E_1(\lambda_1)$ and $E_2(\lambda_2)$ are permutable. By Theorem 8.1 this is the case if and only if $E_2(\lambda_2)$ is permutable with H_1. Now, according to the preceding theorem, $E_2(\lambda_2)$ is permutable with H_1 if and only if it is permutable with U_1; and this is the case if and only if H_2 is permutable with U_1. Finally, the condition just obtained is equivalent to the condition that U_1 and U_2 be permutable.

§ 3. NORMAL TRANSFORMATIONS

By reason of an algebraic analogy with certain concepts of the theory of finite matrices, it was early found desirable to study the class of "normal" transformations: a bounded linear transformation T with domain $\mathfrak{H}$ is said to be normal if it is permutable with its adjoint T^*. Such transformations are of special interest largely because they admit a spectral representation analogous to that developed for self-adjoint transformations and can therefore be subjected to an analysis along the lines developed in Chapters VI and VII. For a complete theory it is evidently important that we be able to include non-bounded transformations in the class of all normal transformations. The first satisfactory definition from this point of view was given by J. v. Neumann.† We shall give a different, but equivalent, definition which is simpler in certain respects than his.

† J. v. Neumann, Mathematische Annalen, 102 (1929), p. 406, Definition 6.

DEFINITION 8.3. *A transformation T is said to be normal if it has the following properties:*

(1) *T^* and T^{**} exist and have the same domain;*

(2) *the transformations $S_1 \equiv \frac{1}{2}(T^{**} + T^*)$ and $S_2 \equiv \frac{1}{2i}(T^{**} - T^*)$ are essentially self-adjoint;*

(3) *the self-adjoint transformations $H_1 \equiv \tilde{S}_1 \equiv S_1^*$, $H_2 \equiv \tilde{S}_2 \equiv S_2^*$ are permutable.*

We shall also make use of the following definition introduced by v. Neumann:†

DEFINITION 8.4. *A normal transformation T is said to be maximal if it possesses no proper normal extension.*

From these definitions, it is clear that the theory of normal transformations is substantially equivalent to a theory of pairs of permutable self-adjoint transformations. Thus we are led to generalize the analysis given for self-adjoint transformations so as to obtain a method appropriate to the study of such pairs. We can outline the program of our development in a few words. For the purpose in view, it is first desirable to extend the results of Chapter V, § 1, and Chapter VI, § 1, to the case of functions of two real variables x and y or of a single complex variable $z = x + iy$; and the exposition in those sections was designed to lend itself readily to this end. We then introduce a family of projections $E(x, y)$ or $E(z)$ to which we assign a rôle similar to that played by the resolution of the identity $E(\lambda)$ in the earlier theory. The extension of the results of Chapters VI and VII to the case in hand is then accomplished simply by paraphrasing the earlier theory.

We commence by defining the class of functions of bounded variation of the complex variable z. It is convenient to adhere to the general terminology of Chapter VI, § 1. We denote by $\varrho(z)$ a complex-valued function of z, by $*\Delta$ the interval or cell

---

† J. v. Neumann, Mathematische Annalen, 102 (1929), p. 406.

$$\alpha_1 < \Re z \leqq \beta_1, \quad \alpha_2 < \Im z \leqq \beta_2,$$
$$-\infty \leqq \alpha_1 \leqq \beta_1 \leqq +\infty, \; -\infty \leqq \alpha_2 \leqq \beta_2 \leqq +\infty,$$

and by $\varrho(*\Delta)$ the quantity

$$\varrho(\beta_1 + i\beta_2) - \varrho(\beta_1 + i\alpha_2) - \varrho(\alpha_1 + i\beta_2) + \varrho(\alpha_1 + i\alpha_2).$$

The variation of $\varrho(z)$ is denoted by $V(\varrho)$ and is defined as the least upper bound of the set of numbers $\sum_{\alpha=1}^{n} |\varrho(*\Delta_\alpha)|$ formed for all finite collections of mutually disjoint finite intervals $*\Delta$. A function $\varrho(z)$ such that $V(\varrho) < +\infty$ is said to be of bounded variation. We shall restrict our attention to a class of functions of bounded variation in normal form, analogous to the class $\mathfrak{B}^*$ of Chapter V, § 1. We shall say that $\varrho(z)$ is a function of bounded variation in normal form or is a function in the class $\mathfrak{B}^*$ if it has the properties

$$\begin{array}{ll} \varrho(z+\zeta) \to \varrho(z) & \text{when } \Re\zeta \geqq 0,\ \Im\zeta \geqq 0,\ \zeta \to 0, \\ \varrho(z) \to 0 & \text{when } \Re z \to \infty, \text{ and when } \Im z \to -\infty. \end{array}$$

It can be shown that when $\Re z \to +\infty$, $\Im z \to +\infty$, $\varrho(z)$ tends to a limit r. It is unnecessary for us to consider the relation of the class $\mathfrak{B}^*$ to the entire class of functions of bounded variation. We may remark, however, that there exists an operation $*$ which, applied to a function $\varrho(z)$ of bounded variation, yields a function $\varrho^*(z)$ in normal form, in a manner similar to that described in the case of functions of a single real variable. A real function $\varrho(z)$ is said to be monotone-increasing if $\varrho(*\Delta) \geqq 0$ for every interval $*\Delta$. Any function in $\mathfrak{B}^*$ can be resolved into real monotone-increasing components in $\mathfrak{B}^*$, according to the equation

$$\varrho(z) = [\varrho_1(z) - \varrho_2(z)] + i[\varrho_3(z) - \varrho_4(z)].$$

Thus from an arbitrary function $\varrho(z)$ in $\mathfrak{B}^*$ we can develop a theory of measure and of integration identical in terminology, symbolism, and logical structure with that described in Chapter VI, § 1, the only difference being that we must now deal with subsets of the collection I consisting of all complex numbers $z = x + iy$, $0 < x \leqq +\infty$, $0 < y \leqq +\infty$. Lemmas 6.1 and 6.2, in particular, hold for the new theory when they are suitably interpreted. By the use of appropriate symbolism we can make a similar assertion concerning Lemmas 6.3–6.6. If we denote by $\int_{-\infty}^{z} F(t)\, d\varrho(t)$ the Radon-Stieltjes integral of $F(t)$ with respect to $\varrho(t)$ over the interval $-\infty < \Re t \leqq \Re z$, $-\infty < \Im t \leqq \Im z$, and retain the various relevant symbols such as $\frac{d\varphi_2}{d\varphi_1}$, $\succ$, $\sim$, then those lemmas hold without formal modification in the new theory.

We next consider the introduction of the family of projections $E(z)$ mentioned above. We make the following formal definition:

DEFINITION 8.5. *A family of projections $E(z)$, where z assumes all complex values, is called a complex resolution of the identity if it has the following properties:*

(1) $E(z_1)E(z_2) \equiv E(z_3)$ *where* $\Re z_3 = \min[\Re z_1, \Re z_2]$ *and* $\Im z_3 = \min[\Im z_1, \Im z_2]$;

(2) $E(z + \zeta) \to E(z)$ *when* $\Re \zeta \geqq 0$, $\Im \zeta \geqq 0$, $\zeta \to 0$;

(3) $E(z) \to O$ *when* $\Re z \to -\infty$ *and when* $\Im z \to -\infty$;

(4) $E(z) \to I$ *when* $\Re z \to +\infty$, $\Im z \to +\infty$.

As usual we shall employ the notation $E(^*\Delta)$, where $^*\Delta$ is the interval $\alpha_1 < \Re z \leqq \beta_1$, $\alpha_2 < \Im z \leqq \beta_2$, to denote the transformation

$$E(\beta_1 + i\beta_2) - E(\beta_1 + i\alpha_2) - E(\alpha_1 + i\beta_2) + E(\alpha_1 + i\alpha_2).$$

It is easily verified that this transformation is a projection. We shall have to deal continually with the functions of z defined by the equation $\varrho(z) = (E(z)f, g)$, where f and g are arbitrary elements in $\mathfrak{H}$. It can be shown, by methods analogous to those given in the proof of Theorem 5.7, that $V(\varrho) \leqq |f||g|$ and that $\varrho(z)$ is a function in $\mathfrak{B}^*$. The resolution of $\varrho(z)$ into its real monotone-increasing components is given by the equation

$$(E(z)f, g) = \left[\left\| E(z)\frac{f+g}{2}\right\|^2 - \left\| E(z)\frac{f-g}{2}\right\|^2\right]$$
$$+ i\left[\left\| E(z)\frac{f+ig}{2}\right\|^2 - \left\| E(z)\frac{f-ig}{2}\right\|^2\right]$$

The resolution, of course, is not unique. We can apply Lemma 6.7 in the new theory as in the old; and we carry over the terminology and symbolism of Definitions 6.1 and 6.2 to the case in hand. We shall state explicitly the definition which corresponds to Definition 6.3.

DEFINITION 8.6. *If $E(z)$ is a complex resolution of the identity, the terms "$E(z)$-measurable", "null set with respect to $E(z)$", "almost everywhere with respect to $E(z)$", and "$E(z)$-equivalent" shall mean respectively "ϱ-measurable", "null set with respect to $\varrho(z)$", "almost everywhere with respect to $\varrho(z)$", and "ϱ-equivalent", for $\varrho(z) = \|E(z)f\|^2$ and every element f in $\mathfrak{H}$.*

We are finally in a position to define a transformation $T(F)$, associated with a given complex resolution of the identity $E(z)$ and an arbitrary complex-valued function $F(z)$, by means of the expressions

$$(T(F)f, g) = \int_{-\infty}^{+\infty} F(z)\, d(E(z)f, g),$$
$$\|T(F)f\|^2 = \int_{-\infty}^{+\infty} |F(z)|^2\, d\|E(z)f\|^2,$$

holding for all elements f such that the second integral exists and for all elements g in $\mathfrak{H}$. We use the symbol $\int_{-\infty}^{+\infty}$ to indicate that the integral is extended over I.

THEOREM 8.7. *Theorems 6.1–6.8 remain valid when the terminology and symbolism are interpreted on the basis of a complex resolution of the identity $E(z)$ and the terms of Definition 6.3 are replaced by those of Definition 8.6.*

We shall leave the detailed examination of this theorem to the reader. It will be observed that the final theorem of Chapter VI has not been included in the preceding statement. The reason is evident as soon as we note that in order to generalize Theorem 6.9 we must associate a complex resolution of the identity with each transformation $T(F)$. We shall be able to do so, and hence to generalize Theorem 6.9, after we have studied the theory of normal transformations.

By appropriate use of the theory which has been outlined above we obtain a complete characterization of the class of all normal transformations. We first show how a given normal transformation can be expressed in terms of a complex resolution of the identity.

THEOREM 8.8. *If T is a normal transformation, then there exists a complex resolution of the identity $E(z)$ such that the relations $T^{**} \equiv T(F)$, $T^* \equiv T(\bar{F})$, $F(z) \equiv z$, are valid. The transformations T^* and T^{**} are maximal normal transformations. The given transformation T possesses no maximal normal extension other than T^{**}.*

We prove this theorem by a consideration of the self-adjoint transformations H_1 and H_2 associated with the given normal transformation T in the manner indicated in Definition 8.3. Corresponding to H_1 and H_2 are the respective resolutions of the identity $E_1(\lambda)$, $E_2(\lambda)$, $-\infty<\lambda<+\infty$. We define the desired complex resolution of the identity $E(z)$ by the relation

$$E(z) \equiv E_1(\mathfrak{R}(z))\, E_2(\mathfrak{I}(z)).$$

When we make use of the fact that $E_1(\lambda_1)$ and $E_2(\lambda_2)$ are permutable in accordance with Definitions 8.1 and 8.2, we can verify that the transformation $E(z)$ thus constructed is actually a complex resolution of the identity, as defined above. In terms of $E(z)$ the relations

$$(H_1 f, g) = \int_{-\infty}^{+\infty} \lambda\, d(E_1(\lambda) f, g),$$

$$\|H_1 f\|^2 = \int_{-\infty}^{+\infty} \lambda^2\, d\|E_1(\lambda) f\|^2,$$

$$(H_2 f, g) = \int_{-\infty}^{+\infty} \lambda\, d(E_2(\lambda) f, g),$$

$$\|H_2 f\|^2 = \int_{-\infty}^{+\infty} \lambda^2\, d\|E_2(\lambda) f\|^2,$$

become

$$(H_1 f, g) = \int_{-\infty}^{+\infty} \mathfrak{R}(z)\, d(E(z) f, g),$$

$$\| H_1 f \|^2 = \int_{-\infty}^{+\infty} (\mathfrak{R}(z))^2\, d\| E(z) f \|^2,$$

$$(H_2 f, g) = \int_{-\infty}^{+\infty} \mathfrak{J}(z)\, d(E(z) f, g),$$

$$\| H_2 f \|^2 = \int_{-\infty}^{+\infty} (\mathfrak{J}(z))^2\, d\| E(z) f \|^2,$$

respectively; in other words

$$H_1 \equiv T(F_1), \qquad H_2 \equiv T(F_2)$$

where $F_1(z) \equiv \mathfrak{R}(z)$, $F_2(z) \equiv \mathfrak{J}(z)$. If we set $F(z) \equiv z$, we can use Theorems 6.1, 6.4 and 8.7 to write

$$\begin{aligned} H_1 + iH_2 &\equiv T(F), & H_1 - iH_2 &\equiv T(\bar{F}), \\ (H_1 + iH_2)^* &\equiv H_1 - iH_2, & (H_1 - iH_2)^* &\equiv H_1 + iH_2. \end{aligned}$$

We can now investigate in detail the transformations T^* and T^{**}. It is evident that both these transformations satisfy the definition of normality. In order to show that

$$T^{**} \equiv H_1 + iH_2 \equiv T(F), \qquad T^* \equiv H_1 - iH_2 \equiv T(\bar{F}),$$

we proceed as follows: from the relations

$$T^{**} \equiv S_1 + iS_2 \subseteq H_1 + iH_2, \qquad T^* \equiv S_1 - iS_2 \subseteq H_1 - iH_2,$$

we see that

$$\begin{aligned} T^{**} &\equiv (T^*)^* \supseteq (H_1 - iH_2)^* \equiv H_1 + iH_2 \supseteq T^{**}, \\ T^* &\equiv (T^{**})^* \supseteq (H_1 + iH_2)^* \equiv (H_1 - iH_2) \supseteq T^*. \end{aligned}$$

We then infer the truth of the desired identities.

Finally we consider normal extensions of T and of T^*. First let T_1 be a normal extension of T^{**}: we shall show that $T_1 \equiv T^{**}$ and, in this way, shall prove that T^{**} is maximal. Without loss of generality we may suppose that $T_1 \equiv T_1^{**}$. From the relation $T_1^{**} \supseteq T^{**}$, it follows that $T^* \supseteq T_1^*$. Now we observe that, by definition, T_1^{**}, T_1^* have the same domain, as do T^{**} and T^*. This means that we must have $T^* \equiv T_1^*$, and hence $T_1 \equiv T_1^{**} \equiv T^{**}$, as we wished to prove. Thus T^{**} is a maximal normal extension of T. Next let T_1 be a second maximal normal extension of T. Since T_1^{**} is a normal extension of T_1, we have by definition $T_1 \equiv T_1^{**}$. Now $T_1 \supseteq T$ implies $T_1^{**} \supseteq T^{**}$. Since T^{**} is a maximal normal transformation, we conclude that $T_1 \equiv T_1^{**} \equiv T^{**}$, and have thus proved that T^{**} is the sole maximal normal extension of T. With regard to the transformation T^*, which we know to be normal, we note that $(T^*)^{**} \equiv T^*$ and that the preceding discussion therefore applies to it.

We can now obtain for maximal normal transformations a theorem analogous to the fundamental theorems for self-adjoint and for unitary transformations. It is

THEOREM 8.9. *Between the class $\mathfrak{N}$ of all maximal normal transformations T and the class $\mathfrak{E}_c$ of all complex resolutions of the identity there exists a one-to-one correspondence such that $T \equiv T(F)$ where $F(z) = z$.*

If T is a given maximal normal transformation, we can apply the preceding theorem to obtain this representation in terms of a complex resolution of the identity on noting that $T \equiv T^{**}$. On the other hand, if $E(z)$ is a given complex resolution of the identity we can form the transformation $T(F)$, $F(z) = z$, and then show with the aid of Theorems 6.1, 6.4, and 8.7 that it is normal. By these theorems, $T^*(F) \equiv T(\bar{F})$ and $T^{**}(F) \equiv T(F)$ exist and have the common domain $\mathfrak{D}(\bar{F}) = \mathfrak{D}(F)$. According to Definition 8.3 we next have to prove that the transformations

$$S_1 \equiv \frac{T(F)+T(\bar{F})}{2}, \quad S_2 \equiv \frac{T(F)-T(\bar{F})}{2i},$$

are essentially self-adjoint. We shall do so by establishing the identities

$$S_1^* \equiv T\left(\frac{F+\bar{F}}{2}\right), \quad S_2^* \equiv T\left(\frac{F-\bar{F}}{2i}\right),$$

in which the right-hand members are self-adjoint transformations in view of Theorems 6.4 and 8.7. We give a detailed demonstration in the case of S_1 alone, that of S_2 being analogous. The domain of S_1 is the common domain $\mathfrak{D}(F) \equiv \mathfrak{D}(\bar{F})$ of $T(F)$ and $T(\bar{F})$, characterized by the existence of the integral $\int_{-\infty}^{+\infty} |F(z)|^2 \, d\|E(z)f\|^2$, and is therefore identical with the domain $\mathfrak{D}(|F|+1)$ of the self-adjoint transformation $T(|F|+1)$. Since the origin in the z-plane is in the resolvent set of $T(|F|+1)$, by Theorems 6.7 and 8.7, we see that every element f of the domain of S_1 is expressible in the form $T((|F|+1)^{-1})\,g$ and that every element so expressible is in the domain of S_1. If h is an element in the domain of S_1^* we have $(S_1 f, h) = (f, S_1^* h)$ for every element f in the domain of S_1, and therefore

$$(S_1 T((|F|+1)^{-1})g, h) = (T((|F|+1)^{-1})\,g, S_1^* h)$$

for every g in $\mathfrak{H}$. Since, by Theorem 6.1 (2) and Theorem 8.7, $S_1 \subseteq T\left(\frac{F+\bar{F}}{2}\right)$ we can apply Theorem 6.1 (6) and Theorem 8.7 to write the last equation in the form

$$\left(T\left(\frac{F+\bar{F}}{2|F|+2}\right)g, h\right) = \left(T\left(\frac{1}{|F|+1}\right)g, S_1^* h\right).$$

This equation can be cast by the use of Theorems 6.4 and 8.7 into the form

$$\left(g, T\left(\frac{F+\bar{F}}{2|F|+2}\right)h\right) = \left(g, T\left(\frac{1}{|F|+1}\right)S_1^* h\right),$$

holding for every g in $\mathfrak{H}$. Consequently, the equation

$$T\left(\frac{F+\bar{F}}{2|F|+2}\right)h = T\left(\frac{1}{|F|+1}\right)S_1^* h$$

is true and the left-hand term follows the right in being an element of $\mathfrak{D}(|F|+1)$. From Theorem 6.1, (6) and (7), we infer that h is in the domain of $T\left(\frac{F+\bar{F}}{2}\right)$ and that

$$T\left(\frac{F+\bar{F}}{2}\right)h = T(|F|+1)\,T\left(\frac{F+\bar{F}}{2|F|+2}\right)h = S_1^* h.$$

We can express this result by the relation $S_1^* \subseteq T\left(\frac{F+\bar{F}}{2}\right)$. Since the previously noted relation $S_1 \subseteq T\left(\frac{F+\bar{F}}{2}\right)$ implies $S_1^* \supseteq T^*\left(\frac{F+\bar{F}}{2}\right) \equiv T\left(\frac{F+\bar{F}}{2}\right)$, we see that the desired identity $S_1^* \equiv T\left(\frac{F+\bar{F}}{2}\right)$ is true. In this manner, we come to the conclusion that the self-adjoint transformations H_1 and H_2 of Definition 8.3 exist and are to be identified with $T\left(\frac{F+\bar{F}}{2}\right)$ and $T\left(\frac{F-\bar{F}}{2i}\right)$ respectively.

Our next step is to show that H_1 and H_2 are permutable in the sense of Definition 8.2. We must therefore construct the resolutions of the identity $E_1(\lambda)$ and $E_2(\lambda)$ corresponding to H_1 and H_2 respectively. To this end, we define the functions

$$F_{1\lambda}(z) = \begin{Bmatrix} 1, & \Re(z) \leqq \lambda \\ 0, & \Re(z) > \lambda \end{Bmatrix}, \qquad F_{2\lambda}(z) = \begin{Bmatrix} 1, & \Im(z) \leqq \lambda \\ 0, & \Im(z) > \lambda \end{Bmatrix},$$

and the transformations $T(F_{1\lambda})$, $T(F_{2\lambda})$. These families of transformations, defined for $-\infty < \lambda < +\infty$, are quickly seen to be resolutions of the identity such that $T(F_{1\lambda})\,T(F_{2\mu})$ and $T(F_{2\mu})\,T(F_{1\lambda})$ are identical. We put $E_1(\lambda) \equiv T(F_{1\lambda})$ and $E_2(\lambda) \equiv T(F_{2\lambda})$. We then verify the identities

$$\begin{aligned}
\int_{-\infty}^{+\infty} \Re(z)\, d(E(z)f, g) &= \int_{-\infty}^{+\infty} \lambda\, d(E_1(\lambda)f, g), \\
\int_{-\infty}^{+\infty} |\Re(z)|^2\, d\|E(z)f\|^2 &= \int_{-\infty}^{+\infty} \lambda^2\, d\|E_1(\lambda)f\|^2, \\
\int_{-\infty}^{+\infty} \Im(z)\, d(E(z)f, g) &= \int_{-\infty}^{+\infty} \lambda\, d(E_2(\lambda)f, g), \\
\int_{-\infty}^{+\infty} |\Im(z)|^2\, d\|E(z)f\|^2 &= \int_{-\infty}^{+\infty} \lambda^2\, d\|E_2(\lambda)f\|^2,
\end{aligned}$$

which show that the resolutions of the identity corresponding to H_1 and H_2 are $E_1(\lambda)$ and $E_2(\lambda)$ respectively. Since $E_1(\lambda)$ and $E_2(\mu)$ are permutable, H_1 and H_2 are permutable. Thus $T(F)$ is a normal transformation. Furthermore, since

$T^*(F) \equiv T(\bar{F})$, $T^{**}(F) \equiv T(F)$, we can apply the preceding theorem to show that $T(F)$ is a maximal normal transformation.

It now appears that a correspondence has been set up between the class of all maximal normal transformations and the class of all complex resolutions of the identity, with the property that corresponding members T and $E(z)$ of these classes are connected by the relation $T \equiv T(F)$ where $F(z) \equiv z$. The work of the preceding paragraph contains the material for showing that this correspondence is a one-to-one correspondence. To a given $E(z)$ there corresponds one and only one maximal normal transformation T, so we have merely to show that two complex resolutions of the identity cannot give rise to the same maximal normal transformation. Now in the preceding paragraph we showed that, if T and $E(z)$ are connected by the relation $T \equiv T(F)$, the self-adjoint transformations H_1 and H_2 together with their corresponding resolutions of the identity $E_1(\lambda)$ and $E_2(\lambda)$ can be constructed directly from $E(z)$ as well as from T itself. Since $E(z)$ is evidently connected with $E_1(\lambda)$ and $E_2(\lambda)$ by means of the identity

$$E(z) \equiv E_1(\mathfrak{R}(z))\, E_2(\mathfrak{I}(z)),$$

we see that $E(z)$ is uniquely determined when $E_1(\lambda)$ and $E_2(\lambda)$ are known; but T determines the latter transformations completely, and therefore fixes $E(z)$. Thus two distinct complex resolutions of the identity must determine distinct maximal normal transformations. This completes the proof of the theorem.

We may now generalize the theorem just proved, as follows:

THEOREM 8.10. *If $E(z)$ is a complex resolution of the identity and $F(z)$ is a complex-valued $E(z)$-measurable function of z defined almost everywhere with respect to $E(z)$, then the associated transformation $T(F)$ is a maximal normal transformation. The complex resolution of the identity $E^{(1)}(z)$ corresponding to $T(F)$ by Theorem* 8.9 *is determined by the relation $E^{(1)}(^*\Delta) \equiv T(\Phi)$, where $^*\Delta$ is an arbitrary interval and $\Phi(z)$ is the* *function which is equal to one or to zero according as $F(z)$ assumes a value in the interval $*\Delta$ or not.*

The proof that the transformation $T(F)$ is normal proceeds at first exactly as in the special case $F(z) \equiv z$ discussed in the preceding theorem. We find that the transformations

$$S_1 \equiv \frac{1}{2}(T(F) + T(\bar{F})), \qquad S_2 \equiv \frac{1}{2i}(T(F) - T(\bar{F}))$$

have the properties

$$S_1^* \equiv H_1 \equiv T\left(\frac{F+\bar{F}}{2}\right) \equiv T(\Re F),$$

$$S_2^* \equiv H_2 \equiv T\left(\frac{F-\bar{F}}{2i}\right) \equiv T(\Im F).$$

It will be noted that up to this point the preceding proof made no use of the relation $F(z) \equiv z$ and can therefore be repeated *verbatim* in the present more general case. We have to modify our procedure, however, when we come to the proof that H_1 and H_2 are permutable in the sense of Definition 8.2. We denote by $E_1^{(1)}(\lambda)$ and $E_2^{(1)}(\lambda)$ the resolutions of the identity corresponding to the self-adjoint transformations H_1 and H_2 respectively. We then calculate $E_1^{(1)}(*\Delta)$ and $E_2^{(1)}(*\Delta)$ for an arbitrary linear interval $*\Delta$, using the methods described in the first part of the proof of Theorem 6.9. By Theorems 6.7 and 8.7 we see that $(H_1 - l)^{-1} \equiv T((\Re F - l)^{-1})$ for all not-real values of l. By the limiting processes analyzed in Theorem 6.9 we find successively

$$\tanh c(H_1 - \mu) \equiv T(\tanh c(\Re F - \mu)),$$

$$\varphi_\mu(H_1) \equiv T(\varphi_\mu(\Re F)), \qquad \psi_{\alpha\beta}(H_1) \equiv T(\psi_{\alpha\beta}(\Im F)).$$

The justification of the limiting processes involved depends upon Theorem 6.2 and its interpretation according to Theorem 8.7. The details will be left to the reader. If $*\Delta$ is the linear interval $\alpha < \lambda \leq \beta$ and $\Phi_1(z) \equiv \psi_{\alpha\beta}(\Re F(z))$ is the function which is equal to one or to zero according as $\Re F(z)$ assumes a value in $*\Delta$ or not, then the last relation can be written $E_2^{(1)}(*\Delta) \equiv T(\Phi_1)$. It is obvious, both directly and by reference to the limiting processes involved above, that $\Phi_1(z)$ is $E(z)$-measurable. If $\Phi_2(z)$ is the function which is equal to one or to zero according as $\mathfrak{J} F(z)$ assumes a value in the linear interval $^*\Delta$ or not, we have similarly

$$E_2^{(1)}(^*\Delta) \equiv T(\Phi_2).$$

We now have

$$E_1^{(1)}(^*\Delta_1)\, E_2(^*\Delta_2) \equiv E_2^{(1)}(^*\Delta_2)\, E_1^{(1)}(^*\Delta_1) \equiv T(\Phi_1 \circ \Phi_2)$$

for arbitrary linear intervals $^*\Delta_1$ and $^*\Delta_2$, by Theorem 6.1, (6) and (7), and Theorem 8.7. If we take $^*\Delta_1$ and $^*\Delta_2$ as intervals with extremities $-\infty$, λ_1 and $-\infty$, λ_2 respectively, we see that

$$E_1^{(1)}(\lambda_1)\, E_2^{(1)}(\lambda_2) \equiv E_2^{(1)}(\lambda_2)\, E_1^{(1)}(\lambda_1)$$

and that H_1 and H_2 are permutable. It is now evident that $T(F)$ has all the properties required of a normal transformation. The relations $T^{**}(F) \equiv T^*(\bar{F}) \equiv T(F)$ show, by virtue of Theorem 8.8, that $T(F)$ is a maximal normal transformation. As we showed in Theorems 8.8 and 8.9, the corresponding complex resolution of the identity $E^{(1)}(z)$ is given by the product $E_1^{(1)}(\mathfrak{R} z)\, E_2^{(1)}(\mathfrak{J} z)$. Thus if $^*\Delta$ is the two-dimensional interval $\alpha_1 < \mathfrak{R} z \leq \beta_1$, $\alpha_2 < \mathfrak{J} z \leq \beta_2$, and if $^*\Delta_1$ and $^*\Delta_2$ are the linear intervals $\alpha_1 < \lambda \leq \beta$, $\alpha_2 < \lambda \leq \beta_2$ respectively, then

$$E^{(1)}(^*\Delta) \equiv E_1^{(1)}(^*\Delta_1)\, E_2^{(1)}(^*\Delta_2) \equiv T(\Phi_1 \circ \Phi_2) \equiv T(\Phi)$$

where $\Phi \equiv \Phi_1 \circ \Phi_2$ is the function described in the statement of the theorem.

We can now establish the theorem which corresponds in the present theory to Theorem 6.9. We have

THEOREM 8.11. *Let T be a maximal normal transformation, $E(z)$ the corresponding complex resolution of the identity, $F(z)$ a complex-valued $E(z)$-measurable function defined almost everywhere with respect to $E(z)$, and $T(F)$ the transformation associated with $F(z)$ and $E(z)$. Then $T(F)$ is a maximal normal transformation which will be denoted by $T^{(1)}$. Let $E^{(1)}(z)$ be the corresponding complex resolution of the identity, $G^{(1)}(z)$ a complex-valued $E^{(1)}(z)$-measurable function defined* *almost everywhere with respect to* $E^{(1)}(z)$, *and* $T^{(1)}(G^{(1)})$ *the maximal normal transformation associated with* $G^{(1)}(z)$ *and* $E^{(1)}(z)$. *If* $\Phi(z) = G^{(1)}(F(z))$, *then* $\Phi(z)$ *is an* $E(z)$*-measurable function defined almost everywhere with respect to* $E(z)$ *and the transformation* $T(\Phi)$ *associated with* $\Phi(z)$ *and* $E(z)$ *is identical with the transformation* $T^{(1)}(G^{(1)})$.

If we take $G^{(1)}(z)$ as the function which is equal to one or to zero according as z is in the interval $*\Delta$ or not, we see that this theorem reduces to the preceding one; for we have

$$T^{(1)}(G^{(1)}) \equiv E^{(1)}(*\Delta) \equiv T(\Phi),$$

where $\Phi(z)$ is the function described in Theorem 8.10. This means that we can establish the theorem for more and more general classes of function $G^{(1)}(z)$ by the processes carried out in the proof of Theorem 6.8 until we attain the generality required by the statement of the theorem. Since no modification in detail is necessary, we do not need to repeat the earlier proof.

Thus we have completed the analogy between the theory of maximal normal transformations and that of self-adjoint transformations, so far as the subject-matter of Chapter VI is concerned. It is convenient to revise our notation so that it is in accord with Definitions 6.3 and 6.4; we are justified in making the change by the results of Theorems 8.7 and 8.9–8.11.

DEFINITION 8.7. *If T is a maximal normal transformation with the corresponding complex resolution of the identity* $E(z)$, *the terms "T-measurable", "null set with respect to T", "almost everywhere with respect to T", and "T-equivalent" shall mean "E(z)-measurable", "null set with respect to E(z)", "almost everywhere with respect to E(z)", and "E(z)-equivalent", respectively. If* $T(F)$ *is the transformation associated with an arbitrary function* $F(z)$ *and with* $E(z)$, *then it may be denoted by* $F(T)$.

We shall turn next to a brief consideration of the reducibility and the spectral theory of normal transformations. We have first

THEOREM 8.12. *Let T be a maximal normal transformation, $E(z)$ the corresponding complex resolution of the identity, and $^*\Delta$ an arbitrary interval. Then the projection $E(^*\Delta)$ is permutable with T and its range is a closed linear manifold which reduces T. If f is an element such that $\|E(^*\Delta)f\| = 1$, then*

$$z = (TE(^*\Delta)f, E(^*\Delta)f)$$

lies in the interval $^*\Delta$, whenever $E(^*\Delta)f$ is in the domain of T.

The assertions concerning permutability and reducibility are already contained in Theorem 6.1 (4) as interpreted according to Theorem 8.7, or are immediately deducible from the facts stated there. The final assertion of the theorem depends upon the relations

$$(TE(^*\Delta)f, E(^*\Delta)f) = \int_{^*\Delta} z\, d\|E(z)f\|^2,$$

$$\|E(^*\Delta)f\|^2 = \int_{^*\Delta} d\|E(z)f\|^2 = 1.$$

They show that $(TE(^*\Delta)f, E(^*\Delta)f)$ is a weighted average of the values of z in $^*\Delta$ and must therefore have a value in that interval.

THEOREM 8.13. *The spectrum of an arbitrary maximal normal transformation has the following properties:*

(1) *the spectrum contains at least one point;*

(2) *the point spectrum is empty or consists of a finite or denumerably infinite set of characteristic values; characteristic elements corresponding to distinct characteristic values are orthogonal;*

(3) *the continuous spectrum may be empty or not empty;*

(4) *the residual spectrum is empty.*

We form the complex resolution of the identity $E(z)$ corresponding to the given maximal normal transformation T and then construct the inverse of $T_l \equiv T - l \cdot I$ as the transformation $T\left(\frac{1}{z-l}\right)$. If the spectrum of T were empty, every point l of the complex z-plane would belong to the resolvent set; to a given l there would correspond a positive constant $C(l)$ such that the set of points defined by the inequality $\left|\frac{1}{z-l}\right| > C(l)$ is a null set with respect to $E(z)$, as we have shown in Theorems 6.7 and 8.7. It is easy to obtain a contradiction from the last assertion. If $*\Delta$ is an arbitrary interval, then each point of $*\Delta$ is the center of a circle $|z-l| < 1/C(l)$ which is a null set with respect to $E(z)$; since, by the Heine-Borel covering theorem, $*\Delta$ can be covered by a finite or denumerably infinite set of such circles, it is evident that $*\Delta$ is likewise a null-set with respect to $E(z)$; in consequence, $E(*\Delta) \equiv O$ for every $*\Delta$, a result in contradiction with the fundamental properties of $E(z)$ as given in Definition 8.5. Thus the spectrum of T must contain at least one point.

The transformation $T\left(\frac{1}{z-l}\right)$ is a closed linear transformation with domain everywhere dense in $\mathfrak{H}$ unless l is a characteristic value of T, as we have already shown in Theorems 6.7 and 8.7. It follows that the residual spectrum of T is empty. If l is a characteristic value of T, then we can determine the corresponding characteristic manifold as follows: if $F(z)$ is the function equal to one when $z = l$ and equal to zero elsewhere, we know by Theorems 6.6 (3), 6.7, and 8.7 that the transformation $T(F)$ is a projection different from O; its range, as we shall prove, is the characteristic manifold in question. If f is in the range of $T(F)$ we have $T(F)f = f$ and hence

$$(T - lI)\,T(F)f = (T - lI)f = 0,$$

by virtue of the identity $(z-l)\,F(z) = 0$. Thus f is a characteristic element of T for the characteristic value l, whenever $f \neq 0$. On the other hand, if f is a characteristic element of T for the characteristic value l, we form the element $E(*\Delta)f$ for an arbitrary interval $*\Delta$ and show that it is equal to f or to zero according as $*\Delta$ contains or does not contain the value $z = l$. If $E(*\Delta)f \neq 0$ we form the element $E(*\Delta)f/\|E(*\Delta)f\|$ and apply to it the last assertion of the preceding theorem. Since we have

$$(TE(^*\Delta)f/\|E(^*\Delta)f\|,\ E(^*\Delta)f/\|E(^*\Delta)f\|) = l$$

by virtue of the equation

$$TE(^*\Delta)f = E(^*\Delta)Tf = lE(^*\Delta)f,$$

we see that l must be contained in $^*\Delta$. Thus $E(^*\Delta)f = 0$ whenever $^*\Delta$ does not contain the value l. If $^*\Delta$ is an interval containing the value $z = l$, we can write

$$^*\bar{\Delta} = I - {}^*\Delta = \sum_{\alpha=1}^{n} {}^*\Delta_\alpha$$

where the intervals $^*\Delta_1, \cdots, {}^*\Delta_n$ are mutually disjoint. We find therefore that

$$E(^*\Delta)f = E(^*\Delta)f + \sum_{\alpha=1}^{n} E(^*\Delta_\alpha)f = E(I)f = f.$$

It is now easy to calculate $(T(F)f, g)$, where g is an arbitrary element in $\mathfrak{H}$, by means of the relations

$$(T(F)f, g) = \int_{-\infty}^{+\infty} F(z)\, d(E(z)f, g) = (f, g),$$

and thus to conclude that $T(F)f = f$. This result shows that f is in the range of $T(F)$. If we consider two distinct characteristic values of T and the corresponding functions $F_1(z)$ and $F_2(z)$, the relation $F_1(z)\,F_2(z) = 0$ shows that the projections $T(F_1)$ and $T(F_2)$ are orthogonal, satisfying the identity $T(F_1)\,T(F_2) \equiv T(F_1\,F_2) \equiv O$. Their ranges are orthogonal characteristic manifolds corresponding to the two characteristic values under consideration. A now familiar argument shows that the point spectrum of T is at most denumerably infinite, since characteristic elements corresponding to distinct characteristic values are known to be orthogonal.

THEOREM 8.14. *Let T be a maximal normal transformation, W the set of values assumed by (Tf, f) when $\|f\| = 1$ and f ranges over the domain of T. The closed set $\bar{W} = W + W'$ is the smallest convex set which includes the spectrum of the transformation T.*

First we shall show that every point of the spectrum of T is either a point of W or a limit point of W. Thus $\bar{W} = W + W'$

certainly includes the spectrum; and, since $\overline{W}$ is a convex point set by Theorem 4.7, $\overline{W}$ also includes the smallest convex point set of which the spectrum is a subset. Let $z = l$ be a point of the spectrum and let $^*\Delta$ be an arbitrarily small interval containing l as an interior point. There exists an element g such that $E(^*\Delta)g \neq 0$, since otherwise the interior points of $^*\Delta$ must belong to the resolvent set of T. We put

$$f = E(^*\Delta)g/\|E(^*\Delta)g\|$$

so that

$$E(^*\Delta)f = f, \quad \|f\| = 1.$$

From Theorem 8.12, we then conclude that

$$z = (TE(^*\Delta)f, E(^*\Delta)f)$$

is a point of $^*\Delta$. Thus l is a point of W or of W', as we wished to prove.

We now construct the smallest convex set which includes the spectrum: it is the set of points $z = \lambda z_1 + (1-\lambda)z_2$ where z_1 and z_2 range over the spectrum and λ over the interval $0 \leqq \lambda \leqq 1$. This set is closed since both the spectrum and the interval are closed sets. If $\overline{W}$ contains a point z_1 not in this set then there exists a uniquely determined point of the set, z_2, nearest to z_1. The straight line in the z-plane through z_2 perpendicular to the segment joining z_1 and z_2 then separates the spectrum from the point z_1. If f is an element of the domain of T such that $\|f\| = 1$, the equation

$$(Tf, f) = \int_{-\infty}^{+\infty} z\, d\|E(z)f\|^2,$$

in which the integral can be thought of as extended only over that half-plane which contains the spectrum, shows that the point $z = (Tf, f)$ must lie in the same half-plane with the spectrum. Since z_1 is a point for which this result is not valid, we must admit that $\overline{W}$ is a subset of the smallest convex set including the spectrum. Since we know that $\overline{W}$ is not a proper subset of the spectrum, we see that the theorem is established.†

† Wintner, Mathematische Zeitschrift, 30 (1929), p. 248, pp. 252–255 has considered this theorem for bounded transformations.

The further examination of the spectral theory of normal transformations would lead us into considerations analogous to those presented for self-adjoint transformations in Chapter VII. We shall not discuss the extension of the various theorems proved there to the case of normal transformations. The reader will see that Theorems 5.13, 7.1–7.7, 7.9 and 7.10, together with Definitions 7.1 and 7.2, can be taken over into the present theory with obvious minor modifications. Other theorems, such as Theorems 7.8, 7.11–7.13, lie much deeper or must be considerably changed before they can be incorporated in the theory of normal transformations. We must also mention the possibility of extending the theory of normal transformations in the direction suggested by the remark that in that theory we have been studying pairs of permutable self-adjoint transformations. It is natural to consider any finite or denumerably infinite collection of mutually permutable self-adjoint transformations. J. v. Neumann has investigated the problems thus indicated in two important papers, to which we shall refer the reader for further information.*

In conclusion we shall discuss briefly the relation of the class of normal transformations to other classes of transformations previously discussed. We first note the following connection with unitary and self-adjoint transformations.

THEOREM 8.15. *Every maximal normal transformation T for which $l = 0$ is not a characteristic value is expressible as the product of a not-negative definite self-adjoint transformation H and a unitary transformation U which is permutable with H.*

We have only to set $H \equiv T(F_1)$, $U \equiv T(F_2)$, where $F_1(z) = |z|$, and $F_2(z) = z/|z|$ except when $z = 0$; since the point $z = 0$ is a null set with respect to T, the definition of $F_2(z)$ at $z = 0$ is immaterial.

In the following theorem we shall consider under what circumstances various special types of transformation are normal. We have

---

* J. v. Neumann, Mathematische Annalen, 102 (1929), pp. 370–427; Annals of Mathematics, (2) 32 (1931), pp. 191–226.

THEOREM 8.16. *A symmetric transformation is normal if and only if it is essentially self-adjoint; and is a maximal normal transformation if and only if it is self-adjoint. A bounded transformation with domain $\mathfrak{H}$ is normal if and only if it is permutable with its adjoint; it is then a maximal normal transformation. A unitary transformation is a maximal normal transformation.*

If T is a symmetric transformation, then

$$S_1 \equiv \frac{T^{**} + T^*}{2} \equiv T^{**}, \quad S_2 \equiv \frac{T^{**} - T^*}{2i} \equiv O$$

in the domain of T^{**} If T is normal, then $S_1^* \equiv S_1^{**}$ and $T^{**} \equiv T^*$, showing that T is essentially self-adjoint. If T is essentially self-adjoint, then $S_1^* \equiv S_1^{**}$ is a consequence of the identity $T^* \equiv T^{**}$; and $S_2^* \equiv S_2^{**} \equiv O$. Thus $H_1 \equiv S_1^{**}$ and $H_2 \equiv O$ exist and are permutable so that T is normal. Finally, T is maximal as well as normal if and only if $T \equiv T^{**} \equiv T^*$; that is, if and only if T is self-adjoint.

If T is a bounded linear transformation with domain $\mathfrak{H}$, the transformations

$$H_1 \equiv \frac{T^{**} + T^*}{2}, \quad H_2 \equiv \frac{T^{**} - T^*}{2i}$$

exist and have $\mathfrak{H}$ as their domain. Since $T^{**} \equiv T$, a necessary and sufficient condition that T be normal is that

$$H_1 H_2 \equiv H_2 H_1,$$
$$T^2 - TT^* + T^*T - (T^*)^2 \equiv T^2 + TT^* - T^*T - (T^*)^2,$$

an identity which is satisfied if and only if T and T^* are permutable. When T is normal, the identity $T^{**} \equiv T$ shows that T is a maximal normal transformation.

Lastly every unitary transformation U is a maximal normal transformation since the identity $U^* \equiv U^{-1}$ renders U and U^* permutable.

An interesting application of the general theory of normal transformations to the case of unitary transformations can be based upon the following theorem:

THEOREM 8.17. *Let T be a maximal normal transformation, $E(z)$ the corresponding complex resolution of the identity, and $F(z)$ a function measurable with respect to $E(z)$. The identity $F(T) \equiv O$ implies that the locus $|F(z)| > 0$ is a null set with respect to $E(z)$ and that, in particular, the points of the locus which belong to the spectrum of T constitute a null set with respect to $E(z)$. The interior points of the locus belong to the resolvent set of T.*

Many of the facts noted in the theorem have been proved in Theorems 6.6 and 8.7. The only new item is the relation of interior points of the locus $|F(z)| > 0$ to the resolvent set of T. An interior point can always be included in an interval $^*\Delta$ consisting of interior points of the locus in question. Since this interval is a null set with respect to $E(z)$ we find that $E(^*\Delta) \equiv O$ and that the entire interval belongs to the resolvent set of T.

Following v. Neumann,† we can apply this theorem to the study of unitary transformations. We have seen that a unitary transformation U is a maximal normal transformation satisfying the identity $F(U) \equiv UU^* - I \equiv O$, where $F(z) = z\bar{z} - 1$. We see therefore that the resolvent set of U includes the loci $|z| > 1$, $|z| < 1$, and that the spectrum of U lies on the unit circle $|z| = 1$. If we define a function $G_\theta(z)$ by the relations

$$G_\theta(re^{2\pi i\varphi}) = 0,\ r \neq 1;\ r = 1,\ \theta < \varphi \leqq 1$$
$$G_\theta(re^{2\pi i\varphi}) = 1,\ r = 1,\ 0 < \varphi \leqq \theta,$$

then the projections $E_0(\theta) \equiv G_\theta(U)$ enable us to express U by means of the integral

$$(Uf, g) = \int_0^1 e^{2\pi i\theta}\, d(E_0(\theta)f, g).$$

This representation is precisely that described in Theorem 8.4.

§ 4. A THEOREM ON FACTORIZATION

We shall now state and prove a generalization of Theorem 8.15 above. We are thus engaged in extending to

† J. v. Neumann, Mathematische Annalen, 102 (1929), pp. 111-122.

Hilbert space a well-known theorem on the factorization of homogeneous linear transformations. We shall consider the extension only in the case of bounded linear transformations with domain $\mathfrak{H}$. The importance of the factorization lies in the fact that it enables us to express every bounded linear transformation in terms of transformations which we have analyzed in detail.

THEOREM 8.18. *If T is a bounded linear transformation which takes $\mathfrak{H}$ in a one-to-one manner into itself, then T can be factored in either of the two forms*

$$T \equiv H_1 U_1, \qquad T \equiv U_2 H_2,$$

where H_1 and H_2 are bounded positive definite self-adjoint transformations and U_1 and U_2 are unitary transformations.

According to Theorems 2.7, 2.26, and 2.29 our hypotheses imply that T^{-1}, T^*, and $(T^*)^{-1}$ all exist and are bounded linear transformations with domain $\mathfrak{H}$. The transformations T^*T and TT^* are evidently self-adjoint transformations with domain $\mathfrak{H}$; their respective inverses $(T^{-1}T^{*-1})$ and $(T^{*-1}T^{-1})$ exist and are self-adjoint transformations with domain $\mathfrak{H}$. From the inequalities

$$(T^*Tf, f) = |Tf|^2 \geqq 0, \qquad (TT^*f, f) = |T^*f|^2 \geqq 0$$

we see that T^*T and TT^* are not-negative definite. Since the point $l = 0$ is a point of the resolvent set for both these transformations, they must be positive definite, their spectra being closed point sets on the positive real axis. We now define the factors of T, to which the theorem refers, by the relations

$$H_1 \equiv \sqrt{TT^*}, \qquad U_1 \equiv (\sqrt{TT^*})^{-1}T,$$
$$H_2 \equiv \sqrt{T^*T}, \qquad U_2 \equiv T(\sqrt{T^*T})^{-1}.$$

We see immediately that H_1 and H_2 are bounded positive definite self-adjoint transformations. The inverses H_1^{-1} and H_2^{-1} exist and are bounded self-adjoint transformations with domain $\mathfrak{H}$. For U_1 we then have the relations

$$U_1 = H_1^{-1}T, \qquad U_1^* = T^* H_1^{-1},$$
$$U_1 U_1^* = H_1^{-1} T T^* H_1^{-1} = H_1^{-1} H_1^2 H_1^{-1} = I,$$
$$U_1^* U_1 = T^* H_1^{-1} H_1^{-1} T = T^* (T T^*)^{-1} T$$
$$= T^* ((T^*)^{-1} T^{-1}) T = I,$$

which show that U_1 is a unitary transformation. We can verify in a similar manner that U_2 is a unitary transformation, and the proof of the theorem is then complete.

THEOREM 8.19. *If T is an arbitrary bounded linear transformation with domain $\mathfrak{H}$ and l is any point of its resolvent set, then there exist bounded positive definite self-adjoint transformations H_1 and H_2 and unitary transformations U_1 and U_2 such that*

$$T = H_1 U_1 + lI, \qquad T = U_2 H_2 + lI.$$

We have only to apply the preceding theorem to the transformation $T - lI$. It will be recalled from Theorem 4.21 that the resolvent set of a bounded linear transformation is never empty.

CHAPTER IX

SYMMETRIC TRANSFORMATIONS

§ 1. The General Theory

In Chapter II, § 2, we have discussed the elementary properties and relations of symmetric transformations. As a result of these preliminary considerations we formulated three important problems concerning transformations of this category:

(1) the determination of all the maximal symmetric extensions of a given symmetric transformation;

(2) the determination of all maximal symmetric transformations;

(3) the determination of all self-adjoint transformations.

The third problem has occupied our attention in preceding chapters, and its various ramifications have been explored in some detail; the theorems of Chapter V, in particular, serve to characterize the class of all self-adjoint transformations and thus provide the solution of the third problem. We shall now turn to the investigation of the more general questions raised by the first two. The results reported here are due to J. v. Neumann,* whose exposition we shall follow with only occasional modifications and additions. After presenting v. Neumann's general theory, we shall examine briefly certain other aspects of the behavior of symmetric transformations and shall indicate applications to the study of integral operators, continued fractions, and related topics.

In view of Theorem 2.15, as we have already remarked, it is sufficient for us to consider closed linear symmetric transformations for the purposes of the present chapter. A transformation of this sort is related in a simple and fun-

* J. v. Neumann, Mathematische Annalen, 102 (1929), pp. 49–131.

damental manner to a certain corresponding isometric transformation which characterizes it. In consequence, the theorems of Chapter II, § 5, become the basis for a solution of the questions now before us. The key theorem runs as follows:

THEOREM 9.1. *Let $\mathfrak{S}$ be the class of all closed linear symmetric transformations H; and let $\mathfrak{F}$ be the class of all closed isometric transformations V such that $V - I$ has range everywhere dense in $\mathfrak{H}$. Then there exists a one-to-one correspondence between $\mathfrak{S}$ and $\mathfrak{F}$ such that, when H and V correspond,*

(1) *if $\mathfrak{D}$, $\mathfrak{R}_{+i}$, and $\mathfrak{R}_{-i}$ denote the domain of H, the range of H_{+i}, and the range of H_{-i}, respectively, then $\mathfrak{R}_{+i}$ and $\mathfrak{R}_{-i}$ are closed linear manifolds in one-to-one correspondence with $\mathfrak{D}$ by the transformations H_{+i} and H_{-i} respectively; V has $\mathfrak{R}_{-i}$ as its domain, $\mathfrak{R}_{+i}$ as its range, and satisfies the relation $V \equiv H_{+i} H_{-i}^{-1}$;*

(2) *the transformation $I - V$ has an inverse $(I - V)^{-1}$, the domain of which is the domain of H; H satisfies the relation $H \equiv i(I + V)(I - V)^{-1}$.*

If H is in $\mathfrak{S}$ we form the corresponding transformation V in $\mathfrak{F}$ as indicated in the statement of the theorem. Theorems 2.5 and 4.14 show that H_{+i}^{-1} and H_{-i}^{-1} exist and are closed bounded linear transformations. By reference to Theorem 2.23 we see that the linear manifolds $\mathfrak{R}_{+i}$ and $\mathfrak{R}_{-i}$, the respective domains of these transformations, must be closed. Thus the transformation $V \equiv H_{+i} H_{-i}^{-1}$ takes $\mathfrak{R}_{-i}$ in a one-to-one manner into $\mathfrak{R}_{+i}$ and is evidently linear. If f and g are arbitrary elements of $\mathfrak{R}_{-i}$, the domain of V, there exist uniquely determined elements f_0, g_0 in the domain of H such that $f = H_{-i} f_0$, $Vf = H_{+i} f_0$, $g = H_{-i} g_0$, $Vg = H_{+i} g_0$. Thus, by recalling the equation $(Hf_0, g_0) = (f_0, Hg_0)$, and carrying out the usual expansions, we obtain

$$
\begin{aligned}
(f, g) &= (H_{-i} f_0, H_{-i} g_0) = (Hf_0, Hg_0) + (f_0, g_0), \\
(Vf, Vg) &= (H_{+i} f_0, H_{+i} g_0) = (Hf_0, Hg_0) + (f_0, g_0), \\
(Vf, Vg) &= (f, g).
\end{aligned}
$$

The last equation holds for every pair of elements in the domain of V and requires that V be isometric. Since the domain of V is a closed linear manifold, we must have $V \equiv \tilde{V}$ according to Theorem 2.23 and can thus conclude that V is closed. Finally, we note that the range of $V-I$ coincides with the domain of H, by virtue of the relations $Vf-f= H_{+i}f_0-H_{-i}f_0=2if_0$, and is therefore everywhere dense in $\mathfrak{H}$.

If V is a member of $\mathfrak{F}$ we construct H as required by the second statement of the theorem. By hypothesis the range of $V-I$ is everywhere dense in $\mathfrak{H}$. We shall show that $I-V$ takes its domain in a one-to-one manner into its range, which, of course, coincides with that of $V-I$. To do so we have to prove that the equation $g-Vg=0$ implies $g=0$. When f is an arbitrary element of the domain of V we have $(Vf, Vg)=(f, g)$ and can therefore write

$$(Vf-f, g) = (Vf, g)-(f, g) = (Vf, g-Vg) = 0.$$

In view of the fact that the range of $V-I$ is everywhere dense in $\mathfrak{H}$, we must have $g=0$. We denote by $\mathfrak{D}$ the range of $I-V$; the transformation $(I-V)^{-1}$ exists, has $\mathfrak{D}$ as its domain, and is closed and linear in accordance with Theorem 2.5. We now define a transformation H by the relation $H \equiv i(I+V)(I-V)^{-1}$. In view of the identity

$$\begin{aligned} i(I+V)(I-V)^{-1} &\equiv i(2I-(I-V))(I-V)^{-1} \\ &\equiv 2i(I-V)^{-1}-iI \end{aligned}$$

it is evident that H is a closed linear transformation with domain everywhere dense in $\mathfrak{H}$. If f and g are arbitrary elements in $\mathfrak{D}$ then there exist uniquely determined elements f_0 and g_0 in the domain of V such that $f=(I-V)f_0$, $Hf=i(I+V)f_0$, $g=(I-V)g_0$, $Hg=i(I+V)g_0$. Thus, by using the equation $(Vf_0, Vg_0)=(f_0, g_0)$ and carrying out the indicated expansions, we obtain

$$\begin{gathered} (Hf, g) = (if_0+iVf_0, g_0-Vg_0) = i\{(Vf_0, g_0)-(f_0, Vg_0)\}, \\ (f, Hg) = (f_0-Vf_0, ig_0+iVg_0) = i\{(Vf_0, g_0)-(f_0, Vg_0)\}, \\ (Hf, g) = (f, Hg). \end{gathered}$$

The last equation, holding for every pair of elements in $\mathfrak{D}$, implies that H is symmetric.

If H is an arbitrary member of $\mathfrak{S}$ and we put $V \equiv H_{+i} H_{-i}^{-1}$, we find immediately that $i(I+V)(I-V)^{-1}$ and H coincide; for we have the relations

$$V \equiv (H_{-i}-2iI)H_{-i}^{-1} \equiv I-2iH_{-i}^{-1},$$
$$I+V \equiv 2(I-iH_{-i}^{-1}),$$
$$(I-V)^{-1} \equiv (2iH_{-i}^{-1})^{-1} \equiv \frac{1}{2i} H_{-i}.$$

On the other hand, if V is an arbitrary member of $\mathfrak{F}$ and we put $H \equiv i(I+V)(I-V)^{-1}$, we find similarly that $H_{+i} H_{-i}^{-1}$ and V coincide; for we have the relations

$$H \equiv i(2I-(I-V))(I-V)^{-1} \equiv i[2(I-V)^{-1}-I],$$
$$H_{+i} \equiv 2i[(I-V)^{-1}-I],$$
$$H_{-i}^{-1} \equiv (2i(I-V)^{-1})^{-1} \equiv \frac{1}{2i}(I-V).$$

It thus appears that the processes described and discussed in the two preceding paragraphs are inverses of one another and therefore determine a one-to-one correspondence between $\mathfrak{S}$ and $\mathfrak{F}$ with the properties described in the theorem.

THEOREM 9.2. *If H_1 and H_2 belong to $\mathfrak{S}$ and if V_1 and V_2 are their respective correspondents in $\mathfrak{F}$, then the relations $H_1 \subset H_2$ and $V_1 \subset V_2$ are equivalent. A transformation H in $\mathfrak{S}$ is maximal if and only if the corresponding transformation V in $\mathfrak{F}$ is maximal. The maximal symmetric extensions of a given closed linear symmetric transformation H constitute a class with the cardinal number* 1 *or with the cardinal number* $\mathfrak{c}$ *of the continuum according as H is maximal or not.*

The identities

$$V_1 \equiv (H_1-iI)(H_1+iI)^{-1}, \quad V_2 \equiv (H_2-iI)(H_2+iI)^{-1},$$
$$H_1 \equiv i(I+V_1)(I-V_1)^{-1}, \qquad H_2 \equiv i(I+V_2)(I-V_2)^{-1}$$

show that the relations $H_1 \subseteq H_2$ and $V_1 \subseteq V_2$ are equivalent. Since $H_1 \equiv H_2$ implies and is implied by $V_1 \equiv V_2$, we conclude that the relations $H_1 \subset H_2$ and $V_1 \subset V_2$ are also equivalent.

If H is a maximal symmetric transformation it belongs to $\mathfrak{S}$ and has a correspondent V in $\mathfrak{F}$. We wish to show that V is a maximal isometric transformation. If V be not maximal, then there exists an isometric extension of V which we denote by V_0. In view of Theorem 2.47 we may suppose that V_0 is closed. Since $V_0 \supset V$, the range of $V_0 - I$ includes the range of $V - I$ and is therefore everywhere dense in $\mathfrak{H}$. We now see that V_0 belongs to $\mathfrak{F}$ and must therefore have a correspondent H_0 in $\mathfrak{S}$ such that $H_0 \supset H$. By hypothesis the latter relation is impossible, so that V must be maximal. By reasoning of the same character we can show that whenever V is a maximal isometric transformation in $\mathfrak{F}$ the corresponding transformation H in $\mathfrak{S}$ is maximal. For if H be not maximal, there exists a closed linear symmetric extension of H which may be denoted by H_0; since $H_0 \supset H$, the corresponding transformation V_0 in $\mathfrak{F}$ satisfies the relation $V_0 \supset V$, which is excluded by the assumption that V is maximal.

If H is a maximal transformation in $\mathfrak{S}$, we know that it has no maximal extension in $\mathfrak{S}$ other than itself. If, on the other hand, H is in $\mathfrak{S}$ but not maximal, its correspondent V in $\mathfrak{F}$ is not maximal. From Theorems 2.48–2.51 we know precisely how all the maximal isometric extensions of V are to be constructed and we know that they constitute a subset of $\mathfrak{F}$ with the cardinal number $\mathfrak{c}$ of the continuum. Corresponding to these maximal extensions of V there are maximal symmetric transformations in $\mathfrak{S}$ constituting a set with the cardinal number $\mathfrak{c}$, each of which is an extension of H. From the preceding paragraph, it is evident that all such extensions of H are obtained in this manner.

Clearly, we can add considerable detail to the description of the relations examined in the preceding theorem by making use of the deficiency-index of the transformation V. On this account we introduce the following definition:

DEFINITION 9.1. *The deficiency-index of a transformation H in $\mathfrak{S}$ is the deficiency-index (m, n) of the corresponding isometric transformation V in $\mathfrak{F}$.*

We recall that m is the dimension-number of the closed linear manifold $\mathfrak{H} \ominus \mathfrak{R}_{-i}$ where $\mathfrak{R}_{-i}$ is the domain of V and the range of H_{-i}; and that n is the dimension-number of $\mathfrak{H} \ominus \mathfrak{R}_{+i}$ where $\mathfrak{R}_{+i}$ is the range of V and of H_{+i}.

We can now state and prove the desired amplification of Theorem 9.2.

THEOREM 9.3. *A transformation H in $\mathfrak{S}$ is maximal if and only if its deficiency-index has the form* $(m, 0)$ *or the form* $(0, n)$; *in particular, it is self-adjoint if and only if its deficiency-index is* $(0, 0)$. *If H is a transformation in $\mathfrak{S}$ with the deficiency-index* (m, n) *where* $\min [m, n] \geqq 1$, *then the transformations H' in $\mathfrak{S}$ with given deficiency-index* (m', n') *such that $H \subset H'$, $m = m' + p$, $n = n' + p$, constitute a class with the cardinal number* $\mathfrak{c}$, *for* $p = 1, \cdots, \min [m, n]$; *every transformation H' in $\mathfrak{S}$ such that $H' \supset H$ belongs to some one of these classes. A transformation H in $\mathfrak{S}$ possesses self-adjoint extensions if and only if its deficiency-index has the form* (m, m): *if this condition is satisfied with m a finite cardinal number, then every maximal extension in $\mathfrak{S}$ is self-adjoint; if this condition is satisfied with $m = \aleph_0$, then the maximal extensions with assigned deficiency-index* $(p, 0)$ *or* $(0, p)$ *constitute a class with the cardinal number* $\mathfrak{c}$, *for* $p = 0, 1, 2, \cdots, \aleph_0$.

Since this theorem is merely a paraphrase of the results of Theorems 2.48–2.51 depending upon Theorems 9.1 and 9.2, we do not need to give a detailed discussion, save with regard to one point. Since we have not yet considered the rôle played by self-adjoint transformations in the general situation studied here, we must elucidate it for the purposes of the present theorem. The necessary information is comprised in the following assertion: if H is a transformation in $\mathfrak{S}$ with the deficiency-index (m, n) and V is its correspondent in $\mathfrak{F}$, then the statements (1) H is self-adjoint, (2) V is unitary, (3) $m = n = 0$, are equivalent. The statements (1) and (2) are equivalent in view of Theorem 4.17, which shows that $H \equiv \tilde{H}$ is self-adjoint if and only if $\mathfrak{R}_{+i} \equiv \mathfrak{R}_{-i} \equiv \mathfrak{H}$ and thus implies that H is self-adjoint if and only if V is unitary, according to Definitions 2.18 and 2.19. In Theorem 2.50 we showed that the statements (2) and (3) are equivalent. We may note that this relation between self-adjoint transformations in $\mathfrak{S}$ and the corresponding unitary transformations in $\mathfrak{F}$ is a special case of the relation discussed in Theorem 8.3, obtained by putting $a = b = -i$, $c = 1$, $d = -1$.

The three theorems which have just been proved thus enable us to master our first problem: the determination of all the maximal symmetric extensions of a given symmetric transformation H. In place of H we consider the closed linear transformation $\tilde{H}$, which belongs to $\mathfrak{S}$, and can at once find and characterize all the maximal extensions of $\tilde{H}$; according to Theorem 2.15, we obtain in this way all the maximal extensions of H. Evidently, if H is not maximal, the maximal extensions of H form a class with cardinal number $\mathfrak{c}$.

Before turning to our second problem—the determination of all maximal transformations in $\mathfrak{S}$—we shall examine certain general properties of $\mathfrak{S}$ which bear on the significance of the deficiency-index. We must commence by introducing several definitions.

DEFINITION 9.2. *The linear manifolds $\mathfrak{M}_1, \cdots, \mathfrak{M}_n$ are said to be linearly independent if the equation $f_1 + \cdots + f_n = 0$, where the element f_k is in $\mathfrak{M}_k$ for $k = 1, \cdots, n$, implies $f_k = 0$ for every k.*

DEFINITION 9.3. *If $\mathfrak{M}$ is a linear manifold, then the elements f and g in are said to be congruent modulo $\mathfrak{M}$ if $f - g$ is an element of $\mathfrak{M}$.*

We shall write $f \equiv g \pmod{\mathfrak{M}}$ to indicate that f and g are congruent modulo $\mathfrak{M}$. Evidently the relation of congruence modulo $\mathfrak{M}$ is reflexive, symmetric, and transitive: for we have $f \equiv f \pmod{\mathfrak{M}}$ for every element f; $f \equiv g \pmod{\mathfrak{M}}$ implies $g \equiv f \pmod{\mathfrak{M}}$; and $f \equiv g \pmod{\mathfrak{M}}$, $g \equiv h \pmod{\mathfrak{M}}$ imply $f \equiv h \pmod{\mathfrak{M}}$. The usual rules for algebraic congruences are found to hold in the present case: if $f_1 \equiv g_2 \pmod{\mathfrak{M}}$ and $f_2 \equiv g_2 \pmod{\mathfrak{M}}$, then $f_1 + f_2 \equiv g_1 + g_2 \pmod{\mathfrak{M}}$; and if $f \equiv g \pmod{\mathfrak{M}}$, then $af = ag \pmod{\mathfrak{M}}$. Thus we may speak of elements in $\mathfrak{H}$ as linearly dependent modulo $\mathfrak{M}$ or linearly independent modulo $\mathfrak{M}$. If $\mathfrak{M}$ is a closed linear manifold, then two elements are congruent modulo $\mathfrak{M}$ if and only if they have the same projection on the orthogonal complement of $\mathfrak{M}$.

In phrasing our next definition, we recall that a linear manifold which contains only a finite number of linearly independent elements is automatically closed.

DEFINITION 9.4. *An arbitrary subset* $\mathfrak{S}$ *of* $\mathfrak{H}$ *is said to determine the dimension number* n *if the set* $\mathfrak{S}+\mathfrak{O}$ *contains at least one* n*-dimensional linear manifold but none of higher dimension number; and it is said to determine the dimension number* $\aleph_0$ *if the set* $\mathfrak{S}+\mathfrak{O}$ *contains a* k*-dimensional linear manifold for* $k=1, 2, 3, \cdots$.

We note that a closed linear manifold with the dimension number n determines the dimension number n, $n=0, 1, 2, \cdots, \aleph_0$, in accord with this definition.

DEFINITION 9.5. *An arbitrary subset* $\mathfrak{S}$ *of* $\mathfrak{H}$ *is said to determine the dimension number* n (mod $\mathfrak{M}$) *if the set* $\mathfrak{S}-\mathfrak{S}\cdot\mathfrak{M}$ *determines the dimension number* n, $n=0, 1, 2, \cdots, \aleph_0$.

Evidently, a linear manifold determines the dimension number n (mod $\mathfrak{M}$) if and only if it contains n, but not $n+1$, elements linearly independent (mod $\mathfrak{M}$), $n=0, 1, 2, \cdots$. Hence, a set $\mathfrak{S}$ determines the dimension number n (mod $\mathfrak{M}$) if and only if $\mathfrak{S}+\mathfrak{O}$ contains a linear manifold which determines the dimension number n (mod $\mathfrak{M}$) but none which determines the dimension number $n+1$ (mod $\mathfrak{M}$); and a set $\mathfrak{S}$ determines the dimension number $\aleph_0$ (mod $\mathfrak{M}$) if and only if $\mathfrak{S}+\mathfrak{O}$ contains a linear manifold which determines the dimension number k (mod $\mathfrak{M}$), for every k.

THEOREM 9.4. *Let* H *denote a transformation in* $\mathfrak{H}$, H^* *its adjoint,* $\mathfrak{D}$ *the domain of* H, $\mathfrak{D}^*$ *the domain of* H^*, $\mathfrak{D}^+$ *the closed linear manifold* $\mathfrak{H}\ominus\mathfrak{R}_{-i}$ *where* $\mathfrak{R}_{-i}$ *is the range of* H_{-i}, *and* $\mathfrak{D}^-$ *the closed linear manifold* $\mathfrak{H}\ominus\mathfrak{R}_{+i}$ *where* $\mathfrak{R}_{+i}$ *is the range of* H_{+i}. *Then* $\mathfrak{D}^+$ *consists of those and only those elements* f *in* $\mathfrak{D}^*$ *such that* $H^*f=if$, $\mathfrak{D}^-$ *of those and only those elements* f *in* $\mathfrak{D}^*$ *such that* $H^*f=-if$; *if* H *has the deficiency-index* (m, n), *then the points* $l=+i$ *and* $l=-i$ *are characteristic values for* H^* *with multiplicities* m *and* n *respectively. The linear manifolds* $\mathfrak{D}$, $\mathfrak{D}^+$, *and* $\mathfrak{D}^-$ *are linearly independent and together determine* $\mathfrak{D}^*$ *according to the relation* $\mathfrak{D}^*=\mathfrak{D}\dotplus\mathfrak{D}^+\dotplus\mathfrak{D}^-$: *in other words, every element in* $\mathfrak{D}^*$ *is* *expressible in just one way as the sum of three elements, one from each of* $\mathfrak{D}$, $\mathfrak{D}^+$, $\mathfrak{D}^-$; *and every such sum is an element of* $\mathfrak{D}^*$. *The transformation* H^{**} *coincides with* H.

We obtain the desired characterization of $\mathfrak{D}^+ \equiv \mathfrak{H} \ominus \mathfrak{R}_{-i}$ as follows: f belongs to $\mathfrak{D}^+$ if and only if it is orthogonal to every element of $\mathfrak{R}_{-i}$, thus satisfying the equation $(H_{-i}f^0, f) = 0$ or the equivalent equation $(Hf^0, f) = (f^0, if)$ for every element f^0 in $\mathfrak{D}$; in view of the definition of H^*, the latter equation holds if and only if f is in $\mathfrak{D}^*$ and $H^*f = if$. By an entirely similar argument we show that $\mathfrak{D}^-$ has the properties described in the theorem. If the deficiency-index of H is (m, n), then $\mathfrak{D}^+$ and $\mathfrak{D}^-$ have the dimension numbers m and n respectively: in other words, the points $l = +i$ and $l = -i$ are characteristic values of H^* with multiplicities m and n respectively, provided that we adopt for the purposes of the theorem the convention that l is to be regarded as a characteristic value of H^* with multiplicity 0 whenever the equation $H^*f = lf$ has no solution other than $f = 0$.

Since H^* is a linear transformation whose domain $\mathfrak{D}^*$ contains $\mathfrak{D}$, $\mathfrak{D}^+$, and $\mathfrak{D}^-$, it is evident that $\mathfrak{D}^*$ is a linear manifold containing $\mathfrak{D} \dotplus \mathfrak{D}^+ \dotplus \mathfrak{D}^-$. We must show that whenever f is in $\mathfrak{D}^*$ there exist uniquely determined elements f^0, f^+, and f^- in $\mathfrak{D}$, $\mathfrak{D}^+$, and $\mathfrak{D}^-$ respectively such that $f = f^0 + f^+ + f^-$. When f is given in $\mathfrak{D}^*$, we form the projections of $H_i^* f$ on the mutually orthogonal closed linear manifolds $\mathfrak{R}_i$ and $\mathfrak{D}^- = \mathfrak{H} \ominus \mathfrak{R}_i$ respectively. The projection on $\mathfrak{R}_i$ is expressible in the form $H_i f^0 = H_i^* f^0$, where f^0 is a uniquely determined element of $\mathfrak{D}$, as we showed in Theorem 9.1; and the projection on $\mathfrak{D}^-$ can be written for convenience in the form $-2if^- = H_i^* f^-$ where f^- is a uniquely determined element of $\mathfrak{D}^-$. We now have $H_i^* f = H_i^* f^0 + H_i^* f^-$; on writing this relation in the form $H_i^* (f - (f^0 + f^-)) = 0$, we see that $f - (f^0 + f^-)$ is in $\mathfrak{D}^+$ and can be denoted by f^+. We have thus proved that $f = f^0 + f^+ + f^-$, as desired. In order to show that this expression for an arbitrary element f in $\mathfrak{D}^*$ is unique, it is sufficient to prove that the null element can be expressed in this form only by putting $f^0 = f^+ = f^- = 0$. If $f^0 + f^+ + f^- = 0$, we apply H_i^* to both sides of this equation, thus obtaining $H_i f^0 - 2if^- = 0$, since $H_i^* f^0 = H_i f^0$, $H_i^* f^+ = 0$, $H_i^* f^- = -2if^-$. Now $H_i f^0$ and f^- belong to the closed linear manifolds $\mathfrak{R}_i$ and $\mathfrak{D}^-$ respectively. In view of the fact that these manifolds are mutually orthogonal and hence linearly independent by Theorem 1.23 we must have $H_i f^0 = 0$, $f^- = 0$. Applying Theorem 9.1, we find $f^0 = 0$. Finally it is evident that f^+ must also reduce to the null element. We have thus proved the uniqueness of our resolution of the given element f in $\mathfrak{D}^*$. According to Definition 9.2, therefore, the linear manifolds $\mathfrak{D}$, $\mathfrak{D}^+$, $\mathfrak{D}^-$ are linearly independent.

If H is a given transformation in $\mathfrak{S}$, then $H_0 \equiv H^{**}$ is an extension of H in $\mathfrak{S}$ according to Theorem 2.11; the corresponding transformations V and V_0 satisfy the relation $V \subseteqq V_0$, as indicated in Theorem 9.2. Since $H_0^* \equiv (H^{**})^* \equiv H^*$, the present theorem shows that $\mathfrak{D}_0^+ = \mathfrak{D}^+$, $\mathfrak{D}_0^- = \mathfrak{D}^-$ and thus requires that V and V_0 should have the same domain and the same range. The relation $V \subset V_0$ is therefore inadmissible, so that we must have $V \equiv V_0$, $H \equiv H_0 \equiv H^{**}$ as we wished to prove.

THEOREM 9.5. *If H is a symmetric transformation, then $\tilde{H}$ and H^{**} are transformations in $\mathfrak{S}$ satisfying the relation $\tilde{H} \equiv H^{**}$.*

From Theorem 2.11 we know that $\tilde{H}$ and H^{**} are in $\mathfrak{S}$. If we apply the operation ** to the relations $H \subseteqq \tilde{H} \subseteqq H^{**}$ we obtain $H^{**} \subseteqq \tilde{H}^{**} \subseteqq H^{**}$ and thus conclude that $\tilde{H}^{**}$ and H^{**} coincide. From the preceding theorem we have $\tilde{H} \equiv \tilde{H}^{**}$ and hence the desired identity $\tilde{H} \equiv H^{**}$.

By making use of Theorem 9.4, we can obtain a characterization of the deficiency-index (m, n) in which the complex numbers $+i$ and $-i$ no longer play any special part. When H is a given transformation in $\mathfrak{S}$, H^* its adjoint, and $\mathfrak{D}$ and $\mathfrak{D}^*$ their respective domains, we commence by resolving $\mathfrak{D}^*$ into three mutually exclusive classes: an element f in $\mathfrak{D}^*$

belongs to $\mathfrak{C}^+$, $\mathfrak{C}^0$, $\mathfrak{C}^-$ according as the real number $\mathfrak{J}(H^*f, f)$ is positive, zero, or negative. Upon applying the concepts introduced in Definitions 9.3—9.5 and computing the dimension numbers (mod $\mathfrak{D}$) of these three sets of elements, we find the numbers m and n appearing in the deficiency-index (m, n) of H; we give the precise result below, after proving an auxiliary theorem.

THEOREM 9.6. *Let f be an element of $\mathfrak{D}^*$ expressed in the form $f = f^0 + f^+ + f^-$, where f^0, f^+, f^- belong respectively to $\mathfrak{D}$, $\mathfrak{D}^+$, $\mathfrak{D}^-$ as described in Theorem 9.4. Then f belongs to $\mathfrak{C}^+$, $\mathfrak{C}^0$, $\mathfrak{C}^-$ according as the difference $|f^+|^2 - |f^-|^2$ is positive, zero, or negative. In particular, the relations*

$$\mathfrak{D} \subseteqq \mathfrak{C}^0, \quad \mathfrak{D}^+ \subseteqq \mathfrak{C}^+ + \mathfrak{O}, \quad \mathfrak{D}^- \subseteqq \mathfrak{C}^- + \mathfrak{O}$$

are satisfied.

The theorem depends upon the equation

$$\mathfrak{J}(H^*f, f) = |f^+|^2 - |f^-|^2,$$

which can be verified by direct computation. Keeping in mind the fundamental relations connecting f^0, f^+, f^-, H and H^* we have

$$\begin{aligned}(H^*f, f) &= (Hf^0 + i(f^+ - f^-),\ f^0 + (f^+ + f^-)) \\ &= (Hf^0, f^0) + \{(Hf^0, f^+ + f^-) + (i(f^+ - f^-), f^0)\} \\ &\qquad + (i(f^+ - f^-), f^+ + f^-) \\ &= (Hf^0, f^0) + \{(f^0, i(f^+ - f^-)) + (i(f^+ - f^-), f^0)\} \\ &\qquad + \{(if^+, f^-) + (f^-, if^+)\} \\ &\qquad + i\{|f^+|^2 - |f^-|^2\}.\end{aligned}$$

It is evident that the first three terms are real, the last pure imaginary: in the case of the first term we have

$$(Hf^0, f^0) = (f^0, Hf^0) = \overline{(Hf^0, f^0)};$$

both the second and third terms are sums of conjugate complex numbers. The evaluation thus yields the result stated.

Since $f = f^0 + f^+ + f^-$ belongs to $\mathfrak{D}$, $\mathfrak{D}^+$, $\mathfrak{D}^-$ respectively if and only if the relations $f^+ = f^- = 0$, $f^0 = f^- = 0$, $f^0 = f^+ = 0$, respectively, are satisfied, it is plain that the relations $\mathfrak{D} \subseteqq \mathfrak{C}^0$, $\mathfrak{D}^+ \subseteqq \mathfrak{C}^+ + \mathfrak{O}$, $\mathfrak{D}^- \subseteqq \mathfrak{C}^- + \mathfrak{O}$ are true. It should be noted that the element 0 belongs to $\mathfrak{C}^0$ and is not a member of $\mathfrak{C}^+$ or of $\mathfrak{C}^-$.

With the aid of this result, we prove

THEOREM 9.7. *If H is a transformation in $\mathfrak{H}$ with domain $\mathfrak{D}$ and deficiency-index (m, n), then the corresponding sets $\mathfrak{C}^+$, $\mathfrak{C}^0$, $\mathfrak{C}^-$ have the following properties:*

(1) *$\mathfrak{C}^+$ and $\mathfrak{C}^+ + \mathfrak{C}^0$ determine the dimension number m* (mod $\mathfrak{D}$);

(2) *$\mathfrak{C}^-$ and $\mathfrak{C}^- + \mathfrak{C}^0$ determine the dimension number n* (mod $\mathfrak{D}$);

(3) *$\mathfrak{C}^0$ determines the dimension number* min $[m, n]$, (mod $\mathfrak{D}$).

Since

$$\mathfrak{C}^+ + \mathfrak{C}^0 \supseteqq \mathfrak{C}^+ + \mathfrak{O} \supseteqq \mathfrak{D}^+,$$

the dimension numbers (mod $\mathfrak{D}$) determined by these sets must satisfy the relations

$$\dim (\mathfrak{C}^+ + \mathfrak{C}^0) \geqq \dim \mathfrak{C}^+ \geqq \dim \mathfrak{D}^+.$$

The inequalities

$$m \leqq \dim \mathfrak{D}^+. \qquad \dim (\mathfrak{C}^+ + \mathfrak{C}^0) \leqq m$$

are therefore sufficient to establish (1). Since $\mathfrak{D}^+$ is a closed linear manifold with the dimension number m which has no element other than 0 in common with $\mathfrak{D}$, $\mathfrak{D}^+$ determines the dimension number m (mod $\mathfrak{D}$): according to Definitions 9.4 and 9.5, we have merely to refer to the identity

$$(\mathfrak{D}^+ - \mathfrak{D}^+ \cdot \mathfrak{D}) + \mathfrak{O} = \mathfrak{D}^+.$$

If $m = \aleph_0$, the inequality $\dim (\mathfrak{C}^+ + \mathfrak{C}^0) \leqq m$ is trivial. On the other hand, if m is finite, this inequality is equivalent to the assertion that the set

$$\mathfrak{S} = [(\mathfrak{C}^+ + \mathfrak{C}^0) - (\mathfrak{C}^+ + \mathfrak{C}^0)\mathfrak{D}] + \mathfrak{O}$$

contains no linear manifold of $m+1$ dimensions. It is evident that $\mathfrak{S}$ has no element in common with $\mathfrak{C}^-$ or with $\mathfrak{D} - \mathfrak{O}$, a fact upon which we base our discussion. Let $f_k = \overset{0}{f_k} + \overset{+}{f_k} + \overset{-}{f_k}$, where $k = 1, \cdots, m+1$, be linearly independent elements of $\mathfrak{S} \subseteqq \mathfrak{D}^*$ expressed in the form described in Theorem 9.4. Since $\mathfrak{D}^+$ has the finite dimension number m, there exist constants $a_1, \cdots, a_{m+1}$ not all zero such that

$$a_1 \overset{+}{f_1} + \cdots + a_{m+1} \overset{+}{f}_{m+1} = 0.$$

The element

$$g = g^0 + g^+ + g^- = a_1 f_1 + \cdots + a_{m+1} f_{m+1}$$

must therefore satisfy the relations $g \neq 0$, $g^+ = 0$. In accordance with Theorems 9.4 and 9.6, we see that g belongs to $\mathfrak{D} - \mathfrak{O}$ when $g^- = 0$, to $\mathfrak{C}^-$ when $g^- \neq 0$, and in either case lies outside $\mathfrak{S}$, contrary to hypothesis. Thus $\mathfrak{S}$ contains no linear manifold of $m+1$ dimensions.

The proof of (2) is analogous to that of (1) and need not be repeated in detail.

From the relations $\mathfrak{C}^0 \subseteqq \mathfrak{C}^+ + \mathfrak{C}^0$, $\mathfrak{C}^0 \subseteqq \mathfrak{C}^- + \mathfrak{C}^0$, and the facts already shown in (1) and (2), we infer that $\dim \mathfrak{C}^0 \leqq \min [m, n]$. In order to demonstrate (3), we have to show that $\dim \mathfrak{C}^0 \geqq p$ for every finite cardinal number $p \leqq \min [m, n]$. If p is such a number then there exist closed linear manifolds $\mathfrak{M}^+ \subseteqq \mathfrak{D}^+$, $\mathfrak{M}^- \subseteqq \mathfrak{D}^-$ with the dimension number p. We select orthonormal sets $\{\varphi_k^+\}$, $\{\varphi_k^-\}$, $k = 1, \cdots, p$, which determine the closed linear manifolds $\mathfrak{M}^+$, $\mathfrak{M}^-$ respectively. We define the elements

$$\varphi_k = \varphi_k^+ + \varphi_k^-, \qquad k = 1, \cdots, p.$$

If

$$\begin{aligned} f = f^+ + f^- &= a_1 \varphi_1 + \cdots + a_p \varphi_p \\ &= (a_1 \varphi_1^+ + \cdots + a_p \varphi_p^+) + (a_1 \varphi_1^- + \cdots + a_p \varphi_p^-), \end{aligned}$$

then f cannot belong to $\mathfrak{D}$ and, in particular, cannot reduce to 0 unless $f^+ = f^- = 0$, a requirement which is satisfied if and only if $a_1 = \cdots = a_p = 0$. Furthermore, we have

$$\begin{aligned} \| f^+ \|^2 &= \| a_1 \varphi_1^+ + \cdots + a_p \varphi_p^+ \|^2 = |a_1|^2 + \cdots + |a_p|^2, \\ \| f^- \|^2 &= \| a_1 \varphi_1^- + \cdots + a_p \varphi_p^- \|^2 = |a_1|^2 + \cdots + |a_p|^2, \end{aligned}$$

so that f belongs to $\mathfrak{C}^0$ in accordance with Theorem 9.6. Thus the p elements $\varphi_1, \cdots, \varphi_p$ are linearly independent and determine a closed linear manifold $\mathfrak{M}$ with the dimension number p such that

$$\mathfrak{M} \subseteqq (\mathfrak{C}^0 - \mathfrak{C}^0 \cdot \mathfrak{D}) + \mathfrak{O}.$$

This result implies $\dim \mathfrak{C}^0 \geqq p$ and thus leads to the truth of (3).

It is interesting to note that some of the results of Theorem 9.3 can be obtained directly on the basis of the new interpretation of the deficiency-index given in the present theorem. We can prove the earlier results: a transformation H in $\mathfrak{H}$ is maximal if and only if its deficiency-index has the form $(m, 0)$ or the form $(0, n)$; it is self-adjoint if and only if it has the deficiency-index $(0, 0)$. First we show that a necessary and sufficient condition that H be maximal is that $\mathfrak{C}^0$ and $\mathfrak{D}$ coincide. We shall consider the equivalent statement: a necessary and sufficient condition that H have a proper symmetric extension is that $\mathfrak{C}^0$ contain some element g not in $\mathfrak{D}$. If H_0 is a proper symmetric extension of H, then the domain of H_0 contains an element g not in the domain of H; such an element g must belong to $\mathfrak{C}^0$ since $H^*g = H_0 g$ exists and $(H^*g, g) = (H_0 g, g)$ is obviously real. On the other hand, if $\mathfrak{C}^0$ contains an element g not in $\mathfrak{D}$ we define a symmetric extension H_0 of H by the relations $H_0 = H$ in $\mathfrak{D}$, $H_0 g = H^* g$. The nature of H_0 is readily determined from the equations

$$(H_0 f, g) = (f, H^* g) = (f, H_0 g), \qquad f \text{ in } \mathfrak{D};$$
$$(H_0 g, g) = \overline{(g, H_0 g)} = (g, H_0 g),$$

Next, a necessary and sufficient condition that $\mathfrak{C}^0 = \mathfrak{D}$ is that $\mathfrak{C}^0$ determine the dimension number 0 (mod $\mathfrak{D}$). The necessity is obvious, and the sufficiency follows as soon as we show that $\mathfrak{C}^0 \supset \mathfrak{D}$ implies that $\mathfrak{C}^0$ determines a dimension number (mod $\mathfrak{D}$) one or greater. If $\mathfrak{C}^0 \supset \mathfrak{D}$ and f is an element of $\mathfrak{C}^0$ not in $\mathfrak{D}$, then the one-dimensional linear manifold consisting of all elements $a \cdot f$ lies in $\mathfrak{C}^0$ but not in $\mathfrak{D} - \mathfrak{O}$; for if (H^*f, f) is real then $(H^*af, af) = |a|^2 (H^*f, f)$ is also real for all values of a. Thus the set

$$(\mathfrak{C}^0 - \mathfrak{C}^0 \cdot \mathfrak{D}) + \mathfrak{O} \equiv \mathfrak{C}^0 - \mathfrak{C}^0(\mathfrak{D} - \mathfrak{O})$$

contains the linear manifold in question: in other words the dimension number (mod $\mathfrak{D}$) determined by $\mathfrak{C}^0$ is one or greater. Thus H is maximal, according to the preceding results, if and only if min $[m, n] = 0$. Finally, H is self-adjoint if and only if $m = n = 0$, since a necessary and sufficient condition that H and H^* be identical is that $\mathfrak{C}^0$, $\mathfrak{D}$, and $\mathfrak{D}^*$ be identical, $\mathfrak{D}^+$ and $\mathfrak{D}^-$ both identical to $\mathfrak{O}$, and $\mathfrak{C}^+$ and $\mathfrak{C}^-$ empty.

The most important consequence of Theorem 9.7 is the light which it throws upon the behavior of the adjoint transformation. The facts are as follows:

THEOREM 9.8. *If H is a transformation in $\mathfrak{S}$ with the deficiency-index (m, n), then l is a characteristic value of multiplicity m or n for the transformation H^* according as $\mathfrak{J}(l) > 0$ or $\mathfrak{J}(l) < 0$. If H is a maximal symmetric transformation with the deficiency-index $(m, 0)$ (the deficiency-index $(0, n)$), then the half-planes $\mathfrak{J}(l) > 0$ and $\mathfrak{J}(l) < 0$ (the half-planes $\mathfrak{J}(l) < 0$ and $\mathfrak{J}(l) > 0$) belong to the resolvent sets of H and of H^* respectively, and $(H_l^{-1})^* \equiv (H^* - \bar{l} I)^{-1}$ for $\mathfrak{J}(l) > 0$ (for $\mathfrak{J}(l) < 0$).*

Let us consider the behavior of the transformation $aH + bI$, where a and b are real, $a \neq 0$. This transformation evidently belongs to $\mathfrak{S}$ and has the same domain as H. Its adjoint $(aH + bI)^*$ reduces to $aH^* + bI$, since a and b are real, and has the same domain $\mathfrak{D}^*$ as H^*. The equation

$$\mathfrak{J}((aH^* + bI)f, f) = \mathfrak{J}a(H^*f, f) + \mathfrak{J}b(f, f) = a\mathfrak{J}(H^*f, f)$$

shows that the class $\mathfrak{C}^0$ for $aH + bI$ is the same as that for H while the classes $\mathfrak{C}^+$, $\mathfrak{C}^-$ for $aH + bI$ and the classes $\mathfrak{C}^+$, $\mathfrak{C}^-$ for H coincide in the same or in contrary order according as a is positive or negative. As a consequence of the preceding theorem, we see that the deficiency-index of $aH + bI$ is (m, n) or (n, m) according as a is positive or negative.

Since the equations

$$H^*f = lf, \quad (aH + bI)^*f = aH^*f + bf = if$$

are equivalent when $-\dfrac{b}{a} = \mathfrak{R}(l)$, $\dfrac{1}{a} = \mathfrak{J}(l)$, we see that l is a characteristic value of multiplicity m or n for H^* according as $\mathfrak{J}(l) > 0$ or $\mathfrak{J}(l) < 0$: for the latter statement is equivalent to the assertion that $+i$ is a characteristic value of multiplicity m or n for $(aH + bI)^*$ according as $a > 0$ or $a < 0$, an assertion known to be true as a result of the preceding paragraph and Theorem 9.4. The statement that l is a characteristic value of multiplicity 0 for H^* is taken to mean that the equation $H^*f = lf$ implies $f = 0$.

When H has the deficiency-index $(m, 0)$ and l is a point of the half-plane $\mathfrak{J}(l) > 0$, l must belong to the resolvent set of H by virtue of Theorem 4.16, since $\bar{l}$ is a characteristic value of multiplicity zero for H^* and hence does not belong to the point spectrum of H^*. In Theorem 2.7 we now put $T \equiv H_l^{-1}$, $T^{-1} \equiv H - l\,I$, $\mathfrak{J}(l) > 0$, and conclude that the adjoints $T^* \equiv (H_l^{-1})^*$ and $(T^{-1})^* \equiv H^* - \bar{l}\,I$ exist and are inverses of one another. Since T is a bounded linear transformation with domain $\mathfrak{H}$, T^* has similar properties by Theorem 2.29; and $\bar{l}$ therefore belongs to the resolvent set of H^*. The case where H has the deficiency-index $(0, n)$ is treated in an analogous manner.

We shall now turn to the analysis of maximal symmetric transformations. From the elementary characteristics of isometric transformations we see immediately certain processes for constructing transformations in $\mathfrak{S}$ which are maximal but not self-adjoint. We shall begin therefore with the examination of one such transformation, with the deficiency-index $(0, 1)$. We shall recognize shortly that this particular transformation appears as a fundamental unit in the structure of the most general maximal transformation which is not self-adjoint.

THEOREM 9.9. *Let V be the isometric transformation with domain $\mathfrak{H}$ defined in terms of the complete orthonormal set $\{\varphi_n\}$, where $n = 0, 1, 2, \cdots$, by the equations $f = \sum_{\alpha=0}^{\infty} a_\alpha \varphi_\alpha$, $Vf = \sum_{\alpha=0}^{\infty} a_\alpha \varphi_{\alpha+1}$. Then V belongs to $\mathfrak{F}$ and has the deficiency-index* $(0, 1)$; *its correspondent H in $\mathfrak{S}$ is a maximal symmetric transformation with the deficiency-index* $(0, 1)$. *The transformations V and H defined in this way for different orthonormal sets $\{\varphi_n^{(1)}\}$ and $\{\varphi_n^{(2)}\}$ are unitary equivalent.*

It is readily verified by reference to Theorems 1.8 and 2.44 that the transformation V is a closed isometric transformation whose domain is $\mathfrak{H}$ and whose range is the closed linear manifold of all elements orthogonal to φ_0. V evidently has the deficiency-index $(0, 1)$. In order to show that V belongs to $\mathfrak{F}$ we have to prove that the range of $V - I$ is everywhere dense in $\mathfrak{H}$. Since the range of $V - I$ is a linear manifold, it is sufficient to show that it contains elements arbitrarily close to the element φ_n for $n = 0, 1, 2, \cdots$. If we put

$$f_{ln} = -\sum_{\alpha=0}^{l-1} \frac{l-\alpha}{l} \varphi_{n+\alpha},$$

we find at once that

$$(V-I) f_{ln} = -\sum_{\alpha=0}^{l-1} \frac{l-\alpha}{l} \varphi_{n+\alpha+1} + \sum_{\alpha=0}^{l-1} \frac{l-\alpha}{l} \varphi_{n+\alpha}$$

since $V\varphi_k = \varphi_{k+1}$ for $k = 0, 1, 2, \cdots$ in accordance with the fundamental definition. On combining the terms in the latter expression, we find

$$g_{ln} = (V-I) f_{ln} = \varphi_n - \frac{1}{l} \sum_{\alpha=1}^{l} \varphi_{n+\alpha}.$$

We now have

$$\|\varphi_n - g_{ln}\|^2 = \left\| \frac{1}{l} \sum_{\alpha=1}^{l} \varphi_{n+\alpha} \right\|^2 = \frac{1}{l^2} \sum_{\alpha=1}^{l} \|\varphi_{n+\alpha}\|^2 = \frac{1}{l}.$$

Thus the distance from the element g_{ln} in the range of $V-I$ to the element φ_n can be made as small as may be desired by a suitable choice of the integer l. Since V thus belongs to $\mathfrak{F}$, it has a correspondent H in $\mathfrak{S}$ with the deficiency-index $(0, 1)$; the form of the index shows that H is a maximal symmetric transformation.

If V_1 and V_2 are two transformations defined in the manner described by means of the complete orthonormal sets $\{\varphi_n^{(1)}\}$ and $\{\varphi_n^{(2)}\}$ respectively, it is evident that they are connected by the relation $V_2 \equiv UV_1U^{-1}$, where U is the unitary transformation determined from the equations

$$f = \sum_{\alpha=0}^{\infty} a_\alpha \varphi_\alpha^{(1)}, \qquad Uf \equiv \sum_{\alpha=0}^{\infty} a_\alpha \varphi_\alpha^{(2)}.$$

We thus find the successive identities

$$V_2 \equiv UV_1U^{-1}, \qquad I \pm V_2 \equiv U(I \pm V_1)U^{-1},$$
$$(I-V_2)^{-1} \equiv U(I-V_1)^{-1}U^{-1},$$
$$H_2 \equiv i(I+V_2)(I-V_2)^{-1} \equiv iU(I+V_1)(I-V_1)^{-1}U^{-1}$$
$$\equiv UH_1U^{-1},$$

so that H_1 and H_2 are equivalent.

DEFINITION 9.6. *A maximal symmetric transformation of the type described in Theorem* 9.9 *is called an elementary symmetric transformation.*

By an analysis whose general outline it is now easy to perceive, we obtain the following result:

THEOREM 9.10. *Every maximal symmetric transformation H is of one of the following types:*

(1) *it has the deficiency index* $(0, 0)$ *and is self-adjoint;*

(2) *it has the deficiency index* $(m, 0)$, $m = 1, 2, 3, \cdots, \aleph_0$; *there exist closed linear manifolds* $\mathfrak{M}_k$, *where k runs through all the finite cardinal numbers* $0, 1, 2, \cdots$ *not exceeding m, with the following properties:*

a) *they are mutually orthogonal, reduce H, and satisfy the relation* $\mathfrak{H} \equiv \mathfrak{M}_0 \oplus \mathfrak{M}_1 \oplus \mathfrak{M}_2 \oplus \cdots$;

b) $\mathfrak{M}_0$ *has dimension number* $N = 0, 1, 2, \cdots, \aleph_0$; *the transformation* $T^{(0)}$ *induced by H in* $\mathfrak{M}_0$ *is an ordinary algebraic (Hermitian) symmetric transformation or a self-adjoint transformation with respect to* $\mathfrak{M}_0$ *according as* $N = 0, 1, 2, \cdots$ *or* $N = \aleph_0$;

c) $\mathfrak{M}_k$, $1 \leqq k \leqq m$, *has the dimension number* $\aleph_0$ *and is therefore an abstract Hilbert space; the transformation* $T^{(k)}$ *induced by H in* $\mathfrak{M}_k$ *is the negative of an elementary symmetric transformation;*

(3) *if its deficiency index is* $(0, n)$, $n = 1, 2, 3, \cdots, \aleph_0$, *there exist closed linear manifolds* $\mathfrak{M}_k$, *where k runs through all the finite cardinal numbers* $0, 1, 2, \cdots$ *not exceeding n, with the same properties as those described under* (2) *save that* $T^{(k)}$, $1 \leqq k \leqq n$, *is an elementary symmetric transformation.*

In the cases (2) *and* (3), *H is completely characterized by its behavior in the manifolds* $\mathfrak{M}_k$.

Case (1) is merely a restatement of part of Theorem 9.3; and case (2) can be deduced from case (3) by an examination of the transformation $-H$ which has the deficiency index $(0, m)$, as we noted in the proof of Theorem 9.8.

In case (3) we consider the transformation V in $\mathfrak{F}$ corresponding to H. We know that, with the notation introduced in Theorem 9.4, the domain and range of V are respectively the closed linear manifolds $\mathfrak{R}_{-i} \equiv \mathfrak{H}$ and $\mathfrak{R}_{+i}$, where $\mathfrak{D}^{-} \equiv \mathfrak{H} \ominus \mathfrak{R}_{+i}$ has the dimension number n. We choose an orthonormal set $\{\varphi_0^{(k)}\}$, $k = 1, 2, 3, \cdots$ which determines the closed linear manifold $\mathfrak{D}^-$ and which therefore contains precisely n elements. We then define

$$\varphi_m^{(k)} = V^m \varphi_0^{(k)}, \quad k = 1, 2, 3, \cdots, \quad m = 0, 1, 2, \cdots.$$

It is evident at once that $\{\varphi_m^{(k)}\}$ is an orthonormal set: for, arranging our notation so that $l \geqq m$, we have

$$(\varphi_l^{(j)}, \varphi_m^{(k)}) = (V^l \varphi_0^{(j)}, V^m \varphi_0^{(k)}) = (V^{(l-m)} \varphi_0^{(j)}, \varphi_0^{(k)});$$

if $l = m$, the last expression reduces to $(\varphi_0^{(j)}, \varphi_0^{(k)}) = \delta_{jk}$, and, if $l > m$, it vanishes by virtue of the fact that $V^{(l-m)} \varphi_0^{(j)}$ is then an element of $\mathfrak{R}_{+i}$, the range of V. The set $\{\varphi_m^{(k)}\}$ may not be complete, but can always be made so by the introduction of an orthonormal set $\{\varphi_m^{(0)}\}$; we shall adapt our notation to the general case where the latter set is infinite, the particular cases where it is infinite or entirely absent being included by the simple convention used before in similar situations.* We define $\mathfrak{M}_k$ as the closed linear manifold determined by the set of elements $\varphi_0^{(k)}, \varphi_1^{(k)}, \varphi_2^{(k)}, \cdots$, with the proviso that when the set $\{\varphi_m^{(0)}\}$ is absent the manifold $\mathfrak{M}_0$ shall be defined by the relation $\mathfrak{M}_0 \equiv \mathfrak{O}$. It is evident at once that the manifolds $\{\mathfrak{M}_k\}$ are mutually orthogonal and together determine $\mathfrak{H}$ according to the identity $\mathfrak{H} \equiv \mathfrak{M}_0 \oplus \mathfrak{M}_1 \oplus \mathfrak{M}_2 \oplus \cdots$. We shall show that each of these manifolds reduces V. First we observe that the projection of $\mathfrak{H}$ on $\mathfrak{M}_k$ is a transformation E_k characterized by the equations

$$f = \sum_{\alpha, \beta = 0}^{\infty} a_\beta^{(\alpha)} \varphi_\beta^{(\alpha)}, \qquad E_k f = \sum_{\beta=0}^{\infty} a_\beta^{(k)} \varphi_\beta^{(k)}.$$

Next we note that by means of the relations used to define the set $\{\varphi_m^{(k)}\}$ we can write

$$V \varphi_m^{(k)} = \varphi_{m+1}^{(k)}, \qquad k = 1, 2, 3, \cdots, \quad m = 0, 1, 2, \cdots$$

and, for an arbitrary element f in $\mathfrak{H}$,

$$f = \sum_{\alpha, \beta = 0}^{\infty} a_\beta^{(\alpha)} \varphi_\beta^{(\alpha)} = E_0 f + \sum_{\alpha=1}^{\infty} \sum_{\beta=0}^{\infty} a_\beta^{(\alpha)} \varphi_\beta^{(\alpha)},$$

$$V f = V E_0 f + \sum_{\alpha=1}^{\infty} \sum_{\beta=0}^{\infty} a_\beta^{(\alpha)} \varphi_{\beta+1}^{(\alpha)}.$$

* Compare the comments on Theorem 1.4.

Thus if we replace f by its projection on $\mathfrak{M}_k$, $E_k f$, for $k = 1, 2, 3, \cdots$, we have

$$VE_k f = \sum_{\beta=0}^{\infty} a_\beta^{(k)} \varphi_{\beta+1}^{(k)}$$

which is evidently identical to $E_k Vf$ provided that $E_k VE_0 f = 0$. Now it is easy to show that $VE_0 f$ is an element of $\mathfrak{M}_0$, so that the equation in question is certainly satisfied: we read the facts directly from the equations

$$(VE_0 f, \varphi_0^{(k)}) = 0, \quad k = 1, 2, 3, \cdots;$$
$$(VE_0 f, \varphi_m^{(k)}) = (VE_0 f, V\varphi_{m-1}^{(k)}) = (E_0 f, \varphi_{m-1}^{(k)}) = 0,$$
$$k = 1, 2, 3, \cdots, \quad m = 1, 2, 3, \cdots,$$

the first set of which depends upon our choice of the elements $\{\varphi_0^{(k)}\}$ while the second set is established by use of the isometric character of V. Since V and E_k are thus known to be permutable, $\mathfrak{M}_k$ reduces V for $k = 1, 2, 3, \cdots$. It is also easy to show that $\mathfrak{M}_0$ reduces V: for the fact that $VE_0 f$ is in $\mathfrak{M}_0$ enables us to write

$$E_0(Vf) = E_0(VE_0 f) = VE_0 f.$$

The behavior of V in each of the closed linear manifolds $\mathfrak{M}_k$ is now readily described. For $k > 0$ the manifold $\mathfrak{M}_k$ evidently has the dimension number $\aleph_0$ and can therefore be treated as an abstract Hilbert space: if $V^{(k)}$ is the transformation with domain $\mathfrak{M}_k$ such that $V^{(k)} = V$ in $\mathfrak{M}_k$, an examination of the formulas given above shows that $V^{(k)}$ is a transformation of the type discussed in Theorem 9.9. The transformation $V^{(0)}$ with domain $\mathfrak{M}_0$ must take $\mathfrak{M}_0$ in a one-to-one isometric fashion into itself: thus $V^{(0)}$ has the character of an algebraic unitary transformation or a unitary transformation in abstract Hilbert space according as the dimension number of $\mathfrak{M}_0$ is less than or equal to $\aleph_0$.

The proof of the present theorem demands nothing more than the translation of the various facts deduced concerning V into corresponding facts concerning H. We recall that an element f belongs to the domain of H if and only if $f = (I - V)g$ and that for such an element the equation $Hf = i(I + V)g$ must hold. Thus if f is in the domain of H the element

$$E_k f = E_k(I-V)g = (I-V)E_k g$$

is also in the domain of H; and

$$HE_k f = i(I+V)E_k g = E_k(i(I+V)g) = E_k Hf.$$

The latter equation implies that the closed linear manifold $\mathfrak{M}_k$ reduces H. From Theorem 4.26 we see that H is completely determined by its behavior in the individual manifolds $\mathfrak{M}_0$, $\mathfrak{M}_1$, $\mathfrak{M}_2, \cdots$. If we denote by $T^{(k)}$ the transformation defined by the relation $T^{(k)} = H$ in $\mathfrak{M}_k$, it is at once evident that

$$V^{(k)} \equiv (T^{(k)}-iI)(T^{(k)}+iI)^{-1}, \quad T^{(k)} \equiv i(I+V^{(k)})(I-V^{(k)})^{-1}.$$

When $k>0$, $T^{(k)}$ is thus a closed linear symmetric transformation with respect to the abstract Hilbert space $\mathfrak{M}_k$, $V^{(k)}$ is the corresponding isometric transformation in $\mathfrak{F}$: from the remarks made above $T^{(k)}$ is identified as an elementary symmetric transformation. For $k = 0$, there are various cases to consider. If $\mathfrak{M}_0$ has the dimension number $\aleph_0$, then $\mathfrak{M}_0$ can be regarded as an abstract Hilbert space and the transformations $T^{(0)}$, $V^{(0)}$ as members of $\mathfrak{S}$ and $\mathfrak{F}$ in correspondence with one another; since $V^{(0)}$ is unitary the transformations $V^{(0)}$ and $T^{(0)}$ have the deficiency-index $(0,0)$ which identifies $T^{(0)}$ as a self-adjoint transformation in $\mathfrak{M}_0$. If the dimension number of $\mathfrak{M}_0$ is finite, it is easily verified that $T^{(0)}$ is defined throughout $\mathfrak{M}_0$ and behaves there like a linear (Hermitian) symmetric transformation in a unitary space. With these remarks, our proof is brought to a close.

It is evident from the facts established in this theorem that we are now in a position to construct maximal symmetric transformations with prescribed structure: in other words, subject to the restrictions noted in the theorem, we can assign the manifolds $\{\mathfrak{M}_k\}$ and the corresponding transformations $\{T^{(k)}\}$ arbitrarily; and the transformation H, defined by the relation $Hf = \sum_{\alpha=0}^{\infty} T^{(\alpha)} E_\alpha f$ whenever the right-hand expression is significant, is readily identified as the desired maximal transformation. More generally, we can construct transformations with prescribed deficiency index (m, n). For example, if we pick m closed linear manifolds $\{\mathfrak{M}_k\}$ and n closed linear manifolds $\{\mathfrak{N}_k\}$, which are mutually orthogonal with dimension number $\aleph_0$, and which together determine the closed linear manifold $\mathfrak{H}$, we can then construct a transformation $S^{(k)}$ in $\mathfrak{M}_k$, identical with the negative of an elementary symmetric transformation, and an elementary symmetric transformation $T^{(k)}$ in $\mathfrak{N}_k$. If we let E_k and F_k denote the projections of $\mathfrak{H}$ on $\mathfrak{M}_k$ and $\mathfrak{N}_k$ respectively, then the transformation H, defined by the equation

$$Hf = \sum_{\alpha=1}^{\infty} S^{(\alpha)} E_\alpha f + \sum_{\alpha=1}^{\infty} T^{(\alpha)} F_\alpha f$$

whenever the right-hand expression is significant, is a closed linear symmetric transformation with the deficiency index (m, n). From these brief indications, we see that every type of symmetric transformation discussed in Chapter II and in the present chapter can be exemplified by means of simple constructions.

From the theorem just proved we can derive an abstract characterization of the elementary symmetric transformations which brings to the fore one of their important properties, not yet noted.

THEOREM 9.11. *A maximal symmetric transformation is irreducible if and only if it is an elementary symmetric transformation or the negative of one.*

The necessity of the condition is an immediate consequence of Theorem 9.10. If a maximal symmetric transformation is irreducible, then one of three cases must arise, corresponding to the three cases of that theorem: the transformation is self-adjoint, is the negative of an elementary symmetric transformation, or is an elementary symmetric transformation. We can exclude the first possibility immediately. If a transformation is self-adjoint and irreducible at once, its resolution of the identity $E(\lambda)$ must coincide for each value of λ either with the identity I or with the null-transformation O; and it results immediately that the self-adjoint transformation in question must have the form $a \cdot I$ where a is a real number. We see, however, that the transformation $a \cdot I$ is reduced by every closed linear manifold and that, in consequence, there is no irreducible self-adjoint transformation.

The truth of the present theorem hinges upon the irreducibility of an elementary transformation H or its negative $-H$. Evidently it is enough to show that H is irreducible. Let $\mathfrak{M}$ be a closed linear manifold which reduces H, $\mathfrak{N} = \mathfrak{H} \ominus \mathfrak{M}$ its orthogonal complement. Both $\mathfrak{M}$ and $\mathfrak{N}$ reduce H according to Theorem 4.23; by an argument analogous to that used in the proof of Theorem 4.27, both these manifolds reduce V, the isometric transformation which corresponds to H. If $\{\varphi_n\}$ is the orthonormal set of Theorem 9.9, we can show that φ_0 belongs either to $\mathfrak{M}$ or to $\mathfrak{N}$. Since $\mathfrak{M}$ reduces V it is evident that V takes $\mathfrak{M}$ in a one-to-one isometric fashion into a closed linear manifold $\mathfrak{M}^*$ which is a subset of $\mathfrak{M}$; the element φ_0 is orthogonal to $\mathfrak{M}^*$ as well as to the entire range of V. By similar reasoning, we see that V takes $\mathfrak{N}$ into a subset $\mathfrak{N}^*$ orthogonal to φ_0. It is easily verified that the closed linear manifold $\mathfrak{M}^* \oplus \mathfrak{N}^*$ is the entire range of V. Since $\mathfrak{M} \ominus \mathfrak{M}^*$ is a closed linear manifold orthogonal to $\mathfrak{M}^*$ and to $\mathfrak{N}^*$, it must be orthogonal to the range of V: thus $\mathfrak{M} \ominus \mathfrak{M}^*$ contains no element other than the null element or it is identical with the closed linear manifold determined by the element φ_0; and if the first possibility is realized the relation $\mathfrak{M} = \mathfrak{M}^*$ holds and requires that φ_0, being orthogonal to $\mathfrak{M}^*$, belong to $\mathfrak{N}$. It is now very easy to show that whichever of the manifolds $\mathfrak{M}$ and $\mathfrak{N}$ contains φ_0 must contain the entire orthonormal set $\{\varphi_n\}$: for the successive images of φ_0 by the transformations

$$V,\ V^2 \equiv V \cdot V,\ \cdots,\ V^{n+1} \equiv V \cdot V^n,\ \cdots$$

all belong to the same manifold and are known to be given by the equation $V^n \varphi_0 = \varphi_n$ for $n = 0, 1, 2, \cdots$. Consequently the manifold which contains φ_0 must be identical with $\mathfrak{H}$ and the other with $\mathfrak{O}$. Finally, therefore, the only manifolds $\mathfrak{M}$ which reduce H are $\mathfrak{H}$ and $\mathfrak{O}$; and H is seen to be irreducible.

In view of the fundamental significance of the deficiency-index of a closed linear symmetric transformation, tests for determining it assume a certain degree of importance. In Chapter II we have already shown that a transformation H in $\mathfrak{H}$ is self-adjoint—that is, has the deficiency index $(0, 0)$—if it is bounded (Theorem 2.24), if it is defined throughout $\mathfrak{H}$ (Theorem 2.17), or if its range is the entire space $\mathfrak{H}$ (Theorem 2.19). We shall obtain various other conditions which restrict the form of the deficiency-index, as we proceed.

§ 2. Real Transformations

Hitherto it has not been necessary for us to pose questions of "reality"; all our developments have referred to "complex" Hilbert space and "complex" transformations. It is now advantageous for us to lay down an abstract definition of what we shall mean by a "real" transformation in "complex" Hilbert space.

First we introduce a particular transformation which, intuitively speaking, may be said to replace each element of $\mathfrak{H}$ by its "conjugate" element with reference to a suitably chosen "real" subspace.

Definition 9.7. *A transformation J with domain $\mathfrak{H}$ is said to be a conjugation if*

$$J^2 \equiv I, \qquad (Jf, Jg) = \overline{(f, g)} = (g, f),$$

for every f and g in $\mathfrak{H}$.

The main properties of conjugations are summarized in the following theorem:

Theorem 9.12. *A conjugation J is a one-to-one transformation of $\mathfrak{H}$ into itself with an inverse J^{-1} such that $J^{-1} \equiv J$. It has the properties*

$$(Jf, g) = \overline{(f, Jg)}, \quad J(f_1 + f_2) = Jf_1 + Jf_2, \quad J(af) = \bar{a}Jf.$$

J is isomorphic with each of the following transformations:

(1) *the transformation J_0 in the concrete Hilbert space $\mathfrak{H}_0$ which takes $(x_1, x_2, x_3, \cdots)$ into $(\bar{x}_1, \bar{x}_2, \bar{x}_3, \cdots)$;*

(2) *the transformation* J_2 *in the concrete Hilbert space* $\mathfrak{L}_2$ (*of Lebesgue-measurable functions* $f(P)$ *such that* $\int_E |f(P)|^2 dP$ *exists*) *which takes* $f(P)$ *into* $\overline{f(P)}$.

Any two conjugations are isomorphic; that is, they are equivalent by a suitably chosen unitary transformation. The class of all conjugations in $\mathfrak{H}$ *has the cardinal number* $\mathfrak{c}$ *of the continuum.*

First we note that J^{-1} must exist since $Jf = 0$ implies $f = 0$ according to the equations

$$\|Jf\|^2 = \overline{(f\ f)} = \|f\|^2 = 0.$$

Next we obtain the identity

$$J^{-1} \equiv J^{-1}(J^2) \equiv (J^{-1}J)\,J \equiv J,$$

which shows that the range of J must be the entire space $\mathfrak{H}$. If we now replace g by Jg in the relation $(Jf, Jg) = \overline{(f, g)}$ we find the corresponding relation $(Jf, g) = \overline{(f, Jg)}$ asserted in the theorem. The latter relation enables us to write

$$\begin{aligned}(J(f_1+f_2), g) &= \overline{(f_1+f_2, Jg)} = \overline{(f_1, Jg)+(f_2, Jg)} \\ &= (Jf_1, g)+(Jf_2, g) = (Jf_1+Jf_2, g)\end{aligned}$$

and hence to conclude that $J(f_1+f_2)$ and Jf_1+Jf_2 are identical. By similar manipulations we establish the equation $J(af) = \bar{a}\,Jf$.

In order to prove that the conjugation J is isomorphic with the transformations J_0 and J_2, which are both conjugations according to Definition 9.7, we construct the set $\mathfrak{S}$ of all elements in $\mathfrak{H}$ such that $Jf = f$. We may refer to $\mathfrak{S}$ as the set of elements real with respect to the conjugation J. This set evidently contains the null element, and, while it is not a linear manifold, it contains together with the elements f_1 and f_2 any linear combination $a_1 f_1 + a_2 f_2$ in which the coefficients a_1 and a_2 are real numbers. Of the numerous other elementary properties of $\mathfrak{S}$ which we might examine, there is only one which we need to state and prove:

the closed linear manifold determined by $\mathfrak{S}$ is the entire space $\mathfrak{H}$. In fact, if g is an arbitrary element we write

$$f_1 = Jg+g, \quad f_2 = i(Jg-g), \quad g = \tfrac{1}{2}(f_1+if_2)$$

and observe that Jf_1 and Jf_2 reduce to f_1 and f_2 respectively; in other words, every element in $\mathfrak{H}$ is a linear combination of elements in $\mathfrak{S}$. On the basis of Theorem 1.18 we select a sequence $\{f_n\}$ which is everywhere dense in $\mathfrak{S}$ and therefore determines the closed linear manifold $\mathfrak{H}$. To this sequence we apply the process described in Theorem 1.13, thus obtaining a complete orthonormal set. By an examination of the process in question we see that this set is a subset of $\mathfrak{S}$; we shall denote it by $\{\varphi_n\}$. It can now be verified without difficulty that J is characterized by the relations

$$f = \sum_{\alpha=1}^{\infty} x_\alpha \varphi_\alpha, \qquad Jf = \sum_{\alpha=1}^{\infty} \bar{x}_\alpha \varphi_\alpha$$

in terms of the set $\{\varphi_n\}$ constructed above. It is plain that J is isomorphic with the transformation J_0 described in the theorem. Since the transformation J_2 is a conjugation it, too, is isomorphic with J_0 and hence also with J. More generally, any two conjugations, being isomorphic with J_0, are isomorphic with each other.

It is clear that in a given space $\mathfrak{H}$ there are many different conjugations. We denote by c the cardinal number of the class of all conjugations in $\mathfrak{H}$ and by $\mathfrak{c}$ the cardinal number of the continuum. We prove that $c = \mathfrak{c}$. The results of the preceding paragraph show that we can associate a conjugation with an arbitrary complete orthonormal set and that by considering all such sets in $\mathfrak{H}$ we obtain each conjugation at least once. Hence c cannot exceed the cardinal number of the class of all complete orthonormal sets in $\mathfrak{H}$. Since the latter class has the cardinal number $\mathfrak{c}$, we see that $c \leqq \mathfrak{c}$. On the other hand, we can construct $\mathfrak{c}$ distinct conjugations and thereby show that $c \geqq \mathfrak{c}$. If $\{\varphi_n\}$ is a complete orthonormal set in $\mathfrak{H}$, then the set $\{\psi_n\}$ defined by the equations $\psi_1 = e^{i\theta}\varphi_1$, $0 \leqq \theta < \pi$, $\psi_n = \varphi_n$ for $n \geqq 2$, is also a complete orthonormal set. The conjugations J_θ defined by the equations

$$f = \sum_{\alpha=1}^{\infty} x_\alpha \psi_\alpha, \qquad J_\theta f = \sum_{\alpha=1}^{\infty} \bar{x}_\alpha \psi_\alpha$$

are distinct for distinct values of θ, $0 \leqq \theta < \pi$, by virtue of the relations

$$\varphi_1 = e^{-i\theta} \psi_1, \qquad J_\theta \varphi_1 = e^{i\theta} \psi_1 = e^{2i\theta} \varphi_1 .$$

The inequalities $c \leqq \mathfrak{c}$, $c \geqq \mathfrak{c}$ show that $c = \mathfrak{c}$, as we wished to prove.

We shall now define the term "real transformation" in a manner which accords with our intuitive concept of its meaning, so far as that is possible. Briefly, we shall say that a transformation is real whenever there exists a conjugation with which it is permutable; with greater precision we state the requirement as follows:

DEFINITION 9.8. *A transformation T and a conjugation J are said to be permutable if whenever f is an element of the domain of T the element Jf is in the domain of T and satisfies the relation $TJf = JTf$. A transformation T is said to be real with respect to the conjugation J if it is permutable with J.*

The importance of this definition for the theory of transformations is emphasized by the elementary theorem which we shall prove next.

THEOREM 9.13. *Let T be a linear transformation real with respect to a conjugation J. Let $A(T)$ be its point spectrum, $B(T)$ its continuous spectrum, $C(T)$ its residual spectrum, $D(T)$ its resolvent set, and $W(T)$ its numerical range. Then each of the five sets A, B, C, D, W is symmetrical in the real axis of the complex l-plane whenever it is not an empty set. If T has an adjoint T^* then T^* is real with respect to J.*

Recalling that J and J^{-1} are identical, we can write $T - \bar{l}I \equiv J(T - lI)J^{-1}$ by virtue of the permutability of T and J. Now J is a one-to-one transformation of $\mathfrak{H}$ into itself which preserves orthogonality relations and distances according to the equations

$$(Jf, Jg) = \overline{(f, g)}, \qquad |Jf - Jg| = |J(f-g)| = |f-g|.$$

In consequence it is a continuous transformation. Because J has these properties we can make use of the same general argument as that on which the proof of Theorem 4.3 depends, to show that a point l and its conjugate $\bar{l}$ must be classified in the same way with regard to the spectrum of T. In particular, we note that when l is a characteristic value of multiplicity n its conjugate $\bar{l}$ is a characteristic value of the same multiplicity.

Next if l is a value assumed by $(Tf, f)/\|f\|^2$, we see that for Jf we have the relations

$$(TJf, Jf)/\|Jf\|^2 = (JTf, Jf)/\|f\|^2 = \overline{(Tf, f)}/\|f\|^2 = \bar{l},$$

so that the set $W(T)$ is symmetrical in the real axis.

Finally, if T has an adjoint T^* we must show that T^* is permutable with J. The element g belongs to the domain of T^* and $T^*g = g^*$ if and only if $(Tf, g) = (f, g^*)$ for every element f in the domain of T. If g is such an element and f is replaced by Jf, the latter equation takes the successive forms

$$(TJf, g) = (Jf, g^*), \qquad (JTf, g) = (Jf, g^*),$$
$$(Tf, Jg) = (f, Jg^*).$$

Thus Jg is in the domain of T^* and $T^*Jg = Jg^* = JT^*g$, as we wished to prove.

We shall apply this result to the theory of symmetric transformations in the following theorem.

THEOREM 9.14. *If a closed linear symmetric transformation H is real with respect to a conjugation J, then its deficiency-index has the form (m, m). The maximal symmetric extensions of H can be described as follows:*

(1) *if $m = 0$, H is self-adjoint and real with respect to J; it possesses no proper symmetric extension;*

(2) *if $m = 1$, every maximal extension of H is self-adjoint and real with respect to J; the class of all maximal extensions has the cardinal number $\mathfrak{c}$ of the continuum;*

(3) *if $1 < m < \aleph_0$, the maximal extensions of H fall into two classes, each with the cardinal number $\mathfrak{c}$ – the class of all*

self-adjoint extensions real with respect to J, and the class of all self-adjoint extensions not real with respect to J;

(4) *if* $m = \aleph_0$ *the maximal extensions of H fall into three classes, each with the cardinal number* $\mathfrak{c}$—*the class of all self-adjoint extensions real with respect to J, the class of all self-adjoint extensions not real with respect to J, and the class of all maximal extensions which are not self-adjoint.*

A maximal symmetric transformation which is real with respect to a conjugation J is self-adjoint; a self-adjoint transformation is real with respect to a conjugation J if and only if its resolution of the identity is real with respect to J.

If H is real with respect to J, then H^* has the same property, as we proved in the preceding theorem; thus the spectrum of H^* is symmetrical in the real axis. By reference to Theorem 9.8 we see that H must have the deficiency-index (m, m).

In order to discuss the maximal extensions of a real transformation H in $\mathfrak{H}$ we must consider the character of the corresponding transformation V in $\mathfrak{F}$. We can show immediately that a transformation H in $\mathfrak{H}$ is real with respect to the conjugation J if and only if the corresponding isometric transformation V in $\mathfrak{F}$ satisfies the relation $VJ \equiv JV^{-1}$. The necessity of the condition follows at once, by use of the identity $J \equiv J^{-1}$, from the relations

$$\begin{aligned} VJ &\equiv H_i(H_{-i})^{-1}J^{-1} \equiv H_i(JH_{-i})^{-1} \equiv H_i(H_iJ)^{-1} \\ &\equiv H_iJ(H_i)^{-1} \equiv JH_{-i}(H_i)^{-1} \equiv JV^{-1}. \end{aligned}$$

The sufficiency of the condition is proved as follows: f is in the domain of H if and only if there exists an element g such that $f = (I-V)g$, $Hf = i(I+V)g$; in these equations we replace g by $g^* = -JVg$, thus obtaining

$$\begin{aligned} f^* &= ((VJ)Vg - JVg) = J(I-V)g = Jf \\ Hf^* &= -i((VJ)Vg + JVg) = -iJ(I+V)g = JHf \end{aligned}$$

and reaching the result that H is permutable with J.

Let H be a transformation in $\mathfrak{H}$ real with respect to a conjugation J. Its deficiency-index is (m, m); and we may suppose that $m > 0$, since the case $m = 0$ is settled at once by the statement in the theorem. The closed linear manifolds $\mathfrak{D}^+$ and $\mathfrak{D}^-$ determined by the sets of characteristic elements of H^* for the characteristic values $+i$ and $-i$ have the common dimension number m. It is evident that J carries $\mathfrak{D}^-$ in a one-to-one continuous manner, with preservation of distance and of orthogonality, into $\mathfrak{D}^+$. If V is the isometric transformation in $\mathfrak{F}$ corresponding to H, then all its unitary extensions and hence all the self-adjoint extensions of H are found by means of the isometric transformations of $\mathfrak{D}^+$ into $\mathfrak{D}^-$. The situation has been discussed in great detail in Theorem 9.3. The class of all unitary extensions of V has the cardinal number $\mathfrak{c}$. In order to determine whether a given unitary extension U yields a real self-adjoint extension of H or not we have merely to verify whether U satisfies the identity $UJ \equiv JU^{-1}$ or not; indeed, we can confine our attention for this purpose to the behavior of U in $\mathfrak{D}^+$. Let U_0 be the isometric transformation of $\mathfrak{D}^+$ into $\mathfrak{D}^-$ which coincides with U in $\mathfrak{D}^+$; the self-adjoint transformation corresponding to U is real with respect to J if and only if $U_0 J \equiv J U_0^{-1}$ in $\mathfrak{D}^-$. We select an orthonormal set $\{\varphi_n\}$, containing precisely m elements, which determines the closed linear manifold $\mathfrak{D}^-$; the orthonormal set $\{\psi_n\}$, where $\psi_n = J\varphi_n$, plays a similar rôle in $\mathfrak{D}^+$. We now prove the following criterion: the relation $U_0 J \equiv J U_0^{-1}$ holds throughout $\mathfrak{D}^-$ if and only if the expression $a_{ik} = (U_0 J \varphi_i, \varphi_k)$ is symmetric in its indices, $a_{ik} = a_{ki}$. If $f = \sum_{\alpha=1}^{\infty} a_\alpha \varphi_\alpha$ is an arbitrary element of $\mathfrak{D}^-$ we can write

$$(U_0 J f, \varphi_i) = \overline{(f, J U_0^{-1} \varphi_i)} = \sum_{\alpha=1}^{\infty} \bar{a}_\alpha (J U_0^{-1} \varphi_i, \varphi_\alpha)$$

$$= \sum_{\alpha=1}^{\infty} \bar{a}_\alpha (U_0 J \varphi_\alpha, \varphi_i) = \sum_{\alpha=1}^{\infty} \bar{a}_\alpha a_{\alpha i},$$

$$(J U_0^{-1} f, \varphi_i) = \overline{(f, U_0 J \varphi_i)} = \sum_{\alpha=1}^{\infty} \bar{a}_\alpha (U_0 J \varphi_i, \varphi_\alpha)$$

$$= \sum_{\alpha=1}^{\infty} \bar{a}_\alpha a_{i\alpha}.$$

A necessary and sufficient condition that $U_0 J f = J U_0^{-1} f$ is that the two expressions evaluated be equal for $i = 1, 2, 3, \cdots$ for all sets of values $\{a_n\}$ such that $\sum_{\alpha=1}^{\infty} |a_\alpha|^2$ converges. Evidently such an equality can hold if and only if $a_{ik} = a_{ki}$. The transformation U_0 is completely determined by the quantities

$$a_{ik} = (U_0 J \varphi_i, \varphi_k) = (U_0 \psi_i, \varphi_k);$$

in order that U_0 be isometric these coefficients must be connected by the relation

$$\sum_{\alpha=1}^{\infty} a_{i\alpha} \bar{a}_{k\alpha} = (U_0 \psi_i, U_0 \psi_k) = (\psi_i, \psi_k) = \delta_{ik}.$$

If $m = 1$ the condition $a_{11} = a_{11}$ is automatically satisfied; hence every maximal transformation of H is self-adjoint and real with respect to J. If $1 < m < \aleph_0$ then every maximal extension of H is known to be self-adjoint, by Theorem 9.3; but there are both real and not-real extensions of H. We can obtain a class of real extensions with cardinal number $\mathfrak{c}$ by choosing U_0 according to the equations $a_{ik} = \alpha \delta_{ik}$, $|\alpha| = 1$. We can likewise obtain a class of extensions not real with respect to J, with the cardinal number $\mathfrak{c}$, by choosing U_0 according to the equations $a_{11} = a_{22} = \cos\theta$, $a_{12} = \sin\theta$, $a_{21} = -\sin\theta$, $a_{ik} = \delta_{ik}$ when $i > 2$ or $k > 2$, $0 < \theta < \pi$. Since all the extensions of H constitute a class with cardinal number $\mathfrak{c}$, it is plain that the assertion of the theorem must hold for the present case. The case $m = \aleph_0$ differs from the preceding case only in the occurrence of $\mathfrak{c}$ maximal extensions which are not self-adjoint, as we have already shown in Theorem 9.3.

If a maximal symmetric transformation is real with respect to a conjugation J, then by the first part of the present theorem and the result of Theorem 9.3 its deficiency-index is $(0, 0)$; it is therefore self-adjoint. If a self-adjoint transformation H is real with respect to a conjugation J, then its resolution of the identity $E(\lambda)$ is real with respect to H: we have merely to apply the argument of Theorem 8.1 in order to show that H is permutable with J if and only if $E(\lambda)$ is permutable with J.

§ 3. Approximation Theorems

In the present section we shall study sequences of symmetric transformations, particularly self-adjoint transformations, with a view to establishing useful generalizations of the results which we have proved and applied in the opening sections of Chapter V. This section is essentially a systematic survey in abstract terms of results already obtained by Carleman,† with occasional extensions.

We first prove an elementary theorem on which our fundamental definition is to be based.

Theorem 9.15. *Let $\{H^{(n)}\}$ be a sequence of linear symmetric transformations, $\mathfrak{S}$ the set of elements in $\mathfrak{H}$ in which the sequence is convergent according to Definition* 2.3. *Then $\mathfrak{S}$ is a linear manifold. If $\mathfrak{S}$ is everywhere dense in $\mathfrak{H}$, the limit transformation $H^{(0)}$ of the sequence is linear and symmetric. If the sequence $\{H^{(n)}\}$ is also uniformly bounded, then $H^{(0)}$ is bounded and essentially self-adjoint; if every transformation of the sequence is real with respect to a conjugation J, then $H^{(0)}$ and $\tilde{H}^{(0)}$ are real with respect to J.*

If $\{H^{(n)}\}$ is an arbitrary sequence of linear symmetric transformations, the set $\mathfrak{S}$ certainly contains the null element. If f_1 and f_2 belong to $\mathfrak{S}$ then $f_3 = a_1 f_1 + a_2 f_2$ also belongs to $\mathfrak{S}$: for

$$H^{(n)} f_1 \to f_1^*, \qquad H^{(n)} f_2 \to f_2^*,$$

$$H^{(n)} f_3 = a_1 H^{(n)} f_1 + a_2 H^{(n)} f_2 \to a_1 f_1^* + a_2 f_2^* = f_3^*.$$

Thus $\mathfrak{S}$ is a linear manifold. If f is in $\mathfrak{S}$ we define the limit transformation $H^{(0)}$ of the sequence for the element f by the relations $H^{(n)} f \to f^*$, $H^{(0)} f = f^*$. From the limiting relations noted above it is evident that $H^{(0)}$ is a linear transformation with the linear manifold $\mathfrak{S}$ as its domain. If f and g are elements of $\mathfrak{S}$, then

† Carleman, *Les équations intégrales singulières à noyau réel et symétrique*, Uppsala 1923, Chapters I and II; and, Annales de l'Institut H. Poincaré, 1 (1931), pp. 401–430.

$$(H^{(0)}f, g) = \lim_{n\to\infty} (H^{(n)}f, g)$$
$$= \lim_{n\to\infty} (f, H^{(n)}g) = (f, H^{(0)}g),$$

so that $H^{(0)}$ is adjoint to itself. When $\mathfrak{S}$ is everywhere dense in $\mathfrak{H}$, the transformation $H^{(0)}$ is therefore symmetric.

If the sequence $\{H^{(n)}\}$ is uniformly bounded, in the sense that there exists a constant C independent of f and of n such that $\|H^{(n)}f\| \leqq C\|f\|$ for f in the domain of $H^{(n)}$, then $H^{(0)}$ is bounded and satisfies the inequality $\|H^{(0)}f\| \leqq C\|f\|$ throughout $\mathfrak{S}$: for if f is in $\mathfrak{S}$, we have

$$\|H^{(0)}f\| = \lim_{n\to\infty} \|H^{(n)}f\| \leqq C\|f\|.$$

Thus when $\mathfrak{S}$ is everywhere dense in $\mathfrak{H}$, the transformation $H^{(0)}$ is essentially self-adjoint, in accordance with Theorem 2.24.

If every transformation $H^{(n)}$ of the sequence is real with respect to the conjugation J, then $H^{(n)}$ is permutable with J for $n = 1, 2, 3, \cdots$; we must show that $H^{(0)}$ is permutable with J. If f is in $\mathfrak{S}$ then $H^{(n)}f \to H^{(0)}f$; thus by using the fact that J is a continuous transformation we find the relations

$$H^{(n)}Jf = JH^{(n)}f \to JH^{(0)}f,$$

which imply that $H^{(0)}Jf$ exists and is equal to $JH^{(0)}f$, as we wished to prove. Thus, when $\mathfrak{S}$ is everywhere dense in $\mathfrak{H}$, $H^{(0)}$ is a symmetric transformation real with respect to J; the transformations $(H^{(0)})^*$ and $(H^{(0)})^{**} \equiv \tilde{H}^{(0)}$ are also real, according to Theorem 9.13.

We now lay down a definition, the usefulness of which will be apparent shortly.

Definition 9.9. *A sequence $\{H^{(n)}\}$ of linear symmetric transformations is said to approximate, or to be an approximating sequence for, the symmetric transformation H if the limit transformation $H^{(0)}$ and the transformation H are connected by the relation $H \subseteq \tilde{H}^{(0)}$.*

In Chapter V we showed how to construct an approximating sequence for a given self-adjoint transformation. The details of that construction can be applied with slight modifications to the case of an arbitrary symmetric transformation, as we shall see immediately.

THEOREM 9.16. *If H is an arbitrary symmetric transformation, then there exists a sequence $\{H^{(n)}\}$ which approximates H and in which each transformation $H^{(n)}$ is a bounded self-adjoint transformation whose spectrum consists of a finite number of characteristic values, each one different from zero having finite multiplicity. In particular, if H is a bounded linear transformation, the sequence $\{H^{(n)}\}$ can be chosen so as to be uniformly bounded; if H is a linear transformation real with respect to a conjugation J, then the sequence $\{H^{(n)}\}$ can be chosen so that $H^{(n)}$ is real with respect to J; and if H is linear and not-negative definite, the sequence $\{H^{(n)}\}$ can be chosen so that $H^{(n)}$ is not-negative definite.*

We may suppose without loss of generality that H is closed and linear, since a sequence which approximates H also approximates $\tilde{H} \equiv H^{**}$. By the process used in the opening paragraphs of Chapter V, § 2, we select a linear manifold $\mathfrak{M}$, everywhere dense in the domain of H and hence in the space $\mathfrak{H}$, such that its images by the transformations $H - iI$, $H + iI$ are everywhere dense in the ranges of those transformations. We then choose an orthonormal subset $\{\varphi_n\}$ which determines the linear manifold $\mathfrak{M}$. We define $\mathfrak{M}_n$, the linear manifold determined by the set of elements $\varphi_1, \cdots, \varphi_n$; and E_n, the projection of $\mathfrak{H}$ on $\mathfrak{M}_n$. The sequence composed of the transformations $H^{(n)} \equiv E_n H E_n$ is then found to have the properties described above. By arguments analogous to those used in Theorems 5.1 and 5.2, we show that $H^{(n)}$ is a bounded self-adjoint transformation of the indicated type and that the sequence $H^{(n)}$ converges in $\mathfrak{M}$ to the limit H. Since $\mathfrak{M}$ was chosen so that the behavior of H is completely determined by its behavior in $\mathfrak{M}$ alone, we see that $\tilde{H}^{(0)}$ must be an extension of H: in other words, $\{H^{(n)}\}$ is an approximating sequence for H in the sense of Definition 9.9.

If H is a bounded transformation satisfying the inequality

$\|Hf\| \leqq C\|f\|$, then the sequence $\{H^{(n)}\}$ is uniformly bounded, by virtue of the relations

$$\|H^{(n)}f\| = \|E_n HE_n f\| \leqq \|HE_n f\| \leqq C\|E_n f\| \leqq C\|f\|,$$

holding for every f in $\mathfrak{H}$.

If H is real with respect to the conjugation J, we proceed as in the general case to the determination of the manifold $\mathfrak{M}$ and the orthonormal set $\{\varphi_n\}$. It is then possible to replace $\{\varphi_n\}$ by a new set $\{\psi_n\}$ which determines a linear manifold $\mathfrak{M}^* \supseteqq \mathfrak{M}$ and which has the property that $J\psi_n = \psi_n$. In order to do so we construct a simple sequence $\{f_n\}$ from the two sequences $\{\varphi_n + J\varphi_n\}$ and $\{i(\varphi_n - J\varphi_n)\}$: evidently f_n satisfies the equation $Jf_n = f_n$; and the sequence $\{f_n\}$ determines a linear manifold $\mathfrak{M}^*$ which contains $\mathfrak{M}$. To the sequence $\{f_n\}$ we apply the process of orthogonalization, to obtain the orthonormal set $\{\psi_n\}$. Once this set is at our disposal we define the linear manifold $\mathfrak{M}_n^*$ determined by the set of elements $\psi_1, \cdots, \psi_n$ and put $H^{(n)} = E_n HE_n$ where E_n is the projection of $\mathfrak{H}$ on $\mathfrak{M}_n^*$. As before we find that the sequence $\{H^{(n)}\}$ has the general properties described in the statement of the theorem; and in addition, we find that $H^{(n)}J = JH^{(n)}$. It is evident from the relations

$$f = \sum_{\alpha=1}^{\infty} a_\alpha \psi_\alpha, \quad Jf = \sum_{\alpha=1}^{\infty} \bar{a}_\alpha \psi_\alpha, \quad E_n f = \sum_{\alpha=1}^{n} a_\alpha \psi_\alpha$$

that $E_n J$ and JE_n are identical. Thus, if we make use of the fact that H is permutable with J, we have

$$\begin{aligned} H^{(n)}J &\equiv E_n HE_n J \equiv E_n HJE_n \\ &\equiv E_n JHE_n \equiv JE_n HE_n \equiv JH^{(n)}. \end{aligned}$$

It follows that $H^{(n)}$ is real with respect to J.

If H is not-negative definite, then $(Hf, f) \geqq 0$ for every element f in the domain of H. Thus $H^{(n)}$ satisfies the relation

$$(H^{(n)}f, f) = (E_n HE_n f, f) = (HE_n f, E_n f) \geqq 0$$

and is likewise not-negative definite.

If we pass from the consideration of a sequence of self-adjoint transformations to the study of the corresponding sequence of resolvents, it is possible for us to obtain a considerable store of information of useful character. We shall develop the consequences of this point of view in the theorems which follow.

THEOREM 9.17. *Let* $\{H^{(n)}\}$ *be a sequence of self-adjoint transformations approximating a closed linear symmetric transformation* H; *let* $\{R_l^{(n)}\}$ *be the corresponding sequence of resolvents; and let* Δ *be the open set of all points* l *such that for some* $N = N(l)$ *and some positive* $\delta = \delta(l)$ *the distance from* l *to the spectrum of* $H^{(n)}$ *exceeds* δ *whenever* n *exceeds* N. *Then there exists a sequence of integers* $\{n(k)\}$, *a family of transformations* X_l, *and a family of functions* $\rho(f, g; \lambda)$ *with the following properties:*

(1) $n(k)$ *tends to infinity with* k;

(2) X_l *is defined when* l *is in* Δ *and is a bounded linear transformation with domain* $\mathfrak{H}$, *satisfying the inequality*

$$\| X_l f \| \leq \| f \| / d(l),$$

where $d(l)$ *is the distance from* l *to the boundary of the set* Δ;

(3) $\rho(f, g; \lambda)$ *is defined for arbitrary* f *and* g *in* $\mathfrak{H}$ *and arbitrary real* λ, $-\infty < \lambda < +\infty$; *for fixed* λ *it is a bilinear function of its arguments* f *and* g; *for fixed* f *and* g *it is a function of bounded variation in* λ *in the prescribed normal form, satisfying the inequality*

$$V(\rho) \leq \| f \| \, \| g \|;$$

when $f = g$, *the function* $\rho(f, g; \lambda)$ *is a real monotone-increasing function of* λ;

(4) *the sequence* $\{(R_l^{(n(k))} f, g)\}$ *converges for fixed* f *and* g *to the limit*

$$(X_l f, g) = \int_{-\infty}^{+\infty} \frac{1}{\lambda - l} \, d\rho(f, g; \lambda)$$

throughout Δ, *the convergence being uniform on any bounded closed subset of* Δ.

The function $\rho(f, g; \lambda)$ *can be represented in the form* $(B(\lambda) f, g)$ *where* $B(\lambda)$ *is a bounded self-adjoint transformation,* $-\infty < \lambda < +\infty$.

The open set Λ, we should notice, contains both the half-planes $\mathfrak{J}(l) > 0$, $\mathfrak{J}(l) < 0$. Our theorem depends fundamentally upon the fact that, in the language of the theory of analytic functions, the family $\{(R_l^{(n)} f, g)\}$ is a "normal family" in the open set Λ. We shall not appeal to the theory of such families for the results we use, but shall prove directly all the assertions made in the statement of the theorem.

For our present purposes, we must note that the inequality of Theorem 4.20 can be extended to the case of an arbitrary self-adjoint transformation T. It is evident from Theorem 5.11 that the real points of the resolvent set of T constitute a set of T-measure zero in accordance with Definition 6.3. Hence, if l is a point at positive distance $d(l)$ from the spectrum of T, the inequality $\left|\dfrac{1}{\lambda - l}\right| \leqq 1/d(l)$ holds almost everywhere with respect to T. Since the resolvent of T is defined according to Theorems 5.7 and 6.1 by the relation $R_l \equiv \dfrac{1}{T-l}$, we conclude that $|(R_l f, g)| \leqq \|R_l f\| \|g\| \leqq \|f\| \|g\| / d(l)$ for every l in the resolvent set of T, in accordance with Theorem 6.5.

We start by choosing a sequence $\{f_i\}$ everywhere dense in $\mathfrak{H}$ and a sequence $\{l_m\}$ everywhere dense in Λ. We then consider the collection of numbers $A_{ijm}^{(n)} = (R_{l_m}^{(n)} f_i, f_j)$, which we classify in simple sequences distinguished by the indices (i, j, m). These simple sequences are ranged in a definite order according to their indices by the following convention: the index (i, j, m) precedes the index (i', j', m') if one of the three mutually exclusive relations

(1) $i+j+m < i'+j'+m'$,

(2) $i+j+m = i'+j'+m', \qquad i+j < i'+j'$,

(3) $i+j+m = i'+j'+m', \qquad i+j = i'+j', \qquad i < i'$,

holds. Each of these simple sequences is bounded, by virtue of the inequality given above:

$$|A_{ijm}^{(n)}| \leqq \|R_{l_m}^{(n)} f_i\| \|f_j\| < \|f_i\| \|f_j\| / \delta_m,$$

which is satisfied for $\delta_m = \delta(l_m)$ and $n > N = N(l_m)$; in consequence every subsequence contains a convergent subsequence. By dealing with each simple sequence in the order of its index and selecting from it a subsequence determined in part by all the preceding selections, we obtain in a familiar manner simple sequences of integers $n(i, j, m, k)$ which are distinguished and ordered according to the indices (i, j, m) and which possess the following properties: when k becomes infinite, $n(i, j, m, k)$ becomes infinite and the corresponding simple sequence $\{A_{ijm}^{(n(i,j,m,k))}\}$ converges to a limit A_{ijm} satisfying the inequality

$$|A_{ijm}| \leqq \|f_i\| \|f_j\|/\delta_m$$

where $\delta_m = \delta(l_m)$; when the sequence $\{n(i, j, m, k)\}$ precedes the sequence $\{n(i', j', m', k)\}$ it contains the latter as a subsequence. We now define $n(k)$ by the equation $n(k) = n(k, k, k, k)$, so that each sequence $\{n(i, j, m, k)\}$ contains all but a finite number of the members of the sequence $\{n(k)\}$. It is evident that $n(k) \to \infty$ and $A_{ijm}^{(n(k))} \to A_{ijm}$ when $k \to \infty$.

From the convergence of the sequence $\{A_{ijm}^{(n(k))}\}$, where $A_{ijm}^{(n(k))} = (R_{l_m}^{(n(k))} f_i, f_j)$ we can deduce the convergence of the sequence $\{(R_{l_m}^{(n(k))} f, g)\}$ for arbitrary f and g in $\mathfrak{H}$. In the inequality

$$\begin{aligned} &|(R_{l_m}^{(n(k))} f, g) - (R_{l_m}^{(n(k))} f_i, f_j)| \\ &\leqq |(R_{l_m}^{(n(k))}(f - f_i), g)| + |(R_{l_m}^{(n(k))} f_i, g - f_j)| \\ &\leqq (\|f - f_i\| \|g\| + \|f_i\| \|g - f_j\|)/\delta_m \end{aligned}$$

we can make the last member small by choosing f_i and f_j sufficiently close to f and to g respectively. If we fix i and j we see that the sequence $\{(R_{l_m}^{(n(k))} f, g)\}$ oscillates in a neighborhood of the value A_{ijm} when k becomes infinite. Since, with a suitable choice of i and j, we can render this neighborhood as small as may be desired, the sequence must converge.

If Δ_0 is a bounded closed subset of Δ and if θ is an arbitrary number in the range $0 < \theta < 1$, then the distance from a point l in Δ_0 to the spectrum of $H^{(n)}$ exceeds $\theta\, d(l)$ for sufficiently large values of n, where $d(l)$ is the distance from l to the boundary of Δ: for if the spectra of infinitely many of the transformations $H^{(n)}$ had points in common with the circle with center at l and radius $\theta\, d(l)$, then Δ would have a boundary point in the circle, in contradiction with the fact that the boundary of Δ is at distance $d(l)$ from the point l. When l is restricted to the bounded closed set Δ_0, the distance $d(l)$ assumes a positive minimum d_0. Hence we can write, by the inequality given above,

$$\|R_l^{(n)} f\| \leqq \|f\|/\theta\, d(l) \leqq \|f\|/\theta\, d_0$$

for all sufficiently large values of n when l is in Δ_0. Now we can select a finite subset of the sequence $\{l_m\}$ with the properties: every point of the subset is within distance $\frac{1}{2} d_0$ of Δ_0; if l is an arbitrary point of Δ_0 there exists a point of the subset within distance ε of l, where ε is a given positive number. The distance from an arbitrary point l_m of this subset to the spectrum of $H^{(n)}$ exceeds $\frac{1}{2}\theta\, d_0$ for all sufficiently large values of n; and for such values of n, the inequality

$$\|R_{l_m}^{(n)} f\| \leqq 2\|f\|/\theta\, d_0$$

is satisfied. If l is an arbitrary point in Δ_0 and l_m is a point of the chosen finite subset of the sequence $\{l_m\}$ such that $|l-l_m|<\varepsilon$, we have

$$\begin{aligned} &|(R_l^{(n(p))} f, g)-(R_l^{(n(q))} f, g)| \\ &\qquad \leqq |((R_l^{(n(p))}-R_{l_m}^{(n(p))})f, g)| + |(R_{l_m}^{(n(p))} f, g)-(R_{l_m}^{(n(q))} f, g)| \\ &\qquad\qquad + |((R_{l_m}^{(n(q))}-R_l^{(n(q))})f, g)|. \end{aligned}$$

The first and last terms of the expression on the right can be modified and appraised on the basis of the functional equation for the resolvent noted in Theorem 4.19; the first term, for instance, satisfies the relations

$$\begin{aligned} &|(R_l^{(n(p))}-R_{l_m}^{(n(p))})f, g)| \\ &\leqq |l-l_m|\,|(R_l^{(n(p))} R_{l_m}^{(n(p))} f, g)| \leqq 2\varepsilon\|f\|\|g\|/\theta^2 d_0^2, \end{aligned}$$

the inequality being a consequence of those stated above. We obtain, therefore, the result

$$\begin{aligned} &|(R_l^{(n(p))} f, g)-(R_l^{(n(q))} f, g)| \\ &\leqq 4\varepsilon\|f\|\|g\|/\theta^2 d_0^2 + |(R_{l_m}^{(n(p))} f, g)-(R_{l_m}^{(n(q))} f, g)|. \end{aligned}$$

When ε and θ have been chosen, the last term can be made uniformly small for the finite number of admitted values l_m by taking p and q sufficiently large. Thus, by arranging these preliminary selections suitably, the entire right-hand member is made small independent of l for all sufficiently large values of p and q. This result establishes the uniform convergence of the sequence $\{(R_l^{(n(k))} f, g)\}$ in the set Δ_0. The limit function $I(f, g; l)$ is defined for all f and g in $\mathfrak{H}$ and all l in the open set Δ; for fixed l, it is a bilinear function of f and g, and, for fixed f and g, it is analytic in l. We note that it satisfies the inequality

$$|I(f, g; l)| \leqq |f||g|/\theta d(l)$$

satisfied by each member of the sequence defining it. The parameter θ can be allowed to tend to 1, with the result that the inequality

$$|I(f, g; l)| \leqq |f||g|/d(l)$$

is found to hold for every f and g in $\mathfrak{H}$ and every l in Δ.

The remaining assertions of the theorem now follow without any difficulty. By the familiar argument based on Theorem 2.28 we construct the transformation X_l satisfying the equation

$$(X_l f, g) = I(f, g; l)$$

and verify that it is a bounded linear transformation with domain $\mathfrak{H}$ for which the inequality

$$|X_l f| \leqq |f|/d(l)$$

holds. From Theorem 5.7 we know that $(R_l^{(n)} f, g)$ admits an integral representation of the form $\int_{-\infty}^{+\infty} \frac{1}{\lambda - l} d(E^{(n)}(\lambda) f, g)$, where $(E^{(n)}(\lambda) f, f)$ is a real, monotone-increasing function of λ; by applying the reasoning used in the proof of Lemma 5.3 we find that $(X_l f, g)$ has an integral representation of the form indicated above and that $\varrho(f, f; \lambda)$ is real and monotone-increasing. The further stated properties of $\varrho(f, g; \lambda)$ can be established by reasoning parallel to that used in the demonstration of Theorem 5.5.

Before attempting to analyze the nature of the transformation X_l in greater detail, we shall prove a preliminary result concerning an important type of equation.

THEOREM 9.18. *If H is a closed linear symmetric transformation, H^* its adjoint, and $\mathfrak{C}^0, \mathfrak{C}^+, \mathfrak{C}^-$ the related sets of elements defined in § 1, then the set of all elements φ satisfying the relation $H^*\varphi - l\varphi = f$, $\mathfrak{I}(l) \neq 0$, has the following properties:*

(1) *for $\psi = 2i\,\mathfrak{I}(l)\,\varphi + f$ the relation*

$$|f|^2 - |\psi|^2 = -4\,\mathfrak{I}(l)\,\mathfrak{I}(H^*\varphi, \varphi)$$

subsists;

(2) *when $\mathfrak{I}(l) > 0$ ($\mathfrak{I}(l) < 0$) the element φ is in $\mathfrak{C}^-, \mathfrak{C}^0, \mathfrak{C}^+$ ($\mathfrak{C}^+, \mathfrak{C}^0, \mathfrak{C}^-$) according as $|f|^2 - |\psi|^2$ is positive, zero, or negative;*

(3) *there exists a unique solution φ_0 with the property that $\varphi \neq \varphi_0$ implies $|\psi| > |\psi_0|$;*

(4) *when, for $\mathfrak{I}(l) > 0$ ($\mathfrak{I}(l) < 0$), the element φ ranges over the set of all solutions in $\mathfrak{C}^- + \mathfrak{C}^0$ (in $\mathfrak{C}^+ + \mathfrak{C}^0$) the complex number (φ, g) ranges over the interior and circumference of the circle with center (φ_0, g) and radius $|g_0|\sqrt{|f|^2 - |\psi_0|^2}/2|\mathfrak{I}(l)|$, where g_0 is the projection of g on the closed linear manifold $\mathfrak{N}(l)$ constituted by the totality of solutions of the homogeneous equation $H^*\chi - l\chi = 0$; in particular, a necessary and sufficient condition that (φ, g) lie on the circumference of this circle is that φ belong to $\mathfrak{C}_0$ and that $\varphi - \varphi_0$ and g_0 be linearly dependent.*

From the definitions of $\mathfrak{C}^+, \mathfrak{C}^0, \mathfrak{C}^-$, we see that (2) follows immediately from (1); and (1) can be verified directly on the basis of the equation $H^*\varphi = l\varphi + f$.

In proving (3), we first show that the equation $H^*\varphi - l\varphi = f$ has at least one solution when $\mathfrak{I}(l) \neq 0$; in other words, that the range of the transformation $H^* - l \cdot I$ is the entire space $\mathfrak{H}$. This range is evidently the same as that of the transformation

$$(aH + bI)^* - iI \equiv aH^* + (b - i)I$$

where

$$a = 1/\mathfrak{I}(l), \qquad b = -\mathfrak{R}(l)/\mathfrak{I}(l).$$

Since $aH + bI$ is a closed linear symmetric transformation, the results of Theorem 9.4 can be used to identify the range in question with $\mathfrak{H}$. Our equation thus has a solution φ_1; a necessary and sufficient condition that an element φ_2 be a solution is that it be congruent (mod $\mathfrak{N}(l)$) to φ_1. We start from an arbitrary solution φ_1 to form the required solution φ_0: using the projections ψ_0 and χ_1 of the element $\psi_1 = 2i\mathfrak{J}(l)\varphi_1 + f$ on $\mathfrak{H} \ominus \mathfrak{N}(l)$ and $\mathfrak{N}(l)$ respectively, we set

$$\varphi_0 = (\psi_0 - f)/2i\mathfrak{J}(l) = \varphi_1 - \chi_1/2i\mathfrak{J}(l).$$

It is evident that φ_0 is a solution. If φ is an arbitrary solution it can be written in the form $\varphi_0 + \chi$ where χ is an element of $\mathfrak{N}(l)$. Thus

$$\psi = 2i\mathfrak{J}(l)\varphi + f = \psi_0 + 2i\mathfrak{J}(l)\chi$$

and

$$\|\psi\|^2 = \|\psi_0\|^2 + 4|\mathfrak{J}(l)|^2 \|\chi\|^2 \geqq \|\psi_0\|^2,$$

the inequality holding unless $\|\chi\| = 0$, $\chi = 0$, $\varphi = \varphi_0$.

It is now comparatively easy to prove the fourth assertion. If φ is an arbitrary solution we write it in the form $\varphi_0 + \chi$ indicated above. The condition that φ lie in $\mathfrak{C}^- + \mathfrak{C}^0$ or $\mathfrak{C}^+ + \mathfrak{C}^0$ according as $\mathfrak{J}(l) > 0$ or $\mathfrak{J}(l) < 0$ is that $\|f\|^2 - \|\psi\|^2$ be positive or zero. Using the evaluation of $\|\psi\|^2$ given in the preceding paragraph, we can write this condition in the equivalent form

$$\|\chi\|^2 \leqq [\|f\|^2 - \|\psi_0\|^2]/4|\mathfrak{J}(l)|^2.$$

The inequality

$$|(\varphi, g) - (\varphi_0, g)| = |(\chi, g)| = |(\chi, g_0)| \leqq \|\chi\| \|g_0\|$$

thus reduces to that demanded by (4). In order to show that (φ, g), in addition to being confined to the circle described in (4), also ranges over the entire circle we can proceed as follows: assuming that the radius does not vanish, we define $\chi_0 = g_0/\|g_0\|$, $\chi = z\chi_0$, $\varphi = \varphi_0 + \chi$ where $|z|$ does not exceed the radius in question; it is evident that

$$(\varphi, g) = (\varphi_0, g) + (\chi, g) = (\varphi_0, g) + z$$

thus describes the circle indicated when z is varied. In order that (φ, g) lie on the circumference of the circle, it is necessary and sufficient that the equality sign hold in the two inequalities above. The result stated in the theorem follows at once.

We return now to the study of the transformation X_l of Theorem 9.17. We employ the notations already introduced.

THEOREM 9.19. *For a fixed value* $l = m$, *the transformation* X_l *has the following properties:*

(1) *the adjoint* X_m^* *exists and is identical with* $X_{\overline{m}}$;

(2) *the element* $X_m f$ *is a solution of the equation* $H^*\varphi - m\varphi = f$ *belonging to* $\mathfrak{C}^- + \mathfrak{C}^0$, $\mathfrak{C}^0$, *or* $\mathfrak{C}^+ + \mathfrak{C}^0$ *according as* $\mathfrak{I}(m) > 0$, $\mathfrak{I}(m) = 0$, *or* $\mathfrak{I}(m) < 0$;

(3) *there exists a closed linear transformation* T, *dependent on* m, *which satisfies the relations* $H \subseteq T \subseteq H^*$ *and* $(T - m\cdot I)^{-1} \equiv X_m$; *when* $\mathfrak{I}(m) \neq 0$, *the spectrum of* T *is confined to that one of the two half-planes* $\mathfrak{I}(l) \geq 0$, $\mathfrak{I}(l) \leq 0$ *which does not contain the point* $l = m$; *and when* $\mathfrak{I}(m) = 0$ *the transformation* T *is self-adjoint.*

If $P_1(\lambda)$ *and* $P_2(\lambda)$ *are polynomials and if* f *and* g *are elements such that* $P_1(H)f$ *and* $P_2(H)g$ *exist, then*

$$(P_1(H)f, P_2(H)g) = \int_{-\infty}^{+\infty} P_1(\lambda)\,\overline{P_2(\lambda)}\,d\varrho(f, g; \lambda)$$

where $\varrho(f, g; \lambda)$ *is the function associated with the transformation* X_l *in the manner described in Theorem* 9.17. *When* f *is an element such that* $H^n f$ *exists, the behavior of the expression* $(X_l f, g)$ *for large values of* $|l|$ *is displayed by the relations*

$$(X_l f, g) = -\sum_{\alpha=0}^{n-1} (H^\alpha f, g)/l^{\alpha+1} + (X_l H^n f, g)/l^n,$$

$$\lim_{|l|\to\infty} (X_l H^n f, g) = 0$$

uniformly in the sectors

$$0 < \varepsilon \leq \arg l \leq \pi - \varepsilon, \quad \pi + \varepsilon \leq \arg l \leq 2\pi - \varepsilon.$$

Since X_m is a bounded linear transformation with domain $\mathfrak{H}$ its adjoint X_m^* exists. We have

$$(X_m^* f, g) = \overline{(X_m g, f)} = \lim_{k\to\infty} \overline{(R_m^{(n(k))} g, f)}$$
$$= \lim_{k\to\infty} (R_{\overline{m}}^{(n(k))} f, g) = (X_{\overline{m}} f, g),$$

and thus conclude that X_m^* and $X_{\overline{m}}$ are identical.

In view of the relations $H \subseteq \tilde{H}^{(0)}$, $H^* \supseteq (\tilde{H}^{(0)})^* \equiv (H^{(0)})^*$, it is sufficient in discussing (2) to prove that $X_m f$ is a solution of the equation $(H^{(0)})^* \varphi - m \varphi = f$. To this end we employ the inequality

$$
\begin{aligned}
|(X_m f, H_{\bar{m}}^{(0)} g) - (f, g)| \\
&\leqq |(X_m f, H_{\bar{m}}^{(0)} g) - (R_m^{(n(k))} f, H_{\bar{m}}^{(0)} g)| \\
&\qquad + |(R_m^{(n(k))} f, H_{\bar{m}}^{(0)} g) - (R_m^{(n(k))} f, H_{\bar{m}}^{(n(k))} g)| \\
&\leqq |(X_m f, H_{\bar{m}}^{(0)} g) - (R_m^{(n(k))} f, H_{\bar{m}}^{(0)} g)| \\
&\qquad + \|f\| \|(H_{\bar{m}}^{(0)} - H_{\bar{m}}^{(n(k))}) g\| / \delta(m).
\end{aligned}
$$

The last expression tends to zero with $1/k$ whenever g is an element of the domain of $H^{(0)}$. Thus the equation $(X_m f, H_{\bar{m}}^{(0)} g) = (f, g)$ must hold; and $X_m f$ belongs to the domain of $(H^{(0)})^*$, satisfying the equation

$$
(H_{\bar{m}}^{(0)})^* X_m f \equiv (H^{(0)})^* X_m f - m X_m f = f.
$$

We must show further that $X_m f$ belongs to the set of elements specified in (2). First, if $\mathfrak{J}(m) = 0$, we have $m = \bar{m}$, $X_m \equiv X_{\bar{m}} \equiv X_m^*$. Consequently

$$
(H^* X_m f, X_m f) = (f + m X_m f, X_m f) = (f, X_m f) + m \|X_m f\|^2
$$

is real and, by definition, $X_m f$ belongs to $\mathfrak{C}^0$. When $\mathfrak{J}(m) \neq 0$, we employ the inequality

$$
\|X_m f\| \leqq \liminf_{k \to \infty} \|R_m^{(n(k))} f\|,
$$

which is an immediate consequence of the relations

$$
\begin{aligned}
\|X_m f\|^2 &= \lim_{k \to \infty} (R_m^{(n(k))} f, X_m f) \\
&\leqq \liminf_{k \to \infty} \|R_m^{(n(k))} f\| \|X_m f\|,
\end{aligned}
$$

in order to establish the inequalities

$$
\begin{aligned}
\mathfrak{J}(m) \mathfrak{J}(H^* X_m f, X_m f) &= |\mathfrak{J}(m)|^2 \|X_m f\|^2 + \mathfrak{J}(m) \mathfrak{J}(f, X_m f) \\
&\leqq \liminf_{k \to \infty} (|\mathfrak{J}(m)|^2 \|R_m^{(n(k))} f\|^2 + \mathfrak{J}(m) \mathfrak{J}(f, R_m^{(n(k))} f)) \\
&\leqq \liminf_{k \to \infty} (\mathfrak{J}(m) \mathfrak{J}(H^{(n(k))} R_m^{(n(k))} f, R_m^{(n(k))} f)) = 0,
\end{aligned}
$$

the last expression vanishing because $H^{(n)}$ is self-adjoint. It follows that $\mathfrak{J}(H^* X_m f, X_m f)$ either vanishes or has sign opposite to that of $\mathfrak{J}(m)$, so that $X_m f$ belongs to the set $\mathfrak{C}^- + \mathfrak{C}^0$ or to the set $\mathfrak{C}^+ + \mathfrak{C}^0$ according as $\mathfrak{J}(m)$ is positive or negative.

Since $X_m f$ satisfies the equation $H^* \varphi - m\varphi = f$, it is evident that $X_m f = 0$ implies $f = 0$. Consequently the inverse X_m^{-1} exists and satisfies the relation $X_m^{-1} \subseteqq H^* - m \cdot I$. The transformation $T \equiv X_m^{-1} + m \cdot I$ is closed and linear, since both X_m and X_m^{-1} have these properties, and evidently satisfies the relation $T \subseteqq H^*$. The range of X_m is everywhere dense in $\mathfrak{H}$, since an element g such that $(X_m f, g) = 0$ for every f has the properties $(f, X_{\bar{m}} g) = 0$, $X_{\bar{m}} g = 0$, $g = 0$. By reference to Theorem 2.7, we now see that X_m^{-1} has an adjoint $(X_m^{-1})^*$. T^* therefore exists and can be computed according to the identities

$$T^* \equiv (X_m^{-1} + m \cdot I)^* \equiv (X_m^{-1})^* + \bar{m} \cdot I \equiv (X_m^*)^{-1} + \bar{m} \cdot I$$
$$\equiv X_{\bar{m}}^{-1} + \bar{m} \cdot I.$$

The relation $T^* \subseteqq H^*$, which has already been established, implies the further relations $T \equiv T^{**} \supseteqq H^{**} \supseteqq H$ which we wished to prove. We may note here that when $\mathfrak{J}(m)$ vanishes T is self-adjoint.

Next we must ascertain some of the major properties of the spectrum of T when $\mathfrak{J}(m) \neq 0$. First we prove that the point spectrum of T lies in that one of the half-planes $\mathfrak{J}(l) \geqq 0$, $\mathfrak{J}(l) \leqq 0$ which does not contain the point $l = m$. If $g = X_m f$ is an element in the domain of T such that $Tg = lg$, $g \neq 0$, we can apply the inequality

$$\mathfrak{J}(m)\mathfrak{J}(l)\|g\|^2 = \mathfrak{J}(m)\mathfrak{J}(H^* X_m f, X_m f) \leqq 0$$

to show that $\mathfrak{J}(l)$ vanishes or has sign opposite to that of $\mathfrak{J}(m)$; the point spectrum of T thus has the location described. The residual spectrum of T has a similar property. If l is in the residual spectrum of T, then $\bar{l}$ is in the point spectrum of T^*, as we proved in Theorem 4.15. Now the point spectrum of $T^* \equiv X_{\bar{m}}^{-1} + \bar{m} \cdot I$ is confined to the half-plane

$\mathfrak{J}(l) \geqq 0$ or $\mathfrak{J}(l) \leqq 0$ which does not contain the point $l = \overline{m}$. It follows that the points of the residual spectrum of T lie in that one of these half-planes which does not contain the point $l = m$. We now see that, if l is any point such that $\mathfrak{J}(m)\mathfrak{J}(l)$ is positive, it belongs either to the continuous spectrum or to the resolvent set of T. In order to show that it is the second alternative which is realized, we discuss first the case in which l satisfies the additional condition $|\mathfrak{J}(l)| \geqq |l-m|$. Under this assumption we prove that the transformation $T_l^{-1} \equiv (T-l\cdot I)^{-1}$, whose existence has been established, is bounded and, in particular, satisfies the inequality

$$\| T_l^{-1} g \| \leqq \| g \| / |\mathfrak{J}(l)|.$$

The restriction on l can then be removed. The inequality to be proved can be expressed in an equivalent form by noting that, since T_l takes $X_m f$ into $f + (m-l) X_m f$ by definition, we can put

$$g = f + (m-l) X_m f, \qquad T_l^{-1} g = X_m f,$$

and can therefore write the inequality as

$$\| X_m f \| \leqq \| f + (m-l) X_m f \| / |\mathfrak{J}(l)|$$

This relation is equivalent to

$$0 \leqq \{|\mathfrak{J}(l)|^2 - |l-m|^2\} \| X_m f \|^2 \leqq \| f \|^2 + 2\mathfrak{R}\{(m-l)(X_m f, f)\}.$$

The latter inequality can be established by a limiting process. We have

$$\begin{aligned} \| R_m^{(n(k))} f \| &= \| R_l^{(n(k))} (f + (m-l) R_m^{(n(k))} f) \| \\ &\leqq \| f + (m-l) R_m^{(n(k))} f \| / |\mathfrak{J}(l)| \end{aligned}$$

so that the inequality

$$\begin{aligned} 0 &\leqq \{|\mathfrak{J}(l)|^2 - |l-m|^2\} \| R_m^{(n(k))} f \|^2 \\ &\leqq \| f \|^2 + 2\mathfrak{R}\{(m-l)(R_m^{(n(k))} f, f)\} \end{aligned}$$

is satisfied. If we allow k to become infinite so that the middle term tends to its inferior limit and then replace $\liminf_{k\to\infty} \| R_m^{(n(k))} f \|^2$ by the smaller term $\| X_m f \|^2$, we obtain the desired inequality in the last of the three forms given above. Thus, whenever l satisfies both conditions $\mathfrak{J}(m)\mathfrak{J}(l) > 0$, $|\mathfrak{J}(l)| \geqq |l-m|$, it belongs to the resolvent set of T; and, according to Theorem 4.11, every point at distance less than $|\mathfrak{J}(l)|$ from l also belongs to the resolvent set of T. The locus defined by the inequality $|\mathfrak{J}(l)| \geqq |l-m|$ is readily identified as that portion of the plane containing the point m and bounded by the parabola with focus at m and directrix along the real axis; for points of the locus the inequality $\mathfrak{J}(m)\mathfrak{J}(l) > 0$ is automatically satisfied. If we now select an arbitrary point l such that $\mathfrak{J}(m)\mathfrak{J}(l)$ is positive, we can determine a real number λ with the same sign as $\mathfrak{J}(l)$ such that the point $l+i\lambda$ belongs to the locus described. Since the point l is at distance $|\lambda| \leqq |\mathfrak{J}(l+i\lambda)|$ from the point $l+i\lambda$, it is evident from the remarks made above that l belongs to the resolvent set of T. If we now use the notation of Theorem 4.11, putting $R_l \equiv T_l^{-1}$ and $R_{l+i\lambda} \equiv T_{l+i\lambda}^{-1}$, we can write

$$\begin{aligned} |(R_l f, g)| &= \left| \sum_{\alpha=0}^{\infty} (-i\lambda)^{\alpha} (R_{l+i\lambda}^{\alpha+1} f, g) \right| \\ &\leqq |f||g| \sum_{\alpha=0}^{\infty} |\lambda^{\alpha}| / |\mathfrak{J}(l+i\lambda)|^{\alpha+1} \\ &\leqq \frac{|f||g|}{|\mathfrak{J}(l+i\lambda)| - |\lambda|} = |f||g| / |\mathfrak{J}(l)| \end{aligned}$$

by virtue of the inequality established above. This result implies that the inequality

$$|T_l^{-1} g| = |R_l g| \leqq |g| / |\mathfrak{J}(l)|$$

holds for every l such that $\mathfrak{J}(m)\mathfrak{J}(l)$ is positive.

We come finally to the examination of the properties of the expressions $(X_l f, g)$ and $\varrho(f, g; \lambda)$ when f and g are appropriately restricted. We shall base our deductions upon the equations

$$\text{(A)} \qquad \begin{aligned} \varrho(H^m f, g; \lambda) &= \int_{-\infty}^{\lambda} \mu^m \, d\varrho(f, g; \mu), \\ (H^m f, g) &= \int_{-\infty}^{+\infty} \mu^m \, d\varrho(f, g; \mu), \end{aligned}$$

where g is arbitrary and f is an element such that $H^m f$ exists. We shall establish them by means of mathematical induction.

When $m = 0$ the first of these relations is trivial while the second requires proof. We begin by showing that the expression $-l(X_l f, g)$ tends to the limit (f, g) when $|l|$ becomes infinite in such a manner that $|l|/|\mathfrak{J}(l)|$ remains bounded. When g is an element in the domain of H we can write

$$(X_l f, Hg) = (H^* X_l f, g) = (f, g) + l(X_l f, g)$$

where the term on the left tends to zero with $1/|\mathfrak{J}(l)|$ because of the inequality

$$|(X_l f, Hg)| \leqq \|f\| \|Hg\|/|\mathfrak{J}(l)|.$$

If g is arbitrary we can select g_0 in the domain of H so close to g that the inequalities

$$|(f, g) - (f, g_0)| \leqq \|f\| \|g - g_0\| < \varepsilon/2,$$
$$|-l(X_l f, g) + l(X_l f, g_0)| = |l|\,|(X_l f, g - g_0)|$$
$$\leqq |l|\, \|f\| \|g - g_0\|/|\mathfrak{J}(l)| < \varepsilon/2$$

are satisfied for fixed positive ε and arbitrary l subject to the restriction noted above. Combining these inequalities and allowing l to become infinite in the prescribed manner, we deduce the inequality

$$\limsup_{|l|\to\infty} |-l(X_l f, g) - (f, g)| \leqq \varepsilon.$$

The desired limiting relation follows at once. On the other hand, we can evaluate $\lim_{|l|\to\infty} - l(X_l f, g)$ by writing

$$-l(X_l f, g) = \int_{-\infty}^{+\infty} \frac{l\, d\varrho(f, g; \mu)}{l - \mu}$$
$$= \int_{-\infty}^{+\infty} d\varrho(f, g; \mu) + \int_{-\infty}^{+\infty} \frac{\mu}{\mu - l} d\varrho(f, g; \mu)$$

and allowing l to become infinite through pure imaginary values, for the second integral on the right then tends to zero by virtue of the fact that its integrand tends boundedly to zero. We equate the two different expressions for the limit of $-l(X_l f, g)$ and thereby obtain the second of equations (A) for $m = 0$.

Next, we assume that the equations (A) hold for $m = 0, \cdots, n-1$, and then prove that they also hold for $m = n$. We use the properties of the transformation X_l to write

$$\begin{aligned}(X_l H^n f, g) &= (H^{n-1} f + l X_l H^{n-1} f, g)\\ &= (H^{n-1} f, g) + l(X_l H^{n-1} f, g)\end{aligned}$$

and then employ the equations (A) for $m = n-1$ to rewrite this equation in the form

$$\begin{aligned}\int_{-\infty}^{+\infty} \frac{d\varrho_n}{\mu - l} &= \int_{-\infty}^{+\infty} d\varrho_{n-1} + \int_{-\infty}^{+\infty} \frac{l}{\mu - l}\, d\varrho_{n-1}\\ &= \int_{-\infty}^{+\infty} \frac{\mu}{\mu - l}\, d\varrho_{n-1}\end{aligned}$$

where we have denoted $\varrho(H^m f, g; \mu)$ by the symbol ϱ_m for $m = 0, \cdots, n$. In order to take advantage of the equality between these integrals, we turn naturally to the range of ideas developed in the proof of Lemma 5.2. We cannot apply the lemma directly to the last integral on the right but are able to circumvent this difficulty. We first select real numbers λ and ν, $\lambda > \nu$, assuming for convenience that the points $\mu = \lambda$ and $\mu = \nu$ are points of continuity for ϱ_{n-1} and ϱ_n. We introduce the contour $C(\lambda, \nu; \alpha, \varepsilon)$ described in the lemma cited above, noting that its length does not exceed $2(\lambda - \nu) + 4\alpha$. Next we choose a negative number A and a positive number B, satisfying the inequalities $A < \nu$ and $B > \lambda$, in such a manner that the inequality

$$\left| \int_{-\infty}^{A} \frac{\mu}{\mu - l}\, d\varrho_{n-1} + \int_{B}^{+\infty} \frac{\mu}{\mu - l}\, d\varrho_{n-1} \right| < \eta$$

holds for given positive η and arbitrarily l on the contour $C(\lambda, \nu; \alpha, \varepsilon)$. Such a choice is possible since there exists a positive constant $K(\delta)$, independent of ε, with the property that $\left|\frac{\mu}{\mu - l}\right| \leqq K(\delta)$ for every μ outside the open interval $(\nu - \delta, \lambda + \delta)$, $\delta > 0$, and for every l on the contour $C(\lambda, \nu; \alpha, \varepsilon)$, and since the variation of ϱ_{n-1} over the combined intervals $-\infty < \mu \leqq A$ and $B \leqq \mu < +\infty$ tends to zero when $|A|$ and B become infinite. By virtue of these various facts we can assert the truth of the inequality

$$\left|\frac{1}{2\pi i}\int_{C(\lambda, \nu;\, \alpha, \varepsilon)} \left\{\int_{-\infty}^{+\infty} \frac{d\varrho_n}{\mu - l} - \int_A^B \frac{\mu\, d\varrho_{n-1}}{\mu - l}\right\} dl\right| \leqq \eta[(\lambda - \nu) + 2\alpha]/\pi.$$

If we define a function σ_n by the equations

$$\begin{aligned} \sigma_n(\mu) &= 0 \qquad , && -\infty < \mu < A, \\ \sigma_n(\mu) &= \int_A^\mu \mu\, d\varrho_{n-1}, && A \leqq \mu < B, \\ \sigma_n(\mu) &= \int_A^B \mu\, d\varrho_{n-1}, && B \leqq \mu < +\infty, \end{aligned}$$

we see that σ_n is a function of bounded variation in the usual normal form with the property that

$$\int_A^B \frac{\mu\, d\varrho_{n-1}}{\mu - l} = \int_{-\infty}^{+\infty} \frac{d\sigma_n}{\mu - l};$$

the points $\mu = \lambda$ and $\mu = \nu$ are points of continuity of σ_n since by hypothesis they are points of continuity of ϱ_{n-1}. Introducing the new function σ_n, we allow ε to tend to the limit zero in the inequality just established; by reference to Lemma 5.2 and to the hypothesis concerning λ and ν, we see that the result is the inequality

$$|(\varrho_n(\lambda) - \varrho_n(\nu)) - (\sigma_n(\lambda) - \sigma_n(\nu))| \leqq \eta[(\lambda - \nu) + 2\alpha]/\pi.$$

Here we can allow η to tend to zero so as to established the equation

$$\varrho_n(\lambda) - \varrho_n(\nu) = \sigma_n(\lambda) - \sigma_n(\nu) = \int_\nu^\lambda \mu\, d\varrho_{n-1}.$$

In the last integral we replace ϱ_{n-1} by its value

$$\int_{-\infty}^{\mu} \mu^{n-1}\, d\varrho(f, g; \mu)$$

and can therefore write

$$\varrho_n(\lambda) - \varrho_n(\nu) = \int_\nu^\lambda \mu^n \, d\varrho(f, g; \mu).$$

When ν becomes negatively infinite, $\varrho_n(\nu)$ tends to zero and this equation reduces to the first of equations (A) for $m = n$, namely

$$\varrho_n(\lambda) = \varrho(H^n f, g; \lambda) = \int_{-\infty}^\lambda \mu^n \, d\varrho(f, g; \mu).$$

The hypothesis that λ is a point of continuity of ϱ_{n-1} and of ϱ_n can now be circumvented. The equation just established holds save for a set of values of λ at most denumerably infinite; but it is an equation connecting two functions of bounded variation in the usual normal form and must therefore hold for every value of λ. The second of equations (A) for $m = n$ now follows without any difficulty: in the equation

$$(f, g) = \int_{-\infty}^{+\infty} d\varrho(f, g; \mu) = \lim_{\mu \to +\infty} \varrho(f, g; \mu),$$

we replace f by $H^m f$, obtaining

$$(H^m f, g) = \lim_{\mu \to +\infty} \varrho(H^m f, g; \mu) = \int_{-\infty}^{+\infty} \mu^m \, d\varrho(f, g; \mu),$$

as we wished to show.

The facts established in the two preceding paragraphs enable us to apply the principle of mathematical induction and to assert the truth of the equations (A) for $m = 0, 1, 2, \cdots$.

The application of these relations to the derivation of the results stated in the theorem is now quite simple. The assertion for general polynomials $P_1(\lambda)$ and $P_2(\lambda)$ is an obvious consequence of the results for the special case $P_1(\lambda) \equiv \lambda^m$ and $P_2(\lambda) \equiv \lambda^n$, so that we confine our attention to this case. We have

$$(H^m f, H^n g) = \int_{-\infty}^{+\infty} \mu^m \, d\varrho(f, H^n g; \mu),$$

$$\varrho(f, H^n g; \mu) = \overline{\varrho(H^n g, f; \mu)} = \int_{-\infty}^{\mu} \mu^n \, d\overline{\varrho(g, f; \mu)}$$

so that the desired equations

$$\begin{aligned}(H^m f, H^n g) &= \lim_{A\to-\infty, B\to+\infty} \int_A^B \mu^m \, d\varrho(f, H^n g; \mu) \\ &= \lim_{A\to-\infty, B\to+\infty} \int_A^B \mu^{m+n} \, d\varrho(f, g; \mu) \\ &= \int_{-\infty}^{+\infty} \mu^{m+n} \, d\varrho(f, g; \mu)\end{aligned}$$

follow immediately.

The behavior of the expression $(X_l f, g)$ for large values of $|l|$ is easily ascertained as follows: when $H^n f$ exists we write

$$\begin{aligned}(X_l f, g) &= \int_{-\infty}^{+\infty} \frac{d\varrho(f, g; \lambda)}{\lambda - l} \\ &= \int_{-\infty}^{+\infty} \left(-\sum_{\alpha=0}^{n-1} \lambda^\alpha / l^{\alpha+1} + \frac{\lambda^n / l^n}{\lambda - l} \right) d\varrho(f, g; \lambda)\end{aligned}$$

and then apply equations (A) to express the last term in the form

$$\begin{aligned}&-\sum_{\alpha=0}^{n-1} (H^\alpha f, g) / l^{\alpha+1} \\ &\qquad + \int_{-\infty}^{+\infty} \frac{1}{\lambda - l} \, d\left(\int_{-\infty}^{\lambda} \mu^n \, d\varrho(f, g; \mu) \right) \Big/ l^n \\ &\quad = -\sum_{\alpha=0}^{n-1} (H^\alpha f, g) / l^{\alpha+1} + (X_l H^n f, g) / l^n.\end{aligned}$$

We obtain further information about the last term by writing down the inequality

$$|(X_l H^n f, g)| \leqq \| H^n f \| \| g \| / |\mathfrak{J}(l)|$$

and noting that on the sectors $\varepsilon \leqq \arg l \leqq \pi - \varepsilon$, $\pi + \varepsilon \leqq \arg l \leqq 2\pi - \varepsilon$ the inequality

$$|\mathfrak{J}(l)| \geqq |l| \sin \varepsilon$$

is satisfied: combining these inequalities we have

$$|(X_l H^n f, g)| \leqq \| H^n f \| \| g \| / |l| \sin \varepsilon$$

and infer that $(X_l H^n f, g)$ tends uniformly to zero when l becomes infinite in the sectors described.

In certain cases the general results described in Theorems 9.17 and 9.19 assume a somewhat simpler form, as we shall show at once.

THEOREM 9.20. *If* $\{H^{(n)}\}$ *is a sequence of self-adjoint transformations approximating a maximal symmetric transformation* H, *then the sequence of integers* $\{n(k)\}$ *of Theorem* 9.17 *can be defined by the equations* $n(k) = k$, $k = 1, 2, 3, \cdots$, *and the transformation* X_l *is identical with* H_l^{-1} *or with* $(H^* - l \cdot I)^{-1}$ *for* $\mathfrak{J}(l) \neq 0$ *according as* l *is in the point spectrum of* H^* *or not. The transformation* $T \equiv X_m^{-1} + m \cdot I$ *coincides with* H *or with* H^* *according as* m, $\mathfrak{J}(m) \neq 0$, *belongs to the point spectrum of* H^* *or not. In particular, if* H *is self-adjoint and* R_l *is its resolvent, then* $X_l \equiv R_l$ *and* $T \equiv H$.

If H is a maximal symmetric transformation, then for a point m in one of the two half-planes $\mathfrak{J}(l) > 0$, $\mathfrak{J}(l) < 0$ the range of the transformation $H_m \equiv H - m \cdot I$ is the entire Hilbert space $\mathfrak{H}$; such a point m is a characteristic value for the adjoint H^* when H is not self-adjoint. If we form the transformation $T \equiv X_m^{-1} + m \cdot I$, discussed in Theorem 9.19, we know that $T - m \cdot I$ is an extension of $H - m \cdot I$; since its range coincides with that of $H - m \cdot I$ and since the equation $Tf = mf$ has no solution other than $f = 0$, we see that $T - m \cdot I$ must coincide with $H - m \cdot I$, T with H. It follows that H_m^{-1} exists and coincides with X_m, for every m in the half-plane indicated. To pass to the consideration of points in the second half-plane, we have merely to replace m by its conjugate $\bar{m}$: $X_{\bar{m}}$ is then found to be determined by the relations

$$X_{\bar{m}} \equiv X_m^* \equiv (H_m^{-1})^* \equiv (H^* - \bar{m} \cdot I)^{-1}$$

and the related transformation $T \equiv X_{\bar{m}}^{-1} + \bar{m} \cdot I$ reduces to H^*.

In order to show that in Theorem 9.17 we can put $n(k) = k$ we have only to observe that the sequences $\{A_{ijm}^{(n)}\}$ constructed in the proof of that theorem can now be chosen so as to converge when n becomes infinite through integral values. On the basis of the assertion that the sequence $\{(R_l^{(n)} f, g)\}$ converges for every possible choice of f, g, and l, $\mathfrak{J}(l) \neq 0$, the desired selection can be effected without difficulty. In proving this assertion it is sufficient for us to confine our attention to values of l situated in just one of the two half-planes $\mathfrak{J}(l)>0$, $\mathfrak{J}(l)<0$: for we can always pass from one to the other by the equation

$$(R_l^{(n)} f, g) = \overline{(R_{\bar{l}}^{(n)} g, f)}.$$

Furthermore, it is sufficient for us to consider elements f and g picked from a sequence $\{g_i\}$ everywhere dense in $\mathfrak{H}$ and not necessarily independent of the number l: the situation for arbitrary f and g can be treated along the lines of the argument used in proving Theorem 9.17. In view of these facts, we shall select l as an arbitrary point such that the range of $H - l \cdot I$ is the entire space $\mathfrak{H}$; remembering that our hypotheses require that $\tilde{H}^{(0)}$ and H be identical, we see that we can select a sequence $\{f_i\}$ in the domain of $H^{(0)} \subseteqq H$ in such wise that the sequence of elements $g_i = H^{(0)} f_i - l f$ is everywhere dense in $\mathfrak{H}$. We now write $(R_l^{(n)} g_i, g_j)$ in the form

$$(R_l^{(n)} H_l^{(n)} f_i, g_j) + (R_l^{(n)} (H_l^{(0)} - H_l^{(n)}) f_i, g_j)$$
$$= (f_i, g_j) + (R_l^{(n)} (H^{(0)} - H^{(n)}) f_i, g_j).$$

The last term satisfies the inequality

$$|(R_l^{(n)} (H^{(0)} - H^{(n)}) f_i, g_j)| \leqq \|(H^{(0)} - H^{(n)}) f_i\| \, \|g_j\| / |\mathfrak{J}(l)|$$

and therefore tends to zero with $1/n$. Thus we have proved that the sequence $\{(R_l^{(n)} g_i, g_j)\}$ is convergent and have brought our demonstration to a close.

We shall continue these investigations with a few remarks concerning the case in which the set Δ of Theorem 9.17 contains points on the real axis. Since Δ is an open set in the l-plane, its intersection with the real axis is either an empty set or an open set on the axis. Thus if Δ contains a point on the real axis it contains an open interval including that point.

THEOREM 9.21. *If the closed linear symmetric transformation H has an approximating sequence of self-adjoint transformations $\{H^{(n)}\}$ such that the set Δ contains an open interval of the* *real axis in the l-plane, then H has the deficiency-index* (m, m) *and possesses a self-adjoint extension whose resolvent set includes the interval in question. In particular, if H is a definite symmetric transformation satisfying an inequality of the form* $(Hf, f) \geqq C\|f\|^2$ *or* $(Hf, f) \leqq C\|f\|^2$, *then H has the deficiency-index* (m, m) *and possesses a self-adjoint extension satisfying the same inequality.*

Let Λ contain the interval $\mu - d < \lambda < \mu + d$ of the real axis. Then the transformation $T \equiv X_\mu^{-1} + \mu \cdot I$ exists and is a self-adjoint extension of H by Theorem 9.19. It is immediately clear that H must have the deficiency-index (m, m). The transformation X_μ is bounded and self-adjoint, satisfying the inequality $\|X_\mu g\| \leqq \|g\|/d$, as we showed in Theorems 9.17 and 9.19. Its spectrum is therefore confined to the interval $-d \leqq \lambda \leqq d$ of the real axis in accordance with Theorem 5.12. Its inverse is a self-adjoint transformation whose spectrum must lie outside the interval $-d < \lambda < d$, as can be verified by reference to the discussion of inverses given in Theorem 5.11. The spectrum of $T \supseteq H$ is therefore outside of the interval $\mu - d < \lambda < \mu + d$, as we wished to prove.

Next let Λ contain the interval $-\infty < \lambda < C$ or the interval $C < \lambda < +\infty$. By simple transformations we can reduce both cases to the particular case in which Λ contains the interval $-\infty < \lambda < 0$. When Λ contains the interval $-\infty < \lambda < C$ we replace H and $H^{(n)}$ by $H - C \cdot I$ and $H^{(n)} - C \cdot I$ respectively and find that the set Λ associated with the sequence $\{H^{(n)} - C \cdot I\}$ contains the interval $-\infty < \lambda < 0$; and when Λ contains the interval $C < \lambda < +\infty$ we replace H and $H^{(n)}$ by $-H + C \cdot I$ and $-H^{(n)} + C \cdot I$ respectively, with a similar result. If $\{H^{(n)}\}$ is a sequence of self-adjoint transformations approximating H in such wise that Λ contains the interval $-\infty < \lambda < 0$, then for sufficiently large n the spectrum of $H^{(n)}$ lies outside an open interval $-\mu_n < \lambda < -\nu_n$, where the positive numbers μ_n and ν_n tend to $+\infty$ and to 0 respectively when n becomes infinite. We choose an arbitrary positive number μ and consider values of n sufficiently large that the inequality $\nu_n < \mu < \mu_n$ is satisfied. By reference to

Theorems 5.11 and 5.12 we verify that the self-adjoint transformation $R^{(n)}_{-\mu} \equiv (H^{(n)} + \mu \cdot I)^{-1}$ has its spectrum located on the interval

$$\frac{1}{\mu - \mu_n} \leqq \lambda \leqq \frac{1}{\mu - \nu_n}$$

and therefore satisfies the inequality

$$\frac{1}{\mu - \mu_n} \| f \|^2 \leqq (R^{(n)}_{-\mu} f, f) \leqq \frac{1}{\mu - \nu_n} \| f \|^2.$$

In the latter relation we allow n to become infinite through values in the sequence $\{n(k)\}$ of Theorem 9.17, thus obtaining the inequality

$$0 \leqq (X_{-\mu} f, f) \leqq \frac{1}{\mu} \| f \|^2,$$

which implies that the spectrum of the self-adjoint transformation $X_{-\mu}$ lies on the interval $0 \leqq \lambda \leqq \frac{1}{\mu}$. Finally, we conclude that the self-adjoint transformation

$$T \equiv X^{-1}_{-\mu} - \mu \cdot I \supseteqq H$$

has its spectrum confined to the interval $0 \leqq \lambda < +\infty$, as we wished to prove. H must have the deficiency-index (m, m) as in the case discussed in the first paragraph.

If H is a given definite symmetric transformation, satisfying an inequality of the form $(Hf, f) \geqq C \| f \|^2$ or of the form $(Hf, f) \leqq C \| f \|^2$, we can study its properties by relating it in the manner indicated in the preceding paragraph to a not-negative definite transformation: for $H - C \cdot I$ or $-H + C \cdot I$ is a not-negative definite transformation according as the first or second inequality holds. Now if we consider the particular case of a not-negative definite transformation H, for which the inequality $(Hf, f) \geqq 0$ holds, we know from Theorem 9.16 that there exists an approximating sequence of not-negative definite self-adjoint transformations. By applying the results of the preceding paragraph we can construct a self-adjoint transformation T such that $T \supseteqq H$, $(Tf, f) \geqq 0$. Thus H has the deficiency-index (m, m) and possesses a not-negative definite self-adjoint extension.

Finally, we shall analyze more carefully the relation between the family of transformations $B(\lambda)$, introduced in Theorem 9.17, and the sequence $\{E^{(n(k))}(\lambda)\}$, where $E^{(n)}(\lambda)$ is the resolution of the identity corresponding to $H^{(n)}$ and $\{n(k)\}$ is the sequence described in that theorem. By confining our attention to the subsequence $\{H^{(n(k))}\}$ and renumbering it in a suitable manner, we can state our result as follows:

THEOREM 9.22. *Let $\{H^{(n)}\}$ be a sequence of self-adjoint transformations approximating a closed linear symmetric transformation H, and let*

$$\lim_{n\to\infty}(R_l^{(n)}f, g) = \int_{-\infty}^{+\infty}\frac{1}{\lambda-l}\,d(B(\lambda)f, g), \qquad \mathfrak{I}(l)\neq 0,$$

where $R_l^{(n)}$ is the resolvent of $H^{(n)}$ and $B(\lambda)$ is the family of transformations described in Theorem 9.17. Then

$$\lim_{n\to\infty}(E^{(n)}(\lambda)f, g) = (B(\lambda)f, g)$$

except possibly on a finite or denumerably infinite set of values λ, independent of the elements f and g, where $E^{(n)}(\lambda)$ is the resolution of the identity corresponding to $H^{(n)}$ The inequality

$$|(B(\lambda)-B(\mu))f|^2 \leqq ((B(\lambda)-B(\mu))f, f)$$

is valid for $\lambda \geqq \mu$ and for all elements f in $\mathfrak{H}$. The transformations $B(\lambda-\varepsilon)$, $B(\lambda+\varepsilon)$ tend to limit transformations $B(\lambda-0)$, $B(\lambda+0)$ respectively when ε tends to zero through positive values; and the transformation $B(\lambda)$ tends to a limit transformation $B(-\infty)$ or $B(+\infty)$ when λ tends to $-\infty$ or to $+\infty$ respectively. The identities $B(\lambda+0)\equiv B(\lambda)$, $B(-\infty)\equiv O$, $B(+\infty)\equiv I$ are valid.

We first consider the relation $\lim_{n\to\infty}(E^{(n)}(\lambda)f,f) = (B(\lambda)f,f)$ for a fixed element f in $\mathfrak{H}$. If $\{n(k)\}$ is an arbitrary sequence such that $n(k)\to\infty$, $\lim_{k\to\infty}(E^{(n(k))}(\lambda)f,f) = \varrho(f,f;\lambda)$ when $k\to\infty$, we can reason as we did in the proof of Lemma 5.3 to show that

$$\begin{aligned}\lim_{k\to\infty} &(R_l^{(n(k))} f, f) \\ &= \lim_{k\to\infty} \int_{-\infty}^{+\infty} \frac{1}{\lambda - l}\, d\,(E^{(n(k))}(\lambda) f, f) \\ &= \int_{-\infty}^{+\infty} \frac{1}{\lambda - l}\, d\,\varrho(f, f;\ \lambda) \\ &= \int_{-\infty}^{+\infty} \frac{1}{\lambda - l}\, d\,(B(\lambda) f, f).\end{aligned}$$

Applying Lemma 5.2, we see that $\varrho^*(f, f;\ \lambda) = (B(\lambda) f, f)$. Now, since $(E^{(n)}(\lambda) f, f)$ is a real monotone-increasing function tending to zero at $\lambda = -\infty$ and to $(f, f) = \|f\|^2$ at $\lambda = +\infty$, the limit function $\varrho(f, f;\ \lambda)$ is a real monotone-increasing function with all its values included in the range $0 \leqq \varrho \leqq \|f\|^2$. On the other hand, we know that $(B(\lambda) f, f)$ is a real monotone-increasing function in normal form, tending to zero at $\lambda = -\infty$; and by the results established in Theorem 9.19 we find that

$$\lim_{\lambda\to+\infty} (B(\lambda) f, f) = \int_{-\infty}^{+\infty} d\,(B(\lambda) f, f) = (f, f) = \|f\|^2.$$

Consequently, we must have

$$\begin{aligned}&\varrho(f, f;\ +\infty) - \varrho(f, f;\ -\infty) \\ &= \varrho^*(f, f;\ +\infty) - \varrho^*(f, f;\ -\infty) = \|f\|^2.\end{aligned}$$

The monotone character of the function $\varrho(f, f;\ \lambda)$ thus permits us to conclude that $\varrho(f, f;\ +\infty) = \|f\|^2$, $\varrho(f, f;\ -\infty) = 0$. Hence we can write $\varrho^*(f, f;\ \lambda) = \varrho(f, f;\ \lambda + 0)$, $\varrho^*(f, f;\ \lambda - 0) = \varrho(f, f;\ \lambda - 0)$ and can conclude that $\varrho^*(f, f;\ \lambda) = (B(\lambda) f, f)$ and $\varrho(f, f;\ \lambda)$ coincide except possibly at their (common) points of discontinuity. It is now easy to show that $\lim_{n\to\infty} (E^{(n)}(\lambda) f, f)$ exists and is equal to $(B(\lambda) f, f)$ at every point where the latter function is continuous. We first select a sequence of integers $\{n(k)\}$ such that, for an arbitrary fixed value $\lambda = \mu$,

$$\lim_{k\to\infty} (E^{(n(k))}(\mu) f, f) = \limsup_{n\to\infty} (E^{(n)}(\mu) f, f),\ n(k) \to \infty,\ k \to \infty.$$

To the sequence $\{(E^{(n(k))}(\lambda)f, f)\}$ we now apply the theorem of Helly to determine a subsequence which converges for every λ. As we pointed out above, the limit function of this subsequence coincides with $(B(\lambda)f, f)$ except possibly at the points of discontinuity of the latter function; and it assumes the value $\limsup_{n\to\infty}(E^{(n)}(\mu)f, f)$ at $\lambda = \mu$. Since μ can be chosen arbitrarily, it follows that $\limsup_{n\to\infty}(E^{(n)}(\lambda)f, f) = (B(\lambda)f, f)$ except possibly at the points of discontinuity of the latter function. A similar argument shows that the equation $\liminf_{n\to\infty}(E^{(n)}(\lambda)f, f) = (B(\lambda)f, f)$ holds in the same sense. Hence we find that

$$\lim_{n\to\infty}(E^{(n)}(\lambda)f, f) = (B(\lambda)f, f)$$

except possibly at the points of discontinuity of $(B(\lambda)f, f)$.

Next we shall prove that the set of all points of discontinuity of all the functions $(B(\lambda)f, f)$ is at most denumerably infinite. If $\{f_n\}$ is a sequence of elements everywhere dense in $\mathfrak{H}$, the set of all points of discontinuity of all the functions $(B(\lambda)f_n, f_n)$ is at most denumerably infinite. Hence it is sufficient to show that any point of discontinuity of the function $(B(\lambda)f, f)$, where f is arbitrary, must be a point of discontinuity for some of the functions $(B(\lambda)f_n, f_n)$. When f is given, we choose from $\{f_n\}$ a subsequence $\{f_n^*\}$ which converges to f; and we show that $(B(\lambda)f_n^*, f_n^*)$ tends uniformly to $(B(\lambda)f, f)$ when n becomes infinite. The inequalities

$$\begin{aligned} &|(B(\lambda)f, f) - (B(\lambda)f_n^*, f_n^*)| \\ &\qquad \leq |(B(\lambda)(f - f_n^*), f_n^*)| + |(B(\lambda)f, f - f_n^*)| \\ &\qquad \leq \|f - f_n^*\| \|f_n^*\| + \|f\| \|f - f_n^*\| \\ &\qquad \leq 2\|f\| \|f - f_n^*\| + \|f - f_n^*\|^2 \end{aligned}$$

are sufficient for this purpose. It is clear, therefore, that a common point of continuity of the functions $(B(\lambda)f_n^*, f_n^*)$ must be a point of continuity of $(B(\lambda)f, f)$. This result implies the desired characterization of the points of discontinuity of the function $(B(\lambda)f, f)$.

Finally, we write

$$(E^{(n)}(\lambda)f, g)$$
$$= \left[\left(E^{(n)}(\lambda)\frac{f+g}{2}, \frac{f+g}{2}\right) - \left(E^{(n)}(\lambda)\frac{f-g}{2}, \frac{f-g}{2}\right)\right]$$
$$+ i\left[\left(E^{(n)}(\lambda)\frac{f+ig}{2}, \frac{f+ig}{2}\right) - \left(E^{(n)}(\lambda)\frac{f-ig}{2}, \frac{f-ig}{2}\right)\right],$$

$$(B(\lambda)f, g)$$
$$= \left[\left(B(\lambda)\frac{f+g}{2}, \frac{f+g}{2}\right) - \left(B(\lambda)\frac{f-g}{2}, \frac{f-g}{2}\right)\right]$$
$$+ i\left[\left(B(\lambda)\frac{f+ig}{2}, \frac{f+ig}{2}\right) - \left(B(\lambda)\frac{f-ig}{2}, \frac{f-ig}{2}\right)\right],$$

and allow n to become infinite in the first equation. Since the right-hand member then tends to the right-hand member of the second equation, except possibly on a finite or denumerably infinite set of values λ independent of f and g, we see that $\lim_{n\to\infty} (E^{(n)}(\lambda)f, g) = (B(\lambda)f, g)$, in the sense stated in the theorem.

If we allow n to become infinite in the relations

$$|((E^{(n)}(\beta) - E^{(n)}(\alpha))f, (B(\lambda) - B(\mu))f)|$$
$$\leqq \|(E^{(n)}(\beta) - E^{(n)}(\alpha))f\| \|(B(\lambda) - B(\mu))f\|$$
$$= ((E^{(n)}(\beta) - E^{(n)}(\alpha))f, f)^{1/2} \|(B(\lambda) - B(\mu))f\|,$$

where $\beta \geqq \alpha$ and $\lambda \geqq \mu$, we find from the preceding results that

$$|((B(\beta) - B(\alpha))f, (B(\lambda) - B(\mu))f)|$$
$$\leqq ((B(\beta) - B(\alpha))f, f)^{1/2} \|(B(\lambda) - B(\mu))f\|,$$

except possibly on a finite or denumerably infinite set of values of α and β. Since the expressions on both sides of this inequality are functions of α and β continuous on the right in accordance with the definition of $B(\lambda)$, we can readily show that the inequality holds without exception. If we put $\alpha = \mu$ and $\beta = \lambda$, we obtain the desired inequality

$$\|(B(\lambda) - B(\mu))f\|^2 \leqq ((B(\lambda) - B(\mu))f, f), \quad \lambda \geqq \mu.$$

Since the term on the right may be expressed in the form $(B(\lambda)f, f) - (B(\mu)f, f)$, it is clear that both terms tend to zero when λ and μ become positively or negatively infinite. Hence there exist transformations $B(-\infty)$ and $B(+\infty)$ such that $B(\lambda) \rightarrow B(-\infty)$ when $\lambda \rightarrow -\infty$, $B(\lambda) \rightarrow B(+\infty)$ when $\lambda \rightarrow +\infty$, throughout $\mathfrak{H}$. From the relations

$$0 = \lim_{\lambda \rightarrow -\infty} (B(\lambda)f, g) = (B(-\infty)f, g),$$

$$(f, g) = \lim_{\lambda \rightarrow +\infty} (B(\lambda)f, g) = (B(+\infty)f, g),$$

the second of which is taken from Theorem 9.19, it is evident that $B(-\infty) \equiv O$, $B(+\infty) \equiv I$. Similarly, from the inequality

$$\|(B(\lambda - \varepsilon_1) - B(\lambda - \varepsilon_2))f\|^2 \leqq ((B(\lambda - \varepsilon_1) - B(\lambda - \varepsilon_2))f, f)$$

where $0 < \varepsilon_1 \leqq \varepsilon_2$, we see that both terms tend to zero with ε_1 and ε_2. Hence we infer the existence of a transformation $B(\lambda - 0)$ such that $B(\lambda - \varepsilon) \rightarrow B(\lambda - 0)$ throughout $\mathfrak{H}$ when ε tends to zero through positive values. Finally, we establish the existence of a transformation $B(\lambda + 0)$ such that $B(\lambda + \varepsilon) \rightarrow B(\lambda + 0)$ throughout $\mathfrak{H}$ when ε tends to zero through positive values; and by virtue of the relations

$$\lim_{\varepsilon \rightarrow 0} \|(B(\lambda + \varepsilon) - B(\lambda))f\|^2 \leqq \lim_{\varepsilon \rightarrow 0} [(B(\lambda + \varepsilon)f, f) - (B(\lambda)f, f)] = 0$$

we can identify $B(\lambda + 0)$ with $B(\lambda)$.

For subsequent use, we shall establish certain results concerning the range of the transformation $B(\mu) - B(\mu - 0)$. It is convenient to retain the simplifying assumptions of the preceding theorem.

THEOREM 9.23. *If Δ is an arbitrary interval $\alpha \leqq \lambda \leqq \beta$ and if $B(\Delta)$ is the transformation $B(\beta) - B(\alpha)$, then $(B(\Delta)Hf, g) = \int_\alpha^\beta \lambda \, d(B(\lambda)f, g)$ for every element f in the domain of H and every element g in $\mathfrak{H}$. Every element different from 0 in the range of $B(\mu) - B(\mu - 0)$ is a characteristic element of H^* corresponding to the characteristic value μ.*

It is evidently sufficient for us to treat the case where α and β are finite values such that $\lim_{n\to\infty}(E^{(n)}(\lambda)f, g) = (B(\lambda)f, g)$ for $\lambda = \alpha, \beta$ and for all f and g in $\mathfrak{H}$, since the general case can then be handled by simple limiting processes. For such values of α and β and for an element f in the domain of $\tilde{H}^{(0)} \supseteqq H$ which has the property $H^{(n)}f \to \tilde{H}^{(0)}f$, we have

$$\begin{aligned}\lim_{n\to\infty}(E^{(n)}(\Delta)\, H^{(n)}f, g) &= \lim_{n\to\infty}\int_\alpha^\beta \lambda\, d(E^{(n)}(\lambda)f, g)\\ &= \lim_{n\to\infty}\left\{[\lambda(E^{(n)}(\lambda)f, g)]_{\lambda=\alpha}^{\lambda=\beta} - \int_\alpha^\beta (E^{(n)}(\lambda)f, g)\, d\lambda\right\}\\ &= [\lambda(B(\lambda)f, g)]_{\lambda=\alpha}^{\lambda=\beta} - \int_\alpha^\beta (B(\lambda)f, g)\, d\lambda\\ &= \int_\alpha^\beta \lambda\, d(B(\lambda)f, g),\end{aligned}$$

the passage to the limit under the sign of integration being justified by the observation that the integrand tends boundedly to its limit almost everywhere on Δ. From the relations

$$(E^{(n)}(\Delta)H^{(n)}f, g) = (E^{(n)}(\Delta)\tilde{H}^{(0)}f, g) - (E^{(n)}(\Delta)(\tilde{H}^{(0)} - H^{(n)})f, g),$$
$$\lim_{n\to\infty}(E^{(n)}(\Delta)\, \tilde{H}^{(0)}f, g) = (B(\Delta)\, \tilde{H}^{(0)}f, g),$$
$$|(E^{(n)}(\Delta)\, (\tilde{H}^{(0)} - H^{(n)})f, g)| \leqq \|(\tilde{H}^{(0)} - H^{(n)})f\|\, \|g\|,$$
$$H^{(n)}f \to \tilde{H}^{(0)}f,$$

we see at once that $\lim_{n\to\infty}(E^{(n)}(\Delta)\, H^{(n)}f, g) = (B(\Delta)\tilde{H}^{(0)}f, g)$. Thus we have $(B(\Delta)\, \tilde{H}^{(0)}f, g) = \int_\alpha^\beta \lambda\, d(B(\lambda)f, g)$, under the indicated restrictions on α, β, and f. If f is an arbitrary element in the domain of $H \subseteqq \tilde{H}^{(0)}$, we can determine a sequence $\{f_n\}$ which satisfies the restriction noted above and which has the property that $f_n \to f$, $\tilde{H}^{(0)}f_n \to Hf$. Such a choice is possible because of the relation between the sequence $\{H^{(n)}\}$ and the transformation H which is demanded by Definition 9.9. If we allow n to become infinite in the equation $(B(\Delta)\, \tilde{H}^{(0)}f_n, g) = \int_\alpha^\beta \lambda\, d(B(\lambda)f_n, g)$, treating the integral on the right by integration by parts just as in the earlier discussion, we obtain the relation $(B(\Delta)Hf,g)=\int_\alpha^\beta \lambda\, d(B(\lambda)f,g)$, which was to be proved.

We shall now show that, if $h=(B(\mu)-B(\mu-0))\,g$, then $H^*h=\mu h$. Setting $\alpha=\mu-\varepsilon$, $\beta=\mu$, $\varepsilon>0$, in the equation just established, we obtain

$$((B(\mu)-B(\mu-\varepsilon))\,Hf,\,g)=\int_{\mu-\varepsilon}^{\mu}\lambda\,d(B(\lambda)\,f,\,g).$$

We then let ε tend to zero, with the result

$$((B(\mu)-B(\mu-0))\,Hf,\,g)=\mu((B(\mu)-B(\mu-0))\,f,\,g).$$

Since the transformation $B(\mu)-B(\mu-0)$ is self-adjoint, we may write this equation in the form $(Hf,\,h)=(f,\,\mu h)$ and thus conclude that $H^*h=\mu h$, as we wished to do.

CHAPTER X

APPLICATIONS

§ 1. INTEGRAL OPERATORS

As a first application of the general theory developed in the preceding chapters, we shall consider integral operators with kernels of Carleman type, which were defined in Chapter III, § 2. We denote by E an arbitrary Lebesgue-measurable set in n-dimensional Euclidean space, with infinite or positive finite measure, by $\mathfrak{L}_2$ the space of all Lebesgue-measurable functions $f(P)$ defined in E such that the Lebesgue integral $\int_E |f(P)|^2 \, dP$ exists. The Lebesgue-measurable set in $2n$-dimensional Euclidean space consisting of all point-pairs (P, Q) where P and Q are in E, is denoted by E_2; and the associated space of functions $F(P, Q)$ is denoted by $\mathfrak{L}_2^{(2)}$. The kernel $K(P, Q)$ is a Lebesgue-measurable function defined almost everywhere in E_2 with the following properties: $K(P, Q)$ is (Hermitian) symmetric, satisfying the equation $K(Q, P) = \overline{K(P, Q)}$ almost everywhere in E_2; and the integral $\int_E |K(P, Q)|^2 \, dQ$ exists for almost every point P in E.

It is convenient to introduce the real not-negative Lebesgue-measurable function $K(P)$ defined by the relations

$$K(P) = \left(\int_E |K(P, Q)|^2 \, dQ\right)^{1/2} \text{ when the integral exists,}$$
$$K(P) = 0 \text{ elsewhere in } E.$$

The kernels of Hilbert-Schmidt type are characterized by the requirement that $K^2(P)$ be integrable over E. Here we abandon any such integrability assumption. Associated with the function $K(P)$ is the set $\mathfrak{D}$ in $\mathfrak{L}_2$ which consists of those functions $f(P)$ in $\mathfrak{L}_2$ such that the integral $\int_E K(P)|f(P)|\,dP$ exists. We shall consider also the set $\mathfrak{D}^*$ which consists of those functions $f(P)$ in $\mathfrak{L}_2$ such that $f^*(P) = \int_E K(P, Q)\, f(Q)\, dQ$ is also a function in $\mathfrak{L}_2$.

We can now prove a theorem which brings all integral operators with kernels of Carleman type under the general theory of Chapter IX.

THEOREM 10.1. *Let H be the integral operator with kernel of Carleman type, $K(P, Q)$, defined for the domain $\mathfrak{D}$, described above, by the equation*

$$Hf(P) = \int_E K(P, Q)\, f(Q)\, dQ, \qquad f(P) \text{ in } \mathfrak{D};$$

and let T be the integral operator defined in the domain $\mathfrak{D}^*$ by the same equation, valid for $f(P)$ in $\mathfrak{D}^*$. Then H is a linear symmetric transformation in the Hilbert space $\mathfrak{L}_2$; and its adjoint H^* coincides with T.

It is evident directly that $\mathfrak{D}$ is a linear manifold. By the customary argument we show that $\mathfrak{D}$ is everywhere dense in $\mathfrak{L}_2$: if $g(P)$ is an element of $\mathfrak{L}_2$ such that $\int_E f(P)\,\overline{g(P)}\,dP = 0$ for every element $f(P)$ in $\mathfrak{D}$, we must have $g(P) = 0$ almost everywhere in E. The simplest proof consists in specializing the function $f(P)$ as follows: if $E(a)$ is the Lebesgue-measurable set specified by the inequality $K(P) \leqq a$ and if e is an arbitrary Lebesgue-measurable subset of $E(a)$ with finite measure, we put $f(P)$ equal to one or to zero according as P is in e or in $E - e$. Thus $f(P)$ is in $\mathfrak{D}$ and

$$\int_E f(P)\,\overline{g(P)}\,dP = \int_e \overline{g(P)}\,dP = 0.$$

The arbitrary character of the set e requires that $g(P) = 0$ almost everywhere in $E(a)$; and, since $\lim_{a\to\infty} E(a) = E$, the relation $g(P) = 0$ must hold almost everywhere in E, as we wished to prove. When $f(P)$ is in $\mathfrak{D}$, the function $Hf(P)$ described in the statement of the theorem is a Lebesgue-measurable function defined almost everywhere in E. We must show that the integral

$$\int_E |Hf(P)|^2 \, dP = \int_E \left(\int_E K(P, Q) f(Q) \, dQ \cdot \overline{\int_E K(P, R) f(R) \, dR} \right) dP$$

exists. For this purpose, it is sufficient to show that the function $K(P, Q)\,\overline{K(P, R)}\, f(Q)\, \overline{f(R)}$ is absolutely integrable with respect to P, Q, and R, the integration being performed in any convenient order. Now, by virtue of the symmetry of the kernel $K(P, Q)$, we have

$$\int_E |K(P, Q)\,\overline{K(P, R)}| \, dP \leqq \left(\int_E |K(P, Q)|^2 dP \int_E |K(P, R)|^2 \, dP \right)^{1/2} = K(Q)\, K(R),$$

for almost every Q and almost every R; and, since $f(P)$ is in $\mathfrak{D}$, we can integrate first with respect to P and then with respect to Q and R, with the desired result. We are now permitted to integrate in any other order; in particular, we may integrate first with respect to Q and R and then with respect to P. Thus H takes its domain into a subset of $\mathfrak{L}_2$, as it should. The linear character of H can easily be verified directly. We pass on to the determination of the adjoint H^*, which exists by virtue of the fact that $\mathfrak{D}$ is everywhere dense in $\mathfrak{L}_2$. A necessary and sufficient condition that an element $g(P)$ in $\mathfrak{L}_2$ belong to the domain of H^* is that there exist an element $g^*(P)$ in $\mathfrak{L}_2$ such that

$$\int_E \left(\int_E K(P, Q) f(Q) \, dQ \right) \overline{g(P)} \, dP = \int_E f(Q)\, \overline{g^*(Q)} \, dQ$$

for every element $f(P)$ in $\mathfrak{D}$; and, when this condition is satisfied, $H^* g = g^*$. In view of the inequality

$$\int_E |K(P, Q)\, \overline{g(P)}| \, dP \leqq \left(\int_E |K(P, Q)|^2 \, dP \int_E |g(P)|^2 \, dP \right)^{1/2} = K(Q) \|g\|$$

and of the fact that $f(P)$ is in $\mathfrak{D}$, the function $K(P, Q) f(Q) \overline{g(P)}$ is absolutely integrable over E_2. We may therefore invert the order of integration in the condition above, writing it in the form

$$\int_E f(Q)\,\overline{h(Q)}\,dQ = 0, \quad h(Q) = \int_E K(Q, P)\,g(P)\,dP - g^*(Q),$$

after taking account of the symmetry of the kernel $K(P, Q)$. A sufficient condition that this relation hold is that $h(Q) = 0$ almost everywhere in E. This condition is also necessary. When g and g^* are given, it is evident that $h(Q)$ is a Lebesgue-measurable function defined almost everywhere in E. If $E(a)$ is the Lebesgue-measurable set specified by the inequalities $K(P) \leqq a$, $|h(P)| \leqq a$, and if e is an arbitrary Lebesgue-measurable subset of $E(a)$ with finite measure, the function $f(P)$ equal to one or to zero according as P is in e or in $E - e$ belongs to $\mathfrak{D}$ and satisfies the equation

$$\int_E f(Q)\,\overline{h(Q)}\,d(Q) = \int_e \overline{h(Q)}\,dQ = 0.$$

Since e is arbitrary, we find that $h(Q) = 0$ almost everywhere in $E(a)$; hence, also, $h(Q) = 0$ almost everywhere in the set $E^* = \lim_{a \to \infty} E(a)$. Evidently, $E - E^*$ is a set of measure zero, so that the desired result is established. We have thus shown that $H^* g(P)$ exists and is equal to $g^*(P)$ if and only if $\int_E K(P, Q)\,g(Q)\,dQ$ is an element of $\mathfrak{L}_2$ equal almost everywhere to $g^*(P)$. This implies that H^* and T coincide. Since we obviously have $H \subseteqq T \equiv H^*$, it is clear that H is a linear symmetric transformation: its domain is a linear manifold everywhere dense in $\mathfrak{L}_2$, and the relation $H \frown H$ is true.

In view of the theorem just proved, we can deduce from the theory developed in Chapter IX, § 1, complete general information concerning the integral operators $\tilde{H} \equiv H^{**}$ and $T \equiv H^*$ with kernels of Carleman type. The special case of a real kernel $K(P, Q) = K(Q, P) = \overline{K(P, Q)}$ is easily treated by the results given in Chapter IX, § 2: we introduce the conjugation J_2 which carries $f(P)$ into $\overline{f(P)}$ and note that $J_2 H f(P) = H J_2 f(P)$, the equation being valid whenever either term is significant. In this way we obtain the general theory of integral equations given by Carleman.*

Many of the results of Carleman depend upon the approximation theory given in Chapter IX, § 3. The application of that theory is made possible by the following result:

THEOREM 10.2. *Let* $K^{(n)}(P, Q)$, $n = 1, 2, 3, \cdots$, *be a (Hermitian) symmetric kernel of Hilbert-Schmidt type defined over* E_2, $H^{(n)}$ *the associated integral operator with domain* $\mathfrak{L}_2$; *let* $K(P, Q)$ *be a kernel of Carleman type defined over* E_2, H *the associated integral operator described in Theorem* 10.1; *and let the relations*

$$|K^{(n)}(P, Q)| \leqq |K(P, Q)|, \quad \lim_{n\to\infty} K^{(n)}(P, Q) = K(P, Q),$$

hold almost everywhere in E_2. *Then the sequence* $\{H^{(n)}\}$ *approximates* H *in the sense of Definition* 9.9. *A particular sequence of kernels* $K^{(n)}(P, Q)$ *satisfying the conditions stated can be formed as follows: let* $E(a)$ *be the set of points* P *such that* $K(P) \leqq a$, $d(OP) \leqq a$, *where* $d(OP)$ *is the distance from a fixed point* O *to the point* P; *let* $\{a_n\}$ *be a sequence of positive numbers such that* a_n *becomes infinite with* n; *we then set* $K^{(n)}(P, Q) = K(P, Q)$ *when* P *and* Q *are in* $E(a_n)$, *and* $K^{(n)}(P, Q) = 0$ *elsewhere.*

It is evident upon inspection that the special sequence constructed above satisfies the general conditions laid down in the first part of the theorem. We shall therefore concern ourselves only with the properties of the general sequences described. It is sufficient to show that $H^{(n)} \to H$ in $\mathfrak{D}$. Now

$$\int_E |Hf(P) - H^{(n)}f(P)|^2 \, dP = \int_E \int_E \int_E F^{(n)}(P, Q, R) \, dP \, dQ \, dR,$$

$$F^{(n)}(P, Q, R) \\ = (K(P, Q) - K^{(n)}(P, Q))(\overline{K(P, R) - K^{(n)}(P, R)}) f(Q) \overline{f(R)}.$$

Since $f(P)$ is in $\mathfrak{D}$ and since, under the hypotheses of the theorem, the inequality $|F^{(n)}(P, Q, R)| \leqq 4\, |K(P, Q)\, K(P, R) f(Q) f(R)|$ is valid, the integrand on the right is absolutely integrable.

---

* Carleman, *Sur les équations intégrales singulières à noyau réel et symétrique*, Uppsala, 1923, Chapter II; and, Annales de l'Institut H. Poincaré, **1** (1931), pp. 401–430.

Furthermore, the inequality just noted can be combined with the relation $\lim_{n\to\infty} F^{(n)}(P, Q, R) = 0$ to yield the result $\lim_{n\to\infty} \int_E \int_E \int_E F^{(n)}(P, Q, R)\, dP\, dQ\, dR = 0$. This requires that $H^{(n)}f(P) \to Hf(P)$ in $\mathfrak{L}_2$ and that $H^{(n)} \to H$ in $\mathfrak{D}$, as we wished to prove.

In order to take advantage of this theorem, we must first examine a few properties of integral operators with (Hermitian) symmetric kernels of Hilbert-Schmidt type. Our analysis is based largely on Theorem 5.14.

THEOREM 10.3. *Let H be a self-adjoint integral operator with (Hermitian) symmetric kernel $K(P, Q)$ of Hilbert-Schmidt type, and let $E(\lambda)$ be the corresponding resolution of the identity. Then the transformation $A(\lambda)$ defined by the relations*

$$A(\lambda) \equiv E(\lambda),\ \lambda < 0; \quad A(0) \equiv O; \quad A(\lambda) \equiv E(\lambda) - I,\ \lambda > 0,$$

is an integral operator with (Hermitian) symmetric kernel $A(P, Q; \lambda)$ of Hilbert-Schmidt type, which satisfies the relations

(1) $\int_E A(P, Q; \lambda)\, A(Q, R; \lambda)\, dQ = -\operatorname{sgn} \lambda \cdot A(P, R; \lambda)$, *where* sgn λ *is equal to* -1, 0, $+1$, *according as* $\lambda < 0$, $\lambda = 0$, $\lambda > 0$;

(2) *for* $P = Q$, $A(P, Q; \lambda)$ *reduces to a function* $A(P, P; \lambda)$ *such that* $|A(P, P; \lambda)| \leq K^2(P)/\lambda^2$ *for* $\lambda \neq 0$;

(3) $V(A(P, Q; \lambda); \Delta) \leq K(P)\, K(Q)/\delta^2$, *where* Δ *is an arbitrary finite or infinite interval at positive distance* δ *from the point* $\lambda = 0$;

(4) $V((A(\lambda) f, g); \Delta) \leq \int_E \int_E K(P)\, K(Q)\, |f(Q)|\, |g(P)|\, dP dQ/\delta^2$ *where* f *and* g *are arbitrary elements in* $\mathfrak{L}_2$.

According to Theorem 5.14, the operator H has a complete orthonormal set $\{\varphi_n\}$ of characteristic elements, only a finite number corresponding to characteristic values outside the interval $-\delta < \lambda < +\delta$, where δ is any positive number. We denote by λ_n the characteristic value associated with φ_n, and introduce the function $A(P, Q; \lambda)$ defined by the equations

$$A(P, Q; \lambda) = \sum_{\lambda_\alpha \leqq \lambda} \varphi_\alpha(P) \overline{\varphi_\alpha(Q)}, \qquad A(P, Q; \lambda) = 0,$$
$$A(P, Q; \lambda) = -\sum_{\lambda_\alpha > \lambda} \varphi_\alpha(P) \overline{\varphi_\alpha(Q)}$$

according as $\lambda < 0$, $\lambda = 0$, or $\lambda > 0$. It is to be noted that the sums which appear in the definition of $A(P, Q; \lambda)$ are finite sums. Since each of the products $\varphi_n(P) \overline{\varphi_n(Q)}$ is a function in $\mathfrak{L}_2^{(2)}$, $A(P, Q; \lambda)$ is evidently an element of $\mathfrak{L}_2^{(2)}$ and, hence, a kernel of Hilbert-Schmidt type. Its (Hermitian) symmetric character can be read directly from the defining equations. By virtue of the relations

$$E(\lambda) f(P) = \sum_{\lambda_\alpha \leqq \lambda} \varphi_\alpha(P) \int_E f(Q) \overline{\varphi_\alpha(Q)}\, dQ,$$
$$f(P) = \sum_{\alpha=1}^{\infty} \varphi_\alpha(P) \int_E f(Q) \overline{\varphi_\alpha(Q)}\, dQ,$$

it is clear that $A(\lambda) f(P) = \int_E A(P, Q; \lambda) f(Q)\, dQ$, for every function $f(P)$ in $\mathfrak{L}_2$. In accordance with Theorem 3.9, relation (1) of the present theorem expresses in terms of kernels the identities

$$A^2(\lambda) \equiv A(\lambda), \quad \lambda \leqq 0;$$
$$A^2(\lambda) \equiv (E(\lambda) - I)^2 \equiv I - E(\lambda) \equiv -A(\lambda), \quad \lambda > 0.$$

The inequality in (2) is established by use of the equation $H\varphi_n = \int_E K(P, Q) \varphi_n(Q)\, dQ = \lambda_n \varphi_n(P)$ and of Bessel's inequality for functions in $\mathfrak{L}_2$. We have

$$\begin{aligned}
|A(P, P; \lambda)| & \\
&= \sum_{\lambda_\alpha \leqq \lambda} \varphi_\alpha(P) \overline{\varphi_\alpha(P)} = \sum_{\lambda_\alpha \leqq \lambda} |\varphi_\alpha(P)|^2 \\
&= \sum_{\lambda_\alpha \leqq \lambda} \left| \int_E K(P, Q) \varphi_\alpha(Q)\, dQ \right|^2 / \lambda_\alpha^2 \\
&\leqq \frac{1}{\lambda^2} \sum_{\lambda_\alpha \leqq \lambda} \left| \int_E K(P, Q) \varphi_\alpha(Q)\, dQ \right|^2 \\
&\leqq \int_E |K(P, Q)|^2\, dQ / \lambda^2 \\
&= K^2(P)/\lambda^2
\end{aligned}$$

when $\lambda < 0$, together with a similar sequence of relations when $\lambda > 0$. If we consider $\Lambda(P, Q; \lambda)$ for fixed (P, Q) as a function of λ alone, we see that it is constant on any interval which contains no characteristic value $\lambda = \lambda_n$ and discontinuous, in general, at each of the points $\lambda = \lambda_n$, provided merely that we avoid the set of points (P, Q) in E_2 for which one or more of the products $\varphi_n(P)\overline{\varphi_n(Q)}$ fails to be defined, a set which obviously has zero measure. Thus $\Lambda(P, Q; \lambda)$ is a function of bounded variation over any interval Δ of the type described in (3). The explicit appraisal of the variation is established by reasoning similar to that used in the proof of (2). It is sufficient, evidently, to consider the intervals $-\infty \leqq \lambda \leqq -\delta$, $\delta \leqq \lambda \leqq +\infty$, where δ is an arbitrary positive number. We shall give the calculations in full only for the first case. We have

$$
\begin{aligned}
&V(\Lambda(P, Q; \lambda); \Delta) \\
&\quad \leqq \sum_{\lambda_\alpha \leqq -\delta} |\varphi_\alpha(P)\overline{\varphi_\alpha(Q)}| \\
&\quad = \sum_{\lambda_\alpha \leqq -\delta} \left| \int_E K(P, R)\,\varphi_\alpha(R)\,dR \cdot \int_E K(Q, R)\,\varphi_\alpha(R)\,dR \right| / \lambda_\alpha^2 \\
&\quad \leqq \frac{1}{\delta^2} \sum_{\lambda_\alpha \leqq -\delta} \left| \int_E K(P, R)\,\varphi_\alpha(R)\,dR \cdot \int_E K(Q, R)\,\varphi_\alpha(R)\,dR \right| \\
&\quad \leqq \Bigg(\sum_{\lambda_\alpha \leqq -\delta} \left| \int_E K(P, R)\,\varphi_\alpha(R)\,dR \right|^2 \\
&\qquad\qquad \times \sum_{\lambda_\alpha \leqq -\delta} \left| \int_E K(Q, R)\,\varphi_\alpha(R)\,dR \right|^2 \Bigg)^{1/2} / \delta^2 \\
&\quad \leqq \left(\int_E |K(P, R)|^2\,dR \cdot \int_E |K(Q, R)|^2\,dR \right)^{1/2} / \delta^2 \\
&\quad = K(P)\,K(Q)/\delta^2.
\end{aligned}
$$

Finally, we can deduce (4) from (3). For any collection $\mathfrak{D}$ of intervals $\Delta_1, \cdots, \Delta_n$, no two of which have interior points in common, contained in a given interval Δ of the type described in (3), we have, in the notation of Chapter V, § 1,

$$\begin{aligned}
V_{\mathfrak{D}}&((A(\lambda)f, g)) \\
&= \sum_{\alpha=1}^{n} |(A(\Delta_\alpha)f, g)| \\
&= \sum_{\alpha=1}^{n} \left| \int_E \int_E A(P, Q; \Delta_\alpha) f(Q) \overline{g(P)}\, dP\, dQ \right| \\
&\leqq \sum_{\alpha=1}^{n} \int_E \int_E |A(P, Q; \Delta_\alpha)|\, |f(Q)|\, |g(P)|\, dP\, dQ \\
&= \int_E \int_E V_{\mathfrak{D}}(A(P, Q; \lambda))\, |f(Q)|\, |g(P)|\, dP\, dQ \\
&\leqq \int_E \int_E V(A(P, Q; \lambda); \Delta)\, |f(Q)|\, |g(P)|\, dP\, dQ \\
&\leqq \int_E \int_E K(P) K(Q)\, |f(Q)|\, |g(P)|\, dP\, dQ/\delta^2.
\end{aligned}$$

The last integral is absolutely convergent since $K(P)$, $f(P)$, and $g(P)$ are all elements of $\mathfrak{L}_2$. This inequality implies the relation given in (4). It is to be observed that (4) supplements the earlier appraisal, according to which

$$\begin{aligned}
V((A(\lambda)f, g); \Delta) &= V\left((E(\lambda)f, g) - \frac{1}{2}(1 + \operatorname{sgn} \lambda)\,(f, g);\ \Delta\right) \\
&= V((E(\lambda)f, g);\ \Delta) \leqq |f| \cdot |g|,
\end{aligned}$$

where Δ is any interval which does not contain the point $\lambda = 0$.

We propose to use the two theorems just proved, together with Theorem 9.17, in order to analyze more thoroughly the nature of integral operators with kernels of Carleman type. If we approximate the integral operator H, described in Theorem 10.1, by a sequence of integral operators $\{H^{(n)}\}$ of the type discussed in Theorem 10.2, we can assert the existence of a sequence of integers $\{n(k)\}$ and of a family of bounded self-adjoint transformations $B(\lambda)$ such that

$$n(k+1) > n(k), \quad \lim_{k \to \infty} (R_l^{(n(k))} f, g) = \int_{-\infty}^{+\infty} \frac{1}{\lambda - l}\, d(B(\lambda)f, g),$$

where $R_l^{(n)}$ is the resolvent of $H^{(n)}$, f and g are arbitrary elements of $\mathfrak{L}_2$, and l is not real. In general, different sequences $\{n(k)\}$ lead to different families $B(\lambda)$, although in the special case where the integral operator $\tilde{H} \equiv H^{**}$

is a maximal symmetric transformation, we know that it is possible to take $n(k) = k$ and that the family $B(\lambda)$ is unique. These facts have been established in Chapter IX, § 3. We shall examine the properties of the family $B(\lambda)$ in some detail. For convenience in stating our results, we shall suppose that the sequence $\{H^{(n)}\}$ has been so chosen that we may take $n(k) = k$: in other words, if the sequence $\{H^{(n)}\}$, as given, leads to more than one family $B(\lambda)$, we select a subsequence which leads to a unique family and confine our attention thenceforth to this subsequence, appropriately renumbered.

THEOREM 10.4. *Let* $K^{(n)}(P, Q)$, $H^{(n)}$, $K(P, Q)$, *and* H *be defined and interrelated in the manner described in Theorem* 10.2; *and let* $B(\lambda)$ *be the family of self-adjoint transformations characterized by the relation*

$$\lim_{n\to\infty} (R_l^{(n)} f, g) = \int_{-\infty}^{+\infty} \frac{1}{\lambda - l}\, d(B(\lambda) f, g), \qquad \mathfrak{J}(l) \neq 0,$$

in accordance with Theorem 9.17 *and the remarks made above, where* $R_l^{(n)}$ *is the resolvent of* $H^{(n)}$. *Then the transformation* $A(\lambda)$ *defined by the identities*

$$A(\lambda) \equiv B(\lambda),\ \lambda < 0; \quad A(0) \equiv O;$$
$$A(\lambda) \equiv B(\lambda) - I,\ \lambda > 0,$$

is an integral operator with kernel $A(P, Q; \lambda)$ *of Carleman type. This kernel has the following properties:*

(1) *if* E_2^* *is a suitably chosen Lebesgue-measurable subset of* E_2 *such that* $E_2 - E_2^*$ *has measure zero, and if* Δ *is an arbitrary interval* $\alpha \leqq \lambda \leqq \beta$ (*the improper values* $\alpha = -\infty$ *and* $\beta = +\infty$ *being admitted*) *at positive distance* δ *from the point* $\lambda = 0$, *then* $A(P, Q; \lambda)$, *considered for fixed* (P, Q) *in* E_2^*, *is a function of* λ *of bounded variation over* Δ, *satisfying the relations*

$$|A(P, Q; \lambda)| \leqq K(P)\, K(Q)/\lambda^2, \quad \lambda \neq 0,$$
$$V(A(P, Q; \lambda);\ \Delta) \leqq K(P)\, K(Q)/\delta^2,$$
$$A(P, Q; \lambda + 0) = A(P, Q; \lambda), \quad \lambda \neq 0,$$
$$A(P, Q; -\infty) = A(P, Q; +\infty) = 0;$$

(2) *if Δ is an interval of the type described above, then*

$$\int_E K(P, R)\, A(R, Q; \Delta)\, dR = \int_\alpha^\beta \lambda\, d A(P, Q; \lambda),$$

where the integral on the right is to be interpreted as an improper integral convergent in the mean (in $\mathfrak{L}_2$) if $\alpha = -\infty$ or $\beta = +\infty$; and

$$K(P, Q) = \int_{-\infty}^{+\infty} \lambda\, d A(P, Q; \lambda),$$

where the integral on the right is to be interpreted as an improper integral at $\lambda = -\infty$, $\lambda = 0$, $\lambda = +\infty$, convergent in the mean (in $\mathfrak{L}_2$);

(3) *if $f(P)$ is an arbitrary element of $\mathfrak{L}_2$, then*

$$\int_E K(P, Q) f(Q)\, dQ = \int_{-\infty}^{+\infty} \lambda\, d\left(\int_E A(P, Q; \lambda) f(Q)\, dQ\right),$$

the integral with respect to λ being improper at $\lambda = -\infty$, $\lambda = 0$, and $\lambda = +\infty$;

(4) *if $f(P)$ is an arbitrary element of $\mathfrak{L}_2$ and $g(P)$ is an arbitrary element of $\mathfrak{D}$, then*

$$\int_E \int_E K(P, Q) f(Q) \overline{g(P)}\, dP\, dQ$$
$$= \int_{-\infty}^{+\infty} \lambda\, d\left(\int_E \int_E A(P, Q; \lambda) f(Q) \overline{g(P)}\, dP\, dQ\right)$$

the integral with respect to λ being improper at $\lambda = -\infty$ and $\lambda = +\infty$;

(5) *if $f(P)$ is an arbitrary element of $\mathfrak{L}_2$ and if μ and ε are positive numbers, then the integral*

$$\left[\int_{-\mu}^{-\varepsilon} + \int_{+\varepsilon}^{+\mu}\right] d\left(\int_E A(P, Q; \lambda) f(Q)\, dQ\right)$$

converges in $\mathfrak{L}_2$ to $f(P) - (B(0) - B(0-0)) f(P)$ when $\mu \to \infty$, $\varepsilon \to 0$.

We begin by establishing the existence of the function $A(P, Q; \lambda)$. Let $E(a)$ denote the set described in Theorem 10.2, for arbitrary positive a; and let $E_2(a, b)$ denote the set of point-pairs (P, Q) where P is in $E(a)$ and Q is in $E(b)$. Let $\mathfrak{M}(a)$ be the set of all functions in $\mathfrak{L}_2$ which vanish on

$E - E(a)$, $\mathfrak{M}_2(a, b)$ the set of all functions in $\mathfrak{L}_2^{(2)}$ which vanish on $E_2 - E_2(a, b)$. It is easily verified that $\mathfrak{M}(a)$ and $\mathfrak{M}_2(a, b)$ are closed linear manifolds. We select an orthonormal set $\{\psi_p^{(a)}(P)\}$ which determines the closed linear manifold $\mathfrak{M}(a)$, for $0 < a < +\infty$. By the reasoning used in the proof of Theorem 3.8, we show that $\{\psi_p^{(a)}(P) \cdot \overline{\psi_q^{(b)}(Q)}\}$ is an orthonormal set which determines the closed linear manifold $\mathfrak{M}_2(a, b)$. In terms of the latter set, we construct a function $A_{ab}(P, Q; \lambda)$ in $\mathfrak{M}_2(a, b)$ according to the equations

$$A_{ab}(P, Q; \lambda) = \sum_{\alpha, \beta = 1}^{\infty} A_{\alpha\beta}^{(a, b)}(\lambda)\, \psi_\alpha^{(a)}(P)\, \overline{\psi_\beta^{(b)}(Q)},$$

$$A_{pq}^{(a, b)}(\lambda) = (A(\lambda)\, \psi_q^{(b)}, \psi_p^{(a)}).$$

To justify this procedure, we must show that the series $\sum_{\alpha, \beta = 1}^{\infty} |A_{\alpha\beta}^{(a, b)}(\lambda)|^2$ is convergent for every value of λ. The case $\lambda = 0$, where $A_{pq}^{(a, b)}(0) = 0$, is trivial and will be disregarded in the sequel. By virtue of Theorem 9.22 we can write, except possibly on a finite or denumerably infinite set of values of λ,

$$\sum_{\alpha, \beta = 1}^{N} |(A(\lambda)\, \psi_\beta^{(b)}, \psi_\alpha^{(a)})|^2 = \lim_{n \to \infty} \sum_{\alpha, \beta = 1}^{N} |(A^{(n)}(\lambda)\, \psi_\beta^{(b)}, \psi_\alpha^{(a)})|^2,$$

where $A^{(n)}(\lambda)$ is associated with $H^{(n)}$ in the manner described in Theorem 10.3. By the use of Bessel's inequality for $\mathfrak{L}_2^{(2)}$, the results of Theorem 10.3, and the inequality $|K^{(n)}(P, Q)| \leqq |K(P, Q)|$, we can appraise the sum appearing on the right, as follows:

$$\begin{aligned} &\sum_{\alpha, \beta = 1}^{N} |(A^{(n)}(\lambda)\, \psi_\beta^{(b)}, \psi_\alpha^{(a)})|^2 \\ &\quad = \sum_{\alpha, \beta = 1}^{N} \left| \int_{E(a)} \int_{E(b)} A^{(n)}(P, Q; \lambda)\, \psi_\beta^{(b)}(Q)\, \overline{\psi_\alpha^{(a)}(P)}\, dQ\, dP \right|^2 \\ &\quad \leqq \int_{E(a)} \int_{E(b)} |A^{(n)}(P, Q; \lambda)|^2\, dQ\, dP \\ &\quad \leqq \int_{E(a)} \int_{E} A^{(n)}(P, Q; \lambda)\, \overline{A^{(n)}(P, Q; \lambda)}\, dQ\, dP \\ &\quad = \int_{E(a)} |A^{(n)}(P, P; \lambda)|\, dP \end{aligned}$$

$$\leqq \int_{E(a)} [K^{(n)}(P)]^2 \, dP/\lambda^2$$
$$\leqq \int_{E(a)} K^2(P) \, dP/\lambda^2.$$

It follows that $\sum_{\alpha,\beta=1}^{N} |A_{\alpha\beta}^{(a,b)}(\lambda)|^2 \leqq \int_{E(a)} K^2(P)\, dP/\lambda^2$ for all values of λ with the possible exception of a finite or denumerably infinite set; but the two terms in this inequality are functions of λ which are continuous on the right when $\lambda \neq 0$, so that no exception other than $\lambda = 0$ need be made. Thus the function $A_{ab}(P, Q; \lambda)$ exists and is an element in $\mathfrak{M}_2(a, b)$ satisfying the inequality

$$\int_E \int_E |A_{ab}(P, Q; \lambda)|^2 \, dQ \, dP \leqq \int_{E(a)} K^2(P) \, dP/\lambda^2.$$

From the defining equations we obtain

$$A_{qp}^{(b,a)}(\lambda) = (A(\lambda)\psi_p^{(a)}, \psi_q^{(b)}) = (\psi_p^{(a)}, A(\lambda)\psi_q^{(b)}) = \overline{A_{pq}^{(a,b)}(\lambda)},$$

$$\begin{aligned} A_{ba}(Q, P; \lambda) &= \sum_{\alpha,\beta=1}^{\infty} A_{\beta\alpha}^{(b,a)}(\lambda)\, \psi_\beta^{(b)}(Q)\, \overline{\psi_\alpha^{(a)}(P)} \\ &= \sum_{\alpha,\beta=1}^{\infty} \overline{A_{\alpha\beta}^{(a,b)}(\lambda)}\, \overline{\psi_\alpha^{(a)}(P)}\, \psi_\beta^{(b)}(Q) \\ &= \overline{A_{ab}(P, Q; \lambda)}, \end{aligned}$$

by virtue of the self-adjoint character of the transformations $B(\lambda)$ and $A(\lambda)$. In particular, the function $A_{aa}(P, Q; \lambda)$ is (Hermitian) symmetric. If $f(P)$ and $g(P)$ are arbitrary elements of $\mathfrak{M}(b)$ and $\mathfrak{M}(a)$ respectively, then the equation

$$(A(\lambda)f, g) = \int_E \int_E A_{ab}(P, Q; \lambda)\, f(Q)\, \overline{g(P)}\, dQ\, dP$$

is satisfied. The term on the left can be written

$$(A(\lambda)f, g) = \sum_{\alpha,\beta=1}^{\infty} A_{\alpha\beta}^{(a,b)}(\lambda)\, (f, \psi_\beta^{(b)})\, (\psi_\alpha^{(a)}, g),$$

as we see by substituting $f = \sum_{\beta=1}^{\infty} (f, \psi_\beta^{(b)})\, \psi_\beta^{(b)}$ and $g = \sum_{\alpha=1}^{\infty} (g, \psi_\alpha^{(a)})\, \psi_\alpha^{(a)}$ and evaluating. Since $\overline{f(Q)}\, g(P)$ is a function in $\mathfrak{M}_2(a, b)$, we have also

$$\int_E \int_E A_{ab}(P, Q; \lambda)\, f(Q)\, \overline{g(P)}\, dQ\, dP$$
$$= \sum_{\alpha, \beta=1}^{\infty} \int_E \int_E A_{ab}(P, Q; \lambda)\, \overline{\psi_\alpha^{(a)}(P)}\, \psi_\beta^{(b)}(Q)\, dQ\, dP$$
$$\times \int_E \int_E \psi_\alpha^{(a)}(P)\, \overline{\psi_\beta^{(b)}(Q)}\, f(Q)\, \overline{g(P)}\, dQ\, dP$$
$$= \sum_{\alpha, \beta=1}^{\infty} A_{\alpha\beta}^{(a,b)}(\lambda)\, (f, \psi_\beta^{(b)})\, (\psi_\alpha^{(a)}, g).$$

Our assertion is established at once by comparison of these results. We are now in a position to show that

$$A_{a'b'}(P, Q; \lambda) = A_{ab}(P, Q; \lambda)$$

almost everywhere in $E_2(a, b)$ whenever $a' \geqq a$, $b' \geqq b$. Since we have $E(a') \supseteqq E(a)$ and $E(b') \supseteqq E(b)$ under these circumstances, we see that $\mathfrak{M}(a') \supseteqq \mathfrak{M}(a)$ and $\mathfrak{M}(b') \supseteqq \mathfrak{M}(b)$. Thus the relation

$$(A(\lambda) f, g) = \int_E \int_E A_{a'b'}(P, Q; \lambda) f(Q) \overline{g(P)}\, dQ\, dP$$
$$= \int_E \int_E A_{ab}(P, Q; \lambda) f(Q) \overline{g(P)}\, dQ\, dP$$

holds for arbitrary functions $f(P)$, $g(P)$ in $\mathfrak{M}(b)$, $\mathfrak{M}(a)$ respectively. We can put this relation in the form

$$\int_{E(a)} \int_{E(b)} [A_{a'b'}(P, Q; \lambda) - A_{ab}(P, Q; \lambda)] f(Q) \overline{g(P)}\, dQ\, dP = 0,$$

and can thus conclude that the desired equation is true. Consequently, if a and b become positively infinite through any prescribed sequences of values, we must have

$$\lim_{a \to \infty} E(a) = E, \quad \lim_{b \to \infty} E(b) = E, \quad \lim_{a \to \infty, b \to \infty} E_2(a, b) = E_2,$$
$$\lim_{a \to \infty, b \to \infty} A_{ab}(P, Q; \lambda) = A(P, Q; \lambda) \text{ almost everywhere in } E_2.$$

The function $A(P, Q; \lambda)$ defined by this limiting relation is, except for modifications affecting its values on a subset of E_2 of zero measure, the function whose existence is asserted in the theorem. Its (Hermitian) symmetric character is obvious by virtue of the equations

$$A(P, Q; \lambda) = \lim_{a\to\infty} A_{aa}(P, Q; \lambda), \quad A_{aa}(Q, P; \lambda) = \overline{A_{aa}(P, Q; \lambda)},$$

which hold almost everywhere in E_2. From the relations

$$\int_{E(a)}\int_E |A(P, Q; \lambda)|^2 dQ\,dP = \lim_{b\to\infty}\int_{E(a)}\int_{E(b)} |A(P, Q; \lambda)|^2 dQ\,dP,$$

$$\int_{E(a)}\int_{E(b)} |A(P, Q; \lambda)|^2 dQ\,dP = \int_E\int_E |A_{ab}(P, Q; \lambda)|^2 dQ\,dP$$

$$\leqq \int_{E(a)} K^2(P)\,dP/\lambda^2,$$

we conclude that the integral $\int_{E(a)}\int_E |A(P, Q; \lambda)|^2 dQ\,dP$ exists and does not exceed $\int_{E(a)} K^2(P)\,dP/\lambda^2$. The theorem of Fubini shows at once that the integral $\int_E |A(P, Q; \lambda)|^2 dQ$ exists for almost every P in $E(a)$, $0 < a < +\infty$; in other words, for almost every P in E. Thus $A(P, Q; \lambda)$ is a kernel of Carleman type for every value of λ, the case $\lambda = 0$ being included because $A(P, Q; 0) = 0$. If $f(P)$ is an arbitrary element in $\mathfrak{L}_2$ and $g(P)$ is an arbitrary element in $\mathfrak{M}(a)$, the function $A(P, Q; \lambda) f(Q) \overline{g(P)}$ is absolutely integrable over the set of point-pairs (P, Q) where P is in $E(a)$ and Q is in E. Thus, if $f_b(P)$ denotes the projection of $f(P)$ on $\mathfrak{M}(b)$, we have

$$\begin{aligned}(A(\lambda) f, g) &= \lim_{b\to\infty} (A(\lambda) f_b, g)\\ &= \lim_{b\to\infty}\int_E\int_E A_{ab}(P, Q; \lambda) f_b(Q)\overline{g(P)}\,dQ\,dP\\ &= \lim_{b\to\infty}\int_{E(a)}\int_{E(b)} A(P, Q; \lambda) f(Q)\overline{g(P)}\,dQ\,dP\\ &= \int_{E(a)}\left(\int_E A(P, Q; \lambda) f(Q)\,dQ\right)\overline{g(P)}\,dP.\end{aligned}$$

It follows that $A(\lambda) f(P) = \int_E A(P, Q; \lambda) f(Q)\,dQ$ almost everywhere in $E(a)$, $0 < a < +\infty$; in other words, almost everywhere in E. Thus the transformation $A(\lambda)$ is an integral operator with kernel of Carleman type, as we wished to show. It is easily verified that the kernel is essentially unique: for if $A(\lambda)$ were defined by two kernels of Carleman type, their difference would be a kernel $A_0(P, Q; \lambda)$ of Carleman type such that $\int_E A_0(P, Q; \lambda) f(Q)\, dQ = 0$ for every element f in $\mathfrak{L}_2$ and almost every P in E, and would therefore vanish almost everywhere in E_2.

We turn now to the study of $A(P, Q; \lambda)$ as a function of λ. We shall first prove that, if Δ is an arbitrary interval at positive distance δ from $\lambda = 0$ and if f and g are arbitrary elements of $\mathfrak{D}$, then

$$V((A(\lambda)f, g); \Delta) \leqq \int_E \int_E K(P)\, K(Q)\, |f(Q)|\, |g(P)|\, dQ\, dP/\delta^2.$$

This inequality follows by the application of suitable limiting processes to the result given in Theorem 10.3 (4). By virtue of the relation $|K^{(n)}(P, Q)| \leqq |K(P, Q)|$ we have

$$\begin{aligned} V((A^{(n)}(\lambda)f, g); \Delta) & \\ \leqq \int_E \int_E & K^{(n)}(P)\, K^{(n)}(Q)\, |f(Q)|\, |g(P)|\, dQ\, dP/\delta^2 \\ \leqq \int_E \int_E & K(P)\, K(Q)\, |f(Q)|\, |g(P)|\, dQ\, dP/\delta^2. \end{aligned}$$

During the remainder of the paragraph, it will be convenient to denote the integral appearing in the last expression by the letter I. We can now apply the theorem of Helly to the sequence $\{(A^{(n)}(\lambda)f, g)\}$, considered on the interval Δ. Thus we find that there exist a sequence $\{n(k)\}$ and a function $\varrho(\lambda)$ defined on Δ such that

$$\lim_{k\to\infty} n(k) = \infty, \quad \lim_{k\to\infty} (A^{(n(k))}(\lambda)f, g) = \varrho(\lambda), \quad V(\varrho; \Delta) \leqq I/\delta^2.$$

We know, however, that $\varrho(\lambda) = (A(\lambda)f, g)$ except possibly on a finite or denumerably infinite set, and that $\varrho(\lambda+0) = (A(\lambda)f, g)$ without exception. Thus if Δ is restricted to be an interval with extremities at which $\varrho(\lambda) = \varrho(\lambda+0)$ we can conclude that

$$V((A(\lambda)f, g); \Delta) = V(\varrho(\lambda+0); \Delta) \leqq V(\varrho(\lambda); \Delta) \leqq I/\delta^2.$$

The details of the argument are similar to those given in the discussion of the relation $V(\varrho^*) \leqq V(\varrho)$ at the beginning of Chapter V, § 1. If Δ is an arbitrary interval, we choose a sequence of intervals $\{\Delta_n\}$ of the restricted type, with the properties $\Delta_n \supseteqq \Delta_{n+1}$, $\lim_{n\to\infty} \Delta_n = \Delta$. If $\delta_n > 0$ is the distance from Δ_n to the point $\lambda = 0$, we have $\lim_{n\to\infty} \delta_n = \delta$. Hence we find that

$$V((A(\lambda)f, g); \Delta) \leqq V((A(\lambda)f, g); \Delta_n) \leqq I/\delta_n^2.$$

When we allow n to become infinite in this relation, we obtain the desired result:

$$V((A(\lambda)f, g); \Delta) \leqq I/\delta^2.$$

Since $(A(\lambda)f, g)$ vanishes at $\lambda = \pm\infty$, in accordance with Theorem 9.22, we can apply this inequality over the interval $(-\infty, \lambda)$ or the interval $(\lambda, +\infty)$, according as $\lambda < 0$ or $\lambda > 0$, and can thus establish the inequality $|(A(\lambda)f, g)| \leqq I/\lambda^2$.

The results of the preceding paragraph enable us to calculate the function $A(P, Q; \lambda)$ from the integral

$$(A(\lambda)f, g) = \int_E \int_E A(P, Q; \lambda) f(Q) \overline{g(P)}\, dQ\, dP$$

by the consideration of appropriately chosen sequences of elements f and g in $\mathfrak{D}$. We introduce the function $\varphi_n(P, Q)$ defined for all point-pairs (P, Q) as follows: if $S_n(Q)$ is the hypersphere consisting of all points P at a distance not exceeding $1/n$ from the center Q, we set $\varphi_n(P, Q)$ equal to $1/m(S_n(Q))$ or to zero according as P is in $S_n(Q)$ or not. We then denote by $\varphi_n^{(a)}(P, Q)$ the function which is equal to $\varphi_n(P, Q)$ when P is in the set $E(a)$ and equal to zero when P is in the set $E - E(a)$. It is evident that $\varphi_n^{(a)}(P, Q)$ is a function in $\mathfrak{M}(a)$, $\mathfrak{D}$, $\mathfrak{L}_2$, for fixed Q. If we form the expressions

$$(A(\lambda)\varphi_n^{(a)}(S, Q), \varphi_n^{(a)}(R, P))$$

$$= \int_E \int_E A(R, S; \lambda)\, \varphi_n^{(a)}(S, Q)\, \overline{\varphi_n^{(a)}(R, P)}\, dS\, dR$$

$$= \int_{E(a)} \int_{E(a)} A(R, S; \lambda)\, \varphi_n(S, Q)\, \overline{\varphi_n(R, P)}\, dS\, dR$$

$$= \int_{S_n(P)E(a)} \int_{S_n(Q)E(a)} A(R, S; \lambda)\, dS\, dR / m(S_n(P))\, m(S_n(Q)),$$

$$\int_E \int_E K(R)\, K(S)\, |\varphi_n^{(a)}(S, Q)|\, |\varphi_n^{(a)}(R, P)|\, dS\, dR$$

$$= \int_{S_n(P)E(a)} \int_{S_n(Q)E(a)} K(R)\, K(S)\, dS\, dR / m(S_n(P))\, m(S_n(Q)),$$

we see by the theory of the generalized derivative† that when n becomes infinite these two expressions tend to the respective limits $A(P, Q; \lambda)$ and $K(P)\,K(Q)$ almost everywhere in $E_2(a, a)$, to the common limit zero almost everywhere in $E_2 - E_2(a, a)$. We shall denote by $E_2^*(a)$ the set of points (P, Q) in $E_2(a, a)$ where the second expression has the limit $K(P)\,K(Q)$. On the other hand, the inequality

$$V((A(\lambda)\, \varphi_n^{(a)}(S, Q), \varphi_n^{(a)}(R, P)); \Delta)$$
$$\leqq \int_E \int_E K(R)\, K(S)\, |\varphi_n^{(a)}(S, Q)|\, |\varphi_n^{(a)}(R, P)|\, dS\, dR / \delta^2$$

enables us to apply the theorem of Helly to the sequence $\{(A(\lambda)\, \varphi_n^{(a)}(S, Q), \varphi_n^{(a)}(R, P))\}$ on any interval Δ, whenever (P, Q) is a point-pair for which the integral on the right is bounded with respect to n—in particular, whenever (P, Q) is in $E_2^*(a)$. For a point in $E_2^*(a)$, therefore, we can select a sequence $\{n(k)\}$ such that

$$\lim_{k \to \infty} n(k) = \infty, \quad \lim_{k \to \infty} (A(\lambda)\, \varphi_{n(k)}^{(a)}(S, Q), \varphi_{n(k)}^{(a)}(R, P)) = \varrho_a(P, Q; \lambda)$$

for all values of λ different from zero. In choosing this sequence we must employ the usual "diagonal process": we first select a sequence $\{n(1, k)\}$ which ensures convergence on the range $-\infty < \lambda \leqq -1$, $+1 \leqq \lambda < +\infty$; when we have obtained a sequence $\{n(i, k)\}$ which ensures convergence on the range $-\infty < \lambda \leqq -1/i$, $+1/i \leqq \lambda < +\infty$, we choose a subsequence $\{n(i+1, k)\}$ which ensures convergence on

---

† Lebesgue, Annales de l'École Normale Supérieure, (3) 27 (1910), pp. 395–401; de la Vallée Poussin, *Intégrales de Lebesgue*, Paris, 1916, Chapters IV and V.

the range $-\infty < \lambda \leqq -1/(i+1)$, $+1/(i+1) \leqq \lambda < +\infty$, for $i = 1, 2, 3, \cdots$; and our final choice is then determined by the equation $n(k) = n(k, k)$. We know by a familiar argument that

$$V(\varrho_a(P, Q; \lambda); \Delta) \leqq K(P)\, K(Q)/\delta^2;$$

and we find, by reference to the inequality

$$|(A(\lambda)\, \varphi_n^{(a)}(S, Q),\ \varphi_n^{(a)}(R, P))|$$

$$\leqq \int_E \int_E K(R)\, K(S)\, |\varphi_n^{(a)}(S, Q)|\, |\varphi_n^{(a)}(R, P)|\, dS\, dR/\lambda^2,$$

that

$$|\varrho_a(P, Q; \lambda)| \leqq K(P)\, K(Q)/\lambda^2.$$

Thus the function $\sigma_a(P, Q; \lambda) = \varrho_a(P, Q; \lambda + 0)$ exists and satisfies the relations

$$V(\sigma_a(P, Q; \lambda); \Delta) \leqq K(P)\, K(Q)/\delta^2,$$
$$|\sigma_a(P, Q; \lambda)| \leqq K(P)\, K(Q)/\lambda^2.$$

By the manner in which $\varrho_a(P, Q; \lambda)$ was defined, it is evident that $\varrho_a(P, Q; \lambda) = A(P, Q; \lambda)$ almost everywhere in $E_2^*(a)$, and hence also in $E_2(a, a)$, for each value of λ. Thus, if $f(P)$ and $g(P)$ are arbitrary functions in $\mathfrak{M}(a)$, we have

$$\lim_{\varepsilon \to 0} \int_{E(a)} \int_{E(a)} \varrho_a(P, Q; \lambda + \varepsilon)\, f(Q)\, \overline{g(P)}\, dQ\, dP$$
$$= \lim_{\varepsilon \to 0} \int_{E(a)} \int_{E(a)} A(P, Q; \lambda + \varepsilon)\, f(Q)\, \overline{g(P)}\, dQ\, dP$$
$$= \lim_{\varepsilon \to 0} (A(\lambda + \varepsilon) f, g) = (A(\lambda) f, g)$$
$$= \int_{E(a)} \int_{E(a)} A(P, Q; \lambda)\, f(Q)\, \overline{g(P)}\, dQ\, dP$$

when ε tends to zero through positive values. In the first expression, we can interchange the limit and integration processes by virtue of the inequality

$$|\varrho_a(P, Q; \lambda + \varepsilon) f(Q) \overline{g(P)}| \leqq K(P)\, K(Q)\, |f(Q)|\, |g(P)|/(|\lambda| - \varepsilon)^2,$$

where the term on the right is integrable over $E_2(a, a)$. We find therefore that

$$\int_{E(a)}\int_{E(a)} [\sigma_a(P, Q; \lambda) - A(P, Q; \lambda)] f(Q)\overline{g(P)}\, dQ\, dP = 0$$

and conclude that $\sigma_a(P, Q; \lambda) = A(P, Q; \lambda)$ almost everywhere in $E_2(a, a)$. We now form the sets $E_{21}^* = E_2^*(1)$, $E_{2n}^* = E_2^*(n) - E_2^*(n-1) E_2^*(n)$, $n = 2, 3, 4, \cdots$, and put $E_2^* = \sum_{\alpha=1}^{\infty} E_{2\alpha}^*$. It is evident that $E_2^* = \sum_{\alpha=1}^{\infty} E_2^*(\alpha)$ differs from the set $E_2 = \sum_{\alpha=1}^{\infty} E_2(\alpha)$ only by a set of zero measure. We modify the function $A(P, Q; \lambda)$ by making it coincide on the set E_{2n}^* with the function $\sigma_n(P, Q; \lambda)$, $n = 1, 2, 3, \cdots$. The modified function differs from the original function only on a set of zero measure and therefore serves equally well as the kernel of the integral operator $A(\lambda)$; but, in addition, it has on the set E_2^* the various properties described under (1) in the statement of the theorem. Our construction is thus brought to an end.

As a first step toward the proof of (2), we note that by Theorem 9.23 we have

$$(A(\Delta) Hf, g) = \int_\alpha^\beta \lambda\, d(A(\lambda) f, g),$$

where Δ is an arbitrary interval $\alpha \leqq \lambda \leqq \beta$ at positive distance δ from the point $\lambda = 0$, f is an arbitrary element in $\mathfrak{D}$, and g is an arbitrary element in $\mathfrak{L}_2$. We next show that

$$\int_E\int_E f(Q)\overline{g(P)} \left(\int_\alpha^\beta \lambda\, d A(P, Q; \lambda)\right) dQ\, dP$$
$$= \int_\alpha^\beta \lambda\, d\left(\int_E\int_E A(P, Q; \lambda) f(Q)\overline{g(P)}\, dQ\, dP\right),$$

where Δ is a finite interval $\alpha \leqq \lambda \leqq \beta$ at positive distance δ from $\lambda = 0$, and $f(P)$ and $g(P)$ are arbitrary functions in $\mathfrak{D}$. If we write each of the Stieltjes integrals involved as a limit of the usual type, we have

$$\int_E\int_E f(Q)\overline{g(P)} \left(\int_\alpha^\beta \lambda\, d A(P, Q; \lambda)\right) dQ\, dP$$
$$= \int_E\int_E f(Q)\overline{g(P)} \left(\lim_{n\to\infty} \sum_{\nu=1}^{n} \lambda_\nu A(P, Q; \Delta_\nu)\right) dQ\, dP,$$

$$\int_{\alpha}^{\beta} \lambda \, d\left(\int_E \int_E A(P, Q; \lambda) f(Q) \overline{g(P)} \, dQ \, dP\right)$$
$$= \lim_{n \to \infty} \int_E \int_E f(Q) \overline{g(P)} \left(\sum_{\nu=1}^{n} \lambda_\nu A(P, Q; \Delta_\nu)\right) dQ \, dP.$$

We have only to show that the resulting expressions are equal. Writing $\gamma = \max[|\alpha|, |\beta|]$, we have

$$\left| f(Q) \overline{g(P)} \sum_{\nu=1}^{n} \lambda_\nu A(P, Q; \Delta_\nu) \right|$$
$$\leqq \gamma \sum_{\nu=1}^{n} |A(P, Q; \Delta_\nu)| \, |f(Q)| \, |g(P)|$$
$$\leqq \gamma V(A(P, Q; \lambda); \Delta) \, |f(Q)| \, |g(P)|$$
$$\leqq \gamma K(Q) |f(Q)| \, K(P) |g(P)| / \delta^2.$$

Since f and g are elements of $\mathfrak{D}$, the last term is integrable over E_2. Thus the processes of integration and of passage to the limit can be applied interchangeably to the sequence $\left\{ f(Q) \overline{g(P)} \sum_{\nu=1}^{n} \lambda_\nu A(P, Q; \Delta_\nu) \right\}$, and the asserted equality is established. By combining the two results so far obtained, we find that

$$(A(\Delta) Hf, g)$$
$$= \int_E \int_E \int_E A(P, R; \Delta) K(R, Q) f(Q) \overline{g(P)} \, dQ \, dR \, dP$$
$$= \int_{\alpha}^{\beta} \lambda \, d(A(\lambda) f, g)$$
$$= \int_E \int_E f(Q) \overline{g(P)} \left(\int_{\alpha}^{\beta} \lambda \, dA(P, Q; \lambda)\right) dQ \, dP,$$

where Δ is a finite interval at positive distance δ from $\lambda = 0$, and f and g are arbitrary elements in $\mathfrak{D}$. If we further require that $g(P)$ belong to some one of the closed linear manifolds $\mathfrak{M}(a)$, $0 < a < +\infty$, described above, we can then perform the integration in the second term in any order that we please. For, by virtue of the inequality

$$\int_{E(a)} \int_E |A(P, Q; \lambda)|^2 \, dQ \, dP \leqq \int_{E(a)} K^2(P) \, dP / \lambda^2$$

established above, we have

$$\int_E |A(P, R; \lambda) K(R, Q)| \, dR$$
$$\leqq K(Q) \left(\int_E |A(P, R; \lambda)|^2 \, dR\right)^{1/2},$$

$$\int_E |g(P)| \left(\int_E |A(P, R; \lambda)|^2 \, dR\right)^{1/2} dP$$
$$= \int_{E(a)} |g(P)| \left(\int_E |A(P, R; \lambda)|^2 \, dR\right)^{1/2} dP$$
$$\leqq \left(\int_{E(a)} |g(P)|^2 \, dP\right)^{1/2} \left(\int_{E(a)} \int_E |A(P, R; \lambda)|^2 \, dR \, dP\right)^{1/2} / \lambda^2$$
$$\leqq \left(\int_E |g(P)|^2 \, dP\right)^{1/2} \left(\int_{E(a)} K^2(P) \, dP\right)^{1/2} / \lambda^2,$$

$$\int_E \int_E \int_E |A(P, R; \lambda) K(R, Q) f(Q) g(P)| \, dR \, dQ \, dP$$
$$\leqq \left(\int_E |g(P)|^2 \, dP\right)^{1/2} \left(\int_{E(a)} K^2(P) \, dP\right)^{1/2} \int_E K(Q) |f(Q)| \, dQ / \lambda^2;$$

and thus conclude that the function $A(P, R; \Delta) K(R, Q) \cdot f(Q) \overline{g(P)}$ is absolutely integrable with respect to P, Q, R. We can therefore write

$$\int_E \int_E f(Q) \overline{g(P)} \left(\int_E A(P, R; \Delta) K(R, Q) \, dR\right) dQ \, dP$$
$$= \int_E \int_E f(Q) \overline{g(P)} \left(\int_\alpha^\beta \lambda \, dA(P, Q; \lambda)\right) dQ \, dP,$$

for f in $\mathfrak{D}$ and g in $\mathfrak{M}(a)$, $0 < a < +\infty$. It follows immediately that

$$\int_E A(P, R; \Delta) K(R, Q) \, dR = \int_\alpha^\beta \lambda \, dA(P, Q; \lambda)$$

almost everywhere in E_2, for an arbitrary finite interval Δ at positive distance from $\lambda = 0$. In order to ascertain the behavior of these expressions when $\alpha \to -\infty$ or $\beta \to +\infty$, we write $K(P, Q) = K_Q(P)$ for fixed Q and observe that for almost every Q in E this is a function of P in $\mathfrak{L}_2$. Thus we have

$$\int_E A(P, R; \Delta) K(R, Q) \, dR = A(\Delta) K_Q(P)$$

and can apply the results of Theorem 9.22. We find that the relation established for a finite interval Δ subsists provided that the integral on the right be treated as an improper integral convergent in the mean—that is, convergent in $\mathfrak{L}_2$ for fixed Q. By interchanging P and Q and then taking conjugates, we obtain the results stated in (2). By similar reasoning, we see that

$$(A(-\varepsilon)-A(-\mu))\,K_Q(P)+(A(\mu)-A(\varepsilon))\,K_Q(P)$$
$$=\left[\int_{-\mu}^{-\varepsilon}+\int_{+\varepsilon}^{+\mu}\right]\lambda\,d\,A(P,Q;\lambda)$$

converges in $\mathfrak{L}_2$ for fixed Q to

$$(A(0-0)-A(-\infty))\,K_Q(P)+(A(+\infty)-A(0+0))\,K_Q(P)$$
$$=K_Q(P)-(B(0)-B(0-0))\,K_Q(P).$$

In order to evaluate the second term on the right, we observe that, for an arbitrary function $f(P)$ in $\mathfrak{L}_2$, we have

$$\begin{aligned}(f,(B(0)-B(0-0))\,K_Q) &= ((B(0)-B(0-0))f,K_Q)\\ &=\int_E(B(0)-B(0-0))f(P)\,\overline{K(P,Q)}\,dP\\ &=\int_E K(Q,P)\,(B(0)-B(0-0))f(P)\,dP\\ &=H^*(B(0)-B(0-0))f(Q)=0,\end{aligned}$$

in accordance with Theorem 9.23. Since $f(P)$ is arbitrary, it follows at once that $(B(0)-B(0-0))\,K_Q(P)=0$. Thus we find that $K(P,Q)=\int_{-\infty}^{+\infty}\lambda\,d\,A(P,Q;\lambda)$, in the sense described in (2).

We turn next to the proof of (3). By an argument similar to that used in establishing Theorem 10.3 (4), we can verify that for almost every P in E the function of λ defined by the integral $\int_E A(P,Q;\lambda)f(Q)\,dQ$ has the property

$$V\left(\int_E A(P,Q;\lambda)f(Q)\,dQ;\Delta\right)\leq K(P)\int_E K(Q)\,|f(Q)|\,dQ/\delta^2,$$

whenever Δ is an interval at positive distance δ from $\lambda=0$ and $f(P)$ is a function in $\mathfrak{D}$. Reasoning analogous to that used in the preceding paragraph shows immediately that

$$\int_E\left(\int_\alpha^\beta \lambda\, d A(P, Q; \lambda)\right) f(Q)\, dQ = \int_\alpha^\beta \lambda\, d\left(\int_E A(P, Q; \lambda) f(Q)\, dQ\right),$$

whenever Δ is a finite interval $\alpha \leqq \lambda \leqq \beta$ at positive distance δ from $\lambda = 0$ and $f(P)$ is a function in $\mathfrak{D}$. If we now apply (2), we find that

$$\begin{aligned}\int_E\left(\int_E \overline{A(Q, R; \Delta)}\, \overline{K(R, P)}\, dR\right) f(Q)\, dQ \\ = \int_E\left(\int_E K(P, R)\, A(R, Q; \Delta)\, dR\right) f(Q)\, dQ \\ = \int_\alpha^\beta \lambda\, d\left(\int_E A(P, Q; \lambda) f(Q)\, dQ\right).\end{aligned}$$

If for fixed P we write $K(Q, P) = K_P(Q)$, then this function of Q belongs to $\mathfrak{L}_2$ for almost every P in E; and we can put the equation just established into the more convenient form

$$(f, A(\Delta) K_P) = \int_\alpha^\beta \lambda\, d\left(\int_E A(P, Q; \lambda) f(Q)\, dQ\right).$$

Thus we can obtain the relation

$$\begin{aligned}\int_E A(P, Q; \Delta) f(Q)\, dQ \\ = \int_\alpha^\beta 1/\lambda\, d\left(\int_\alpha^\lambda \mu\, d\left(\int_E A(P, Q; \mu) f(Q)\, dQ\right)\right) \\ = \int_\alpha^\beta 1/\lambda\, d(f, A(\lambda) K_P)\end{aligned}$$

by an obvious application of Lemma 5.1(6). We may now remove the restrictions previously imposed upon $f(P)$. If $f(P)$ is an arbitrary function in $\mathfrak{L}_2$ and $\{f_n(P)\}$ is a sequence in $\mathfrak{D}$ which converges in $\mathfrak{L}_2$ to the limit $f(P)$, then $(f_n, A(\lambda) K_P)$ converges boundedly to $(f, A(\lambda) K_P)$ for $-\infty < \lambda < +\infty$. Hence, by a familiar argument, based on the formula for integration by parts given in Lemma 5.1(9), we find that

$$\lim_{n\to\infty} \int_\alpha^\beta 1/\lambda\, d(f_n, A(\lambda) K_P) = \int_\alpha^\beta 1/\lambda\, d(f, A(\lambda) K_P).$$

Since

$$\lim_{n\to\infty} \int_E A(P, Q; \Delta) f_n(Q)\, dQ = \int_E A(P, Q; \Delta) f(Q)\, dQ,$$

we find that

$$\int_E A(P, Q; \Delta) f(Q)\, dQ = \int_\alpha^\beta 1/\lambda\, d(f, A(\lambda) K_P)$$

for every function $f(P)$ in $\mathfrak{L}_2$. Thus, by using Lemma 5.1(5), we see that, considered as a function of λ, the integral $\int_E A(P, Q; \lambda) f(Q)\, dQ$ is of bounded variation over any interval which is at positive distance from $\lambda = 0$; and, by using Lemma 5.1 (6) in the same manner as above, we obtain the equation

$$(f, A(\Delta) K_P) = \int_\alpha^\beta \lambda\, d\left(\int_E A(P, Q; \lambda) f(Q)\, dQ\right)$$

for every finite interval Δ at positive distance from $\lambda = 0$ and for every function $f(P)$ in $\mathfrak{L}_2$. Thus it follows that

$$\begin{aligned} &\int_{-\infty}^{+\infty} \lambda\, d\left(\int_E A(P, Q; \lambda) f(Q)\, dQ\right) \\ &= \lim_{\varepsilon \to 0, \mu \to \infty} \left(\int_{-\mu}^{-\varepsilon} + \int_{+\varepsilon}^{+\mu}\right) \lambda\, d\left(\int_E A(P, Q; \lambda) f(Q)\, dQ\right) \\ &= (f, (A(0-0) - A(-\infty)) K_P) + (f, (A(+\infty) - A(0+0)) K_P) \\ &= (f, K_P) - (f, (B(0) - B(0-0)) K_P), \qquad \mu > 0,\ \varepsilon > 0. \end{aligned}$$

The first term in this result can be written in the form

$$(f, K_P) = \int_E f(Q)\, \overline{K(Q, P)}\, dQ = \int_E K(P, Q) f(Q)\, dQ.$$

The second vanishes, by the argument used in the preceding paragraph. Thus the proof of (3) is completed.

The proof of (4) is immediate. If $f(P)$ is an arbitrary element in $\mathfrak{L}_2$ and $g(P)$ an arbitrary element in $\mathfrak{D}$, then

$$\begin{aligned} &\int_E \int_E K(P, Q) f(Q)\, \overline{g(P)}\, dQ\, dP \\ &= \int_E f(Q) \overline{\left(\int_E K(Q, P) g(P)\, dP\right)}\, dQ, \end{aligned}$$

as we showed in the proof of Theorem 10.1. The second integral can be written

$$\begin{aligned}(f, Hg) &= \int_{-\infty}^{+\infty} \lambda\, d(B(\lambda) f, g)\\ &= \int_{-\infty}^{+\infty} \lambda\, d(A(\lambda) f, g)\\ &= \int_{-\infty}^{+\infty} \lambda\, d\left(\int_E \int_E A(P, Q; \lambda) f(Q) \overline{g(P)}\, dQ\, dP\right),\end{aligned}$$

by virtue of the formulas given in Theorem 9.19 and the representation of $A(\lambda)$ as an integral operator.

Finally, we may observe that (5) is merely the concrete form of the abstract relation

$$\begin{aligned}&(A(-\varepsilon) - A(-\mu)) f + (A(\mu) - A(\varepsilon)) f\\ &\to (A(0-0) - A(-\infty)) f + (A(+\infty) - A(0+0)) f\end{aligned}$$

where the second term can be written as $f - (B(0) - B(0-0))f$. This appears at once by reference to Theorem 9.22.

In conclusion we shall make a few remarks concerning the cases where the integral operator H has the property that $\tilde{H}$ is maximal or self-adjoint. We have*

THEOREM 10.5. *If $\tilde{H}$ is a maximal symmetric transformation, then the kernel $\mathcal{A}(P, Q; \lambda)$ constructed in Theorem 10.4 is independent of the sequence $\{H^{(n)}\}$ used in the construction. If $\tilde{H}$ is self-adjoint, then the kernel $\mathcal{A}(P, Q; \lambda)$ has the property that*

$$\int_E \mathcal{A}(P, R; \lambda) \mathcal{A}(R, Q; \mu)\, dR = \begin{cases} \mathcal{A}(P, Q; \mu), & \mu \leqq \lambda < 0\\ 0, & \mu \leqq 0 \leqq \lambda\\ -\mathcal{A}(P, Q; \lambda), & 0 < \mu \leqq \lambda\end{cases}$$

The results of Theorem 9.20 show that, whenever $\tilde{H}$ is maximal, then $B(\lambda)$, $A(\lambda)$, and $\mathcal{A}(P, Q; \lambda)$ are unique, in the sense that they are independent of the approximating sequence $\{H^{(n)}\}$ in terms of which they are constructed. The particular relations satisfied by the kernel $\mathcal{A}(P, Q; \lambda)$ in case $\tilde{H}$ is self-adjoint merely reflect the relations

* Carleman, *Sur les équations intégrales singulières à noyau réel et symètrique,* Uppsala, 1923, Chapter III. Carleman considers real operators so that $\tilde{H}$ is maximal if and only if it is self-adjoint. A real kernel such that the associated operator $\tilde{H}$ is self-adjoint is called by Carleman a kernel of Class I; all other real kernels constitute his Class II.

$$A(\lambda)\,A(\mu) \equiv \left\{\begin{array}{ll} A(\mu), & \mu \leqq \lambda < 0 \\ O, & \mu \leqq 0 \leqq \lambda \\ -A(\lambda), & 0 < \mu \leqq \lambda \end{array}\right\},$$

which are valid because $B(\lambda) \equiv E(\lambda)$ is the resolution of the identity associated with $\tilde{H}$. We shall give the proof in detail for the case $\mu \leqq \lambda < 0$; the other cases can then be treated in an analogous manner. In the equation $(A(\lambda)\,A(\mu)f, g) = (A(\mu)f, g)$, $\mu \leqq \lambda < 0$, we take f and g as arbitrary elements in the closed linear manifold $\mathfrak{M}(a)$, described in the opening paragraph of the proof of Theorem 10.4. The equation can then be put in the form

$$\int_E \int_E \int_E A(P, R; \lambda)\,A(R, Q; \mu) f(Q) \overline{g(P)}\, dQ\, dP\, dR$$
$$= \int_E \int_E A(P, Q; \mu) f(Q) \overline{g(P)}\, dQ\, dP,$$

where both integrals are absolutely convergent. The absolute convergence of the first integral can be established by means of the inequalities

$$\int_E |A(P, R; \lambda)\,A(R, Q; \mu)|\, dR$$
$$\leqq \left(\int_E |A(P, R; \lambda)|^2\, dR\right)^{1/2} \left(\int_E |A(R, Q; \mu)|^2\, dR\right)^{1/2},$$

$$\int_E \left(\int_E |A(R, Q; \mu)|^2\, dR\right)^{1/2} |f(Q)|\, dQ$$
$$\leqq \left(\int_{E(a)} \int_E |A(Q, R; \mu)|^2\, dR\, dQ\right)^{1/2} \left(\int_E |f(Q)|^2\, dQ\right)^{1/2},$$

$$\int_E \left(\int_E |A(P, R; \lambda)|^2\, dR\right)^{1/2} |\overline{g(P)}|\, dP$$
$$\leqq \left(\int_{E(a)} \int_E |A(P, R; \lambda)|^2\, dR\, dQ\right)^{1/2} \left(\int_E |g(P)|^2\, dP\right)^{1/2};$$

and that of the second integral can be proved in a similar manner. We can therefore write

$$\int_E \int_E \left(\int_E A(P, R; \lambda)\,A(R, Q; \mu)\, dR\right) f(Q) \overline{g(P)}\, dQ\, dP$$
$$= \int_E \int_E A(P, Q; \mu) f(Q) \overline{g(P)}\, dQ\, dP$$

for all f and g in $\mathfrak{M}(a)$, $0<a<\infty$. It follows that

$$\int_E A(P, R; \lambda) A(R, Q; \mu)\, dR = A(P, Q; \mu), \qquad \mu \leqq \lambda < 0,$$

as we wished to prove.

§ 2. Ordinary Differential Operators of the First Order

We shall now apply the general theory of Chapter IX to the study of symmetric transformations associated with formal differential operators. It is not our object to develop a complete theory; we wish rather to illustrate the abstract results of Chapter IX by a detailed discussion of some special cases of recognized importance. In this section we shall examine ordinary linear differential operators of the first order, and in the next we shall consider those of the second order.

In order that the formal differential operator $P\dfrac{d}{dx}+Q\cdot$ coincide with its formal or Lagrange adjoint $-\dfrac{d}{dx}(\bar{P}\cdot)+\bar{Q}\cdot$, we require that $P=-\bar{P}$ and $Q=-\bar{P}'+\bar{Q}$, where the dash indicates differentiation with respect to x. These conditions imply that $P = p/i$, $Q = -p'/2i+q$, where p and q are real functions. We shall therefore consider the formal differential operator $p\dfrac{1}{i}\dfrac{d}{dx}+(-p'/2i+q)\cdot$ on an arbitrary finite or infinite interval (a, b), where the coefficients p and q have the properties:

(1) $p(x)$ is a positive absolutely continuous function* on the open interval $a<x<b$;

(2) $q(x)$ is Lebesgue-integrable over every closed interval interior to (a, b). It is possible to show that the study of this formal differential operator is equivalent to the study of the operator $\dfrac{1}{i}\dfrac{d}{dx}$ on an appropriately chosen interval (a', b').

---

* We shall say that a function is absolutely continuous on an open interval if it is absolutely continuous on every closed subinterval. For the properties of absolutely continuous functions which we use here we may refer to Carathéodory, *Vorlesungen über reelle Funktionen*, second edition, Leipzig, 1927, Chapters IX and X.

The indicated reduction depends upon the following theorem.

THEOREM 10.6. *Let* $\alpha(x) = A\int_c^x 1/p(\xi)\,d\xi + B$ *and* $\beta(x) = (|A|/p(x))^{1/2}\exp\left(i\int_c^x q(\xi)/p(\xi)\,d\xi\right)$, *where* $A \neq 0$ *and* B *are arbitrary real numbers,* $x = c$ *is an arbitrary interior point of the interval* (a, b), *and* $\exp z = e^z$; *let* a' *and* b' *be respectively the lesser and the greater of the numbers* $\alpha(a)$ *and* $\alpha(b)$; *let* $\mathfrak{L}_2(a, b)$ *be the Hilbert space of all complex-valued Lebesgue-measurable functions* $f(x)$, $a < x < b$, *such that the Lebesgue integral* $\int_a^b |f(x)|^2\,dx$ *exists; and let* $\mathfrak{L}_2(a', b')$ *be the Hilbert space similarly defined over the interval* (a', b'). *Then* $\mathfrak{L}_2(a, b)$ *and* $\mathfrak{L}_2(a', b')$ *can be put in one-to-one isometric correspondence by defining* $f(x) = g(\alpha(x))/\beta(x)$ *as the correspondent in* $\mathfrak{L}_2(a, b)$ *of* $g(x)$ *in* $\mathfrak{L}_2(a', b')$. *If* $f(x)$ *in* $\mathfrak{L}_2(a, b)$ *and* $g(x)$ *in* $\mathfrak{L}_2(a', b')$ *are corresponding functions, then a necessary and sufficient condition that* $g(x)$ *be absolutely continuous on the open interval* (a', b') *and have a derivative* $g'(x)$ *in* $\mathfrak{L}_2(a', b')$ *is that* $f(x)$ *be absolutely continuous on the open interval* (a, b) *and have a derivative* $f'(x)$ *such that* $pf'/i + (-p'/2i + q)f$ *is a function in* $\mathfrak{L}_2(a, b)$. *When these conditions are satisfied, the functions* Ag'/i *and* $pf'/i + (-p'/2i + q)f$ *are in correspondence. By appropriate choice of the constants* A *and* B, *the interval* (a', b') *can be reduced to one of the three intervals* $(0, 1)$, $(0, +\infty)$, $(-\infty, +\infty)$, *according as both, just one, or neither of the two integrals* $\int_a^c 1/p(\xi)\,d\xi$, $\int_c^b 1/p(\xi)\,d\xi$ *is convergent. In the second and third cases, we may restrict* A *to the values* ± 1.

Our assumptions concerning the function p show that α is a real monotone function of x with no interval of constancy, increasing or decreasing according as A is positive or negative. The relation $y = \alpha(x)$ therefore determines a one-to-one correspondence between the open intervals $a < x < b$ and $a' < y < b'$, so that x can be expressed in terms of y by the relation $x = \gamma(y)$, where γ is a real monotone function of y. It is easily shown that γ is absolutely continuous on the open interval (a', b'): for if two sets on the intervals (a, b) and (a', b') respectively are in correspondence by the relation $x = \gamma(y)$, $y = \gamma(x)$, then one is a Borel set if and only if the other is a Borel set; and, by virtue of the inequality $|A|/p(x) > 0$, one is a Borel set of zero measure if and only if the other is a Borel set of zero measure. The function $\beta(x)$ is also absolutely continuous on the open interval (a, b), since it is built from the absolutely continuous functions $p(x) > 0$ and $\int_c^x q(\xi)/p(\xi)\,d\xi$ by operations, such as multiplication and exponentiation, which are known to be applicable within the class of absolutely continuous functions. Finally, we observe that $1/\beta = \overline{\beta}/\beta\overline{\beta} = \overline{\beta}p/|A|$ is absolutely continuous on the open interval (a, b).

Between the class of all functions $f(x)$ defined on the open interval (a, b) and the class of all functions $g(x)$ defined on the open interval (a', b'), we now determine a one-to-one correspondence by the relations

$$f(x) = g(\alpha(x))/\beta(x), \qquad g(x) = \beta(\gamma(x))\,f(\gamma(x)).$$

The absolute continuity of the monotone functions $y = \alpha(x)$, $x = \gamma(y)$ implies that $f(x)$ is Lebesgue-measurable if and only if its correspondent $g(x)$ is Lebesgue-measurable. By virtue of the relations $1/|\beta(x)|^2 = p/|A| = |\alpha'(x)|$, we find that

$$\begin{aligned}\int_a^b |f(x)|^2\,dx &= \int_a^b |g(\alpha(x))|^2/|\beta(x)|^2\,dx\\ &= \int_a^b |g(\alpha(x))|^2\,|\alpha'(x)|\,dx\\ &= \int_{a'}^{b'} |g(x)|^2\,dx,\end{aligned}$$

where the existence of the first integral implies and is implied by that of the last. Hence, $f(x)$ is in $\mathfrak{L}_2(a, b)$ if and only if its correspondent $g(x)$ is in $\mathfrak{L}_2(a', b')$. The isometric character of the correspondence thus determined between $\mathfrak{L}_2(a, b)$ and $\mathfrak{L}_2(a', b')$ is a consequence of the relation

$$\begin{aligned}\int_a^b f_1(x)\,\overline{f_2(x)}\,dx &= \int_a^b (g_1(\alpha(x))/\beta(x))\,\overline{(g_2(\alpha(x))/\beta(x))}\,dx \\ &= \int_a^b g_1(\alpha(x))\,\overline{g_2(\alpha(x))}\,|\alpha'(x)|\,dx \\ &= \int_{a'}^{b'} g_1(x)\,\overline{g_2(x)}\,dx,\end{aligned}$$

where g_1 and g_2 are the correspondents of f_1 and f_2 respectively. An important property of the correspondence is that $f(x)$ is absolutely continuous if and only if its correspondent $g(x)$ is absolutely continuous. If, for example, $f(x)$ is absolutely continuous, then $f(\gamma(x))$ and $\beta(\gamma(x))$ are absolutely continuous, since each is an absolutely continuous function of a monotone absolutely continuous function; and their product $g(x)$ is likewise absolutely continuous. The converse result is obtained by similar reasoning. If now $f(x)$ and $g(x)$ are corresponding functions in $\mathfrak{L}_2(a, b)$ and $\mathfrak{L}_2(a', b')$ respectively, and if both functions are absolutely continuous, we find that the function $f^*(x)$, where

$$\begin{aligned}f^*(x) &= pf'/i+(-p'/2i+q)f \\ &= (p/i)\,(g(\alpha(x))/\beta(x))'+(-p'/2i+q)\,g(\alpha(x))/\beta(x) \\ &= A\,g'(\alpha(x))/i\beta(x),\end{aligned}$$

and the function $g^*(x) = Ag'(x)/i$ are in correspondence with each other; and we know that the first belongs to $\mathfrak{L}_2(a, b)$ if and only if the second belongs to $\mathfrak{L}_2(a', b')$. With this result we have established all the assertions of the theorem concerning the correspondence under discussion.

The final remarks bearing on the reduction of the interval (a', b') to a normal form require no detailed examination.

As a result of the theorem just proved, it is clear that every linear transformation in $\mathfrak{L}_2(a, b)$ associated with the formal differential operator $p(x)\dfrac{1}{i}\dfrac{d}{dx}+(-p'(x)/2i+q(x))$, $a<x<b$, is isomorphic with a linear transformation in $\mathfrak{L}_2(a', b')$ associated with the formal differential operator $A\dfrac{1}{i}\dfrac{d}{dx}$,

$a' < x < b'$, where (a', b') is one of the three typical intervals $(0, 1)$, $(0, +\infty)$, $(-\infty, +\infty)$. Since the constant multiplier $A \neq 0$ affects spectral and other properties in an obvious and quite trivial manner, it is sufficient for us to examine the formal differential operator $\frac{1}{i}\frac{d}{dx}$ on each of the three typical intervals. We describe the facts in the three theorems which follow.

THEOREM 10.7. *Let* $\mathfrak{D}^*$ *be the class of all functions* $f(x)$ *in* $\mathfrak{L}_2(0, 1)$ *such that* $f(x)$ *is absolutely continuous for* $0 < x < 1$ *and has a derivative* $f'(x)$ *belonging to* $\mathfrak{L}_2(0, 1)$*—in other words, the class of all functions* $f(x)$ *expressible in the form* $f(x) = c + \int_0^x g(\xi)\, d\xi$, *where* c *is an arbitrary constant and* $g(x)$ *is an arbitrary function in* $\mathfrak{L}_2(0, 1)$*; let* $\mathfrak{D}(\varphi)$ *be the class of all functions* $f(x)$ *in* $\mathfrak{D}^*$ *such that* $f(0) - e^{i\varphi} f(1) = 0$, $0 \leqq \varphi < 2\pi$*—in other words, the class of all functions* $f(x)$ *expressible in the form* $f(x) = c\, e^{-i\varphi x} + \int_0^x g(\xi)\, d\xi$, *where* c *is an arbitrary constant and* $g(x)$ *is an arbitrary function in* $\mathfrak{L}_2(0, 1)$ *orthogonal to the function* $h(x) \equiv 1$, *so that* $\int_0^1 g(\xi)\, d\xi = 0$*; and let* $\mathfrak{D}$ *be the class of all functions* $f(x)$ *in* $\mathfrak{D}^*$ *such that* $f(0) = f(1) = 0$*—in other words, the class of all functions* $f(x)$ *expressible in the form* $f(x) = \int_0^x g(\xi)\, d\xi$, *where* $g(x)$ *is an arbitrary function in* $\mathfrak{L}_2(0, 1)$ *orthogonal to the function* $h(x) \equiv 1$. *Let* H, $H(\varphi)$, *and* T *be transformations with the domains* $\mathfrak{D}$, $\mathfrak{D}(\varphi)$, *and* $\mathfrak{D}^*$ *respectively, each of which takes an arbitrary function* $f(x)$ *in its domain into the function* $\frac{1}{i} f'(x)$. *Then* H *is a closed linear symmetric transformation with the adjoint* $H^* \equiv T$ *and the deficiency-index* $(1, 1)$*; and the transformation* $H(\varphi)$ *is a self-adjoint extension of* H. *If* S *is an arbitrary maximal symmetric extension of* H, *then* S *coincides with* $H(\varphi)$ *for exactly one value of* φ *on the range* $0 \leqq \varphi < 2\pi$. *The transformations* $H(\varphi + \psi)$ *and* $H(\varphi) - \psi \cdot I$ *are equivalent by the unitary transformation* $U(\psi)$ *which takes an arbitrary function*

$f(x)$ in $\mathfrak{L}_2(0, 1)$ into $e^{i\psi x}f(x)$, $-\infty<\varphi<+\infty$, $-\infty<\psi<+\infty$. The transformation $H(\varphi)$ has a simple point spectrum, consisting of the characteristic values $2n\pi-\varphi$, $n=0, \pm 1, \pm 2, \cdots$, with corresponding characteristic functions $e^{i(2n\pi-\varphi)x}$.

We must first establish the equivalence of the two characterizations given for each of the sets $\mathfrak{D}^*$, $\mathfrak{D}(\varphi)$, $\mathfrak{D}$. In the case of $\mathfrak{D}^*$, it is evident that any function $f(x)$ expressible in the indicated form is absolutely continuous on the closed interval $(0, 1)$, belongs to $\mathfrak{L}_2(0, 1)$ (by virtue of the fact that it is a bounded continuous function on the closed interval), and has a derivative $f'(x) = g(x)$ in $\mathfrak{L}_2(0, 1)$. It is to be observed that, whenever $|g(x)|^2$ is integrable over the finite interval $(0, 1)$, $g(x)$ is also integrable over that interval. Thus every function $f(x)$ of the indicated type belongs to $\mathfrak{D}^*$. On the other hand, if $f(x)$ is an absolutely continuous function on the open interval $(0, 1)$ and if its derivative $f'(x) = g(x)$ is in $\mathfrak{L}_2(0, 1)$, we can write $f(x) = c+\int_0^x g(\xi)\, d\xi$, $0<x<1$, where the constant c is determined by the substitution of some particular value of x, such as $x=\frac{1}{2}$. The validity of this equation at the points $x=0$ and $x=1$ remains in question. Since two functions in the Hilbert space $\mathfrak{L}_2(0, 1)$ are regarded as identical if they coincide almost everywhere on the interval $(0, 1)$, we may suppose that $f(x)$ is modified, if necessary, so that the representation is valid for $0 \leqq x \leqq 1$. The modified function is then absolutely continuous on the closed interval $(0, 1)$. We shall therefore introduce the simplifying convention that, whenever we consider a function in $\mathfrak{D}^*$, it shall appear in the modified, absolutely continuous form capable of the integral representation on the closed interval $0 \leqq x \leqq 1$. The integral representations for functions in $\mathfrak{D}$ and $\mathfrak{D}(\varphi)$ are now readily obtained, by applying the boundary conditions $f(0)=f(1)=0$ and $f(0)-e^{i\varphi}f(1)=0$ to the integral representation of functions in $\mathfrak{D}^*$. Thus, in the case of $\mathfrak{D}$, we find that the equations $f(x) = c+\int_0^x g(\xi)\, d\xi$, $f(0)=f(1)=0$, are satisfied if and only if $c=0$, $\int_0^1 g(\xi)\, d\xi=0$, as asserted in the theorem. The integral representation of functions in $\mathfrak{D}(\varphi)$ results immediately from the observation that $f(x)$ is in $\mathfrak{D}(\varphi)$ if and only if $f(x)-f(0)\,e^{-i\varphi x}$ is in $\mathfrak{D}$.

The facts just established show that the sets $\mathfrak{D}$, $\mathfrak{D}(\varphi)$, $\mathfrak{D}^*$ are linear manifolds in $\mathfrak{L}_2(0, 1)$ satisfying the relations $\mathfrak{D} \subseteqq \mathfrak{D}(\varphi) \subseteqq \mathfrak{D}^*$. It is evident that for distinct numbers φ_1 and φ_2 on the range $0 \leqq \varphi < 2\pi$ the sets $\mathfrak{D}(\varphi_1)$ and $\mathfrak{D}(\varphi_2)$ have only the elements of $\mathfrak{D}$ in common. We shall now show that $\mathfrak{D}$ is everywhere dense in $\mathfrak{L}_2(0, 1)$; that is, that the only function $f(x)$ in $\mathfrak{L}_2(0, 1)$ which is orthogonal to all functions $g(x)$ in $\mathfrak{D}$ is the function $f(x) = 0$. By an integration by parts, we can write

$$0 = \int_0^1 f(x)\,\overline{g(x)}\,dx = -\int_0^1 F(x)\,\overline{g'(x)}\,dx$$

where $F(x) = \int_0^x f(\xi)\,d\xi$; the boundary conditions $g(0) = g(1) = 0$ must be employed in the calculation. By the characterization of $\mathfrak{D}$ given above, we see that g' is an arbitrary function in $\mathfrak{L}_2(0, 1)$ orthogonal to the function $h(x) \equiv 1$. We conclude that $F(x)$ must be a constant multiple of $h(x)$ and that $f(x) = F'(x)$ must vanish identically, as we wished to prove. Since the linear manifolds $\mathfrak{D}^*$ and $\mathfrak{D}(\varphi)$ contain $\mathfrak{D}$, they must also be everywhere dense in $\mathfrak{L}_2(0, 1)$.

The transformations H, $H(\varphi)$, and T, described in the statement of the theorem, are obviously linear transformations satisfying the relations $H \subseteqq H(\varphi) \subseteqq T$. Since the domain of H is everywhere dense in $\mathfrak{L}_2(0, 1)$, its adjoint H^* exists. A necessary and sufficient condition that $g(x)$ belong to the domain of H^* and that H^*g coincide with g^* is that the relation

$$\int_0^1 \frac{1}{i} f'(x)\,\overline{g(x)}\,dx = \int_0^1 f(x)\,\overline{g^*(x)}\,dx$$

be satisfied for some function $g^*(x)$ in $\mathfrak{L}_2(0, 1)$ and all functions $f(x)$ in $\mathfrak{D}$. By an integration by parts this condition becomes

$$\int_0^1 \frac{1}{i} f'(x)\,\overline{g(x)}\,dx = -\int_0^1 f'(x)\,\overline{G^*(x)}\,dx,$$

where $G^*(x) = \int_0^x g^*(\xi)\, d\xi$, and is therefore equivalent to the assertion that the function $G^*(x) - g(x)/i$ is orthogonal to $f'(x)$ whenever $f(x)$ is in $\mathfrak{D}$. By the argument used above, we see that this condition is satisfied if and only if $G^*(x) - g(x)/i$ reduces to a constant a; in other words, if and only if $g(x) = a + \int_0^x i g^*(\xi)\, d\xi$ is a function in $\mathfrak{D}^*$ and $g'(x)/i = g^*(x)$. Thus the transformations H^* and T are identical. Now from the relation $H \subseteq T$ we conclude that $T^* \subseteq H^* \equiv T$. This means that the domain of T^* consists of functions in $\mathfrak{D}^*$ and that T^* carries a function $g(x)$ in its domain into $g'(x)/i$. Hence we see that $g(x)$ is in the domain of T^* if and only if it satisfies the additional condition $\int_0^1 \frac{1}{i} f'(x)\, \overline{g(x)}\, dx = \int_0^1 f(x) \overline{\frac{1}{i} g'(x)}\, dx$ for every function $f(x)$ in $\mathfrak{D}^*$, the domain of T. The formula for integration by parts shows that this condition is equivalent to the equation $f(1)\, \overline{g(1)} - f(0)\, \overline{g(0)} = 0$. This equation is satisfied if and only if $g(0) = g(1) = 0$, as we see by setting $f(x) = ax + b$, where a and b are arbitrary constants. Thus the domain of T^* is the set $\mathfrak{D}$, and T^* is identical with H. The relations $H \subseteq T \equiv H^*$, $T^* \equiv H$ show us immediately that H is a closed linear symmetric transformation. The deficiency-index of H is easily calculated by determining the characteristic functions of the transformation $H^* \equiv T$, as we have shown in Theorem 9.8. These functions are the solutions of the differential equation $\frac{1}{i} f' = l f$ which belong to $\mathfrak{D}^*$. It is easily ascertained that the desired solutions are constant multiples of e^{ilx}. Every complex number l is therefore a characteristic value of multiplicity one for H^*; and the deficiency-index of H is $(1, 1)$, as we wished to prove.

We shall now determine all the maximal symmetric extensions of H, by the methods described in Chapter IX, § 1. According to the general theory, we must first find all the isometric transformations of $\mathfrak{D}^+ = \mathfrak{L}_2(0, 1) \ominus \mathfrak{R}_{-i}$ into

$\mathfrak{D}^- = \mathfrak{L}_2(0,1) \ominus \mathfrak{R}_{+i}$ where $\mathfrak{R}_{+i}$ and $\mathfrak{R}_{-i}$ are the ranges of the transformations H_{+i} and H_{-i} respectively. By Theorem 9.4, $\mathfrak{D}^+$ consists of all the functions $f(x) = ae^{-x}$, $\mathfrak{D}^-$ of all the functions $f(x) = ae^{+x}$, where a is an arbitrary constant. Hence the desired isometric transformations are those which carry ae^{-x} into cae^{+x}, where c is an arbitrary complex constant subject to the condition $\int_0^1 e^{-2x}\,dx = |c|^2 \int_0^1 e^{+2x}\,dx$. It is found that $c = e^{-1+i\theta}$, where $0 \leqq \theta < 2\pi$. Corresponding to each value of θ, there exists a maximal symmetric extension of H which we denote by $S(\theta)$. The process given in Theorems 9.1–9.3 for the construction of $S(\theta)$ shows that a function $f(x)$ is in the domain of $S(\theta)$ if and only if it is expressible in the form $f(x) = g(x) + a(e^{-x} - e^{-1+i\theta}e^{x})$ where $g(x)$ is a function in the domain of H. Every such function evidently belongs to $\mathfrak{D}^*$ and satisfies the boundary condition $f(0) - \dfrac{e - e^{i\theta}}{1 - ee^{i\theta}} f(1) = 0$, as we see by simple calculations depending upon the fact that $g(0) = g(1) = 0$. Conversely, any function $f(x)$ in $\mathfrak{D}^*$ which satisfies this boundary condition is in the domain of $S(\theta)$; for the function $g(x) = f(x) - a(e^{-x} - e^{-1+i\theta}e^{x})$, where $a = ef(1)/(1 - ee^{i\theta})$, is a function in $\mathfrak{D}^*$ satisfying the boundary conditions $g(0) = g(1) = 0$ and hence belongs to $\mathfrak{D}$. If it be observed that the linear fractional transformation $w = \dfrac{e - z}{1 - ez}$ takes the unit circle $z = e^{i\theta}$ into the unit circle $w = e^{i\varphi}$, it is immediately clear that there is a one-to-one correspondence between the ranges $0 \leqq \theta < 2\pi$ and $0 \leqq \varphi < 2\pi$ such that $\dfrac{e - e^{i\theta}}{1 - ee^{i\theta}} = e^{i\varphi}$. Thus the boundary condition which characterizes the domain of $S(\theta)$ can be put in the form $f(0) - e^{i\varphi}f(1) = 0$. Since $S(\theta) \subset H^*$, we see that $S(\theta)$ and $H(\varphi)$ are identical for corresponding values θ and φ. By reference to the results of Theorem 9.3, we find that all the maximal symmetric extensions of the transformation H, with deficiency-index $(1, 1)$, are self-adjoint. Hence $S(\theta)$ and $H(\varphi)$ are self-adjoint transformations.

We shall now examine the connection between the transformations $H(\varphi+\psi)$, $H_\psi(\varphi) \equiv H(\varphi) - \psi \cdot I$, and $U(\psi)$. The unitary character of $U(\psi)$ is evident from the defining relations given in the statement of the theorem. The family of transformations $U(\psi)$, $-\infty < \psi < +\infty$, is seen at once to be a group, by virtue of the identities $U(0) \equiv I$, $U^{-1}(\psi) \equiv U(-\psi)$, $U(\psi_1)\,U(\psi_2) \equiv U(\psi_1+\psi_2)$. We first show that $U(\psi)$ takes $\mathfrak{D}^*$ in a one-to-one manner into itself. If $f(x)$ is absolutely continuous, then $U(\psi)f(x) = e^{i\psi x}f(x)$ is also absolutely continuous; and if $f(x)$ further has a derivative $f'(x)$ in $\mathfrak{L}_2(0, 1)$, then $e^{i\psi x}f(x)$ also has a derivative $e^{i\psi x}f'(x) + i\psi e^{i\psi x}f(x)$ in $\mathfrak{L}_2(0, 1)$. Hence $U(\psi)$ takes $\mathfrak{D}^*$ in a one-to-one manner into a subset of itself. Since this statement is valid for $U^{-1}(\psi) \equiv U(-\psi)$, we see that the image of $\mathfrak{D}^*$ by the transformation $U(\psi)$ is $\mathfrak{D}^*$ itself. Incidentally, we have shown that

$$TU(\psi)f(x) = U(\psi)\,Tf(x) + \psi \cdot U(\psi)f(x)$$

whenever $f(x)$ is in $\mathfrak{D}^*$. This relation can be written in the form

$$T_\psi \equiv U(\psi)\,T\,U^{-1}(\psi).$$

We show next that $U(\psi)$ takes $\mathfrak{D}(\varphi)$ in a one-to-one manner into $\mathfrak{D}(\varphi-\psi)$. For this purpose it is sufficient to observe that, if $f(x)$ is a function in $\mathfrak{D}^*$ satisfying the boundary condition $f(0) - e^{i\varphi}f(1) = 0$, then

$$f^*(x) = U(\psi)f(x) = e^{i\psi x}f(x)$$

is a function in $\mathfrak{D}^*$ satisfying the boundary condition $f^*(0) - e^{i(\varphi-\psi)}f^*(1) = 0$, and conversely. By virtue of this result and of the equation $H(\varphi) = T$, holding in $\mathfrak{D}(\varphi)$, we see that the identity $T_\psi \equiv U(\psi)\,T\,U^{-1}(\psi)$ established above can be written as a relation holding in $\mathfrak{D}(\varphi)$ and can thus be put in the form

$$H_\psi(\varphi) \equiv U(\psi)\,H(\varphi+\psi)\,U^{-1}(\psi).$$

Hence the transformations $H_\psi(\varphi)$ and $H(\varphi+\psi)$ are unitary equivalent, as asserted in the theorem.

Finally we shall determine the spectral properties of the transformation $H(\varphi)$. The characteristic values and characteristic functions for $H(\varphi)$ are easily found by applying to the characteristic functions e^{ilx} of $H^* \equiv T$ the boundary condition which characterizes the domain $\mathfrak{D}(\varphi)$ of $H(\varphi)$. The admissible values of l are calculated from the equation $1 - e^{i\varphi} e^{il} = 0$ and are thus found to be $l = 2n\pi - \varphi$, $n = 0, \pm 1, \pm 2, \cdots$. Each of these values is a characteristic value for $H(\varphi)$; and the characteristic functions associated with $l = 2n\pi - \varphi$ are constant multiples of $e^{i(2n\pi-\varphi)x}$, so that each of these characteristic values has multiplicity one. We shall now show that any number l different from these characteristic values is in the resolvent set of $H(\varphi)$. For this it is sufficient to show that the equation $H(\varphi)f - lf = g$, where l is not a characteristic value and g is an arbitrary function in $\mathfrak{L}_2(0, 1)$, has a unique solution, as we see by reference to Theorem 4.18. The equation to be solved is equivalent to the differential system $\frac{1}{i} f'(x) - lf(x) = g(x)$, $f(0) - e^{i\varphi} f(1) = 0$, where $g(x)$ is given and $f(x)$ is to be determined as a function in $\mathfrak{D}^*$. The general solution of the differential equation $\frac{1}{i} f' - lf = g$ is the function $f(x) = a e^{ilx} + i e^{ilx} \int_0^x e^{-il\xi} g(\xi)\, d\xi$, where a is an arbitrary constant; and this function is clearly an element of $\mathfrak{D}^*$. In order that $f(x)$ satisfy the boundary condition, it is necessary and sufficient that a have the value

$$i e^{i(l+\varphi)} \int_0^1 e^{-il\xi} g(\xi)\, d\xi / (1 - e^{i(l+\varphi)}),$$

as we see by direct calculation; the denominator here cannot vanish because of our restriction on l. With this value for a, we find that $f(x) = \int_0^1 G(x, \xi; l)\, g(\xi)\, d\xi$, where $G(x, \xi; l)$ is a kernel of Hilbert-Schmidt type given by the equations

$$G(x, \xi; l) = \frac{-ie^{il(x-\xi)}}{1-e^{-i(l+\varphi)}}, \qquad 0 \leqq \xi < x \leqq 1,$$

$$G(x, \xi; l) = \frac{+ie^{il(x-\xi)}}{1-e^{+i(l+\varphi)}}, \qquad 0 \leqq x < \xi \leqq 1,$$

$$G(x, x; l) = 0, \qquad 0 \leqq x \leqq 1.$$

Thus the transformation $H_l(\varphi)$ has an inverse which is a bounded integral operator of Hilbert-Schmidt type with domain $\mathfrak{L}_2(0, 1)$, whenever l is not a characteristic value of $H(\varphi)$. This result shows that all such values l belong to the resolvent set of $H(\varphi)$. Hence $H(\varphi)$ has a simple point spectrum, as we wished to prove.

We turn now to the study of the differential operator $\frac{1}{i}\frac{d}{dx}$ on the interval $(0, +\infty)$. It is convenient to state our results in terms of the negative of this operator, namely, the operator $i\frac{d}{dx}$.

THEOREM 10.8. *Let* $\mathfrak{D}^*$ *be the class of all functions* $f(x)$ *in* $\mathfrak{L}_2(0, +\infty)$ *such that* $f(x)$ *is absolutely continuous for* $0<x<+\infty$ *and has a derivative belonging to* $\mathfrak{L}_2(0, +\infty)$*—in other words, the class of all functions* $f(x)$ *in* $\mathfrak{L}_2(0, \infty)$ *expressible in the form* $f(x) = ce^{-x} + \int_0^x g(\xi)\,d\xi$*, where* c *is an arbitrary constant and* $g(x)$ *is a function in* $\mathfrak{L}_2(0, \infty)$*; let* $\mathfrak{D}$ *be the class of all functions* $f(x)$ *in* $\mathfrak{D}^*$ *such that* $f(0) = 0$*—in other words, the class of all functions* $f(x)$ *in* $\mathfrak{L}_2(0, \infty)$ *expressible in the form* $f(x) = \int_0^x g(\xi)\,d\xi$ *where* $g(x)$ *is in* $\mathfrak{L}_2(0, \infty)$*; and let* H *and* T *be transformations with the domains* $\mathfrak{D}$ *and* $\mathfrak{D}^*$ *respectively, each of which takes an arbitrary function* $f(x)$ *in its domain into the function* $if'(x)$*. Then* H *is an elementary symmetric transformation in the sense of Definition* 9.6*; and the transformations* T *and* H^* *coincide.*†

If $f(x)$ is a function in $\mathfrak{D}^*$, we first examine its behavior on a finite interval $(0, a)$: since $f(x)$ is absolutely continuous

† This theorem is due to J. v. Neumann, Journal für Mathematik, 161 (1929), pp. 234–236.

for $0 < x \leqq a$ and since $f'(x)$, considered on the interval $(0, a)$, is a function in $\mathfrak{L}_2(0, a)$, we can proceed as in the first paragraph of the proof of Theorem 10.7 to show that $f(x)$ is an absolutely continuous function on the closed interval $(0, a)$, expressible in the form $f(x) = f(0) + \int_0^x g(\xi)\, d\xi$, $g(x) = f'(x)$. Since a is arbitrary, this expression is valid over the entire interval $(0, +\infty)$. When $f(0) = 0$, so that $f(x)$ is in $\mathfrak{D}$, this representation is entirely satisfactory; but, when $f(0) \neq 0$, it causes some inconvenience due to the fact that the constant $f(0)$ is no longer a function in $\mathfrak{L}_2(0, +\infty)$. In the latter case, however, we observe that the function $f(x) - f(0)\, e^{-x}$ is in $\mathfrak{D}$ and can therefore be represented as the integral $\int_0^x g(\xi)\, d\xi$, where $g(x) = f'(x) + f(0) e^{-x}$; we thus obtain the more satisfactory representation for the function $f(x)$ in $\mathfrak{D}^*$ given by the relation

$$f(x) = f(0)\, e^{-x} + \int_0^x g(\xi)\, d\xi,$$

where $g(x)$ is in $\mathfrak{L}_2(0, \infty)$. We must call attention to the fact that in these representations for functions in $\mathfrak{D}$ and in $\mathfrak{D}^*$ the function $g(x)$ is not an entirely arbitrary element of $\mathfrak{L}_2(0, \infty)$: it must have the property that the indefinite integral $\int_0^x g(\xi)\, d\xi$ is also a function in $\mathfrak{L}_2(0, \infty)$. Under this condition on $g(x)$, the indicated representations always lead to functions in $\mathfrak{D}$ and $\mathfrak{D}^*$ respectively. It is evident that $\mathfrak{D}$ and $\mathfrak{D}^*$ are linear manifolds satisfying the relation $\mathfrak{D} \subseteqq \mathfrak{D}^*$. We can now show that $\mathfrak{D}$, and hence $\mathfrak{D}^*$ also, is everywhere dense in $\mathfrak{L}_2(0, \infty)$. To this end, we prove that the only function $f(x)$ in $\mathfrak{L}_2(0, \infty)$ which satisfies the equation

$$\int_0^{+\infty} f(x)\, \overline{g(x)}\, dx = 0$$

for every function $g(x)$ in $\mathfrak{D}$ is the function $f(x) = 0$. If $h(x)$ is an arbitrary function in $\mathfrak{L}_2(0, \infty)$ with the properties $h(x) = 0$ for $x \geqq a$ and $\int_0^a h(x)\, dx = 0$, then $g(x) = \int_0^x h(\xi)\, d\xi$

is a function in $\mathfrak{D}$ vanishing for $x \geqq a$. For such a function $g(x)$, the relation above takes the form

$$-\int_0^a F(x)\,\overline{h(x)}\,dx = \int_0^a f(x)\,\overline{g(x)}\,dx = \int_0^\infty f(x)\,\overline{g(x)}\,dx = 0,$$

where

$$F(x) = \int_0^x f(\xi)\,d\xi.$$

By reasoning like that used in the proof of the preceding theorem, we conclude that $F(x)$ is constant on the interval $(0, a)$ and that $f(x) = F'(x)$ vanishes there. Since a is arbitrary, we must have $f(x) = 0$, as we wished to prove.

The transformations H and T are evidently linear transformations with domains everywhere dense in $\mathfrak{L}_2(0, \infty)$. We shall give a direct proof of the symmetric character of H. By an integration by parts, we find that

$$\begin{aligned}(Hf, g) - (f, Hg) &= \lim_{a\to+\infty}\left[\int_0^a i f'(x)\,\overline{g(x)}\,dx - \int_0^a f(x)\,\overline{i g'(x)}\,dx\right] \\ &= \lim_{a\to+\infty} i f(a)\,\overline{g(a)}\end{aligned}$$

for all functions $f(x)$ and $g(x)$ in $\mathfrak{D}$. Now $f(x)\,\overline{g(x)}$ is integrable over $(0, \infty)$ since each of its factors belongs to $\mathfrak{L}_2(0, \infty)$; and it tends to a limit when x becomes infinite, as we have just seen. These two properties are compatible only if the limit in question has the value zero. Hence we find that $(Hf, g) = (f, Hg)$ and that H is a linear symmetric transformation. Next we shall show that the transformation H_{-i}^{-1}, the existence of which follows by Theorem 4.14 from the symmetric character of H, has $\mathfrak{L}_2(0, \infty)$ as its domain and is expressed by the equation

$$f(x) = H_{-i}^{-1} g = -i e^{-x} \int_0^x e^{\xi} g(\xi)\,d\xi.$$

For this purpose we must determine all the functions $f(x)$ in $\mathfrak{D}$ such that $i f'(x) + i f(x) = g(x)$, where $g(x)$ is a given function in $\mathfrak{L}_2(0, \infty)$. The general solution of this differential equation is seen to be

$$f(x) = c\,e^{-x} - i\,e^{-x}\int_0^x e^{\xi}\,g(\xi)\,d\xi.$$

In order that this solution belong to $\mathfrak{D}$ it is necessary that $f(0) = 0$, $c = 0$. We thus have to determine whether or not the function

$$f(x) = i\,e^{-x}\int_0^x e^{\xi}\,g(\xi)\,d\xi$$

belongs to $\mathfrak{D}$. It is evident that this is an absolutely continuous function on the interval $0 \leqq x < \infty$, which vanishes at $x = 0$ and has the derivative $f'(x) = -f(x) - ig(x)$; but it is not certain that $f(x)$ or $f'(x)$ belongs to $\mathfrak{L}_2(0, \infty)$. We observe, however, that when $f(x)$ belongs to $\mathfrak{L}_2(0, \infty)$ then $f'(x)$ also belongs to $\mathfrak{L}_2(0, \infty)$. Thus, in order to prove that $f(x)$ is in $\mathfrak{D}$, it is sufficient for us to show that it is in $\mathfrak{L}_2(0, \infty)$. If we form the function $g_n(x)$ equal to $g(x)$ or to zero according as $0 \leqq x \leqq n$ or $x > n$, $n = 1, 2, 3, \cdots$. we see that the corresponding function

$$f_n(x) = -i\,e^{-x}\int_0^x e^{\xi}\,g_n(\xi)\,d\xi$$

is equal to $f(x)$ for $0 \leqq x \leqq n$ and equal to $C_n e^{-x}$, where

$$C_n = -i\int_0^n e^{\xi}\,g(\xi)\,d\xi,$$

for $x > n$. It is obvious at once that $f_n(x)$ belongs to $\mathfrak{L}_2(0, \infty)$ and is therefore in $\mathfrak{D}$ in accordance with the remarks made above. By applying Theorem 4.14, we find that $\| f_n \| = \| H_{-i}^{-1} g_n \| \leqq \| g_n \|$. We can therefore write

$$\int_0^n |f(x)|^2\,dx = \int_0^n |f_n(x)|^2\,dx \leqq \int_0^\infty |f_n(x)|^2\,dx$$
$$\leqq \int_0^\infty |g_n(x)|^2\,dx \leqq \int_0^\infty |g(x)|^2\,dx,$$

and can conclude that

$$\int_0^\infty |f(x)|^2\,dx \leqq \int_0^\infty |g(x)|^2\,dx.$$

Thus $f(x)$ is in $\mathfrak{L}_2(0, \infty)$ and hence in $\mathfrak{D}$. This result shows that H_{-i}^{-1} is a bounded linear transformation with domain

$\mathfrak{L}_2(0, \infty)$, as we wished to prove. By reference to Theorems 2.23 and 2.5 we see that H_{-i}^{-1}, H_{-i}, and H are all closed linear transformations. Theorems 9.1–9.3 can now be applied to the transformation H; they show that H is a maximal symmetric transformation with the deficiency-index $(0, n)$. We can complete our analysis by means of the associated isometric transformation $V \equiv H_{+i} H_{-i}^{-1}$ discussed in Theorem 9.1. The range of V coincides with the range of H_{+i}, and consists of all functions

$$g(x) = i f'(x) - i f(x) \text{ in } \mathfrak{L}_2(0, \infty)$$

where $f(x)$ is in $\mathfrak{D}$. This relation between $f(x)$ and $g(x)$ assumes the form

$$f(x) = -i e^{+x} \int_0^x e^{-\xi} g(\xi) \, d\xi$$

when it is solved under the boundary condition $f(0) = 0$. Since e^{-x}, $f(x)$, and $g(x)$ are functions in $\mathfrak{L}_2(0, \infty)$, the products $e^{-x} f(x)$ and $e^{-x} g(x)$ are integrable over $(0, \infty)$. The equation

$$e^{-x} f(x) = -i \int_0^x e^{-\xi} g(\xi) \, d\xi$$

shows that $e^{-x} f(x)$ tends to the limit $-i \int_0^\infty e^{-\xi} g(\xi) \, d(\xi)$ when x becomes infinite; and the integrability of $e^{-x} f(x)$ requires that this limit have the value zero. Hence the closed linear manifold which is the common range of V and H_{+i} consists of functions $g(x)$ which are orthogonal to the function e^{-x}. In consequence, the deficiency-index of H is subject to the restriction $n \geqq 1$, and H is certainly not self-adjoint. By direct calculation from the formulas given above it is found that

$$V g(x) = g(x) - 2 e^{-x} \int_0^x e^{\xi} g(\xi) \, d\xi$$

for all functions $g(x)$ in $\mathfrak{L}_2(0, \infty)$. We can now determine without difficulty the sequence of functions which can be obtained be repeated applications of V to the function $\varphi_0(x) = \sqrt{2}\, e^{-x}$. By the argument used in the proof of

Theorem 9.10 we see that the set $\{\varphi_n(x)\}$, where $\varphi_{n+1}(x) = V\varphi_n(x) = \varphi_n(x) - 2e^{-x}\int_0^x e^{\xi}\varphi_n(\xi)\,d\xi$ for $n = 0, 1, 2, \cdots$, is an orthonormal set in $\mathfrak{L}_2(0, \infty)$. If we write $\varphi_n(x) = e^{-x}p_n(x)$, we have

$$p_0(x) = \sqrt{2},\ p_{n+1}(x) = p_n(x) - 2\int_0^x p_n(\xi)\,d\xi,$$

and therefore conclude that $p_n(x)$ is a polynomial of degree n. Since these polynomials satisfy the relations

$$\int_0^\infty e^{-2x}p_m(x)\overline{p_n(x)}\,dx = \delta_{mn}, \qquad m, n = 0, 1, 2, \cdots,$$

they are essentially the Laguerre polynomials. Thus the orthonormal set $\{\varphi_n\}$ is known to be complete.† In terms of this set we have

$$g = \sum_{\alpha=0}^{\infty} a_\alpha \varphi_\alpha, \quad Vg = \sum_{\alpha=0}^{\infty} a_\alpha \varphi_{\alpha+1},$$

where $a_n = (g, \varphi_n)$, for every function $g(x)$ in $\mathfrak{L}_2(0, \infty)$. This result identifies V and shows that H is an elementary symmetric transformation, in accordance with Theorem 9.9 and Definition 9.6. We may add that the range of V is the closed linear manifold of all functions in $\mathfrak{L}_2(0, \infty)$ orthogonal to $\varphi_0(x) = \sqrt{2}\,e^{-x}$, and that H has the deficiency-index (0,1).

Theorem 9.4 now enables us to determine the transformation H^*. The linear manifolds $\mathfrak{D}$, $\mathfrak{D}^+$, and $\mathfrak{D}^-$ associated with H are found to be, respectively, the set $\mathfrak{D}$ described in the statement of the present theorem, the set $\mathfrak{O}$ which contains only the function $f(x) = 0$, and the set of all functions ce^{-x} where c is an arbitrary constant. It follows that the domain of H^* consists of those and only those functions $f(x)$ which are expressible in the form $f(x) = ce^{-x} + h(x)$ where $h(x)$ is in $\mathfrak{D}$ — in other words, of those functions which belong to the set $\mathfrak{D}^*$ described in the statement of the theorem. For such a function we have

$$H^*f(x) = H^*ce^{-x} + H^*h(x) = -ice^{-x} + ih'(x) = if'(x).$$

† Courant-Hilbert, *Methoden der Mathematischen Physik*, vol. 1, second edition, 1931, pp. 81–82.

Hence H^* coincides with the transformation T defined above. This result completes the proof of the theorem.

THEOREM 10.9. *Let* $\mathfrak{D}$ *be the class of all functions* $f(x)$ *in* $\mathfrak{L}_2(-\infty, +\infty)$ *such that* $f(x)$ *is absolutely continuous for* $-\infty < x < +\infty$ *and has a derivative* $f'(x)$ *in* $\mathfrak{L}_2(-\infty, +\infty)$; *and let* H *be the transformation with domain* $\mathfrak{D}$ *which takes* $f(x)$ *in* $\mathfrak{D}$ *into* $\frac{1}{i} f(x)$. *Then* H *is a self-adjoint transformation with a simple continuous spectrum. If* U *is the unitary transformation which takes* $f(x)$ *into* $\frac{1}{\sqrt{2\pi}} \int_{-\infty}^{+\infty} e^{ix\xi} f(\xi)\, d\xi$, *in accordance with Theorem* 3.10; *and if* S *is the self-adjoint transformation which takes* $f(x)$ *into* $x f(x)$, *whenever* $f(x)$ *is a function in* $\mathfrak{L}_2(-\infty, +\infty)$ *such that the integral* $\int_{-\infty}^{+\infty} x^2 |f(x)|^2\, dx$ *exists — then* $H \equiv USU^{-1}$.

The set $\mathfrak{D}$ is evidently a linear manifold in $\mathfrak{L}_2(-\infty, +\infty)$. As in the two preceding theorems, we find that $\mathfrak{D}$ is everywhere dense in $\mathfrak{L}_2(-\infty, +\infty)$: if $f(x)$ is a function in this space such that $\int_{-\infty}^{+\infty} f(x) \overline{g(x)}\, dx = 0$ for every function $g(x)$ in $\mathfrak{D}$, we can write

$$-\int_a^b F(x) \overline{h(x)}\, dx = \int_a^b f(x) \overline{g(x)}\, dx = \int_{-\infty}^{+\infty} f(x) \overline{g(x)}\, dx = 0$$

where

$$F(x) = \int_a^x f(\xi)\, d\xi, \quad g(x) = \int_a^x h(\xi)\, d\xi,$$

and $h(x)$ is an arbitrary function in $\mathfrak{L}_2(-\infty, +\infty)$ with the properties $h(x) = 0$ for $x \leqq a$, $x \geqq b$, $\int_a^b h(x)\, dx = 0$; and we can therefore conclude that $F(x)$ is constant on the arbitrary interval (a, b) and that $f(x) = F'(x)$ vanishes identically.

The transformation H is therefore a linear transformation with domain everywhere dense in $\mathfrak{L}_2(-\infty, +\infty)$. We can show by an argument similar to the one used in the proof of Theorem 10.8 that it is symmetric. We have

$$(Hf, g)-(f, Hg)$$
$$= \lim_{a\to-\infty,\, b\to+\infty}\left[\int_a^b \frac{1}{i} f'(x)\overline{g(x)}\,dx - \int_a^b f(x)\overline{\frac{1}{i} g'(x)}\,dx\right]$$
$$= \lim_{b\to+\infty} \frac{1}{i} f(b)\overline{g(b)} - \lim_{a\to-\infty} \frac{1}{i} f(a)\overline{g(a)},$$

for arbitrary functions $f(x)$ and $g(x)$ in $\mathfrak{D}$. Since the function $f(x)\overline{g(x)}$ is integrable over $(-\infty, +\infty)$ and tends to a limit when x becomes either positively or negatively infinite, each of these limits must have the value zero. Thus the equation $(Hf, g) = (f, Hg)$ is valid and implies that H is symmetric. The transformation $H_l \equiv H - l \cdot I$ has, in accordance with Theorem 4.14, an inverse H_l^{-1} whenever l is not real. We shall prove that the domain of H_l^{-1} is the entire space $\mathfrak{L}_2(-\infty, +\infty)$. Let us consider first the case $\mathfrak{J}(l) > 0$. The general solution of the differential equation $\frac{1}{i} f' - lf = g$, where g is an arbitrary function in $\mathfrak{L}_2(-\infty, +\infty)$, is given by

$$f(x) = ce^{ilx} + ie^{ilx}\int_{-\infty}^{x} e^{-il\xi} g(\xi)\,d\xi, \quad \mathfrak{J}(l) > 0,$$

where c is an arbitrary constant. The integral which occurs in this expression is absolutely convergent since $|e^{-il\xi}|^2$ and $|g(\xi)|^2$ are both integrable over $(-\infty, x)$. For large negative values of x, the function $|f(x)|$ is of the order of magnitude of $|c|\,|e^{ilx}|$ and hence cannot belong to $\mathfrak{L}_2(-\infty, +\infty)$ unless $c = 0$. We shall therefore examine the solution

$$f(x) = ie^{ilx}\int_{-\infty}^{x} e^{-il\xi} g(\xi)\,d\xi,$$

with the intention of showing that it belongs to $\mathfrak{L}_2(-\infty, +\infty)$. If $g_n(x)$ is that function in $\mathfrak{L}_2(-\infty, +\infty)$ which is equal to $g(x)$ for $-n \le x \le +n$ and equal to zero elsewhere, $n = 1, 2, 3, \cdots$, then the function

$$f_n(x) = ie^{ilx}\int_{-\infty}^{x} e^{-il\xi} g_n(\xi)\,d\xi$$

is an absolutely continuous function for $-\infty < x < +\infty$ with the properties

$$f_n(x) = 0, \qquad x < -n,$$
$$f_n(x) = f(x) - ie^{ilx}\int_{-\infty}^{-n} e^{-il\xi}\, g(\xi)\, d\xi, \qquad -n \leqq x \leqq +n,$$
$$f_n(x) = C_n e^{ilx}, \qquad C_n = i\int_{-n}^{+n} e^{-il\xi}\, g(\xi)\, d\xi, \qquad x > +n.$$

From the behavior of $f_n(x)$ for large values of $|x|$, it is evidently a function in $\mathfrak{L}_2(-\infty, +\infty)$. Hence $f_n' = ilf_n + ig_n$ is also a function in $\mathfrak{L}_2(-\infty, +\infty)$ and $f_n(x)$ must belong to $\mathfrak{D}$. By applying Theorem 4.14, we obtain the inequality

$$\|f_n\| = \|H_l^{-1} g_n\| \leqq \|g_n\|/|\mathfrak{J}(l)|.$$

Now the expressions given above show that, on any finite closed interval, $f_n(x)$ tends uniformly to the limit $f(x)$ as n becomes infinite. By combining the relations

$$\lim_{n\to\infty}\int_a^b |f_n(x)|^2\, dx = \int_a^b |f(x)|^2\, dx,$$
$$\|g_n\|^2 = \int_{-n}^{+n} |g(x)|^2\, dx \leqq \int_{-\infty}^{+\infty} |g(x)|^2\, dx,$$

with the inequality given above, we find that

$$\int_a^b |f(x)|^2\, dx \leqq \int_{-\infty}^{+\infty} |g(x)|^2\, dx/|\mathfrak{J}(l)|^2,$$

and thus conclude that $f(x)$ belongs to $\mathfrak{L}_2(-\infty, +\infty)$. It is now apparent that $f' = ilf + ig$ must also belong to $\mathfrak{L}_2(-\infty, +\infty)$, $f(x)$ to $\mathfrak{D}$. Thus H_l^{-1} carries $\mathfrak{L}_2(-\infty, +\infty)$ in a one-to-one manner into $\mathfrak{D}$, according to the relation

$$f(x) = H_l^{-1} g(x) = \int_{-\infty}^{x} i e^{il(x-\xi)}\, g(\xi)\, d\xi.$$

In the case $\mathfrak{J}(l) < 0$, which presents no essential novelty, we find that H_l^{-1} carries $\mathfrak{L}_2(-\infty, +\infty)$ in a one-to-one manner into $\mathfrak{D}$, according to the relation

$$f(x) = H_l^{-1} g(x) = \int_{+\infty}^{x} i e^{il(x-\xi)}\, g(\xi)\, d\xi.$$

By introducing the function $\psi(z; l)$ defined by the equations

$$\begin{aligned} \psi(z; l) &= 0, & z<0, \quad & \mathfrak{J}(l)>0; \\ \psi(z; l) &= i e^{ilz}, & z>0, \quad & \mathfrak{J}(l)>0; \\ \psi(z; l) &= i e^{ilz}, & z<0, \quad & \mathfrak{J}(l)<0; \\ \psi(z; l) &= 0, & z>0, \quad & \mathfrak{J}(l)<0, \end{aligned}$$

we can express both these results by the single formula

$$f(x) = H_l^{-1} g(x) = \int_{-\infty}^{+\infty} \psi(x-\xi; l)\, g(\xi)\, d\xi.$$

The symmetric transformation H is therefore self-adjoint in accordance with Theorems 9.1–9.3; and its resolvent $R_l \equiv H_l^{-1}$, $\mathfrak{J}(l) \neq 0$, is given by the relation above.

For fixed x and l, we may consider $\psi(x-\xi; l)$ as a function of ξ belonging to $\mathfrak{L}_2(-\infty, +\infty)$. By direct calculation we find that

$$\begin{aligned} U\psi(x-\xi; l) &= \frac{1}{\sqrt{2\pi}} \int_{-\infty}^{+\infty} e^{i\xi\eta}\, \psi(x-\eta; l)\, d\eta \\ &= e^{ix\xi}/\sqrt{2\pi}\,(\xi - l), \end{aligned}$$

and can therefore write

$$\begin{aligned} \psi(x-\xi; l) &= U^{-1}\left(e^{ix\xi}/\sqrt{2\pi}\,(\xi-l)\right) \\ &= \frac{1}{\sqrt{2\pi}} \int_{-\infty}^{+\infty} e^{-i\xi\eta}\, e^{ix\eta}/(\eta - l)\, d\eta \\ &= \overline{U(e^{-ix\xi}/\sqrt{2\pi}\,(\xi-\bar{l}))}. \end{aligned}$$

We therefore have

$$\begin{aligned} H_l^{-1} g(x) &= \int_{-\infty}^{+\infty} g(\xi)\, \overline{U(e^{-ix\xi}/\sqrt{2\pi}\,(\xi-\bar{l}))}\, d\xi \\ &= \frac{1}{\sqrt{2\pi}} \int_{-\infty}^{+\infty} \left(U^{-1} g(\xi)\right) \overline{\left(\frac{e^{-ix\xi}}{\xi - \bar{l}}\right)}\, d\xi \\ &= \frac{1}{\sqrt{2\pi}} \int_{-\infty}^{+\infty} e^{ix\xi} \left(\frac{1}{\xi - l}\, U^{-1} g(\xi)\right) d\xi \\ &= U\left(\frac{1}{x-l}\, U^{-1} g(x)\right), \end{aligned}$$

by virtue of the unitary character of the transformation U. We have thus shown that H_l^{-1} is equivalent by the unitary transformation U to the transformation which carries an arbitrary function $f(x)$ in $\mathfrak{L}_2(-\infty, +\infty)$ into the function $\frac{1}{x-l} f(x)$ in $\mathfrak{L}_2(-\infty, +\infty)$. The latter transformation is clearly the inverse of $S_l \equiv S - l \cdot I$, where S is the transformation described in the statement of the theorem. By Theorem 2.54, we see that $H_l \equiv U S_l U^{-1}$ and hence that $H \equiv U S U^{-1}$. We note that the transformation S, being unitary-equivalent to the self-adjoint transformation H, is self-adjoint by Theorem 2.55.

We can now determine the spectral properties of S and hence those of H. Since the equation $S_l f(x) = (x-l) f(x) = 0$ implies $f(x) = 0$ almost everywhere $-\infty < x < +\infty$, S has no characteristic values. Its spectrum is therefore a pure continuous spectrum. The resolution of the identity associated with S can be calculated by the use of Theorem 5.10, as we indicated earlier in the proof of Theorem 7.12, or by the use of limiting processes such as those described in the proof of Theorem 6.9. Denoting it by $E(\lambda)$, we find that $E(\lambda) f(x)$ is the function which is equal to $f(x)$ or to zero according as $x \leqq \lambda$ or $x > \lambda$. If $f(x)$ is any function in $\mathfrak{L}_2(-\infty, +\infty)$ which vanishes only on a set of zero measure, then the family of functions $E(\lambda) f(x)$ determines a closed linear manifold $\mathfrak{M}(f)$ which coincides with $\mathfrak{L}_2(-\infty, +\infty)$: for if $g(x)$ is a function in $\mathfrak{L}_2(-\infty, +\infty)$ with the property that

$$\int_{-\infty}^{+\infty} E(\lambda) f(x) \overline{g(x)}\, dx = 0$$

for all values of λ, we have successively

$$\int_{-\infty}^{\lambda} f(x) \overline{g(x)}\, dx = 0, \qquad f(x) \overline{g(x)} = 0, \qquad g(x) = 0.$$

According to Theorem 7.9, the transformation S has a simple spectrum. Thus H likewise has a simple continuous spectrum, as we wished to prove.

On the basis of Theorem 10.9, it is possible to develop a sound theory of the Heaviside operational calculus, which has shown itself so useful in certain branches of applied mathematics. For various reasons we shall confine ourselves to brief remarks of a general nature concerning this topic. As is well known, the Heaviside calculus is a body of rules, largely formal in character, for manipulating the differential operator $\frac{d}{dx}$; its usefulness is due to its heuristic transparency, its flexibility, and its simplifying power. An obvious method for discussing the Heaviside calculus from a mathematical point of view is to apply the operational calculus of Chapter VI to the self-adjoint transformation H defined by the formal differential operator $\frac{1}{i}\frac{d}{dx}$ in accordance with Theorem 10.9. By virtue of the connection established there between the transformations H, S, and U, it is easily shown that $F(H) \equiv UF(S)U^{-1}$ for an arbitrary function $F(\lambda)$. It can be proved that the function $F(\lambda)$ is H-measurable if and only if it is S-measurable and that it is S-measurable if and only if it is Lebesgue-measurable. Now, if $F(\lambda)$ is a Lebesgue-measurable function, the transformation $F(S)$ is characterized as follows: its domain consists of those functions $f(x)$ in $\mathfrak{L}_2(-\infty, +\infty)$ for which the integral $\int_{-\infty}^{+\infty} |F(x)f(x)|^2\,dx$ exists; and $F(S)$ carries a function $f(x)$ belonging to its domain into the function $F(x)f(x)$. Hence, if $F(\lambda)$ is a Lebesgue-measurable function, the transformation $F(H)$ is determined as follows: its domain consists of those functions $f(x)$ in $\mathfrak{L}_2(-\infty, +\infty)$ such that

$$g(x) = U^{-1}f(x) = \frac{1}{\sqrt{2\pi}}\int_{-\infty}^{+\infty} e^{-ix\xi} f(\xi)\,d\xi$$

is in the domain of $F(S)$; and, for a function $f(x)$ in the domain of $F(H)$,

$$F(H)f(x) = \frac{1}{2\pi}\int_{-\infty}^{+\infty} e^{ix\xi} F(\xi) \int_{-\infty}^{+\infty} e^{-i\xi\eta} f(\eta)\,d\eta\,d\xi.$$

This formula appears as the defining equation for the operator $F(H)$ in many treatments of the Heaviside calculus, notably that due to Wiener.* For practical purposes it is not suf-

* N. Wiener, Mathematische Annalen, 95 (1925–6), pp. 557–584.

ficient to deal with operations confined to the Hilbert space $\mathfrak{L}_2(-\infty, +\infty)$. The definitions and theorems of Chapter VI must therefore be supplemented by others which apply to operands not contained in this space. Furthermore, an adequate theory must not be limited to a statement of rules as general as those given in Chapter VI; it must contain in addition many special rules for the convenient handling of typical problems which occur frequently in practice. For a detailed development of the program indicated, we refer to the paper of Wiener just cited and the book of Bush,† where a good bibliography is to be found.

In connection with differential operators of the first order, we give the following curious illustration of some aspects of the theory of adjoint operators.

THEOREM 10.10. *Let* $\mathfrak{B}$ *be the set of all continuous functions of bounded variation on the closed interval* $(0, 1)$ *with derivatives in* $\mathfrak{L}_2(0, 1)$; *let* $\mathfrak{B}_0$ *be the set of all functions* f *in* $\mathfrak{B}$ *such that* $f' = 0$ *almost everywhere on* $(0, 1)$; *and let* A *be the transformation with domain* $\mathfrak{B}$ *which carries* f *into* f'/i. *Then* $\mathfrak{B}$ *and* $\mathfrak{B}_0$ *are linear manifolds everywhere dense in* $\mathfrak{L}_2(0, 1)$; *and* A *is a linear transformation which is non-bounded and which possesses no closed linear extension. Every function in* $\mathfrak{B}_0$ *is a characteristic function of* A *for the characteristic value zero. The adjoint* A^* *exists and has as its domain the set* $\mathfrak{O}$ *which consists of the null element alone.*

It is clear that both $\mathfrak{B}$ and $\mathfrak{B}_0$ are linear manifolds in $\mathfrak{L}_2(0, 1)$. Since $\mathfrak{B}$ contains the sets $\mathfrak{D}$ and $\mathfrak{D}^*$ of Theorem 10.7, it is also everywhere dense in $\mathfrak{L}_2(0, 1)$. We now prove that $\mathfrak{B}_0$ likewise is everywhere dense in $\mathfrak{L}_2(0, 1)$. Let $\varphi_{ab}(x)$ be the function defined by the equations

$$\varphi_{ab}(x) = 1, \quad a \leqq x \leqq b;$$
$$\varphi_{ab}(x) = 0, \quad 0 \leqq x < a, \quad b < x \leqq 1,$$

where $0 < a < b < 1$. We then construct an approximating function $\varphi_{ab}(x; \epsilon)$ in $\mathfrak{B}_0$ as follows: we divide the interval $(0, 1)$

† Vannevar Bush, *Operational Circuit Analysis*, New York, 1929.

into the five intervals $(0, a(1-\varepsilon))$, $(a(1-\varepsilon), a)$, (a, b), $(b, b+(1-b)\varepsilon)$, $(b+(1-b)\varepsilon, 1)$, where $0<\varepsilon<1$; on the first and last we put $\varphi_{ab}(x;\varepsilon)=0$; on the second we arrange that $\varphi_{ab}(x;\varepsilon)$ increase continuously and monotonely from zero to one and have vanishing derivative almost everywhere; on the third we put $\varphi_{ab}(x;\varepsilon)=1$; and on the fourth we arrange that $\varphi_{ab}(x;\varepsilon)$ decrease continuously and monotonely from one to zero and have vanishing derivative almost everywhere. It is easily verified that $\varphi_{ab}(x;\varepsilon)\to\varphi_{ab}(x)$ in $\mathfrak{L}_2(0, 1)$ as $\varepsilon\to 0$. Since the set of all functions $\varphi_{ab}(x)$ determines the closed linear manifold $\mathfrak{L}_2(0, 1)$, we see that $\mathfrak{B}_0$ is everywhere dense in $\mathfrak{L}_2(0, 1)$.

The transformation A is a linear transformation with domain everywhere dense in $\mathfrak{L}_2(0, 1)$, as we see by inspection; and, in consequence of these properties, the adjoint transformation A^* exists. If H is the transformation defined in Theorem 10.7, if O is the transformation which carries every element of $\mathfrak{L}_2(0, 1)$ into the null element, and if T_0 is the transformation with domain $\mathfrak{B}_0$ defined by the equation $T_0 = O$ in $\mathfrak{B}_0$, then the relations $A \supseteqq H^*$, $A \supseteqq T_0$, are valid and imply the relations $A^* \subseteqq H$, $A^* \subseteqq T_0^* \equiv O$. Thus the domain of A^* consists of those functions f in the domain $\mathfrak{D}$ of H which satisfy the equation $Hf = A^*f = T_0 f = 0$. Obviously, the only function with these properties is the function $f=0$. Hence the domain of A^* is the set $\mathfrak{O}$, as we wished to prove. Theorem 2.29 shows immediately that the transformation A is non-bounded; this result also follows from the relation $A \supseteqq H^*$. We show finally that A possesses no closed linear extension: for if there were such an extension, the transformation $\tilde{A}$ defined in Theorem 2.10 would exist and would satisfy the contradictory relations $\tilde{A} \supseteqq \tilde{T}_0 \equiv O$, $H^* \subseteqq \tilde{A} \equiv O$.

§ 3. Ordinary Differential Operators of the Second Order

In the present section we shall give a detailed examination of the formal differential operator $-\dfrac{d}{dx}\,p(x)\,\dfrac{d}{dx}+r(x)\cdot$,

$a<x<b$, where (a, b) is an arbitrary finite or infinite interval, and where $p(x)$ and $r(x)$ are Lebesgue-measurable real functions such that $1/p(x)$ and $r(x)$ are Lebesgue-integrable over every closed interval interior to (a, b). We commence with a few introductory remarks of more or less heuristic nature, with a view to justifying our choice of this operator.

The most general formal linear ordinary differential operator of the second order can be written as the sum of a finite number of terms of the form

$$p_1(x)\frac{d}{i\,dx}p_2(x)\frac{d}{i\,dx}p_3(x)\cdot \quad +q_1(x)\frac{d}{i\,dx}q_2(x)\cdot \quad +r_1(x)\cdot \quad ,$$

where the functions p, q, and r are complex-valued. Unless these functions are appropriately restricted, neither the sum nor its individual terms can be reduced to simpler form. We shall therefore assume at the outset that each of them is differentiable to an order sufficiently high to ensure the possibility of reducing each term to the form

$$-p(x)\frac{d^2}{dx^2}+q(x)\frac{d}{i\,dx}+r(x)\cdot \quad ,$$

where $p = p_1p_2p_3$, $q = -ip_1p_2'p_3-2ip_1p_2p_3'+q_1q_2$, and $r = -p_1p_2'p_3'-p_1p_2p_3''-iq_1q_2'+r_1$; the given formal operator, as the sum of a finite number of such terms, then has the same form. In discussing symmetry properties of the transformations associated with the formal operator $-p\dfrac{d^2}{dx^2}+q\dfrac{d}{i\,dx}+r\cdot$, we must consider the formal or Lagrange adjoint $-\dfrac{d^2}{dx^2}(\overline{p}\cdot \quad)+\dfrac{d}{i\,dx}(\overline{q}\cdot \quad)+\overline{r}\cdot$. By assuming the existence of the appropriate derivatives, we can reduce the Lagrange adjoint to the form

$$-\overline{p}\frac{d^2}{dx^2}+(-i\overline{p}'+\overline{q})\frac{d}{i\,dx}+(-\overline{p}''-i\overline{q}'+\overline{r})\cdot$$

In order that the given operator and its Lagrange adjoint coincide, it is therefore necessary and sufficient that $p=\overline{p}$, $q=-2i\overline{p}'+\overline{q}$, $r=-\overline{p}''-i\overline{q}'+\overline{r}$ or that $\mathfrak{I}p = 0$,

$\mathfrak{J} q = -p'$, $\mathfrak{J} r = -\frac{1}{2}\mathfrak{R} q'$. Similarly, in discussing reality with respect to the conjugation J_2 which replaces a function by its complex-conjugate, we are led to consider the formal conjugate operator

$$-\overline{p(x)}\frac{d^2}{dx^2} - \overline{q(x)}\frac{d}{idx} + \overline{r(x)}.$$

In order to obtain transformations which are real with respect to J_2, we require that the given operator coincide with its formal conjugate. For this purpose it is necessary and sufficient to require that $\mathfrak{J} p = 0$, $\mathfrak{R} q = 0$, $\mathfrak{J} r = 0$. Thus the formal differential operator $-p\frac{d^2}{dx^2} + q\frac{d}{idx} + r\cdot$ coincides both with its Lagrange adjoint and with its formal conjugate if and only if $\mathfrak{J} p = 0$, $\mathfrak{R} q = 0$, $\mathfrak{J} q = -p'$, $\mathfrak{J} r = 0$; in other words, if and only if it can be written as

$$-\frac{d}{dx}p(x)\frac{d}{dx} + r(x).$$

where the functions p and r are real-valued.

By means of the transformation-theory of Liouville, we can classify formal differential operators of the second order into various types. The result is similar to that obtained for differential operators of the first order in Theorem 10.6; but it is less satisfactory because it leads to no corresponding simplification. First, we may consider the unitary transformation in $\mathfrak{L}_2(a, b)$ which takes $f(x)$ into $f(x)/\delta(x)$, where $|\delta(x)| = 1$. The image or transform of the formal operator $-p\frac{d^2}{dx^2} + q\frac{d}{idx} + r\cdot\ $, $a \leq x \leq b$, by this unitary transformation is the formal operator

$$-P\frac{d^2}{dx^2} + Q\frac{d}{idx} + R\cdot$$

$$= \frac{1}{\delta}\left\{-p\frac{d^2}{dx^2}(\delta\cdot\ ) + q\frac{d}{idx}(\delta\cdot\ ) + r\delta\cdot\ \right\};$$

it is evident that

$$P = p, \quad Q = q + 2p\delta'/i\delta, \quad R = r - p\delta''/\delta + q\delta'/i\delta.$$

If the original operator coincides with its Lagrange adjoint and if $\mathfrak{R}(q/p)$ is integrable over (a, b), we may determine δ by the equation $\delta'/\delta = -\frac{1}{2} i \mathfrak{R}(q/p)$ so that $P = p$, $Q = -ip'$, $R = \mathfrak{R} r - (\mathfrak{R} q)^2/4p$, where the function p is real-valued. Thus the transformed operator coincides not only with its Lagrange adjoint but also with its formal conjugate, in accordance with the conditions indicated in the preceding paragraph. Secondly, we shall consider changes of variable analogous to those discussed in Theorem 10.6. Let $\beta(x)$ be a real function different from zero almost everywhere on (a, b) with the property that $\beta^2(x)$ is Lebesgue-integrable over every closed interval interior to (a, b); let $x = c$ be an arbitrary interior point of the interval (a, b); and let C be an arbitrary real constant. We define a monotone real function with no interval of constancy by the equation

$$\alpha(x) = \pm \left[\int_c^x \beta^2(\xi)\, d\xi + C \right].$$

If we denote the lesser and the greater of the numbers $\alpha(a)$ and $\alpha(b)$ by a' and b' respectively, then the relation $y = \alpha(x)$ defines a one-to-one correspondence between the ranges $a \leqq x \leqq b$ and $a' \leqq y \leqq b'$, with the inverse $x = \gamma(y)$. The equations $g(x) = \beta(x) f(\alpha(x))$, $f(y) = g(\gamma(y))/\beta(\gamma(y))$ define a one-to-one isometric correspondence between functions $f(y)$ in $\mathfrak{L}_2(a', b')$ and $g(x)$ in $\mathfrak{L}_2(a, b)$. We shall consider the result of transforming the operator $-\frac{d}{dx} p \frac{d}{dx} + r \cdot$, where p and r are real functions, by means of this isometric correspondence. Starting with a function $f(y)$ in $\mathfrak{L}_2(a', b')$, we must pass to its correspondent $g(x)$ in $\mathfrak{L}_2(a, b)$, apply the given operator to the latter function, and then determine from the resulting function $g^*(x)$ the corresponding function in $\mathfrak{L}_2(a', b')$. Formally, the result of the first two steps is the function

$$\begin{aligned} g^*(x) &= -\frac{d}{dx} p(x) \frac{d}{dx} (\beta(x) f(\alpha(x))) + r(x)\, \beta(x) f(\alpha(x)) \\ &= -p\beta\alpha'^2 f''(\alpha(x)) - ((p\beta\alpha')' + p\beta'\alpha') f'(\alpha(x)) \\ &\qquad + (-(p\beta')' + r\beta) f(\alpha(x)). \end{aligned}$$

The correspondent of this function is then seen to be

$$f^*(y) = -P(y)f''(y) - Q(y)f'(y) + R(y)f(y),$$

where P, Q, and R are obtained by setting $x = \gamma(y)$ in the functions $p\alpha'^2$, $((p\beta\alpha')' + p\beta'\alpha')/\beta$, and $(-(p\beta')' + r\beta)/\beta$, respectively. By means of the relation $\alpha'(x) = \pm\beta^2(x)$, it is easily verified by direct calculation that $Q(y) = \frac{dP}{dy}$; in other words, that $((p\beta\alpha')' + p\beta'\alpha')/\beta = (p\alpha'^2)'/\alpha'$. It is therefore apparent that the transformed operator can be written $-\frac{d}{dy}P(y)\frac{d}{dy} + R(y)\cdot\quad$, $a' \leqq y \leqq b'$, where the functions P and R are real-valued. By selection of the functions α and β we can control the interval (a', b'), the function P, or the function R; but it is not possible for us to assign all three arbitrarily. In order to prescribe $P(y)$, for example, we must choose α as a solution of the differential equation $p(x)\alpha'^2 = P(\alpha)$. Similarly, if we wish to prescribe $R(y)$, we must choose α and β as solutions of the simultaneous differential equations $-(p\beta')' + r\beta = R(\alpha)\beta$, $\alpha' = \pm\beta^2$. Some particular cases here are not without interest. If $p(x) \geqq 0$ and $1/p(x)$ is Lebesgue-integrable over every closed interval interior to (a, b), we can take $P(y) = 1$ by setting $\alpha'^2 = 1/p(x)$, $\beta = (p(x))^{-1/4}$; the function $R(y)$ is then given by the equations

$$R(y) = r(x) + \frac{1}{4}p''(x) - \frac{1}{16}p'^2(x)/p(x), \quad y = \alpha(x).$$

Similarly, if the equation $-(p\beta')' + r\beta = 0$ admits a solution which does not vanish on the open interval (a, b), we can take $R(y) = 0$ by choosing β as the solution in question. Finally, we remark that we may combine transformations of the indicated types to reduce a given formal differential operator to its simplest and most convenient equivalent form.

The developments sketched in the two preceding paragraphs show that under certain conditions we may treat the differential operator $-\frac{d}{dx}p(x)\frac{d}{dx} + r(x)\cdot\quad$, where p and r

are real functions, as typical of the entire class of formal ordinary linear differential operators of the second order which coincide with their Lagrange adjoints. Furthermore, it is a familiar fact that differential operators of this special form occur with great frequency in various problems of pure and applied mathematics. In particular, the Schrödinger equation of the quantum mechanics involves an operator of this type both in one-dimensional problems and in problems which can be reduced to one-dimensional problems by separation of variables or by an idealization of the physical situation under consideration. Thus we have both theoretical and practical grounds for considering the study of the indicated special operator as an important illustration of the general theory of Chapter IX.

We must now recall some of the fundamental existence theorems for linear differential equations, as they apply to the operator which we wish to examine. Since we impose upon the functions p and r conditions somewhat lighter than those assumed in the more familiar treatments, we shall indicate the proof of the results which we need here. It is convenient to state these results in the following form:

LEMMA 10.1. *Let $P(x)$, $R(x)$, and $f^*(x)$ be complex-valued Lebesgue-measurable functions defined on a finite closed interval (a', b'), the functions $1/P$, R, and f^* being Lebesgue-integrable over (a', b'); let $x = c$ be an arbitrary point of the interval (a', b'); and let c_1 and c_2 be arbitrary complex constants. Then there exists a unique function $f(x)$ with the properties*

(1) *$f(x)$ is absolutely continuous on the closed interval (a', b');*

(2) *$P(x) f'(x)$ is absolutely continuous on the closed interval (a', b');*

(3) *$-(Pf')' + Rf = f^*$ almost everywhere on (a', b');*

(4) *$f(c) = c_1$, $P(c) f'(c) = c_2$;*

and the function $f(x)$ is real whenever P, R, f^*, c_1, and c_2 are real. If $f_1(x)$ and $f_2(x)$ are two functions with the properties (1), (2), *and* (3) *for $f^*(x) = 0$, then the function*

$$W(f_1, f_2) = P(x)(f_1(x) f_2'(x) - f_1'(x) f_2(x))$$

is a constant which vanishes if and only if $f_1(x)$ *and* $f_2(x)$ *are linearly dependent; and if* $f_1(x)$, $f_2(x)$, *and* $f_3(x)$ *are any three functions with the properties* (1), (2), *and* (3) *for* $f^*(x) = 0$, *then they are linearly dependent. A function* $f(x)$ *has properties* (1)—(4) *if and only if it can be expressed in the form*

$$f(x) = a_1 f_1(x) + a_2 f_2(x) + \int_c^x \frac{f_1(x) f_2(\xi) - f_1(\xi) f_2(x)}{W(f_1, f_2)} f^*(\xi) d\xi,$$

where $f_1(x)$ *and* $f_2(x)$ *are two functions satisfying conditions* (1)-(3) *with* $f^*(x) \doteq 0$ *and* $W(f_1, f_2) \neq 0$, *and where* a_1 *and* a_2 *are constants determined by the equations*

$$\begin{aligned} a_1 f_1(c) + a_2 f_2(c) &= f(c) = c_1, \\ a_1 P(c) f_1'(c) + a_2 P(c) f_2'(c) &= P(c) f'(c) = c_2 \end{aligned}$$

with determinant $W(f_1, f_2) \neq 0$.

Before sketching the proof of this lemma, we note, in connection with condition (2), that the statement must be interpreted to mean that $P(x) f'(x)$ is equal almost everywhere to a function which is absolutely continuous on the closed interval (a', b').

Our first step is to treat the special case where $R(x) = 0$. By direct integration of the differential equation $-(Pf')' = f^*$, we established the existence and uniqueness of the function $f(x)$ satisfying conditions (1)-(4) and we obtain for this function the explicit formula

$$f(x) = c_1 + c_2 \int_c^x 1/P(\xi)\, d\xi - \int_c^x (1/P(\xi)) \left(\int_c^\xi f^*(\eta)\, d\eta \right) d\xi.$$

On integrating by parts in the last term of this formula, we find that

$$\begin{aligned} f(x) &= c_1 + c_2 \int_c^x 1/P(\xi)\, d\xi \\ &\quad + \int_c^x \left(\int_c^\xi 1/P(\eta)\, d\eta - \int_c^x 1/P(\eta)\, d\eta \right) f^*(\xi)\, d\xi \\ &= c_1 + c_2 u(x) + \int_c^x (u(\xi) - u(x)) f^*(\xi)\, d\xi, \end{aligned}$$

where $u(x) = \int_c^x 1/P(\xi)\,d\xi$; and we recognize this result as a special case of the last statement of the lemma, with $R(x) = 0$, $f_1(x) = 1$, $f_2(x) = u(x)$, and $W(f_1, f_2) = 1$.

We now see that in the general case a function $f(x)$ satisfies conditions (1)-(4) if and only if it satisfies the equation

$$f(x) = c_1 + c_2 u(x) + \int_c^x (u(\xi) - u(x))\,(f^*(\xi) - R(\xi) f(\xi))\,d\xi.$$

For the results of the preceding paragraph show that this relation is equivalent to conditions (1)-(4) with (3) written in the form $-(Pf')' = f^* - Rf$. To solve this equation by the method of successive approximations, we introduce the sequences $\{f_n(x)\}$ and $\{g_n(x)\}$ defined by the relations

$$f_0(x) = 0,$$
$$f_{n+1}(x) = c_1 + c_2 u(x) + \int_c^x (u(\xi) - u(x))\,(f^*(\xi) - R(\xi) f_n(\xi))\,d\xi,$$
$$g_0(x) = f_0(x) = 0, \qquad g_{n+1}(x) = f_{n+1}(x) - f_n(x),$$

where $n = 0, 1, 2, \cdots$. We then have

$$f_n(x) = \sum_{\alpha=0}^{n} g_\alpha(x)$$

for all n, and

$$g_{n+1}(x) = \int_c^x (u(x) - u(\xi))\,R(\xi)\,g_n(\xi)\,d\xi$$

for all n except $n = 0$. We now show that on a certain closed interval, described presently, the sequence $\{f_n(x)\}$ tends uniformly to a limit $f(x)$; in other words, that on this interval the series $\sum_{\alpha=0}^{\infty} g_\alpha(x)$ converges uniformly to the sum $f(x)$. Since the function $u(x)$ is continuous on the closed interval (a', b'), there exists a constant M such that $|u(x) - u(\xi)| \leqq M$ for all x and all ξ on this interval; and since the function $R(x)$ is Lebesgue-integrable on (a', b'), there corresponds to each positive number ε a positive $\eta = \eta(\varepsilon)$ such that $\int_\Delta |R(\xi)|\,d\xi \leqq \varepsilon$ for every subinterval Δ of (a', b') with length not exceeding η. We denote by ϱ some number on the open interval (0, 1) and set $\delta = \eta(\varrho/M)$. We then introduce the interval Δ which comprises those values of x on the interval (a', b') satisfying the inequality $c-\delta \leqq x \leqq c+\delta$. For the present we restrict our attention to the interval Δ. If γ_n is the maximum of the continuous function $|g_n(x)|$ on Δ, we have

$$\begin{aligned}\gamma_{n+1} &= \max \left| \int_c^x (u(x)-u(\xi)) R(\xi) g_n(\xi) d\xi \right| \\ &\leqq M\gamma_n \max \left| \int_c^x |R(\xi)| d\xi \right| \leqq \varrho\gamma_n\end{aligned}$$

for x in Δ and $n = 1, 2, 3, \cdots$. Thus the series $\sum_{\alpha=0}^{\infty} g_\alpha(x)$ converges uniformly on the interval Δ, as we wished to prove. It is easily verified that the sum of this series is a solution of the equation under consideration, for all x in Δ. We can now show that there is no other solution on any subinterval Δ_0 of Δ. In fact, if $f(x)$ is the difference of two solutions valid on the interval Δ_0 and if γ is the maximum of the continuous function $|f(x)|$ on Δ_0, we have

$$f(x) = \int_c^x (u(x)-u(\xi)) R(\xi) f(\xi) d\xi, \quad \gamma \leqq \varrho\gamma,$$

and infer, by virtue of the inequality $0<\varrho<1$, that γ and $f(x)$ vanish. Thus we have established the existence of a unique function $f(x)$ satisfying conditions (1)–(4) on the interval Δ. In this result, the position of the point $x=c$ on the interval (a', b') is entirely unrestricted, the constants c_1 and c_2 are arbitrary, and the length of the interval Δ is never less than the smaller of the two fixed numbers δ and $b'-a'$. We are therefore enabled to construct, by a finite number of applications of this result, a function $f(x)$ satisfying conditions (1)–(4) on the entire interval (a', b'). We begin by covering the interval (a', b') by a finite set of overlapping intervals $\Delta_1, \cdots, \Delta_n$, each of which has length equal to the lesser of the numbers δ and $b'-a'$. From all the functions satisfying conditions (1)–(3) on an arbitrary interval Δ_k, we can select a unique function $f_k(x)$ by assigning a point in Δ_k and the values of $f_k(x)$ and $P(x)f_k'(x)$

at this point. When k has a value such that Δ_k contains the point $x = c$, we determine $f_k(x)$ by the requirement that $f_k(c) = c_1$, $P(c) f_k'(c) = c_2$. When $f_k(x)$ has been determined for a given value of k, we choose $f_l(x)$ on any interval Δ_l which overlaps Δ_k by the requirement that $f_k(x) = f_l(x)$ on $\Delta_k \Delta_l$, observing that the desired relation is satisfied if and only if

$$f_k(x_0) = f_l(x_0), \qquad P(x_0) f_k'(x_0) = P(x_0) f_l'(x_0)$$

at some point $x = x_0$ in $\Delta_k \Delta_l$. It is evident that this set of requirements leads to a uniquely determined function $f_k(x)$ for $k = 1, \cdots, n$. The function $f(x)$ which is defined as equal to $f_k(x)$ on Δ_k, $k = 1, \cdots, n$, is obviously a single-valued function satisfying conditions (1)–(4) on the entire interval (a', b'). The method employed in the construction of this function shows that it is unique.

The arguments necessary to establish the remaining statements of the lemma are sufficiently familiar that we do not repeat them here; they are strictly analogous to those given in the usual discussions where P and Q are assumed to be continuous.*

As we pointed out in Chapter III, § 3, the formula of Lagrange is essential in the discussion of the symmetry of transformations associated with the formal operator $-\frac{d}{dx} p \frac{d}{dx} + r \cdot$, which we wish to study. In the present instance, the formula is

$$\int_{a'}^{b'} [(-(pf')'\overline{g} + rf\overline{g}) - (-f\overline{(pg')'} + rf\overline{g})]\, dx = [-pf'\overline{g} + f\overline{pg'}]_{x=a'}^{x=b'},$$

where $a < a' < b' < b$. In using this formula it is convenient to denote the expression $-p(x) f'(x) \overline{g(x)} + f(x) \overline{p(x) g'(x)}$, for values of x where it is significant, by the symbol $B_x(f, g)$. We shall write

* See, for example, Bôcher, *Leçons sur les méthodes de Sturm*, Paris, 1917, Chapter I and Chapter V, § 20.

$$B_a(f, g) = \lim_{x \to a} B_x(f, g) \quad \text{and} \quad B_b(f, g) = \lim_{x \to b} B_x(f, g),$$

whenever the indicated limits exist; and we shall also write

$$B(f, g) = B_b(f, g) - B_a(f, g),$$

when the two terms on the right are significant.

After these preliminary remarks, we can proceed to the analysis of the various transformations associated with the formal differential operator under consideration.

THEOREM 10.11. *Let $p(x)$ and $r(x)$ be real Lebesgue-measurable functions defined on a finite or infinite interval (a, b), where $1/p(x)$ and $r(x)$ are Lebesgue-integrable over every finite closed interval (a', b') interior to (a, b). Let $\mathfrak{D}^*$ be the set of all functions in $\mathfrak{L}_2(a, b)$ such that*

(1) $f(x)$ is absolutely continuous on the open interval (a, b);

(2) $p(x)f'(x)$ is absolutely continuous on the open interval (a, b);

(3) $-(pf')' + rf$ is a function in $\mathfrak{L}_2(a, b)$;

and let T be the linear transformation with domain $\mathfrak{D}^*$ which takes f into $-(pf')' + rf$.

Then $B_a(f, g)$, $B_b(f, g)$, and $B(f, g) = (Tf, g) - (f, Tg)$ are complex-valued bilinear functions of f and g defined for all f and g in $\mathfrak{D}^*$. Let $\mathfrak{D}$ be the set of all functions f in $\mathfrak{D}^*$ such that $B(f, g) = 0$ for every function g in $\mathfrak{D}^*$; and let H be the linear transformation with domain $\mathfrak{D}$ which takes f into $-(pf')' + rf$. Then H is a closed linear symmetric transformation with the adjoint $H^* \equiv T$ and the deficiency-index (m, m), where $m = 0, 1,$ or 2. Both H and T are real with respect to the conjugation J_2 which takes $f(x)$ into $\overline{f(x)}$.

In connection with the descriptive properties of the class $\mathfrak{D}^*$, we point out that a function $f(x)$ in $\mathfrak{L}_2(a, b)$ is considered to be absolutely continuous for $a < x < b$ if it is equal almost everywhere to a function which has this property in the strict sense, and that the derivative of such a function is to be calculated by operating with the equivalent function which is absolutely continuous in the strict sense.

The formula of Lagrange shows us immediately that

$$B(f,g)=(Tf,g)-(f,Tg)=\lim_{a'\to a,\, b'\to b}\int_{a'}^{b'}(Tf\bar{g}-f\overline{Tg})\,dx$$
$$=\lim_{b'\to b}B_{b'}(f,g)-\lim_{a'\to a}B_{a'}(f,g),$$

the existence of the limits $B_a(f,g)$, $B_b(f,g)$, and $B(f,g)$ being established by the convergence of the integral of $Tf\bar{g}-f\overline{Tg}$ over the interval (a, b), provided that f and g are functions in $\mathfrak{D}^*$. The fact that $B(f, g)$ is thus significant for all f and g in $\mathfrak{D}^*$ allows us to define the set $\mathfrak{D}$ in the manner indicated in the theorem.

We now determine all pairs of functions g and g^* in $\mathfrak{L}_2(a, b)$ which satisfy the relation $(Hf, g) = (f, g^*)$ for every function f in $\mathfrak{D}$. For this purpose it is convenient to restrict the function f to belong to a suitably chosen subset of $\mathfrak{D}$, described below. Let (a', b') be an arbitrary closed interval interior to (a, b). Then Lemma 10.1 establishes the existence of solutions of the differential equation $-(pu')'+ru=0$ on the interval (a', b'). In fact, we can construct two real solutions u_1 and u_2 with $W(u_1, u_2) = 1$ and we can express every other solution u in the form $u = a_1u_1 + a_2u_2$ where a_1 and a_2 are complex constants. Since every solution u is continuous on (a', b'), we see that the totality of solutions is a two-dimensional closed linear manifold in $\mathfrak{L}_2(a', b')$. Let f^* be an arbitrary function in $\mathfrak{L}_2(a, b)$ which vanishes identically outside the interval (a', b') and which has the property that

$$\int_{a'}^{b'} f^*(x)\,\overline{u(x)}\,dx = 0$$

for every solution u of the differential equation $-(pu')'+ru=0$ on (a', b'); and let $f(x)$ be the function defined by the relations

$$f(x) = 0, \quad a<x<a', \quad b'<x<b;$$
$$f(x) = \int_{a'}^{x}(u_1(x)\,u_2(\xi)-u_1(\xi)\,u_2(x))\,f^*(\xi)\,d\xi, \quad a'\leqq x\leqq b'.$$

The conditions imposed upon f^* demand that $f(x)$ and $p(x)f'(x)$, where

$$p(x)f'(x) = 0, \quad a<x<a', \quad b'<x<b;$$

$$p(x)f'(x) = \int_{a'}^{x} (p(x)u_1'(x)u_2(\xi) - u_1(\xi)p(x)u_2'(x))f^*(\xi)\,d\xi,$$
$$a' \leqq x \leqq b',$$

be continuous for $a<x<b$, vanishing at $x=a'$ and $x=b'$. It is easily verified, with the help of Lemma 10.1, that f is in $\mathfrak{D}$ and that $Hf = -(pf')'+rf=f^*$. Since both f and f^* vanish identically outside (a', b'), the relation $(Hf, g) = (f, g^*)$, which refers to $\mathfrak{L}_2(a, b)$, can be written in the form

$$\int_{a'}^{b'} f^*\bar{g}\,dx = \int_{a'}^{b'} f\overline{g^*}\,dx,$$

which can be treated as a relation holding in $\mathfrak{L}_2(a', b')$. Now the function $h(x)$ defined for $a' \leqq x \leqq b'$ by the equation

$$h(x) = \int_{a'}^{x} (u_1(x)u_2(\xi) - u_1(\xi)u_2(x))g^*(\xi)\,d\xi$$

belongs to $\mathfrak{L}_2(a', b')$ and satisfies the equation $-(ph')'+rh=g^*$, in accordance with Lemma 10.1. The formula of Lagrange shows that

$$\int_{a'}^{b'} (f\overline{g^*} - f^*\bar{h})\,dx = \int_{a'}^{b'} (-f\overline{(ph')'} + (pf')'\bar{h})\,dx$$
$$= (pf'\bar{h} - f\overline{ph'})_{x=a'}^{x=b'} = 0$$

and thus enables us to write the relation $(Hf, g) = (f, g^*)$ in the form

$$\int_{a'}^{b'} f^*\overline{(g-h)}\,dx = 0.$$

Since f^*, considered as a function on the interval (a', b'), is arbitrary save for the condition that it be a function in $\mathfrak{L}_2(a', b')$ orthogonal to every solution u of the equation $-(pu')'+ru=0$, $a' \leqq x \leqq b'$, we conclude that $g-h = a_1u_1 + a_2u_2$ or

$$g(x) = a_1u_1(x) + a_2u_2(x) + \int_{a'}^{x} (u_1(x)u_2(\xi) - u_1(\xi)u_2(x))g^*(\xi)\,d\xi$$

for $a' \leqq x \leqq b'$. By Lemma 10.1, this relation implies that g is a solution of the equation $-(pg')'+rg = g^*$ on (a', b').

Since (a', b') is an arbitrary closed interval interior to (a, b), and since g and g^* belong to $\mathfrak{L}_2(a, b)$, this result means that the equation $(Hf, g) = (f, g^*)$ holds for every f in $\mathfrak{D}$ only if g is in $\mathfrak{D}^*$ and $g^* = Tg$. On the other hand, we have $(Hf, g) - (f, Tg) = (Tf, g) - (f, Tg) = B(f, g) = 0$ for all f in $\mathfrak{D}$ and all g in $\mathfrak{D}^*$. We conclude therefore that the adjoint of H exists and is identical with T. From Theorem 2.6 we infer that $\mathfrak{D}$ determines the closed linear manifold $\mathfrak{L}_2(a, b)$. Since $\mathfrak{D}^*$ contains $\mathfrak{D}$ as a subset, it also determines the closed linear manifold $\mathfrak{L}_2(a, b)$. In view of this fact, the transformation T^* exists. The relations $H \subseteqq T$, $H^* \equiv T$ imply the relations $H \frown T$ and $H \subseteqq T^* \subseteqq T$. Thus a function f belongs to the domain of T^* if and only if it is in $\mathfrak{D}^*$ and satisfies the relation

$$0 = (T^* f, g) - (f, Tg) = (Tf, g) - (f, Tg) = B(f, g)$$

for every g in $\mathfrak{D}^*$; in other words, if and only if it is in $\mathfrak{D}$. Thus T^* and H must coincide. The relations $H \frown T$, $H \subseteqq T$, $H \equiv T^*$ now imply that H is a closed linear symmetric transformation. We note that $\mathfrak{D}$ and $\mathfrak{D}^*$ are both linear manifolds everywhere dense in $\mathfrak{L}_2(a, b)$, in accordance with the relations $H \equiv T^*$, $T \equiv H^*$.

If f is a function in $\mathfrak{D}^*$, then $\bar{f}$ is also in $\mathfrak{D}^*$; and

$$T\bar{f} = -(p\bar{f}')' + r\bar{f} = \overline{-(pf')' + rf} = \overline{Tf}.$$

Thus the transformation T is permutable with the conjugation J_2, and T is real with respect to J_2 in accordance with Definition 9.8. Theorem 9.13 shows that $H \equiv T^*$ is also real with respect to J_2.

By Theorem 9.14, the deficiency-index of the transformation H must have the form (m, m). We know from the general theory of Chapter IX, § 1, that m is the dimension number of the characteristic manifold of the transformation $T \equiv H^*$ corresponding to an arbitrary not-real characteristic value l. To determine m, we solve the differential equation $-(pf')' + (r - l)f = 0$ on the open interval (a, b). Lemma 10.1, with $P = p$ and $R = r - l$, establishes the existence of solutions on an arbitrary closed interval (a', b') interior to (a, b); each solution can be extended to the entire open interval (a, b) by allowing a' and b' to tend to a and to b, respectively. The lemma shows that there are exactly two linearly independent solutions of the differential equation. Hence the number of linearly independent solutions belonging to $\mathfrak{L}_2(a, b)$ is 0, 1, or 2. The solutions which belong to $\mathfrak{L}_2(a, b)$ are precisely the solutions of the equation $Tf - lf = 0$. It follows, therefore, that m has the value 0, 1, or 2.

We shall now examine the three cases which may arise according as $m = 0, 1$, or 2. We first obtain the following immediate result, the discussion of which may be left to the reader.

THEOREM 10.12. *In order that the transformation H of Theorem* 10.10 *be self-adjoint, it is necessary and sufficient that one of the four following conditions be satisfied:*

(1) *the sets $\mathfrak{D}$ and $\mathfrak{D}^*$ coincide;*

(2) $B(f, g) = 0$ *for all f and g in $\mathfrak{D}^*$;*

(3) *m has the value* 0;

(4) *the differential equation* $-(pf')' + (r - l)f = 0$, $\mathfrak{J}(l) \neq 0$, *has no solution in $\mathfrak{L}_2(a, b)$ other than $f = 0$.*

Whenever these conditions are satisfied, the differential equation $-(pf') + (r - l)f = f^*$, $\mathfrak{J}(l) \neq 0$, *has a unique solution in $\mathfrak{L}_2(a, b)$ for each function f^* in $\mathfrak{L}_2(a, b)$.*

We have next

THEOREM 10.13. *A necessary and sufficient condition that the transformation H of Theorem* 10.11 *have the deficiency-index* (1, 1) *is that the solutions in $\mathfrak{L}_2(a, b)$ of the differential equation* $-(pf')' + (r - l)f = 0$, $\mathfrak{J}(l) \neq 0$, *constitute a one-dimensional closed linear manifold. When H has the deficiency-index* (1, 1), *we can select a normalized characteristic element u of the transformation $H^* \equiv T$ corresponding to the characteristic value $+i$—in other words, a function satisfying the relations* $\int_a^b |u|^2 dx = 1$ *and* $-(pu')' + (r - i)u = 0$—*and we set* $v_\theta = (e^{i\theta/2}u - e^{-i\theta/2}\bar{u})/2i$ *where* $0 \leq \theta < 2\pi$. *The domain $\mathfrak{D}$ of H and the associated sets $\mathfrak{D}^+$ and $\mathfrak{D}^-$ described in*

Theorem 9.4 *can then be identified as follows:* $\mathfrak{D}$ *is the set of all functions* f *in* $\mathfrak{D}^*$ *such that* $B(f, u) = B(f, \bar{u}) = 0$; *the set* $\mathfrak{D}^+$ *is the closed linear manifold determined by* u; *and the set* $\mathfrak{D}^-$ *is the closed linear manifold determined by* $\bar{u}$. *If* $\mathfrak{D}(\theta)$ *is the set of all functions* f *in* $\mathfrak{D}^*$ *such that* $B(f, v_\theta) = 0$, *and if* $H(\theta)$ *is the transformation with domain* $\mathfrak{D}(\theta)$ *which takes* f *into* $-(pf')' + rf$, *then* $H(\theta)$ *is a self-adjoint extension of* H *real with respect to the conjugation* J_2; *and if* S *is an arbitrary maximal symmetric extension of* H, *then* S *coincides with* $H(\theta)$ *for exactly one value of* θ *on the range* $0 \leqq \theta < 2\pi$. *The differential system* $-(pf')' + (r - l)f = f^*$, $\mathfrak{I}(l) \neq 0$, $B(f, v_\theta) = 0$, *has just one solution in* $\mathfrak{L}_2(a, b)$ *for each function* f^* *in* $\mathfrak{L}_2(a, b)$.

The condition that H have the deficiency-index $(1, 1)$ has already been discussed in Theorem 10.11. Thus, if H has the deficiency-index $(1, 1)$, we can choose a normalized solution u of the equation $Tf = if$. Since T is real with respect to the conjugation J_2, we find that $\bar{u} = J_2 u$ is a normalized solution of the equation $Tf = -if$, in accordance with the relations $T\bar{u} = TJ_2 u = J_2 Tu = J_2 iu = -iJ_2 u = -i\bar{u}$. By direct computation we find that $B(u, u) = 2i$, $B(\bar{u}, \bar{u}) = -2i$, $B(u, \bar{u}) = B(\bar{u}, u) = 0$. According to Theorem 9.4, $\mathfrak{D}^+$ is the closed linear manifold determined by u; $\mathfrak{D}^-$ is the closed linear manifold determined by $\bar{u}$; and $\mathfrak{D}^*$ consists of those and only those functions f in $\mathfrak{L}_2(a, b)$ which are expressible in the form $f = f_0 + a_1 u + a_2 \bar{u}$, where f_0 is in $\mathfrak{D}$ and a_1 and a_2 are complex constants. The condition that $B(f, g) = 0$ for every g in $\mathfrak{D}^*$ characterizes those functions f in $\mathfrak{D}^*$ which belong to $\mathfrak{D}$. This condition is now seen to be fulfilled if and only if

$$B(f, g_0) + \bar{a}_1 B(f, u) + \bar{a}_2 B(f, \bar{u}) = B(f, g_0 + a_1 u + a_2 \bar{u}) = 0$$

for every g_0 in $\mathfrak{D}$ and every pair of constants a_1 and a_2. Since $B(f, g_0) = 0$ whenever f is in $\mathfrak{D}^*$ and since a_1 and a_2 are arbitrary, this condition is satisfied if and only if $B(f, u) = B(f, \bar{u}) = 0$. Hence $\mathfrak{D}$ is characterized in the manner stated in the theorem.

We shall now determine all the maximal symmetric extensions of H, by the methods described in Chapter IX, § 1. According to the general theory, we must first determine all the isometric transformations of $\mathfrak{D}^+ = \mathfrak{L}_2(a, b) \ominus \mathfrak{R}_{-i}$ into $\mathfrak{D}^- = \mathfrak{L}_2(a, b) \ominus \mathfrak{R}_{+i}$, where $\mathfrak{R}_{+i}$ and $\mathfrak{R}_{-i}$ are the ranges of H_{+i} and H_{-i} respectively. Hence the desired isometric transformations are those which carry αu into $e^{-i\theta}\alpha \bar{u}$, $0 \leqq \theta < 2\pi$. Corresponding to each value of θ, there is a distinct maximal symmetric extension of H, which we denote by $S(\theta)$. The process given in Theorems 9.1–9.3 for the construction of $S(\theta)$ shows that a function f is in the domain of $S(\theta)$ if and only if it is expressible in the form $f = g + \alpha(u - e^{-i\theta}\bar{u})$ where g is a function in $\mathfrak{D}$ and α is a complex constant. By setting $\alpha = e^{i\theta/2}c/2i$ and $(e^{i\theta/2}u - e^{-i\theta/2}\bar{u})/2i = v_\theta$, we can write this relation as $f = g + cv_\theta$. A necessary and sufficient condition that f have this form is that it belong to the set $\mathfrak{D}(\theta)$ described in the statement of the theorem, as we shall now prove. When $f = g + cv_\theta$, we see that it is a function in $\mathfrak{D}^*$ satisfying the condition

$$B(f, v_\theta) = B(g, v_\theta) + cB(v_\theta, v_\theta) = 0,$$

since $B(g, v_\theta)$ vanishes by virtue of the fact that g and v_θ belong to $\mathfrak{D}$ and to $\mathfrak{D}^*$ respectively and since $B(v_\theta, v_\theta)$ vanishes by direct calculation. Hence f belongs to $\mathfrak{D}(\theta)$. On the other hand, if f is in $\mathfrak{D}(\theta)$ it belongs to $\mathfrak{D}^*$ and can therefore be written in the form $f = g + \alpha_1 u + \alpha_2 \bar{u}$ where g is in $\mathfrak{D}$; and it satisfies the condition $B(f, v_\theta) = 0$ which implies the relations

$$\begin{aligned} B(\alpha_1 u + \alpha_2 \bar{u}, v_\theta) &= \alpha_1 B(u, v_\theta) + \alpha_2 B(\bar{u}, v_\theta) \\ &= -e^{-i\theta/2}\alpha_1 - e^{i\theta/2}\alpha_2 = 0 \end{aligned}$$

and

$$\alpha_1 = ce^{i\theta/2}/2i, \qquad \alpha_2 = -ce^{-i\theta/2}/2i.$$

Hence we see that $f = g + cv_\theta$, as we wished to prove. This result shows that $S(\theta)$ has $\mathfrak{D}(\theta)$ as its domain. Since we know that the transformation $H^* \equiv T$ is an extension of $S(\theta)$, we conclude that $S(\theta) \equiv H(\theta)$, as stated in the theorem. The fact that $H(\theta)$ is self-adjoint is an immediate consequence of Theorem 9.3; and the fact that $H(\theta)$ is real with respect to the conjugation J_2 follows from Theorem 9.14. We can also deduce the reality of $H(\theta)$ from the observation that the boundary condition $B(f, v_\theta) = 0$ is expressed in terms of the real-valued function v_θ. Since the resolvent set of $H(\theta)$ contains all not-real values l, the equation $H(\theta)f - lf = f^*$, $\mathfrak{I}(l) \neq 0$, has a unique solution f in $\mathfrak{L}_2(a, b)$ whenever f^* is in $\mathfrak{L}_2(a, b)$. This equation is obviously equivalent to the differential system $-(pf')' + (r - l)f = f^*$, $\mathfrak{I}(l) \neq 0$, $B(f, v_\theta) = 0$, where the given function f^* and the solution f are required to belong to $\mathfrak{L}_2(a, b)$.

THEOREM 10.14. *A necessary and sufficient condition that the transformation H of Theorem 10.11 have the deficiency-index (2, 2) is that every solution of the differential equation $-(pu')' + (r - l)u = 0$, $\mathfrak{I}(l) \neq 0$, belong to $\mathfrak{L}_2(a, b)$. When H has the deficiency-index (2, 2), we can introduce characteristic functions u_1 and u_2 of $H^* \equiv T$ corresponding to the characteristic value $+i$ and constituting an orthonormal set,*

$$(u_j, u_k) = \int_a^b u_j \bar{u}_k \, dx = \delta_{jk};$$

and we set

$$v_{k\mathfrak{a}} = u_k - \sum_{\alpha=1}^{2} a_{k\alpha} \bar{u}_\alpha, \qquad k = 1, 2,$$

where $\mathfrak{a} = \{a_{jk}\}$ is a unitary matrix,

$$\sum_{\alpha=1}^{2} a_{j\alpha} \bar{a}_{k\alpha} = \delta_{jk}, \qquad j, k = 1, 2.$$

The domain $\mathfrak{D}$ of H and the associated sets $\mathfrak{D}^+$ and $\mathfrak{D}^-$ described in Theorem 9.4 can then be characterized as follows: $\mathfrak{D}$ is the set of all functions f in $\mathfrak{D}^*$ such that

$$B(f, u_1) = B(f, u_2) = B(f, \bar{u}_1) = B(f, \bar{u}_2) = 0;$$

$\mathfrak{D}^+$ is the closed linear manifold determined by the orthonormal set $\{u_k\}$; and $\mathfrak{D}^-$ is the closed linear manifold determined by the orthonormal set $\{\bar{u}_k\}$. If $\mathfrak{D}(\mathfrak{a})$ is the set of all functions f

in $\mathfrak{D}^*$ *such that* $B(f, v_{1\mathfrak{a}}) = B(f, v_{2\mathfrak{a}}) = 0$ *and if* $H(\mathfrak{a})$ *is the transformation with domain* $\mathfrak{D}(\mathfrak{a})$ *which takes* f *into* $-(pf')' + rf$, *then* $H(\mathfrak{a})$ *is a self-adjoint extension of* H; *and if* S *is an arbitrary maximal symmetric extension of* H, *there exists a unique matrix* $\mathfrak{a}$ *such that* $S \equiv H(\mathfrak{a})$. *The differential system* $-(pf')' + (r - l)f = f^*$, $\mathfrak{J}(l) \neq 0$, $B(f, v_{1\mathfrak{a}}) = B(f, v_{2\mathfrak{a}}) = 0$, *has just one solution in* $\mathfrak{L}_2(a, b)$ *for each function* f^* *in* $\mathfrak{L}_2(a, b)$. *The transformation* $H(\mathfrak{a})$ *is real with respect to the conjugation* J_2 *if and only if the matrix* $\mathfrak{a}$ *satisfies the condition* $a_{21} = a_{12}$. *When this condition is satisfied, there exist two real functions* $v_{\mathfrak{a}}^{(1)}$ *and* $v_{\mathfrak{a}}^{(2)}$ *such that* $v_{\mathfrak{a}}^{(j)} = \sum_{\beta=1}^{2} b_{j\beta} v_{\beta\mathfrak{a}}$, *where* $\det\{b_{jk}\} \neq 0$; *and the set* $\mathfrak{D}(\mathfrak{a})$ *is the set of all functions* f *in* $\mathfrak{D}^*$ *such that*

$$B(f, v_{\mathfrak{a}}^{(1)}) = B(f, v_{\mathfrak{a}}^{(2)}) = 0.$$

The condition that H have the deficiency-index (2,2) has already been discussed in Theorem 10.11. Thus, if H has the index (2,2), we can select an orthonormal set which determines the two-dimensional closed linear manifold $\mathfrak{D}^+$ of all solutions of the equation $Tf = if$ or $-(pf')' + (r-i)f = 0$. This set contains two elements, which we denote by u_1 and u_2. Since T is real with respect to the conjugation J_2, we find that $\{\bar{u}_k\}$ is an orthonormal set which determines the two-dimensional closed linear manifold $\mathfrak{D}^-$ of all solutions of the equation $Tf = -if$. By direct calculation we have

$$B(u_j, u_k) = 2i\delta_{jk}, \quad B(u_j, \bar{u}_k) = B(\bar{u}_j, u_k) = 0,$$
$$B(\bar{u}_j, \bar{u}_k) = -2i\delta_{jk}, \qquad j, k = 1, 2.$$

Since we have characterized the manifolds $\mathfrak{D}^+$ and $\mathfrak{D}^-$, we can apply Theorem 9.4 to show that $\mathfrak{D}^*$, the domain of $H^* \equiv T$, comprises those and only those functions f in $\mathfrak{L}_2(a, b)$ which are expressible in the form $f = f_0 + a_1 u_1 + a_2 u_2 + a_3 \bar{u}_1 + a_4 \bar{u}_2$ where f_0 is in $\mathfrak{D}$ and a_1, a_2, a_3, a_4 are complex constants. The condition that $B(f, g) = 0$ for every g in $\mathfrak{D}^*$ characterizes those functions f in $\mathfrak{D}^*$ which belong to $\mathfrak{D}$. This condition is fulfilled if and only if

$$B(f, g_0)+\bar{a}_1 B(f, u_1)+\bar{a}_2 B(f, u_2)+\bar{a}_3 B(f, \bar{u}_1)+\bar{a}_4 B(f, \bar{u}_2)$$
$$= B(f, g_0 + a_1 u_1 + a_2 u_2 + a_3 \bar{u}_1 + a_4 \bar{u}_2) = 0$$

for every g_0 in $\mathfrak{D}$ and every set of complex constants a_1, a_2, a_3, a_4. Since $B(f, g_0) = 0$ for arbitrary f in $\mathfrak{D}^*$ and since the constants in question are arbitrary, this condition is satisfied if and only if

$$B(f, u_1) = B(f, u_2) = B(f, \bar{u}_1) = B(f, \bar{u}_2) = 0,$$

as we wished to prove. $\mathfrak{D}$ is thus determined in the manner stated in the theorem.

We shall now determine all the maximal symmetric extensions of H, by the methods of Chapter IX, § 1. According to the general theory, we must first determine all the isometric transformations of $\mathfrak{D}^+ = \mathfrak{L}_2(a, b) \ominus \mathfrak{R}_{-i}$ into $\mathfrak{D}^- = \mathfrak{L}_2(a, b) \ominus \mathfrak{R}_{+i}$, where $\mathfrak{R}_{+i}$ and $\mathfrak{R}_{-i}$ are the ranges of H_{+i} and H_{-i} respectively. Clearly, the desired isometric transformations are those which carry $f = \sum_{\alpha=1}^{2} a_\alpha u_\alpha$ into $f^* = \sum_{\alpha,\beta=1}^{2} a_\alpha \mathfrak{a}_{\alpha\beta} \bar{u}_\beta$, where $\mathfrak{a} = \{\mathfrak{a}_{jk}\}$ is a unitary matrix. Corresponding to each matrix $\mathfrak{a}$, there is a distinct maximal symmetric extension of H which we denote by $S(\mathfrak{a})$. The process given in Theorems 9.1–9.3 for the construction of $S(\mathfrak{a})$ shows that a function f is in the domain of $S(\mathfrak{a})$ if and only if it is expressible in the form

$$f = g + \sum_{\alpha,\beta=1}^{2} a_\alpha (\delta_{\alpha\beta} u_\beta - \mathfrak{a}_{\alpha\beta} \bar{u}_\beta) = g + \sum_{\alpha=1}^{2} a_\alpha v_{\alpha\mathfrak{a}},$$

where g is a function in $\mathfrak{D}$, a_1 and a_2 are complex constants, and $v_{1\mathfrak{a}}$ and $v_{2\mathfrak{a}}$ are the functions defined in the statement of the theorem. When f is expressed in this form, we see that it is a function in $\mathfrak{D}^*$ satisfying the relations

$$B(f, v_{k\mathfrak{a}}) = B(g, v_{k\mathfrak{a}}) + a_1 B(v_{1\mathfrak{a}}, v_{k\mathfrak{a}}) + a_2 B(v_{2\mathfrak{a}}, v_{k\mathfrak{a}}) = 0$$

for $k = 1, 2$, since $B(g, v_{k\mathfrak{a}})$ vanishes by virtue of the fact that g and $v_{k\mathfrak{a}}$ are in $\mathfrak{D}$ and in $\mathfrak{D}^*$ respectively and since

$$B(v_{j\mathfrak{a}}, v_{k\mathfrak{a}}) = 2i\left(\delta_{jk} - \sum_{\alpha=1}^{2} \mathfrak{a}_{j\alpha} \bar{\mathfrak{a}}_{k\alpha}\right) = 0 \quad \text{for} \quad j, k = 1, 2,$$

by direct calculation. On the other hand, if f is in $\mathfrak{D}(\mathfrak{a})$ it belongs to $\mathfrak{D}^*$ and can therefore be written in the form $f = g + a_1 u + a_2 u_2 + a_3 \bar{u}_1 + a_4 \bar{u}_2$, where g is in $\mathfrak{D}$. The conditions $B(f, v_{k\mathfrak{a}}) = 0$, $k = 1, 2$, reduce by direct calculation to

$$a_1 B(u_1, v_{k\mathfrak{a}}) + a_2 B(u_2, v_{k\mathfrak{a}}) + a_3 B(\bar{u}_1, v_{k\mathfrak{a}}) + a_4 B(\bar{u}_2, v_{k\mathfrak{a}}) \\ = 2i(a_1 \delta_{1k} + a_2 \delta_{2k} + a_3 \bar{a}_{k1} + a_4 \bar{a}_{k2}) = 0.$$

The relations $a_3 \bar{a}_{11} + a_4 \bar{a}_{12} = -a_1$, $a_3 \bar{a}_{21} + a_4 \bar{a}_{22} = -a_2$ are thus seen to be valid; and they imply that $a_3 = -a_1 a_{21} - a_2 a_{22}$, $a_4 = -a_1 a_{21} - a_2 a_{22}$. In consequence the function f has the form $f = g + a_1 v_{1\mathfrak{a}} + a_2 v_{2\mathfrak{a}}$. Hence the transformation $S(\mathfrak{a})$ has $\mathfrak{D}(\mathfrak{a})$ as its domain. Since the transformation $H^* \equiv T$ is an extension of $S(\mathfrak{a})$, we conclude that $S(\mathfrak{a}) \equiv H(\mathfrak{a})$, as stated in the theorem. The fact that $H(\mathfrak{a})$ is self-adjoint is an immediate consequence of Theorem 9.3. Since the resolvent set of $H(\mathfrak{a})$ contains all not-real values l, the equation $H(\mathfrak{a})f - lf = f^*$, $\mathfrak{I}(l) \neq 0$, has a unique solution f in $\mathfrak{L}_2(a, b)$ whenever f^* is in $\mathfrak{L}_2(a, b)$. This equation is obviously equivalent to the differential system $-(pf')' + (r - l)f = f^*$, $\mathfrak{I}(l) \neq 0$, $B(f, v_{1\mathfrak{a}}) = B(f, v_{2\mathfrak{a}}) = 0$, where the solution f and the given function f^* are required to belong to $\mathfrak{L}_2(a, b)$. In order to determine under what conditions the transformation $H(\mathfrak{a})$ is real with respect to the conjugation J_2, we have merely to examine the discussion under Theorem 9.14 with reference to the special case before us. We find that $H(\mathfrak{a})$ is real with respect to J_2 if and only if the matrix $\mathfrak{a}$ has the property that $a_{kj} = a_{jk}$ for $j, k = 1, 2$. Obviously, these conditions reduce to the single relation $a_{21} = a_{12}$, the others being trivial. We may also obtain this result by examining the conditions under which J_2 carries $\mathfrak{D}(\mathfrak{a})$ into itself. Since J_2 carries $\mathfrak{D}$ into itself, we have only to determine the conditions under which $\bar{v}_{1\mathfrak{a}}$ and $\bar{v}_{2\mathfrak{a}}$ belong to $\mathfrak{D}(\mathfrak{a})$. It is evident that the equations

$$B(\bar{v}_{j\mathfrak{a}}, v_{k\mathfrak{a}}) = 2i(\bar{a}_{kj} - \bar{a}_{jk}) = 0, \qquad j, k = 1, 2,$$

furnish the desired criterion. When $H(\mathfrak{a})$ is real with respect to J_2 we find by direct calculation that for $k = 1, 2$ the relations

$$\bar{v}_{k\mathfrak{a}} = -\sum_{\beta=1}^{2} \bar{a}_{k\beta}\, v_{\beta\mathfrak{a}},$$

$$v_{k\mathfrak{a}} = -\sum_{\beta=1}^{2} a_{k\beta}\, \bar{v}_{\beta\mathfrak{a}},$$

$$\mathfrak{R}\, v_{k\mathfrak{a}} = \tfrac{1}{2} \sum_{\beta=1}^{2} (\delta_{k\beta} - \bar{a}_{k\beta})\, v_{\beta\mathfrak{a}},$$

$$\mathfrak{J}\, v_{k\mathfrak{a}} = \frac{1}{2i} \sum_{\beta=1}^{2} (\delta_{k\beta} + \bar{a}_{k\beta})\, v_{\beta\mathfrak{a}},$$

are satisfied. Since the linear manifolds $\mathfrak{D}^+$ and $\mathfrak{D}^-$ are linearly independent in accordance with Theorem 9.4, we see easily that the functions $v_{1\mathfrak{a}}$ and $v_{2\mathfrak{a}}$ are linearly independent. We therefore conclude from the relations indicated above that we can choose a pair of linearly independent real functions $v_{\mathfrak{a}}^{(1)}$ and $v_{\mathfrak{a}}^{(2)}$ from the set $\mathfrak{R}\, v_{1\mathfrak{a}}$, $\mathfrak{R}\, v_{2\mathfrak{a}}$, $\mathfrak{J}\, v_{1\mathfrak{a}}$, $\mathfrak{J}\, v_{2\mathfrak{a}}$, and that we can express the two remaining functions in the set as linear combinations of them. It is now clear that we can express $v_{1\mathfrak{a}}$ and $v_{2\mathfrak{a}}$ as linear combinations of $v_{\mathfrak{a}}^{(1)}$ and $v_{\mathfrak{a}}^{(2)}$, and conversely. Thus we have $v_{\mathfrak{a}}^{(j)} = \sum_{\beta=1}^{2} b_{j\beta}\, v_{\beta\mathfrak{a}}$, $\det \{b_{jk}\} \neq 0$. The boundary conditions $B(f, v_{1\mathfrak{a}}) = B(f, v_{2\mathfrak{a}}) = 0$ which characterize $\mathfrak{D}(\mathfrak{a})$ are obviously equivalent to the real conditions $B(f, v_{\mathfrak{a}}^{(1)}) = B(f, v_{\mathfrak{a}}^{(2)}) = 0$. This result completes our discussion of the theorem.

In order to make further progress in the study of the self-adjoint transformations $H(\theta)$ and $H(\mathfrak{a})$ described above, we must first discuss certain special cases. In addition to providing tools for the treatment of the general problem, these special cases are of interest in themselves. The proposed specialization consists in restricting the behavior of the functions $p(x)$ and $r(x)$ at one or at both ends of the interval (a, b). There are essentially two different cases, one corresponding to the differential systems discussed by Weyl,* the other to the

---

* Weyl, Göttinger Nachrichten, 1909, pp. 37–63; Mathematische Annalen, 68 (1909), pp. 220–269.

Sturm-Liouville differential systems.† It is the first case which will enable us to gain a deeper insight into the general problem. We shall use without further comment the notations introduced in the preceding theorems.

THEOREM 10.15. *Let* (a, b) *be an arbitrary finite or infinite interval with* $a > -\infty$; *and let the functions* $p(x)$ *and* $r(x)$ *satisfy, in addition to the conditions stated in Theorem* 10.11, *the condition that* $1/p(x)$ *and* $r(x)$ *be absolutely integrable over every interval* (a, c), $a < c < b$. *The set* $\mathfrak{D}$ *is then the set of all functions* f *in* $\mathfrak{D}^*$ *such that* $f(a) = p(a)f'(a) = 0$ *and* $B_b(f, g) = 0$ *for every* g *in* $\mathfrak{D}^*$. *Let* $\mathfrak{G}^*(\varphi)$ *be the set of all functions* f *in* $\mathfrak{D}^*$ *such that* $\sin\varphi . f(a) + \cos\varphi \cdot p(a)f'(a) = 0$, $0 \leqq \varphi < \pi$; *and let* $T(\varphi)$ *be the transformation with domain* $\mathfrak{G}^*(\varphi)$ *which takes* f *into* $-(pf')' + rf$. *Let* $\mathfrak{G}(\varphi)$ *be the set of all functions* f *in* $\mathfrak{G}^*(\varphi)$ *such that* $B_b(f, g) = 0$ *for every* g *in* $\mathfrak{G}^*(\varphi)$; *and let* $G(\varphi)$ *be the transformation with domain* $\mathfrak{G}(\varphi)$ *which takes* f *into* $-(pf')' + rf$. *Then* $G(\varphi)$ *is a proper closed linear symmetric extension of* H *with the adjoint* $G^*(\varphi) \equiv T(\varphi)$. *The deficiency-index of* H *is either* $(1, 1)$ *or* $(2, 2)$. *Both the transformations* $G(\varphi)$ *and* $G^*(\varphi) \equiv T(\varphi)$ *are real with respect to the conjugation* J_2.

By applying Lemma 10.1 to the differential equation $-(pf')' + rf = f^*$, $a \leqq x \leqq c$, $a < c < b$, where f^* is in $\mathfrak{L}_2(a, b)$, we see that each of its solutions f is absolutely continuous on the closed interval (a, c). The set $\mathfrak{D}^*$ consists of functions equivalent to such solutions. Hence, if f is a function in $\mathfrak{D}^*$, we calculate $f(a)$ by replacing f by the equivalent function which is absolutely continuous on every closed interval (a, c). The expression $p(a)f'(a)$ is to be determined similarly.

We first construct a function f_φ in $\mathfrak{G}(\varphi)$ with the properties

$$\begin{aligned} f_\varphi(a) &= \cos\varphi, & p(a)f'_\varphi(a) &= -\sin\varphi; \\ f_\varphi(x) &= 0, & b' < x &< b, \end{aligned}$$

† We use this expression to denote a self-adjoint differential system of the second order, the cases actually considered by Sturm and Liouville being obtained from the general system by specialization of the boundary conditions. For an elementary discussion of these systems, see, for example, Bôcher, *Leçons sur les méthodes de Sturm*, Paris, 1917, pp. 69–91.

where $x = b'$ is a fixed point interior to (a, b). We begin by selecting two linearly independent real solutions, u_1 and u_2, of the differential equation $-(pf')' + rf = 0$ which satisfy the conditions $\int_a^{b'} u_j \bar{u}_k \, dx = \int_a^{b'} u_j u_k \, dx = \delta_{jk}$. We then put

$$\begin{aligned} f_\varphi^*(x) &= a_1(\varphi) u_1(x) + a_2(\varphi) u_2(x), \qquad & a \leqq x \leqq b', \\ f_\varphi^*(x) &= 0, & b' < x < b, \end{aligned}$$

where $a_1(\varphi)$ and $a_2(\varphi)$ are the solutions of the linear equations

$$\begin{aligned} (u_2(a) a_1 - u_1(a) a_2) / W(u_1, u_2) &= \cos \varphi, \\ (p(a) u_2'(a) a_1 - p(a) u_1'(a) a_2) / W(u_1, u_2) &= -\sin \varphi, \end{aligned}$$

with the determinant $+1$. The function f_φ^* is obviously in $\mathfrak{L}_2(a, b)$. We now define the desired function f_φ by the formula

$$f_\varphi(x) = \int_{b'}^{x} \frac{u_1(x) u_2(\xi) - u_1(\xi) u_2(x)}{W(u_1, u_2)} f_\varphi^*(\xi) \, d\xi.$$

Since, according to Lemma 10.1, f_φ is a solution of the equation $-(pf')' + rf = f_\varphi^*$ and since f_φ vanishes on the interval (b', b), it is clear that f_φ is in $\mathfrak{D}^*$. By direct calculation we find that $f_\varphi(a) = \cos\varphi$, $p(a) f'(a) = -\sin\varphi$, so that f_φ satisfies the boundary condition

$$\sin\varphi \, f_\varphi(a) + \cos\varphi \, p(a) f_\varphi'(a) = 0$$

and therefore belongs to $\mathfrak{G}^*(\varphi)$. Since f_φ and pf_φ' vanish on (b', b), the boundary condition $B_b(f_\varphi, g) = 0$ is valid for all g in $\mathfrak{G}^*(\varphi)$, and f_φ thus belongs to $\mathfrak{G}(\varphi)$.

By making use of the function f_φ we can characterize the set $\mathfrak{D}$. A function f in $\mathfrak{D}$ must evidently satisfy the relation

$$B(f, f_\varphi) = -p(a) f'(a) \sin\varphi - f(a) \cos\varphi = 0$$

for every value of φ, and must therefore have the property that $f(a) = p(a) f'(a) = 0$. Thus a function f in $\mathfrak{D}^*$ belongs to $\mathfrak{D}$ if and only if it satisfies the relations $f(a) = 0$, $p(a) f'(a) = 0$, and $B_b(f, g) = B_b(f, g) - B_a(f, g) = B(f, g) = 0$ for every g in $\mathfrak{D}^*$. This is the result asserted in the theorem.

We are now prepared to discuss the transformations $G^*(\varphi)$ and $T^*(\varphi)$. The relations $\mathfrak{D} \subseteq \mathfrak{G}(\varphi) \subseteq \mathfrak{G}^*(\varphi) \subseteq \mathfrak{D}^*$ and $H \subseteq G(\varphi) \subseteq T(\varphi) \subseteq T$ are obviously true; and they imply, by virtue of the identities $H^* \equiv T$ and $T^* \equiv H$, that $T \supseteq G^*(\varphi) \supseteq T^*(\varphi) \supseteq H$. Hence, in order to determine the transformations $G^*(\varphi)$ and $T^*(\varphi)$, we have only to ascertain their respective domains as subsets of $\mathfrak{D}^*$. If f is in the domain of $G^*(\varphi)$, it is a function in $\mathfrak{D}^*$ such that

$$(f, G(\varphi) f_\varphi) = (G^*(\varphi) f, f_\varphi).$$

If we write this relation in the form $(f, Tf_\varphi) = (Tf, f_\varphi)$ and make use of the properties of the function f_φ, we find that

$$\begin{aligned} \sin \varphi f(a) + \cos \varphi p(a) f'(a) &= B(f, f_\varphi) \\ &= (Tf, f_\varphi) - (f, Tf_\varphi) = 0. \end{aligned}$$

Consequently f is in $\mathfrak{G}^*(\varphi)$; and the domain of $G^*(\varphi)$ is a subset of $\mathfrak{G}^*(\varphi)$. On the other hand, if f and g are arbitrary functions in $\mathfrak{G}(\varphi)$ and in $\mathfrak{G}^*(\varphi)$ respectively, we have

$$\begin{aligned} (G(\varphi) f, g) - (f, Tg) &= (Tf, g) - (f, Tg) \\ &= B(f, g) = B_b(f, g) - B_a(f, g). \end{aligned}$$

We know that $B_b(f, g)$ vanishes in accordance with the definition of $\mathfrak{G}(\varphi)$; and we can infer from the equations

$$\sin \varphi f(a) + \cos \varphi p(a) f'(a) = 0,$$

$$\sin \varphi \overline{g(a)} + \cos \varphi \overline{p(a) g'(a)} = \overline{\sin \varphi g(a) + \cos \varphi p(a) g'(a)} = 0$$

that the expression

$$B_a(f, g) = -p(a) f'(a) \overline{g(a)} + f(a) \overline{p(a) g'(a)}$$

also vanishes. Hence $(G(\varphi) f, g) = (f, Tg)$ for all f in $\mathfrak{G}(\varphi)$ and all g in $\mathfrak{G}^*(\varphi)$. The domain of $G^*(\varphi)$ therefore includes the set $\mathfrak{G}^*(\varphi)$. From the relations which have now been established we conclude that $G^*(\varphi)$ and $T(\varphi)$ coincide. It is at once obvious that $G(\varphi)$ is a symmetric transformation, by virtue of the relations

$$H \subseteq G(\varphi) \subseteq T(\varphi) \equiv G^*(\varphi);$$

and it follows that

$$G(\varphi) \subseteq G^{**}(\varphi) \equiv T^*(\varphi) \subseteq G^*(\varphi) \equiv T(\varphi).$$

If f is in the domain of $T^*(\varphi)$, it must therefore belong to $\mathfrak{G}^*(\varphi)$; and for all g in $\mathfrak{G}^*(\varphi)$ it must satisfy the equation

$$\begin{aligned} B_b(f,g) &= B(f,g) = (Tf, g) - (f, Tg) \\ &= (T^*(\varphi)f, g) - (f, T(\varphi)g) = 0, \end{aligned}$$

since $B_a(f,g)$ vanishes in accordance with the earlier analysis. Hence f is in $\mathfrak{G}(\varphi)$, and the transformations $T^*(\varphi)$ and $G(\varphi)$ must coincide. We conclude from the relations $G(\varphi) \subseteq T(\varphi)$, $G^*(\varphi) \equiv T(\varphi)$, $T^*(\varphi) \equiv G(\varphi)$, that $G(\varphi)$ is a closed linear symmetric transformation. It is easily verified that $T(\varphi)$ is real with respect to the conjugation J_2; and it then follows by Theorem 9.13 that $G(\varphi) \equiv T^*(\varphi)$ is also real with respect to J_2.

Since the function f_φ belongs to $\mathfrak{G}(\varphi)$ but not to $\mathfrak{D}$, it is evident that $G(\varphi)$ is a proper symmetric extension of H. Theorem 2.13 therefore shows that H is not self-adjoint; according to Theorems 10.11 and 10.12, H must have the deficiency-index (1, 1) or (2, 2).

THEOREM 10.16. *Let the hypotheses of Theorem* 10.15 *be satisfied, and let* H *have the deficiency-index* (1, 1). *Then the relation* $B_b(f, g) = 0$ *is satisfied for all* f *and* g *in* $\mathfrak{D}^*$; *the set* $\mathfrak{D}$ *consists of all those functions* f *in* $\mathfrak{D}^*$ *such that* $f(a) = p(a)f'(a) = 0$. *The transformation* $G(\varphi)$ *is self-adjoint; its domain consists of all those functions* f *in* $\mathfrak{D}^*$ *such that* $\sin\varphi f(a) + \cos\varphi\, p(a) f'(a) = 0$. *The relation* $\sin\varphi\, v_\theta(a) + \cos\varphi\, p(a)\, v_\theta'(a) = 0$ *determines a one-to-one correspondence between the ranges* $0 \leqq \varphi < \pi$ *and* $0 \leqq \theta < 2\pi$ *such that* $\mathfrak{G}(\varphi) \equiv \mathfrak{D}(\theta)$ *and* $G(\varphi) \equiv H(\theta)$ *if and only if* φ *and* θ *are corresponding values.*

According to Theorem 9.3, H can have no proper closed linear symmetric extensions save those with the deficiency-index (0, 0). Hence $G(\varphi)$ has the deficiency-index (0,0) and is therefore self-adjoint. Thus we see that $\mathfrak{G}(\varphi) \equiv \mathfrak{G}^*(\varphi)$ and that the domain of $G(\varphi)$ may be described in the manner indicated above.

If f and g are arbitrary functions in $\mathfrak{D}^*$ and if φ_1 and φ_2 are values such that $\varphi_1 \neq \varphi_2$, $0 \leqq \varphi_1 < \pi$, $0 \leqq \varphi_2 < \pi$, we can determine the linear combinations $f_0 = f - a_1 f_{\varphi_1} - a_2 f_{\varphi_2}$ and $g_0 = g - b_1 f_{\varphi_1} - b_2 f_{\varphi_2}$ so that they belong to $\mathfrak{G}(\varphi) \equiv \mathfrak{G}^*(\varphi)$ for every φ. In the case of f, for example, we choose a_1 and a_2 as the solutions of the linear equations

$$\begin{aligned} f_0(a) &= f(a) - a_1 \cos \varphi_1 - a_2 \cos \varphi_2 &= 0, \\ p(a) f_0'(a) &= p(a) f'(a) + a_1 \sin \varphi_1 + a_2 \sin \varphi_2 &= 0, \end{aligned}$$

with determinant $\sin(\varphi_1 - \varphi_2) \neq 0$. If we treat f_0 as a function in $\mathfrak{G}(\varphi)$ and g_0 as a function in $\mathfrak{G}^*(\varphi)$, we see that $B_b(f_0, g_0)$ vanishes by virtue of the definition of $\mathfrak{G}(\varphi)$. It is immediately evident that $B_b(f, g) = B_b(f_0, g_0)$ since the functions f_{φ_1} and f_{φ_2} vanish on the interval (b', b). We have thus proved that $B_b(f, g)$ vanishes for all f and g in $\mathfrak{D}^*$. The characterization of the set $\mathfrak{D}$ given in the preceding theorem may now be specialized to the case under consideration; it appears at once that the set $\mathfrak{D}$ here consists of those functions f in $\mathfrak{D}^*$ such that $f(a) = p(a) f'(a) = 0$.

By means of the preceding results, we can show that the transformations $G(\varphi)$, $0 \leqq \varphi < \pi$, exhaust the maximal symmetric extensions of H. The central feature of the proof is the observation that φ and θ satisfy the relation $G(\varphi) \equiv H(\theta)$ if and only if they are connected by the equation $\sin \varphi\, v_\theta(a) + \cos \varphi\, p(a) v_\theta'(a) = 0$. When $G(\varphi) \equiv H(\theta)$, it is evident that v_θ, as a function in the domain of $H(\theta)$, must belong to the domain of $G(\varphi)$ and therefore satisfies the indicated boundary condition. On the other hand, when this condition is satisfied, v_θ is a function in $\mathfrak{G}(\varphi)$. By reference to the description of $\mathfrak{D}(\theta)$ given in Theorem 10.13 and to the characterization of $\mathfrak{D}$ just obtained for the present case, we see that $\mathfrak{D}(\theta)$ must be contained in $\mathfrak{G}(\varphi)$ and that $H(\theta)$ must have $G(\varphi)$ as an extension. Since $H(\theta)$ and $G(\varphi)$ are both self-adjoint transformations, we must have $H(\theta) \equiv G(\varphi)$. As we have already shown in Theorem 10.13, there corresponds to a given value of φ on the range $0 \leqq \varphi < \pi$ just one value of θ on the range $0 \leqq \theta < 2\pi$ such that

$G(\varphi) \equiv H(\theta)$. On the other hand, if θ is a given value on the range $0 \leqq \theta < 2\pi$, there exists at least one solution φ, $0 \leqq \varphi < \pi$, of the equation $\sin\varphi\, v_\theta(a) - \cos\varphi\, p(a)\, v'_\theta(a) = 0$, since the coefficients $v_\theta(a)$ and $p(a)\, v'_\theta(a)$ are real numbers; and this solution is unique unless $v_\theta(a)$ and $p(a)\, v'_\theta(a)$ both vanish. We shall now show that the latter situation never arises. The equations $v_\theta(a) = p(a)\, v'_\theta(a) = 0$ imply that v_θ belongs to $\mathfrak{D}$. By definition, however, v_θ is a linear combination with non-vanishing coefficients of functions in $\mathfrak{D}^+$ and $\mathfrak{D}^-$; and, by Theorem 9.4, v_θ cannot belong to $\mathfrak{D}$. This contradiction shows that the exceptional case does not occur; and a given value of θ leads therefore to a unique value φ. With this result, the proof of the theorem is complete.

THEOREM 10.17. *Let the hypotheses of Theorem* 10.15 *be satisfied, and let H have the deficiency-index* (2, 2). *Then $\mathfrak{D}$ is the set of all functions f in $\mathfrak{D}^*$ such that*

$$\begin{aligned} f(a) = p(a) f'(a) = B_b(f, u_1) &= B_b(f, u_2) \\ &= B_b(\bar{f}, u_1) = B_b(\bar{f}, u_2) = 0, \end{aligned}$$

where u_1 and u_2 are the functions described in Theorem 10.14; *and $\mathfrak{G}(\varphi)$ is the set of all functions f in $\mathfrak{D}^*$ such that*

$$\begin{gathered} \sin\varphi\, f(a) + \cos\varphi\, p(a) f'(a) \\ = B_b(f, u_1) = B_b(f, u_2) = B_b(\bar{f}, u_1) = B_b(\bar{f}, u_2) = 0. \end{gathered}$$

The closed linear symmetric transformation $G(\varphi)$ has the deficiency-index (1, 1); *in other words, the solutions in $\mathfrak{L}_2(a, b)$ of the differential system $-(pf')' + (r - l) f = 0$, $\mathfrak{J}(l) \neq 0$, $\sin\varphi\, f(a) + \cos\varphi\, p(a) f'(a) = 0$, constitute a one-dimensional closed linear manifold. Let u_φ be a normalized function in this manifold; let $v_{\varphi,\theta}$ denote the real function $(e^{i\theta/2} u_\varphi - e^{-i\theta/2} \bar{u}_\varphi)/2i$; let $\mathfrak{G}(\varphi, \theta)$ be the set of all functions f in $\mathfrak{D}^*$ such that*

$$\sin\varphi\, f(a) + \cos\varphi\, p(a) f'(a) = 0, \quad B_b(f, v_{\varphi,\theta}) = 0;$$

and let $G(\varphi, \theta)$ be the transformation with domain $\mathfrak{G}(\varphi, \theta)$ which takes f into $-(pf')' + rf$. Then $G(\varphi, \theta)$ is a self-adjoint extension of $G(\varphi)$ real with respect to the conjugation J_2;

and, if S is an arbitrary maximal symmetric extension of $G(\varphi)$, then S coincides with $G(\varphi, \theta)$ for exactly one value of θ on the range $0 \leqq \theta < 2\pi$. The transformations $G(\varphi, \theta)$ constitute a proper subclass of the class of all transformations $H(\mathfrak{a})$ which are real with respect to the conjugation J_2.

The description of $\mathfrak{D}$ given in Theorems 10.14 and 10.15 shows that in the present case $\mathfrak{D}$ is the set of all functions f in $\mathfrak{D}^*$ such that

$$f(a) = p(a)f'(a)$$
$$= B_b(f, u_1) = B_b(f, u_2) = B_b(f, \bar{u}_1) = B_b(f, \bar{u}_2) = 0.$$

By means of the identity $B(\bar{f}, g) = \overline{B(f, \bar{g})}$, we can write the last two equations in the alternative form given above. In order to characterize $\mathfrak{G}(\varphi)$, we proceed, as in the proof of Theorem 10.16, to form linear combinations $u_{k,0} = u_k - a_{k1} f_{\varphi_1} - a_{k2} f_{\varphi_2}$ such that

$$u_{k,0}(a) = p(a)\, u'_{k,0}(a) = 0 \quad \text{for } k = 1, 2.$$

It is obvious that $u_{10}, u_{20}, \bar{u}_{10}, \bar{u}_{20}$ are functions in $\mathfrak{G}^*(\varphi)$ for every φ. Thus a function f in $\mathfrak{G}(\varphi)$ must belong to $\mathfrak{D}^*$ and must satisfy the conditions

$$\sin\varphi\, f(a) + \cos\varphi\, p(a)f'(a) = B_b(f, u_{10}) = B_b(f, u_{20})$$
$$= B_b(f, \bar{u}_{10}) = B_b(f, \bar{u}_{20}) = 0.$$

The fact that f_{φ_1} and f_{φ_2} vanish identically on the interval (b', b) enables us to write $B_b(f, u_k) = B_b(f, u_{k,0})$, $B_b(f, \bar{u}_k) = B_b(f, \bar{u}_{k,0})$, for $k = 1, 2$, and hence to obtain the conditions stated in the theorem. Let us suppose, on the other hand, that f is a function in $\mathfrak{D}^*$ satisfying the indicated conditions. By virtue of the equation $\sin\varphi\, f(a) + \cos\varphi\, p(a)f'(a) = 0$, we can choose a constant c so that the function $f_0 = f - cf_\varphi$ has the property $f_0(a) = p(a)f_0'(a) = 0$. Since f_φ vanishes identically on the interval (b', b), the four remaining conditions on f imply that

$$B_b(f_0, u_1) = B_b(f_0, u_2) = B_b(\bar{f}_0, u_1) = B_b(\bar{f}_0, u_2) = 0.$$

Thus f_0 must belong to $\mathfrak{D}$ and $f = f_0 + cf_\varphi$ to $\mathfrak{G}(\varphi)$, as we wished to prove.

In order to determine the deficiency-index of $G(\varphi)$, we must ascertain the multiplicity of the not-real characteristic values of the transformation $G^*(\varphi)$—in other words, we must determine the number of linearly independent solutions in $\mathfrak{L}_2(a, b)$ of the differential system $-(pf')' + (r-l)f = 0$, $\mathfrak{J}(l) \neq 0$, $\sin\varphi f(a) + \cos\varphi\, p(a) f'(0) = 0$. Since H has the deficiency-index (2, 2) by hypothesis, the differential equation has two linearly independent solutions in $\mathfrak{L}_2(a, b)$; and the solutions which satisfy the boundary condition therefore constitute a one-dimensional closed linear manifold in $\mathfrak{L}_2(a, b)$. Thus the transformation $G(\varphi)$ has the deficiency-index (1, 1).

The relations between $G(\varphi)$, u_φ, $v_{\varphi,\theta}$, $G(\varphi, \theta)$ can be discussed by the same methods as those applied to the treatment of $H, u, v_\theta, H(\theta)$ in Theorem 10.13, with the results stated above. It is evident that each transformation $G(\varphi, \theta)$ coincides with just one of the transformations $H(\mathfrak{a})$ described in Theorem 10.14; and it is found that there exist some real transformations $H(\mathfrak{a})$ which coincide with no transformation $G(\varphi, \theta)$. In order to survey these relations in detail, we shall indicate the determination of the matrix $\mathfrak{a}$ such that $H(\mathfrak{a}) \equiv G(\varphi, \theta)$. According to Theorem 10.14, an arbitrary function f in $\mathfrak{D}^*$ can be expressed in the form $f = f_0 + X_1 u_1 + X_2 u_2 + X_3 \bar{u}_1 + X_4 \bar{u}_2$, where the X's are complex variables and f_0 is in $\mathfrak{D}$. In order that a function f in $\mathfrak{D}^*$ belong to $\mathfrak{G}(\varphi, \theta)$, it must satisfy conditions which reduce to linear relations between these complex variables. If we denote by $W_\varphi(f)$ the expression $\sin\varphi f(a) + \cos\varphi\, p(a) f'(a)$ and write $W_\varphi^{(1)}$, $W_\varphi^{(2)}$, $\overline{W_\varphi^{(1)}}$, $\overline{W_\varphi^{(2)}}$ in place of $W_\varphi(u_1)$, $W_\varphi(u_2)$, $W_\varphi(\bar{u}_1)$, $W_\varphi(\bar{u}_2)$ respectively, we can calculate u_φ and $v_{\varphi,\theta}$ in terms of u_1 and u_2 by means of the formulas

$$u_\varphi = A_\varphi (W_\varphi^{(2)} u_1 - W_\varphi^{(1)} u_2),$$

$$v_{\varphi,\theta} = A_\varphi (e^{i\theta/2} W_\varphi^{(2)} u_1 - e^{i\theta/2} W_\varphi^{(1)} u_2 - e^{-i\theta/2} \overline{W_\varphi^{(2)}}\, \bar{u}_1 - e^{-i\theta/2} \overline{W_\varphi^{(1)}}\, \bar{u}_2)/2i,$$

where the constant A_φ is chosen so that $\int_a^b |u_\varphi|^2 \, dx = 1$. In order that a function f in $\mathfrak{D}^*$ belong to $\mathfrak{G}(\varphi, \theta)$, it is necessary and sufficient that $W_\varphi(f) = B(f, v_{\varphi,\theta}) = 0$. By direct calculation we find that these conditions are equivalent to the relations

$$W_\varphi^{(1)} X_1 + W_\varphi^{(2)} X_2 + \overline{W_\varphi^{(1)}} X_3 + \overline{W_\varphi^{(2)}} X_4 = 0,$$
$$\overline{W_\varphi^{(2)}} X_1 - \overline{W_\varphi^{(1)}} X_2 + e^{i\theta} W_\varphi^{(2)} X_3 - e^{i\theta} W_\varphi^{(1)} X_4 = 0,$$

respectively. On the other hand, a function f in $\mathfrak{D}^*$ belongs to $\mathfrak{D}(\mathfrak{a})$ if and only if $B(f, v_{1\mathfrak{a}}) = B(f, v_{2\mathfrak{a}}) = 0$; that is, if and only if

$$X_1 - \bar{a}_{11} X_3 - \bar{a}_{12} X_4 = 0,$$
$$X_2 - \bar{a}_{21} X_3 - \bar{a}_{22} X_4 = 0.$$

In order that $G(\varphi, \theta) \equiv H(\mathfrak{a})$, $\mathfrak{G}(\varphi, \theta) \equiv \mathfrak{D}(\mathfrak{a})$, it is thus necessary and sufficient that the first pair of linear forms introduced above be linearly dependent upon the second. This requirement is easily found to be equivalent to the following set of relations between the coefficients of the four forms:

$$-W_\varphi^{(1)} \bar{a}_{11} - W_\varphi^{(2)} \bar{a}_{21} = \overline{W_\varphi^{(1)}},$$
$$-W_\varphi^{(1)} \bar{a}_{12} - W_\varphi^{(2)} \bar{a}_{22} = \overline{W_\varphi^{(2)}},$$
$$-\overline{W_\varphi^{(2)}} \bar{a}_{11} + \overline{W_\varphi^{(1)}} \bar{a}_{21} = e^{i\theta} W_\varphi^{(2)},$$
$$-\overline{W_\varphi^{(2)}} \bar{a}_{12} + \overline{W_\varphi^{(1)}} \bar{a}_{22} = -e^{i\theta} W_\varphi^{(1)}.$$

By solving these equations, we obtain

$$a_{11} = \frac{-(W_\varphi^{(1)})^2 - e^{-i\theta} (\overline{W_\varphi^{(2)}})^2}{|W_\varphi^{(1)}|^2 + |W_\varphi^{(2)}|^2},$$

$$a_{12} = a_{21} = \frac{-W_\varphi^{(1)} W_\varphi^{(2)} + e^{-i\theta} \overline{W_\varphi^{(1)}}\, \overline{W_\varphi^{(2)}}}{|W_\varphi^{(1)}|^2 + |W_\varphi^{(2)}|^2},$$

$$a_{22} = \frac{-e^{-i\theta} (\overline{W_\varphi^{(1)}})^2 - (W_\varphi^{(2)})^2}{|W_\varphi^{(1)}|^2 + |W_\varphi^{(2)}|^2},$$

where $|W_\varphi^{(1)}|^2 + |W_\varphi^{(2)}|^2 > 0$. The matrix $\mathfrak{a} = \{a_{jk}\}$ can be shown by direct calculation to be a unitary matrix which satisfies the conditions $\bar{a}_{22} = e^{i\theta} a_{11}$, $\bar{a}_{12} = -e^{i\theta} a_{12}$, as well as the condition $a_{12} = a_{21}$. It is obvious that the elements of this matrix are rational functions of $e^{i\theta}$, $\sin\varphi$, and $\cos\varphi$. Our determination of this matrix leads, of course, to the identity $H(\mathfrak{a}) \equiv G(\varphi, \theta)$, and thus shows that distinct pairs (φ, θ) on the range $0 \leqq \varphi < \pi$, $0 \leqq \theta < 2\pi$ yield distinct matrices $\mathfrak{a}$. On the other hand, we find without difficulty that the most general unitary matrix $\mathfrak{a}$ which satisfies the condition $a_{12} = a_{21}$ can be written in the form

$$a_{11} = e^{i\alpha}\cos\gamma, \quad a_{12} = \pm e^{i(\alpha+\beta)/2}\sin\gamma, \quad a_{22} = -e^{i\beta}\cos\gamma,$$

where $0 \leqq \alpha < 2\pi$, $0 \leqq \beta < 2\pi$, $0 \leqq \gamma \leqq \frac{\pi}{2}$. Distinct choices of the arbitrary sign in a_{12} and of the numbers α, β, γ on the range $0 \leqq \alpha < 2\pi$, $0 \leqq \beta < 2\pi$, $0 < \gamma < \frac{\pi}{2}$ yield distinct matrices; but when γ has the value 0 or the value $\frac{\pi}{2}$, distinct choices of the sign in a_{12} and of the numbers α, β may lead to the same matrix. We see therefore that the transformations $H(\mathfrak{a})$ which are real with respect J_2 depend upon three essential parameters, while those which satisfy a relation of the form $H(\mathfrak{a}) \equiv G(\varphi, \theta)$ depend upon only two. Since in each case the dependence is expressed by analytic relations which are in general non-singular, there exist real transformations $H(\mathfrak{a})$ which are not of the type $G(\varphi, \theta)$; in fact, such transformations $H(\mathfrak{a})$ form a class with the cardinal number $\mathfrak{c}$ of the continuum.

In the three preceding theorems we have considered the case where the functions $p(x)$ and $r(x)$ are restricted at $x = a$. The corresponding case where $b < +\infty$ and the functions $1/p(x)$ and $r(x)$ are integrable over (c, b), $a < c < b$, can obviously be reduced to the one just considered by putting $y = -x$, $g(y) = f(x)$ and examining the formal differential operator $-\frac{d}{dy}p(-y)\frac{d}{dy} + r(-y)\cdot$, as applied to functions $g(y)$ in $\mathfrak{L}_2(-b, -a)$. Hence we shall not study this case in detail.

We shall now discuss the case where restrictions are imposed at both extremities of the interval (a, b). We have the following result.

THEOREM 10.18. *Let (a, b) be an arbitrary finite interval; and let the functions $p(x)$ and $r(x)$ have the property that $1/p(x)$ and $r(x)$ are Lebesgue-integrable over (a, b). Then $\mathfrak{D}^*$ is the set of all functions f such that*

(1) $f(x)$ is absolutely continuous on the closed interval (a, b);

(2) $p(x)f'(x)$ is absolutely continuous on the closed interval (a, b);

(3) $-(pf')' + rf$ is a function in $\mathfrak{L}_2(a, b)$;

$\mathfrak{D}$ is the set of all functions f in $\mathfrak{D}^*$ such that

$$f(a) = p(a)f'(a) = f(b) = p(b)f'(b) = 0;$$

and the transformation H has the deficiency-index $(2, 2)$. Between the class of all self-adjoint transformations $H(\mathfrak{a})$ and the class of all self-adjoint differential systems of the type

$$-(pf')' + (r - l)f = f^*,$$

$$W_1(f) = a_1 f(a) + b_1 p(a) f'(a) + c_1 f(b) + d_1 p(b) f'(b) = 0,$$

$$W_2(f) = a_2 f(a) + b_2 p(a) f'(a) + c_2 f(b) + d_2 p(b) f'(b) = 0,$$

where systems with equivalent boundary conditions are regarded as identical, there is a one-to-one correspondence such that the self-adjoint transformation defined by a system of this type coincides with the corresponding transformation $H(\mathfrak{a})$; in other words, between the class of all unitary matrices $\mathfrak{a}$ and the class of all pairs of self-adjoint boundary conditions $W_1(f)$, $W_2(f)$ associated with the formal differential operator $-\frac{d}{dx}p(x)\frac{d}{dx} + r(x)\cdot$, equivalent pairs being regarded as identical, there is a one-to-one correspondence such that corresponding elements of the two classes are connected by the relations $B(f, v_{1\mathfrak{a}}) = b_{11} W_1(f) + b_{12} W_2(f)$, $B(f, v_{2\mathfrak{a}}) = b_{21} W_1(f) + b_{22} W_2(f)$, $\det \{b_{jk}\} \neq 0$, holding for every function f in $\mathfrak{D}^*$.

In the present case we can apply Lemma 10.1 directly to the entire interval (a, b). All solutions of the differential equation $-(pf')' + (r - l)f = f^*$, where f^* is Lebesgue-integrable over (a, b) are absolutely continuous on the closed interval (a, b) and therefore belong to $\mathfrak{L}_2(a, b)$. We see at once that $\mathfrak{D}^*$ can be described in the manner indicated above; and we can apply the results of Theorems 10.16 and 10.17, which are relevant both at $x = a$ and at $x = b$, to show that $\mathfrak{D}$ is the subset of $\mathfrak{D}^*$ determined by the boundary conditions $f(a) = p(a)f'(a) = f(b) = p(b)f'(b) = 0$. Since all the solutions of the differential equation $-(pf')' + (r-l)f = 0$ belong to $\mathfrak{L}_2(a, b)$, we see that H has the deficiency-index $(2, 2)$, in accordance with Theorem 10.12.

By definition, $\mathfrak{D}(\mathfrak{a})$ is the subset of $\mathfrak{D}^*$ determined by the boundary conditions $B(f, v_{1\mathfrak{a}}) = B(f, v_{2\mathfrak{a}}) = 0$, which can be written in the form $W_1(f) = W_2(f) = 0$ by a suitable choice of the constant coefficients in W_1 and W_2; and, since $H(\mathfrak{a})$ is self-adjoint, we see that these boundary conditions must be self-adjoint with respect to the formal differential operator $-\frac{d}{dx}p(x)\frac{d}{dx} + r(x)\cdot$, as we have already pointed out in Chapter III, § 3. On the other hand, if we determine a transformation S by using self-adjoint boundary conditions $W_1(f) = W_2(f) = 0$ to assign the domain of this formal operator as a subset of $\mathfrak{D}^*$, we can show by the arguments sketched in Chapter III, §3, that S is self-adjoint and hence coincides with one of the transformations $H(\mathfrak{a})$. We can give a detailed analysis of the situation by writing the functions belonging to $\mathfrak{D}^*$ in the form $f = f_0 + X_1 u_1 + X_2 u_2 + X_3 \bar{u}_1 + X_4 \bar{u}_2$ and expressing the boundary conditions as linear relations between the complex variables X_1, X_2, X_3, X_4. The substitution of the X's for the variables $\xi_1 = f(a)$, $\xi_2 = p(a)f'(a)$, $\xi_3 = f(b)$, $\xi_4 = p(b)f'(b)$, which appear initially in the boundary conditions, is justified by the observation that the ξ's are directly expressible as linear combinations of the X's and that, conversely, the X's can be similarly expressed in terms of the ξ's by virtue of the relations

$$B(f, u_1) = 2iX_1, \qquad B(f, u_2) = 2iX_2,$$
$$B(f, \bar{u}_1) = -2iX_3, \qquad B(f, \bar{u}_2) = -2iX_4.$$

In order to define self-adjointness for boundary conditions expressed in terms of the X's, we note that

$$B(f, g) = 2i(X_1 \bar{Y}_1 + X_2 \bar{Y}_2 - X_3 \bar{Y}_3 - X_4 \bar{Y}_4)$$

where

$$f = f_0 + X_1 u_1 + X_2 u_2 + X_3 \bar{u}_1 + X_4 \bar{u}_2 \ .$$

and

$$g = g_0 + Y_1 u_1 + Y_2 u_2 + Y_3 \bar{u}_1 + Y_4 \bar{u}_2$$

are arbitrary functions in $\mathfrak{D}^*$. Let W_k and V_k, $k = 1, 2, 3, 4$, be any linearly independent linear forms in the X's and in the Y's respectively, connected by the relation

$$2i(X_1 \bar{Y}_1 + X_2 \bar{Y}_2 - X_3 \bar{Y}_3 - X_4 \bar{Y}_4) = \sum_{\alpha=1}^{4} W_\alpha \bar{V}_{5-\alpha}.$$

Then the boundary conditions $V_1 = V_2 = 0$ are said to be adjoint to the boundary conditions $W_1 = W_2 = 0$; and the latter conditions are said to be self-adjoint if the forms V_1 and V_2, when written in terms of the X's, are linearly dependent upon the forms W_1 and W_2. It is now easy to show that the boundary conditions $B(f, v_{1\mathfrak{a}}) = B(f, v_{2\mathfrak{a}}) = 0$, associated with the transformation $H(\mathfrak{a})$, are self-adjoint in the indicated sense. We first introduce the forms

$$W_j = \sum_{\alpha=1}^{4} A_{j\alpha} X_\alpha, \qquad j = 1, 2, 3, 4,$$

where

$$\begin{aligned} B(f, v_{1\mathfrak{a}}) = W_1 &= X_1 \phantom{+ a_{21} X_2} - \bar{a}_{11} X_3 - \bar{a}_{12} X_4, \\ B(f, v_{2\mathfrak{a}}) = W_2 &= X_2 - \bar{a}_{21} X_3 - \bar{a}_{22} X_4, \\ W_3 &= a_{11} X_1 + a_{21} X_2 + X_3 \quad , \\ W_4 &= a_{12} X_1 + a_{22} X_2 + X_4. \end{aligned}$$

It is readily verified that the matrix $\{A_{jk}\}$ has the properties $\sum_{\alpha=1}^{4} A_{\alpha j} \bar{A}_{\alpha k} = 2\delta_{jk}$, $j, k = 1, 2, 3, 4$. Thus, if we set $Y_1^* = -iY_1$, $Y_2^* = -iY_2$, $Y_3^* = iY_3$, $Y_4^* = iY_4$, $V_{5-j} = \sum_{\alpha=1}^{4} A_{j\alpha} Y_\alpha^*$, we obtain

$$\sum_{\alpha=1}^{4} W_\alpha \bar{V}_{5-\alpha} = \sum_{\alpha,\beta,\gamma=1}^{4} A_{\alpha\beta} \bar{A}_{\alpha\gamma} X_\beta \bar{Y}_\gamma^* = 2 \sum_{\beta=1}^{4} X_\beta \bar{Y}_\beta^*$$
$$= 2i(X_1 \bar{Y}_1 + X_2 \bar{Y}_2 - X_3 \bar{Y}_3 - X_4 \bar{Y}_4) = B(f, g).$$

Hence the boundary conditions adjoint to $B(f, v_{1a}) = B(f, v_{2a}) = 0$ are obtained by equating to zero the forms

$$V_1 = -a_{12}\,i\,Y_1 - a_{22}\,i\,Y_2 \qquad + i\,Y_4,$$
$$V_2 = -a_{11}\,i\,Y_1 - a_{21}\,i\,Y_2 + i\,Y_3$$

If we replace the Y's by the X's in these forms, we see directly that

$$V_1 = -i\,a_{12}\,W_1 - i\,a_{22}\,W_2, \quad V_2 = -i\,a_{11}\,W_1 - i\,a_{21}\,W_2.$$

Hence the boundary conditions $B(f, v_{1a}) = B(f, v_{2a}) = 0$ are self-adjoint. On the other hand, if W_1 and W_2 are linearly independent linear forms in the X's, we can define a transformation S which has as its domain the set of all functions $f = f_0 + X_1 u_1 + X_2 u_2 + X_3 \bar{u}_1 + X_4 \bar{u}_2$ in $\mathfrak{D}^*$ such that $W_1 = W_2 = 0$, and which carries f into $-(pf')' + rf$. From the obvious relations $H \subseteq S \subseteq H^* \equiv T$ we conclude that $T \equiv H^* \supseteq S^* \supseteq H \equiv T^*$. Hence the domain of S^* consists of those functions $g = g_0 + Y_1 u_1 + Y_2 u_2 + Y_3 \bar{u}_1 + Y_4 \bar{u}_2$ in $\mathfrak{D}^*$ such that

$$B(f, g) = (Tf, g) - (f, Tg) = (Sf, g) - (f, Tg) = 0$$

for all functions f in the domain of S; and S^* carries g into $-(pg')' + rg$. We introduce the linear forms W_3, W_4, V_1, V_2, V_3, V_4 connected by the relation $B(f, g) = \sum_{\alpha=1}^{4} W_\alpha \bar{V}_{5-\alpha}$, as described above. The condition on the domain of S^* is then expressed by the equation $W_3 \bar{V}_2 + W_4 \bar{V}_1 = 0$. Since the values of W_3 and W_4 are independent and unrestricted when $W_1 = W_2 = 0$, we conclude that the domain of S^* consists of those functions g in $\mathfrak{D}^*$ which satisfy the boundary conditions $V_1 = V_2 = 0$. If, in particular, the boundary conditions $W_1 = W_2 = 0$ are self-adjoint, the transformations S and S^* have the same domain and are therefore identical. In this case, S is a self-adjoint extension of H and must therefore coincide with just one of the transformations $H(a)$. The condition that two sets of boundary conditions $W_1 = W_2 = 0$ and $W_1' = W_2' = 0$ determine the same subset of $\mathfrak{D}^*$ is that they be equivalent in the sense that $W_1' = c_{11} W_1 + c_{12} W_2$,

$W_2' = c_{21} W_1 + c_{22} W_2$, $\det \{c_{jk}\} \neq 0$. The results obtained in this paragraph serve to establish the statements made in the theorem.

Since the theory developed in Theorem 10.17 can be applied here, it is of some interest to indicate the significance of $\mathfrak{G}(\varphi)$, $\mathfrak{G}^*(\varphi)$, $\mathfrak{G}(\varphi, \theta)$, $G(\varphi)$, $G^*(\varphi)$, and $G(\varphi, \theta)$ in the present instance. We shall state the facts without proof. The sets $\mathfrak{G}(\varphi)$ and $\mathfrak{G}^*(\varphi)$ are readily identified as the subsets of $\mathfrak{D}^*$ determined by the boundary conditions

$$\sin \varphi\, f(a) + \cos \varphi\, p(a) f'(a) = f(b) = p(b) f'(b) = 0$$

and

$$\sin \varphi\, f(a) + \cos \varphi\, p(a) f'(a) = 0,$$

respectively. For fixed φ, the relation

$$\sin \chi\, v_{\varphi,\theta}(b) + \cos \chi\, p(b)\, v'_{\varphi,\theta}(b) = 0$$

determines a one-to-one correspondence between the ranges $0 \leqq \theta < 2\pi$ and $0 \leqq \chi < \pi$; if θ and χ are in correspondence, then $\mathfrak{G}(\varphi, \theta)$ is the subset of $\mathfrak{D}^*$ determined by the self-adjoint boundary conditions

$$\sin \varphi f(a) + \cos \varphi\, p(a) f'(a) = \sin \chi f(b) + \cos \chi\, p(b) f'(b) = 0.$$

The transformations $G(\varphi)$, $G^*(\varphi) \equiv T(\varphi)$, and $G(\varphi, \theta)$ are to be described accordingly.

We shall now return to the general situation discussed under Theorems 10.11–10.14. It will be assumed henceforth that $p(x)$ and $r(x)$ are subject merely to the conditions stated there. With the aid of the special results obtained in Theorems 10.15–10.17, we are able to obtain much more exact knowledge concerning the general case. Both the methods and the results are suggested by a paper of Weyl.† We begin by treating the case where H has the deficiency-index (2, 2), since this case throws light on the others.

THEOREM 10.19. *A necessary and sufficient condition that the transformation H have the deficiency-index* (2, 2) *is that the solutions of the differential equation* $-(pf')' + (r - l) f = 0$

---

† Weyl, Göttinger Nachrichten, 1910, pp. 442–46ı.

all belong to $\mathfrak{L}_2(a, b)$, *even when* l *is real. When* H *has the deficiency-index* $(2, 2)$, *the resolvent* $R_l(\mathfrak{a})$ *of the self-adjoint transformation* $H(\mathfrak{a})$ *is an integral operator with kernel* $G(x, y; l; \mathfrak{a})$ *of Hilbert-Schmidt type. If* $w_1(x; l)$ *and* $w_2(x; l)$ *are arbitrary linearly independent solutions of the differential equation* $-(pf')' + (r - l)f = 0$, *where* l *is a point in the resolvent set of* $H(\mathfrak{a})$, *then* $G(x, y; l; \mathfrak{a})$ *is expressed as the quotient*

$$\frac{\begin{vmatrix} w_1(x; l) & w_2(x; l) & g(x, y; l) \\ B(w_1, v_{1\mathfrak{a}}) & B(w_2, v_{1\mathfrak{a}}) & B(g, v_{1\mathfrak{a}}) \\ B(w_1, v_{2\mathfrak{a}}) & B(w_2, v_{2\mathfrak{a}}) & B(g, v_{2\mathfrak{a}}) \end{vmatrix}}{\begin{vmatrix} B(w_1, v_{1\mathfrak{a}}) & B(w_2, v_{1\mathfrak{a}}) \\ B(w_1, v_{2\mathfrak{a}}) & B(w_2, v_{2\mathfrak{a}}) \end{vmatrix}}$$

where

$$g(x, y; l) = \operatorname{sgn}(x - y)(w_1(x; l) w_2(y; l) - w_1(y; l) w_2(x; l))/2W(w_1, w_2)$$

and where $B(g, v_{k\mathfrak{a}})$, $k = 1, 2$, *is calculated with* g *as a function of* x. *The point spectrum of* $H(\mathfrak{a})$ *is a denumerably infinite set of isolated characteristic values, none of which has multiplicity greater than two; and the continuous spectrum of* $H(\mathfrak{a})$ *is empty.*

We first assume tnat l is a point in the resolvent set of the self-adjoint transformation $H(\mathfrak{a})$ and is also a characteristic value of multiplicity two for the transformation $T \equiv H^*$. According to Theorems 4.18 and 10.14 both conditions are satisfied whenever l is not real; and we shall show subsequently that the second condition is always satisfied and hence implies no restriction on l. The determination of the resolvent of $H(\mathfrak{a})$ is accomplished by solving the differential equation $-(pf')' + (r - l)f = f^*$, with the boundary conditions $B(f, v_{1\mathfrak{a}}) = B(f, v_{2\mathfrak{a}}) = 0$, where f^* is an arbitrary function in $\mathfrak{L}_2(a, b)$. The general solution of the differential equation can be written as

$$f(x) = a_1 w_1(x; l) + a_2 w_2(x; l) + \int_a^x \frac{w_1(x; l)\, w_2(\xi; l) - w_1(\xi; l)\, w_2(x; l)}{W(w_1, w_2)} f^*(\xi)\, d\xi,$$

where a_1 and a_2 are complex constants and $w_1(x; l)$ and $w_2(x; l)$ are linearly independent solutions of the homogeneous differential equation obtained by setting $f^* = 0$. By hypothesis w_1 and w_2 are functions in $\mathfrak{L}_2(a, b)$ so that the integral appearing in this formula is convergent. Furthermore, it is evident that $f(x)$ is expressed by this formula as a linear combination of w_1 and w_2 with coefficients which are bounded continuous functions of x on (a, b) and must therefore be a function in $\mathfrak{L}_2(a, b)$. The verification of the assertion that f is in $\mathfrak{D}^*$ and is a solution of the differential equation offers no further difficulty. We can also write the general solution in the form

$$f(x) = b_1 w_1(x; l) + b_2 w_2(x; l) + \int_b^x \frac{w_1(x; l)\, w_2(\xi; l) - w_1(\xi; l)\, w_2(x; l)}{W(w_1, w_2)} f^*(\xi)\, d\xi.$$

Hence, if we multiply each of the two formulas by $\frac{1}{2}$ and add, we obtain a third expression for the general solution,

$$f(x) = c_1 w_1(x; l) + c_2 w_2(x; l) + \int_a^b g(x, \xi; l) f^*(\xi)\, d\xi,$$

where $g(x, y; l)$ is the function introduced in the statement of the theorem. We must now choose the constants c_1 and c_2 so that f satisfies the boundary conditions. By direct calculation we find that

$$B_x(f, v_{ka}) = c_1 B_x(w_1, v_{ka}) + c_2 B_x(w_2, v_{ka}) + \int_a^b B_x(g(x, \xi; l), v_{ka}) f^*(\xi)\, d\xi,$$

$$B_x(g(x, y; l), v_{ka}) = \operatorname{sgn}(x - y) \times (B_x(w_1, v_{ka})\, w_2(y; l) - B_x(w_2, v_{ka})\, w_1(y; l))/2W(w_1, w_2)$$

for $k = 1, 2$. We see from the form of the expression for $B_x(g, v_{ka})$ that the limits

$$B_a(g, v_{ka}) = -(B_a(w_1, v_{ka})\, w_2(y; l) - B_a(w_2, v_{ka})\, w_1(y; l))/2W(w_1, w_2),$$

$$B_b(g, v_{ka}) = (B_b(w_1, v_{ka})\, w_2(y; l) - B_b(w_2, v_{ka})\, w_1(y; l))/2W(w_1, w_2),$$

exist for $a<y<b$, and that

$$|B_x(g, v_{k\mathfrak{a}})| \leqq K(|w_1(y; l)|+|w_2(y; l)|)$$

for a suitably chosen positive constant K. In the expression for $B_x(f, v_{k\mathfrak{a}})$ we allow x to tend to a and to b respectively. The inequality just noted enables us to effect the passage to the limit under the sign of integration in each case. We thus find that

$$B_a(f, v_{k\mathfrak{a}}) = c_1 B_a(w_1, v_{k\mathfrak{a}}) + c_2 B_a(w_2, v_{k\mathfrak{a}}) + \int_a^b B_a(g(x, \xi; l), v_{k\mathfrak{a}}) f^*(\xi)\, d\xi,$$

$$B_b(f, v_{k\mathfrak{a}}) = c_1 B_b(w_1, v_{k\mathfrak{a}}) + c_2 B_b(w_2, v_{k\mathfrak{a}}) + \int_a^b B_b(g(x, \xi; l), v_{k\mathfrak{a}}) f^*(\xi)\, d\xi.$$

On subtracting the first equation from the second, we find that

$$B(f, v_{k\mathfrak{a}}) = c_1 B(w_1, v_{k\mathfrak{a}}) + c_2 B(w_2, v_{k\mathfrak{a}}) + \int_a^b B(g(x, \xi; l), v_{k\mathfrak{a}}) f^*(\xi)\, d\xi.$$

In order that f satisfy the conditions $B(f, v_{1\mathfrak{a}}) = B(f, v_{2\mathfrak{a}}) = 0$, the constants c_1 and c_2 must be chosen as solutions of the linear equations

$$c_1 B(w_1, v_{k\mathfrak{a}}) + c_2 B(w_2, v_{k\mathfrak{a}}) = -\int_a^b B(g; v_{k\mathfrak{a}}) f^*(\xi)\, d\xi, \qquad k = 1, 2.$$

If we solve these equations and substitute the resulting values for c_1 and c_2 in the expression for f, we obtain

$$f(x) = \int_a^b G(x, \xi; l; \mathfrak{a}) f^*(\xi)\, d\xi,$$

where G is the function introduced in the statement of the theorem. This procedure is made possible by the fact that the determinant

$$\begin{vmatrix} B(w_1, v_{1\mathfrak{a}}) & B(w_2, v_{1\mathfrak{a}}) \\ B(w_1, v_{2\mathfrak{a}}) & B(w_2, v_{2\mathfrak{a}}) \end{vmatrix}$$

cannot vanish when l is in the resolvent set of $H(\mathfrak{a})$: for the vanishing of this determinant would imply the existence of a function $f = c_1 w_1 + c_2 w_2$ which satisfies the boundary conditions $B(f, v_{1\mathfrak{a}}) = B(f, v_{2\mathfrak{a}}) = 0$ without being identically zero, and which is therefore a solution of the equation $H(\mathfrak{a})f - lf = 0$. Thus the resolvent $R_l(\mathfrak{a})$ of $H(\mathfrak{a})$ is an integral operator with the kernel $G(x, y; l; \mathfrak{a})$ for the values of l under consideration. Since this kernel can be written in the forms

$$\sum_{\alpha,\beta=1}^{2} A_{\alpha\beta}(l; \mathfrak{a})\, w_\alpha(x; l)\, w_\beta(y; l), \quad x > y;$$

$$\sum_{\alpha,\beta=1}^{2} B_{\alpha\beta}(l; \mathfrak{a})\, w_\alpha(x; l)\, w_\beta(y; l), \quad x < y,$$

the integral $\int_a^b \int_a^b |G(x, y; l; \mathfrak{a})|^2\, dx\, dy$ exists and the kernel is of Hilbert-Schmidt type.

Whenever l is not real, the resolvent $R_l(\mathfrak{a})$ exists and has the properties indicated above. By Theorem 3.8 it is a bounded linear transformation of finite norm; it is also a normal transformation in accordance with Theorem 8.16. Thus we can discuss its spectrum by methods analogous to those used in the proof of Theorem 5.14. As a result we find that the spectrum of $R_l(\mathfrak{a})$ consists of the origin together with a denumerably infinite set of characteristic values having the origin as sole limit point; since the equation $R_l(\mathfrak{a})f = 0$ implies $f = 0$, by Theorem 4.10, the origin must belong to the continuous spectrum of $R_l(\mathfrak{a})$. By virtue of the identity $H(\mathfrak{a}) \equiv R_l^{-1}(\mathfrak{a}) + lI$, the spectrum of $H(\mathfrak{a})$ can be determined according to Theorems 6.7 and 8.7. We find that $H(\mathfrak{a})$ has a pure point spectrum of isolated characteristic values, as stated in the theorem. It is evident that the equation $H(\mathfrak{a})f = lf$ never has more than two linearly independent solutions—in other words, that no characteristic value of $H(\mathfrak{a})$ has multiplicity greater than two.

We shall now prove that H has the deficiency-index (2, 2) if and only if all the solutions of the differential equation $-(pf')' + (r - l)f = 0$ belong to $\mathfrak{L}_2(a, b)$. The case where l is not real has already been treated in Theorem 10.14, so that we may suppose henceforth that l is real. We first consider the situation where the differential equation has two linearly independent solutions, $w_1(x; l)$ and $w_2(x; l)$, in $\mathfrak{L}_2(a, b)$ for real l. Then the function $g(x, y; l)$ defined above is a kernel of Hilbert-Schmidt type, and therefore defines a bounded linear transformation with domain $\mathfrak{L}_2(a, b)$. By Theorem 4.21, the equation

$$\int_a^b g(x, \xi; l) f(\xi)\, d\xi - m f(x) = f^*(x)$$

has a unique solution in $\mathfrak{L}_2(a, b)$, whenever f^* is in $\mathfrak{L}_2(a, b)$ and m is sufficiently remote from the origin. We denote by $w_k(x; l, m)$ the solution obtained for $f^*(x) = w_k(x; l)$, $k = 1, 2$. The relation

$$m w_k(x; l, m) = -w_k(x; l) + \int_a^b g(x, \xi; l) w_k(\xi; l, m)\, d\xi$$

shows that $w_k(x; l, m)$ is in $\mathfrak{D}^*$ and that

$$m(H^* w_k(x; l, m) - l w_k(x; l, m)) = w_k(x; l, m).$$

Thus $w_1(x; l, m)$ and $w_2(x; l, m)$ are linearly independent solutions in $\mathfrak{L}_2(a, b)$ of the differential equation $-(pf')' + (r - (l - 1/m))f = 0$. Since we can assign to m not-real values of large modulus, we see by Theorem 10.14 that H has the deficiency-index (2,2). When H has the deficiency-index (2,2), we wish to prove that every solution of the differential equation $-(pf')' + (r - l)f = 0$ belongs to $\mathfrak{L}_2(a, b)$. We first consider the special case where the hypotheses of Theorem 10.17 are valid. For arbitrary not-real l the equation has two linearly independent solutions, $w_1(x; l)$ and $w_2(x; l)$, in $\mathfrak{L}_2(a, b)$. If $R_l(\mathfrak{a})$ is the resolvent of $H(\mathfrak{a})$, the equation $R_l(\mathfrak{a}) f - m f = w_k$ has a unique solution $w_k(x; l, m)$ in $\mathfrak{L}_2(a, b)$ for $k = 1, 2$, whenever m is different from zero and from the characteristic values of $R_l(\mathfrak{a})$. It is now easy to show that

$$H^* w_k(x; l, m) - (l - 1/m)\, w_k(x; l, m) = 0,$$

and hence that every solution of the differential equation $-(pf')' + (r - (l - 1/m))f = 0$ belongs to $\mathfrak{L}_2(a, b)$, unless m assumes one of the forbidden values—in other words, unless $l-1/m$ is a characteristic value of $H(\mathfrak{a})$. Since we can choose $H(\mathfrak{a})$ at pleasure, we see that every solution of the differential equation $-(pf')'+(r-l)f=0$ belongs to $\mathfrak{L}_2(a, b)$ unless l is a characteristic value for all the transformations $H(\mathfrak{a})$. If l is such an exceptional value, the solutions of the equation which belong to $\mathfrak{L}_2(a, b)$ constitute a closed linear manifold with dimension number two or one. We exclude the second alternative here by showing that it leads to a contradiction. If this alternative were realized, there would exist a function $w(x; l)$ in $\mathfrak{L}_2(a, b)$ which would satisfy the equation $H(\mathfrak{a})f-lf=0$ for every transformation $H(\mathfrak{a})$ without vanishing identically. In particular, $w(x; l)$ would belong to the domain $\mathfrak{G}(\varphi, \theta)$ of the transformation $G(\varphi, \theta)$ described in Theorem 10.17, and would therefore satisfy the relation

$$\sin\varphi\, w(a; l)+\cos\varphi\, p(a)\, w'(a; l) = 0$$

for every φ. Thus w would satisfy the equations

$$-(pw')'+(r-l)w = 0, \qquad w(a; l) = p(a)w'(a; l) = 0.$$

According to Lemma 10.1, this is possible only if $w(x; l)$ vanishes identically, contrary to hypothesis. We have thus established the desired result for the particular case under discussion. We next remark, as at the close of Theorem 10.17, that the case where $p(x)$ and $r(x)$ are restricted at $x = b$ instead of at $x = a$ leads to a similar result. Finally, in the general case, we consider the behavior of the solutions of the differential equation $-(pf')'+(r-l)f=0$ on each of the intervals (a, b), (a, c), (c, b), where $a<c<b$. For not-real l, every solution on (a, b) belongs to $\mathfrak{L}_2(a, b)$, by hypothesis; and hence every solution on (a, c) belongs to $\mathfrak{L}_2(a, c)$, every solution on (c, b) to $\mathfrak{L}_2(c, b)$. The behavior of the solutions for arbitrary l on each of the intervals (a, c) (c, b) is therefore governed by the results proved for the two special cases treated above. It follows that every solution of the equation on (a, c) belongs to $\mathfrak{L}_2(a, c)$, every solution on (c, b) to $\mathfrak{L}_2(c, b)$, even in case l is real. Hence we find

that every solution on (a, b) belongs to $\mathfrak{L}_2(a, b)$, as we wished to prove.

THEOREM 10.20. *Let n_a be the number of linearly independent solutions in $\mathfrak{L}_2(a, c)$ of the differential equation $-(pf')'+(r-l)f = 0$, $a<x<c$, where $a<c<b$; and let n_b be the number of linearly independent solutions in $\mathfrak{L}_2(c, b)$ of the differential equation $-(pf')'+(r-l)f = 0$, $c<x<b$. Then the transformation H has the deficiency-index $(1, 1)$ if and only if one of the following two situations occurs:*

(1) *$n_a = 2$ for all l and all c; $n_b = 1$ for all not-real l and all c;*

(2) *$n_a = 1$ for all not-real l and all c; $n_b = 2$ for all l and all c. If the first condition is satisfied, then $B_b(f, g) = 0$ for all f and g in $\mathfrak{D}^*$, $\mathfrak{D}$ is the set of all functions f in $\mathfrak{D}^*$ such that $B_a(f, u) = B_a(f, \bar{u}) = 0$, and $\mathfrak{D}(\theta)$ is the set of all functions f in $\mathfrak{D}^*$ such that $B_a(f, v_\theta) = 0$. For not-real l the resolvent $R_l(\theta)$ of the self-adjoint transformation $H(\theta)$ is an integral operator with the kernel*

$$G(x, y; l; \theta) = \begin{Bmatrix} w(x; l)\, w_\theta(y; l)/W(w, w_\theta), & x>y \\ w(y; l)\, w_\theta(x; l)/W(w, w_\theta), & x<y \end{Bmatrix},$$

where $w(x; l)$ is a non-trivial solution in $\mathfrak{L}_2(a, b)$ of the differential equation $-(pf')'+(r-l)f = 0$, and $w_\theta(x; l)$ is a non-trivial solution of this equation satisfying the boundary condition $B_a(w_\theta, v_\theta) = 0$; the real and imaginary parts of $G(x, y; l; \theta)$ are real kernels of Carleman type. The transformation $H(\theta)$ has no characteristic values with multiplicity greater than one. If the second condition is satisfied, analogous results are valid, the rôles played by the points $x = a$ and $x = b$ being interchanged.

We first observe that the numbers n_a and n_b are obviously independent of c; and that they are also independent of l, when l is not real, by virtue of their significance in connection with the application of Theorems 10.11–10.17 over the intervals (a, c) and (c, b). According to Theorem 10.15, each of the numbers n_a and n_b must have the value 1 or the value 2, when l is not real. If $n_a = n_b = 2$, then the differential equation $-(pf')'+(r-l)f=0$ has two linearly independent solutions in $\mathfrak{L}_2(a, b)$, and H has the deficiency-index (2, 2) in accordance with Theorems 10.14 and 10.19. If $n_a=2$ and $n_b=1$ or if $n_a=1$ and $n_b=2$, for some not-real l, then the solutions of the differential equation which belong to $\mathfrak{L}_2(a, b)$ constitute a one-dimensional closed linear manifold, and H has the deficiency-index (1, 1) in accordance with Theorem 10.13. In these cases it should be noted that if $n_a=2$ for a single value of l, real or not, then $n_a=2$ for all values of l by Theorem 10.19; and that if $n_a=1$ for a single value of l, real or not, then $n_a=1$ for all not-real l and $n_a=0$ or $n_a=1$ for each real l, by Theorems 10.13 and 10.19. Similar remarks apply to n_b. Finally, if $n_a=n_b=1$, we can show that H has the deficiency-index (0, 0). If f and g are functions in $\mathfrak{L}_2(a, b)$ which, considered on (a, c), belong to the associated set $\mathfrak{D}^* \subseteqq \mathfrak{L}_2(a, c)$ and which, considered on (c, b), belong to the associated set $\mathfrak{D}^* \subseteqq \mathfrak{L}_2(c, b)$, then $B_a(f, g) = B_b(f, g) = 0$, as we have proved in Theorem 10.16. Since all functions in the set $\mathfrak{D}^* \subseteqq \mathfrak{L}_2(a, b)$ associated with the interval (a, b) are functions of the type just described, we conclude that the equation

$$B(f, g) = B_b(f, g) - B_a(f, g) = 0$$

holds for all f and g in $\mathfrak{D}^*$, and that the set $\mathfrak{D}$ therefore coincides with $\mathfrak{D}^*$. This result implies that H has the deficiency-index (0, 0) in accordance with Theorems 10.12–10.14. It is important to remark in this case that the solutions of the differential equation $-(pf')'+(r-l)f=0$, $\mathfrak{J}(l) \neq 0$, which are of integrable absolute square over (a, c) and (c, b) respectively, in accordance with the relations $n_a=n_b=1$, must be linearly independent: for otherwise the equation would have a non-trivial solution in $\mathfrak{L}_2(a, b)$ and H would have the deficiency-index (1, 1), contrary to fact. We may also add that if the equation $n_a=n_b=1$ holds for a single value of l, real or not, then it holds for all not-real l, and n_a, n_b assume the values 0 or 1 for each real l. From the preceding analysis of the various possibilities, the statement given in the theorem follows at once.

We shall now discuss in detail the case where $n_a = 2$, $n_b = 1$ for not-real l. As we pointed out above, we have $B_b(f, g) = 0$ for all f and g in $\mathfrak{D}^*$, by virtue of the application of Theorem 10.16 over the interval (c, b). In consequence, the descriptions of the sets $\mathfrak{D}$ and $\mathfrak{D}(\theta)$ given in Theorem 10.13 reduce to those stated in the present theorem. We next consider the existence and properties of the functions $w(x; l)$ and $w_\theta(x; l)$ described above. The existence of $w(x; l)$ follows immediately from the fact that $n_a = 2$, $n_b = 1$, as we indicated in the preceding paragraph. It is easily seen that the quantity $B_a(w, v_\theta)$ cannot vanish: for otherwise $w(x; l)$, which is obviously a function in $\mathfrak{D}^*$, would belong to $\mathfrak{D}(\theta)$ and would be a characteristic function for the self-adjoint transformation $H(\theta)$ corresponding to the not-real characteristic value l. Since we have $n_a = 2$, there exist two linearly independent solutions, $w_1(x; l)$ and $w_2(x; l)$, of the differential equation $-(pf')' + (r - l)f = 0$ which, considered on the interval (a, c), belong to $\mathfrak{L}_2(a, c)$. At least one of the two quantities $B_a(w_1, v_\theta)$ and $B_a(w_2, v_\theta)$ is different from zero: for otherwise $B_a(w, v_\theta)$ would vanish by virtue of the fact that w is a linear combination of w_1 and w_2. Thus we can define $w_\theta(x; l)$ by the equation

$$w_\theta(x; l) = B_a(w_2, v_\theta)\, w_1(x; l) - B_a(w_1, v_\theta)\, w_2(x; l).$$

It is to be noted that w and w_θ are linearly independent because of the relations $B_a(w, v_\theta) \neq 0$ and $B_a(w_\theta, v_\theta) = 0$, and that the quantity $W(w, w_\theta)$ is therefore different from zero. It may also be remarked that the functions w and w_θ are uniquely determined, save for constant multipliers different from zero. We can now introduce the function $G(x, y; l; \theta)$, $\mathfrak{I}(l) \neq 0$, by means of the formula given in the theorem, and can verify directly that the integral $\int_a^b |G(x, \xi; l; \theta)|^2 d\xi$ exists. Since $G(x, y; l; \theta) = G(y, x; l; \theta)$, it is clear that the real and imaginary parts of this function are real symmetric functions of x and y which belong to $\mathfrak{L}_2(a, b)$ when treated as functions of x or of y alone—in other words, these functions are kernels of Carleman type, as defined in Chapter III, § 2. We can now form the integral $f(x) = \int_a^b G(x, \xi; l; \theta) f^*(\xi)\, d\xi$ for an arbitrary function $f^*(x)$ in $\mathfrak{L}_2(a, b)$. From the formula

$$f(x) = w(x; l) \int_a^x \frac{w_\Theta(\xi; l)}{W(w, w_\Theta)} f^*(\xi)\, d\xi$$
$$+ w_\Theta(x; l) \int_x^b \frac{w(\xi; l)}{W(w, w_\Theta)} f^*(\xi)\, d\xi,$$

it is clear that f is an absolutely continuous function on the open interval (a, b). Direct calculation now shows that

$$p(x) f'(x) = \int_a^b p(x) \frac{\partial G(x, \xi; l; \theta)}{\partial x} f^*(\xi)\, d\xi$$

where

$$\frac{\partial G(x, y; l; \theta)}{\partial x} = \left\{\begin{matrix} w'(x; l)\, w_\Theta(y; l)/W(w, w_\Theta), & x>y \\ w(y; l)\, w'_\Theta(x; l)/W(w, w_\Theta), & x<y \end{matrix}\right\}.$$

In the same manner as above we find that $p(x) f'(x)$ is absolutely continuous on the open interval (a, b). Further calculation shows that $-(pf')' + (r-l)f = f^*$. Our next step is to prove that $B_a(f, v_\Theta) = 0$. For $a<x<b$, we find without difficulty that

$$B_x(f, v_\Theta) = \int_a^b B_x(G(x, \xi; l; \theta), v_\Theta) f^*(\xi)\, d\xi,$$

where

$$B_x(G(x, y; l; \theta), v_\Theta(x))$$
$$= \left\{\begin{matrix} B_x(w, v_\Theta)\, w_\Theta(y; l)/W(w, w_\Theta), & x>y \\ B_x(w_\Theta, v_\Theta)\, w(y; l)/W(w, w_\Theta), & x<y \end{matrix}\right\}.$$

Since the limits $B_a(w, w_\Theta)$, $B_a(w_\Theta, w_\Theta)$ exist, there is a positive constant K such that for $a<x\leqq c$

$$|B_x(G, v_\Theta)| \leqq K(|w(y; l)| + |w_\Theta(y; l)|), \quad a<y\leqq c,$$
$$|B_x(G, v_\Theta)| \leqq K|w(y; l)|, \quad c<y<b.$$

Hence we can allow x to tend to a under the sign of integration in the expression for $B_x(f, v_\Theta)$. Since we have

$$\lim_{x\to a} B_x(G, v_\theta) = \lim_{\substack{x\to a\\ x<y}} B_x(w_\theta, v_\theta)\, w(y; l)/W(w, w_\theta) = 0,$$

for every y on the open interval (a, b), we conclude that $B_a(f, v_\theta) = 0$. Finally we shall show that $f(x)$ belongs to $\mathfrak{L}_2(a, b)$; because of the other properties of f which have already been established, this result implies that f belongs to $\mathfrak{D}(\theta)$ and that $H(\theta)f - lf = f^*$. The proof will be modelled on the discussion of similar situations encountered in Theorems 10.8 and 10.9. We introduce the function $f_c^*(x)$ equal to $f^*(x)$ for $a < x \leqq c$ and equal to zero for $c < x < b$. This function is obviously in $\mathfrak{L}_2(a, b)$. The corresponding function $f_c(x) = \int_a^b G(x, \xi; l; \theta) f_c^*(\xi)\, d\xi$ thus has the form

$$f_c(x) = w(x; l) \int_a^x \frac{w_\theta(\xi; l)}{W(w, w_\theta)} f^*(\xi)\, d\xi$$
$$+ w_\theta(x; l) \int_x^c \frac{w(\xi; l)}{W(w, w_\theta)} f^*(\xi)\, d\xi, \quad a < x \leqq c,$$

$$f_c(x) = w(x; l) \int_a^c \frac{w_\theta(\xi; l)}{W(w, w_\theta)} f^*(\xi)\, d\xi, \quad c < x < b.$$

The fact that w belongs to $\mathfrak{L}_2(a, b)$ and w_θ, considered on (a, c), to $\mathfrak{L}_2(a, c)$, shows that $f_c(x)$ belongs to $\mathfrak{L}_2(a, b)$. As we pointed out above, we conclude that f_c is in $\mathfrak{D}(\theta)$ and that $H(\theta)f_c - lf_c = f_c^*$. Theorem 4.14 now shows that

$$\| f_c \|^2 \leqq \| f_c^* \|^2/|\mathfrak{J}(l)|^2 \quad \text{or} \quad \int_a^b |f_c|^2\, dx \leqq \int_a^b |f_c^*|^2\, dx/|\mathfrak{J}(l)|^2.$$

The expressions given above for f_c show that on any closed interval (a', b') interior to (a, b), f_c tends uniformly to the limit f as c tends to b. By combining the relations

$$\lim_{c\to b} \int_{a'}^{b'} |f_c|^2\, dx = \int_{a'}^{b'} |f|^2\, dx, \qquad \int_a^b |f_c^*|^2\, dx \leqq \int_a^b |f^*|^2\, dx$$

with the inequality given above, we find that

$$\int_{a'}^{b'} |f(x)|^2\, dx \leqq \int_a^b |f^*(x)|^2\, dx/|\mathfrak{J}(l)|^2,$$

and thus conclude that f is in $\mathfrak{L}_2(a, b)$. According to remarks made previously, f is a solution of the equation $H(\theta)f - lf = f^*$.

Since this equation is known to have a unique solution, we have

$$R_l(\theta)f^* = f(x) = \int_a^b G(x, \xi; l; \theta) f^*(\xi)\, d\xi$$

as the explicit form of the resolvent $R_l(\theta)$ of $H(\theta)$.

The assertion of the theorem concerning the characteristic values of $H(\theta)$, if any exist, is an obvious consequence of Theorem 10.19.

The alternative case $n_a = 1$, $n_b = 2$ can be reduced to the one just discussed by simple transformations discussed at the close of Theorem 10.17. We shall therefore omit the treatment of this case.

THEOREM 10.21. *The transformation H has the deficiency-index* $(0, 0)$ *if and only if the numbers n_a and n_b introduced in Theorem* 10.20 *are both equal to one, for all not-real l and all c. For all not-real l the resolvent R_l of the self-adjoint transformation H is an integral operator with the kernel*

$$G(x, y; l) = \begin{cases} w_1(x; l)\, w_2(y; l)/W(w_1, w_2), & x > y \\ w_1(y; l)\, w_2(x; l)/W(w_1, w_2), & x < y \end{cases}$$

where $w_1(x; l)$ and $w_2(x; l)$ are linearly independent solutions of the differential equation $-(pf')' + (r - l)f = 0$ *such that*

(1) *$w_1(x; l)$, considered on the interval (c, b), belongs to* $\mathfrak{L}_2(c, b)$;

(2) *$w_2(x; l)$, considered on the interval (a, c), belongs to* $\mathfrak{L}_2(a, c)$.

The real and imaginary parts of $G(x, y; l)$ are real kernels of Carleman type. No characteristic value of the transformation H has multiplicity greater than one.

The significance of the condition $n_a = n_b = 1$ has already been established in connection with the proof of Theorem 10.20. The existence and linear independence of the functions w_1 and w_2 were also proved there. We may add the remark that these functions are uniquely determined, save for constant multipliers different from zero. We can therefore proceed immediately to the discussion of the resolvent R_l of H. We find directly that the integral $\int_a^b |G(x,\xi;l)|^2\,d\xi$ exists, and show, as in the proof of the preceding theorem, that the real and imaginary parts of $G(x,y;l)$ are real kernels of Carleman type. We then form the function

$$f(x)=\int_a^b G(x,\xi;l)f^*(\xi)\,d\xi,$$

where $f^*(x)$ is an arbitrary function in $\mathfrak{L}_2(a,b)$. By following the argument used in the proof of Theorem 10.20, we find that f is absolutely continuous on the open interval (a,b), that

$$p(x)f'(x)=\int_a^b p(x)\frac{\partial G(x,\xi;l)}{\partial x}f^*(\xi)\,d\xi$$

is absolutely continuous on the open interval (a,b), and that $-(pf')'+(r-l)f=f^*$ almost everywhere on (a,b). We now show that $f(x)$ belongs to $\mathfrak{L}_2(a,b)$; the properties of this function already established then demand that $f(x)$ belong to $\mathfrak{D}\equiv\mathfrak{D}^*$ and that $Hf-lf=f^*$. We introduce the function $f^*_{a'b'}(x)$ equal to $f^*(x)$ for $a<a'\leqq x\leqq b'<b$ and equal to zero for $a<x<a'$ and $b'<x<b$. This function evidently belongs to $\mathfrak{L}_2(a,b)$. The corresponding function $f_{a'b'}(x)=\int_a^b G(x,\xi;l)f^*_{a'b'}(\xi)\,d\xi$ then has the form

$$f_{a'b'}(x)=w_2(x;l)\int_{a'}^{b'}w_1(\xi;l)f^*(\xi)\,d\xi/W(w_1,w_2),\quad a<x<a',$$

$$f_{a'b'}(x)=w_1(x;l)\int_{a'}^{x}w_2(\xi;l)f^*(\xi)\,d\xi/W(w_1,w_2)$$
$$+w_2(x;l)\int_{x}^{b'}w_1(\xi;l)f^*(\xi)\,d\xi/W(w_1,w_2),$$
$$a'\leqq x\leqq b',$$

$$f_{a'b'}(x)=w_1(x;l)\int_{a'}^{b'}w_2(\xi;l)f^*(\xi)\,d\xi/W(w_1,w_2),\quad b'<x<b.$$

The properties (1) and (2) of the functions $w_1(x;l)$ and $w_2(x;l)$ show that $f_{a'b'}(x)$ belongs to $\mathfrak{L}_2(a,b)$ and is therefore a function in $\mathfrak{D}$ such that $Hf_{a'b'}-lf_{a'b'}=f^*_{a'b'}$, as we pointed out above. Theorem 4.14 yields the inequality $\|f_{a'b'}\|^2\leqq\|f^*_{a'b'}\|^2/|\mathfrak{I}(l)|^2$, which may be written

$$\int_a^b|f_{a'b'}(x)|^2\,dx\leqq\int_{a'}^{b'}|f^*(x)|^2\,dx/|\mathfrak{I}(l)|^2.$$

It is evident that on any closed interval (a'', b'') interior to (a, b) the function $f_{a'b'}(x)$ tends uniformly to the limit $f(x)$ when a' and b' tend to a and b respectively. As in the proof of Theorem 10.20, we conclude that

$$\int_{a''}^{b''} |f(x)|^2\, dx \leqq \int_a^b |f^*(x)|^2\, dx / |\mathfrak{J}(l)|^2$$

and hence that $f(x)$ belongs to $\mathfrak{L}_2(a, b)$. The argument employed in the proof of Theorem 10.20 leads finally to the desired result that

$$R_l f^*(x) = \int_a^b G(x, \xi; l) f^*(\xi)\, d\xi.$$

The closing remark of the theorem with regard to the characteristic values of H, if any exist, is an obvious consequence of Theorem 10.19.

We propose to apply the preceding theorems concerning the resolvents of self-adjoint extensions of H to the characterization of the corresponding resolutions of the identity. At certain points in the analysis we require information concerning the solutions of the homogeneous equation $-(pf')' + (r - l)f = 0$ somewhat more precise than that given in Lemma 10.1. We therefore state

LEMMA 10.2. *If $u(x, c; c_1, c_2; l)$ is the solution of the differential equation* $-\dfrac{d}{dx}\, p(x) \dfrac{d}{dx} f(x) + (r(x) - l) f(x) = 0$, $a < x < b$, *determined by the initial conditions* $u(c, c; c_1, c_2; l) = c_1$, $p(x) \dfrac{\partial}{\partial x} u(x, c; c_1, c_2; l)\Big|_{x=c} = c_2$, *where $p(x)$ and $r(x)$ are subject to the hypotheses of Theorem* 10.10, *$x = c$ is an interior point of the interval (a, b), and c_1, c_2, and l are complex parameters,—then $u(x, c; c_1, c_2; l)$ and $p(x) \dfrac{\partial}{\partial x} u(x, c; c_1, c_2; l)$ are continuous functions of their five arguments. When c_1, c_2, and l are arbitrary fixed values such that $|c_1|^2 + |c_2|^2 > 0$, the set of points (x, c) determined by the equation $u(x, c; c_1, c_2; l) = 0$ is a null set on the range $a < x < b$, $a < c < b$.*

We first show that for fixed c_1, c_2, and for $l = 0$, the functions $f(x, c) = u(x, c; c_1, c_2; 0)$ and $p(x) \dfrac{\partial}{\partial x} f(x, c)$ are continuous in (x, c) in the neighborhood of any point (c_0, c_0) where $a < c_0 < b$. As we saw in the proof of Lemma 10.1, $f(x, c)$ is the unique solution of the integral equation

$$f(x, c) = c_1 + c_2 \int_c^x 1/p(\eta)\, d\eta + \int_c^x \left(\int_\xi^x 1/p(\eta)\, d\eta \right) r(\xi) f(\xi, c)\, d\xi$$

and can be calculated by the method of successive approximations. In the earlier discussion, we put $f_0(x, c) = 0$ and denoted by $f_{n+1}(x, c)$ the function obtained by substituting $f_n(x, c)$ for $f(x, c)$ in the right-hand member of the integral equation; we then defined the functions

$$g_0(x, c) = f_0(x, c) = 0, \qquad g_{n+1}(x, c) = f_{n+1}(x, c) - f_n(x, c);$$

and we showed that

$$f(x, c) = \lim_{n \to \infty} f_n(x, c) = \sum_{\alpha=0}^{\infty} g_\alpha(x, c),$$

the series being uniformly convergent when c is fixed and x is confined to a sufficiently small neighborhood of the point $x = c$. By minor modifications of the method of proof used in the case where c is fixed, we can show that the series converges uniformly for all (x, c) in a sufficiently small neighborhood of the point (c_0, c_0) where c_0 is any value between a and b. Since the functions f_n and g_n are obviously continuous in (x, c), we conclude that $f(x, c)$ is a continuous function of (x, c) in a sufficiently small neighborhood of any point (c_0, c_0). The details of this demonstration may be left to the reader. The corresponding property of the function $p(x) \frac{\partial}{\partial x} f(x, c)$ can be read at once from the relation

$$p(x) \frac{\partial}{\partial x} f(x, c) = c_2 + \int_c^x r(\xi) f(\xi, c)\, d\xi.$$

We next prove that the functions $f(x, c)$, $p(x) \frac{\partial}{\partial x} f(x, c)$ are continuous in the neighborhood of an arbitrary point (x_0, c_0) of the range $a < x < b$, $a < c < b$, and are therefore continuous throughout this range. We introduce the functions

$$\varphi_1(x, c; l) = u(x, c; 1, 0; l), \qquad \varphi_2(x, c; l) = u(x, c; 0, 1; l)$$

and employ them to write

$$f(x, c) = f(c_0, c)\varphi_1(x, c_0; 0) + p(c_0)\frac{\partial}{\partial c_0} f(c_0, c)\varphi_2(x, c_0; 0),$$

$$p(x)\frac{\partial}{\partial x}f(x, c) = f(c_0, c)p(x)\frac{\partial}{\partial x}\varphi_1(x, c_0; 0)$$

$$+ p(c_0)\frac{\partial}{\partial c_0}f(c_0, c)\,p(x)\frac{\partial}{\partial x}\varphi_2(x, c_0; 0),$$

in accordance with Lemma 10.1. On the right of each of these equations, each term is the product of a function of x which we know to be continuous on the open interval (a, b), by a function of c which we have just proved to be continuous in the neighborhood of the point $c = c_0$. Hence the functions on the left are continuous in the neighborhood of the arbitrary point (x_0, c_0), as we wished to prove.

Finally, we can show that the functions under consideration are continuous in all five arguments. We shall sketch the proof, leaving all details to the reader. By Lemma 10.1, we see that $u(x, c; c_1, c_2; l)$ is the unique solution of the integral equation

$$u(x, c; c_1, c_2; l) = \Psi(x, c; c_1, c_2) + l\int_c^x \Phi(x, \xi; c)u(\xi, c; c_1, c_2; l)\,d\xi,$$

where

$$\Psi = c_1\varphi_1(x, c; 0) + c_2\varphi_2(x, c; 0)$$

and

$$\Phi = \varphi_1(x, c; 0)\varphi_2(\xi, c; 0) - \varphi_2(x, c; 0)\varphi_1(\xi, c; 0)$$

are continuous functions of their arguments by virtue of the result established in the preceding paragraph. We can calculate the solution of this integral equation by the method of successive approximations, starting with a first approximating function which vanishes identically. It is then easy to show that the successive approximating functions are continuous in the variables x, c, c_1, c_2, l and constitute a sequence which converges uniformly to the solution u on any range of the type $a' \leqq x \leqq b'$, $a' \leqq c \leqq b'$, $|c_1| \leqq C_1$, $|c_2| \leqq C_2$, $|l| \leqq L$, where $a < a' < b' < b$. The continuity of the solution u is thus established. We can now deduce the continuity of the function $p(x)\dfrac{\partial}{\partial x}u(x, c; c_1, c_2; l)$ directly from the relation

$$p(x)\frac{\partial}{\partial x}u(x, c; c_1, c_2; l) = p(x)\frac{\partial}{\partial x}\Psi(x, c; c_1, c_2)$$
$$+ l\int_c^x p(x)\frac{\partial}{\partial x}\Phi(x, \xi; c)\,u(\xi, c; c_1, c_2; l)\,d\xi,$$

since all the functions appearing on the right are continuous in their respective arguments in accordance with the earlier results. We may add the obvious remark that this method shows u and $p\dfrac{\partial u}{\partial x}$ to be analytic in c_1, c_2, and l.

When c_1, c_2, and l have fixed values with $|c_1|^2+|c_2|^2>0$, the equation $u(x, c; c_1, c_2; l) = 0$ determines a measurable subset of the range $a<x<b$, $a<c<b$. The intersection of this set with an arbitrary line $c = c_0$ is the set of points x, $a<x<b$, where the function $u(x, c_0; c_1, c_2; l)$ vanishes and must therefore be a null set or the entire range $a<x<b$. The second possibility is excluded by the inequality $|c_1|^2+|c_2|^2>0$, which does not permit u to vanish identically in x. The theorem of Fubini now shows that the set of points (x, c) under consideration is a null set, as we wished to prove.

THEOREM 10.22. *Let A be an arbitrary self-adjoint extension of the symmetric transformation H of Theorem* 10.11; *let $E(\lambda)$ be the corresponding resolution of the identity; let Δ be an arbitrary finite interval $\alpha \leqq \lambda \leqq \beta$; and let $\varphi_1(x, c; l)$ and $\varphi_2(x, c; l)$ be the solutions of the differential equation $-\dfrac{d}{dx}p(x)\dfrac{d}{dx}f(x)+(r(x)-l)f(x) = 0$, $a<x<b$, determined by the initial conditions*

$$\varphi_1(x, c; l) = p(x)\frac{\partial}{\partial x}\varphi_2(x, c; l) = 1,$$

$$\varphi_2(x, c; l) = p(x)\frac{\partial}{\partial x}\varphi_1(x, c; l) = 0, \quad x = c,$$

where $a<c<b$. Then the transformation $E(\Delta) \equiv E(\beta)-E(\alpha)$ is an integral operator with kernel $\mathbf{E}(x, y; \Delta)$ of Carleman type. This kernel can be expressed as a Stieltjes integral

$$E(x, y; \Delta) = \sum_{\mu,\nu=1}^{2} \int_{\alpha}^{\beta} \varphi_\mu(x, c; \lambda)\, \varphi_\nu(y, c; \lambda)\, d\varrho_{\mu\nu}(\lambda),$$

where $\varrho_{jk}(\lambda)$, $j, k = 1, 2$, *is a function which is of bounded variation over every finite interval and which is continuous on the right. For a fixed value* c, *the functions* $\varrho_{jk}(\lambda)$ *are uniquely determined apart from additive constants. If these constants are suitably determined, the matrix* $\{\varrho_{jk}(\lambda)\}$ *is (Hermitian) symmetric; this matrix can be taken as real if and only if the transformation* A *is real with respect to the conjugation* J_2. *For fixed* Δ *and almost all* y *on the interval* (a, b), $E(x, y; \Delta)$ *is a function of* x *in the domain of* A, *and*

$$A E(x, y; \Delta) = -\frac{\partial}{\partial x} p(x) \frac{\partial}{\partial x} E(x, y; \Delta) + r(x) E(x, y; \Delta)$$
$$= \int_{\alpha}^{\beta} \mu \, d E(x, y; \Delta_\mu),$$

where Δ_μ *is the interval with extremities* 0 *and* μ.

In developing the rather lengthy proof of this theorem, we shall actually establish much more than we have asserted in the formal statement. The additional material can be found below, chiefly in the various numbered formulas. The facts which we encounter here occupy a position in the theory of differential operators roughly analogous to that of Theorems 10.4–10.5 in the theory of integral operators. In the formal statement, however, the points of divergence are emphasized at the expense of the points of similarity. Thus, the integral representation of the kernel E is evidently peculiar to the theory of differential operators, while the last assertion of the theorem and also many of the related facts established in the course of the proof are strict analogues of the results of Theorems 10.4–10.5.

Throughout the demonstration of the present theorem, we shall find it convenient to make use of a fixed terminology, to which we now devote a few explanatory remarks. The first two variables appearing in the kernel E shall be denoted by the letters x, y, z, c, ξ, η and shall be restricted without comment to the open interval (a, b). Thus the assertion that a specified property holds for almost all $(x_1, \cdots, x_n)$, or for all $(x_1, \cdots, x_n)$ outside a null set, is understood to mean that the property holds for all points $(x_1, \cdots, x_n)$, where $a < x_k < b$ for $k = 1, \cdots, n$, with the exception of a set of points of Lebesgue measure zero. In this connection we recall that a set of points $(x_1, \cdots, x_n)$ is of measure zero, according to the theorem of Fubini, if and only if its intersection with the hyperplane $x_1 = c_1, \cdots, x_k = c_k$ is a set of points $(x_{k+1}, \cdots, x_n)$ of zero measure, for almost all $(c_1, \cdots, c_k)$. In a similar way, the variables $\alpha, \beta, \lambda, \mu, \nu$ are to be associated with the interval $(-\infty, +\infty)$, unless some further restriction is explicitly noted. At certain points of the proof, we shall have occasion to confine these variables to a fixed denumerably infinite set Λ which is everywhere dense in the interval $(-\infty, +\infty)$ and which contains the point $\lambda = 0$. In this connection it is convenient to introduce the class $\mathfrak{D}_0(\Lambda)$ of all intervals Δ which have extremities α and β in the set Λ and which contain the point $\lambda = 0$. We shall say that a function $F(x_1, \cdots, x_n; \lambda)$ belongs to the class $\mathfrak{B}^+(x_1, \cdots, x_n)$ if it has the properties

(1) $F(x_1, \cdots, x_n; \lambda)$ is of bounded variation in λ over every finite interval Δ;

(2) $F(x_1, \cdots, x_n; \lambda)$ is continuous on the right in λ,

for all $(x_1, \cdots, x_n)$ outside a fixed null set independent of Δ, λ. Naturally, we may include the case where F is independent of some or all of the variables $x_1, \cdots, x_n$ simply by dropping the variables in question from our notation. The properties of the Stieltjes integral summarized in Lemma 5.1 are easily carried over to the case of a Stieltjes integral formed with such a function $F(x_1, \cdots, x_n; \lambda)$. We shall therefore refer directly to Lemma 5.1 whenever we need such properties.

We begin the proof of the present theorem by considering the bounded self-adjoint transformation

$$B(l) \equiv \frac{1}{2i}(A_l^{-1} - (A_l^{-1})^*) \equiv \frac{1}{2i}\left(\frac{1}{A - l} - \frac{1}{A - \bar{l}}\right)$$

$$\equiv \frac{\nu}{(A - \mu)^2 + \nu^2},$$

where $l=\mu+i\nu$, $\nu>0$. On the basis of Theorems 10.19–10.21, it is easily shown that $B(l)$ is an integral operator with kernel $K(x, y; l)$ of Carleman type. We have, according as $A\equiv H$, $A\equiv H(\theta)$, or $A\equiv H(\mathfrak{a})$, the following explicit formulas for this kernel:

$$\begin{aligned}
K(x,y;l) &= (G(x,y;l)-G(x,y;\bar{l}))/2i \\
&= (G(x,y;l)-\overline{G(y,x;l)})/2i \\
&= (G(x,y;l)-\overline{G(x,y;l)})/2i = \mathfrak{I}G(x,y;l), \\
K(x,y;l) &= (G(x,y;l;\theta)-G(x,y;\bar{l};\theta))/2i \\
&= (G(x,y;l;\theta)-\overline{G(y,x;l;\theta)})/2i \\
&= (G(x,y;l;\theta)-\overline{G(x,y;l;\theta)})/2i = \mathfrak{I}G(x,y;l;\theta), \\
K(x,y;l) &= (G(x,y;l;\mathfrak{a})-G(x,y;\bar{l};\mathfrak{a}))/2i \\
&= (G(x,y;l;\mathfrak{a})-\overline{G(y,x;l;\mathfrak{a})})/2i.
\end{aligned}$$

The first two formulas can be established by direct manipulation of the formulas given in the two preceding theorems, while the last is a special instance of Theorems 3.8–3.9. The asserted properties of the kernel $K(x, y; l)$ are obvious from these formulas. The resolution of the identity $F(\varrho)$ associated with the transformation $B(l)$ can be expressed by means of the operational calculus of Chapter VI in the form $F(\varrho)\equiv F_\varrho(B(l))$, where the function $F_\varrho(\lambda)$ is equal to one or to zero according as $\lambda\leqq\varrho$ or $\lambda>\varrho$. Theorem 10.4 shows that the transformation $I-F_\varrho(B(l))$, $\varrho>0$, is an integral operator with kernel $F(x, y; \varrho; l)$ of Carleman type. By examining the details of the construction of the kernel given in the first paragraph of the proof of Theorem 10.4, we find that the inequality

$$\int_{a'}^{b'}\int_a^b |F(x,y;\varrho;l)|^2\,dy\,dx \leqq \int_{a'}^{b'}\int_a^b |K(x,y;l)|^2\,dy\,dx/\varrho^2$$

is satisfied. In fact, the explicit formulas of Theorems 10.19–10.21 show that the function

$$K(x;l) = \left(\int_a^b |K(x,\xi;l)|^2\,d\xi\right)^{1/2}$$

is continuous on the open interval $a<x<b$; it is therefore permissible to replace the sets $E(a)$ used in the proof of

Theorem 10.4 by intervals (a', b'), $a < a' < b' < b$, without modifying the reasoning in any essential respect. The desired relation then appears as a special case of the inequality given in the first paragraph of the proof of Theorem 10.4. By the use of Theorem 6.9, we can now write $I - F_\varrho(B(l)) \equiv G_\varrho(A)$, where $G_\varrho(\lambda)$ is equal to one or to zero according as $\dfrac{\nu}{(\lambda-\mu)^2+\nu^2} > \varrho$ or $\dfrac{\nu}{(\lambda-\mu)^2+\nu^2} \leqq \varrho$. For values of ϱ such that $0 < \varrho < 1/\nu$, $\nu > 0$, we have $G_\varrho(A) \equiv E(\beta - 0) - E(\alpha)$, where $\alpha = \mu - (\nu/\varrho - \nu^2)^{1/2}$ and $\beta = \mu + (\nu/\varrho - \nu^2)^{1/2}$. It is clear that, by assigning suitable values to μ, ν, and ϱ, we can give α and $\beta > \alpha$ arbitrary finite values. Hence we see that $E(\beta - 0) - E(\alpha)$ is an integral operator with kernel of Carleman type—namely, the function $F(x, y; \varrho; l)$ with appropriate values of ϱ and l. It is easy to show that $E(\beta) - E(\beta - 0)$ is an integral operator with kernel of Hilbert-Schmidt type. The range of the projection $E(\beta) - E(\beta - 0)$ is the closed linear manifold $\mathfrak{M}_\beta$ consisting of all solutions of the equation $Af = \beta f$. As we have already pointed out in Theorems 10.19–10.21, the dimension number of this manifold is at most equal to two. When it is equal to zero, we have $E(\beta) - E(\beta - 0) \equiv O$ and note that the kernel of this operator vanishes identically; and when it is equal to N, $N = 1, 2$, the kernel of the operator $E(\beta) - E(\beta - 0)$ is the function $\sum_{\nu=1}^{N} u_\nu(x)\overline{u_\nu(y)}$ where $\{u_n\}$ is an orthonormal set which determines $\mathfrak{M}_\beta$. Thus we can assert that

$$E(\Delta) \equiv E(\beta) - E(\alpha) \equiv (E(\beta) - E(\beta - 0)) + (E(\beta - 0) - E(\alpha))$$

is an integral operator with kernel $E(x, y; \Delta)$ of Carleman type. In particular, we know that the integral $\int_{a'}^{b'} \int_a^b |E(x, y; \Delta)|^2 \, dy \, dx$ exists whenever $a < a' < b' < b$. Since we have used a family of kernels, $K(x, y; l)$, in deducing the existence and elementary properties of the kernel $E(x, y; \Delta)$, we cannot apply in its entirety the theory of Chapter X, § 1. We must therefore appeal to direct methods in order to establish further properties of the kernel E.

The fact that E is a kernel of Carleman type requires that $E(y, x; \Delta) = \overline{E(x, y; \Delta)}$ for fixed Δ and almost all (x, y); and also that $E(\xi, x; \Delta)$ and $E(\xi, y; \Delta)$ be functions of ξ belonging to $\mathfrak{L}_2(a, b)$ for fixed Δ and almost all (x, y). The relation $E(\Delta_1 \Delta_2) \equiv E(\Delta_1) E(\Delta_2)$ implies, by the reasoning given in the proof of Theorem 10.5, that

$$E(x, y; \Delta_1 \Delta_2) = \int_a^b E(x, \xi; \Delta_1) E(\xi, y; \Delta_2) \, d\xi$$

for fixed Δ_1, Δ_2 and for almost all (x, y). It is evident that these properties hold for all (x, y) outside a null set independent of the intervals Δ, Δ_1, and Δ_2, provided that the extremities of these intervals be confined to the denumerably infinite set $\varLambda$. In fact, when this restriction is made, we can determine a set $E_2(\Delta, \Delta_1, \Delta_2)$ of points (x, y) such that the properties in question are valid for all (x, y) in this set and such that the neglected set of points (x, y) is a null set; and we then see that the intersection, E_2^*, of the denumerably infinite collection of sets $E_2(\Delta, \Delta_1, \Delta_2)$ is a set of points (x, y) such that the properties in question are valid upon it for all intervals Δ, Δ_1, Δ_2 with extremities in $\varLambda$, and such that the set of neglected points (x, y) is a null set. We shall now consider the special function $E_0(x, y; \lambda) = \operatorname{sgn} \lambda \, E(x, y; \Delta_\lambda)$, where Δ_λ is the interval with extremities 0 and λ. It is evident that this function is the kernel for the operator $E(\lambda) - E(0) \equiv \operatorname{sgn} \lambda \, E(\Delta_\lambda)$, and that it determines the kernel $E(x, y; \Delta)$ through the relation $E(x, y; \Delta) = E_0(x, y; \beta) - E_0(x, y; \alpha)$. We therefore carry out our investigation in terms of the kernel E_0. Let Δ be a fixed interval (α, β) which contains the point $\lambda = 0$. Since $E(\xi, x; \Delta)$ and $E(\xi, y; \Delta)$ are functions of ξ belonging to $\mathfrak{L}_2(a, b)$ for almost all (x, y), we can form the expression

$$((E(\lambda) - E(0)) E(\xi, y; \Delta), \; E(\xi, x; \Delta))$$
$$= \int_a^b \left(\int_a^b \operatorname{sgn} \lambda \, E(\xi, \eta; \Delta_\lambda) \, E(\eta, y; \Delta) \, d\eta \right) \cdot \overline{E(\xi, x; \Delta)} \, d\xi$$

for arbitrary λ and almost all (x, y). The integral on the right can be reduced to the form

$$\int_a^b \operatorname{sgn}\lambda\, E(\xi, y; \Delta_\lambda \Delta)\, E(x, \xi; \Delta)\, d\xi = \operatorname{sgn}\lambda\, E(x, y; \Delta_\lambda \Delta) = E_0(x, y; \lambda)$$

whenever λ is in the interval Δ, by virtue of the properties indicated above. Hence we can write

$$((E(\lambda) - E(0))\, E(\xi, y; \Delta),\ E(\xi, x; \Delta)) = E_0(x, y; \lambda),$$

where the equality sign holds, first, for fixed α, β, and λ, $\alpha \leqq \lambda \leqq \beta$, and for almost all (x, y), and, secondly, for all α, β, and λ in the set Λ, $\alpha \leqq \lambda \leqq \beta$, and all (x, y) in E_2^*. For fixed α and β in Λ and fixed (x, y) in E_2^*, the left-hand term is a function of λ which is continuous on the right on the interval $(-\infty, +\infty)$ and whose variation over this infinite interval is bounded by the quantity

$$\left(\int_a^b |E(\xi, y; \Delta)|^2\, d\xi \int_a^b |E(\xi, x; \Delta)|^2\, d\xi\right)^{1/2}.$$

In order to obtain these results, it is sufficient to apply the known properties of the resolution of the identity $E(\lambda)$, as indicated in Theorem 5.7 and its proof, to the function $(E(\lambda)\, E(\xi, y; \Delta),\ E(\xi, x; \Delta))$, which differs from the function of λ actually under consideration by the additive constant $-(E(0)\, E(\xi, y; \Delta),\ E(\xi, x; \Delta))$. Thus if α and β are fixed points in the set Λ, if λ is an arbitrary point in the interval Δ, if $\{\lambda + \epsilon_k\}$ is any sequence of points in the set Λ such that $\epsilon_k > 0$ and $\lim_{k\to\infty} \epsilon_k = 0$, and if (x, y) is an arbitrary point in E_2^*, then the limit $\lim_{k\to\infty} E_0(x, y; \lambda + \epsilon_k)$ exists and is equal to $((E(\lambda) - E(0))\, E(\xi, y; \Delta),\ E(\xi, x; \Delta))$. The latter expression is in turn equal to $E_0(x, y; \lambda)$ for almost all (x, y). If for each value of λ we replace $E_0(x, y; \lambda)$ by the limit $\lim_{k\to\infty} E_0(x, y; \lambda + \epsilon_k)$, formed for an arbitrary sequence $\{\lambda + \epsilon_k\}$ in the set Λ such that $\epsilon_k > 0$ and $\lim_{k\to\infty} \epsilon_k = 0$, we thereby change the values of $E_0(x, y; \lambda)$ only on a null set which may depend upon λ; and we obtain a new function which is defined for all λ and all (x, y) in E_2^* and which is evidently a member of the class $\mathfrak{B}^+(x, y)$. The modified function can obviously be used as the kernel of the integral operator $E(\lambda) - E(0)$, since it differs from the kernel as first constructed only on a set of zero measure. We shall therefore suppose that, until further notice, the original kernel is replaced by the modified function and that the symbol $E_0(x, y; \lambda)$ denotes the new kernel. It is clear that the kernel of the integral operator $E(\Delta)$ may be replaced by the expression

$$E(x, y; \Delta) = E_0(x, y; \beta) - E_0(x, y; \alpha),$$

formed in terms of the new function E_0. The modified kernels E and E_0 evidently enjoy all the properties of the original kernels which were set forth above. We can even state some of these properties in more precise form. Thus we see that $E_0(y, x; \lambda) = \overline{E_0(x, y; \lambda)}$ and $E(y, x; \Delta) = \overline{E(x, y; \Delta)}$ for all λ, all Δ, and all (x, y) in E_2^*: the first equation obviously holds for all λ in Λ and all (x, y) in E_2^* and must therefore persist for arbitrary λ by virtue of the fact that $E_0(x, y; \lambda)$ is continuous on the right in λ for arbitrary (x, y) in E_2^*; and the second equation follows directly from the first. Similarly, we note the inequality

$$\text{(A)} \quad V(E_0(x,y;\lambda); \Delta) \leqq \left(\int_a^b |E(\xi,y;\Delta)|^2 d\xi \cdot \int_a^b |E(\xi,x;\Delta)|^2 d\xi\right)^{1/2},$$

which is valid for all (x, y) in E_2^* and all Δ in the class $\mathfrak{D}_0(\Lambda)$.

For fixed λ and almost all y, the function $E_0(x, y; \lambda)$ is a function of x which belongs to $\mathfrak{L}_2(a, b)$ and which has the property that

$$\begin{aligned} E(\Delta_\lambda) E_0(x, y; \lambda) &= \int_a^b E(x, \xi; \Delta_\lambda) \operatorname{sgn} \lambda \, E(\xi, y; \Delta_\lambda) \, d\xi \\ &= \operatorname{sgn} \lambda \, E(x, y; \Delta_\lambda) = E_0(x, y; \lambda) \end{aligned}$$

for almost all x. This relation implies that $E_0(x, y; \lambda)$, considered as a function of x, belongs to the range of the projection $E(\Delta_\lambda)$ and hence to the domain of the self-adjoint transformation A in accordance with Theorem 5.9. We may now calculate the function $AE_0(x, y; \lambda)$ by means of the formulas given in that theorem. We start from the equation

$$(AE_0(\xi, y; \lambda), E_0(\xi, x; \lambda)) = \int_{-\infty}^{+\infty} \mu \, d\,(E(\mu) E_0(\xi, y; \lambda), E_0(\xi, x; \lambda))$$

which is valid for fixed λ and almost all (x, y). The term on the left becomes

$$\begin{aligned}\int_a^b & AE_0(\xi, y; \lambda) \cdot \overline{E_0(\xi, x; \lambda)} \, d\xi \\ &= \int_a^b E_0(x, \xi; \lambda) AE_0(\xi, y; \lambda) \, d\xi \\ &= E(\Delta_\lambda) AE(x, y; \Delta_\lambda) = AE(\Delta_\lambda) E(x, y; \Delta_\lambda) \\ &= \operatorname{sgn} \lambda \, AE_0(x, y; \lambda),\end{aligned}$$

for fixed λ and almost all (x, y). The integral on the right is unchanged in value if the function of bounded variation appearing under the differential sign is replaced by

$$\varrho(x, y; \lambda; \mu) = ((E(\mu) - E(0)) E_0(\xi, y; \lambda), E_0(\xi, x; \lambda)).$$

For fixed λ and almost all (x, y), the functions $E_0(\xi, y; \lambda)$ and $E_0(\xi, x; \lambda)$ are functions of ξ belonging to $\mathfrak{L}_2(a, b)$, so that $\varrho(x, y; \lambda; \mu)$ is defined for all such values (x, y) and for all μ and is a function of μ which is of bounded variation over the interval $(-\infty, +\infty)$ and which is continuous on the right. In short, $\varrho(x, y; \lambda; \mu)$ is a function in the class $\mathfrak{B}^+(x, y)$ for fixed λ. For fixed λ and μ, we find by easy calculations that

$$\varrho(x, y; \lambda; \mu) = \sigma(x, y; \lambda; \mu)$$

for almost all (x, y), where σ is defined for all λ, all μ, and all (x, y) in E_2^* by the equations

$$\begin{aligned}\sigma(x, y; \lambda; \mu) &= 0, & \lambda\mu &\leqq 0, \\ \sigma(x, y; \lambda; \mu) &= E_0(x, y; \mu), & \Delta_\mu &\subseteqq \Delta_\lambda, \\ \sigma(x, y; \lambda; \mu) &= E_0(x, y; \lambda), & \Delta_\lambda &\subseteqq \Delta_\mu.\end{aligned}$$

For fixed λ, the function $\sigma(x, y; \lambda; \mu)$ belongs to $\mathfrak{B}^+(x, y)$, by the results of the preceding paragraph. It is evident that, when μ is confined to the set Λ and λ is fixed, the equation

$\varrho(x, y; \lambda; \mu) = \sigma(x, y; \lambda; \mu)$ holds for all (x, y) outside a null set independent of μ. Since both ϱ and σ are functions of μ which are continuous on the right, this equation must persist for arbitrary values of μ without the introduction of further exceptional points (x, y). We see therefore that the integral under consideration reduces to

$$\int_{\Delta_\lambda} \mu \, d E_0(x, y; \mu) = \operatorname{sgn} \lambda \int_0^\lambda \mu \, d E_0(x, y; \mu)$$

for fixed λ and almost all (x, y). By combining the preceding results, we find that

$$\begin{aligned} A E_0(x, y; \lambda) &= -\frac{\partial}{\partial x} p(x) \frac{\partial}{\partial x} E_0(x, y; \lambda) + r(x) E_0(x, y; \lambda) \\ &= \int_0^\lambda \mu \, d E_0(x, y; \mu), \end{aligned}$$

for fixed λ and almost all (x, y). This formula obviously leads without difficulty to the result asserted in the last sentence of the statement of the present theorem. It is to be observed that the derivatives involved are calculated for fixed λ and y by treating $E_0(x, y; \lambda)$ and $p(x) \frac{\partial}{\partial x} E_0(x, y; \lambda)$ as equivalent to functions of x which are absolutely continuous on the open interval (a, b), while the function $E_0(x, y; \mu)$ which appears in the Stieltjes integral is considered as a member of the class $\mathfrak{B}^+(x, y)$. Now Lemma 10.1 enables us to cast this relation into the equivalent integral form

$$\begin{aligned} E_0(x, y; \lambda) = {} & E_0(z, y; \lambda) \varphi_1(x, z; 0) \\ & + p(z) \frac{\partial}{\partial z} E_0(z, y; \lambda) \varphi_2(x, z; 0) \\ & + \int_z^x \Phi(x, \xi; z) \left(\int_0^\lambda \mu \, d E_0(\xi, y; \mu) \right) d\xi, \end{aligned} \tag{1}$$

where φ_1 and φ_2 are the functions described in the statement of the theorem and $\Phi(x, y; z)$ denotes the expression $\varphi_1(x, z; 0) \varphi_2(y, z; 0) - \varphi_2(x, z; 0) \varphi_1(y, z; 0)$. This relation is valid for fixed λ, for fixed y outside a set of zero measure, and for all (x, z), provided that the functions $E_0(x, y; \lambda)$,

$E_0(z, y; \lambda)$, and $p(z)\frac{\partial}{\partial z}E_0(z, y; \lambda)$ in the first three terms are treated as absolutely continuous functions of x or of z on the open interval (a, b) while the function $E_0(\xi, y; \mu)$ in the integral term is treated as a member of the class $\mathfrak{B}^+(\xi, y)$. It is convenient, however, to interpret this relation and the expressions which it connects in a somewhat different fashion. For this reason, we examine the integral term more carefully, with a view to showing that it is a function in the class $\mathfrak{B}^+(x, y, z)$. In the first place, it is clear that, if Δ is an arbitrary interval (α, β) in $\mathfrak{D}_0(\Lambda)$ and if γ is the greater of $|\alpha|$ and $|\beta|$, then the inequality given in Lemma 5.1 (1) and the inequality (A) noted above enable us to write

$$
\begin{aligned}
V\left(\int_z^x \Phi(x, \xi; z)\left(\int_0^\lambda \mu\, dE_0(\xi, y; \mu)\right) d\xi; \Delta\right)
&\leqq \left|\int_z^x |\Phi(x, \xi; z)|\, V\left(\int_0^\lambda \mu\, dE_0(\xi, y; \mu); \Delta\right) d\xi\right| \\
&\leqq \gamma\left|\int_z^x |\Phi(x, \xi; z)|\, V(E_0(\xi, y; \mu); \Delta)\, d\xi\right| \\
&\leqq \gamma\left|\int_z^x |\Phi(x, \xi; z)|^2\, d\xi\right|^{1/2}\left|\int_z^x V^2(E_0(\xi, y; \mu); \Delta)\, d\xi\right|^{1/2} \\
&\leqq \gamma\left|\int_z^x |\Phi(x, \xi; z)|^2\, d\xi\right|^{1/2}\left|\int_z^x \int_a^b |E(\eta, \xi; \Delta)|^2\, d\eta\, d\xi\right|^{1/2} \\
&\qquad \times\left|\int_a^b |E(\eta, y; \Delta)|^2\, d\eta\right|^{1/2}
\end{aligned}
$$

for all (x, z) and all y outside a null set independent of Δ. In the final expression, the first integral exists for all (x, z) by virtue of Lemma 10.2, the second exists for all (x, z) in accordance with remarks made above, and the third exists for all y outside a null set independent of Δ as a result of the properties established in the preceding paragraph. Thus the integral term is a function of λ which is of bounded variation over every finite interval Δ, for all (x, y, z) outside a null set independent of Δ. In the second place, we observe that the Stieltjes integral $\int_0^\lambda \mu\, dE_0(x, y; \mu)$ is continuous on the right in λ for all (x, y) in E_2^*, a property analogous to that stated in Lemma 5.1 (8); and that the inequality

$$\left| \Phi(x, \xi; z) \int_0^\lambda \mu \, d E_0(\xi, y; \mu) \right|$$
$$\leqq |\Phi(x, \xi; z)| \, V\left(\int_0^\lambda \mu \, d E_0(\xi, y; \mu); \Delta\right)$$
$$\leqq \gamma |\Phi(x, \xi; z)| \left(\int_a^b |E(\eta, \xi; \Delta)|^2 d\eta \cdot \int_a^b |E(\eta, y; \Delta)|^2 d\eta\right)^{1/2}$$

is satisfied for all (x, z) and all (ξ, y) in E_2^*. By the inequality of Schwarz, the final expression is integrable with respect to ξ over the interval with extremities z and x. Hence we can combine the two properties just noted with a well-known theorem† concerning passage to the limit under the sign of integration in a Lebesgue integral to establish the following result: if Δ is an arbitrary interval in the class $\mathfrak{D}_0(\Lambda)$ and if λ and $\lambda + \varepsilon$, $\varepsilon > 0$, are points contained in Δ, then

$$\lim_{\varepsilon \to 0} \int_z^x \Phi(x, \xi; z) \left(\int_0^{\lambda+\varepsilon} \mu \, d E_0(\xi, y; \mu)\right) d\xi$$
$$= \int_z^x \Phi(x, \xi; z) \left(\lim_{\varepsilon \to 0} \int_0^{\lambda+\varepsilon} \mu \, d E_0(\xi, y; \mu)\right) d\xi$$
$$= \int_z^x \Phi(x, \xi; z) \left(\int_0^\lambda \mu \, d E_0(\xi, y; \mu)\right) d\xi$$

for all (x, z) and for all y outside a null set independent of Δ and of λ. Thus the integral term is continuous on the right in λ for all (x, y, z) outside a null set independent of λ. It now appears that the integral term is a function in $\mathfrak{B}^+(x, y, z)$, as we wished to prove. We return to the consideration of the relation (1). If we regard the expressions $E_0(x, y; \lambda)$ and $E_0(z, y; \lambda)$ and the integral term as functions in $\mathfrak{B}^+(x, y, z)$, then (1) is still valid for fixed λ and almost all (x, y, z). Hence we can solve (1) for the second term on the right, finding that $p(z) \frac{\partial}{\partial z} E_0(z, y; \lambda) \, \varphi_2(x, z; 0)$ is equal for fixed λ and almost all (x, y, z) to a function of x, y, z, and λ which belongs to the class $\mathfrak{B}^+(x, y, z)$. By virtue of this result, we see that we can calculate the function $p(z) \frac{\partial}{\partial z} E_0(z, y; \lambda) \, \varphi_2(x, z; 0)$

---

† de la Vallée Poussin, *Intégrales de Lebesgue,* Paris, 1916, p. 49.

in the manner indicated above and then modify its values on a set of points (x, y, z) of zero measure, dependent upon λ so as to obtain a function in $\mathfrak{B}^+(x, y, z)$. After this modification has been effected, the relation (1) obviously holds for all λ and all (x, y, z) outside a null set independent of λ. Furthermore, since the factor $\varphi_2(x, z; 0)$ vanishes only on a null set of points (x, z) by virtue of Lemma 10.2, the modification reduces $p(z)\frac{\partial}{\partial z}E_0(z, y; \lambda)$ to an equivalent function in $\mathfrak{B}^+(y, z)$. Until further notice, we shall suppose that the functions $p(z)\frac{\partial}{\partial z}E_0(z, y; \lambda)$ and $p(z)\frac{\partial}{\partial z}E_0(z, y; \lambda)\, \varphi_2(x, z; 0)$ and the relation (1) are interpreted in harmony with these remarks.

The equation $E_0(y, x; \lambda) = \overline{E_0(x, y; \lambda)}$ holds for all λ and all (x, y) outside a null set independent of λ. In consequence, we find that $p(y)\frac{\partial}{\partial y}E_0(y, x; \lambda) = \overline{p(y)\frac{\partial}{\partial y}E_0(x, y; \lambda)}$, at least for fixed λ and almost all (x, y). If we apply these relations, together with the fact that φ_1, φ_2, and Φ are real functions, we can easily deduce from (1) the relation

$$
\begin{aligned}
E_0(x, y; \lambda) = {} & E_0(x, z; \lambda)\, \varphi_1(y, z; 0) \\
& + p(z)\frac{\partial}{\partial z}E_0(x, z; \lambda)\, \varphi_2(y, z; 0) \\
& + \int_z^y \Phi(y, \xi; z) \left(\int_0^\lambda \mu \, dE_0(x, \xi; \mu)\right) d\xi .
\end{aligned} \tag{2}
$$

Since this relation holds for fixed λ and almost all (x, y, z), we can subject it to the same treatment as that applied to (1) in the preceding paragraph. Thus we shall regard (2) as an equation connecting functions in $\mathfrak{B}^+(x, y, z)$ and holding for all λ and all (x, y, z) outside a null set independent of λ. In particular, we shall suppose, until further notice, that the function $p(z)\frac{\partial}{\partial z}E_0(x, z; \lambda)$ has been so modified as to belong to the class $\mathfrak{B}^+(x, z)$.

We must next investigate the integral term on the right of (2). Its significant properties, as will appear immediately, are not exhausted by the statement that it is a function in $\mathfrak{B}^+(x, y, z)$. We first prove that for all λ and all (y, z) the integral term is a function of x belonging to $\mathfrak{L}_2(a, b)$. By the formula for integration by parts given in Lemma 5.1 (9), we have

$$\int_z^y \Phi(y, \xi; z)\left(\int_0^\lambda \mu\, dE_0(x, \xi; \mu)\right) d\xi$$

$$= \lambda \int_z^y \Phi(y, \xi; z) E_0(x, \xi; \lambda)\, d\xi - \int_z^y \int_0^\lambda \Phi(y, \xi; z) E_0(x, \xi; \mu)\, d\mu\, d\xi.$$

If we denote by $f(x)$ the function which is equal to $\Phi(y, x; z)$ or to zero according as x is in the interval with extremities y and z or not, then $f(x)$ is obviously a function in $\mathfrak{L}_2(a, b)$. The first term of the difference above can be written as

$$\lambda \int_a^b E_0(x, \xi; \lambda) f(\xi)\, d\xi = \lambda(E(\lambda) - E(0)) f(x)$$

for fixed (y, z), and is therefore a function of x in $\mathfrak{L}_2(a, b)$. The second term must be treated by more direct methods. We have to prove that the function

$$\Phi(y, \xi; z) E_0(x, \xi; \mu) \Phi(y, \eta; z) E_0(x, \eta; \nu)$$

is absolutely integrable with respect to ξ, η, μ, ν, x over a range which is specified by the inequalities $z \leqq \xi \leqq y$, $z \leqq \eta \leqq y$, $0 \leqq \mu \leqq \lambda$, $0 \leqq \nu \leqq \lambda$, $a < x < b$, when $z < y$ and $\lambda > 0$, and by similar inequalities in the other cases. Since Φ is bounded and continuous on this range in accordance with Lemma 10.2, it is sufficient for us to consider the product $E_0(x, \xi; \mu) E_0(x, \eta; \nu)$. By Schwarz's inequality, we find that the integral of the absolute value of this product with respect to ξ, η, and x is dominated by the quantity

$$\left|\int_z^y \int_a^b |E_0(x, \xi; \mu)|^2\, dx\, d\xi\right|^{1/2} \left|\int_z^y \int_a^b |E_0(x, \eta; \nu)|^2\, dx\, d\eta\right|^{1/2}$$

We have still to show that this expression is integrable with respect to μ and ν. We shall confine our attention to the integration with respect to μ, which is evidently typical. We can prove that the function

$$\varrho(\mu) = \operatorname{sgn} \mu \cdot \left|\int_z^y \int_a^b |E_0(x, \xi; \mu)|^2\, dx\, d\xi\right|$$

is monotone-increasing and hence that $|\varrho(\mu)|^{1/2}$ is integrable over every finite interval. If Δ is an arbitrary finite interval containing the point $\mu = 0$, we have

$$\begin{aligned}((E(\mu)-E(0))\,E(x, \xi; \Delta),\ E(x, \xi; \Delta)) \\ = \operatorname{sgn}\mu\,((E(\mu)-E(0))\,E(x, \xi; \Delta), (E(\mu)-E(0))\,E(x, \xi; \Delta)) \\ = \operatorname{sgn}\mu\,(E_0(x, \xi; \mu),\ E_0(x, \xi; \mu)) \\ = \operatorname{sgn}\mu \int_a^b |E_0(x, \xi; \mu)|^2\,dx\end{aligned}$$

for fixed Δ, fixed μ in Δ, and almost all ξ. Since the first term is a real monotone-increasing function of μ for almost all ξ, we have

$$\operatorname{sgn}\mu_1 \int_a^b |E_0(x, \xi; \mu_1)|^2\,dx \leqq \operatorname{sgn}\mu_2 \int_a^b |E_0(x, \xi; \mu_2)|^2\,dx, \qquad \mu_1 \leqq \mu_2,$$

for fixed μ_1 and μ_2 and almost all ξ. It follows at once, by integration of both terms in this inequality with respect to ξ, that $\varrho(\mu_1) \leqq \varrho(\mu_2)$ for $\mu_1 \leqq \mu_2$. This completes the proof of the assertion that the integral term in (2) is a function of x in $\mathfrak{L}_2(a, b)$. Secondly, we shall prove that

$$E(\Delta_\nu) \int_z^y \Phi(y, \xi; z) \left(\int_0^\lambda \mu\, dE_0(x, \xi; \mu)\right) d\xi = \left\{\begin{array}{c} 0 \\ \displaystyle\int_z^y \Phi(y, \xi; z) \left(\int_0^\nu \mu\, dE_0(x, \xi; \mu)\right) d\xi \\ \displaystyle\int_z^y \Phi(y, \xi; z) \left(\int_0^\lambda \mu\, dE_0(x, \xi; \mu)\right) d\xi \end{array}\right\}$$

for fixed λ and ν, all (y, z), and almost all x, according as $\lambda\nu \leqq 0$, $\Delta_\nu \subseteqq \Delta_\lambda$, or $\Delta_\nu \supseteqq \Delta_\lambda$. This result is based upon the relations

$$\int_a^b E(x, \eta; \Delta_\nu)\, E_0(\eta, \xi; \mu)\, d\eta = \operatorname{sgn}\mu\, E(x, \xi; \Delta_\mu \Delta_\nu) = \left\{\begin{array}{ll} 0\ , & \mu\nu \leqq 0 \\ E_0(x, \xi; \nu), & \Delta_\nu \subseteqq \Delta_\mu \\ E_0(x, \xi; \mu), & \Delta_\nu \supseteqq \Delta_\mu \end{array}\right\},$$

which hold for fixed μ, fixed ν, and almost all (x, ξ). We shall discuss in detail only the case where $\Delta_\nu \subseteqq \Delta_\lambda$. In the integral term we apply the formula for integration by parts given in Lemma 5.1 (9) just as we did above. We then replace the variable x by a new variable η, multiply by $E(x, \eta; \Delta_\nu)$, and integrate with respect to η over the interval (a, b). The result of these operations is an explicit formula for the function obtained by applying the projection $E(\Delta_\nu)$ to the integral term, regarded as a function of x in $\mathfrak{L}_2(a, b)$. This formula involves integrations with respect to the three variables ξ, η, μ. By inequalities similar to those used above, we can show that the integral is absolutely convergent and hence that the integration can be performed in any desired order. The details may be left to the reader. If we integrate in the order η, μ, ξ, we obtain

$$\begin{aligned}
&\int_z^y \Phi(y, \xi; z)\left(\lambda \int_a^b E(x, \eta; \Delta_\nu)\, E_0(\eta, \xi; \lambda)\, d\eta\right) d\xi \\
&\qquad - \int_z^y \Phi(y, \xi; z) \int_0^\lambda \left(\int_a^b E(x, \eta; \Delta_\nu)\, E_0(\eta, \xi; \mu)\, d\eta\right) d\mu\, d\xi \\
&\quad = \int_z^y \Phi(y, \xi; z)\, \lambda\, E_0(x, \xi; \nu)\, d\xi \\
&\qquad - \int_z^y \Phi(y, \xi; z) \left(\int_0^\nu E_0(x, \xi; \mu)\, d\mu + \int_\nu^\lambda E_0(x, \xi; \nu)\, d\mu\right) d\xi \\
&\quad = \int_z^y \Phi(y, \xi; z) \left(\nu E_0(x, \xi; \nu) - \int_0^\nu E_0(x, \xi; \mu)\, d\mu\right) d\xi \\
&\quad = \int_z^y \Phi(y, \xi; z) \left(\int_0^\nu \mu\, d E_0(x, \xi; \mu)\right) d\xi,
\end{aligned}$$

for fixed λ and ν subject to the condition $\Delta_\nu \subseteqq \Delta_\lambda$, for fixed (y, z), and for almost all x. The two remaining cases can be discussed in the same manner.

In order to apply the information obtained in the preceding paragraph, we solve the relation (2) for the function $p(z) \dfrac{\partial}{\partial z} E_0(x, z; \lambda)$, finding that

$$\begin{aligned}
&p(z) \frac{\partial}{\partial z} E_0(x, z; \lambda) \\
&\quad = E_0(x, y; \lambda)/\varphi_2(y, z; 0) - E_0(x, z; \lambda)\, \varphi_1(y, z; 0)/\varphi_2(y, z; 0) \\
&\qquad - \int_z^y \Phi(y, \xi; z) \left(\int_0^\lambda \mu\, d E_0(x, \xi; \mu)\right) d\xi/\varphi_2(y, z; 0)
\end{aligned}$$

for all (x, y, z) outside a null set independent of λ. As we have already indicated, the function $p(z)\frac{\partial}{\partial z}E_0(x, z; \lambda)$ is in $\mathfrak{B}^+(x, z)$ as a consequence of these relations. For future reference, we shall give an actual appraisal of the variation of this function of λ. If Δ is an arbitrary interval (α, β) in the class $\mathfrak{D}_0(\Delta)$ and if γ is the greater of $|\alpha|$ and $|\beta|$, we find with the help of inequalities already noted that

$$
\begin{aligned}
&V\left(p(z)\frac{\partial}{\partial z}E_0(x, z; \lambda); \Delta\right) \\
&\quad \leqq V(E_0(x, y; \lambda); \Delta)/|\varphi_2(y, z; 0)| \\
&\qquad + V(E_0(x, z; \lambda); \Delta)\,|\varphi_1(y, z; 0)|/|\varphi_2(y, z; 0)| \\
&\qquad + V\left(\int_z^y \Phi(y, \xi; z)\left(\int_0^\lambda \mu\, dE_0(\xi, x; \mu)\right) d\xi; \Delta\right)\Big/|\varphi_2(y, z; 0)| \\
&\quad \leqq \left(\int_a^b |E(\xi, x; \Delta)|^2 d\xi \cdot \int_a^b |E(\xi, y; \Delta)|^2 d\xi\right)^{1/2}\Big/|\varphi_2(y, z; 0)| \\
&\qquad + \left(\int_a^b |E(\xi, x; \Delta)|^2 d\xi \cdot \int_a^b |E(\xi, z; \Delta)|^2 d\xi\right)^{1/2} \\
&\qquad\qquad \times |\varphi_1(y, z; 0)|/|\varphi_2(y, z; 0)| \\
&\qquad + \gamma \left|\int_z^y |\Phi(y, \xi; z)|^2 d\xi\right|^{1/2} \left|\int_z^y \int_a^b |E(\eta, \xi; \Delta)|^2 d\eta\, d\xi\right|^{1/2} \\
&\qquad\qquad \times \left|\int_a^b |E(\eta, x; \Delta)|^2 d\eta\right|^{1/2} /|\varphi_2(y, z; 0)|
\end{aligned}
$$

for all (x, y, z) outside a null set independent of Δ. For fixed Δ and fixed (y, z) outside a null set which does not depend upon Δ, this inequality assumes the form

$$
\text{(B)} \quad V\left(p(z)\frac{\partial}{\partial z}E_0(x, z; \lambda); \Delta\right) \leqq K\left(\int_a^b |E(\eta, x; \Delta)|^2 d\eta\right)^{1/2}
$$

valid for all x outside a null set which is independent of Δ but may depend upon (y, z) and for some constant K independent of x. If we apply the results of the preceding paragraph and the known properties of the kernel E_0 to the expression for $p(z)\frac{\partial}{\partial z}E_0(x, z; \lambda)$ given above, we find at once that $p(z)\frac{\partial}{\partial z}E_0(x, z; \lambda)$, considered as a function of x, belongs to $\mathfrak{L}_2(a, b)$ for fixed λ and almost all z, and satisfies the relations

$$E(\Delta_\mu)\, p(z) \frac{\partial}{\partial z} E_0(x, z; \lambda) = \begin{cases} 0 & , \ \lambda\mu \leqq 0 \\ p(z) \dfrac{\partial}{\partial z} E_0(x, z; \mu), & \Delta_\mu \subseteqq \Delta_\lambda \\ p(z) \dfrac{\partial}{\partial z} E_0(x, z; \lambda), & \Delta_\mu \supseteqq \Delta_\lambda \end{cases}$$

for fixed λ, fixed μ, and almost all (x, z). Thus, for fixed λ and almost all z, the function $p(z) \dfrac{\partial}{\partial z} E_0(x, z; \lambda)$ is a function of x in the range of the projection $E(\Delta_\lambda)$ and therefore belongs to the domain of the self-adjoint transformation A. We are now in a position to treat this function by methods analogous to those which led to the relation (1). From the equation

$$\begin{aligned} &\left(A p(y) \frac{\partial}{\partial y} E_0(\xi, y; \lambda),\ E_0(\xi, x; \lambda)\right) \\ &\qquad = \int_{-\infty}^{+\infty} \mu \, d\left(E(\mu)\, p(y) \frac{\partial}{\partial y} E_0(\xi, y; \lambda),\ E_0(\xi, x; \lambda)\right), \end{aligned}$$

where we have written y in place of z, we find that

$$\begin{aligned} &-\frac{\partial}{\partial x} p(x) \frac{\partial}{\partial x}\left(p(y) \frac{\partial}{\partial y} E_0(x, y; \lambda)\right) \\ &\quad + r(x)\left(p(y) \frac{\partial}{\partial y} E_0(x, y; \lambda)\right) \\ &\qquad\qquad = \int_0^\lambda \mu \, d\left(p(y) \frac{\partial}{\partial y} E_0(x, y; \mu)\right) \end{aligned}$$

for fixed λ and almost all (x, y). Lemma 10.1 shows that

$$\begin{aligned} p(y) \frac{\partial}{\partial y} E_0(x, y; \lambda) &= p(y) \frac{\partial}{\partial y} E_0(z, y; \lambda)\, \varphi_1(x, z; 0) \\ &\quad + p(z) \frac{\partial}{\partial z}\left(p(y) \frac{\partial}{\partial y} E_0(z, y; \lambda)\right) \varphi_2(x, z; 0) \qquad (3) \\ &\quad + \int_z^x \Phi(x, \xi; z) \left(\int_0^\lambda \mu \, d\left(p(y) \frac{\partial}{\partial y} E_0(\xi, y; \mu)\right)\right) d\xi, \end{aligned}$$

valid for fixed λ, almost all y, and all (x, z), provided that the first three terms involving the kernel E_0 are treated as absolutely continuous functions of x or of z on the open interval (a, b) and that the function $p(y)\frac{\partial}{\partial y}E_0(\xi, y; \mu)$ in the integral term is treated as a function in $\mathfrak{B}^+(\xi, y)$. Just as in the case of relation (1), we are able to prove that the integral term in (3) is a function in $\mathfrak{B}^+(x, y, z)$. If Δ is an arbitrary interval (α, β) in the class $\mathfrak{D}_0(\varDelta)$ and if γ is the greater of $|\alpha|$ and $|\beta|$, we find that

$$V\left(\int_z^x \Phi(x, \xi; z)\int_0^\lambda \mu\, d\left(p(y)\frac{\partial}{\partial y}E_0(\xi, y; \mu)\right) d\xi; \Delta\right)$$

$$\leqq \gamma\left|\int_z^x |\Phi(x, \xi; z)|^2\, d\xi\right|^{1/2}\left|\int_z^x V^2\left(p(y)\frac{\partial}{\partial y}(E_0(\xi, y; \mu); \Delta\right) d\xi\right|^{1/2}$$

for all (x, y, z) outside a null set independent of Δ. By virtue of the inequality (B) or of the more extended inequality which precedes it, we see that in the relation under consideration the last integral exists for all (x, y, z) outside a null set independent of Δ. Hence the integral term in (3) is a function of λ which is of bounded variation over every finite interval for all (x, y, z) outside a fixed null set. We see next that the function $\Phi(x, \xi; z)\int_0^\lambda \mu\, d\left(p(y)\frac{\partial}{\partial y}E_0(\xi, y; \mu)\right)$ is continuous on the right in λ for all (x, y, z, ξ) outside a null set independent of λ and that it is dominated by the function $\gamma\,|\Phi(x, \xi; z)|\; V\left(p(y)\frac{\partial}{\partial y}E_0(\xi, y; \lambda); \Delta\right)$ for all λ in the interval Δ and for all (x, y, z, ξ) outside a null set independent of λ and of Δ. With the help of Schwarz's inequality and the inequality (B), we find that the dominating function is integrable with respect to ξ over the interval with extremities z and x. It follows directly that the integral term in (3) is continuous on the right in λ for all (x, y, z) outside a null set independent of λ. This completes the proof of the assertion that the integral term is a function in $\mathfrak{B}^+(x, y, z)$. These properties of the integral term enable us to interpret (3) as a relation which connects functions in $\mathfrak{B}^+(x, y, z)$ and which is valid for all λ and all (x, y, z) outside a null set independent of λ. In particular the expression $p(z)\dfrac{\partial}{\partial z}\left(p(y)\dfrac{\partial}{\partial y}E_0(z, y; \lambda)\right)$ is to be modified in the usual manner so that it becomes a function in $\mathfrak{B}^+(y, z)$; and we shall suppose until further notice that the function appears in this modified form. The detailed discussion of the necessary modifications may be left to the reader, since they are analogous to those carried out explicitly in the case of relation (1).

At this point it is desirable that we summarize the preceding results in a form slightly more suitable for the applications which we intend to make. We have shown that the functions

$$E_0(x, y; \lambda),\ E_0(x, z_2; \lambda),\ E_0(z_1, z_2; \lambda),\ p(z_2)\frac{\partial}{\partial z_2}E_0(x, z_2; \lambda),$$

$$p(z_1)\frac{\partial}{\partial z_1}E_0(z_1, z_2; \lambda),\ p(z_1)\frac{\partial}{\partial z_1}\left(p(z_2)\frac{\partial}{\partial z_2}E_0(z_1, z_2; \lambda)\right)$$

are in the class $\mathfrak{B}^+(x, y, z_1, z_2)$. In particular, we have shown that, if Δ is an interval in the class $\mathfrak{D}_0(\varLambda)$, then

$$\text{(A}'\text{)}\qquad V(E_0(x, y; \lambda); \Delta) \leqq K\left(\int_a^b |E(\xi, y; \Delta)|^2\, d\xi\right)^{1/2}$$

for fixed Δ, fixed x outside a null set independent of Δ, all y outside a null set which is independent of Δ but may be dependent upon x, and some constant K independent of y;

$$\text{(A}''\text{)}\qquad V(E_0(x, z_2; \lambda); \Delta) \leqq K\left(\int_a^b |E(\xi, x; \Delta)|^2\, d\xi\right)^{1/2}$$

for fixed Δ, fixed z_2 outside a null set independent of Δ, all x outside a null set which is independent of Δ but may be dependent upon z_2, and some constant K independent of x; and

$$\text{(B}'\text{)}\qquad V\left(p(z_2)\frac{\partial}{\partial z_2}E_0(x, z_2; \lambda); \Delta\right) \leqq K\left(\int_a^b |E(\xi, x; \Delta)|^2\, d\xi\right)^{1/2}$$

for fixed Δ, fixed z_2 outside a null set independent of Δ, all x outside a null set which is independent of Δ but may depend upon z_2, and some constant K independent of x. We have shown that the various functions under consideration are connected by the relations

$$(2') \quad E_0(x, y; \lambda) = E_0(x, z_2; \lambda)\,\varphi_1(y, z_2; 0) + p(z_2)\frac{\partial}{\partial z_2} E_0(x, z_2; \lambda)\,\varphi_2(y, z_2; 0) + \int_{z_2}^{y} \Phi(y, \xi; z_2)\left(\int_0^{\lambda} \mu\, d E_0(x, \xi; \mu)\right) d\xi,$$

$$(1') \quad E_0(x, z_2; \lambda) = E_0(z_1, z_2; \lambda)\,\varphi_1(x, z_1; 0) + p(z_1)\frac{\partial}{\partial z_1} E_0(z_1, z_2; \lambda)\,\varphi_2(x, z_1; 0) + \int_{z_1}^{x} \Phi(x, \xi; z_1)\left(\int_0^{\lambda} \mu\, d E_0(\xi, z_2; \mu)\right) d\xi,$$

$$(3') \quad p(z_2)\frac{\partial}{\partial z_2} E_0(x, z_2; \lambda) = p(z_2)\frac{\partial}{\partial z_2} E_0(z_1, z_2; \lambda)\,\varphi_1(x, z_1; 0) + p(z_1)\frac{\partial}{\partial z_1}\left(p(z_2)\frac{\partial}{\partial z_2} E_0(z_1, z_2; \lambda)\right)\varphi_2(x, z_1; 0) + \int_{z_1}^{x} \Phi(x, \xi; z_1)\left(\int_0^{\lambda} \mu\, d\left(p(z_2)\frac{\partial}{\partial z_2} E_0(\xi, z_2; \mu)\right)\right) d\xi,$$

valid for all λ and for all (x, y, z_1, z_2) outside a null set independent of λ. We may now assign to z_1 and z_2 fixed values c_1 and c_2 respectively such that the various properties and relations enumerated above are true for all λ, all Δ in $\mathfrak{D}_0(\Lambda)$, and all (x, y) outside a null set independent of λ and of Δ. In the sequel it is sufficient for our purposes to deal with the case where $z_1 = c_1$ and $z_2 = c_2$ have been so chosen.

We shall now prove that (2′), with $z_2 = c_2$, implies

$$(4) \quad E_0(x, y; \lambda) = \int_0^{\lambda} \varphi_1(y, c_2; \mu)\, d E_0(x, c_2; \mu) + \int_0^{\lambda} \varphi_2(y, c_2; \mu)\, d\!\left(p(c_2)\frac{\partial}{\partial c_2} E_0(x, c_2; \mu)\right)$$

for all λ and all (x, y) outside a null set independent of λ. We first examine the term on the right of (4), denoting it for convenience by the symbol $H_0(x, y; \lambda)$. Since $E_0(x, c_2; \mu)$ and $p(c_2)\frac{\partial}{\partial c_2} E_0(x, c_2; \mu)$ are functions in $\mathfrak{B}^+(x)$, the function $H_0(x, y; \lambda)$ is defined as a Stieltjes integral for all λ, all y, and all x outside a null set independent of λ and of y. It is evident that $H_0(x, y; \lambda)$ is a function in $\mathfrak{B}^+(x, y)$. If Δ is an arbitrary interval (α, β) and if (a', b') is a finite closed interval interior to (a, b), then Lemma 10.2 shows that there exists a constant M such that $|\varphi_k(y, c_2; \mu)| \leqq M$, $k = 1, 2$, for all μ in Δ and all y in (a', b'). Hence we find that, for fixed Δ in the class $\mathfrak{D}_0(\varDelta)$ and fixed x outside a null set independent of λ, of Δ, and of y, the expression $V(H_0(x,y;\lambda); \Delta)$ is a bounded function of y on the interval (a', b') in accordance with the inequality

$$V(H_0(x, y; \lambda); \Delta) \leqq M(V(E_0(x, c_2; \mu); \Delta) + V(p(c_2)\frac{\partial}{\partial c_2} E_0(x, c_2; \mu); \Delta)).$$

Since $V(H_0(x, y; \lambda); \Delta)$ is a measurable function of y for fixed x, it must be integrable with respect to y over every finite interval interior to (a, b). If we note that

$$\varphi_k(y, c_2; \mu) = \varphi_k(y, c_2; 0) + \mu \int_{c_2}^{y} \Phi(y, \xi; c_2)\, \varphi_k(\xi, c_2; \mu)\, d\xi$$

for $k = 1, 2$, by virtue of Lemma 10.1, we can show by an easily justified change of order of integration that $H_0(x, y; \lambda)$ satisfies the relation

$$\begin{aligned} H_0&(x, y; \lambda) \\ &= E_0(x, c_2; \lambda)\, \varphi_1(y, c_2; 0) \\ &\quad + p(c_2)\frac{\partial}{\partial c_2} E_0(x, c_2; \lambda)\, \varphi_2(y, c_2; 0) \\ &\quad + \int_0^{\lambda} \mu \left(\int_{c_2}^{y} \Phi(y, \xi; c_2)\, \varphi_1(\xi, c_2; \mu)\, d\xi\right) d E_0(x, c_2; \mu) \\ &\quad + \int_0^{\lambda} \mu \left(\int_{c_2}^{y} \Phi(y, \xi; c_2)\, \varphi_2(\xi, c_2; \mu)\, d\xi\right) \\ &\qquad\qquad \times d\left(p(c_2)\frac{\partial}{\partial c_2} E_0(x, c_2; \mu)\right) \\ &= E_0(x, c_2; \lambda)\, \varphi_1(y, c_2; 0) \\ &\quad + p(c_2)\frac{\partial}{\partial c_2} E_0(x, c_2; \lambda)\, \varphi_2(y, c_2; 0) \\ &\quad + \int_{c_2}^{y} \Phi(y, \xi; c_2) \left(\int_0^{\lambda} \mu\, d H_0(x, \xi; \mu)\right) d\xi \end{aligned}$$

for all λ, all y, and all x outside a null set independent of λ and of y. We are now prepared to study the difference $\Gamma_0(x, y; \lambda) = E_0(x, y; \lambda) - H_0(x, y; \lambda)$. From the fact that $E_0(x, y; \lambda)$ and $H_0(x, y; \lambda)$ are in $\mathfrak{B}^+(x, y)$, we conclude that $\Gamma_0(x, y; \lambda)$ is also in $\mathfrak{B}^+(x, y)$. The inequality (A') shows that the function $V(E_0(x, y; \lambda); \Delta)$ is integrable with respect to y over every finite closed interval interior to (a, b), for fixed Δ in $\mathfrak{D}_0(A)$ and fixed x outside a null set independent of Δ; for this function satisfies the inequality

$$\begin{aligned}\int_{a'}^{b'} & V(E_0(x, y; \lambda); \Delta)\, dy \\ & \leqq K\int_{a'}^{b'}\left(\int_a^b |E(\xi, y; \Delta)|^2\, d\xi\right)^{1/2} dy \\ & \leqq K(b'-a')^{1/2}\left(\int_{a'}^{b'}\int_a^b |E(\xi, y; \Delta)|^2\, d\xi\, dy\right)^{1/2}\end{aligned}$$

under the circumstances indicated. On combining this property of E_0 with the corresponding property of H_0, we see that $V(\Gamma_0(x, y; \lambda); \Delta)$ is integrable with respect to y over every finite closed interval interior to (a, b), for fixed Δ in $\mathfrak{D}_0(A)$ and fixed x outside a null set independent of Δ. In fact, we may even remove the restriction on the interval Δ; for, if Δ_1 is an arbitrary interval and Δ_2 is any interval in $\mathfrak{D}_0(A)$ such that $\Delta_2 \supseteqq \Delta_1$, we have

$$V(\Gamma_0(x, y; \lambda); \Delta_1) \leqq V(\Gamma_0(x, y; \lambda); \Delta_2),$$

the first term being significant whenever the second is defined and the relation holding whenever both terms are significant. Finally, we see that (2′) and the corresponding relation established for H_0 lead to the equation

$$\Gamma_0(x, y; \lambda) = \int_{c_2}^{y} \Phi(y, \xi; c_2)\left(\int_0^\lambda \mu\, d\Gamma_0(x, \xi; \mu)\right) d\xi,$$

holding for all λ and all (x, y) outside a null set independent of λ. We note in particular that $\Gamma_0(x, y; 0) = 0$ for almost all (x, y). It is now easy to prove that $\Gamma_0(x, y; \lambda) = 0$ for all λ and all (x, y) outside a null set independent of λ, and thus to establish the validity of (4). Let (a', b') be an arbitrary interval which contains c_2 and which is completely interior to (a, b); and let Δ be an arbitrary interval (α, β). Then it is clear that, for all (x, y) outside a null set independent of Δ, we have

$$V(\Gamma_0(x, y; \lambda); \Delta) \leqq \gamma \left| \int_{c_2}^{y} |\Phi(y, \xi; c_2)| \, V(\Gamma_0(x, \xi; \lambda); \Delta)\, d\xi \right|$$
$$\leqq N\gamma \left| \int_{c_2}^{y} V(\Gamma_0(x, \xi; \lambda); \Delta)\, d\xi \right|,$$

where γ is the greater of $|\alpha|$ and $|\beta|$ and N is the least upper bound of $|\Phi(y, \xi; c_2)|$ for $a' \leqq y \leqq b'$, $a' \leqq \xi \leqq b'$. Now we know that

$$\left| \int_{c_2}^{y} V(\Gamma_0(x, \xi; \lambda); \Delta)\, d\xi \right| \leqq \int_{a'}^{b'} V(\Gamma_0(x, \xi; \lambda); \Delta)\, d\xi$$

for fixed x outside a null set independent of Δ, the integral on the right being convergent. We denote the value of the latter expression by C. We conclude that

$$V(\Gamma_0(x, y; \lambda); \Delta) \leqq CN\gamma,$$
$$V(\Gamma_0(x, y; \lambda); \Delta) \leqq N\gamma \left| \int_{c_2}^{y} V(\Gamma_0(x, \xi; \lambda); \Delta)\, d\xi \right|$$

for fixed x outside a null set independent of Δ and all y on (a', b') outside a null set which is independent of Δ but may depend upon x. By an obvious inductive argument, we can obtain from these inequalities the relation

$$V(\Gamma_0(x, y; \lambda); \Delta) \leqq CN^{k+1} \gamma^{k+1} |y - c_2|^k / k!, \quad k = 0, 1, 2, \cdots,$$

which holds for the indicated values (x, y): for we have already established this relation in the case $k = 0$; and, if it holds for $k = n$, we can establish it for $k = n + 1$ by means of the inequalities

$$V(\Gamma_0(x, y; \lambda); \Delta) \leqq N\gamma \left| \int_{c_2}^{y} V(\Gamma_0(x, \xi; \lambda); \Delta)\, d\xi \right|$$
$$\leqq CN^{n+1} \gamma^{n+1} \left| \int_{c_2}^{y} |\xi - c_2|^n \, d\xi \right| \Big/ n!$$
$$\leqq CN^{n+1} \gamma^{n+1} |y - c_2|^{n+1} / (n + 1)!.$$

Evidently the right-hand member of this relation tends to zero with $1/k$ for arbitrary y in (a', b'). Hence we find that

$V(\Gamma_0(x, y; \lambda); \Delta) = 0$ for fixed x outside a null set independent of Δ and all y on (a', b') outside a null set which is independent of Δ but may depend upon x. Since the intervals (a', b') and Δ are arbitrary, and since $\Gamma_0(x, y; 0)$ vanishes for almost all (x, y), this result implies that $\Gamma_0(x, y; \lambda) = 0$ for all λ and all (x, y) outside a null set independent of λ. This completes the proof of (4).

It is now clear that, by reasoning like that just carried through in detail, we can deduce from (1′) and (3′) the respective corresponding equations

$$(5) \qquad E_0(x, c_2; \lambda) = \int_0^\lambda \varphi_1(x, c_1; \mu)\, d E_0(c_1, c_2; \mu) + \int_0^\lambda \varphi_2(x, c_1; \mu)\, d\left(p(c_1) \frac{\partial}{\partial c_1} E_0(c_1, c_2; \mu)\right),$$

$$(6) \quad p(c_2) \frac{\partial}{\partial c_2} E_0(x, c_2; \lambda) = \int_0^\lambda \varphi_1(x, c_1; \mu)\, d\left(p(c_2) \frac{\partial}{\partial c_2} E_0(c_1, c_2; \mu)\right) + \int_0^\lambda \varphi_2(x, c_1; \mu)\, d\left(p(c_1) \frac{\partial}{\partial c_1}\left(p(c_2) \frac{\partial}{\partial c_2} E_0(c_1, c_2; \mu)\right)\right),$$

valid for all λ and all x outside a null set independent of λ. In fact, if we denote by $\Gamma(x; \lambda)$ the difference of the two terms involved in (5) or in (6), we find that

$$\Gamma(x; \lambda) = \int_{c_1}^x \Phi(x, \xi; c_1) \left(\int_0^\lambda \mu\, d\Gamma(\xi; \mu)\right) d\xi$$

and can show by the argument used above that $\Gamma(x; \lambda)$ vanishes in the indicated sense. It is to be noted that the variable x now plays a rôle similar to that of the variable y in the situation discussed in the preceding paragraph.

By substituting from (5) and (6) in the right-hand term of (4) and applying Lemma 5.1 (6), we obtain the equation

$$(7) \quad E_0(x, y; \lambda) = \sum_{\alpha, \beta = 1}^{2} \int_0^\lambda \varphi_\alpha(x, c_1; \mu)\, \varphi_\beta(y, c_2; \mu)\, d\sigma_{\alpha\beta}(\mu),$$

holding for all λ and all (x, y) outside a null set independent of λ, where

$$\sigma_{11}(\mu) = E_0(c_1, c_2; \mu), \quad \sigma_{12}(\mu) = p(c_2)\frac{\partial}{\partial c_2} E_0(c_1, c_2; \mu),$$
$$\sigma_{21}(\mu) = p(c_1)\frac{\partial}{\partial c_1} E_0(c_1, c_2; \mu),$$
$$\sigma_{22}(\mu) = p(c_1)\frac{\partial}{\partial c_1}\left(p(c_2)\frac{\partial}{\partial c_2} E_0(c_1, c_2; \mu)\right).$$

The sum on the right of this equation is obviously a function of λ which is continuous on the right and which is of bounded variation over every finite interval, for all (x, y) without exception. Hence we may effect a definitive modification of the kernel $E_0(x, y; \lambda)$ which makes it identical with the right-hand term of (7) for all λ and all (x, y). It is of interest to observe that the kernel thereby becomes a continuous function of (x, y) for fixed λ. We may now transform the expression on the right of (7), so as to obtain the result actually asserted in the theorem. If c is an arbitrary point of the interval (a, b), we have

$$\varphi_j(x, c_1; \mu) = \sum_{\gamma=1}^{2} a_{j\gamma}(\mu)\,\varphi_\gamma(x, c; \mu),$$
$$\varphi_k(y, c_2; \mu) = \sum_{\delta=1}^{2} b_{k\delta}(\mu)\,\varphi_\delta(y, c; \mu),$$

where

$$a_{j1} = \varphi_j(c, c_1; \mu), \quad a_{j2} = p(c)\frac{\partial}{\partial c}\varphi_j(c, c_1; \mu),$$
$$b_{k1} = \varphi_k(c, c_2; \mu), \quad b_{k2} = p(c)\frac{\partial}{\partial c}\varphi_k(c, c_2; \mu), \qquad j, k = 1, 2.$$

It follows immediately that

$$\begin{aligned} E_0(x, y; \lambda) &\\ &= \sum_{\alpha,\beta,\gamma,\delta=1}^{2}\int_0^\lambda a_{\alpha\gamma}(\mu)\, b_{\beta\delta}(\mu)\,\varphi_\gamma(x, c; \mu) \\ (8) \qquad &\qquad\qquad \times \varphi_\delta(y, c; \mu)\, d\sigma_{\alpha\beta}(\mu) \\ &= \sum_{\gamma,\delta=1}^{2}\int_0^\lambda \varphi_\gamma(x, c; \mu)\,\varphi_\delta(y, c; \mu)\, d\varrho_{\gamma\delta}(\mu) \end{aligned}$$

where

$$\varrho_{jk}(\mu) = \sum_{\alpha,\beta=1}^{2}\int_0^\mu a_{\alpha j}(\nu)\, b_{\beta k}(\nu)\, d\sigma_{\alpha\beta}(\nu), \qquad j, k = 1, 2.$$

The integral expression for $E(x, y; \Delta)$ given in the theorem is an obvious consequence of this result. We can now calculate the various derivatives of $E_0(x, y; \lambda)$ by means of (8), obtaining

$$p(x)\frac{\partial}{\partial x}E_0(x, y; \lambda) = \sum_{\gamma,\delta=1}^{2}\int_0^{\lambda} p(x)\frac{\partial}{\partial x}\varphi_\gamma(x, c; \mu)\,\varphi_\delta(y, c; \mu)\,d\varrho_{\gamma\delta}(\mu), \tag{9}$$

$$p(y)\frac{\partial}{\partial y}E_0(x, y; \lambda) = \sum_{\gamma,\delta=1}^{2}\int_0^{\lambda} \varphi_\gamma(x, c; \mu)\,p(y)\frac{\partial}{\partial y}\varphi_\delta(y, c; \mu)\,d\varrho_{\gamma\delta}(\mu), \tag{10}$$

$$\begin{aligned} p(x)\frac{\partial}{\partial x}\left(p(y)\frac{\partial}{\partial y}E_0(x, y; \lambda)\right) &= p(y)\frac{\partial}{\partial y}\left(p(x)\frac{\partial}{\partial x}E_0(x, y; \lambda)\right) \\ &= \sum_{\gamma,\delta=1}^{2}\int_0^{\lambda} p(x)\frac{\partial}{\partial x}\varphi_\gamma(x, c; \mu) \\ &\qquad\times p(y)\frac{\partial}{\partial y}\varphi_\delta(y, c; \mu)\,d\varrho_{\gamma\delta}(\mu), \end{aligned} \tag{11}$$

for all λ and all (x, y) without exception. We shall indicate briefly the deduction of (9). The integral which appears on the right is obviously defined for all λ and all (x, y), when $p(x)\frac{\partial}{\partial x}\varphi_k(x, c; \mu)$ is regarded as an absolutely continuous function of x on the open interval (a, b) for $k = 1, 2$. We can show that for fixed λ and fixed (x, y) we have

$$\begin{aligned} \int_c^x 1/p(\xi)\left(\sum_{\gamma,\delta=1}^{2}\int_0^{\lambda} p(\xi)\frac{\partial}{\partial \xi}\varphi_\gamma(\xi, c; \mu)\right. \\ \left.\times\,\varphi_\delta(y, c; \mu)\,d\varrho_{\gamma\delta}(\mu)\right) d\xi \\ = \sum_{\gamma,\delta=1}^{2}\int_0^{\lambda}\left(\int_c^x (1/p(\xi))\,p(\xi)\frac{\partial}{\partial \xi}\varphi_\gamma(\xi, c; \mu)\right. \\ \left.\times\,\varphi_\delta(y, c; \mu)\,d\xi\right) d\varrho_{\gamma\delta}(\mu). \end{aligned}$$

The necessary interchange of order of integration is easily justified by the argument used in the discussion of Theorem 10.4 (2): we express each Stieltjes integral as the limit of a sum of the usual type and show, by virtue of the inequality

$$\left| \sum_{\gamma,\delta=1}^{2} \sum_{\nu=1}^{n} (1/p(\xi))\, p(\xi) \frac{\partial}{\partial \xi} \varphi_\gamma(\xi, c; \mu_\nu) \times \varphi_\delta(y, c; \mu_\nu)\, \varrho_{\gamma\delta}(\Delta^{(\nu)}) \right| \leqq M/|p(\xi)|$$

which is valid for fixed λ, fixed (x, y), and all ξ and μ on the range of integration, that the operations of integrating with respect to ξ and of passing to the limit, $n \to \infty$, can be applied interchangeably to such sums. Obviously, this result can be written as

$$\begin{aligned} \int_c^x 1/p(\xi) \left(\sum_{\gamma,\delta=1}^{2} \int_0^\lambda p(\xi) \frac{\partial}{\partial \xi} \varphi_\gamma(\xi, c; \mu) \times \varphi_\delta(y, c; \mu)\, d\varrho_{\gamma\delta}(\mu) \right) d\xi \\ = \sum_{\gamma,\delta=1}^{2} \int_0^\lambda \varphi_\gamma(x, c; \mu)\, \varphi_\delta(y, c; \mu)\, d\varrho_{\gamma\delta}(\mu) \\ - \sum_{\gamma,\delta=1}^{2} \int_0^\lambda \varphi_\gamma(c, c; \mu)\, \varphi_\delta(y, c; \mu)\, d\varrho_{\gamma\delta}(\mu) \\ = E_0(x, y; \lambda) - E_0(c, y; \lambda) \end{aligned}$$

for all λ and all (x, y). By applying the operator $p(x) \frac{\partial}{\partial x}$ to both members of this equation and interpreting the resulting functions as absolutely continuous functions of x on the open interval (a, b), we finally obtain (9). The relations (10) and (11) can be established by reasoning of the same character. If we put $x = y = c$ in these equations, we obtain

(8′) $$E_0(c, c; \lambda) = \varrho_{11}(\lambda) - \varrho_{11}(0),$$

(9′) $$p(c) \frac{\partial}{\partial c} E_0(c, y; \lambda) \Big|_{y=c} = \varrho_{21}(\lambda) - \varrho_{21}(0),$$

(10′) $$p(c) \frac{\partial}{\partial c} E_0(x, c; \lambda) \Big|_{x=c} = \varrho_{12}(\lambda) - \varrho_{12}(0),$$

$$
\begin{aligned}
(11') \qquad & p(x)\frac{\partial}{\partial x}\left(p(y)\frac{\partial}{\partial y}E_0(x, y; \lambda)\right)\Big|_{x=y=c} \\
&= p(y)\frac{\partial}{\partial y}\left(p(x)\frac{\partial}{\partial x}E_0(x, y; \lambda)\right)\Big|_{x=y=c} \\
&= \varrho_{22}(\lambda) - \varrho_{22}(0).
\end{aligned}
$$

The essential uniqueness of the functions $\varrho_{jk}(\lambda)$, $j, k = 1, 2$, as stated in the theorem, follows immediately from these relations.

It is now a simple matter to establish the remaining assertions of the theorem concerning the matrix $\{\varrho_{jk}(\lambda)\}$. Since $E_0(x, y; \lambda)$ is now continuous in (x, y), we see that the equation $E_0(x, y; \lambda) = \overline{E_0(y, x; \lambda)}$ must hold for all λ and all (x, y) without exception. This equation becomes

$$
\begin{aligned}
&\sum_{\gamma, \delta=1}^{2} \int_0^\lambda \varphi_\gamma(x, c; \mu)\, \varphi_\delta(y, c; \mu)\, d\varrho_{\gamma\delta}(\mu) \\
&= \sum_{\gamma, \delta=1}^{2} \int_0^\lambda \varphi_\gamma(y, c; \mu)\, \varphi_\delta(x, c; \mu)\, d\overline{\varrho_{\gamma\delta}(\mu)}
\end{aligned}
$$

and thus implies that $\varrho_{jk}(\lambda) - \overline{\varrho_{kj}(\lambda)} = c_{jk}$ is a constant such that $\overline{c_{jk}} = -c_{kj}$, for $j, k = 1, 2$. Evidently we can replace the matrix $\{\varrho_{jk}(\lambda)\}$ by the (Hermitian) symmetric matrix $\{\varrho_{jk}(\lambda) - \frac{1}{2}c_{jk}\}$, as stated in the theorem. A similar argument shows that $E_0(x, y; \lambda)$ is a real function if and only if the matrix $\{\varrho_{jk}(\lambda)\}$ can be so determined that its elements are real functions. Now it is found that $E_0(x, y; \lambda)$ is real if and only if the resolution of the identity $E(\lambda)$ is real with respect to the conjugation J_2; and, by Theorem 9.14, the latter condition is satisfied if and only if the corresponding self-adjoint transformation A is real with respect to J_2. We may remark that, if the matrix $\{\varrho_{jk}(\lambda)\}$ is real, then it can be replaced by a real symmetric matrix since the constant c_{jk}, on which the replacement may be made to depend in the manner indicated above, is evidently real.

We have thus brought the proof of Theorem 10.22 to a close. With this theorem we shall terminate our discussion of differential operators of the second order, although we have left unanswered many questions which are immediately suggested by the results given above. It is worthwhile to point out some of the more interesting details which we are unable to touch here. Further properties and interrelations of the functions $\varrho_{jk}(\lambda)$ should be explored in a complete theory, as should the connections between the matrix $\{\varrho_{jk}(\lambda)\}$ and the spectrum of the self-adjoint transformation A. It is also of importance to determine the influence of the boundary conditions associated with A upon the integral representation of the kernel $E(x, y; \Delta)$. The solutions of such problems would seem to require fairly powerful methods of analysis.

We must refrain also from discussing some of the more familiar aspects of the theory of differential operators. We may mention in particular the examination of the further spectral properties of the self-adjoint transformation A. Many more or less general results in this direction are to be found in the literature, to which we refer the reader for further information.* A number of instructive examples are scattered through the papers cited.

§ 4. Jacobi Matrices and Allied Topics

In Chapter VII, § 3, we examined in some detail the connection between self-adjoint transformations with simple spectra and Jacobi matrices. We are now in a position to study, with the aid of the theory of Chapter IX, the most general Jacobi matrix. By removing the restriction imposed in Theorem 7.14, we not only complete the investigations

* Hilb, Mathematische Annalen, 66 (1908), pp. 1–66;—Plancherel, Mathematische Annalen, 67 (1909), pp. 519–534;—Weyl, Göttinger Nachrichten, 1909, pp. 37–63; Mathematische Annalen, 68 (1909), pp. 220-269; Göttinger Nachrichten, 1910, pp. 442–467;—Gray; American Journal of Mathematics, 50 (1928), pp. 431–458;—Stone, Mathematische Zeitschrift, 28 (1928), pp. 654–676;—Milne, Transactions of the American Mathematical Society, 30 (1928), pp. 797–802. These papers deal with actual extensions of the Sturm-Liouville differential systems; that is, with cases more general than the one discussed in Theorem 10.18.

initiated in that section, but also obtain the means for solving important problems in the theory of continued fractions and the theory of moments.

An infinite matrix $A = \{a_{mn}\}$ satisfying the conditions

$$a_{mn} = \bar{a}_{nm}; \qquad a_{mn} = 0, \qquad m < n - 1;$$
$$a_{nn} = a_n; \qquad a_{n,n+1} = b_n \neq 0$$

will be called a Jacobi matrix. The matrix $A_p = \{a_{mn}(p)\}$ obtained from a Jacobi matrix $A = \{a_{mn}\}$ by the defining equations $a_{mn}(p) = a_{mn}$, $m \leqq p$, $n \leqq p$ and $a_{mn}(p) = 0$, $m > p$ or $n > p$, will be referred to as a reduced Jacobi matrix of order p. While matrices of the latter type can evidently be defined and studied without reference to any prescribed Jacobi matrix, it is their relation to the theory of Jacobi matrices which concerns us here and which makes it convenient to think of them in terms of the definition given above. In Theorem 7.14 we introduced the polynomials $G_n(l)$ constructed from a given Jacobi matrix A by the recursion formulas

$$G_1(l) = 1, \qquad G_2(l) = \frac{l - a_1}{\bar{b}_1}$$

$$G_n(l) = \frac{(l - a_{n-1})\, G_{n-1}(l) - b_{n-2}\, G_{n-2}(l)}{\bar{b}_{n-1}};$$

we shall find these polynomials fundamental in the study of a general Jacobi matrix A and its reduced matrices. It will be observed that the elements of the reduced Jacobi matrix A_p suffice to determine the polynomials $G_1, \cdots, G_p$, but not their successors.

THEOREM 10.23. *The transformation $T_1(A_p)$ associated by the complete orthonormal set $\{g_n\}$ with the reduced Jacobi matrix A_p is an essentially self-adjoint transformation; and the self-adjoint transformation $H^{(p)} \equiv \tilde{T}_1(A_p) \equiv T_1^*(A_p) \equiv T_1^{**}(A_p)$ is bounded. The function $\varrho_p(\lambda) = \| E^{(p)}(\lambda)\, g_1 \|^2$, where $E^{(p)}(\lambda)$ is the resolution of the identity corresponding to $H^{(p)}$, has the following properties:*

(1) $\varrho_p(\lambda)$ *is a real monotone-increasing function in* $\mathfrak{B}^*$ *which assumes exactly* $p+1$ *distinct values; its* p *distinct points of discontinuity are located at the roots of the polynomial* $G_{p+1}(l)$;

$$(2) \int_{-\infty}^{+\infty} d\varrho_p(\lambda) = 1, \quad \int_{-\infty}^{+\infty} G_m(\lambda)\overline{G_n(\lambda)}\, d\varrho_p(\lambda) = \delta_{mn},$$

$$\int_{-\infty}^{+\infty} \lambda\, G_n(\lambda)\overline{G_m(\lambda)}\, d\varrho_p(\lambda) = a_{mn}(p) = a_{mn}, \qquad m, n = 1, \cdots, p;$$

(3) *the integral* $I(l;\varrho_p) = \int_{-\infty}^{+\infty} \frac{1}{\lambda - l}\, d\varrho_p(\lambda)$ *is a rational function of* l *which is analytic at infinity and developable there in the convergent series*

$$-\sum_{\alpha=0}^{\infty} c_\alpha(p)/l^{\alpha+1}, \qquad c_n(p) = \int_{-\infty}^{+\infty} \lambda^n\, d\varrho_p(\lambda);$$

(4) *the rational function* $I(l;\varrho_p)$ *can be expressed as the continued fraction*†

$$\frac{1\,|}{|\,a_1 - l} - \frac{|\,b_1|^2\,|}{|\,a_2 - l} - \cdots - \frac{|\,b_{p-2}|^2\,|}{|\,a_{p-1} - l} - \frac{|\,b_{p-1}|^2\,|}{|\,a_p - l}.$$

The constants $c_n(p)$ *which appear in* (3) *have the property that* $c_n(p) = c_n(q)$ *for* $0 \leqq n \leqq 2p-1$, $q \geqq p$.

Since the matrix is symmetric and contains only a finite number of elements different from zero, Theorems 3.2, 3.4 and 3.5 are applicable to it; they show that $T_1(A_p)$ is an essentially self-adjoint transformation and that $H^{(p)} \equiv T_2(A_p) \equiv T_1^*(A_p)$ is a bounded self-adjoint transformation which coincides with $\tilde{T}_1(A_p)$ and $T_1^{**}(A_p)$. We see at once that the relations $H^{(p)} g_n = \sum_{\alpha=1}^{\infty} a_{\alpha n}(p)\, g_\alpha$ are satisfied and reduce to the equations

$$\begin{aligned}
H^{(p)} g_1 &= a_1 g_1 + \bar{b}_1 g_2,\\
H^{(p)} g_{n-1} &= b_{n-2} g_{n-2} + a_{n-1} g_{n-1} + \bar{b}_{n-1} g_n, \quad 3 \leqq n \leqq p,\\
H^{(p)} g_p &= b_{p-1} g_{p-1} + a_p g_p,\\
H^{(p)} g_n &= 0, \qquad n > p.
\end{aligned}$$

† We use the notation of Perron, *Die Lehre von den Kettenbrüchen*, second edition, Leipzig, 1929, pp. 3–4.

By manipulations like those indicated in the proof of Theorem 7.14 we conclude that $g_n = G_n(H^{(p)})\,g_1$ for $1 \leqq n \leqq p$. Obviously, the elements g_n, $n > p$, are characteristic elements of $H^{(p)}$ for the characteristic value $l = 0$. Thus, if $E^{(p)}(\lambda)$ is the resolution of the identity corresponding to $H^{(p)}$, the element g_n, $n > p$, satisfies the relations $E^{(p)}(\lambda)\,g_n = 0$, $\lambda < 0$, and $E^{(p)}(\lambda)\,g_n = g_n$, $\lambda \geqq 0$. In consequence, we see that every element expressible in the form $E^{(p)}(\lambda)\,g_1$ must be orthogonal to g_n, $n > p$, by virtue of the equation $(E^{(p)}(\lambda)g_1, g_n) = (g_1, E^{(p)}(\lambda)\,g_n) = 0$, and that $E^{(p)}(\lambda)\,g_1$ is an element of the linear manifold determined by the elements $g_1, \cdots, g_p$. On the other hand, we have noted that $g_n = G_n(H^{(p)})\,g_1$, $1 \leqq n \leqq p$. From Theorem 7.2 it follows that the closed linear manifold $\mathfrak{M}_p(g_1)$ determined by the set of elements $E^{(p)}(\lambda)g_1$, $-\infty < \lambda < +\infty$, coincides with the linear manifold determined by the set of elements $g_1, \cdots, g_p$. In order that $\mathfrak{M}_p(g_1)$ have the correct dimension number p, it is necessary that the function $\varrho_p(\lambda) = \|E^{(p)}(\lambda)g_1\|^2$ assume precisely $p+1$ distinct values and have precisely p distinct points of discontinuity, $\lambda_{p1} < \lambda_{p2} < \cdots < \lambda_{p,p-1} < \lambda_{pp}$, as we see by reference to Lemma 6.2, Theorem 6.2, and Theorem 7.2. We now locate these points of discontinuity. From the equation for $H^{(p)} g_p$ given above we see that

$$G_{p+1}(H^{(p)})g_1 = ((H^{(p)} - a_p)g_p - b_{p-1}\,g_{p-1})/\bar{b}_p = 0.$$

If Δ is an interval containing the point $\lambda = \lambda_{pk}$ but no other point of discontinuity of $\varrho_p(\lambda)$, then

$$\begin{aligned} G_{p+1}(\lambda_{pk})\|E^{(p)}(\Delta)g_1\|^2 &= \int_\Delta G_{p+1}(\lambda)\,d\varrho_p(\lambda) \\ &= (E^{(p)}(\Delta)\,G_{p+1}(H^{(p)})g_1, g_1) = 0. \end{aligned}$$

Since

$$\|E^{(p)}(\Delta)g_1\|^2 = \varrho_p(\lambda_{pk}+0) - \varrho_p(\lambda_{pk}-0) \neq 0,$$

we conclude that $G_{p+1}(\lambda_{pk}) = 0$. Thus the polynomial $G_{p+1}(l)$ vanishes at the p points of discontinuity of $\varrho_p(\lambda)$; and the fact that this polynomial is of degree p shows that it has no other roots.

The relations given in (2) are easily established. We have

$$\int_{-\infty}^{+\infty} d\varrho_p(\lambda) = \|g_1\|^2 = 1,$$

$$\int_{-\infty}^{+\infty} G_m(\lambda)\overline{G_n(\lambda)}\,d\varrho_p(\lambda) = (G_m(H^{(p)})g_1, G_n(H^{(p)})g_1) = (g_m, g_n) = \delta_{mn},$$

$$\int_{-\infty}^{+\infty} \lambda G_n(\lambda)\overline{G_m(\lambda)}\,d\varrho_p(\lambda) = (H^{(p)} G_n(H^{(p)})g_1, G_m(H^{(p)})g_1) = (H^{(p)} g_n, g_m) = a_{mn}(p) = a_{mn},$$

$$m, n = 1, \cdots, p.$$

If $R_l^{(p)}$ is the resolvent of the self-adjoint transformation $H^{(p)}$, then the integral $I(l; \varrho_p)$ gives the value of $(R_l^{(p)} g_1, g_1)$. From (1) it is evident that this integral is a rational function of l representable in the partial fraction form $\sum_{\alpha=1}^{p} \frac{\pi_{p\alpha}}{\lambda_{p\alpha} - l}$ and therefore analytic at infinity. The series development at infinity for this function follows directly from Theorem 4.21. Since the spectrum of $H^{(p)}$ consists of the characteristic values $0, \lambda_{p1}, \cdots, \lambda_{pp}$, we have $\|H^{(p)} f\| \leq C\|f\|$ where C is the greater of $|\lambda_{p1}|$ and $|\lambda_{pp}|$. Hence, Theorem 4.21 shows that

$$I(l; \varrho_p) = (R_l^{(p)} g_1, g_1) = -\sum_{\alpha=0}^{\infty} ((H^{(p)})^\alpha g_1, g_1)/l^{\alpha+1} = -\sum_{\alpha=0}^{\infty} c_\alpha(p)/l^{\alpha+1},$$

$$c_n(p) = \int_{-\infty}^{+\infty} \lambda^n\,d\varrho_p(\lambda),$$

the series being convergent for $|l| > C$.

In order to obtain the expansion of $I(l; \varrho_p)$ as a continued fraction, we first note that $R_l^{(p)} g_1$ is an element of $\mathfrak{M}_p(g_1)$ and is therefore expressible in the form $\sum_{\alpha=1}^{p} x_\alpha g_\alpha$. We therefore have

$$g_1 = \sum_{\alpha=1}^{p} x_\alpha (H^{(p)} g_\alpha - l g_\alpha).$$

By substituting this expression for g_1 in the relations $(g_1, g_n) = \delta_{1n}$, we obtain a system of p equations satisfied by the p coefficients $x_1, \cdots, x_p$:

$$\begin{aligned} x_1(a_1-l)+x_2 b_1 &= 1, \\ x_{n-1}\bar{b}_{n-1}+x_n(a_n-l)+x_{n+1}b_n &= 0,\ 2\leqq n\leqq p-1, \\ x_{p-1}\bar{b}_{p-1}+x_p(a_p-l) &= 0. \end{aligned}$$

This system can be written in the equivalent form

$$\begin{aligned} x_1 &= \frac{1}{a_1-l+b_1 x_2/x_1}, \\ x_n/x_{n-1} &= \frac{-\bar{b}_{n-1}}{a_n-l+b_n x_{n+1}/x_n}, \quad 2\leqq n\leqq p-1, \\ x_p/x_{p-1} &= \frac{-\bar{b}_p}{a_p-l}. \end{aligned}$$

By substituting from each equation of this system into the preceding one, we obtain for $x_1=(R_l^{(p)}g_1, g_1)=I(l;\varrho_p)$ the expression given in (4).

Since $G_n(l)$ is a polynomial of degree $n-1$, there exists a unique set of constants C_{mn} such that

$$l^m=\sum_{\alpha=1}^{m+1} C_{m\alpha}G_\alpha(l), \qquad m=0,1,2,\cdots.$$

Hence, if n is any one of the integers $0,\cdots,2p-1$, we can write

$$\lambda^n=\lambda\cdot\lambda^j\cdot\bar{\lambda}^k=\sum_{\alpha=1}^{j+1}\sum_{\beta=1}^{k+1}C_{j\alpha}\bar{C}_{k\beta}\,\lambda\,G_\alpha(\lambda)\overline{G_\beta(\lambda)}$$

where $0\leqq j\leqq p-1$, $0\leqq k\leqq p-1$, and can therefore conclude that

$$\begin{aligned} c_n(q) &= \int_{-\infty}^{+\infty}\lambda^n\,d\varrho_q(\lambda) \\ &= \sum_{\alpha=1}^{j+1}\sum_{\beta=1}^{k+1}C_{j\alpha}\bar{C}_{k\beta}\int_{-\infty}^{+\infty}\lambda\,G_\alpha(\lambda)\overline{G_\beta(\lambda)}\,d\varrho_q(\lambda) \\ &= \sum_{\alpha=1}^{j+1}\sum_{\beta=1}^{k+1}C_{j\alpha}\bar{C}_{k\beta}\,a_{\beta\alpha}(q) \\ &= \sum_{\alpha=1}^{j+1}\sum_{\beta=1}^{k+1}C_{j\alpha}\bar{C}_{k\beta}\,a_{\beta\alpha}, \qquad q\geqq p, \end{aligned}$$

is independent of q for $q\geqq p$. This completes the proof of the theorem.

With the aid of Theorem 10.23 we can establish some of the properties of various systems of polynomials associated with a given Jacobi matrix, among them the polynomials $G_n(l)$ already defined. In preparation for our first theorem in this connection, we observe that the final assertion of Theorem 10.23 suggests the introduction of a sequence of constants $\{c_n\}$ associated with the Jacobi matrix A: we set $c_0 = 1$; and, if n is a positive integer, we denote by p_n the least integer such that $2p_n - 1 \geqq n$ and set $c_n = c_n(p_n)$. It is evident that $c_n(p) = c_n(p_n) = c_n$ for $p \geqq p_n$ and that

$$c_n = \int_{-\infty}^{+\infty} \lambda \cdot \lambda^{p_n-1} \cdot \bar{\lambda}^{p_n-1}\, d\varrho_{p_n}(\lambda) = \sum_{\alpha=1}^{p_n} \sum_{\beta=1}^{p_n} C_{p_n-1,\alpha}\, \bar{C}_{p_n-1,\beta}\, a_{\beta\alpha},$$

$$c_n = \int_{-\infty}^{+\infty} \lambda \cdot \lambda^{p_n-2} \cdot \bar{\lambda}^{p_n-1}\, d\varrho_{p_n}(\lambda) = \sum_{\alpha=1}^{p_n-1} \sum_{\beta=1}^{p_n} C_{p_n-2,\alpha}\, \bar{C}_{p_n-1,\beta}\, a_{\beta\alpha},$$

according as n is odd or even. The latter relations, taken together with the definition of the constants C_{jk}, show that c_n is determined by the reduced Jacobi matrix A_{p_n}.

THEOREM 10.24. *Let A be a Jacobi matrix, $\{G_n(l)\}$ the associated sequence of polynomials, and $\{c_n\}$ the associated sequence of constants. Then*

(1) *if we define $a_0 = b_{-1} = b_0 = 1$, $G_{-1}(l) = -1$, $G_0(l) = 0$, the recursion formula*

$$G_n(l) = \frac{(l - a_{n-1})\,G_{n-1}(l) - b_{n-2}\,G_{n-2}(l)}{\bar{b}_{n-1}}$$

is valid for $n \geqq 1$;

(2) $(l - \bar{m}) \sum_{\alpha=1}^{n} G_\alpha(l)\overline{G_\alpha(m)} = G_{n+1}(l)\overline{b_n G_n(m)} - b_n G_n(l)\overline{G_{n+1}(m)}$, $n \geqq 1$;

(3) $G_n(l) = (-1)^{n-1} \det_{j,k=1,\cdots,n-1} \{a_{jk} - l\delta_{jk}\} / \bar{b}_0 \cdots \bar{b}_{n-1}$, $n \geqq 2$;

(4) $G_n(l) = (-1)^{n-1}\gamma_n \det_{j,k=0,\cdots,n-2} \{c_{j+k+1} - l c_{j+k}\}/D_{n-2}^{1/2}\overline{D_{n-2}^{1/2}}$, $n \geqq 2$, *where* $\gamma_n = b_0 \cdots b_{n-1}/|b_0 \cdots b_{n-1}|$, $D_0 = 1$, *and* $D_n = \det_{j,k=0,\cdots,n-1} \{c_{j+k}\}$, $n \geqq 1$;

(5) $G_n(l)$ *is a polynomial of degree* $n-1$ *with* $n-1$ *distinct real roots;*

(6) *the roots of* $G_n(l)$ *separate the roots of* $G_{n+1}(l)$ *for* $n \geqq 2$;

(7) *the inequality* $|G_n(\lambda + i\mu_1)| \leqq |G_n(\lambda + i\mu_2)|$ *is satisfied for* $|\mu_1| \leqq |\mu_2|$ *whenever* λ, μ_1, *and* μ_2 *are real numbers.*

The artificial scheme given in (1) for extending the recursion formula to the values $n = 1, 2$ is readily justified; while the extension is trivial, it often simplifies applications of the recursion formula. In proving (2), we have only to write

$$l G_n(l) = \bar{b}_n G_{n+1}(l) + a_n G_n(l) + b_{n-1} G_{n-1}(l),$$
$$\overline{m G_n(m)} = b_n \overline{G_{n+1}(m)} + a_n \overline{G_n(m)} + \bar{b}_{n-1} \overline{G_{n-1}(m)},$$

and substitute in the left-hand member of the equation to be verified: the right-hand member is obtained by virtue of obvious cancellations on the left. We establish (3) by an inductive process based on (1). First, it is easy to calculate and verify (3) for $n = 2, 3$. We next suppose that the equation has been proved for $n = 2, \cdots, p$ and show that it is also true for $n = p+1$, $p \geqq 3$. By expanding the determinant in an obvious manner, we find

$$\begin{aligned}(-1)^{p+1} &\det_{j,k=1,\cdots,p} \{a_{jk} - l\delta_{jk}\}/\bar{b}_0 \cdots \bar{b}_p \\ &= (-1)^{p+1}(a_p - l) \det_{j,k=1,\cdots,p-1} \{a_{jk} - l\delta_{jk}\}/\bar{b}_0 \cdots \bar{b}_p \\ &\qquad - (-1)^{p+1}\bar{b}_{p-1} b_{p-2} \det_{j,k=1,\cdots,p-2} \{a_{jk} - l\delta_{jk}\}/\bar{b}_0 \cdots \bar{b}_p \\ &= (l - a_p) G_p(l)/\bar{b}_p - b_{p-2} G_{p-1}(l)/\bar{b}_p = G_{p+1}(l).\end{aligned}$$

Hence (3) is valid for $n \geqq 2$. The proof of (4) depends upon the evaluation of $G_n(l)$ just obtained. We consider the linear transformation $y_j = \sum_{\alpha=1}^{n-1} d_{j\alpha} x_\alpha$, $1 \leqq j \leqq n-1$, defined by the relation $\sum_{\alpha=1}^{n-1} y_\alpha l^{\alpha-1} = \sum_{\beta=1}^{n-1} x_\beta G_\beta(l)$. It is easily verified that d_{jk} is the coefficient of l^{j-1} in the polynomial $G_k(l)$. Hence $d_{jk} = 0$ for $j > k$; and $d_{kk} = 1/\bar{b}_0 \cdots \bar{b}_{k-1}$, $b_0 = 1$, by virtue of (1) and (3). It follows that

$$\det_{j,k=1,\cdots,n-1} \{d_{jk}\} = 1/\bar{b}_0^{n-1} \cdots \bar{b}_{n-2}.$$

We now use the equation

$$\int_{-\infty}^{+\infty} (\lambda - l)\left(\sum_{\beta=1}^{n-1} y_\beta \lambda^{\beta-1}\right)\left(\sum_{\alpha=1}^{n-1} \bar{y}_\alpha \bar{\lambda}^{\alpha-1}\right) d\varrho_n(\lambda)$$
$$= \int_{-\infty}^{+\infty} (\lambda - l)\left(\sum_{\nu=1}^{n-1} x_\nu G_\nu(\lambda)\right)\left(\sum_{\mu=1}^{n-1} \bar{x}_\mu \overline{G_\mu(\lambda)}\right) d\varrho_\mu(\lambda)$$

to show that

$$\sum_{\alpha,\beta=1}^{n-1} (c_{\alpha+\beta-1} - l c_{\alpha+\beta-2})\, y_\beta \bar{y}_\alpha = \sum_{\mu,\nu=1}^{n-1} (a_{\mu\nu} - l\delta_{\mu\nu})\, x_\nu \bar{x}_\mu$$

when the variables x and y are connected by the indicated relation. If we replace the y's by their expressions in terms of the x's and equate the coefficients of $x_k \bar{x}_j$ in the resulting identity, we obtain the matrix equation

$$\{d^*_{j\alpha}\}\,\{c_{\alpha+\beta-1} - l c_{\alpha+\beta-2}\}\,\{d_{\beta k}\} = \{a_{jk} - l\delta_{jk}\}, \quad d^*_{jk} = \bar{d}_{kj}.$$

We conclude that

$$\det_{j,k=0,\cdots,n-2} \{c_{j+k+1} - l c_{j+k}\} / |b_0|^{2n-2} \cdots |b_{n-2}|^2$$
$$= \det_{j,k=1,\cdots,n-1} \{d^*_{jk}\} \det_{j,k=1,\cdots,n-1} \{c_{j+k-1} - l c_{j+k-2}\} \det_{j,k=1,\cdots,n-1} \{d_{jk}\}$$
$$= \det_{j,k=1,\cdots,n-1} \{a_{jk} - l\delta_{jk}\}.$$

By comparing the coefficients of l^{n-1} in the two extreme terms of this equation, we find that

$$D_{n-2} / |b_0|^{2n-2} \cdots |b_{n-2}|^2 = 1$$

and hence that

$$D_{n-1}/D_{n-2} = |b_0|^2 \cdots |b_{n-1}|^2, \quad D_{n-2} D_n / D_{n-1}^2 = |b_n|^2.$$

On substituting these results in (3), we obtain the formula stated in (4). We have already proved (5), which is merely a modification of Theorem 10.23 (1). In order to prove that the roots of $G_n(l)$ separate those of $G_{n+1}(l)$, we shall calculate the quantities $\overline{G_n(\lambda_{nk})}$ where $\lambda_{n1} < \lambda_{n2} < \cdots < \lambda_{n,n-1} < \lambda_{nn}$ are the roots of $G_{n+1}(l)$. The equations $\int_{-\infty}^{+\infty} \lambda^p \overline{G_n(\lambda)}\, d\varrho_n(\lambda) = 0$ are valid for $0 \leqq p \leqq n-2$ and can obviously be written in the form $\sum_{\alpha=1}^{n} \lambda_{n\alpha}^p \overline{G_n(\lambda_{n\alpha})} = 0$. Thus there exists a constant C such that $\overline{G_n(\lambda_{nk})} = (-1)^k C \Delta_{nk}$, where Δ_{nk} is the Vandermonde determinant obtained by deleting the row $p = n-2$ and the column $q = k$ from $\det_{p+1, q=1,\cdots,n-1} \{\lambda_{nq}^p\}$. It is clear that the inequalities $\lambda_{n1} < \lambda_{n2} < \cdots < \lambda_{n,n-1} < \lambda_{nn}$ imply $\Delta_{nk} > 0$, and it is also clear that C is different from zero since $G_n(l)$ cannot have n roots. Thus the polynomial $\overline{G_n(\lambda)}/C$ in the real variable λ assumes real values of alternate sign as λ runs through the values $\lambda_{n1} < \lambda_{n2} < \cdots < \lambda_{n,n-1} < \lambda_{nn}$. We therefore conclude that this polynomial is real and that it vanishes at least once on the interval $\lambda_{nk} < \lambda < \lambda_{n,k+1}$, $k = 1, \cdots, n-1$. This result is equivalent to that asserted in (6). Finally, we obtain (7) by writing $|G_n(\lambda + i\mu)|$ in factored form as $\prod_{\alpha=1}^{n-1} |(\lambda - \lambda_{n-1,\alpha}) + i\mu| / D_{n-2}^{1/2} D_{n-1}^{1/2}$ and then noting that the expression $|(\lambda - \lambda_0) + i\mu|$ increases with $|\mu|$ when λ, λ_0 and μ are real numbers. The trivial case $n = 1$ is not included under this method, since $|G_1(l)| = 1$.

THEOREM 10.25. *Let $A = \{a_{mn}\}$ and $A^{(1)} = \{a_{mn}^{(1)}\}$ be Jacobi matrices such that $a_{mn}^{(1)} = a_{m+1,n+1}$; let $\{G_n(l)\}$ and $\{G_n^{(1)}(l)\}$ be the sequences of polynomials associated with A and with $A^{(1)}$ respectively; and let $H_n(l)$ be the polynomial of degree $n-2$ defined by the equation $H_n(l) = -G_{n-1}^{(1)}(l)/\overline{b_1}$ for $n \geqq 2$. Then*

(1) *if we define $b_0 = 1$, $H_0 = 1$, $H_1 = 0$, the recursion formula*

$$H_n(l) = \frac{(l - a_{n-1})\, H_{n-1}(l) - b_{n-2}\, H_{n-2}(l)}{\overline{b}_{n-1}}$$

is valid for $n \geqq 2$;

(2)
$$1 + (l - \overline{m}) \sum_{\alpha=2}^{n} G_\alpha(l) \, \overline{H_\alpha(m)}$$
$$= G_{n+1}(l) \, \overline{b_n \, H_n(m)} - b_n \, G_n(l) \, \overline{H_{n+1}(m)}, \quad n \geq 2;$$

(3) *the polynomial* $H_{n+1}(l)$ *has* $n-1$ *real roots which separate those of* $G_{n+1}(l)$, *for* $n \geq 2$;

(4) $\quad |H_n(l)/G_n(l)| \leq 1/|\mathfrak{J}(l)|, \quad n \geq 2$;

(5) *the continued fraction*

$$F_n(l, t) = \frac{|b_0|^2 \,|}{|a_1 - l} - \frac{|b_1|^2 \,|}{|a_2 - l} - \cdots - \frac{|b_{n-2}|^2 \,|}{|a_{n-1} - l} - \frac{|b_{n-1}|^2 \,|}{|a_n - l - t}$$

satisfies the relation

$$F_n(l, t) = \frac{\overline{b_n} \, H_{n+1}(l) + H_n(l) \, t}{\overline{b_n} \, G_{n+1}(l) + G_n(l) \, t}$$

for $n \geq 1$;

(6) *when* l *is a fixed number,* $\mathfrak{J}(l) \neq 0$, *and* t *describes the half-plane* $\mathfrak{J}(l) \, \mathfrak{J}(t) \geq 0$, *then* $z = F_n(l, t)$ *describes a locus* $C_n(l)$ *consisting of the interior and circumference of the circle with center at the point*

$$\frac{\overline{b_n} \, H_{n+1}(l) \, \overline{G_n(l)} - b_n \, H_n(l) \, \overline{G_{n+1}(l)}}{\overline{b_n} \, G_{n+1}(l) \, \overline{G_n(l)} - b_n \, G_n(l) \, \overline{G_{n+1}(l)}}$$

$$= \left(-1 + 2i \, \mathfrak{J}(l) \sum_{\alpha=1}^{n} \overline{G_\alpha(l)} \, H_\alpha(l) \right) \Big/ 2i \, \mathfrak{J}(l) \sum_{\alpha=1}^{n} |G_\alpha(l)|^2$$

and radius

$$\frac{|b_n \, G_{n+1}(l) \, H_n(l) - b_n \, G_n(l) \, H_{n+1}(l)|}{|\overline{b_n} \, G_{n+1}(l) \, \overline{G_n(l)} - b_n \, G_n(l) \, \overline{G_{n+1}(l)}|} = 1/2 |\mathfrak{J}(l)| \sum_{\alpha=1}^{n} |G_\alpha(l)|^2;$$

the locus $C_n(l)$ *lies in the half-plane* $\mathfrak{J}(l) \, \mathfrak{J}(z) > 0$ *and contains* $C_{n+1}(l)$ *in its interior.*

For $n = 2, 3$, the recursion formula given in (1) can be verified directly; and for $n \geq 4$, it follows immediately from the equation

$$G_{n-1}^{(1)}(l) = \frac{(l - a_{n-2}^{(1)}) \, G_{n-2}^{(1)}(l) - b_{n-3}^{(1)} \, G_{n-3}^{(1)}(l)}{\overline{b}_{n-2}^{(1)}}$$

$$= \frac{(l - a_{n-1}) \, G_{n-2}^{(1)}(l) - b_{n-2} \, G_{n-3}^{(1)}(l)}{\overline{b}_{n-1}},$$

which is satisfied by the polynomials $\{G_n^{(1)}(l)\}$ associated with the Jacobi matrix $A^{(1)}$. This recursion formula leads to the result stated in (2) by a series of manipulations analogous to those used in the proof of Theorem 10.24 (2).

We now pass to the consideration of (5). It is easy to verify that the formula $F_n(l, t) = \dfrac{\overline{b}_n H_{n+1} + H_n t}{\overline{b}_n G_{n+1} + G_n t}$ is valid for $n = 1$. If it is valid for $n = p \geqq 1$, we have

$$\begin{aligned} F_{p+1}(l, t) &= F_p\left(l, \frac{b_p \overline{b}_p}{a_{p+1} - l - t}\right) \\ &= \frac{\overline{b}_p(a_{p+1} - l) H_{p+1} + b_p \overline{b}_p H_p - \overline{b}_p H_{p+1} t}{\overline{b}_p(a_{p+1} - l) G_{p+1} + b_p \overline{b}_p G_p - \overline{b}_p G_{p+1} t} \\ &= \frac{\overline{b}_{p+1} H_{p+2} + H_{p+1} t}{\overline{b}_{p+1} G_{p+2} + G_{p+1} t}, \end{aligned}$$

by virtue of the recursion formulas. Hence the formula is true for all $n \geqq 1$.

By the use of (5) and of Theorem 10.23, we can now establish the assertions made in (3) and (4). Since it is true that

$$H_{n+1}/G_{n+1} = F_n(l, 0) = I(l; \varrho_n) = \sum_{\alpha=1}^{n} \frac{\pi_{n\alpha}}{\lambda_{n\alpha} - l},$$

where $\lambda_{n1} < \lambda_{n2} < \cdots < \lambda_{n, n-1} < \lambda_{nn}$ are the roots of $G_{n+1}(l)$ and $\pi_{n1}, \cdots, \pi_{nn}$ are positive constants, we see that as l increases through real values from λ_{nk} to $\lambda_{n, k+1}$ the rational function H_{n+1}/G_{n+1} assumes all real values from $-\infty$ to $+\infty$. Thus $H_{n+1}(l)$ vanishes at least once on the open interval $(\lambda_{nk}, \lambda_{n, k+1})$, $1 \leqq k \leqq n - 1$. Since $H_{n+1}(l)$ is a polynomial of degree $n - 1$, it has exactly one root on each of these intervals, as we wished to prove. From the familiar inequality $|I(l; \varrho_{n-1})| \leqq 1/|\mathfrak{J}(l)|$ we conclude that (4) is satisfied.

For fixed l, the relation $z = F_n(l, t)$ defines a linear fractional transformation of the t-plane into the z-plane. The determinant of this transformation has the value

$\bar{b}_n G_{n+1}(l) H_n(l) - \bar{b}_n G_n(l) H_{n+1}(l)$. By means of the recursion formulas we can easily show that

$$\begin{aligned} | b_n | \, | G_{n+1}(l) H_n(l) - G_n(l) H_{n+1}(l) | \\ = | b_{n-1} | \, | G_n(l) H_{n-1}(l) - G_{n-1}(l) H_n(l) | = \cdots \\ = | b_0 | \, | G_1(l) H_0(l) - G_0(l) H_1(l) | = 1. \end{aligned}$$

Hence the transformation is non-singular and carries the half-plane $\mathfrak{J}(l)\mathfrak{J}(t) \geqq 0$, $\mathfrak{J}(l) \neq 0$, into a locus $C_n(l)$ in the z-plane which is either a circle together with all its interior or all its exterior points, or another half-plane. In the case $n = 1$, we can verify directly that the transformation $z = F_1(l, t)$ $= \dfrac{|b_0|^2}{a_1 - l - t}$ defines $C_1(l)$ as the circumference and interior of a circle lying on the half-plane $\mathfrak{J}(l)\mathfrak{J}(z) > 0$. Let us suppose now that the locus $C_n(l)$ has been shown to consist of the circumference and interior of a circle lying on this half-plane. Since the transformation $z = F_{n+1}(l, t)$ is the product of the two transformations $z = F_n(l, t')$ and $t' = \dfrac{|b_n|^2}{a_{n+1} - l - t}$, it is then clear that $C_{n+1}(l)$ must consist of interior points of $C_n(l)$ and is therefore composed of the circumference and interior points of a circle lying on the half-plane $\mathfrak{J}(l)\mathfrak{J}(z) > 0$. By induction we conclude that the qualitative description of $C_n(l)$ given in (6) is valid for all $n \geqq 1$. In order to determine the center and radius of the circular region $C_n(l)$, we observe that the transformation $z = F_n(l, t)$ carries the points 0, $-\bar{b}_n G_{n+1}/G_n$, $-b_n \bar{G}_{n+1}/\bar{G}_n$ of the t-plane into the points H_{n+1}/G_{n+1}, ∞, $\dfrac{\bar{b}_n H_{n+1} \bar{G}_n - b_n H_n \bar{G}_{n+1}}{\bar{b}_n G_{n+1} \bar{G}_n - b_n G_n \bar{G}_{n+1}}$, respectively, of the extended z-plane. Since the points $-\bar{b}_n G_{n+1}/G_n$ and $-b_n \bar{G}_{n+1}/\bar{G}_n$ are symmetric in the real axis, $\mathfrak{J}(t) = 0$, their images in the z-plane must be inverse with respect to the circle which is the boundary of $C_n(l)$. Thus the center of this circle is the point $\dfrac{\bar{b}_n H_{n+1} \bar{G}_n - b_n H_n \bar{G}_{n+1}}{\bar{b}_n G_{n+1} \bar{G}_n - b_n G_n \bar{G}_{n+1}}$, and its radius has the value

$$\left| \frac{H_{n+1}}{G_{n+1}} - \frac{\bar{b}_n H_{n+1} \bar{G}_n - b_n H_n \bar{G}_{n+1}}{\bar{b}_n G_{n+1} \bar{G}_n - b_n G_n \bar{G}_{n+1}} \right| = \frac{| b_n G_{n+1} H_n - b_n G_n H_{n+1} |}{| \bar{b}_n G_{n+1} \bar{G}_n - b_n G_n \bar{G}_{n+1} |}.$$

The expression for the center can be reduced to the second form given in (6) by use of (2) and of Theorem 10.24 (2); and the expression for the radius can be evaluated by use of the equation $|b_n| \, |G_{n+1} H_n - G_n H_{n+1}| = 1$, established above, and of Theorem 10.24 (2).

It is to be observed that the polynomials $H_n(l)$ have many properties analogous to those of the polynomials $G_n(l)$, by virtue of their relation to the Jacobi matrix $A^{(1)}$. Since such properties can be obtained by means of Theorem 10.24, we have confined our attention in Theorem 10.25 to relations which involve both sequences $\{G_n\}$ and $\{H_n\}$.

In our next theorem we shall give results which become significant only in connection with certain facts which are established later. It is convenient to state them at this point because of their intimate algebraic relations with the formulas of Theorems 10.24 and 10.25.

THEOREM 10.26. *The linear fractional transformation defined by the equations*

$$t = \frac{-b_n \overline{G_{n+1}(0)} - b_n \overline{H_{n+1}(0)}\, s}{\overline{G_n(0)} + \overline{H_n(0)}\, s},$$

$$s = \frac{-b_n \overline{G_{n+1}(0)} - \overline{G_n(0)}\, t}{b_n \overline{H_{n+1}(0)} + \overline{H_n(0)}\, t}$$

for $n \geqq 1$ *takes the half-planes* $\mathfrak{J}(s) \geqq 0$ *and* $\mathfrak{J}(s) \leqq 0$ *into the half-planes* $\mathfrak{J}(t) \geqq 0$ *and* $\mathfrak{J}(t) \leqq 0$ *respectively; and it carries the function* $F_n(l, t) = \dfrac{\bar{b}_n H_{n+1} + H_n t}{\bar{b}_n G_{n+1} + G_n t}$ *into the function* $\Phi_n(l, s) = \dfrac{P_n(l) + U_n(l)\, s}{Q_n(l) + V_n(l)\, s}$, *where*

$$P_n(l) = H_{n+1}(l) \overline{b_n G_n(0)} - b_n H_n(l) \overline{G_{n+1}(0)}$$
$$= -1 + l \sum_{\alpha=1}^{n} \overline{G_\alpha(0)} H_\alpha(l),$$

$$\begin{aligned} Q_n(l) &= G_{n+1}(l)\overline{b_n G_n(0)} - b_n G_n(l)\overline{G_{n+1}(0)} \\ &= l\sum_{\alpha=1}^{n} \overline{G_\alpha(0)}\, G_\alpha(l), \\ U_n(l) &= H_{n+1}(l)\overline{b_n H_n(0)} - b_n H_n(l)\overline{H_{n+1}(0)} \\ &= l\sum_{\alpha=1}^{n} \overline{H_\alpha(0)}\, H_\alpha(l), \\ V_n(l) &= G_{n+1}(l)\overline{b_n H_n(0)} - b_n G_n(l)\overline{H_{n+1}(0)} \\ &= 1 + l\sum_{\alpha=1}^{n} \overline{H_\alpha(0)}\, G_\alpha(l). \end{aligned}$$

When l is a fixed number, $\mathfrak{J}(l) \neq 0$, and s describes the half-plane $\mathfrak{J}(l)\mathfrak{J}(s) \geqq 0$, then $z = \Phi_n(l, s)$ describes the locus $C_n(l)$ of Theorem 10.25 (6).

We shall write the transformation under consideration in the equivalent form $t = \dfrac{A+Bs}{C+Ds}$, where $A = -\gamma_n b_n \overline{G_{n+1}(0)}$, $B = -\gamma_n \overline{b_n H_{n+1}(0)}$, $C = \gamma_n \overline{G_n(0)}$, $D = \gamma_n \overline{H_n(0)}$, and $\gamma_n = b_0 \cdots b_{n-1}/|b_0 \cdots b_{n-1}|$. This form has the advantage that the coefficients are real numbers, as we shall now show. From Theorem 10.24, (3) and (4), it is clear that the polynomial $G_n(l)$ is the product of a polynomial with real coefficients and the complex constant γ_n. By virtue of the fact that the polynomial $H_n(l)$ has only real roots, it is also clear that $H_n(l)$ is the product of a polynomial with real coefficients and a complex constant η_n. The relation $H_n/G_n = I(l; \varrho_{n-1})$, $n \geqq 2$, shows that we may take $\eta_n = \gamma_n$, since $I(l; \varrho_{n-1})$ can obviously be written as the quotient of two polynomials with real coefficients. The case $n = 1$, which is not included under this method of approach, is trivial since $H_1 = 0$, $G_1 = 1$, $\gamma_1 = 1$. From these facts we see at once that the coefficients C and D are real numbers. To discuss the two remaining coefficients, we put them in the new forms

$$\begin{aligned} A &= -(b_n\gamma_n/|b_n|\gamma_{n+1})\,|b_n|\gamma_{n+1}\overline{G_{n+1}(0)}, \\ B &= -(b_n\gamma_n/|b_n|\gamma_{n+1})\,|b_n|\gamma_{n+1}\overline{H_{n+1}(0)}. \end{aligned}$$

By direct evaluation, we find that $b_n \gamma_n / |b_n| \gamma_{n+1} = 1$ and hence conclude that A and B are also real constants. In order to determine the effect of the transformation $t = \dfrac{A + Bs}{C + Ds}$, we make use of the relation $\mathfrak{I}(t) = (BC - AD)\mathfrak{I}(s)/|C + Ds|^2$. By use of the recursion formulas of Theorem 10.24 (2) and Theorem 10.25 (2), we find that

$$\begin{aligned}
BC&-AD \\
&= \gamma_n^2 b_n (\overline{G_{n+1}(0)}\, \overline{H_n(0)} - \overline{G_n(0)}\, \overline{H_{n+1}(0)}) \\
&= \gamma_n^2 \overline{b}_{n-1} (\overline{G_n(0)}\, \overline{H_{n-1}(0)} - \overline{G_{n-1}(0)}\, \overline{H_n(0)}) \\
&= (\gamma_n^2 \overline{b}_{n-1}/b_{n-1}) b_{n-1} (\overline{G_n(0)}\, \overline{H_{n-1}(0)} - \overline{G_{n-1}(0)}\, \overline{H_n(0)}) = \cdots \\
&= (\gamma_n^2 \overline{b}_{n-1} \cdots \overline{b}_0 / b_{n-1} \cdots b_0) b_0 (\overline{G_1(0)}\, \overline{H_0(0)} - \overline{G_0(0)}\, \overline{H_1(0)}) \\
&= \gamma_n^2 \overline{b}_{n-1} \cdots \overline{b}_0 / b_{n-1} \cdots b_0 = 1.
\end{aligned}$$

Hence the transformation is non-singular and takes the half-planes $\mathfrak{I}(s) \geqq 0$ and $\mathfrak{I}(s) \leqq 0$ into the half-planes $\mathfrak{I}(t) \geqq 0$ and $\mathfrak{I}(t) \leqq 0$ respectively. On substituting the expression $t = \dfrac{-b_n \overline{G_{n+1}(0)} - b_n \overline{H_{n+1}(0)} s}{\overline{G_n(0)} + \overline{H_n(0)} s}$ in the function $F_n(l, t)$ we obtain the function $\Phi_n(l, s)$ described in the theorem. This substitution yields the formulas for P_n, Q_n, U_n, V_n in terms of G_n, H_n, G_{n+1}, H_{n+1} without any difficulty. The final expressions are obtained from these by use of Theorem 10.24 (2) and Theorem 10.25 (2). The calculation of U_n in this manner depends upon the closing remarks in the proof of Theorem 10.25. The transformation defined by the equation $z = \Phi_n(l, s)$, where l is a fixed number, is evidently the product of the two transformations $z = F_n(l, t)$, $t = \dfrac{-b_n \overline{G_{n+1}(0)} - b_n \overline{H_{n+1}(0)} s}{\overline{G_n(0)} + \overline{H_n(0)} s}$, and therefore behaves in the manner described in the theorem.

We turn now to the contemplated generalization of Theorem 7.14. We have

THEOREM 10.27. *Let A be a Jacobi matrix, $\{G_n(l)\}$ the associated sequence of polynomials, and $\{c_n\}$ the associated* *sequence of constants; let $T_1(A)$ be the symmetric transformation associated with the matrix A by the complete orthonormal set $\{g_n\}$; let T be the closed linear symmetric transformation $\tilde{T}_1(A) \equiv T_1^{**}(A)$; and let J be the transformation which carries $f = \sum_{\alpha=1}^{\infty} x_\alpha g_\alpha$ into $Jf = \sum_{\alpha=1}^{\infty} \omega_\alpha \bar{x}_\alpha g_\alpha$, where $\omega_n = \bar{\gamma}_n/\gamma_n = \bar{b}_0 \cdots \bar{b}_{n-1}/b_0 \cdots b_{n-1}$. Then J is a conjugation, and T is real with respect to J. The transformation T has the deficiency-index (m, m) where $m = 0$ or $m = 1$; the two possible cases are distinguished by the following properties:*

(1) *when $m = 0$, the series $\sum_{\alpha=1}^{\infty} |G_\alpha(l)|^2$ diverges for all not-real l, and for all real l with the possible exception of a finite or denumerably infinite set of values; the set where this series converges is the point spectrum of the self-adjoint transformation T;*

(2) *when $m = 1$, the series $\sum_{\alpha=1}^{\infty} |G_\alpha(l)|^2$ converges for all l.*

If the transformation X_l and the function $\varrho(f, g; \lambda)$ are associated with the symmetric transformation T by an approximating sequence of self-adjoint transformations in the manner specified in Theorem 9.17, then

(1) *$\varrho(\lambda) = \varrho(g_1, g_1; \lambda)$ is a real monotone-increasing function in the class $\mathfrak{V}^*$ which assumes infinitely many distinct values between its greatest lower bound 0 and its least upper bound $\|g_1\|^2 = 1$;*

(2)
$$\int_{-\infty}^{+\infty} G_m(\lambda)\overline{G_n(\lambda)}\, d\varrho(\lambda) = \delta_{mn},$$
$$\int_{-\infty}^{+\infty} \lambda\, G_n(\lambda)\overline{G_m(\lambda)}\, d\varrho(\lambda) = a_{mn},$$
$$(X_l g_m, g_n) = \int_{-\infty}^{+\infty} \frac{G_m(\lambda)\overline{G_n(\lambda)}}{\lambda - l}\, d\varrho(\lambda), \quad m, n = 1, 2, 3, \cdots;$$

(3) *the function $(X_l g_1, g_1)$ is represented in the sectors $0 < \varepsilon \leqq \arg l \leqq \pi - \varepsilon$, $\pi + \varepsilon \leqq \arg l \leqq 2\pi - \varepsilon$, by the asymptotic series*
$$-\sum_{\alpha=0}^{\infty} c_\alpha / l^{\alpha+1}, \qquad c_n = \int_{-\infty}^{+\infty} \lambda^n\, d\varrho(\lambda).$$

If $X_l^{(1)}$ *and* $X_l^{(2)}$ *are two such transformations and if* $\varrho^{(1)}(\lambda)$ *and* $\varrho^{(2)}(\lambda)$ *are the associated functions, the identity* $X_l^{(1)} \equiv X_l^{(2)}$, *for all not-real* l, *is equivalent to the identity* $\varrho^{(1)}(\lambda) \equiv \varrho^{(2)}(\lambda)$. *When* T *has the deficiency-index* $(0, 0)$, *it is a self-adjoint transformation with simple spectrum; the transformation* X_l *coincides with the resolvent of* T *for all not-real* l *and the function* $\varrho(\lambda)$ *associated with it is equal to* $(E(\lambda)g_1, g_1)$, *where* $E(\lambda)$ *is the resolution of the identity corresponding to* T. *When* T *has the deficiency-index* $(1, 1)$, *it has* $\mathfrak{c}$ *self-adjoint extensions, each of which is real with respect to the conjugation* J *and has a simple spectrum.*

As in the proof of Theorem 7.14, we find that

$$\begin{aligned} Tg_1 &= a_1 g_1 + \bar{b}_1 g_2, \\ Tg_{n-1} &= b_{n-2} g_{n-2} + a_{n-1} g_{n-1} + \bar{b}_{n-1} g_n, \qquad n \geqq 3, \end{aligned}$$

and hence conclude that $g_n = G_n(T) g_1$ for $n \geqq 1$. With the aid of these relations we proceed to examine the connection between the transformations T and J. It is easily verified that J is a conjugation in the sense of Definition 9.7, when the equation $|\omega_n| = 1$ is brought to bear. Theorem 9.13 shows that T is real with respect to J if the equation $JTg_n = TJg_n$ is satisfied for $n \geqq 1$. By virtue of the relation $\omega_n = \bar{b}_{n-1}\omega_{n-1}/b_{n-1}$, we have

$$\begin{aligned} JTg_1 &= a_1 g_1 + b_1 \omega_2 g_2 = a_1 g_1 + \bar{b}_1 g_2 = Tg_1 = TJg_1, \\ JTg_{n-1} &= \bar{b}_{n-2}\omega_{n-2} g_{n-2} + a_{n-1}\omega_{n-1} g_{n-1} + b_{n-1}\omega_n g_n \\ &= \omega_{n-1}(b_{n-2} g_{n-2} + a_{n-1} g_{n-1} + \bar{b}_{n-1} g_n) \\ &= T\omega_{n-1} g_{n-1} = TJg_{n-1}, \qquad n \geqq 3. \end{aligned}$$

Thus T is real with respect to J, as asserted in the theorem. This result implies that if the complete orthonormal set $\{g_n\}$ be replaced by the set $\{\bar{\gamma}_n g_n\}$, with the property $J\bar{\gamma}_n g_n = \omega_n \gamma_n g_n = \bar{\gamma}_n g_n$, then the transformation T will be described in terms of the new set by a matrix of real elements; in consequence, there would be no loss of generality in restricting our attention to real Jacobi matrices, and even to Jacobi matrices with $b_n > 0$, if we so desired. Another consequence of this result is that the deficiency-index of T must have the form (m, m) in accordance with Theorem 9.14.

We shall now determine the deficiency-index of T by finding all the characteristic values and characteristic elements of the adjoint transformation T^*. The number l is a characteristic value of $T^* \equiv T_1^*(A)$ if and only if there exists an element $g \neq 0$ such that

$$(T^* g, g_n) = l(g, g_n) = (g, Tg_n), \qquad n \geqq 1.$$

Such an element g exists and is expressible in the form $g = \sum_{\alpha=1}^{\infty} x_\alpha g_\alpha$ if and only if there exists a sequence of complex numbers $\{x_n\}$ satisfying the equations

$$l x_1 = l(g, g_1) = \sum_{\alpha=1}^{\infty} x_\alpha (g_\alpha, Tg_1) = \sum_{\alpha=1}^{\infty} a_{1\alpha} x_\alpha = a_1 x_1 + b_1 x_2,$$

$$\begin{aligned} l x_{n-1} &= l(g, g_{n-1}) = \sum_{\alpha=1}^{\infty} x_\alpha (g_\alpha, Tg_{n-1}) = \sum_{\alpha=1}^{\infty} a_{n-1,\alpha} x_\alpha \\ &= \bar{b}_{n-2} x_{n-2} + a_{n-1} x_{n-1} + b_{n-1} x_n, \qquad n \geqq 3, \end{aligned}$$

and conforming to the further condition that the series $\sum_{\alpha=1}^{\infty} |x_\alpha|^2$ converge to a sum different from zero. The equations just derived can be written in the successive equivalent forms

$$\bar{l}\bar{x}_1 = a_1 \bar{x}_1 + \bar{b}_1 \bar{x}_2;$$

$$\bar{l}\bar{x}_{n-1} = b_{n-2} \bar{x}_{n-2} + a_{n-1} \bar{x}_{n-1} + \bar{b}_{n-1} \bar{x}_n, \quad n \geqq 3;$$

$$\bar{x}_n = G_n(\bar{l})\bar{x}_1, \ n \geqq 1; \qquad x_n = \overline{G_n(\bar{l})}\, x_1, \ n \geqq 1;$$

if we assign to x_1 an arbitrary complex value ξ, the further condition on the sequence $\{x_n\}$ can be written as

$$0 < |\xi|^2 \sum_{\alpha=1}^{\infty} |G_\alpha(\bar{l})|^2 < +\infty.$$

It is now clear that l is a characteristic value for T^* if and only if the series $\sum_{\alpha=1}^{\infty} |G_\alpha(\bar{l})|^2$ is convergent, and that all the characteristic elements corresponding to a characteristic value l are given by the expression $g = \xi \sum_{\alpha=1}^{\infty} \overline{G_\alpha(\bar{l})\, g_\alpha}$, where ξ is an arbitrary complex number different from zero. From Theorem 9.8 we see that the series $\sum_{\alpha=1}^{\infty} |G_\alpha(\bar{l})|^2$ must converge for all not-real l or must diverge for all not-real l, its behavior throughout the half-planes $\mathfrak{J}(l) > 0$, $\mathfrak{J}(l) < 0$ being determined by its behavior at a single point in one of them. When this series diverges for not-real l, the transformation T has the deficiency-index (0, 0) and must therefore be self-adjoint; the identity $T^* \equiv T$ shows that the series $\sum_{\alpha=1}^{\infty} |G_\alpha(\bar{l})|^2$ converges if and only if l belongs to the point-spectrum of T. When the series converges for all not-real l, the inequality of Theorem 10.24 (7) shows that it converges also for all real l. Later we shall prove that the series, known to be convergent over the whole l-plane in this case, converges uniformly over any bounded closed set. Now we can conclude from the facts established above that the convergence of the series $\sum_{\alpha=1}^{\infty} |G_\alpha(\bar{l})|^2$ for a single not-real l implies that every l is a characteristic value of multiplicity one for the transformation T^*; the deficiency-index of T is seen therefore to be (1, 1) in this case. Since $G_n(l)$ is a polynomial with real roots, we may replace $|G_n(\bar{l})|$ by $|G_n(l)|$ in the series $\sum_{\alpha=1}^{\infty} |G_\alpha(\bar{l})|^2$.

If the transformation X_l and the corresponding function $\varrho(f, g; \lambda)$ are taken as specified in the statement of the theorem, we can deduce their chief properties from the general theorems given in Chapter IX, § 3. According to Theorem 9.17 (3), $\varrho(g_1, g_1; \lambda)$ is a real monotone-increasing function of λ in the usual normal form. From Theorem 9.19, we have

$$\int_{-\infty}^{+\infty} G_m(\lambda)\, \overline{G_n(\lambda)}\, d\varrho(\lambda)$$
$$= (G_m(T) g_1, G_n(T) g_1) = (g_m, g_n) = \delta_{mn},$$

$$\int_{-\infty}^{+\infty} \lambda\, G_n(\lambda)\, \overline{G_m(\lambda)}\, d\varrho(\lambda)$$
$$= (T G_n(T) g_1, G_m(T) g_1) = (T g_n, g_m) = a_{mn}.$$

In particular, we observe that

$$\varrho(+\infty) - \varrho(-\infty) = \int_{-\infty}^{+\infty} d\varrho(\lambda) = 1$$

and hence that $\varrho(+\infty) = 1$ because of the equation $\varrho(-\infty) = 0$. It is easy to see that $\varrho(\lambda)$ must assume infinitely many distinct values. If we suppose that it has exactly $p+1$ values and hence p points of discontinuity, we obtain a contradiction at once: there exist constants $C_1, \cdots, C_p$ such that the equation $G_{p+1}(l) = \sum_{\alpha=1}^{p} C_\alpha G_\alpha(l)$ holds at each of the points of discontinuity of $\varrho(\lambda)$; we therefore have

$$\int_{-\infty}^{+\infty} |G_{p+1}(\lambda)|^2\, d\varrho(\lambda) = \sum_{\alpha=1}^{p} C_\alpha \int_{-\infty}^{+\infty} G_\alpha(\lambda)\, \overline{G_{p+1}(\lambda)}\, d\varrho(\lambda) = 0$$

in contradiction with the result just established above. We proceed now to the proof of the remaining formulas stated in (2). We start from the known equation

$$(X_l g_1, g_1) = \int_{-\infty}^{+\infty} \frac{1}{\lambda - l}\, d\varrho(\lambda),$$

valid for all not-real l, and make use of a double induction to obtain the desired general result. If we set $b_0 = 1$, $G_0(l) = 0$, $g_0 = 0$, we have

$$g_n = G_n(T) g_1, \qquad n \geqq 0$$

and

$$g_{n+1} = G_{n+1}(T) g_1 = \left(\frac{(T - a_n) G_n(T) - b_{n-1} G_{n-1}(T)}{\overline{b}_n} \right) g_1$$
$$= ((T - a_n) g_n - b_{n-1} g_{n-1}) / \overline{b}_n, \qquad n \geqq 1.$$

Hence we can apply the equation $X_l(T - lI) = I$, which holds in the domain of the symmetric transformation T according to Theorem 9.19 (3), to write

$$(X_l g_{m+1}, g_n) \\ = ((g_m, g_n) + (l - a_m)(X_l g_m, g_n) - b_{m-1}(X_l g_{m-1}, g_n))/\bar{b}_m$$

for $m \geqq 1$ and $n \geqq 1$. Assuming that

$$(X_l g_k, g_n) = \int_{-\infty}^{+\infty} \frac{G_k(\lambda)\,\overline{G_n(\lambda)}}{\lambda - l}\, d\varrho(\lambda)$$

for all not-real l, for all $n \geqq 1$, and for $k = 1, \cdots, m$, we can show with the help of the formula just proved that

$$(X_l g_{m+1}, g_n) \\ = \int_{-\infty}^{+\infty} \left(G_m(\lambda)\,\overline{G_n(\lambda)} + (l - a_m) \frac{G_m(\lambda)\,\overline{G_n(\lambda)}}{\lambda - l} \right. \\ \left. - b_{m-1} \frac{G_{m-1}(\lambda)\,\overline{G_n(\lambda)}}{\lambda - l} \right) d\varrho(\lambda)/\bar{b}_m \\ = \int_{-\infty}^{+\infty} \frac{G_{m+1}(\lambda)\,\overline{G_n(\lambda)}}{\lambda - l}\, d\varrho(\lambda)$$

for all not-real l and for all $n \geqq 1$. Thus we find that the desired formula holds for all not-real l, for all $n \geqq 1$, and for all $m \geqq 1$, if it holds for all not-real l, for all $n \geqq 1$, and for $m = 1$. Since we know by Theorem 9.19 (1) that

$$(X_{\bar{l}} g_n, g_1) = (g_n, X_l g_1) = \overline{(X_l g_1, g_n)},$$

we see that the equations

$$(X_l g_1, g_n) = \int_{-\infty}^{+\infty} \frac{\overline{G_n(\lambda)}}{\lambda - l}\, d\varrho(\lambda)$$

and

$$(X_{\bar{l}} g_n, g_1) = \int_{-\infty}^{+\infty} \frac{G_n(\lambda)}{\lambda - \bar{l}}\, d\varrho(\lambda)$$

are equivalent. The inductive argument just sketched shows that the latter equation is true for all not-real l and for $n \geqq 1$ if it is true for all not-real l and for $n = 1$. Since the case $n = 1$ has already been settled, we see that the desired formula is true in all cases.

As we have already pointed out in Theorem 9.19, the function $(X_l g_1, g_1)$ is represented in the sectors described above by the asymptotic series $-\sum_{\alpha=0}^{\infty}(T^{\alpha}g_1, g_1)/l^{\alpha+1}$. We can write

$$(T^n g_1, g_1) = \int_{-\infty}^{+\infty} \lambda^n \, d\varrho(g_1, g_1; \lambda) = \int_{-\infty}^{+\infty} \lambda^n \, d\varrho(\lambda)$$

and then prove by means of the formulas

$$\begin{aligned}\int_{-\infty}^{+\infty} \lambda^n \, d\varrho(\lambda) &= \int_{-\infty}^{+\infty} \lambda \cdot \lambda^j \cdot \bar{\lambda}^k \, d\varrho(\lambda) \\ &= \sum_{\alpha=1}^{j+1}\sum_{\beta=1}^{k+1} C_{j\alpha}\bar{C}_{k\beta}\int_{-\infty}^{+\infty} \lambda G_\alpha(\lambda)\overline{G_\beta(\lambda)}\, d\varrho(\lambda) \\ &= \sum_{\alpha=1}^{j+1}\sum_{\beta=1}^{k+1} C_{j\alpha}\bar{C}_{k\beta}\, a_{\beta\alpha}\end{aligned}$$

that $\int_{-\infty}^{+\infty} \lambda^n \, d\varrho(\lambda)$ is equal to the constant c_n defined by the Jacobi matrix A.

The equivalence of the relations $X_l^{(1)} \equiv X_l^{(2)}$, $\varrho^{(1)}(\lambda) \equiv \varrho^{(2)}(\lambda)$, in the sense made precise in the statement of the theorem, is easy to establish. If $X_l^{(1)} \equiv X_l^{(2)}$ for all not-real l. we have

$$I(l; \varrho^{(1)}) = (X_l^{(1)} g_1, g_1) = (X_l^{(2)} g_1, g_1) = I(l; \varrho^{(2)})$$

for all not-real l and can conclude from Lemma 5.2 that $\varrho^{(1)}(\lambda) \equiv \varrho^{(2)}(\lambda)$. If $\varrho^{(1)}(\lambda) \equiv \varrho^{(2)}(\lambda)$, we have

$$\begin{aligned}(X_l^{(1)} g_m, g_n) &= \int_{-\infty}^{+\infty} \frac{G_m(\lambda)\overline{G_n(\lambda)}}{\lambda - l}\, d\varrho^{(1)}(\lambda) \\ &= \int_{-\infty}^{+\infty} \frac{G_m(\lambda)\overline{G_n(\lambda)}}{\lambda - l}\, d\varrho^{(2)}(\lambda) = (X_l^{(2)} g_m, g_n)\end{aligned}$$

for all not-real l, for all $m \geqq 1$, and for all $n \geqq 1$, by virtue of (2); since $\{g_n\}$ is a complete orthonormal set and since both $X_l^{(1)}$ and $X_l^{(2)}$ are bounded linear transformations with domain $\mathfrak{H}$, we conclude that $X_l^{(1)} \equiv X_l^{(2)}$ for all not-real l.

When T has the deficiency-index (0, 0), Theorem 9.3 shows that it is a self-adjoint transformation and Theorem 7.14 that it has a simple spectrum. According to Theorem 9.20, we have $X_l \equiv R_l$, where R_l is the resolvent of T, and

$\varrho(\lambda) = (E(\lambda)g_1, g_1)$, where $E(\lambda)$ is the resolution of the identity corresponding to T.

When T has the deficiency-index $(1, 1)$, it has $\mathfrak{c}$ maximal symmetric extensions, each of which must be self-adjoint, by virtue of Theorem 9.3. According to Theorem 9.14 every self-adjoint extension of T is real with respect to the conjugation J. If H is a self-adjoint extension of T, the closed linear manifold $\mathfrak{M}(g_1)$ determined by the set of elements $E(\lambda)g_1$, $-\infty < \lambda < +\infty$, where $E(\lambda)$ is the resolution of the identity corresponding to H, contains the elements

$$G_n(H)\,g_1 = G_n(T)\,g_1 = g_n, \qquad n \geqq 1,$$

and therefore coincides with $\mathfrak{H}$. The relation $\mathfrak{M}(g_1) = \mathfrak{H}$ implies that H has a simple spectrum, in accordance with Theorem 7.9.

We shall now study in greater detail the effective methods of constructing all the transformations X_l and all the associated functions $\varrho(\lambda) = \varrho(g_1, g_1; \lambda)$ defined by a given Jacobi matrix A. Since X_l is determined by $\varrho(\lambda)$ and conversely, we have to describe and justify methods of approximation which will yield all possible functions $\varrho(\lambda)$. The results which are given in the four following theorems are sufficient for this purpose; they are due to Carleman.* Subsequently we shall be able to obtain more precise information concerning the case where T has the deficiency-index $(1, 1)$.

THEOREM 10.28. *Let $A = \{a_{mn}\}$ be a Jacobi matrix and $A^{(k)} = \{a_{mn}^{(k)}\}$ a reduced Jacobi matrix of order $p(k)$, $k = 1, 2, 3, \cdots$, such that $\lim_{k\to\infty} p(k) = \infty$ and $\lim_{k\to\infty} a_{mn}^{(k)} = a_{mn}$ for $m, n = 1, 2, 3, \cdots$. Let T be the closed linear symmetric transformation associated with A by the complete orthonormal set $\{g_n\}$ in the manner indicated in Theorem* 10.27; *and let $H^{(k)}$ be the self-adjoint transformation associated with $A^{(k)}$ by the same orthonormal set in the manner indicated in Theorem* 10.23. *Then the sequence $\{H^{(k)}\}$ is an approximating sequence of self-*

---

* Carleman, *Les équations intégrales singulières à noyau réel et symétrique*, Uppsala, 1923, pp. 189–220.

adjoint transformations for the closed linear symmetric transformation T. If $R_l^{(k)}$ is the resolvent of $H^{(k)}$, $E^{(k)}(\lambda)$ the resolution of the identity corresponding to $H^{(k)}$, and $\varrho^{(k)}(\lambda)$ the function $(E^{(k)}(\lambda)g_1, g_1)$; if X_l is a transformation of the type described in Theorem 10.27 and $\varrho(\lambda) = \varrho(g_1, g_1; \lambda)$ is the associated function; and if there exists a sequence of integers $\{k(j)\}$ such that $\lim_{j\to\infty} k(j) = \infty$ and $\lim_{j\to\infty}(R_l^{(k(j))}g_1, g_1) = (X_l g_1, g_1)$ for all not-real l;—then $\lim_{j\to\infty}\varrho^{(k(j))}(\lambda)$ exists and is equal to $\varrho(\lambda)$ except possibly at the points of discontinuity of the latter function. When T has the deficiency-index $(0, 0)$, X_l and $\varrho(\lambda)$ are uniquely determined and can be calculated by means of the relations

$$\lim_{k\to\infty}(R_l^{(k)}g_1, g_1) = (X_l g_1, g_1), \qquad \lim_{k\to\infty}\varrho^{(k)}(\lambda) = \varrho(\lambda);$$

and, furthermore, the sequence $\{A^{(k)}\}$ can be so specialized that $A^{(k)}$ is the kth reduced matrix of the Jacobi matrix A.

Once we have proved that $\{H^{(k)}\}$ is an approximating sequence for T in the sense of Definition 9.9, the other assertions of the theorem are consequences or obvious analogues of assertions proved in Theorems 9.20 and 9.22. In order to show that the sequence $\{H^{(k)}\}$ has the indicated property, we have only to show that $H^{(k)}g_n \to Tg_n$ as $k\to\infty$, for all $n \geqq 1$. Since all the elements of the matrices A and $A^{(k)}$ save those along or adjacent to the principal diagonal are equal to zero, we have

$$\begin{aligned} H^{(k)}g_1 &= \sum_{\alpha=1}^{\infty} a_{\alpha,1}^{(k)} g_\alpha = a_{11}^{(k)} g_1 + a_{21}^{(k)} g_2 \to a_{11} g_1 + a_{21} g_2 \\ &= \sum_{\alpha=1}^{\infty} a_{\alpha,1} g_\alpha = Tg_1, \\ H^{(k)}g_n &= \sum_{\alpha=1}^{\infty} a_{\alpha,n}^{(k)} g_\alpha = \sum_{\alpha=n-1}^{n+1} a_{\alpha,n}^{(k)} g_\alpha \to \sum_{\alpha=n-1}^{n+1} a_{\alpha n} g_\alpha \\ &= \sum_{\alpha=1}^{\infty} a_{\alpha n} g_\alpha = Tg_n, \end{aligned}$$

for $n \geqq 2$, as we wished to prove.

THEOREM 10.29. *Let A be a Jacobi matrix, $\{c_n\}$ the associated sequence of constants, and $\varrho(\lambda)$ an arbitrary real monotone-increasing function in the class $\mathfrak{B}^*$ such that the integral $\int_{-\infty}^{+\infty} \lambda^n \, d\varrho(\lambda)$ exists and is equal to c_n for all $n \geqq 0$. Then there exists a sequence $\{A^{(k)}\}$ of reduced Jacobi matrices of the respective orders $p(k) \leqq k$ which is related to the matrix A in the manner described in Theorem 10.28 and which has the further property that the sequence $\{\varrho^{(k)}(\lambda)\}$ of associated functions converges to $\varrho(\lambda)$ except possibly at the points of discontinuity of the latter function.*

The first step in the proof is to construct a suitable function $\varrho^{(k)}(\lambda)$ with the properties

$$\lim_{k\to\infty} \varrho^{(k)}(\lambda) = \varrho(\lambda), \qquad \lim_{k\to\infty} \int_{-\infty}^{+\infty} \lambda^n \, d\varrho^{(k)}(\lambda) = c_n.$$

We assign to $\varrho^{(1)}(\lambda)$ the value 0 or 1 according as $\lambda < 0$ or $\lambda \geqq 0$; and, for $k \geqq 2$, we introduce the numbers

$$\lambda_p^{(k)} = -k^{1/2} + (p-1)\frac{2k^{1/2}}{k-1}, \quad p = 1, \cdots, k,$$

and define $\varrho^{(k)}(\lambda)$ by means of the equations

$$\begin{aligned} \varrho^{(k)}(\lambda) &= 0, & -\infty < \lambda < \lambda_1^{(k)}, \\ \varrho^{(k)}(\lambda) &= \varrho(\lambda_{p+1}^{(k)}), & \lambda_p^{(k)} \leqq \lambda < \lambda_{p+1}^{(k)}, \quad p = 1, \cdots, k-1, \\ \varrho^{(k)}(\lambda) &= 1, & \lambda_k^{(k)} \leqq \lambda < +\infty. \end{aligned}$$

It is immediately evident that $\lim_{k\to\infty} \varrho^{(k)}(\lambda) = \varrho(\lambda)$ at all the points of continuity of $\varrho(\lambda)$. In order to show that $c_n^{(k)} = \int_{-\infty}^{+\infty} \lambda^n \, d\varrho^{(k)}(\lambda)$ tends to the limit $c_n = \int_{-\infty}^{+\infty} \lambda^n \, d\varrho(\lambda)$, we assume that $k \geqq 2$ and introduce the function $F_n^{(k)}(\lambda)$ defined by the equations

$$\begin{aligned} F_n^{(k)}(\lambda) &= (\lambda_1^{(k)})^n, & -\infty < \lambda \leqq \lambda_1^{(k)}, \\ F_n^{(k)}(\lambda) &= (\lambda_p^{(k)})^n, & \lambda_p^{(k)} < \lambda \leqq \lambda_{p+1}^{(k)}, \quad p = 1, \cdots, k-1, \\ F_n^{(k)}(\lambda) &= (\lambda_k^{(k)})^n, & \lambda_k^{(k)} < \lambda < +\infty, \end{aligned}$$

observing that $c_n^{(k)} = \int_{-\infty}^{+\infty} F_n^{(k)}(\lambda)\, d\varrho(\lambda)$. It is clear that we have also

$$\lim_{k\to\infty} F_n^{(k)}(\lambda) = \lambda^n,$$

$$|F_n^{(k)}(\lambda)| \leqq \left(|\lambda| + \frac{2k^{1/2}}{k-1}\right)^n \leqq (|\lambda| + 2^{3/2})^n$$

for $k \geqq 2$. Since $(|\lambda| + 2^{3/2})^n$ is integrable with respect to $\varrho(\lambda)$, by hypothesis, we can apply Lemma 6.1 (6) to show that

$$\lim_{k\to\infty} c_n^{(k)} = \lim_{k\to\infty} \int_{-\infty}^{+\infty} F_n^{(k)}(\lambda)\, d\varrho(\lambda) = \int_{-\infty}^{+\infty} \lambda^n\, d\varrho(\lambda) = c_n,$$

for $n \geqq 0$. The function $\varrho^{(k)}(\lambda)$ assumes $p(k)+1$ distinct values and has $p(k)$ points of discontinuity, where $p(k) \leqq k$ and $\lim_{k\to\infty} p(k) = \infty$.

The second step is to determine a reduced Jacobi matrix $A^{(k)}$ of order $p = p(k)$ so that the monotone-increasing function associated with it in the manner described in Theorem 10.23 coincides with the function $\varrho^{(k)}(\lambda)$ just defined. We begin by defining a symmetric transformation $T^{(k)}$ in the p-dimensional unitary space $\mathfrak{L}_2(\varrho^{(k)})$ by means of the relation $T^{(k)}F(\lambda) = \lambda F(\lambda)$, where $F(\lambda)$ is an arbitrary function in $\mathfrak{L}_2(\varrho^{(k)})$. This transformation can obviously be studied by methods analogous to those used in the proofs of Theorems 7.12 and 7.13. We determine an orthonormal set $\{G_n^{(k)}(\lambda)\}$, $n = 1, \cdots, p(k)$, in $\mathfrak{L}_2(\varrho^{(k)})$ by the requirement that $G_n^{(k)}(\lambda)$ be a polynomial of degree $n-1$. This set can be constructed by applying the process of Theorem 1.13 to the set of functions $1, \lambda, \cdots, \lambda^{p-2}, \lambda^{p-1}$, which are obviously linearly independent elements of $\mathfrak{L}_2(\varrho^{(k)})$ by virtue of the fact that no polynomial of degree less than p can vanish at the p points of discontinuity of $\varrho^{(k)}(\lambda)$ without vanishing identically. This process determines the polynomial $G_n^{(k)}(\lambda)$ uniquely save for a multiplicative constant of absolute value 1, which can be assigned arbitrarily. Since two functions in $\mathfrak{L}_2(\varrho^{(k)})$ are equivalent if and only if they coincide at the p points of discontinuity of $\varrho^{(k)}(\lambda)$, every function in $\mathfrak{L}_2(\varrho^{(k)})$ is equivalent to a polynomial of degree less than p and can therefore be expressed as a linear combination of the polynomials $G_n^{(k)}(\lambda)$. Hence the transformation $T^{(k)}$ is completely determined by the matrix $\{a_{mn}^{(k)}\}$, $m, n = 1, \cdots, p$, where

$$a_{mn}^{(k)} = \int_{-\infty}^{+\infty} \lambda G_n^{(k)}(\lambda) \overline{G_m^{(k)}(\lambda)} \, d\varrho^{(k)}(\lambda).$$

It is easily verified that this matrix is Hermitian symmetric, that all its elements except those along or adjacent to the principal diagonal vanish, and that none of the elements adjacent to the principal diagonal has the value zero. We may now define the reduced Jacobi matrix $A^{(k)} = \{a_{mn}^{(k)}\}$, by assigning to $a_{mn}^{(k)}$ the value indicated above when $m, n = 1, \cdots, p$ and the value zero when $m > p$ or $n > p$. We denote by $H^{(k)}$ the self-adjoint transformation associated with $A^{(k)}$ by the complete orthonormal set $\{g_n\}$ and by $\mathfrak{M}^{(k)}(g_1)$ the closed linear manifold determined by the set of all elements $E^{(k)}(\lambda) g_1$, $-\infty < \lambda < +\infty$, where $E^{(k)}(\lambda)$ is the resolution of the identity associated with $H^{(k)}$. As we showed in the proof of Theorem 10.23, $\mathfrak{M}^{(k)}(g_1)$ is the closed linear manifold determined by the set of elements $g_1, \cdots, g_p$. Furthermore, the transformation $H^{(k)}$ is completely determined in $\mathfrak{M}^{(k)}(g_1)$ by the matrix $\{(H^{(k)} g_n, g_m)\} = \{a_{mn}^{(k)}\}$, $m, n = 1, \cdots, p$. Hence if we set up an isomorphism between $\mathfrak{L}_2(\varrho^{(k)})$ and $\mathfrak{M}^{(k)}(g_1)$ by putting $G_n^{(k)}(\lambda)$ and g_n, $1 \leqq n \leqq p$, in correspondence, the image of $T^{(k)}$ in $\mathfrak{M}^{(k)}(g_1)$ coincides there with $H^{(k)}$. We conclude that the image of the resolvent $R_l^{(k)}$ of $H^{(k)}$ in the passage from $\mathfrak{M}^{(k)}(g_1)$ to $\mathfrak{L}_2(\varrho^{(k)})$ is the transformation which takes $F(\lambda)$ into $\dfrac{1}{\lambda - l} F(\lambda)$. This result shows that we can write

$$\begin{aligned} \int_{-\infty}^{+\infty} \frac{1}{\lambda - l} \, d(E^{(k)}(\lambda) g_1, g_1) &= (R_l^{(k)} g_1, g_1) \\ &= \int_{-\infty}^{+\infty} \frac{1}{\lambda - l} G_1^{(k)}(\lambda) \overline{G_1^{(k)}(\lambda)} \, d\varrho^{(k)}(\lambda) \\ &= \int_{-\infty}^{+\infty} \frac{1}{\lambda - l} \, d\varrho^{(k)}(\lambda). \end{aligned}$$

Lemma 5.2 shows that $\varrho^{(k)}(\lambda) = (E^{(k)}(\lambda) g_1, g_1)$, as we wished to prove. Thus the matrix $A^{(k)}$ which we have constructed is related to the function $\varrho^{(k)}(\lambda)$ in the desired manner.

The third and final step is to show that the sequence $\{A^{(k)}\}$ is connected with the given Jacobi matrix A by the relations $\lim_{k\to\infty} a_{mn}^{(k)} = a_{mn}$, for $m \geqq 1$ and $n \geqq 1$. In order to secure this property we must take advantage of a remark made in the preceding paragraph, to the effect that the polynomial $G_n^{(k)}(\lambda)$ defined there is determinate save for a multiplicative constant of absolute value 1. We shall, in fact, assign a suitable value to the constant in question. According to Theorem 10.24, we have

$$G_n^{(k)}(\lambda)$$
$$= (-1)^{n-1} \gamma_n^{(k)} \det_{p,q=0,\cdots,n-2} \{c_{p+q+1}^{(k)} - \lambda\, c_{p+q}^{(k)}\} / (D_{n-2}^{(k)})^{1/2} (D_{n-1}^{(k)})^{1/2},$$

for $n = 1, \cdots, p(k)$, where

$$c_n^{(k)} = \int_{-\infty}^{+\infty} \lambda^n \, d\varrho^{(k)}(\lambda), \quad D_n^{(k)} = \det_{p,q=0,\cdots,n-1} \{c_{p+q}^{(k)}\},$$

and $\gamma_n^{(k)}$ is a complex constant of absolute value 1. It is therefore clear that the only indeterminacy in the polynomial $G_n^{(k)}(\lambda)$ is due to the constant $\gamma_n^{(k)}$. If $\{G_n(\lambda)\}$ is the sequence of polynomials associated with the Jacobi matrix A, we have similarly

$$G_n(\lambda) = (-1)^{n-1} \gamma_n \det_{p,q=0,\cdots,n-2} \{c_{p+q+1} - \lambda\, c_{p+q}\} / D_{n-2}^{1/2} D_{n-1}^{1/2},$$

for $n \geqq 1$, where

$$c_n = \int_{-\infty}^{+\infty} \lambda^n \, d\varrho(\lambda), \qquad D_n = \det_{p,q=0,\cdots,n-1} \{c_{p+q}\},$$

and γ_n is a complex constant of absolute value 1 determined by the matrix A. We require that $\gamma_n^{(k)} = \gamma_n$ for $n = 1, \cdots, p(k)$ and $k = 1, 2, 3, \cdots$. If we now substitute these expressions for $G_n^{(k)}(\lambda)$ and $G_n(\lambda)$ in the formulas

$$a_{mn}^{(k)} = \int_{-\infty}^{+\infty} \lambda\, G_n^{(k)}(\lambda) \overline{G_m^{(k)}(\lambda)} \, d\varrho^{(k)}(\lambda),$$
$$a_{mn} = \int_{-\infty}^{+\infty} \lambda\, G_n(\lambda) \overline{G_m(\lambda)} \, d\varrho(\lambda),$$

for fixed m and n and $p(k) \geqq \max[m, n]$, it is easy to prove with the help of the relation $\lim_{k\to\infty} c_n^{(k)} = c_n$ that $\lim_{k\to\infty} a_{mn}^{(k)} = a_{mn}$. Thus we see that the sequence $\{A^{(k)}\}$ is related to the matrix A in the desired manner.

We may now combine the preceding results in an obvious manner to obtain the following characterization of the transformations X_l and the associated functions $\varrho(\lambda)$ defined by a given Jacobi matrix A. The evaluation of the cardinal number of the class of associated functions is actually carried out in the proof of Theorem 10.31.

THEOREM 10.30. *Let A be a Jacobi matrix and $\{c_n\}$ the associated sequence of constants. Then the totality of functions $\varrho(\lambda) = \varrho(g_1, g_1; \lambda)$ associated with the matrix A in the manner indicated in Theorem* 10.27 *coincides with the totality of real monotone-increasing functions $\varrho(\lambda)$ in the class $\mathfrak{B}^*$ with the property that the integral $\int_{-\infty}^{+\infty} \lambda^n d\varrho(\lambda)$ exists and is equal to c_n for all $n \geqq 0$. In case the transformation T associated with the matrix A by the complete orthonormal set $\{g_n\}$ has the deficiency-index* (0,0), *there is just one such function; but if T has the deficiency-index* (1,1) *there are $\mathfrak{c}$ such functions. The transformation X_l corresponding to a particular function $\varrho(\lambda)$ is determined by the relations*

$$(X_l g_m, g_n) = \int_{-\infty}^{+\infty} \frac{G_m(\lambda)\overline{G_n(\lambda)}}{\lambda - l}\, d\varrho(\lambda)$$

given in Theorem 10.27 (2).

We may present the facts just established in a somewhat different light by indicating their relation to the theory of infinite continued fractions. From a given Jacobi matrix A we can form the infinite continued fraction

$$\frac{|b_0|^2}{|a_1 - l} - \frac{|b_1|^2}{|a_2 - l} - \cdots - \frac{|b_n|^2}{|a_{n+1} - l} - \cdots$$

where $b_0 = 1$; and, conversely, any such continued fraction determines infinitely many Jacobi matrices with which it may be associated. The meaning of this formal expression is to be sought in the behavior as $k \to \infty$ of the approximant

$$F_k(l) = F_k(l, 0)$$
$$= \frac{|b_0|^2|}{|a_1 - l} - \frac{|b_1|^2|}{|a_2 - l} - \cdots - \frac{|b_{k-2}|^2|}{|a_{k-1} - l} - \frac{|b_{k-1}|^2|}{|a_k - l}$$

or of the generalized approximant

$$F_k^*(l)$$
$$= \frac{|b_0^{(k)}|^2}{|a_1^{(k)} - l} - \frac{|b_2^{(k)}|^2|}{|a_2^{(k)} - l} - \cdots - \frac{|b_{k-2}^{(k)}|^2|}{|a_{k-1}^{(k)} - l} - \frac{|b_{k-1}^{(k)}|^2|}{|a_k^{(k)} - l},$$

where $b_0^{(k)} = 1$, $\lim_{k\to\infty} a_n^{(k)} = a_n$, and $\lim_{k\to\infty} b_n^{(k)} = b_n$. We have

THEOREM 10.31. *Let A be a Jacobi matrix, T the closed linear symmetric transformation associated with A by the complete orthonormal set $\{g_n\}$, and $\{c_n\}$ the sequence of constants associated with A. A necessary and sufficient condition that there exist a sequence of integers $\{k(j)\}$ and a sequence $\{F_k^*(l)\}$ of generalized approximants, related to A and the associated infinite continued fraction in the manner just described, with the properties $\lim_{j\to\infty} k(j) = \infty$ and $\lim_{j\to\infty} F_{k(j)}^*(l) = F(l)$ for all not-real l, is that the function $F(l)$ be expressible in the form*

$$F(l) = I(l; \varrho) = \int_{-\infty}^{+\infty} \frac{1}{\lambda - l} d\varrho(\lambda)$$

where $\varrho(\lambda)$ is an arbitrary real monotone-increasing function in the class $\mathfrak{B}^*$ for which the relation $\int_{-\infty}^{+\infty} \lambda^n d\varrho(\lambda) = c_n$ is valid. A necessary and sufficient condition that all possible sequences $\{F_k^*(l)\}$ of generalized approximants converge and have a common limit is that T have the deficiency-index $(0, 0)$.

To a given sequence of generalized approximants associated with the matrix A, we order a sequence of reduced Jacobi matrices $\{A^{(k)}\}$ defined by the equations

$$a_{nn}^{(k)} = a_n^{(k)}, \quad a_{n,n+1}^{(k)} = b_n^{(k)}, \quad a_{mn}^{(k)} = 0,$$
$$|m - n| > 1, \; m, n = 1, \cdots, k;$$
$$a_{mn}^{(k)} = 0, \qquad m > k \text{ or } n > k.$$

It is evident that the sequence $\{A^{(k)}\}$ is related to A in the manner described in Theorem 10.28. On the other hand, any such sequence determines a sequence of generalized approximants for the infinite continued fraction formed from the matrix A, as we see by reference to Theorem 10.23 (4). If $H^{(k)}$ is the self-adjoint transformation associated with the matrix $A^{(k)}$ by the complete orthonormal set $\{g_n\}$ and $R_l^{(k)}$ is its resolvent, the sequence of generalized approximants is specified by the formula $F_k^*(l) = (R_l^{(k)} g_1, g_1)$. According to Theorems 10.28, 9.17, and 10.27, we can choose a sequence of integers $\{k(j)\}$ such that

$$\lim_{j\to\infty} k(j) = \infty, \qquad \lim_{j\to\infty} F_{k(j)}^*(l) = \int_{-\infty}^{+\infty} \frac{1}{\lambda - l}\, d\varrho(\lambda),$$

where $\varrho(\lambda)$ is a real monotone-increasing function in $\mathfrak{B}^*$ with the property that $\int_{-\infty}^{+\infty} \lambda^n\, d\varrho(\lambda) = c_n$. According to Theorems 10.29 and 10.30, we may adjust the sequence $\{F_k^*(l)\}$ so as to obtain an arbitrary function $\varrho(\lambda)$ satisfying the indicated conditions. When T has the deficiency-index (0, 0), every sequence of generalized approximants converges to the limit $(R_l g_1, g_1)$, where R_l is the resolvent of the self-adjoint transformation T, as we have shown in Theorem 9.20. When T has the deficiency-index (1, 1), it has $\mathfrak{c}$ distinct self-adjoint extensions. Since the transformation X_l described in Theorem 10.27 can be taken as the resolvent of any one of these self-adjoint extensions, it is clear that there are at least $\mathfrak{c}$ distinct functions $\varrho(\lambda)$ satisfying the conditions of the theorem and hence that there are sequences of generalized approximants which converge to distinct limits. Since the class $\mathfrak{B}^*$ has the cardinal number $\mathfrak{c}$, there are precisely $\mathfrak{c}$ such functions. The final assertion of the theorem is obvious.

In the case where the transformation T associated with the Jacobi matrix A has the deficiency-index (1, 1), Theorems 10.28–10.31 leave much to be desired because they fail to indicate a unique method for determining a given associated transformation X_l and the corresponding function

$\varrho(\lambda)$. We shall therefore investigate this case in greater detail with a view to obtaining the theory developed by R. Nevanlinna.*

We shall commence by applying Theorem 9.18 and Theorem 9.19, (1), (2), and (3), to the transformations T and X_l associated with a Jacobi matrix A.

THEOREM 10.32. *Let A be a Jacobi matrix, $\{G_n(l)\}$ the associated sequence of polynomials, T and X_l the transformations associated with A by the complete orthonormal set $\{g_n\}$, and $\varrho(\lambda) = \varrho(g_1, g_1; \lambda)$ the function corresponding to X_l. Then the values assumed by the expression*

$$z = (X_l g_1, g_1) = \int_{-\infty}^{+\infty} \frac{1}{\lambda - l}\, d\varrho(\lambda), \qquad \mathfrak{J}(l) \neq 0,$$

constitute a set $C(l)$ which consists of the interior and circumference of a circle of radius $1/2\,|\mathfrak{J}(l)|\sum_{\alpha=1}^{\infty}|G_\alpha(l)|^2$ lying on the half-plane $\mathfrak{J}(l)\,\mathfrak{J}(z) > 0$, and which varies continuously with l. The radius of $C(l)$ is positive or zero according as the transformation T has the deficiency-index $(1, 1)$ or $(0, 0)$.

The relation $\mathfrak{J}\left(\int_{-\infty}^{+\infty} \frac{1}{\lambda - l}\, d\varrho(\lambda)\right) = \mathfrak{J}(l) \int_{-\infty}^{+\infty} \frac{1}{|\lambda - l|^2}\, d\varrho(\lambda)$ shows at once that the set $C(l)$ must lie on the half-plane $\mathfrak{J}(l)\,\mathfrak{J}(z) > 0$.

When T has the deficiency-index $(0, 0)$, the set $C(l)$ reduces to a single point since the transformation X_l and the function $\varrho(\lambda)$ are unique. By virtue of the fact that the series $\sum_{\alpha=1}^{\infty}|G_\alpha(l)|^2$ is divergent in this case, according to Theorem 10.27, we may state the facts in the form given above.

When T has the deficiency-index $(1, 1)$, we may construct all its maximal symmetric extensions by means of the theory developed in Chapter IX. It is convenient for our present purposes to work with a general not-real value l rather than with the particular value $l = i$, as we did in Theorems 9.1–9.3. This can be accomplished by means of the

* R. Nevanlinna, Annales Academiae Scientiarum Fennicae, (A) 18 (1922), no. 5; (A) 32 (1929), no. 7.

device used in the proof of Theorem 9.8: we consider the transformation $aT+bI$, where $-\frac{b}{a}=\Re(l)$ and $\frac{1}{a}=\Im(l)$, with the adjoint aT^*+bI in place of the transformation T with the adjoint T^*. We thus find that every maximal symmetric extension coincides with some one of the self-adjoint transformations $H(l, \theta) \supseteqq T$, $0 \leqq \theta < 2\pi$, defined as follows: the domain $\mathfrak{D}(l, \theta)$ of $H(l, \theta)$ comprises those and only those elements which are expressible in the form $f+c(\varphi_l - e^{i\theta}\varphi_{\bar{l}})$, where f is an element in the domain of T, c is a complex constant, and

$$\varphi_l = \sum_{\alpha=1}^{\infty} \overline{G_\alpha(\bar{l})g_\alpha} \Big/ \left(\sum_{\alpha=1}^{\infty} |G_\alpha(l)|^2\right)^{1/2}$$

and

$$\varphi_{\bar{l}} = \sum_{\alpha=1}^{\infty} \overline{G_\alpha(l)g_\alpha} \Big/ \left(\sum_{\alpha=1}^{\infty} |G_\alpha(l)|^2\right)^{1/2}$$

are normalized characteristic elements of T^* for the characteristic values l and $\bar{l}$ respectively; and $H(l, \theta)$ is defined in this domain by the equation $H(l, \theta) = T^*$. The details of the investigation will be left to the reader. We shall now compute the expression $(R_l(l, \theta)g_1, g_1)$, where $R_l(l, \theta)$ is the inverse of the transformation $H(l, \theta) - lI$. Since $R_l(l, \theta)g_1$ is in $\mathfrak{D}(l, \theta)$, we can write

$$R_l(l, \theta)g_1 = f + c(\varphi_l - e^{i\theta}\varphi_{\bar{l}})$$

where f is an element in the domain of T. We then have

$$\begin{aligned} g_1 &= (H(l, \theta) - lI)f + c(H(l, \theta) - lI)(\varphi_l - e^{i\theta}\varphi_{\bar{l}}) \\ &= (Tf - lf) + c(T^*\varphi_l - l\varphi_l) - ce^{i\theta}(T^*\varphi_{\bar{l}} - l\varphi_{\bar{l}}) \\ &= (Tf - lf) + 2ce^{i\theta}\Im(l)\varphi_{\bar{l}}. \end{aligned}$$

By virtue of the relations

$$(Tf - lf, \varphi_{\bar{l}}) = (f, T^*\varphi_{\bar{l}} - \bar{l}\varphi_{\bar{l}}) = 0, \quad (\varphi_{\bar{l}}, \varphi_{\bar{l}}) = 1,$$

we see that $(g_1, \varphi_{\bar{l}}) = 2ce^{i\theta}\Im(l)$. On the other hand, the expression given for $\varphi_{\bar{l}}$ above allows us to obtain the direct evaluation

36*

$$(g_1, \varphi_{\bar{l}}) = G_1(l) \Big/ \left(\sum_{\alpha=1}^{\infty} |G_\alpha(l)|^2 \right)^{1/2} = 1 \Big/ \left(\sum_{\alpha=1}^{\infty} |G_\alpha(l)|^2 \right)^{1/2}.$$

By comparison we find that

$$c = e^{-i\theta}/2\mathfrak{J}(l) \left(\sum_{\alpha=1}^{\infty} |G_\alpha(l)|^2 \right)^{1/2}.$$

With this value of c we now have

$$(R_l(l, \theta) g_1, g_1) = (f, g_1) + c(\varphi_l, g_1) - c e^{i\theta} (\varphi_{\bar{l}}, g_1)$$
$$= (f, g_1) + (e^{-i\theta} - 1)/2\mathfrak{J}(l) \sum_{\alpha=1}^{\infty} |G_\alpha(l)|^2.$$

Thus when θ varies from 0 to 2π, the variable $z = (R_l(l, \theta) g_1, g_1)$ describes a circle with radius $1/2|\mathfrak{J}(l)| \sum_{\alpha=1}^{\infty} |G_\alpha(l)|^2$. We shall prove that $C(l)$ consists of this circle together with its interior points. Since we may take $X_l \equiv R_l(l, \theta)$, we conclude that $C(l)$ contains the circle described. In order to show that $C(l)$ also contains the interior of this circle, we observe that $C(l)$ is a convex set. If z_1 and z_2 are points of $C(l)$, then there exist functions $\varrho^{(1)}(\lambda)$ and $\varrho^{(2)}(\lambda)$ such that

$$z_1 = \int_{-\infty}^{+\infty} \frac{1}{\lambda - l} \, d\varrho^{(1)}(\lambda), \quad z_2 = \int_{-\infty}^{+\infty} \frac{1}{\lambda - l} \, d\varrho^{(2)}(\lambda),$$
$$\int_{-\infty}^{+\infty} \lambda^n \, d\varrho^{(1)}(\lambda) = \int_{-\infty}^{+\infty} \lambda^n \, d\varrho^{(2)}(\lambda) = c_n,$$

where $\{c_n\}$ is the sequence of constants associated with the Jacobi matrix A. If α_1 and α_2 are real numbers such that $\alpha_1 \geqq 0$, $\alpha_2 \geqq 0$, $\alpha_1 + \alpha_2 = 1$, the function $\varrho(\lambda) = \alpha_1 \varrho^{(1)}(\lambda) + \alpha_2 \varrho^{(2)}(\lambda)$ is a real monotone-increasing function in the class $\mathfrak{B}^*$ with the property

$$\int_{-\infty}^{+\infty} \lambda^n \, d\varrho(\lambda) = \alpha_1 \int_{-\infty}^{+\infty} \lambda^n \, d\varrho^{(1)}(\lambda) + \alpha_2 \int_{-\infty}^{+\infty} \lambda^n \, d\varrho^{(2)}(\lambda) = c_n.$$

Thus $\varrho(\lambda)$ is a function associated with the Jacobi matrix A in the desired manner, as we proved above in Theorem 10.31; and the point $z = \alpha_1 z_1 + \alpha_2 z_2 = \int_{-\infty}^{+\infty} \frac{1}{\lambda - l} \, d\varrho(\lambda)$ belongs to $C(l)$. This result shows that $C(l)$ is a convex set, as we wished to prove. It now remains to exclude the possibility that $C(l)$ contain points outside this circle. According to Theorem 9.19, the element $X_l g_1$ is a solution of the equation $T^* \varphi - l\varphi = g_1$ belonging to $\mathfrak{C}^- + \mathfrak{C}^0$ or to $\mathfrak{C}^+ + \mathfrak{C}^0$ according as $\mathfrak{J}(l) > 0$ or $\mathfrak{J}(l) < 0$. Theorem 9.18 (4) shows that $(X_l g_1, g_1)$ is confined to the circumference and interior of a certain circle and gives the precise conditions under which it lies on the circumference. In the present instance, the second condition is automatically satisfied since $\mathfrak{N}(l)$ is a one-dimensional closed linear manifold. Hence we see that $(X_l g_1, g_1)$ is a point of the circumference if and only if the element $X_l g_1$ belongs to the set $\mathfrak{C}^0$. This condition is satisfied whenever X_l is identical with $R_l(l, \theta)$, as we see by reference to Theorem 9.6 and the paragraph preceding it. Hence the circle of Theorem 9.18 is precisely the circle described above; and $C(l)$ can contain no point exterior to this circle. By reference to Theorem 10.27, we see that the radius of $C(l)$ is always positive in the present case, which is characterized by the convergence of the infinite series $\sum_{\alpha=1}^{\infty} |G_\alpha(l)|^2$.

The fact that $C(l)$ depends continuously on l, $\mathfrak{J}(l) \neq 0$, is an obvious consequence of the inequality

$$\begin{aligned} \left| \int_{-\infty}^{+\infty} \frac{1}{\lambda - l'} d\varrho(\lambda) - \int_{-\infty}^{+\infty} \frac{1}{\lambda - l} d\varrho(\lambda) \right| \\ \leqq \int_{-\infty}^{+\infty} \frac{|l - l'|}{|\lambda - l'|\,|\lambda - l|} d\varrho(\lambda) \\ \leqq |l' - l| / |\mathfrak{J}(l')|\,|\mathfrak{J}(l)| < \varepsilon, \end{aligned}$$

which holds whenever $|l' - l| < \delta = \delta(l, \varepsilon)$. The fact that the radius of $C(l)$ is a continuous function of l is particularly useful in the next theorem.

THEOREM 10.33. *Let A be a Jacobi matrix such that the associated transformation T has the deficiency-index* (1, 1); *let $\{G_n(l)\}$ be the associated sequence of polynomials; and let $\{H_n(l)\}$ be the sequence of polynomials defined in Theorem* 10.25.

Then the series $\sum_{\alpha=1}^{\infty} |G_\alpha(l)|^2$ *and* $\sum_{\alpha=2}^{\infty} |H_\alpha(l)|^2$ *converge for all* l *and converge uniformly on any bounded closed point set in the* l*-plane. If* $\{a_n\}$ *is any sequence of constants such that the series* $\sum_{\alpha=1}^{\infty} |a_\alpha|^2$ *is convergent, the series* $\sum_{\alpha=1}^{\infty} a_\alpha G_\alpha(l)$ *and* $\sum_{\alpha=2}^{\infty} a_\alpha H_\alpha(l)$ *converge for all* l *and converge uniformly on any bounded closed point set in the* l*-plane; their sums are entire functions of the complex variable* l. *In particular, the polynomials* $P_n(l)$, $Q_n(l)$, $U_n(l)$, $V_n(l)$ *defined in Theorem* 10.26 *converge for all* l, *when* n *becomes infinite, to entire functions* $P(l)$, $Q(l)$, $U(l)$, $V(l)$ *respectively, the convergence being uniform on any bounded closed set in the* l*-plane.*

The convergence of the series $\sum_{\alpha=1}^{\infty} |G_\alpha(l)|^2$ has already been established in Theorem 10.27. Since this is a series of positive terms, we have

$$\sum_{\alpha=1}^{n} |G_\alpha(l)|^2 \leqq \sum_{\alpha=1}^{m} |G_\alpha(l)|^2$$

for $m \geqq n$. According to the preceding theorem, the series $\sum_{\alpha=1}^{\infty} |G_\alpha(l)|^2$ has as its sum the quantity $1/2 |\mathfrak{J}(l)| r(l)$, $\mathfrak{J}(l) \neq 0$, where $r(l)$ is the radius of the circular region $C(l)$, and the sum is therefore a continuous function of l, $\mathfrak{J}(l) \neq 0$. Hence we can apply a well-known theorem* to show that the series converges uniformly on any bounded closed set at positive distance from the real axis. The inequality given in Theorem 10.24 (7) then enables us, in an obvious manner, to assert that the series converges uniformly on any bounded closed set whatsoever.

We can now show, by virtue of the inequality

$$\left| \sum_{\alpha=n}^{m} a_\alpha G_\alpha(l) \right|^2 \leqq \sum_{\alpha=n}^{m} |a_\alpha|^2 \sum_{\alpha=n}^{m} |G_\alpha(l)|^2,$$

---

* A monotone-increasing sequence of continuous functions which converges to a continuous limit, converges uniformly on any bounded closed subset of the domain of convergence. See Carathéodory, *Vorlesungen über reelle Funktionen*, second edition, Leipzig, 1927, p. 176.

that the series $\sum_{\alpha=1}^{\infty} a_\alpha G_\alpha(l)$ converges for all l and converges uniformly on any bounded closed set. The sum of this series is evidently an entire function.

The remark made at the close of Theorem 10.25 shows that the series $\sum_{\alpha=2}^{\infty} |H_\alpha(l)|^2$ and $\sum_{\alpha=2}^{\infty} a_\alpha H_\alpha(l)$ must behave in a similar manner, since the polynomials $H_n(l)$ are associated with a certain Jacobi matrix in essentially the same way as the polynomials $G_n(l)$ are associated with the matrix A. Thus we have only to show that the series $\sum_{\alpha=2}^{\infty} |H_\alpha(l)|^2$ converges for some not-real l and we can then deduce all the results asserted above. The convergence of this series for all not-real l follows directly from the inequality given in Theorem 10.25 (4).

The application of the preceding results to the polynomials P_n, Q_n, U_n, V_n is obvious in view of the explicit formulas given in Theorem 10.26.

THEOREM 10.34. *Let* $C_n(l)$ *and* $C(l)$ *be the circular regions associated with a Jacobi matrix* A *in the manner indicated in Theorems* 10.25 *and* 10.32 *respectively; and let* $P(l)$, $Q(l)$, $U(l)$, $V(l)$ *be the entire functions defined in Theorem* 10.33 *in the case where the transformation* T *associated with* A *has the deficiency-index* (1, 1). *Then the sequence of point sets* $\{C_n(l)\}$ *has a limit set which coincides with* $C(l)$. *The transformation* $z = \dfrac{P(l) + U(l)s}{Q(l) + V(l)s}$, $\mathfrak{I}(l) \neq 0$, *is non-singular and takes the half-plane* $\mathfrak{I}(l)\mathfrak{I}(s) \geqq 0$ *into the circular region* $C(l)$.

The relation $C_{n+1}(l) \subseteq C_n(l)$ established in Theorem 10.25 shows that the sequence $\{C_n(l)\}$ has a limit set consisting of all the points common to the sets of the sequence. This limit set is therefore a circular region composed of the interior and circumference of a circle with radius $1/2\,|\mathfrak{I}(l)|\sum_{\alpha=1}^{\infty}|G_\alpha(l)|^2$ and center

$$(-1 + 2i\mathfrak{I}(l)\sum_{\alpha=1}^{\infty}\overline{G_\alpha(l)}H_\alpha(l))/2i\mathfrak{I}(l)\sum_{\alpha=1}^{\infty}|G_\alpha(l)|^2,$$

calculated by allowing n to become infinite in the expressions given for the radius and center of $C_n(l)$ in Theorem 10.25 (6). Since the radius of this circular region is the same as that of $C(l)$, we can identify the two regions by showing that the boundary of the limit set belongs to $C(l)$. The boundary of the region $C_n(l)$ is the circle described by the variable $z = F_n(l, t)$ when t describes the real axis. Hence we can determine a real sequence $\{t_n\}$ so that the sequence $\{z_n\}$ defined by the equation $z_n = F_n(l, t_n)$ converges to a prescribed point on the circumference of the limit set of the sequence $\{C_n(l)\}$. Now it is evident that $\{F_n(l, t_n)\}$ is a sequence of generalized approximants for the infinite continued fraction associated with the Jacobi matrix A: it is obtained by setting $a_n^{(n)} = a_n - t_n$, $a_k^{(n)} = a_k$, $b_k^{(n)} = b_k$, for $1 \leqq k \leqq n-1$, in the generalized approximant $F_n^*(l)$. Theorem 10.31 shows at once that there exists at least one function $\varrho(\lambda)$, associated with the Jacobi matrix A in the manner indicated in Theorem 10.27, with the property that

$$\lim_{n\to\infty} z_n = \lim_{n\to\infty} F_n(l, t_n) = \int_{-\infty}^{+\infty} \frac{1}{\lambda - l}\, d\varrho(\lambda),$$

for a fixed not-real value l. It must be observed that the sequence $\{t_n\}$ depends upon l and that the sequence $\{F_n(l', t_n)\}$ need not, so far as we have shown, converge for $l' \neq l$; but Theorem 10.31 establishes the existence of a function $\varrho(\lambda)$ and of a subsequence of $\{F_n(l', t_n)\}$ which converges for all not-real l' to the limit $\int_{-\infty}^{+\infty} \frac{1}{\lambda - l'}\, d\varrho(\lambda)$. From these facts we conclude that every point on the circumference of the limit set belongs to $C(l)$ and hence that this set coincides with $C(l)$. We note that the center of the circular region $C(l)$ can now be calculated by means of the formula given above. We remark also that the set $\lim_{n\to\infty} C_n(l)$ reduces to a single point if and only if the transformation T associated with the Jacobi matrix A has the deficiency-index (0,0). In this case, the formula for the center of $C(l)$ must evidently be interpreted as a symbolic expression for the limit of the center of $C_n(l)$ as n becomes infinite.

In case the associated transformation T has the deficiency-index (1,1), it is of interest to consider the behavior of the transformation $z = \dfrac{P_n(l) + U_n(l)s}{Q_n(l) + V_n(l)s}$ as n becomes infinite. This transformation, according to Theorem 10.26, takes the half-plane $\mathfrak{J}(l)\,\mathfrak{J}(s) \geqq 0$ into the circular region $C_n(l)$ when $\mathfrak{J}(l) \neq 0$. When n becomes infinite, the formula for the transformation tends to the limit relation $z = \dfrac{P(l) + U(l)s}{Q(l) + V(l)s}$ and the circular region $C_n(l)$ contracts to the circular region $C(l)$ with positive radius. It is easily seen that these facts imply that the transformation $z = \dfrac{P + Us}{Q + Vs}$ is a non-singular transformation which takes the half-plane $\mathfrak{J}(l)\,\mathfrak{J}(s) \geqq 0$ into the circular region $C(l)$ when $\mathfrak{J}(l) \neq 0$.

Theorems 10.32—10.34 point the way to the further study of the case where the transformation T associated with a given Jacobi matrix A has the deficiency-index (1,1). Let us consider such a matrix together with the associated functions $\varrho(\lambda)$, $P(l)$, $Q(l)$, $U(l)$, $V(l)$. We can solve the relation

$$\frac{P(l) + U(l)s}{Q(l) + V(l)s} = \int_{-\infty}^{+\infty} \frac{1}{\lambda - l}\, d\varrho(\lambda), \qquad \mathfrak{J}(l) \neq 0,$$

for s, by virtue of Theorem 10.34, obtaining

$$s = \varphi(l) = \frac{-P(l) + Q(l)\displaystyle\int_{-\infty}^{+\infty} \frac{1}{\lambda - l}\, d\varrho(\lambda)}{U(l) - V(l)\displaystyle\int_{-\infty}^{+\infty} \frac{1}{\lambda - l}\, d\varrho(\lambda)}, \quad \mathfrak{J}(l) \neq 0.$$

We shall allow the function $s = \varphi(l)$ to assume values in the extended s-plane in order to include all the cases which may arise. Since, by Theorem 10.32, the integral term assumes a value in the set $C(l)$, we infer from Theorem 10.34 that the inequality $\mathfrak{J}(l)\,\mathfrak{J}(\varphi(l)) \geqq 0$, $\mathfrak{J}(l) \neq 0$, is valid. Furthermore, we find that $\varphi(\bar{l}) = \overline{\varphi(l)}$ when $\mathfrak{J}(l) \neq 0$. To establish this equation, we first note the relation $\bar{\gamma}_n G_n(\bar{l}) = \gamma_n \overline{G_n(l)}$, which follows directly from the formula given in Theorem 10.24(4). From the remarks made in the first paragraph of the proof of Theorem 10.26, we see also that $\bar{\gamma}_n H_n(\bar{l}) = \gamma_n \overline{H_n(l)}$. Hence we must have

$$P(\bar{l}) = -1 + \bar{l} \sum_{\alpha=1}^{\infty} \overline{G_\alpha(0)} H_\alpha(\bar{l}) = -1 + \bar{l} \sum_{\alpha=1}^{\infty} G_\alpha(0) \overline{H_\alpha(l)} = \overline{P(l)},$$

$$Q(\bar{l}) = \overline{Q(l)}, \quad U(\bar{l}) = \overline{U(l)}, \quad V(\bar{l}) = \overline{V(l)}.$$

By combining these equations with the relation $I(\bar{l}; \varrho) = \overline{I(l; \varrho)}$, where $I(l; \varrho)$ denotes the integral term, we obtain the desired result. Now $\varphi(l)$ is the quotient of the functions $\varphi_1(l) = -P(l) + Q(l) I(l; \varrho)$, $\varphi_2(l) = U(l) - V(l) I(l; \varrho)$, both single-valued and analytic for $\mathfrak{J}(l) \neq 0$. Since $\varphi_2(\bar{l}) = \overline{\varphi_2(l)}$, the latter function cannot vanish identically in one of the half-planes $\mathfrak{J}(l) > 0$, $\mathfrak{J}(l) < 0$ without vanishing identically in the other also. Hence the quotient $\varphi(l) = \varphi_1(l)/\varphi_2(l)$ must be a single-valued meromorphic function in the half-planes $\mathfrak{J}(l) > 0$, $\mathfrak{J}(l) < 0$, unless $\varphi_2(l) \equiv 0$ and $\varphi(l) = \infty$ for $\mathfrak{J}(l) \neq 0$. The inequality $\mathfrak{J}(l) \mathfrak{J}(\varphi(l)) \geqq 0$ established above shows that, when $\varphi(l)$ is meromorphic for $\mathfrak{J}(l) \neq 0$, it can have no poles. Thus the function $\varphi(l)$ is either a single-valued analytic function with the special properties noted above, for $\mathfrak{J}(l) \neq 0$, or is the singular function $\varphi(l) = \infty$. It is our aim to prove that every such function leads to a unique corresponding function $\varrho(\lambda)$ associated with the Jacobi matrix A in the requisite manner. In order to do so, we must first investigate and characterize the class of functions with which we have to deal. The necessary information is summarized in the three following theorems.

THEOREM 10.35. *A necessary and sufficient condition that a function $f(z)$, single-valued and analytic in the circle $|z| < 1$, satisfy the condition $\Re f(z) \geqq 0$ for $|z| < 1$ is that $f(z)$ be representable as an integral*

$$f(z) = \int_{-\pi}^{+\pi} \frac{e^{i\psi} + z}{e^{i\psi} - z} \, d\tau(\psi) + ic,$$

where $\tau(\psi)$ *is a real monotone-increasing function,* $-\pi \leqq \psi \leqq +\pi$, *with the properties*

$$\tau(-\pi) = 0; \quad \tau(\psi + 0) = \tau(\psi), \quad -\pi \leqq \psi < +\pi,$$

and c is a real constant, equal to $\mathfrak{I} f(0)$. *When this representation is possible, it is unique.**

A function $f(z)$ defined by an integral of the type described in the theorem is obviously single-valued and analytic in the circle $|z| < 1$. If we put $z = re^{i\theta}$, where $r = |z|$, we see at once that

$$\mathfrak{R} f(z) = \int_{-\pi}^{+\pi} \frac{1 - r^2}{1 - 2r\cos(\theta - \psi) + r^2} \, d\tau(\psi) \geqq 0$$

for $r < 1$, since the integrand is positive under this condition. Hence the integral representation is sufficient for $f(z)$ to have the indicated properties.

If $f(z)$ is a single-valued function analytic in the circle $|z| < 1$ and if $\mathfrak{R} f(z) \geqq 0$ for $|z| < 1$, we consider the function $f_n(z) = f\left(\frac{n-1}{n} z\right)$. The function $\mathfrak{R} f_n(z)$ is harmonic and not-negative in the closed region $|z| \leqq 1$ and assumes continuous not-negative boundary values $\mu_n(\theta) = \mathfrak{R} f_n(z)$ on the circumference $z = e^{i\theta}$. By Poisson's integral formula we have

$$\mathfrak{R} f_n(z) = \int_{-\pi}^{+\pi} \frac{1 - r^2}{1 - 2r\cos(\theta - \psi) + r^2} \, d\tau_n(\psi)$$

where

$$\tau_n(\psi) = \frac{1}{2\pi} \int_{-\pi}^{\psi} \mu_n(\theta) \, d\theta$$

is a continuous real monotone-increasing function for $-\pi \leqq \psi \leqq +\pi$. When $z = 0$, this formula becomes

$$\mathfrak{R} f(0) = \mathfrak{R} f_n(0) = \int_{-\pi}^{+\pi} d\tau_n(\psi) = \tau_n(\pi) - \tau_n(-\pi) = \tau_n(\pi).$$

---

* This theorem is due to Herglotz, Berichte der Sächsischen Gesellschaft der Wissenschaften zu Leipzig, Mathematisch-Physikalische Klasse 63 (1911), pp. 501–511.

Hence the sequence $\{\tau_n(\psi)\}$ satisfies the conditions of Helly's theorem. We can therefore determine a sequence of integers $\{n(k)\}$ and a real monotone-increasing function $\tau(\psi)$ such that

$$\lim_{k\to\infty} n(k) = \infty, \quad \lim_{k\to\infty} \tau_{n(k)}(\psi) = \tau(\psi), \quad 0 \leqq \tau(\psi) \leqq \mathfrak{R} f(0),$$

for $-\pi \leqq \psi \leqq +\pi$. If we allow n to become infinite through values in the sequence $\{n(k)\}$ in the formula for $\mathfrak{R} f_n(z)$ given above, we find by virtue of the relations just noted and of the further relations

$$\lim_{n\to\infty} f_n(z) = f(z), \quad \lim_{n\to\infty} \mathfrak{R} f_n(z) = \mathfrak{R} f(z), \quad |z| < 1,$$

that

$$f(z) = \int_{-\pi}^{+\pi} \frac{1-r^2}{1-2r\cos(\theta-\psi)+r^2} \, d\tau(\psi)$$

for $|z| < 1$. The function $\mathfrak{I} f(z)$, harmonic in the open region $|z| < 1$, is now determined by the fact that it is conjugate to $\mathfrak{R} f(z)$ and assumes the value $\mathfrak{I} f(0)$ for $z = 0$. Hence it is given by the formula

$$\mathfrak{I} f(z) = \int_{-\pi}^{+\pi} \frac{2r\sin(\theta-\psi)}{1-2r\cos(\theta-\psi)+r^2} \, d\tau(\psi) + \mathfrak{I} f(0).$$

By combining the expressions for $\mathfrak{R} f(z)$ and $\mathfrak{I} f(z)$, we find that

$$f(z) = \int_{-\pi}^{+\pi} \frac{e^{i\psi}+z}{e^{i\psi}-z} \, d\tau(\psi) + i\,\mathfrak{I} f(0)$$

for $|z| < 1$. The function $\tau(\psi)$ may not have all the properties demanded by the theorem; but it can be replaced by the function $\tau^*(\psi)$ defined by the equations

$$\tau^*(-\pi) = 0; \qquad \tau^*(\pi) = \tau(\pi) - \tau(-\pi);$$
$$\tau^*(\psi) = \tau(\psi+0) - \tau(-\pi+0), \qquad -\pi < \psi < +\pi,$$

a function which obviously has the desired properties.

Since we make little use of the uniqueness of the integral representation under discussion, we shall confine ourselves to a mere sketch of the proof. When $f(z)$ is represented by an integral of the indicated type, we can obtain its power series development at the origin, $f(z)=\sum_{\alpha=0}^{\infty} a_\alpha z^\alpha$, by writing the integrand as a power series and integrating term by term. We thus find that

$$a_n = 2\int_{-\pi}^{+\pi} e^{-ni\psi}\, d\tau(\psi), \qquad n \geqq 1,$$

while

$$a_0 = \int_{-\pi}^{+\pi} d\tau(\psi) + ic.$$

We must therefore ascertain whether or not the function $\tau(\psi)$, with the properties described in the statement of the theorem, is uniquely determined by the further conditions

$$\int_{-\pi}^{+\pi} d\tau(\psi) = \mathfrak{R}\, a_0; \qquad \int_{-\pi}^{+\pi} e^{ni\psi}\, d\tau(\psi) = \tfrac{1}{2}\bar{a}_n;$$
$$\int_{-\pi}^{+\pi} e^{-ni\psi}\, d\tau(\psi) = \tfrac{1}{2} a_n$$

where $n = 1, 2, 3, \cdots$. From the familiar theory of Fourier coefficients, it is found that the function $\tau(\psi)$ is unique. The method used in the last paragraph of the proof of Theorem 8.4 can also be applied to the present problem, with the desired result.

THEOREM 10.36. *A necessary and sufficient condition that a function $\varphi(l)$, single-valued and analytic in the half-planes $\mathfrak{J}(l)>0$, $\mathfrak{J}(l)<0$, satisfy the conditions $\mathfrak{J}(l)\,\mathfrak{J}(\varphi(l)) \geqq 0$, $\varphi(\bar{l}) = \overline{\varphi(l)}$ for $\mathfrak{J}(l) \neq 0$ is that $\varphi(l)$ be representable by the integral formula*

$$\varphi(l) = \alpha l + \beta + \int_{-\infty}^{+\infty} \frac{\lambda l + 1}{\lambda - l}\, d\sigma(\lambda),$$

where α, β are real constants with $\alpha \geqq 0$, and $\sigma(\lambda)$ is a real monotone-increasing function in the class $\mathfrak{B}^*$. When this representation is possible it is unique. The singular function $\varphi(l) = \infty$ may be considered as representable in this form, with $\alpha = 0$, $\beta = \pm\infty$, $\sigma(\lambda) = 0$.

The transformation given by the equations $z = \dfrac{i-l}{i+l}$, $l = i\dfrac{1-z}{1+z}$ maps the half-plane $\mathfrak{J}(l) \geqq 0$ on the circle

$|z| \leqq 1$, in such a manner that, when l increases through real values from $-\infty$ to $+\infty$, the variable $\theta = \arg z$ increases from $-\pi$ to $+\pi$. When z and l are connected by these relations, the equation $if(z) = \varphi(l)$ sets up a one-to-one correspondence between the class of all functions $f(z)$ which are single-valued and analytic in the circle $|z| < 1$ and have not-negative real parts in that region, and the class of all functions $\varphi(l)$ which are single-valued and analytic in the half-plane $\mathfrak{J}(l) > 0$ and have not-negative imaginary parts there. Similarly, when z and l are connected by these relations, we can write

$$-c + i\int_{-\pi}^{+\pi} \frac{e^{i\psi}+z}{e^{i\psi}-z}\, d\tau(\psi) = \alpha l + \beta + \int_{-\infty}^{+\infty} \frac{\lambda l + 1}{\lambda - l}\, d\sigma(\lambda)$$

where

$$\psi = 2 \arctan \lambda, \quad \lambda = i\frac{1-e^{i\psi}}{1+e^{i\psi}} = \tan(\psi/2), \quad \tau(\psi) = \sigma(\lambda)$$

for $-\pi < \psi < +\pi$ and $-\infty < \lambda < +\infty$, $\alpha = \tau(\pi) - \tau(\pi - 0)$, and $\beta = -c$; the functions $\tau(\psi)$ and $\sigma(\lambda)$ are real monotone-increasing functions subject to the conditions stated in Theorem 10.35 and in the present theorem respectively, and the equation holds in the sense that, if the integral on either side exists, the integral on the other also exists and is equal to it. By combining these facts with the result stated in Theorem 10.35, we see that a function $\varphi(l)$ single-valued and analytic in the half-plane $\mathfrak{J}(l) > 0$ satisfies the condition $\mathfrak{J}\varphi(l) \geqq 0$ for $\mathfrak{J}(l) > 0$ if and only if $\varphi(l)$ is representable by the integral formula

$$\varphi(l) = \alpha l + \beta + \int_{-\infty}^{+\infty} \frac{\lambda l + 1}{\lambda - l}\, d\sigma(\lambda)$$

where α, β, and $\sigma(\lambda)$ are subject to the conditions indicated in the statement of the present theorem; and we see, further, that this representation is unique.

When $\varphi(l)$ is a function single-valued and analytic in the half-planes $\mathfrak{J}(l) > 0$, $\mathfrak{J}(l) < 0$, it satisfies the condition $\mathfrak{J}\varphi(l) \geqq 0$ for $\mathfrak{J}(l) > 0$ if and only if it is representable in the half-plane $\mathfrak{J}(l) > 0$ by the indicated integral formula.

This formula is significant in the half-plane $\mathfrak{J}(l) < 0$ also; but it represents $\varphi(l)$ in this half-plane if and only if $\varphi(l)$ satisfies the functional relation $\varphi(\bar{l}) = \overline{\varphi(l)}$. When the latter relation is valid, the inequality $\mathfrak{J}\varphi(l) \geqq 0$, $\mathfrak{J}(l) > 0$ implies the inequality $\mathfrak{J}\varphi(l) \leqq 0$, $\mathfrak{J}(l) < 0$. This completes the proof of the theorem.

THEOREM 10.37. *Any function $\varphi(l)$ which is single-valued and analytic in the half-planes $\mathfrak{J}(l) > 0$, $\mathfrak{J}(l) < 0$ and which satisfies the conditions $\mathfrak{J}(l)\,\mathfrak{J}(\varphi(l)) \geqq 0$, $\varphi(\bar{l}) = \overline{\varphi(l)}$ for $\mathfrak{J}(l) \neq 0$ can be constructed as the limit of a sequence $\{\varphi_n(l)\}$ of rational functions representable in the form*

$$\varphi_n(l) = \beta_n + \int_{-\infty}^{+\infty} \frac{\lambda l + 1}{\lambda - l}\, d\sigma_n(\lambda),$$

where β_n is a real constant and $\sigma_n(\lambda)$ is a real monotone-increasing function in the class $\mathfrak{B}^*$ which assumes precisely $n+1$ distinct values and has its n points of discontinuity located in an arbitrary prescribed denumerably infinite set Λ everywhere dense on the range $-\infty < \lambda < +\infty$. The singular function $\varphi(l) = \infty$ may also be considered as the limit of such a sequence.

By the preceding theorem, we can represent $\varphi(l)$ by the integral formula

$$\varphi(l) = \alpha l + \beta + \int_{-\infty}^{+\infty} \frac{\lambda l + 1}{\lambda - l}\, d\sigma(\lambda)$$

for all not-real l. We choose $\{\beta_n\}$ as any sequence of real numbers such that $\lim_{n \to \infty} \beta_n = \beta$. We select at pleasure an array $\{\lambda_{nk}\}$, $k = 1, \cdots, n$, $n = 1, 2, 3, \cdots$, of numbers which belong to the prescribed set Λ and which have the properties $\lambda_{n1} < \lambda_{n2} < \cdots < \lambda_{n,n-1} < \lambda_{nn}$, $\lim_{n \to \infty} \lambda_{n1} = -\infty$, $\lim_{n \to \infty} \lambda_{nn} = +\infty$, $\lim_{n \to \infty} \max_{k=1,\cdots,n-1} (\lambda_{n,k+1} - \lambda_{nk}) = 0$. We then define two real monotone-increasing functions $\sigma_{n1}(\lambda)$ and $\sigma_{n2}(\lambda)$ in the class $\mathfrak{B}^*$ as follows: we put $\sigma_{n1}(\lambda)$ equal to 0 or to 1 according as $\lambda < \lambda_{nn}$ or $\lambda \geqq \lambda_{nn}$; we put $\sigma_{12}(\lambda)$ equal to 0 or to 1 according as $\lambda < \lambda_{11}$ or $\lambda \geqq \lambda_{11}$; and we determine $\sigma_{n2}(\lambda)$ for $n \geqq 2$ by the equations

$$\sigma_{n2}(\lambda) = 0, \qquad -\infty < \lambda < \lambda_{n1},$$
$$\sigma_{n2}(\lambda) = \sigma(\lambda_{nk}) + k/n^2, \quad \lambda_{nk} \leqq \lambda < \lambda_{n,k+1}, \quad k = 1, \cdots, n-1,$$
$$\sigma_{n2}(\lambda) = \sigma(\lambda_{nn}) + 1/n, \quad \lambda_{nn} \leqq \lambda < +\infty.$$

Finally, we form the function $\sigma_n(\lambda) = \alpha_n \sigma_{n1}(\lambda) + \sigma_{n2}(\lambda)$, where $\{\alpha_n\}$ is an arbitrary sequence of not-negative real numbers such that $\lim_{n\to\infty} \alpha_n = \alpha$. We now have

$$\begin{aligned}
\lim_{n\to\infty} \left(\beta_n + \int_{-\infty}^{+\infty} \frac{\lambda l + 1}{\lambda - l}\, d\sigma_n(\lambda)\right)
&= \lim_{n\to\infty} \beta_n + \lim_{n\to\infty} \alpha_n \int_{-\infty}^{+\infty} \frac{\lambda l + 1}{\lambda - l}\, d\sigma_{n1}(\lambda) \\
&\quad + \lim_{n\to\infty} \int_{-\infty}^{+\infty} \frac{\lambda l + 1}{\lambda - l}\, d\sigma_{n2}(\lambda) \\
&= \beta + \lim_{n\to\infty} \alpha_n \frac{\lambda_{nn} l + 1}{\lambda_{nn} - l} + \lim_{n\to\infty} \int_{-\infty}^{+\infty} \frac{\lambda l + 1}{\lambda - l}\, d\sigma_{n2}(\lambda) \\
&= \beta + \alpha l + \int_{-\infty}^{+\infty} \frac{\lambda l + 1}{\lambda - l}\, d\sigma(\lambda) = \varphi(l)
\end{aligned}$$

for all not-real l, by virtue of the relations

$$\lim_{n\to\infty} \alpha_n = \alpha, \quad \lim_{n\to\infty} \beta_n = \beta, \quad \lim_{n\to\infty} \lambda_{nn} = +\infty,$$
$$\lim_{n\to\infty} \sigma_{n2}(\lambda) = \sigma(\lambda),$$

the last of which is valid for every value of λ which is a point of continuity of $\sigma(\lambda)$. The integral involving $\sigma_{n2}(\lambda)$ is most easily treated by writing it in the form

$$\begin{aligned}
\int_{-\infty}^{+\infty} \left(l + \frac{l^2+1}{\lambda - l}\right) d\sigma_{n2}(\lambda)
&= l\left(\sigma(\lambda_{nn}) + \frac{1}{n}\right) + (l^2+1) \int_{-\infty}^{+\infty} \frac{1}{\lambda - l}\, d\sigma_{n2}(\lambda) \\
&= l\left(\sigma(\lambda_{nn}) + \frac{1}{n}\right) + (l^2+1) \int_{-\infty}^{+\infty} \frac{\sigma_{n2}(\lambda)}{(\lambda - l)^2}\, d\lambda.
\end{aligned}$$

Since the function $\sigma_n(\lambda)$ obviously has the properties demanded by the theorem, the proof is complete.

THEOREM 10.38. *Let A be a Jacobi matrix such that the associated transformation T has the deficiency-index* $(1, 1)$, *and let $P(l)$, $Q(l)$, $U(l)$, $V(l)$ be the associated entire functions introduced in Theorem* 10.33. *Then the relation*

$$\frac{P(l)+U(l)\varphi(l)}{Q(l)+V(l)\varphi(l)}=\int_{-\infty}^{+\infty}\frac{1}{\lambda-l}\,d\varrho(\lambda),\qquad \mathfrak{I}(l)\neq 0,$$

determines a one-to-one correspondence between the class consisting of the singular function $\varphi(l)=\infty$ and all functions $\varphi(l)$ which are single-valued and analytic in the half-planes $\mathfrak{I}(l)>0$, $\mathfrak{I}(l)<0$ and which satisfy the relations $\mathfrak{I}(l)\,\mathfrak{I}(\varphi(l))\geqq 0$, $\varphi(\overline{l})=\overline{\varphi(l)}$ for $\mathfrak{I}(l)\neq 0$, and the class of all real monotone-increasing functions $\varrho(\lambda)$ in $\mathfrak{B}^*$ which are associated with the Jacobi matrix A in the manner indicated in Theorem 10.27. *When $\varphi(l)$ is given, its correspondent $\varrho(\lambda)$ is determined by means of the contour integral described in Lemma* 5.2; *and when $\varrho(\lambda)$ is given, its correspondent $\varphi(l)$ is found by direct solution of the relation above.*

In the preliminary discussion designed to motivate the developments summarized in Theorems 10.35–10.38, we showed that the relation now under consideration sets up a correspondence between the class of all functions $\varrho(\lambda)$ and a subclass of the class of all functions $\varphi(l)$. It is evident that the correspondence is a one-to-one correspondence. We must now show that every function $\varphi(l)$ corresponds to some function $\varrho(\lambda)$.

If $\varphi(l)$ is an arbitrary function of the class described above, we first introduce a sequence of rational functions $\{\varphi_k(l)\}$ which has the properties indicated in Theorem 10.37 and which converges to $\varphi(l)$ for all not-real l. The sequence $\{\beta_k\}$ involved in the construction of the sequence $\{\varphi_k(l)\}$ will be specified presently. We next consider the function

$$F_k^*(l)=\frac{P_k(l)+U_k(l)\,\varphi_k(l)}{Q_k(l)+V_k(l)\,\varphi_k(l)},$$

where P_k, Q_k, U_k, V_k are the polynomials associated with the Jacobi matrix A in the manner indicated in Theorem 10.26. In accordance with that theorem, we have

$$F_{2k}^*(l) = \frac{\bar{b}_k H_{k+1}(l) + H_k(l)\,\psi_k(l)}{\bar{b}_k G_{k+1}(l) + G_k(l)\,\psi_k(l)},$$

$$\psi_k(l) = \frac{-\,b_k\,\overline{G_{k+1}(0)} - b_k\,\overline{H_{k+1}(0)}\,\varphi_k(l)}{\overline{G_k(0)} + \overline{H_k(0)}\,\varphi_k(l)}.$$

It is necessary for our purposes to consider in some detail the properties of the functions $\varphi_k(l)$ and $\psi_k(l)$. From the integral representation

$$\varphi_k(l) = \beta_k + \int_{-\infty}^{+\infty} \frac{\lambda l + 1}{\lambda - l}\, d\sigma_k(\lambda)$$

we conclude that $\varphi_k(l)$ is a rational function of degree k with k distinct simple poles located on the real axis in the finite l-plane and that, for real values of l different from these poles, $\varphi_k(l)$ assumes real values while its derivative

$$\varphi_k'(l) = \int_{-\infty}^{+\infty} \frac{\lambda^2 + 1}{(\lambda - l)^2}\, d\sigma_k(\lambda)$$

assumes positive real values. Hence the equation $\varphi_k(l) = c$ has k distinct real roots when c is any real constant different from

$$\varphi_k(\infty) = \beta_k - \int_{-\infty}^{+\infty} \lambda\, d\sigma_k(\lambda).$$

By reference to the properties of the transformation

$$t = \frac{-\,b_k\,\overline{G_{k+1}(0)} - b_k\,\overline{H_{k+1}(0)}\,s}{\overline{G_k(0)} + \overline{H_k(0)}\,s}$$

given in Theorem 10.26, we can now ascertain the nature of the function $\psi_k(l)$. This function is obviously a rational function of degree k with poles at the roots of the equation $\varphi_k(l) = s_k$, where $s = s_k$ is the point in the extended s-plane which is carried into the point $t = \infty$ by the indicated transformation. If $s_k = \infty$, the poles of $\psi_k(l)$ coincide with those of $\varphi_k(l)$. Otherwise, s_k is a real number and $\psi_k(l)$ has k distinct poles on the finite real axis except in the special case where

$$s_k = \varphi_k(\infty) = \beta_k - \int_{-\infty}^{+\infty} \lambda\, d\sigma_k(\lambda).$$

We suppose henceforth that the sequence $\{\beta_k\}$ has been so chosen that this special case is avoided for every k; by reference to the proof of Theorem 10.37, it is clear that such a choice is possible. We now see that $\psi_k(l)$ has k distinct poles on the real axis in the finite l-plane, each of which must be a simple pole since the degree of $\psi_k(l)$ is equal to k. If we write the equation connecting $\varphi_k(l)$ and $\psi_k(l)$ in the form

$$\psi_k(l) = \frac{A + B\varphi_k(l)}{C + D\varphi_k(l)}$$

obtained in the proof of Theorem 10.26, where A, B, C, D are real numbers such that $BC - AD = 1$, we find at once that, for real values of l distinct from the poles of $\psi_k(l)$, this function assumes real values while its derivative

$$\psi_k'(l) = \frac{BC - AD}{(C + D\varphi_k(l))^2}\varphi_k'(l)$$

assumes positive real values. The properties of the rational function $\psi_k(l)$ which have now been established lead to the conclusion that the development in partial fractions for $\psi_k(l)$ must have the form

$$\psi_k(l) = \pi_{k0} + \sum_{\alpha=1}^{k} \frac{\pi_{k\alpha}}{\lambda_{k\alpha} - l},$$

where $\pi_{kj} > 0$ for $1 \leq j \leq k$. Hence, $\psi_k(l)$ can be represented by the integral formula

$$\psi_k(l) = \pi_{k0} + \delta_k \int_{-\infty}^{+\infty} \frac{1}{\lambda - l}\, d\varrho_k(\lambda)$$

where π_{k0} and δ_k are real constants, $\delta_k > 0$, and $\varrho_k(\lambda)$ is a real monotone-increasing function in $\mathfrak{B}^*$ which assumes precisely $k+1$ distinct values, including the value 0 for large negative λ and the value 1 for large positive λ. As in the proof of Theorem 10.29, we can determine a reduced Jacobi matrix of order k so that the monotone-increasing function associated with it in the manner indicated in Theorem 10.29 coincides with the function $\varrho_k(\lambda)$ just obtained. It follows that $\psi_k(l)$ can be expressed as a continued fraction

$$\psi_k(l) = \pi_{k0} + \frac{|\beta_0^{(k)}|^2|}{|\alpha_1^{(k)} - l} - \frac{|\beta_1^{(k)}|^2|}{|\alpha_2^{(k)} - l} - \cdots$$

$$\cdots - \frac{|\beta_{k-2}^{(k)}|^2|}{|\alpha_{k-1}^{(k)} - l} - \frac{|\beta_{k-1}^{(k)}|^2|}{|\alpha_k^{(k)} - l}, \; |\beta_0^{(k)}|^2 = \delta_k,$$

in accordance with Theorem 10.23 (4). We are now prepared to take the final step in the proof of the present theorem.

By Theorem 10.25 (5) we have for

$$F_{2k}^*(l) = \frac{\bar{b}_k H_{k+1} + H_k \psi_k}{\bar{b}_k G_{k+1} + G_k \psi_k}$$

the expression

$$F_{2k}^*(l) = \frac{|b_0|^2|}{|a_1 - l} - \frac{|b_1|^2|}{|a_2 - l} - \cdots$$

$$\cdots - \frac{|b_{k-2}|^2|}{|a_{k-1} - l} - \frac{|b_{k-1}|^2|}{|a_k - l - \psi_k(l)}, \; b_0 = 1.$$

If we substitute the continued fraction which represents $\psi_k(l)$, we obtain the result

$$F_{2k}^*(l) = \frac{|b_0|^2|}{|a_1 - l} - \cdots - \frac{|b_{k-1}|^2|}{|(a_k - \pi_{k0}) - l}$$

$$- \frac{|\beta_0^{(k)}|^2|}{|\alpha_1^{(k)} - l} - \cdots - \frac{|\beta_{k-1}^{(k)}|^2|}{|\alpha_k^{(k)} - l}.$$

It is immediately evident that $\{F_{2k}^*(l)\}$ is a sequence of generalized approximants for the infinite continued fraction associated with the Jacobi matrix A. By virtue of Theorem 10.31, we can select a subsequence wich converges for all not-real l to a limit

$$F(l) = \int_{-\infty}^{+\infty} \frac{1}{\lambda - l} \, d\varrho(\lambda),$$

where $\varrho(\lambda)$ is a real monotone-increasing function in $\mathfrak{B}^*$ which is associated with the Jacobi matrix A in the manner indicated in Theorem 10.27. On the other hand, the original expression for $F_{2k}^*(l)$ in terms of P_k, Q_k, U_k, V_k, and φ_k shows that the sequence $\{F_{2k}^*(l)\}$ converges to the limit

$\dfrac{P+U\varphi}{Q+V\varphi}$. Thus we have found a function $\varrho(\lambda)$ of the desired type which corresponds to the given function $\varphi(l)$ in accordance with the relation

$$\frac{P(l)+U(l)\varphi(l)}{Q(l)+V(l)\varphi(l)}=\int_{-\infty}^{+\infty}\frac{1}{\lambda-l}\,d\varrho(\lambda).$$

This completes our proof.

With the aid of Theorem 10.38 we can obtain further information concerning the self-adjoint extensions of the transformation T associated with a given Jacobi matrix in the case where T has the deficiency-index (1,1).

THEOREM 10.39. *Let A be a Jacobi matrix, T the closed linear symmetric transformation associated with A by the complete orthonormal set $\{g_n\}$, and $C(l)$ the circular region defined in Theorem* 10.32. *When T has the deficiency-index* (1,1), *the real monotone-increasing functions $\varrho(\lambda)$ associated with A have the following properties:*

(1) *for fixed $\varrho(\lambda)$, the point $z=\int_{-\infty}^{+\infty}\frac{1}{\lambda-l}\,d\varrho(\lambda)$ is always interior to $C(l)$ or always on the boundary of $C(l)$, $\mathfrak{J}(l)\neq 0$;*

(2) *if z is a given point on the boundary of $C(l)$ for fixed l, there is a unique function $\varrho(\lambda)$ such that $\int_{-\infty}^{+\infty}\frac{1}{\lambda-l}\,d\varrho(\lambda)=z$;*

(3) *if z is a given point interior to $C(l)$ for fixed l, the class of all functions $\varrho(\lambda)$ such that $\int_{-\infty}^{+\infty}\frac{1}{\lambda-l}\,d\varrho(\lambda)=z$ has the cardinal number $\mathfrak{c}$;*

(4) *the relation $\dfrac{P(l)+U(l)\beta}{Q(l)+V(l)\beta}=\int_{-\infty}^{+\infty}\frac{1}{\lambda-l}\,d\varrho(\lambda)$ determines a one-to-one correspondence between the class of all real numbers β such that $-\infty<\beta\leqq+\infty$ and the class of all functions $\varrho(\lambda)$ such that the point $z=\int_{-\infty}^{+\infty}\frac{1}{\lambda-l}\,d\varrho(\lambda)$ lies on the boundary of $C(l)$, $\mathfrak{J}(l)\neq 0$;*

(5) *between the class of all functions $\varrho(\lambda)$ described in* (4) *and the class of all self-adjoint extensions H of the symmetric*

transformation T there is a one-to-one correspondence such that corresponding members $\varrho(\lambda)$ and H in the two classes are connected by the relation $\varrho(\lambda) = (E(\lambda)g_1, g_1)$ where $E(\lambda)$ is the resolution of the identity for H.

For a fixed value l, $\mathfrak{J}(l) \neq 0$, we consider the relation

$$z = \frac{P(l) + U(l)s}{Q(l) + V(l)s}, \quad z = \int_{-\infty}^{+\infty} \frac{1}{\lambda - l}\, d\varrho(\lambda), \quad s = \varphi(l),$$

established in Theorem 10.38, in the light of Theorem 10.34. In order that z lie on the boundary of $C(l)$, it is necessary and sufficient that the corresponding point in the s-plane be either the point $s = \infty$ or a point on the real axis. In the class of functions $\varphi(l)$ under discussion, $\varphi(l)$ can assume the value $s = \infty$ for given not-real l if and only if it is the singular function $\varphi(l) \equiv \infty$, $\mathfrak{J}(l) \neq 0$. Similarly, the function $\varphi(l)$ can assume a real value $s = \beta$ for given not-real l if and only if it is identically equal to β for all not-real l: for the not-negative function $\mathfrak{J}\varphi(l)$ is harmonic at every point of the open set $\mathfrak{J}(l) > 0$, and can therefore assume its minimum value 0 in this region if and only if it vanishes identically; and the relation $\mathfrak{J}(\varphi(\bar{l})) = -\mathfrak{J}(\varphi(l))$ requires that the function $\mathfrak{J}\varphi(l)$ vanish for an arbitrary not-real l if and only if it vanishes also for the conjugate value $\bar{l}$. From this result, we can derive (1), (2), and (4) without further difficulty; and we can also establish (3) by observing that, when z is interior to $C(l)$ for fixed not-real l, there exist $\mathfrak{c}$ functions $\varphi(l)$ which assume for the given value l the not-real value s corresponding to z.

If H is a self-adjoint extension of T with the resolvent R_l and the resolution of the identity $E(\lambda)$, the point $z = (R_l g_1, g_1)$ $= \int_{-\infty}^{+\infty} \frac{1}{\lambda - l}\, d\varrho(\lambda)$, where $\varrho(\lambda) = (E(\lambda)g_1, g_1)$, lies on the circumference of the circular region $C(l)$, as we showed in the proof of Theorem 10.32; and, furthermore, when H ranges in a suitable manner over the class of all self-adjoint extensions of T, the corresponding point z passes once and only once through every point of the circumference. This fact, taken in conjunction with (1), (2), and (4) shows that the correspondence described in (5) exists.

THEOREM 10.40. *Let A be a Jacobi matrix, T the transformation associated with A by the complete orthonormal set $\{g_n\}$, and $\varrho(\lambda)$ an associated real monotone-increasing function in $\mathfrak{B}^*$. The sequence of polynomials $\{G_n(\lambda)\}$ associated with the Jacobi matrix A is an orthonormal set in the Hilbert space $\mathfrak{L}_2(\varrho)$; it is complete if and only if one of the following equivalent conditions is satisfied:*
(1) *$\varrho(\lambda)$ is expressible in the form $\varrho(\lambda) = (E(\lambda)g_1, g_1)$, where $E(\lambda)$ is the resolution of the identity corresponding to a self-adjoint extension H of T;*
(2) *$\varrho(\lambda)$ has the property that the point $z = \int_{-\infty}^{+\infty} \frac{1}{\lambda - l}\, d\varrho(\lambda)$ lies on the boundary of the set $C(l)$.*
When (1) *is satisfied, the transformation H is isomorphic with the self-adjoint transformation in $\mathfrak{L}_2(\varrho)$ which has domain consisting of those and only those functions $F(\lambda)$ in $\mathfrak{L}_2(\varrho)$ such that $\lambda F(\lambda)$ belongs to $\mathfrak{L}_2(\varrho)$ and which carries a function $F(\lambda)$ in its domain into the function $\lambda F(\lambda)$. In particular, the conditions* (1) *and* (2) *are always satisfied when T has the deficiency-index* (0, 0).†

The fact that the sequence $\{G_n(\lambda)\}$ is an orthonormal set in the Hilbert space $\mathfrak{L}_2(\varrho)$, which consists of all those ϱ-measurable functions $F(\lambda)$ such that the integral $\int_{-\infty}^{+\infty} |F(\lambda)|^2\, d\varrho(\lambda)$ exists, is an immediate consequence of Theorem 10.27 (2). When the condition (1) is satisfied, we know from the last two paragraphs of the proof of Theorem 10.27 that the closed linear manifold $\mathfrak{M}(g_1)$ determined by the set of elements $E(\lambda)g_1$, $-\infty < \lambda < +\infty$, coincides with $\mathfrak{H}$. Hence we can apply Theorems 6.2 and 7.2 to set up an isomorphism between $\mathfrak{H}$ and $\mathfrak{L}_2(\varrho)$ so that corresponding elements f and $F(\lambda)$ are connected by the relation $f = F(H)g_1$. Since g_n and $G_n(\lambda)$ are in correspondence, by virtue of the equation

† This theorem is due essentially to M. Riesz, Acta Litterarum ac Scientiarum, Sectio Mathematicarum, Szeged, 1 (1922–23), pp. 209–227.

$G_n(H)g_1 = G_n(T)g_1 = g_n$, we see immediately that $\{G_n(\lambda)\}$ must be a complete orthonormal set in $\mathfrak{L}_2(\varrho)$. The relations $f = F(H)g_1$ and $Hf = HF(H)g_1$ show furthermore that the image of H under this isomorphism is the transformation in $\mathfrak{L}_2(\varrho)$ described in the statement of the theorem. When $\{G_n(\lambda)\}$ is a complete orthonormal set in $\mathfrak{L}_2(\varrho)$, we set up an isomorphism between $\mathfrak{H}$ and $\mathfrak{L}_2(\varrho)$ by putting g_n and $G_n(\lambda)$ in correspondence. We then determine the image of the bounded linear transformation X_l under this isomorphism, where X_l denotes the transformation associated with the function $\varrho(\lambda)$ in the manner indicated in Theorem 10.27. The equations $(X_l g_m, g_n) = \int_{-\infty}^{+\infty} \frac{G_m(\lambda)\overline{G_n(\lambda)}}{\lambda - l}\, d\varrho(\lambda)$, established in Theorem 10.27 (2), show at once that the image of X_l is the transformation in $\mathfrak{L}_2(\varrho)$ which takes $F(\lambda)$ into $\frac{1}{\lambda - l} F(\lambda)$, $\mathfrak{I}(l) \neq 0$. The latter transformation is obviously the resolvent of the self-adjoint transformation which takes $F(\lambda)$ into $\lambda F(\lambda)$ whenever both these functions belong to $\mathfrak{L}_2(\varrho)$. Thus X_l is the resolvent of a self-adjoint transformation H in $\mathfrak{H}$; and Theorem 9.19 (3) shows that H is an extension of the symmetric transformation T. By the methods used in the proof of Theorem 7.12, we can now determine the resolution of the identity $E(\lambda)$ corresponding to H; in particular, we can prove that $\varrho(\lambda) = (E(\lambda)g_1, g_1)$. We have thus shown that (1) is a necessary and sufficient condition for the completeness of the set $\{G_n(\lambda)\}$. The equivalence of conditions (1) and (2) is trivial in case T has the deficiency-index (0, 0), since $\varrho(\lambda)$ is unique and $C(l)$ reduces to a single point. In this case both conditions are always satisfied. The equivalence of the two conditions in case T has the deficiency-index (1, 1) has been demonstrated in Theorem 10.39.

THEOREM 10.41. *Let A be a Jacobi matrix such that the transformation T associated with it by the complete orthonormal set $\{g_n\}$ has the deficiency-index (1, 1); let $P(l)$, $Q(l)$, $U(l)$, $V(l)$ be the associated entire functions; let H be a self-adjoint extension of T with the corresponding resolution of the identity*

$E(\lambda)$; and let $\varrho(\lambda)$ be a real monotone-increasing function in $\mathfrak{B}^*$ which is associated with A in the manner indicated in Theorem 10.27 *and which has the property that*

$$z = \int_{-\infty}^{+\infty} \frac{1}{\lambda - l}\, d\varrho(\lambda)$$

is a point of the boundary of the region $C(l)$ for not-real l. Then the functions

$$R(l, \theta) = \cos\theta\, P(l) + \sin\theta\, U(l),$$
$$S(l, \theta) = \cos\theta\, Q(l) + \sin\theta\, V(l),$$

where $-\frac{\pi}{2} < \theta \leqq +\frac{\pi}{2}$, are entire functions with the following properties:

(1) *$R(l, \theta)$ and $S(l, \theta)$ both have infinitely many roots, all real and simple;*

(2) *the roots of $R(l, \theta)$ separate the roots of $S(l, \theta)$ in the sense that every root of $R(l, \theta)$ lies between two roots of $S(l, \theta)$ while there is just one root of $R(l, \theta)$ between any two adjacent roots of $S(l, \theta)$;*

(3) *if $\theta_1 \neq \theta_2$, the roots of $S(l, \theta_1)$ separate the roots of $S(l, \theta_2)$ in the sense that there is just one root of $S(l, \theta_1)$ between any two adjacent roots of $S(l, \theta_2)$;*

(4) *if l is an arbitrary real number, there exists a unique value θ, $-\frac{\pi}{2} < \theta \leqq +\frac{\pi}{2}$, such that l is a root of $S(l, \theta)$.*

If the transformation H, the function $\varrho(\lambda)$, and the real numbers β and θ are connected by the relations

$$\varrho(\lambda) = (E(\lambda) g_1, g_1),$$
$$\frac{R(l, \theta)}{S(l, \theta)} = \frac{P(l) + U(l)\beta}{Q(l) + V(l)\beta} = \int_{-\infty}^{+\infty} \frac{1}{\lambda - l}\, d\varrho(\lambda),$$
$$\beta = \tan\theta,$$

in accordance with Theorem 10.39, *then*

(1) *$\varrho(\lambda)$ is constant on any interval which contains no root of $S(l, \theta)$ and increases if and only if λ passes through a root of $S(l, \theta)$;*

(2) *H has a simple point spectrum, each root of $S(l, \theta)$ being a simple characteristic value and each real number l such that $S(l, \theta) \neq 0$ being a point of the resolvent set.*

If μ is an arbitrary real number, H and $\varrho(\lambda)$ can be chosen in just one way so that μ is a characteristic value of H and a point of discontinuity of $\varrho(\lambda)$. Two distinct self-adjoint extensions of T have no common characteristic value.

We shall first prove that $U(l)\,Q(l) - P(l)\,V(l) = 1$ for all l. By virtue of the relations

$$\frac{P_n(l) + U_n(l)\,s}{Q_n(l) + V_n(l)\,s} = \frac{\bar{b}_n\,H_{n+1}(l) + H_n(l)\,t}{\bar{b}_n\,G_{n+1}(l) + G_n(l)\,t},$$

$$t = \frac{-b_n\,\overline{G_{n+1}(0)} - b_n\,\overline{H_{n+1}(0)}\,s}{\overline{G_n(0)} + \overline{H_n(0)}\,s},$$

we see that

$$\begin{aligned} |U_n(l)\,Q_n(l) - P_n(l)\,V_n(l)| &= |\bar{b}_n\,H_n(l)\,G_{n+1}(l) - \bar{b}_n\,H_{n+1}(l)\,G_n(l)| \\ &\quad\times |-b_n\,\overline{G_n(0)}\,\overline{H_{n+1}(0)} + b_n\,\overline{G_{n+1}(0)}\,\overline{H_n(0)}|. \end{aligned}$$

In establishing Theorems 10.25 and 10.26, we proved that each of the factors on the right of this equation has the absolute value 1. Hence we have

$$|U_n(l)\,Q_n(l) - P_n(l)\,V_n(l)| = 1;$$

and we conclude by passing to the limit that

$$|U(l)\,Q(l) - P(l)\,V(l)| = 1.$$

Thus the entire function $U(l)\,Q(l) - P(l)\,V(l)$ must be a constant. It is easily seen, by virtue of the series developments for the functions P, Q, U, V obtained from Theorem 10.26, that this function assumes the value 1 for $l = 0$ and that the constant is therefore equal to 1. By using the identity just established, we obtain the relation

$$R(l, \theta_1)\,S(l, \theta_2) - R(l, \theta_2)\,S(l, \theta_1) = \sin(\theta_1 - \theta_2),$$

holding for all l, $-\frac{\pi}{2} < \theta_1 \leqq +\frac{\pi}{2}$, $-\frac{\pi}{2} < \theta_2 \leqq +\frac{\pi}{2}$.

If we take $\theta_1 \neq \theta_2$ in the relation just established, the term on the right has a value different from zero. It is therefore obvious that $R(l, \theta_1)$ and $S(l, \theta_1)$ have no common root and, likewise, that $S(l, \theta_1)$ and $S(l, \theta_2)$ have no common root. If we put $\beta = \tan\theta$, we know by Theorem 10.39 that there exists a unique function $\varrho(\lambda)$ which has the properties described above and, which satisfies the equation

$$\frac{R(l, \theta)}{S(l, \theta)} = \frac{P(l) + U(l)\beta}{Q(l) + V(l)\beta} = \int_{-\infty}^{+\infty} \frac{1}{\lambda - l} \delta\varrho(\lambda)$$

for all not-real l. Since the function $R(l, \theta)/S(l, \theta)$ is meromorphic and since the integral term is analytic for not-real l, we conclude that the poles of this function are confined to the real axis, where they constitute a set of isolated points without finite limit point. If we calculate the function $\varrho(\lambda)$ by means of the contour integral of Lemma 5.2, we find that it is constant on any interval which contains no pole of the function $R(l, \theta)/S(l, \theta)$ and increases if and only if λ passes through one of these poles. Now $\varrho(\lambda)$, according to Theorem 10.27, assumes infinitely many distinct values and must therefore have infinitely many points of discontinuity. Hence $R(l, \theta)/S(l, \theta)$ has infinitely many poles. By virtue of the fact that $R(l, \theta)$ and $S(l, \theta)$ have no common root, the poles of $R(l, \theta)/S(l, \theta)$ coincide with the roots of $S(l, \theta)$. Hence $S(l, \theta)$ has infinitely many real roots. If μ is a root of $S(l, \theta)$, if $\varepsilon > 0$ is small enough that the interval $\mu - \varepsilon \leqq \lambda \leqq \mu + \varepsilon$ contains no other root of $S(l, \theta)$, and if π is the positive number $\varrho(\mu + 0) - \varrho(\mu - 0)$, we can write

$$\begin{aligned} \frac{R(l, \theta)}{S(l, \theta)} &= \int_{-\infty}^{+\infty} \frac{1}{\lambda - l} d\varrho(\lambda) \\ &= \frac{\pi}{\mu - l} + \left(\int_{-\infty}^{\mu-\varepsilon} + \int_{\mu+\varepsilon}^{+\infty}\right) \frac{1}{\lambda - l} d\varrho(\lambda), \end{aligned}$$

where the final integral term is clearly analytic at $l = \mu$. Hence we see that the poles of $R(l, \theta)/S(l, \theta)$ are all simple poles, the roots of $S(l, \theta)$ all simple roots. Since

$$\mathfrak{J}\left(\int_{-\infty}^{+\infty} \frac{1}{\lambda - l}\, d\varrho(\lambda)\right) = \mathfrak{J}(l) \int_{-\infty}^{+\infty} \frac{1}{|\lambda - l|^2}\, d\varrho(\lambda)$$

vanishes if and only if $\mathfrak{J}(l) = 0$, we see that the roots of $R(l, \theta)/S(l, \theta)$ are all real. The roots of this function obviously coincide with those of the function $R(l, \theta)$, which therefore has only real roots. For real values of l different from the roots of $S(l, \theta)$, the derivative

$$\frac{d}{dl} \frac{R(l, \theta)}{S(l, \theta)} = \int_{-\infty}^{+\infty} \frac{1}{(\lambda - l)^2}\, d\varrho(\lambda)$$

evidently has positive values. Hence if l increases through real values from one root of $S(l, \theta)$ to the next greater root, the function $R(l, \theta)/S(l, \theta)$ increases through real values, vanishing just once. It follows that $R(l, \theta)$ has just one simple root between any two adjacent roots of $S(l, \theta)$. If $S(l, \theta)$ has a smallest real root μ, then the integral $\int_{-\infty}^{+\infty} \frac{1}{\lambda - l}\, d\varrho(\lambda)$ does not vanish when l is real and less than μ. We conclude that the roots of $R(l, \theta)$ must all exceed μ in this case. Similarly, when $S(l, \theta)$ has a greatest real root, this root exceeds all the roots of $R(l, \theta)$. Hence every root of $R(l, \theta)$ lies between two roots of $S(l, \theta)$. On arranging the results so far established, we may state them as properties (1) and (2) of the functions $R(l, \theta)$, $S(l, \theta)$ and property (1) of the function $\varrho(\lambda)$.

We know from previous results that the functions $S(l, \theta_1)$ and $S(l, \theta_2)$, $\theta_1 \neq \theta_2$, have no common root. If l_1 and l_2 are adjacent roots of $S(l, \theta_2)$, we must have $R(l_1, \theta_2)\, R(l_2, \theta_2) < 0$, by virtue of the relative position of the roots of the functions $R(l, \theta_2)$ and $S(l, \theta_2)$. We have further the equations

$$-R(l_1, \theta_2)\, S(l_1, \theta_1) = -R(l_2, \theta_2)\, S(l_2, \theta_1) = \sin(\theta_1 - \theta_2).$$

Since the final term does not vanish, we conclude that $S(l_1, \theta_1)\, S(l_2, \theta_1) < 0$ and that $S(l, \theta_1)$ vanishes at least once between the given adjacent roots of $S(l, \theta_2)$. If we interchange the rôles of the numbers θ_1 and θ_2, we see that

$S(l, \theta_2)$ also vanishes at least once between any two adjacent roots of $S(l, \theta_1)$. We can now infer that the roots of $S(l, \theta_1)$ and $S(l, \theta_2)$ are located in the manner described in (3) above.

We can show that every real number is a root of at least one function $S(l, \theta)$; that it cannot be a root of more than one we have already proved. If we apply the implicit function theorem to the relation $S(l, \theta) = 0$, we find that each real number l in a sufficiently small neighborhood of a known root must satisfy this relation for a suitably determined θ. Hence the set of all real numbers which are roots of at least one function $S(l, \theta)$ is an open set. On the other hand, it is easily verified that this set must be closed. Hence the set, being non-vacuous and both open and closed, must be the set of all real numbers, as we wished to prove. We confine ourselves to the outline of the proof, since we shall discuss the question from another angle immediately.

When H is the self-adjoint transformation corresponding to $\varrho(\lambda)$ in the manner indicated in Theorem 10.39, we can determine its properties most simply by reference to Theorem 10.40. Since H is isomorphic with the self-adjoint transformation which takes the function $F(\lambda)$ into the function $\lambda F(\lambda)$ whenever both functions belong to the space $\mathfrak{L}_2(\varrho)$, and since $\varrho(\lambda)$ has the special properties already established, we see at once that the points of discontinuity of $\varrho(\lambda)$ are simple characteristic values of H, the remaining points of the real axis points of the resolvent set of H. The relation of the spectrum of H to the corresponding function $S(l, \theta)$ is then obvious. We can now give the second proof of the result discussed in the preceding paragraph. If μ is an arbitrary real number, it is a simple characteristic value of T^* by Theorem 10.27. If φ is a corresponding characteristic element, we can define a symmetric extension T_0 of the transformation T by assigning to it the domain $\mathfrak{D}_0 = \mathfrak{D} + \{\varphi\}$, where $\mathfrak{D}$ is the domain of T, and setting $T_0 = T^*$ in $\mathfrak{D}_0$. The details of the discussion have been given in the paragraph following the proof of Theorem 9.6. Then there exists a self-adjoint extension H of T_0, which obviously must satisfy the relations

$H \supseteqq T$, $H \subseteqq T^*$, $H\varphi = T^*\varphi = \mu\varphi$. The function $S(l, \theta)$ corresponding to H by virtue of the earlier results must vanish for $l = \mu$. As we have already seen, no two self-adjoint extensions of T can have a characteristic value in common.

It is clear that the methods used above can be applied to the study of the relations between the roots of $R(l, \theta_1)$ and those of $R(l, \theta_2)$. Since the question is of little interest in connection with subsequent developments, we shall not enter into it.

We shall next examine certain relations between a given Jacobi matrix A, the spectral properties of the associated transformation T, and the behavior of the associated functions $\varrho(\lambda)$. It will be observed that the results obtained below indicate that in the case where T has the deficiency-index (1, 1) these relations become extremely vague.

THEOREM 10.42. *Let A be a Jacobi matrix, T the transformation associated with A by the complete orthonormal set $\{g_n\}$, $\varrho(\lambda)$ an arbitrary real monotone-increasing function in $\mathfrak{B}^*$ associated with A in the manner indicated in Theorem* 10.27, *and $\{G_n(l)\}$ the associated sequence of polynomials. We introduce the following numbers and sets of points:*

(1) *Δ is the closed set of all real numbers λ in every neighborhood of which lie roots of infinitely many of the polynomials $G_n(l)$, $-\infty \leqq \lambda \leqq +\infty$;*

(2) *λ_{n1} and λ_{nn} are the least and the greatest roots respectively of $G_n(l)$;*

(3) *Δ_1 is the set of all real numbers λ such that $\bar{\gamma}_n G_n(\lambda) > 0$ for all $n \geqq 2$, the number $\lambda = +\infty$ being counted as one which always satisfies the condition;*

(4) *Δ_2 is the set of all real numbers λ such that $(-1)^{n-1}\bar{\gamma}_n G_n(\lambda) > 0$ for all $n \geqq 2$, the number $\lambda = -\infty$ being counted as one which always satisfies the condition;*

(5) *P is the set of all points of increase of the function $\varrho(\lambda)$— in other words, the complement of the sum of all open intervals on which $\varrho(\lambda)$ remains constant;*

(6) *C, C_1, C_2 are respectively the bound, upper bound, and lower bound of the closed linear symmetric transformation T,*

as defined in Definition 2.14, *the values* $C = +\infty$, $C_1 = +\infty$, $C_2 = -\infty$ *being admitted;*

(7) $\mathfrak{a}$ *is the least upper bound of the set of numbers* $|a_{mn}|$, *where* a_{mn} *is the general element of the Jacobi matrix* A.

If E *is an arbitrary set we denote by* $\inf E$, $\sup E$, $\operatorname{ext} E$ *respectively the greatest lower bound, the least upper bound, and the greater of the two numbers* $|\inf E|$ *and* $|\sup E|$, *the values* $-\infty$ *and* $+\infty$ *being admitted. The relations*

$$\sup \Lambda_2 = \inf \Lambda = \lim_{n\to\infty} \lambda_{n1} < \lim_{n\to\infty} \lambda_{nn} = \sup \Lambda = \inf \Lambda_1,$$
$$\inf P \leqq \inf \Lambda = C_2 < C_1 = \sup \Lambda \leqq \sup P,$$
$$C \leqq \operatorname{ext} P, \quad \mathfrak{a} \leqq \operatorname{ext} P \leqq 3\mathfrak{a}, \quad C_2 < a_{nn} < C_1$$

are satisfied regardless of whether the deficiency-index of T *is* (0, 0) *or* (1,1). *When the index is* (0,0), *we have the additional relations*

$$S(T) = P \subseteqq \Lambda, \quad C_2 = \inf P = \inf \Lambda, \quad C_1 = \sup P = \sup \Lambda,$$

where $S(T)$ *denotes the spectrum of the transformation* T. *When the index is* (1,1), *it is necessary that* $C = \operatorname{ext} P = \mathfrak{a} = +\infty$. *If the relations*

$$\frac{P(l) + U(l)\varphi(l)}{Q(l) + V(l)\varphi(l)} = \int_{-\infty}^{+\infty} \frac{1}{\lambda - l}\, d\varrho(\lambda),$$
$$\varphi(l) = \alpha l + \beta + \int_{-\infty}^{+\infty} \frac{\lambda l + 1}{\lambda - l}\, d\sigma(\lambda)$$

hold for all not-real l *in accordance with Theorems* 10.36 *and* 10.38, *and if* Σ *is the set of all points of increase of* $\sigma(\lambda)$, *then the derived set of* P *coincides with the derived set of* Σ; *only the isolated points of the set* P *are restricted in any way by the Jacobi matrix* A. *In the special case where* $C_2 = -\infty$ *and* $C_1 = +\infty$, *we have*

$$C_2 = \inf P = \inf \Lambda, \quad C_1 = \sup P = \sup \Lambda.$$

In the special case where $C_2 > -\infty$, $C_1 = +\infty$, *we have* $C_1 = \sup P = \sup \Lambda = +\infty$. *There exist* $\mathfrak{c}$ *functions* $\varrho(\lambda)$ *for which* $\inf P$ *assumes an arbitrary value* μ *on the range* $-\infty \leqq \mu < C_2$ *and* *for which the derived set P' is an arbitrarily prescribed closed set on the range $\mu \leqq \lambda < +\infty$; but there is just one function $\varrho(\lambda)$ for which $\inf P = C_2$. In this special case, furthermore, every self-adjoint extension of T has exactly one characteristic value on the range $-\infty < \lambda \leqq C_2$, and every value on this range is a characteristic value for exactly one self-adjoint extension of T. In the special case where $C_2 = -\infty$, $C_1 < +\infty$, analogous statements may be made.*

When $n \geqq 2$, the numbers λ_{n1}, λ_{nn} described in (2) exist and have the property

$$\cdots < \lambda_{n+1,1} < \lambda_{n1} < \cdots < \lambda_{31} < \lambda_{21}$$
$$= \lambda_{22} < \lambda_{33} < \cdots < \lambda_{nn} < \lambda_{n+1,n+1} < \cdots.$$

Hence the limits $\lim\limits_{n\to\infty} \lambda_{n1}$, $\lim\limits_{n\to\infty} \lambda_{nn}$ exist and belong to the set Λ. Since all the roots of $G_n(l)$, $n \geqq 2$, lie between these two numbers, we have

$$\inf \Lambda = \lim_{n\to\infty} \lambda_{n1} < \lim_{n\to\infty} \lambda_{nn} = \sup \Lambda.$$

By virtue of Theorem 10.24 (4), the function $(-1)^{n-1}\bar{\gamma}_n G_n(\lambda)$ assumes real values; and, if its least root and its next greater root be denoted by λ_{n1} and λ_{n2} respectively, it satisfies the inequalities

$$(-1)^{n-1}\bar{\gamma}_n G_n(\lambda) > 0, \quad -\infty \leqq \lambda < \lambda_{n1};$$
$$(-1)^{n-1}\bar{\gamma}_n G_n(\lambda) \leqq 0, \quad \lambda_{n1} \leqq \lambda \leqq \lambda_{n2},$$

for $n \geqq 3$. From Theorem 10.24 (6), we recall that $\lambda_{n1} < \lambda_{n-1,1} < \lambda_{n2}$. It is now clear that the set Λ_2 contains every point of the range $-\infty \leqq \lambda \leqq \lim\limits_{n\to\infty} \lambda_{n1}$, but no point of the range $\lim\limits_{n\to\infty} \lambda_{n1} < \lambda \leqq \lambda_{21}$, which is the sum of the intervals $\lambda_{n1} \leqq \lambda \leqq \lambda_{n-1,1}$ for $n \geqq 3$. Since the polynomial $-\bar{\gamma}_2 G_2(\lambda)$ is not-positive for $\lambda \geqq \lambda_{21} = \lambda_{22}$, we see that Λ_2 is the set $-\infty \leqq \lambda \leqq \lim\limits_{n\to\infty} \lambda_{n1}$ and that $\sup \Lambda_2 = \inf \Lambda$. In a similar manner, it is found that Λ_1 is the set $\lim\limits_{n\to\infty} \lambda_{nn} \leqq \lambda \leqq +\infty$ and that $\inf \Lambda_1 = \sup \Lambda$.

The transformation T is constructed by forming the transformation $T_1(A)$ with the complete orthonormal set $\{g_n\}$ as its domain in accordance with Theorems 3.2 and 3.4 and by then setting $T \equiv \tilde{T}_1(A) \equiv T_1^{**}(A)$. It is easily verified that the linear transformation $\hat{T}_1(A)$, with the linear manifold determined by the set $\{g_n\}$ as its domain, and the closed linear transformation $T \equiv \tilde{T}_1(A)$ have the same bounds, since, in accordance with Theorem 2.10, every element f in the domain of T can be approximated in $\mathfrak{H}$ by a sequence $\{f_n\}$ in the domain of $\hat{T}_1(A)$ with the properties $f_n \to f$, $\hat{T}_1(A)f_n \to Tf$. If g is an arbitrary element in the domain of $\hat{T}_1(A)$, it can be expressed as a linear combination of the elements $g_n = G_n(T)g_1$ and hence can be written in the form $g = F(T)g_1$, where $F(\lambda)$ is a polynomial. Using Theorem 9.19 and recalling that integrals with respect to the function $\varrho(\lambda)$ may be treated as integrals over the range $\inf P \leqq \lambda \leqq \sup P$, we find that

$$(\hat{T}_1(A)g, g) = (TF(T)g_1, F(T)g_1) = \int_{-\infty}^{+\infty} \lambda\, |F(\lambda)|^2\, d\varrho(\lambda)$$

$$\geqq \inf P \int_{-\infty}^{+\infty} |F(\lambda)|^2\, d\varrho(\lambda) = \inf P(g, g),$$

$$(\hat{T}_1(A)g, g) \leqq \sup P \int_{-\infty}^{+\infty} |F(\lambda)|^2\, d\varrho(\lambda) = \sup P(g, g).$$

Thus the symmetric transformations T and $\hat{T}_1(A)$ have common upper and lower bounds, C_1 and C_2 respectively, which satisfy the inequalities

$$\inf P \leqq C_2 < C_1 \leqq \sup P.$$

In order to show that $\inf \varLambda = C_2$, we apply Theorems 10.27 and 9.21 to construct a self-adjoint transformation $H \supseteqq T$ with lower bound equal to C_2. If $E(\lambda)$ is the resolution of the identity corresponding to H and $\varrho(\lambda)$ is the function $(E(\lambda)g_1, g_1)$, we find that $C_2 = \inf P$. This result is most readily verified by means of the characterization of H given in Theorems 10.27, 7.14, and 10.40, according to which H is isomorphic with the transformation in $\mathfrak{L}_2(\varrho)$ which carries $F(\lambda)$ into $\lambda F(\lambda)$ whenever both functions are in $\mathfrak{L}_2(\varrho)$. We now observe that whenever $n \geqq 2$ and $\mu \leqq C_2 = \inf P$, the Hermitian form

$$\int_{-\infty}^{+\infty} (\lambda - \mu) \sum_{\alpha=0}^{n-1} x_\alpha \lambda^\alpha \cdot \sum_{\beta=0}^{n-1} \bar{x}_\beta \lambda^\beta \, d\varrho(\lambda)$$
$$= \sum_{\alpha,\beta=0}^{n-1} (c_{\alpha+\beta+1} - \mu c_{\alpha+\beta}) x_\alpha \bar{x}_\beta,$$

where $\{c_n\}$ is the sequence of constants associated with the Jacobi matrix A, is positive definite. In consequence, the determinant of this form, which can be written

$$\det_{j,k=0,\cdots,n-1} \{c_{j+k+1} - \mu c_{j+k}\}$$
$$= D_{n-2}^{1/2} D_{n-1}^{1/2} (-1)^{n-1} \bar{\gamma}_n G_n(\mu)$$

by virtue of Theorem 10.24 (4), must be positive for the indicated values of μ. Since the constants D_n are all positive, we conclude that the interval $-\infty \leqq \lambda \leqq C_2$ is contained in the set Λ_2 and hence that $C_2 \leqq \sup \Lambda_2 = \inf \Lambda$. On the other hand, we know from Theorem 10.28 that, if $H^{(p)}$ is the self-adjoint transformation defined by the reduced Jacobi matrix A_p of order p, then $H^{(p)} \to \hat{T}_1(A)$ in the domain of $\hat{T}_1(A)$; and from Theorem 10.23 that, when g is an element of the closed linear manifold $\mathfrak{M}_p(g_1)$, the inequality

$$\lambda_{p1}(g, g) \leqq (H^{(p)} g, g) \leqq \lambda_{pp}(g, g)$$

is satisfied. If g is an arbitrary element in the domain of $\hat{T}_1(A)$, it is a linear combination of the elements of the set $\{g_n\}$ and therefore belongs to $\mathfrak{M}_p(g_1)$ for all sufficiently large values of p. Since

$$\inf \Lambda < \lambda_{p1} \leqq \lambda_{pp} < \sup \Lambda, \qquad \lim_{p\to\infty} (H^{(p)} g, g) = (\hat{T}_1(A) g, g),$$

we conclude that

$$\inf \Lambda (g, g) \leqq (\hat{T}_1(A) g, g) \leqq \sup \Lambda (g, g)$$

throughout the domain of $\hat{T}_1(A)$ and hence that $C_2 \geqq \inf \Lambda$. The equation $C_2 = \inf \Lambda$ follows at once. In similar fashion we show that $C_1 = \sup \Lambda$. From these results it is evident that $C = \operatorname{ext} \Lambda$.

From the relations

$$a_{mn} = \int_{-\infty}^{+\infty} \lambda\, G_n(\lambda)\, \overline{G_m(\lambda)}\, d\varrho(\lambda),$$

$$\int_{-\infty}^{+\infty} |G_n(\lambda)|^2\, d\varrho(\lambda) = 1$$

established in Theorem 10.27 (2), we conclude that

$$\begin{aligned} |a_{mn}| &\leqq \int_{-\infty}^{+\infty} |\lambda|\, |G_n(\lambda)|\, |G_m(\lambda)|\, d\varrho(\lambda) \\ &\leqq \operatorname{ext} P \int_{-\infty}^{+\infty} |G_n(\lambda)|\, |G_m(\lambda)|\, d\varrho(\lambda) \\ &\leqq \operatorname{ext} P \Big(\int_{-\infty}^{+\infty} |G_n(\lambda)|^2\, d\varrho(\lambda) \\ &\qquad \times \int_{-\infty}^{+\infty} |G_m(\lambda)|^2\, d\varrho(\lambda) \Big)^{1/2} = \operatorname{ext} P, \end{aligned}$$

and hence that $\alpha \leqq \operatorname{ext} P$. It is easy to show that for a suitably chosen Jacobi matrix A the difference $\operatorname{ext} P - \alpha$ does not exceed a preassigned positive ε. In fact, if $\varrho(\lambda)$ is any real monotone-increasing function in $\mathfrak{V}^*$ for which $\varrho(-\infty) = 0$, $\varrho(+\infty) = 1$, $\inf P = 1 - \varepsilon$, $\sup P = 1$, the transformation in $\mathfrak{L}_2(\varrho)$ which takes $F(\lambda)$ into $\lambda F(\lambda)$ is bounded and self-adjoint provided that $\varrho(\lambda)$ assumes infinitely many distinct values; this transformation is isomorphic with the transformation T associated with the Jacobi matrix $\{a_{mn}\}$,

$$a_{mn} = \int_{-\infty}^{+\infty} \lambda\, G_n(\lambda)\, \overline{G_m(\lambda)}\, d\varrho(\lambda),$$

where $\{G_n(\lambda)\}$ is a complete orthonormal set in $\mathfrak{L}_2(\varrho)$ determined subject to the requirement that $G_n(\lambda)$ be a polynomial of degree $n-1$; and the real number a_{nn} obviously satisfies the inequality $1 - \varepsilon \leqq a_{nn} \leqq 1$. A closer examination of the inequality given above shows that $|a_{mn}| < \operatorname{ext} P$ in all cases; but the question of the existence of a case in which $\alpha = \operatorname{ext} P$ remains open. The inequality $\operatorname{ext} P \leqq 3\alpha$ is trivial unless the right-hand member is finite. We shall therefore confine our attention to this case. If $g = \sum_{\alpha=1}^{n} x_\alpha g_\alpha$

is an arbitrary element in the domain of $\hat{T}_1(A)$ and if $|x_k| = \xi_k$, we have

$$|(\hat{T}_1(A)g, g)|$$
$$= \left|\sum_{\alpha,\beta=1}^{n} (Tg_\alpha, g_\beta) x_\alpha \bar{x}_\beta\right| = \left|\sum_{\alpha,\beta=1}^{n} a_{\beta\alpha} x_\alpha \bar{x}_\beta\right|$$
$$\leqq \sum_{\alpha,\beta=1}^{n} |a_{\beta\alpha}| \xi_\alpha \xi_\beta$$
$$\leqq \mathfrak{a}\left[\xi_1(\xi_1+\xi_2) + \sum_{\alpha=2}^{n-1} \xi_\alpha(\xi_{\alpha-1}+\xi_\alpha+\xi_{\alpha+1}) + \xi_n(\xi_{n-1}+\xi_n)\right]$$
$$\leqq \mathfrak{a}\left[2\xi_1^2 + 3\sum_{\alpha=2}^{n-1} \xi_\alpha^2 + 2\xi_n^2\right]$$
$$\leqq 3\mathfrak{a}\sum_{\alpha-1}^{n} \xi_\alpha^2 = 3\mathfrak{a}(g, g).$$

Hence the inequality $C \leqq 3\mathfrak{a}$ is valid. The closed linear symmetric transformation T is now seen to be bounded and hence self-adjoint, by virtue of Theorems 2.23 and 2.24. The argument used above shows that in these circumstances we have $C_2 = \inf P$, $C_1 = \sup P$, $C = \operatorname{ext} P$. Thus the inequality $\operatorname{ext} P \leqq 3\mathfrak{a}$ is established. It is easy to show that $\operatorname{ext} P = C = 3\mathfrak{a}$ for a suitably chosen Jacobi matrix A. In fact, if we put $a_n = 1$, $b_n = 1$, $g = \sum_{\alpha=1}^{n} g_\alpha$, we find that $(Tg, g) = (3-2/n)(g, g)$, $3-2/n \leqq C \leqq 3 = 3\mathfrak{a}$ and infer that $\operatorname{ext} P = C = 3\mathfrak{a}$. In order to prove that, in the general case, the real number a_{nn} exceeds C_2, we choose a self-adjoint transformation $H \supseteq T$ with the lower bound C_2. The associated function $\varrho(\lambda)$ then has the property that $\inf P = C_2$. We therefore have

$$a_{nn} = \int_{-\infty}^{+\infty} \lambda |G_n(\lambda)|^2 d\varrho(\lambda) > \inf P \int_{-\infty}^{+\infty} |G_n(\lambda)|^2 d\varrho(\lambda) = \inf P = C_2,$$

where the strict inequality must hold because $\varrho(\lambda)$ assumes infinitely many distinct values. In a similar manner, we prove that $a_{nn} < C_1$.

When T has the deficiency-index $(0, 0)$, it is a self-adjoint transformation and the associated function $\varrho(\lambda)$ is unique, as we pointed out in Theorem 10.27. The equations $S(T) = P$, $C_2 = \inf P$, $C_1 = \sup P$ are established by considering the image of T in the space $\mathfrak{L}_2(\varrho)$, as in similar situations discussed earlier. By reference to Theorem 10.28, we see that $\lim_{p\to\infty} \varrho_p(\lambda) = \varrho(\lambda)$ at the points of continuity of the latter function, where $\varrho_p(\lambda)$ is the function associated with the reduced Jacobi matrix A_p in the manner indicated in Theorem 10.23. Since the points of discontinuity of $\varrho_p(\lambda)$ are located at the roots of the polynomial $G_{p+1}(\lambda)$, we see that, if Δ is any closed interval at positive distance from the set Λ, then $\varrho_p(\lambda)$ is constant on Δ for all sufficiently large p. Hence $\varrho(\lambda)$ is constant on any such interval and the relation $P \subseteqq \Lambda$ is satisfied.

We turn now to the consideration of the case where T has the deficiency-index (1, 1). The general inequalities which we have established above, show that the three quantities C, $\mathfrak{a}$, ext P are all finite, or all infinite except C. If C were finite, the closed linear symmetric transformation T would be bounded and hence self-adjoint, contrary to hypothesis. We must therefore have $C = \mathfrak{a} = \text{ext } P = +\infty$. The analysis of this case depends upon the formulas

$$\frac{P(l) + U(l)\varphi(l)}{Q(l) + V(l)\varphi(l)} = \int_{-\infty}^{+\infty} \frac{1}{\lambda - l}\, d\varrho(\lambda),$$

$$\varphi(l) = \alpha l + \beta + \int_{-\infty}^{+\infty} \frac{\lambda l + 1}{\lambda - l}\, d\sigma(\lambda)$$

established in Theorems 10.36 and 10.38. If Δ is any closed interval at positive distance from the closed set Σ consisting of the points of increase of $\sigma(\lambda)$, the integral formula shows that the function $\varphi(l)$ is single-valued and analytic in the connected set obtained by adjoining to the half-planes $\mathfrak{I}(l) > 0$, $\mathfrak{I}(l) < 0$ the interval Δ on the real axis. Hence the function $I(l; \varrho) = \int_{-\infty}^{+\infty} \frac{1}{\lambda - l}\, d\varrho(\lambda)$ can be defined on Δ so that it is single-valued and analytic in this connected set except possibly for a finite number of poles on the interval Δ; if there are any poles on Δ they occur at the roots of the function $Q(l)+V(l)\varphi(l)$ and must therefore be distinct from the roots of $V(l)$, by virtue of the fact that $Q(l)$ and $V(l)$ have no common roots. The behavior of $I(l;\varrho)$ in the neighborhood of the interval Δ on the real axis shows that $\varrho(\lambda)$ increases at only a finite number of points on Δ. We can therefore conclude that the derived set P' is contained in Σ. If $\lambda=\mu$ is an isolated point of Σ, $\varphi(l)$ is single-valued and analytic in the neighborhood of $l=\mu$, where it has a simple pole. It is easily verified from the formula for $I(l;\varrho)$ that the latter function is analytic at $l=\mu$ when $V(\mu)\neq 0$ and has a simple pole at $l=\mu$ when $V(\mu)=0$. Thus the point $\lambda=\mu$ is either a point of constancy or an isolated point of increase for $\varrho(\lambda)$. We may therefore replace the relation $P'\subseteqq\Sigma$ by the sharper relation $P'\subseteqq\Sigma'$. It is to be noted that the case of the singular function $\varphi(l)=\infty$ must be reserved for separate treatment. In this case we take $\sigma(\lambda)=0$ and recall from Theorem 10.41 that $\varrho(\lambda)$ increases only as λ increases through a root of $V(l)$; we infer that both P' and Σ' are vacuous. If we write

$$\varphi(l)=\frac{-P(l)+Q(l)\,I(l;\varrho)}{U(l)-V(l)\,I(l;\varrho)},$$

we can discuss the influence of $\varrho(\lambda)$ upon $\sigma(\lambda)$ in a similar manner. We have to exclude the special case where the denominator $U(l)-V(l)\,I(l;\varrho)$ vanishes identically and $\varphi(l)$ is the singular function $\varphi(l)=\infty$. It is helpful, furthermore, to put the relation connecting $\varphi(l)$ and $\sigma(\lambda)$ in the form

$$I(l;\sigma)=\int_{-\infty}^{+\infty}\frac{1}{\lambda-l}\,d\sigma(\lambda)=\frac{\varphi(l)-\alpha' l-\beta}{l^2+1},$$

$$\alpha'=\alpha+\int_{-\infty}^{+\infty}d\sigma(\lambda),$$

obtained by the use of the identity $\dfrac{\lambda l+1}{\lambda-l}=l+\dfrac{l^2+1}{\lambda-l}$. If Δ is a closed interval at positive distance from the closed set P, both $\varphi(l)$ and $I(l;\sigma)$ are found to be analytic on Δ, except possibly for a finite number of simple poles which they share; if there are any poles, they occur at the roots of the function $U(l) - V(l)\, I(l; \varrho)$ and must therefore be distinct from the roots of $V(l)$, by virtue of the fact that $U(l)$ and $V(l)$ have no common roots. Thus $\sigma(\lambda)$ has only a finite number of points of increase on Δ. We conclude that $\Sigma' \subseteqq P$. If $\lambda = \mu$ is an isolated point of P, $I(l; \varrho)$ has a simple pole at $l = \mu$. It is easily verified from the formulas for $\varphi(l)$ and $I(l; \sigma)$, that these functions are analytic at $l = \mu$ when $V(\mu) \neq 0$ and have simple poles at $l = \mu$ when $V(\mu) = 0$. Thus the point $\lambda = \mu$ is either a point of constancy or an isolated point of increase for $\sigma(\lambda)$. We conclude that $\Sigma' \subseteqq P'$. By combining the results already established, we see that $P' = \Sigma'$, even in the case of the singular function $\varphi(l) = \infty$, where both sets are vacuous. Since the function $\sigma(\lambda)$ can be taken as any real monotone-increasing function in the class $\mathfrak{B}^*$, the closed set Σ' can be assigned at pleasure; and the closed set P' is therefore completely unrestricted.

We shall next discuss the particular case where T has the index $(1, 1)$ and the bounds $C_2 = 0$, $C_1 = +\infty$. Later we shall reduce the special cases described in the statement of the theorem to this one. We wish to ascertain some of the more significant properties of the class of associated functions $\varrho(\lambda)$ for which $\inf P$ has an assigned value. The inequality $\inf P \leqq C_2 = 0$ excludes all positive values from consideration. By applying the facts established in the preceding paragraph, we see that the equation $\inf P = \mu$, $-\infty < \mu \leqq 0$, holds if and only if the following conditions are satisfied:

(1) Σ' is an unrestricted closed set on the interval $\mu \leqq \lambda < +\infty$;

(2) the isolated points of Σ on the interval $-\infty < \lambda < \mu$ are distinct from the roots of $V(l)$;

(3) the function $Q(l) + V(l)\, \varphi(l)$ has no roots on the interval $-\infty < \lambda < \mu$;

(4) the point $\lambda = \mu$ has one of the following three incompatible properties:

(a) it is a point of Σ';
(b) it is both an isolated point of Σ and a root of $V(l)$;
(c) it is both a point in the complement of Σ and a root of the function $Q(l)+V(l)\varphi(l)$.

The case of the singular function $\varphi(l)=\infty$ may be included if we let Σ be vacuous and interpret the term "root of $Q(l)+V(l)\varphi(l)$" to mean "root of $V(l)$". The detailed examination of these conditions and their significance may be left to the reader. When $\sigma(\lambda)$ has been chosen so that (1) is satisfied, the function

$$\varphi(l)=\alpha l+\beta+\int_{-\infty}^{+\infty}\frac{\lambda l+1}{\lambda-l}\,d\sigma(\lambda)$$

is analytic except for simple poles on the entire open interval $-\infty<\lambda<\mu$ on the real axis. The function $-\dfrac{Q(l)}{V(l)}$ behaves in a similar manner. Hence the investigation of the remaining conditions depends upon the relative positions of the curves $y=\varphi(l)$, $y=-\dfrac{Q(l)}{V(l)}$, where l is a real variable on the interval $(-\infty,\mu)$. Since the derivative

$$\varphi'(l)=\alpha+\int_{-\infty}^{+\infty}\frac{\lambda^2+1}{(\lambda-l)^2}\,d\sigma(\lambda),\qquad \alpha\geqq 0,$$

is never negative for $\mathfrak{I}(l)=0$, the curve $y=\varphi(l)$ increases monotonely on any interval which contains no pole of $\varphi(l)$ and which is itself contained in the interval $(-\infty,\mu)$. The curve $y=-\dfrac{Q(l)}{V(l)}$ will be investigated in greater detail. First we observe that $Q(0)=0$ by virtue of the series expansion obtained from Theorems 10.26 and 10.33. According to Theorem 10.41, T has a unique self-adjoint extension with the characteristic value 0, the associated function $\varrho(\lambda)$ being determined from the relation $\displaystyle\int_{-\infty}^{+\infty}\frac{1}{\lambda-l}\,d\varrho(\lambda)=\frac{P(l)}{Q(l)}$. On the other hand, we know that T has a self-adjoint extension with lower bound equal to zero. With the help of Theorem 10.41 it is easy to identify the two self-adjoint extensions just described, and hence to show that the function $Q(l)$ has no roots on the negative real axis. The results given in that theorem concerning the roots of the function $S(l, \theta) = \cos\theta\, Q(l) + \sin\theta\, V(l)$, $-\frac{\pi}{2} < \theta \leqq +\frac{\pi}{2}$ can now be applied to prove that $S(l, \theta)$ has exactly one root on the negative real axis when $\theta \neq 0$, and that every point on the negative real axis is a root of $S(l, \theta)$ for some value of θ. It is evident that the existence of two such roots for $S(l, \theta)$ would require the existence of a separating root of $S(l, 0) = Q(l)$, in contradiction with what has just been proved. On the other hand, if $S(l, \theta)$ has no negative real root, its smallest root on the real axis is different from that of $S(l, 0) = Q(l)$ and is therefore positive. The roots of $S(l, \theta)$ form the spectrum of a self-adjoint extension of T, so that T must be positive definite rather than not-negative definite, as we assumed above. Hence $S(l, \theta)$ has just one negative real root when $\theta \neq 0$. In particular, $V(l) = S\left(l, \frac{\pi}{2}\right)$ has such a root, which we denote henceforth by ν. The fact that every negative real number is a root of some function $S(l, \theta)$ has been proved in Theorem 10.41. These properties of the functions $S(l, \theta)$ show that the quotient $-\frac{Q(l)}{V(l)}$ is analytic on the negative real axis save for a simple pole at $l = \nu$ and assumes the value $y = \tan\theta$, $-\frac{\pi}{2} < \theta < \frac{\pi}{2}$, for just one not-positive value of l. In order to ascertain the behavior of $-\frac{Q(l)}{V(l)}$ in the neighborhood of the point $l = \nu$ we employ the identity $U(l)\,Q(l) - P(l)\,V(l) = 1$ established in the proof of Theorem 10.41, writing it in the form $-\frac{Q}{V} \cdot V^2\left(\frac{U}{V} - \frac{P}{Q}\right) = -1$. Since we have

$$\frac{U(l)}{V(l)} = \int_{-\infty}^{+\infty} \frac{1}{\lambda - l}\, d\varrho(\lambda)$$

in accordance with Theorem 10.38, $\frac{U(l)}{V(l)}$ becomes positively infinite when l approaches ν through real values less than ν. At the same time, $\frac{P(l)}{Q(l)}$ remains bounded, being analytic at $l = \nu$. Hence the product $V^2\left(\frac{U}{V} - \frac{P}{Q}\right)$ assumes positive values whenever l assumes real values less than ν in the neighborhood of the point $l = \nu$; and the function $-\frac{Q(l)}{V(l)}$ therefore becomes negatively infinite as l approaches ν through real values less than ν. The information now at our disposal shows that the curve $y = -\frac{Q(l)}{V(l)}$ decreases monotonely from 0 to $-\infty$ as l increases through real values from $-\infty$ to ν and decreases monotonely from $+\infty$ to 0 as l increases through real values from ν to 0.

We are now prepared to construct functions $\varrho(\lambda)$ for which $\inf \boldsymbol{P}$ is a preassigned negative number and $\boldsymbol{P}'$ a preassigned closed set on the interval $\inf \boldsymbol{P} \leqq \lambda < +\infty$. We denote the given number and the given closed set by μ and $\boldsymbol{\Pi}$ respectively. We begin by choosing a real monotone-increasing function $\sigma_0(\lambda)$ in the class $\mathfrak{B}^*$ for which $\boldsymbol{\Sigma}_0$ is a subset of the interval $\mu \leqq \lambda < +\infty$ with the property $\boldsymbol{\Sigma}_0' = \boldsymbol{\Pi}$. We then form the function

$$\sigma(\lambda) = \int_{-\infty}^{\lambda} \frac{\lambda - \mu}{\lambda^2 + 1} d\sigma_0(\lambda),$$

which is obviously a real monotone-increasing function in $\mathfrak{B}^*$. It is easily verified that the set $\boldsymbol{\Sigma}$ is contained in the interval $\mu \leqq \lambda < +\infty$ and has the property that $\boldsymbol{\Sigma}' = \boldsymbol{\Pi}$. By virtue of the relations

$$\int_{-\infty}^{+\infty} \frac{\lambda l + 1}{\lambda - l} d\sigma(\lambda)$$
$$= -\int_{-\infty}^{+\infty} \lambda \, d\sigma(\lambda) + \int_{-\infty}^{+\infty} \frac{\lambda^2 + 1}{\lambda - l} d\sigma(\lambda)$$

$$= -\int_{-\infty}^{+\infty} \frac{\lambda(\lambda-\mu)}{\lambda^2+1}\, d\sigma_0(\lambda) + \int_{-\infty}^{+\infty} \frac{\lambda-\mu}{\lambda-l}\, d\sigma_0(\lambda)$$

$$= -\int_{\mu}^{+\infty} \frac{\lambda(\lambda-\mu)}{\lambda^2+1}\, d\sigma_0(\lambda) + \int_{\mu}^{+\infty} \frac{\lambda-\mu}{\lambda-l}\, d\sigma_0(\lambda),$$

we see that the function

$$\varphi(l) = \beta + \int_{-\infty}^{+\infty} \frac{\lambda l+1}{\lambda-l}\, d\sigma(\lambda)$$

has the properties

$$\lim_{l\to-\infty} \varphi(l) = \beta - \int_{-\infty}^{+\infty} \frac{\lambda(\lambda-\mu)}{\lambda^2+1}\, d\sigma_0(\lambda) = \varphi_1,\ \mathfrak{J}(l) = 0,$$

$$\lim_{l\to\mu} \varphi(l) = \beta - \int_{-\infty}^{+\infty} \frac{\lambda(\lambda-\mu)}{\lambda^2+1}\, d\sigma_0(\lambda) + \int_{\mu+0}^{+\infty} d\sigma_0(\lambda) = \varphi_2,$$

$$\mathfrak{J}(l) = 0,$$

and is analytic on the interval $-\infty < l < \mu$, $\mathfrak{J}(l) = 0$. Obviously, we may arrange that $\varphi_2 - \varphi_1 = \int_{\mu+0}^{+\infty} d\sigma_0(\lambda)$ have any desired positive value without imposing any essential restriction on $\sigma_0(\lambda)$; and we may arrange that φ_1 have any desired real value merely by selecting β in an appropriate manner. Since the conditions imposed upon $\sigma_0(\lambda)$ allow us to choose this function in $\mathfrak{c}$ ways and hence to construct $\mathfrak{c}$ corresponding functions $\sigma(\lambda)$, we now have at our disposal $\mathfrak{c}$ functions $\varphi(l)$ such that the curve $y = \varphi(l)$ increases monotonely from a preassigned value φ_1 to a preassigned value $\varphi_2 > \varphi_1$, when l increases through real values from $-\infty$ to μ. In the special case where $\mu = \nu$ and the point $\lambda = \nu$ does not belong to $\Sigma' = \Pi$, we must be able to take $\varphi_2 = +\infty$. We can do so by replacing the function $\sigma(\lambda)$ by a new function obtained from the old by the addition of a function which is equal to 0 or to 1 according as $\lambda < \nu$ or $\lambda \geqq \nu$; the new function $\varphi(l)$ so obtained differs from the old by the additive term $\dfrac{\nu l+1}{\nu-l}$ and therefore behaves in the desired manner. When $\nu < \mu < 0$, we take $\varphi_1 = 0$, $\varphi_2 = -\dfrac{Q(\mu)}{V(\mu)}$, so that the curves $y = \varphi(l)$

and $y = -\frac{Q(l)}{V(l)}$ do not intersect on the interval $-\infty < l < \mu$. It is then evident that our construction satisfies the first three conditions laid down in the preceding paragraph. The three possibilities stated in the fourth condition must be examined in detail. The condition 4 (a) is self-explanatory and requires no discussion. The condition 4 (b) cannot be realized in the present instance, since $V(\mu) \neq 0$. When $\lambda = \mu$ is not a point of $\Sigma' = \Pi$, we must therefore prove that condition 4 (c) is satisfied. First, it is clear that the point $\lambda = \mu$ is not a point of Σ: it does not belong to Σ', by hypothesis; and it cannot be an isolated point of Σ since the integral formula connecting $\sigma(\lambda)$ and $\sigma_0(\lambda)$ implies that $\sigma(\mu + 0) = \sigma(\mu - 0) = 0$. Secondly, the curves $y = \varphi(l)$ and $y = -\frac{Q(l)}{V(l)}$ are now defined on an open interval containing the interval $(-\infty, \mu)$ and, by construction, intersect when $l = \mu$. Hence all four conditions are satisfied and there exist $\mathfrak{c}$ functions $\varrho(\lambda)$, obtained from the functions $\varphi(l)$ which we have constructed above, with the properties $\inf P = \mu$, $P' = \Pi$, when $\nu < \mu < 0$. The case where $-\infty < \mu < \nu$ is treated in a similar manner, by taking $\varphi_2 = -\frac{Q(\mu)}{V(\mu)}$, $\varphi_1 < \varphi_2$. The case $\mu = \nu$ is most easily discussed by putting $\varphi_1 = 0$, $\varphi_2 = +\infty$ in the manner indicated above. Conditions (1)–(3) are then satisfied, as before. Since $\lambda = \nu$ is now a point of Σ, we must show that, when 4 (a) is not satisfied, 4 (b) is valid. It is evident upon inspection that such is the case. The case $\mu = -\infty$ does not require such cautious treatment: it is easily verified that it is sufficient to take $\Sigma' = \Pi$, $\inf \Sigma = -\infty$.

We have still to investigate the situation which arises when we require that $\inf P = 0$. In order that condition (3) may be satisfied, the curves $y = \varphi(l)$ and $y = -\frac{Q(l)}{V(l)}$ must not intersect on the open intervals $(-\infty, \nu)$, $(\nu, 0)$. Hence we must have $\varphi(l) \geqq 0$ on the first interval, $\varphi(l) \leqq 0$ on the second. These inequalities can be satisfied in just two ways: either $\varphi(l)$ increases monotonely from φ_1 to $+\infty$ on the first interval and from $-\infty$ to φ_2 on the second, with $\varphi_1 \geqq 0$ and $\varphi_2 \leqq 0$; or $\varphi(l) = 0$ for all negative real l. The first alternative is in contradiction with condition (2) and must therefore be discarded: it requires that $\varphi(l)$ have a pole for $l = \nu$ and hence that ν be an isolated point of Σ. The second alternative requires that $\varphi(l)$ vanish identically, since it is analytic on the negative real axis. Hence there is just one function $\varrho(\lambda)$ such that $\inf P = 0$, namely, the function defined by the relation $\dfrac{P(l)}{Q(l)} = \displaystyle\int_{-\infty}^{+\infty} \frac{1}{\lambda - l}\, d\varrho(\lambda)$.

The self-adjoint extensions of T and the corresponding functions $\varrho(\lambda)$ are readily discussed by applying to the general statements of Theorem 10.41 the special properties of the functions $S(l, \theta)$ which we have established above for the case where T has the lower bound zero.

Finally, we shall reduce the cases $C_2 > -\infty$, $C_1 < +\infty$ to the particular case $C_2 = 0$ which we have just studied. When $C_2 > -\infty$, we consider the Jacobi matrix $A_0 = \{a_{mn} - C_2 \delta_{mn}\}$ in connection with the given Jacobi matrix A. The transformations T_0 and T associated with A_0 and A respectively by the complete orthonormal set $\{g_n\}$ are connected by the relation $T_0 \equiv T - C_2 I$ and therefore have the lower bounds 0 and C_2 respectively. The self-adjoint extensions of T_0 and T are in one-to-one correspondence by virtue of the relation $H_0 \equiv H - C_2 I$. The constants $c_n^{(0)}$ and c_n associated with A_0 and A respectively are related by the equations

$$\begin{aligned} c_n^{(0)} &= ((T_0)^n g_1, g_1) = ((T - C_2 I)^n g_1, g_1) \\ &= \sum_{\nu=0}^{n} (-1)^\nu \frac{n!}{\nu!\,(n-\nu)!} C_2^\nu c_{n-\nu}, \\ c_n &= (T^n g_1, g_1) = ((T_0 + C_2 I)^n g_1, g_1) \\ &= \sum_{\nu=0}^{n} \frac{n!}{\nu!\,(n-\nu)!} C_2^\nu c_{n-\nu}^{(0)}, \end{aligned}$$

by virtue of Theorem 10.27 (3) and Theorem 9.19. These equations show that if $\varrho_0(\lambda)$ and $\varrho(\lambda)$ are real monotone-increasing functions in $\mathfrak{B}^*$ such that $\varrho_0(\lambda) = \varrho(\lambda + C_2)$ the equations $\int_{-\infty}^{+\infty} \lambda^n \, d\varrho_0(\lambda) = c_n^{(0)}$, $n \geqq 0$, are equivalent to the equations $\int_{-\infty}^{+\infty} \lambda^n \, d\varrho(\lambda) = c_n$, $n \geqq 0$. Theorem 10.30 now shows that the functions $\varrho_0(\lambda)$ associated with A_0 and the functions $\varrho(\lambda)$ associated with A are in one-to-one correspondence by virtue of the relation $\varrho_0(\lambda) = \varrho(\lambda + C_2)$. The reduction of the case $C_2 > -\infty$ is thus completed. When $C_1 < +\infty$, we first consider the Jacobi matrix $A_0 = -A$. As analogues of the relations just discussed in the case $C_2 > -\infty$, we now have $T_0 \equiv -T$, $C_2^{(0)} = -C_1$, $H_0 \equiv -H$, $c_n^{(0)} = (-1)^n c_n$, $\varrho_0(\lambda) = 1 - \varrho(-\lambda - 0)$. Thus we can reduce the case $C_1 < +\infty$ to the case $C_2 > -\infty$ and hence to the case $C_2 = 0$.

In Theorem 10.30 we showed that the theory of Jacobi matrices is intimately related to the following problem, known as the moment problem: when a sequence $\{c_n\}$, $n \geqq 0$, of real constants is given, it is required to determine all real monotone-increasing functions $\varrho(\lambda)$ in the class $\mathfrak{B}^*$ such that $\int_{-\infty}^{+\infty} \lambda^n \, d\varrho(\lambda) = c_n$. If we interpret such a function $\varrho(\lambda)$ in physical terms as the mass of that portion of matter lying on the interval $(-\infty, \lambda)$ in the case of a linear material distribution, the constant c_n is known as the nth moment of the distribution or of the function $\varrho(\lambda)$. In the theory of probability it is useful to interpret $\varrho(\lambda)$ as a cumulative probability (or frequency) function, $\varrho(\lambda)$ being equal to the probability that a given statistical variable assume a value on the interval $(-\infty, \lambda)$. The problem stated above may therefore be formulated in more concrete terms, as follows: it is required to determine all the material or probability distributions with preassigned moments. This formulation has caused the problem to be called the moment problem. We may observe that the equation $c_0 = 0$ leads to the trivial solution $\varrho(\lambda) = 0$ and that all the higher moments must vanish. We shall therefore discard this case. If $c_0 \neq 0$, we must have $c_0 > 0$. When this inequality is satisfied, we may normalize the problem so that $c_0 = 1$: we have only to replace $\varrho(\lambda)$ and c_n by $\varrho(\lambda)/c_0$ and c_n/c_0 respectively. In all subsequent discussions we shall assume that the constant c_0 has the value 1. Any solution of the normalized problem evidently has the property $\lim_{\lambda\to+\infty} \varrho(\lambda) = 1$. We shall apply the theory of Jacobi matrices to solve the moment problem in this normalized form.

We first study more carefully the connection between a Jacobi matrix A and the associated sequence of constants. We have

THEOREM 10.43. *Let $\{c_n\}$ be a sequence of real constants, $n \geqq 0$, with $c_0 = 1$; and let D_n be the determinant $\det_{j,k=0,\cdots,n-1} \{c_{j+k}\}$. A necessary and sufficient condition that there exist a reduced Jacobi matrix A_p of order p with the associated constant $c_n(p)$ equal to the given constant c_n for $0 \leqq n \leqq 2p-1$, $p \geqq 1$, is that D_n be positive for $1 \leqq n \leqq p$. A necessary and sufficient condition that the sequence $\{c_n\}$ be the sequence of constants associated with a Jacobi matrix A is that $D_n > 0$ for $n \geqq 1$. Two Jacobi matrices $A^{(1)} = \{a^{(1)}_{mn}\}$ and $A^{(2)} = \{a^{(2)}_{mn}\}$ have the same sequence of associated constants if and only if $|a^{(1)}_{mn}| = |a^{(2)}_{mn}|$ for $m \geqq 1$, $n \geqq 1$.*

When A_p is a reduced Jacobi matrix of order p such that $c_n(p) = c_n$ for $0 \leqq n \leqq 2p-1$, we regard A_p as obtained from a Jacobi matrix A. The first $2p$ constants of the sequence associated with A then coincide, in order, with the constants $c_n(p) = c_n$, $n = 0, \cdots, 2p-1$. Hence the determinants D_n, $1 \leqq n \leqq p$, coincide, in order, with the analogous determinants formed from the sequence of constants associated with A. As we saw in the proof of Theorem 10.24 (4), the latter determinants are expressible as products of absolute values of non-vanishing elements of the matrix A. Thus it is necessary that D_n be positive for $1 \leqq n \leqq p$. Similarly, when $\{c_n\}$ is the sequence of constants associated with a Jacobi matrix A, the determinants D_n must be positive for all n.

When the sequence $\{c_n\}$ is given, we can define a linear operation L in the class of all complex-valued polynomials

$P(\lambda)$, as follows: if $P(\lambda) = \sum_{\alpha=0}^{k} x_\alpha \lambda^\alpha$, we set $L(P(\lambda)) = \sum_{\alpha=0}^{k} x_\alpha c_\alpha$. It is easily verified that $L(a_1 P_1 + a_2 P_2) = a_1 L(P_1) + a_2 L(P_2)$. If we wish to consider only the first $2p$ constants, we must confine our attention to the class of all polynomials $P(\lambda)$ of degree less than or equal to $2p-1$. If the determinants $D_1, \cdots, D_k$, $k \geqq 1$, are positive, there exists a polynomial $G_k(\lambda)$ of degree $k-1$ with the properties $L(\lambda^i G_k) = 0$, $0 \leqq i \leqq k-2$, $L(|G_k|^2) = 1$ In the case $k = 1$, we take $G_k(\lambda) = 1$. When $k \geqq 2$, we substitute a polynomial $P(\lambda) = \sum_{\alpha=0}^{k-1} x_\alpha \lambda^\alpha$, with undetermined coefficients, in the relations which are to be satisfied; we thus obtain the conditions

$$L(\lambda^i P) = \sum_{\alpha=0}^{k-1} c_{i+\alpha} x_\alpha = 0, \ 0 \leqq i \leqq k-2;$$

$$L(|P|^2) = \sum_{\alpha,\beta=0}^{k-1} c_{\alpha+\beta} x_\alpha \bar{x}_\beta = 1.$$

If we assign to x_{k-1} an arbitrary value $\xi \neq 0$, the remaining coefficients $x_0, \cdots, x_{k-2}$ are uniquely determined from the equations

$$\sum_{\alpha=0}^{k-2} c_{i+\alpha} x_\alpha = -c_{i+k-1}\xi, \ 0 \leqq i \leqq k-2,$$

since $D_{k-1} > 0$; they are proportional to ξ. The Hermitian form $\sum_{\alpha,\beta=0}^{k-1} c_{\alpha+\beta} x_\alpha \bar{x}_\beta$ is positive definite by virtue of the inequalities $D_1 > 0, \cdots, D_k > 0$. Hence if we assign to $x_0, \cdots, x_{k-1}$ the values just determined, the form assumes the value $C|\xi|^2$, where C is a positive real constant dependent upon the first $2k$ constants of the sequence $\{c_n\}$. We may therefore complete the determination of the polynomial $G_k(\lambda)$ by putting $\xi = \gamma_k C^{-1/2}$, where γ_k is any complex number of absolute value 1. Thus the polynomial $G_k(\lambda)$ exists and is uniquely determined save for the constant factor γ_k. If $D_k > 0$ for $k = 1, \cdots, p$, we may therefore construct the polynomials $G_1(\lambda), \cdots, G_p(\lambda)$, and, if $D_k > 0$

for all $k \geqq 1$, we obtain similarly an infinite sequence of polynomials $\{G_k(\lambda)\}$. In either case, these polynomials clearly have the property that $L(G_i \overline{G}_k) = \delta_{ik}$. Similarly, the constants $a_{mn} = L(\lambda G_n \overline{G}_m)$ satisfy the relations $a_{mn} = \overline{a}_{nm}$, $a_{mn} = 0$ when $|m-n| > 1$. We shall write $a_n = a_{nn}$, $b_n = a_{n,n+1}$. It is then possible to show that the recursion formulas

$$\begin{aligned} \overline{b}_1 G_2 &= (\lambda - a_1) G_1 = \lambda - a_1, \\ \overline{b}_{n-1} G_n &= (\lambda - a_{n-1}) G_{n-1} - b_{n-1} G_{n-2}, \quad n \geqq 3, \end{aligned}$$

are valid for the finite or infinite sequence of polynomials at our disposal. In particular, we see that the constants b_n appearing on the left must be different from zero for the indicated values of n. The proof depends upon the fact that any polynomial $P(\lambda)$ of degree less than or equal to $n-1$ which satisfies the equations $L(P\overline{G}_m) = 0$, $1 \leqq m \leqq n$, must vanish identically. If we write $P(\lambda) = \sum_{\alpha=0}^{n-1} x_\alpha \lambda^\alpha$ and replace the indicated conditions by the equivalent set $L(P\lambda^i) = 0$, $0 \leqq i \leqq n-1$, we find that the coefficients of $P(\lambda)$ must satisfy the equations $\sum_{\alpha=0}^{n-1} c_{i+\alpha} x_\alpha = 0$, $0 \leqq i \leqq n-1$; since the determinant D_n is positive, these coefficients must all vanish. Putting $b_0 = 1$, $G_0 = 0$, we now observe that the polynomial $\lambda G_{n-1} - \overline{b}_{n-1} G_n - a_{n-1} G_{n-1} - b_{n-2} G_{n-2}$ is of degree less than or equal to $n-1$ and satisfies the equations

$$\begin{aligned} L((\lambda G_{n-1} &- \overline{b}_{n-1} G_n - a_{n-1} G_{n-1} - b_{n-2} G_{n-2}) \overline{G}_m) \\ &= a_{m,n-1} - \overline{b}_{n-1} \delta_{nm} - a_{n-1} \delta_{n-1,m} - b_{n-2} \delta_{n-2,m} = 0 \end{aligned}$$

for $1 \leqq m \leqq n$, $n \geqq 2$. Hence this polynomial must vanish identically and the indicated recursion formulas are valid. When $D_k > 0$ for $k = 1, \cdots, p$, we define a reduced Jacobi matrix $A_p = \{a_{mn}^{(p)}\}$ by setting $a_{mn}^{(p)} = a_{mn}$, when $1 \leqq m \leqq p$ and $1 \leqq n \leqq p$, and $a_{mn}^{(p)} = 0$, when $m > p$ or $n > p$. The polynomials $G_1(\lambda), \cdots, G_p(\lambda)$ employed in the construction of this matrix must coincide with the associated polynomials by virtue of the recursion formulas just established. Hence the constants $c_n(p)$, $0 \leqq n \leqq 2p-1$, associated with A_p can be calculated as follows:

$$c_n(p) = \sum_{\alpha=1}^{j+1} \sum_{\beta=1}^{k+1} C_{j\alpha} \bar{C}_{k\beta}\, a_{\beta\alpha}$$

$$= L\left(\lambda\left(\sum_{\alpha=1}^{j+1} C_{j\alpha} G_\alpha\right)\left(\sum_{\beta=1}^{k+1} \bar{C}_{k\beta} \bar{G}_\beta\right)\right) = L(\lambda \cdot \lambda^j \cdot \lambda^k) = c_n,$$

$$0 \leqq n = 1+j+k \leqq 2p-1, \quad 0 \leqq j \leqq p-1, \quad 0 \leqq k \leqq p-1,$$

$$\lambda^m = \sum_{\alpha=1}^{m+1} C_{m\alpha} G_\alpha(\lambda), \quad 0 \leqq m \leqq p-1.$$

Thus the first $2p$ constants of the given sequence $\{c_n\}$ are coincident, in order, with the constants $c_n(p)$, $0 \leqq n \leqq 2p-1$, associated with the reduced Jacobi matrix A_p which we have constructed on the basis of the assumption that $D_k > 0$ for $1 \leqq k \leqq p$. Similarly, when the determinants D_k are positive for $k \geqq 1$, we consider the Jacobi matrix $A = \{a_{mn}\}$, where a_{mn} is the constant defined above. We find that the sequence of polynomials $\{G_k(\lambda)\}$ used in the construction is the sequence associated with the matrix A and that the given sequence of constants is identical with the sequence associated with A.

The analysis of the preceding paragraph leads at once to the necessary and sufficient conditions that two Jacobi matrices have the same sequence $\{c_n\}$ of associated constants. In fact, if we start from the sequence $\{c_n\}$, known to be associated with one Jacobi matrix, the construction given above evidently yields all possible Jacobi matrices with the same associated sequence, distinct matrices being obtained by different choices of the sequence $\{\gamma_k\}$ encountered in the definition of the sequence of polynomials $\{G_k(\lambda)\}$. The conditions stated in the theorem then follow without difficulty.

THEOREM 10.44. *Let* $\{c_n\}$, $n \geqq 0$, *be a sequence of real constants with* $c_0 = 1$. *In order that there exist a real monotone-increasing function* $\varrho(\lambda)$ *in the class* $\mathfrak{B}^*$ *which assumes exactly* $p+1$ *distinct values and which has the first* $2p$ *constants of the given sequence as its first* $2p$ *moments, it is necessary and sufficient that one of the following two equivalent conditions be satisfied:*

(1) *the first* $2p$ *constants of the given sequence coincide, in order, with the constants* $c_n(p)$, $0 \leqq n \leqq 2p-1$, *associated with a reduced Jacobi matrix matrix* A_p *of order* p;

(2) *the determinants* D_n *are positive for* $1 \leqq n \leqq p$.

When these conditions are satisfied, the function $\varrho(\lambda)$ *is unique; and its moments of order* $n \geqq 2p$ *are functions of the constants* $c_0, \cdots, c_{2p-1}$. *In order that there exist a real monotone-increasing function* $\varrho(\lambda)$ *in the class* $\mathfrak{B}^*$ *which assumes infinitely many distinct values and which has the given sequence as its sequence of moments, it is necessary and sufficient that one of the following two equivalent conditions be satisfied:*

(1) *the given sequence is the sequence of constants associated with a Jacobi matrix* A;

(2) *the determinants* D_n *are positive for* $n \geqq 1$.

When these conditions are satisfied, the totality of functions $\varrho(\lambda)$ *with the indicated properties has the cardinal number* 1 *or the cardinal number* $\mathfrak{c}$ *according as the Jacobi matrix* A *of condition* (1) *determines an associated closed linear symmetric transformation* T *with the deficiency-index* (0,0) *or* (1,1). *If the sequence* $\{c_n\}$ *is the sequence of moments for some real monotone-increasing function* $\varrho(\lambda)$ *in the class* $\mathfrak{B}^*$, *then*

(1) $\varrho(\lambda)$ *assumes exactly* $p+1$ *distinct values if and only if* $D_n > 0$ *for* $1 \leqq n \leqq p$ *and* $D_n = 0$ *for* $n > p$;

(2) $\varrho(\lambda)$ *assumes infinitely many distinct values if and only if* $D_n > 0$ *for all* $n \geqq 1$.

If $\varrho(\lambda)$ is a real monotone-increasing function in $\mathfrak{B}^*$ with the moments $\{c_n\}$, $c_n = \int_{-\infty}^{+\infty} \lambda^n \, d\varrho(\lambda)$, the Hermitian form $\sum_{\alpha,\beta=0}^{n-1} c_{\alpha+\beta} x_\alpha \bar{x}_\beta = \int_{-\infty}^{+\infty} \left| \sum_{\alpha=0}^{n-1} x_\alpha \lambda^\alpha \right|^2 d\varrho(\lambda)$ is never negative. In order that $\varrho(\lambda)$ assume infinitely many distinct values it is necessary and sufficient that this form be positive definite for $n \geqq 1$, and thus necessary that the determinant D_n be positive for $n \geqq 1$. In order that $\varrho(\lambda)$ assume exactly $p+1$ distinct values it is necessary and sufficient that this form be positive definite for $1 \leqq n \leqq p$ and not-negative definite for $n > p$, and thus necessary that D_n be positive for $1 \leqq n$

$\leqq p$ and vanish for $n > p$. These results follow at once from the fact that a polynomial $P(\lambda) = \sum_{\alpha=0}^{n-1} x_\alpha \lambda^\alpha$ has the property $\int_{-\infty}^{+\infty} |P(\lambda)|^2 d\varrho(\lambda) = 0$ if and only if all the points of increase of $\varrho(\lambda)$ are roots of $P(\lambda)$. All the necessary conditions on the determinants D_n stated in the theorem now appear as consequences of these results; and the sufficiency of the conditions in the final assertion of the theorem can be deduced from them by an elementary argument, which we leave to the reader.

We now consider the two conditions for the existence of a function $\varrho(\lambda)$ with exactly $p+1$ values and assigned moments of order $n = 0, \cdots, 2p-1$. The equivalence of the two conditions was proved in the preceding theorem. Since we have just shown the second condition to be necessary, the first condition is also necessary. Since the first condition is sufficient, by virtue of Theorem 10.23, the second condition is also sufficient. The uniqueness of the function $\varrho(\lambda)$ can be established most easily by calculating its points of discontinuity $\lambda_1, \cdots, \lambda_p$ and the quantities $\pi_k = \varrho(\lambda_k) - \varrho(\lambda_k - 0)$, $1 \leqq k \leqq p$, in terms of the constants $c_0, \cdots, c_{2p-1}$. We determine the polynomial $P(\lambda) = \sum_{\alpha=0}^{p} x_\alpha \lambda^\alpha$ so that

$$\sum_{\alpha=0}^{p} c_{i+\alpha} x_\alpha = \int_{-\infty}^{+\infty} \lambda^i P(\lambda)\, d\varrho(\lambda) = 0$$

for $0 \leqq i \leqq p-1$, by setting $x_p = 1$ and choosing the remaining coefficients as solutions of the equations

$$\sum_{\alpha=0}^{p-1} c_{i+\alpha} x_\alpha = -c_{i+p}, \qquad 0 \leqq i \leqq p-1,$$

with the determinant $D_p > 0$. It is evident that $P(\lambda)$ is uniquely determined, save for a constant factor, by the constants $c_0, \cdots, c_{2p-1}$. With this polynomial we now have

$$\int_{-\infty}^{+\infty} \lambda^i P(\lambda)\, d\varrho(\lambda) = \sum_{\alpha=1}^{p} \lambda_\alpha^i P(\lambda_\alpha) \pi_\alpha = 0, \quad 0 \leqq i \leqq p-1.$$

Since the Vandermonde determinant $\det_{i+1,k=1,\cdots,p} \{\lambda_k^i\}$ does not vanish, we conclude that $\pi_k P(\lambda_k) = 0$ and hence that $\lambda_1, \cdots, \lambda_p$ are roots of the polynomial $P(\lambda)$. Thus the points of discontinuity of $\varrho(\lambda)$ are uniquely determined by $c_0, \cdots, c_{2p-1}$. The relations

$$\int_{-\infty}^{+\infty} \lambda^i \, d\varrho(\lambda) = \sum_{\alpha=1}^{p} \lambda_\alpha^i \pi_\alpha = c_i, \quad 0 \leqq i \leqq p-1,$$

show in a similar manner that the quantities $\pi_1, \cdots, \pi_p$ are likewise uniquely determined. Thus the function $\varrho(\lambda)$ is unique under the stated conditions.

The conditions for the existence of a function $\varrho(\lambda)$ with infinitely many values and assigned moments are equivalent by the preceding theorem. The second condition, and hence also the first, is necessary by the results proved above. The first condition, and hence also the second, is sufficient by Theorems 10.27 and 10.30. The cardinal number of the totality of functions satisfying the indicated conditions has been determined in Theorem 10.30.

Theorem 10.44 evidently reduces the study of the moment problem to the translation of the results obtained in the preceding theory of Jacobi matrices. Thus Theorems 10.38–10.42 contain fairly complete information about the indeterminate case, in which the function $\varrho(\lambda)$ is not uniquely determined by its moments. Theorem 10.42 gives special information concerning the set of points of increase of the function $\varrho(\lambda)$ in the general case. It should be observed that the sets Λ_2 and Λ_1 which play so important a rôle in that theorem can be characterized by means of the expressions

$$\det_{j,k=0,\cdots,n-1} \{c_{j+k+1} - \lambda c_{j+k}\}, \quad (-1)^{n-1} \det_{j,k=0,\cdots,n-1} \{c_{j+k+1} - \lambda c_{j+k}\},$$

formed from the moments $\{c_n\}$. This observation depends upon the formula of Theorem 10.24 (4).

For further information on the questions discussed in this section, we shall refer to the literature,* where many different

* Stieltjes, Annales de la Faculté des Sciences de Toulouse, (1) 8 (1894), pp. J1–J122, (1) 9 (1895), pp. A5–A47;—F. Riesz, Annales de l'École

methods for treating the theory of continued fractions, the moment problem, and related topics will be found. In particular, we call attention to the fact that examples of the indeterminate case of the moment problem are given in some of the references.

Normale Supérieure, (3) 28 (1911), pp. 33–62;—Hamburger, Mathematische Annalen, 81 (1920), pp. 235–319; 82 (1920), pp. 120–164; pp. 168–187;—R. Nevanlinna, Annales Academiae Scientiarum Fennicae, (A) 18 (1922), no. 5; (A) 32 (1929), no. 7;—M. Riesz, Arkiv för Matematik, Astronomi och Fysik, 16 (1922), nos. 12 and 19; 17 (1922–23), no. 16; Acta Litterarum ac Scientiarum, Sectio Mathematicarum, Szeged, 1 (1922–23), pp. 209–227;—Hellinger and Toeplitz, Journal für Mathematik, 144 (1914), pp. 212–238;—Carleman, *Les équations intégrales singulières à noyau réel et symétrique*, Uppsala, 1923, pp. 189–220;—Hausdorff, Mathematische Zeitschrift, 16 (1923), pp. 220–248;—Perron, *Die Lehre von den Kettenbrüchen*, second edition, Leipzig, 1929, Chapter IX;—Fréchet and Shohat, Transactions of the American Mathematical Society, 33 (1932), pp. 533–543. Examples of the indeterminate case of the moment problem are given by Stieltjes, Perron, and Fréchet and Shohat.

INDEX

Numbers refer to the appropriate pages.
Under the heading Symbols will be found references to those places where notations which are continually employed are first introduced.

COLLOQUIUM PUBLICATIONS

Volume I. H. S. WHITE, Linear Systems of Curves on Algebraic Surfaces; F. S. WOODS, Forms of Non-Euclidean Space; E. B. VAN VLECK, Topics in the Theory of Divergent Series and of Continued Fractions. 1905. 12 + 187 pp. $2.75.

Volume II. Out of Print.

Volume III. Out of Print.

Volume IV. L. E. DICKSON, On Invariants and the Theory of Numbers; W. F. OSGOOD, Topics in the Theory of Functions of Several Complex Variables. 1914. 4 + 230 pp. $2.50.

Volume V, PART I. Out of print.

Volume V, PART II. OSWALD VEBLEN, Analysis Situs. Second edition. 1931. 10 + 194 pp. $2.00.

Volume VI. G. C. EVANS, The Logarithmic Potential. Discontinuous Dirichlet and Neumann Problems. 1927. 8 + 150 pp. $2.00.

Volume VII. E. T. BELL, Algebraic Arithmetic. 1927. 4 + 180 pp. $2.50.

Volume VIII. L. P. EISENHART, Non-Riemannian Geometry. 1927. 8 + 184 pp. $2.50.

Volume IX. G. D. BIRKHOFF, Dynamical Systems. 1927. 8+295 pp. $3.00.

Volume X. A. B. COBLE, Algebraic Geometry and Theta Functions. 1929. 8 + 282 pp. $3.00.

Volume XI. DUNHAM JACKSON, The Theory of Approximation. 1930. 8 + 178 pp. $2.50.

Volume XII. S. LEFSCHETZ, Topology. 1930. 10 + 410 pp. $4.50.

Volume XIII. R. L. MOORE, Foundations of Point Set Theory. 1932. 7 + 486 pp. $5.00.

Volume XIV. J. F. RITT, Differential Equations from the Algebraic Standpoint. 1932. 10 + 172 pp. $2.50.

Volume XV. M. H. STONE, Linear Transformations in Hilbert Space. 1932. 8 + 622 pp. $6.50.

Orders may be addressed to

AMERICAN MATHEMATICAL SOCIETY
501 West 116th Street
NEW YORK CITY

or to any of the following official agents of the Society:

BOWES & BOWES, 1 and 2 Trinity Street, Cambridge, England;

HIRSCHWALDSCHE BUCHHANDLUNG, Unter den Linden 68, Berlin NW 7, Germany;

LIBRAIRIE SCIENTIFIQUE ALBERT BLANCHARD, 3 Place de la Sorbonne, Paris 5, France;

LIBRERIA EDITRICE NICOLA ZANICHELLI, 12, Via A. de Togni, Milano (116), Italie.

Closed symmetric operators in a Hilbert space H.

Let A be a closed symmetric operator in H.

1) $A = A^{**} \subseteq A^{*}$.

2) $\overline{\mathcal{R}(A)}^{\perp} \cap \mathcal{D}(A) = \mathcal{N}(A)$.

3) If $\mathcal{D}(A) = H$ then A is self-adjoint and bounded.

4) If $\mathcal{R}(A) = H$ then A is self-adjoint, $\mathcal{N}(A) = \{0\}$, A^{-1} is self-adjoint and bounded.